Electrochemical Polymer Electrolyte Membranes

ELECTROCHEMICAL ENERGY STORAGE AND CONVERSION

Series Editor: Jiujun Zhang
National Research Council Institute for Fuel Cell Innovation
Vancouver, British Columbia, Canada

Published Titles

Electrochemical Supercapacitors for Energy Storage and Delivery: Fundamentals and Applications
Aiping Yu, Victor Chabot, and Jiujun Zhang

Proton Exchange Membrane Fuel Cells
Zhigang Qi

Graphene: Energy Storage and Conversion Applications
Zhaoping Liu and Xufeng Zhou

Electrochemical Polymer Electrolyte Membranes
Jianhua Fang, Jinli Qiao, David P. Wilkinson, and Jiujun Zhang

Forthcoming Titles

Lithium-Ion Batteries: Fundamentals and Applications
Yuping Wu

Lead-Acid Battery Technologies: Fundamentals, Materials, and Applications
Joey Jung, Lei Zhang, and Jiujun Zhang

Electrochemical Energy: Advanced Materials and Technologies
Pei Kang Shen, Chao-Yang Wang, San Ping Jiang, Xueliang Sun, and Jiujun Zhang

Solid Oxide Fuel Cells: From Fundamental Principles to Complete Systems
Radenka Maric

ELECTROCHEMICAL ENERGY STORAGE AND CONVERSION

Electrochemical Polymer Electrolyte Membranes

Edited by
Jianhua Fang • Jinli Qiao
David P. Wilkinson • Jiujun Zhang

CRC Press
Taylor & Francis Group
Boca Raton London New York

CRC Press is an imprint of the
Taylor & Francis Group, an **informa** business

CRC Press
Taylor & Francis Group
6000 Broken Sound Parkway NW, Suite 300
Boca Raton, FL 33487-2742

First issued in paperback 2017

© 2015 by Taylor & Francis Group, LLC
CRC Press is an imprint of Taylor & Francis Group, an Informa business

No claim to original U.S. Government works

ISBN-13: 978-1-4665-8146-3 (hbk)
ISBN-13: 978-1-138-74897-2 (pbk)

Visit the Taylor & Francis Web site at
http://www.taylorandfrancis.com

and the CRC Press Web site at
http://www.crcpress.com

Contents

Preface

Clean energy technologies, which include storage and conversion, play an important role in the sustainable development of human society. Among clean energy technologies, electrochemical and solar technologies are considered the most promising, environmentally friendly, and sustainable technologies. Polymer electrolyte membrane (PEM) is one of the most important and necessary components in these energy technologies, which include fuel cells, batteries, electrolyzers, and solar energy conversion devices. Therefore, a comprehensive book with a focus on PEM science and technology, and associated major energy applications, seemed useful and necessary in order to support this technology's advancements.

The book chapters contributed by various experts in the field give a comprehensive description of the science and technology of various types of PEMs, including material selection, synthesis, characterization, and applications in various energy devices. The authors of all chapters in this book have both strong academic and industrial expertise; they provide fundamentals as well as up-to-date technical knowledge and information. Furthermore, this book emphasizes and systematically summarizes in detail PEM science and technologies, current achievements, challenges, and future directions, making the book attractive to researchers working in the fields of both polymer materials and clean energy.

It is also our belief that this book will serve as a resource benefiting researchers, students, industrial professionals, and manufacturers by providing a systematic overview of the materials, system design, and related aspects of the development of PEMs for various energy applications. In particular, information on selecting existing materials/technologies and developing new materials/technologies to improve the performance of energy devices should be very useful.

As the editors of this book, we acknowledge and express our sincere thanks to all the chapter contributors for their effective work in preparing and developing their respective chapters. We also express our appreciation to CRC Press for inviting us to lead this book project and to Allison Shatkin for her guidance and support in the book preparation process.

Dr. Jianhua Fang
Shanghai, People's Republic of China

Dr. Jinli Qiao
Shanghai, People's Republic of China

Dr. David P. Wilkinson
Vancouver, British Columbia, Canada

Dr. Jiujun Zhang
Vancouver, British Columbia, Canada

Editors

Dr. Jianhua Fang is an associate professor at the School of Chemistry and Chemical Engineering, Shanghai Jiao Tong University (SJTU), China. Dr. Fang received his BSc in carbon materials from Hunan University, his MSc in electric materials and insulation technologies from SJTU in 1990, and his PhD in materials science from Yamaguchi University, Japan, in 2000. Starting in 2000, he carried out two terms of postdoctoral research at Yamaguchi University and Case Western Reserve University, United States. Dr. Fang has more than 20 years of R&D experience in high-performance polymeric materials (polyimides, polybenzimidazoles, polysulfones, etc.), including more than 5 years of membrane materials for gas separation and pervaporation and 13 years of polymer electrolyte membranes for fuel cells. Dr. Fang has coauthored approximately 100 publications, including more than 70 refereed journal papers and approximately 30 conference and invited oral presentations. He also holds more than 20 Chinese patents and 5 Japanese patent publications. Dr. Fang serves as the editorial board member for the journal *Scientific Reports*.

Dr. Jinli Qiao is a professor, PhD supervisor, and core competency leader of the College of Environmental Science and Engineering, Donghua University, China. She is the vice chairman and vice president of the International Academy of Electrochemical Energy Science (IAOEES). She received her PhD in electrochemistry from Yamaguchi University, Japan, in 2004. After that, she joined the National Institute of Advanced Industrial Science and Technology (AIST), Japan, as a research scientist working on both acidic/alkaline polymer electrolyte membranes (PEMs) and nonnoble metal catalysts for PEM fuel cells. From 2004 to 2008, as a principal investigator, she carried out seven fuel cell–related projects, including two New Energy and Industrial Technology Development Organization (NEDO) projects of Japan, and starting from 2008, she carried out and has been carrying out a total of eleven projects funded by the Chinese government. Her research areas include PEM fuel cell catalysts/membranes, CO_2 electroreduction catalysts, supercapacitors, and metal–air batteries. As the first author and corresponding author, Dr. Qiao has published more than 100 research papers in peer-reviewed journals, 40 conference and invited oral presentations, 4 coauthored books/book chapters, and 20 patent publications. Dr. Qiao is a referee for the project assessment of the National Natural Science Foundation of China, Scientific Research Foundation of the State Education Ministry of China, Foundation of Shanghai Science and Technology Committee, and Shanghai "Twelfth Five-Year Plan" for clean energy, as well as more than 15 high-impacting international journals. Dr. Qiao is a board committee member of the IAOEES and an active member of the Electrochemical Society (ECS), the Electrochemical Society of Japan (ESJ), and China Association of Hydrogen Energy (CAHE).

Dr. David P. Wilkinson is a professor at the Department of Chemical and Biological Engineering at the University of British Columbia (UBC) and Canada Research Chair (Tier 1). He is a fellow of several organizations, including the Engineering Institute of Canada, the Canadian Academy of Engineering, the Chemical Institute of Canada, and the Royal Society of Canada. Dr. Wilkinson received his BASc in chemical engineering from UBC in 1978 and his PhD in chemistry from the University of Ottawa in 1987, where his graduate work was done with Professor Brian Conway. He has more than 20 years of industrial experience in the areas of fuel cells and advanced lithium batteries. He also maintained a joint appointment with the National Research Council Institute for Fuel Cell Innovation for several years where he was a principal research officer and a senior advisor. In addition to being a professor at UBC, Dr. Wilkinson was the director of UBC's Clean Energy Research Center (CERC) for four years. Prior to his university appointment, Dr. Wilkinson was the director and then vice president of research at Ballard Power Systems Inc., involved in the research, development, and application of fuel cell technology. Prior to joining Ballard in 1990, he was the group leader for chemistry and electrochemistry at Moli Energy and part of the team that developed the world's first commercial rechargeable lithium AA battery. Dr. Wilkinson received a number of awards for his work, including the R.A. McLachlan Award, the highest award for professional engineering in British Columbia; the Electrochemical Society Battery Division Technology Award; a Lifetime Achievement Award from the Canadian Hydrogen and Fuel Cells Association; the international Grove Medal award for contributions to fuel cell technology; and the R.S. Jane Memorial Award, the highest award for chemical engineering in Canada. Dr. Wilkinson's main research interests are in electrochemical and photochemical power devices and processes to create clean and sustainable energy. He has more than 80 patents and more than 150 peer-reviewed publications covering innovative research in these fields. He is a board committee steering member of the International Academy of Electrochemical Energy Science (IAOEES) and a board member of the Canadian Hydrogen and Fuel Cells Association.

Dr. Jiujun Zhang is a principal research officer at the National Research Council (NRC) of Canada, a fellow of the International Society of Electrochemistry (ISE), and the chairman and president of the International Academy of Electrochemical Energy Science (IAOEES). His areas of technical expertise are electrochemistry, photoelectrochemistry, spectroelectrochemistry, electrocatalysis, fuel cells (PEMFC [polymer electrolyte membrane fuel cell], SOFC [solid oxide fuel cell], and DMFC [direct methanol fuel cell]), batteries, and supercapacitors. Dr. Zhang received his BSc and MSc in electrochemistry from Peking University in 1982 and 1985, respectively, and his PhD in electrochemistry from Wuhan University in 1988. Starting in 1990, he carried out three terms of postdoctoral research at the California Institute of Technology, York University, and the University of British Columbia (UBC). Dr. Zhang holds more than 10 adjunct professorships, including one at the University of Waterloo, one at UBC, and one at Peking University. He has approximately 400 publications with more than 12,000 citations, including 230 refereed journal papers with an *H-index* of 48, 13 edited/coauthored books, 32 book chapters, 110 conference oral and keynote/invited presentations, and 10 U.S./EU/WO/JP/CA patents, and has produced in excess of 90 industrial technical reports. Dr. Zhang serves as the editor and editorial board member for several international journals as well as editor for a book series (Electrochemical Energy Storage and Conversion, CRC Press). Dr. Zhang is the cofounder and board committee member of the IAOEES and an active member of the Electrochemical Society (ECS), the ISE, the American Chemical Society (ACS), and the Canadian Institute of Chemistry (CIC).

Contributors

Shouwen Chen
School of Environmental & Biological
  Engineering
Nanjing University of Science and
  Technology
Nanjing, Jiangsu, People's Republic of
  China

Xinbing Chen
Key Laboratory of Applied Surface and
  Colloid Chemistry
School of Materials Science and
  Engineering
and
Scientific Research Department
Shaanxi Normal University
Xi'an, Shaanxi, People's Republic of
  China

Jeffrey De Lile
School of Automotive Studies
Tongji University
Shanghai, People's Republic of China

Jianhua Fang
Shanghai Key Lab of Electrical
  Insulation and Thermal Aging
School of Chemistry and Chemical
  Engineering
Shanghai Jiao Tong University
Shanghai, People's Republic of China

Xiaoxia Guo
Shanghai Key Lab of Electrical
  Insulation and Thermal Aging
School of Chemistry and Chemical
  Engineering
Shanghai Jiao Tong University
Shanghai, People's Republic of China

Shaojian He
School of Renewable Energy
North China Electric Power University
Beijing, People's Republic of China

Tomoya Higashihara
Department of Polymer Science and
  Engineering
Graduate School of Science and
  Engineering
Yamagata University
Yonezawa, Japan

Zhaoxia Hu
School of Environmental and Biological
  Engineering
Nanjing University of Science and
  Technology
Nanjing, Jiangsu, People's Republic of
  China

Hong Li
Shanghai Electrochemical Energy
  Devices Research Center
School of Chemistry and Chemical
  Engineering
Shanghai Jiao Tong University
Shanghai, People's Republic of China

Jun Lin
School of Renewable Energy
North China Electric Power University
Beijing, People's Republic of China

Yuyu Liu
Graduate School of Environmental
  Studies
Tohoku University
Sendai, Miyagi, Japan

Hui Na
Alan G. MacDiarmid Institute
College of Chemistry
Jilin University
Changchun, Jilin, People's Republic of
    China

Fenglai Pei
School of Automotive Studies
Tongji University
Shanghai, People's Republic of China

Jinli Qiao
College of Environmental Science and
    Engineering
Donghua University
Shanghai, People's Republic of China
and
Key Laboratory of Green Chemical
    Media and Reactions
School of Chemistry and Chemical
    Engineering
Henan Normal University
Xinxiang, People's Republic of China

Junkun Tang
School Material Science and
    Engineering
East China University of Science and
    Technology
Shanghai, People's Republic of China

Mitsuru Ueda
Department of Polymer Science and
    Engineering
Graduate School of Science and
    Engineering
Yamagata University
Yonezawa, Japan

Yan Yin
State Key Laboratory of Engines
Tianjin University
Tianjin, People's Republic of China

T. Leon Yu
Department of Chemical Engineering
    and Materials Science
Yuan Ze University
Toayuan, Taiwan, Republic of China

Gang Zhang
Alan G. MacDiarmid Institute
College of Chemistry
Jilin University
Changchun, Jilin, People's Republic of
    China

Xuan Zhang
Jiangsu Key laboratory of Chemical
    Pollution Control and Resources
    Reuse
Nanjing University of Science and
    Technology
Nanjing, Jiangsu, People's Republic of
    China

Yongming Zhang
Shanghai Electrochemical Energy
    Devices Research Center
School of Chemistry and Chemical
    Engineering
Shanghai Jiao Tong University
Shanghai, People's Republic of China

Chengji Zhao
Alan G. MacDiarmid Institute
College of Chemistry
Jilin University
Changchun, Jilin, People's Republic of
    China

Su Zhou
School of Automotive Studies &
    Sino-German College for Graduate
    Studies & Zhejiang College
Tongji University
Shanghai, People's Republic of China

1 Overview of Electrochemical Polymer Electrolyte Membranes

T. Leon Yu

CONTENTS

1.1 INTRODUCTION

The growing concern over the shortage of fossil fuel sources and environmental pollution in the past decades has attracted several of researchers to devote in searching of new power generators and energy storage systems. Compared to the conventional liquid electrolytes, polymer electrolytes exhibit many advantages, such as less leakage of harmful liquids and gases, simpler fabrication technique, and easier production of miniaturized structures for both in the high-energy-density batteries and fuel cell power generators. Polymer electrolyte membranes (PEMs) had shown potential components of the energy storage (i.e., Li batteries) and energy conversion devices (i.e., polyelectrolyte membrane fuel cells [PEMFCs]). In these devices, the PEM acts as a separator and a charged ion carrier (Li$^+$ ions in Li batteries, H$^+$ ion in PEMFCs). The mobility of ions in the PEMs determines the conductivity σ of the PEMs and the performance of the energy storage and energy conversion devices.

1.2   POLYMER ELECTROLYTE CLASSIFICATION

Polymer electrolytes can be divided into two classes: (1) the neutral polymers doped with ionic salt or inorganic acid (the Lewis acid–base complex electrolytes) [1,2] and (2) the polymers composed of charged ionic groups that are covalently bonded either on the polymer backbones or on the side chains grafted on the polymer backbones [3–6]. Some examples of class 1 polymer electrolytes are polyethylene oxide (PEO) doped with $LiClO_4$ or NaI for Li-polymer batteries and polybenzimidazole (PBI) doped with H_3PO_4 for high-temperature PEMFCs (HT-PEMFCs; operating temperature ~120°C–200°C) and PBI doped with KOH for alkaline membrane fuel cells (AMFCs) (Figure 1.1). The class 2 polymer electrolytes are also called *polyelectrolytes* or *polyions*. These are polymers composed of covalently linked anionic or cationic groups and small-molecule *counterions* balancing the electroneutrality. These polymers possess properties of both electrolytes and polymers. The polyelectrolytes composed of charged ionic groups less than 15% of the total number of repeat units or those with the dissociated charged ionic groups in low-polarity organic solvents, such as toluene, hexane, and benzene, less than 15% of their total number of repeat units are also called *ionomers* [6].

According to the composed ionic functional groups, the class 2 polymer electrolytes are classified into polyanions, polycations, and polyampholytes. Most of polyanions consist of acidic functional groups, such as $-SO_3H$, $-COOH$, $-PO_3H_2$, $-SO_3^-M^+$, $-COO^-M^+$, and $-PO_3^{2-}M_2^{2+}$ (where M is a metal), which can be dissociated into anionic groups and positively charged metal or proton counterions in high-polarity solvents, such as water and alcohols. Some examples of polyanions are poly(methacrylate), poly(styrene sulfonate), and poly(isopropylene phosphonate), as shown in Figure 1.2. Polycations consist of the Lewis base functional groups, which can be bound with the Lewis acids and form anionic counterions in the aqueous solutions. The most common polycations consist of amines, sulfonium, and phosphonium functional groups. Some examples of polycations are poly(diallyl dimethyl

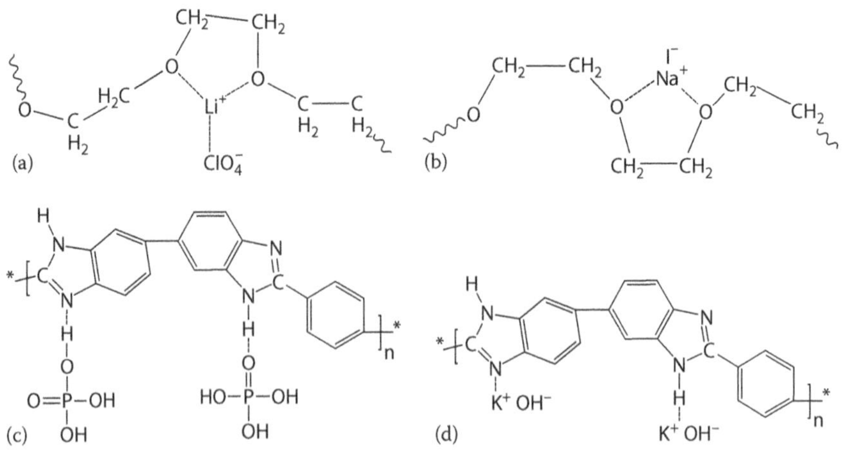

FIGURE 1.1   Chemical structures of (a) PEO doped with $LiClO_4$, (b) PEO doped with NaI, (c) PBI doped with H_3PO_4, and (d) PBI doped with KOH.

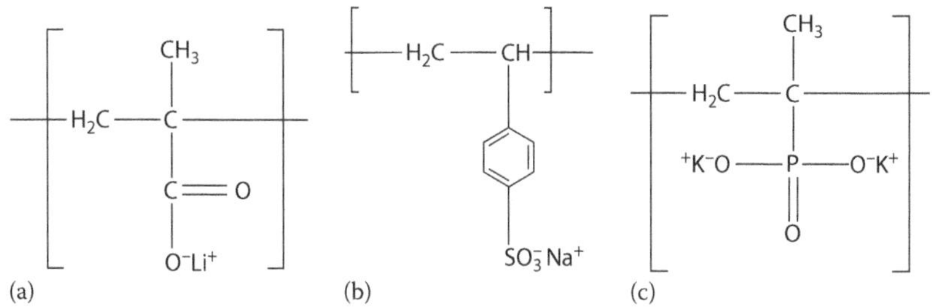

FIGURE 1.2 Examples of polyanions: (a) polymethacrylate, (b) poly(styrene sulfonate), and (c) poly(isopropylene phosphonate).

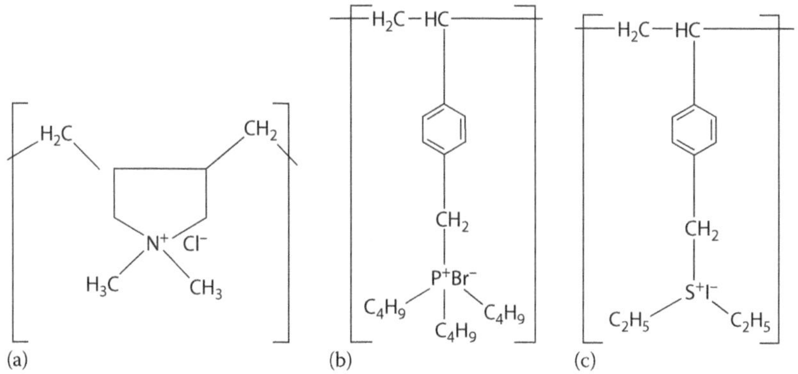

FIGURE 1.3 Examples of polycations: (a) poly(diallyl dimethyl ammonium chloride), (b) poly(vinyl benzyl phosphonium bromide), and (c) poly(vinyl benzyl diethyl sulfonium iodide).

ammonium chloride), poly(vinyl benzyl tributylphosphonium, and poly(vinyl benzyl diethyl sulfonium) as shown in Figure 1.3. The metal ions Li^+, Na^+, and K^+ (Figure 1.2) and the Cl^-, Br^-, and I^- ions (Figure 1.3) are the counterions of the polyelectrolytes. Polyampholytes carry both anionic and cationic groups covalently bound to the polymers. Most of polyampholytes are natural proteins, but some are synthetic polymers. One example of polyampholytes is shown in Figure 1.4.

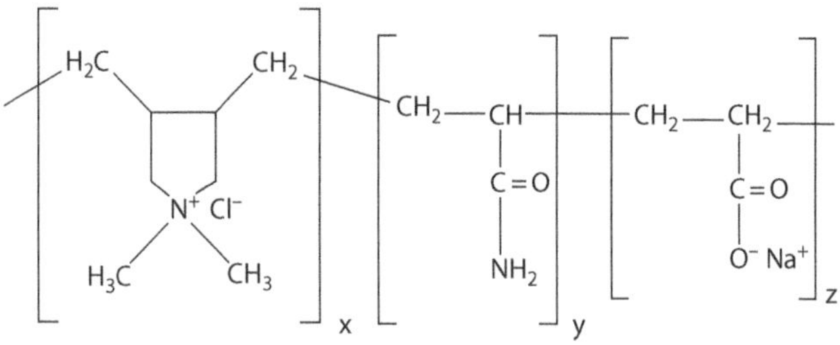

FIGURE 1.4 Example of a polyampholyte.

1.3 POLYMER ELECTROLYTE MEMBRANE STRUCTURE AND DESIGN

The PEMs are the key components of the energy storage (i.e., Li batteries) and energy conversion devices (i.e., PEMFCs). In this section, we discuss the main functions, structures and designs, conducting mechanism, and recent developments of PEMs in these devices.

1.3.1 POLYMER ELECTROLYTE MEMBRANES FOR LI-BATTERY APPLICATIONS

1.3.1.1 Main Functions and Required Properties of the Polymer Electrolyte Membranes for Li Batteries

The ionic conduction of polyether–alkali-metal salt (PEO–alkali-metal salt) complexes to electrochemical devices was first reported by Fenton in 1973 [7]. Since then, lots of research reports on these polymer electrolytes, such as increasing the conductivity of the dry PEMs and reducing the cost of manufacturing devices, have appeared in the literature [8–20]. Although lots of polymers satisfy the criteria of these devices, a few of them have shown advantages over the PEO–Li-salt system. Figure 1.5 illustrates the typical configuration of a polymer electrolyte battery. It consists of a PEM (i.e., PEO–Li-salt membrane) located at the center of the battery with an anode and a composite cathode attached on both the surfaces of the PEM. The PEM is a liquid-free system and is formed by dissolving salts in a

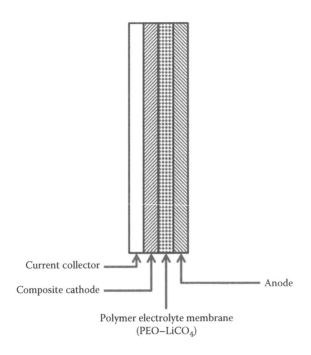

Current collector

Composite cathode

Anode

Polymer electrolyte membrane
(PEO–LiCO$_4$)

FIGURE 1.5 Configuration for PEM batteries.

high-molecular-weight polar polymer matrix. Its conducting phase in the PEM is the combination of the dissolved salts and the polar polymer matrix. The main functions of PEO–Li-salt complex PEM in the PEM batteries are to act as (1) a Li^+ ion carrier, (2) an electrode spacer, and (3) a binder of the anode and cathode. The basic required properties for a PEM used in a battery are (1) good ionic conductivity, (2) low electronic conductivity, (3) good mechanical properties, (4) excellent electrochemical and chemical stability, and (5) easy processing. The excellent mechanical properties of the high-molecular-weight amorphous polymers allow the electrochemical cell to have a solid-state construction. On a macroscopic level, the amorphous polymers behave like a solid, but at an atomic level, the local relaxations at temperatures above T_g lead the polymers to behave like liquid and possessing degrees of freedom similar to those of liquid molecules. The flexibility of amorphous polymer leads the solid PEMs to have good interfacial contacts with both the anode and cathode electrodes and thus leads the interfacial adhesion to be remained for a long time under stress when the cells are charged and discharged.

In the polymer doped with alkali-metal salt complexes, the polymers act like a solvent and provide the Lewis basic sites to coordinate the Lewis acid alkali-metal salts. The polymers are usually polyethers, polyesters, polyimines, and polythiols, which have strong Lewis base coordinating groups (e.g., –O–, >C=O, –NH–, and –S–); and the alkali-metal salts are usually $LiClO_4$, $LiCF_3SO_3$, $LiBF_4$, LiI, etc. The polymer chain should have suitable Lewis base coordinating sites for wrapping around the cations without excess strain. Thus, the polymer chain structure and the salt–polymer solvation properties are important for a PEM possessing good battery performance. For example, the ethylene oxide $-(CH_2CH_2O)_n-$ provides suitable spacing for solvating and wrapping cations; however, the methylene oxide $-(CH_2-O)_n-$ and propylene oxide $-(CH_2CH_2CH_2-O)_n-$ are not suitable for wrapping cations. The salvation of a salt in a polymer solvent depends on the cation–polymer interaction, which follows the hard/soft acid–base interaction principle suggested by Pearson [21]. Hard acids and bases are relatively small, compact, and nonpolarizable compounds; soft acids and bases are larger and more polarizable compounds. The hard acids are cations with large positive charge or those whose d-electrons are relatively unavailable for π-bonding (e.g., Li^+, Na^+, Be^{2+}, Al^{3+}, Cr^{3+}, Fe^{3+}, and Co^{3+}); soft acids are those whose d-electrons or orbitals are available for π-bonding (e.g., Tl^+, Cu^+, Ag^+, Au^+, Pd^{2+}, Pt^{2+}, Br_2, and I_2). The hard bases are nonpolarizable ligands of high electronegativity (e.g., –O– in ether); soft bases are more polarizable (e.g., –S– in thioether). Acid–base interactions are more favorable for hard–hard and soft–soft interactions than for a mixed hard–soft interactions. The hard Lewis base polyether has strong solvation stability with the hard cations, for example, Li^+, Na^+, and Mg^{2+}. The relative stability for the hard Lewis acid cations, such as Li^+, with the Lewis base donors is –O– > –NH– >> –S–. The relative stability for the soft Lewis acid cations, such as Ag^+, with the Lewis base donors is –NH– > –S– > –O–. The PEO is a good polymer solvent for Li^+ ion.

1.3.1.2 Conducting Mechanism of PEO–Li-Salt Membranes

In PEO–Li-salt PEMs, the high-molecular-weight PEO is a nonliquid solvent of the Li salt. PEO has a linear polymer chain possessing helical conformation and a high

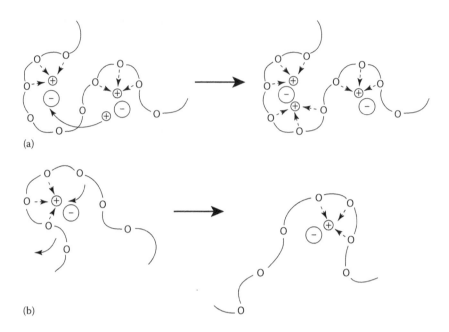

FIGURE 1.6   Cation motion in a PEO–Li-salt PEM. (a) Cation moves from one ionic cluster to another ionic cluster via positive–negative charge interaction. (b) Intrachain polymer segmental motion promotes cation mobility. The "O" represents the ether oxygen atom and the curve line represents the CH_2–CH_2 in PEO. (Modified with permission from Gary, F.M., *Polymer Electrolytes*, The Royal Society of Chemistry, Cambridge, U.K., 1997.)

degree of crystallinity (70%–85%) [22]. The melting point of its crystalline phase is T_m = ~70°C, and the glass transition temperature is T_g = −60°C to −30°C [22]. The conductivity of the PEO–Li-salt PEM results from the hopping of the cations upon the vacant coordinating sites inside the PEO matrices. At temperatures above T_g, the polymer segmental relaxation motion promotes ion mobility by forming and breaking the coordinations between the cation and the O–C–C–O ligand and providing free volume for ions to move. The cation moves from one ionic cluster to another ionic cluster via the positive–negative charge interactions (Figure 1.6a) or moves under the assistance of the polymer segmental motion around the C–C and C–O bonds (Figure 1.6b). In Figure 1.6a, the PEO provides coordinating sites that act only as anchor points for wrapping cations. In Figure 1.6b, the PEO not only provides coordinating sites but also promotes cation mobility by segmental motion.

1.3.2   Polymer Electrolyte Membranes for Fuel Cell Applications

PEMFCs employ a thin PEM for electrolytes. Currently, there are five main types of PEMFCs, differing from each other by their electrolytes (i.e., H+ ion and OH− ion), input fuels (i.e., H_2 gas fuel and methanol/water fuel), and operating temperatures: (1) low-temperature PEMFCs (LT-PEMFCs; H_2/O_2 [or air] fuel,

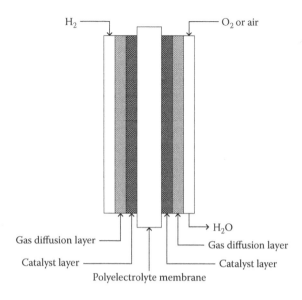

FIGURE 1.7 The structure of an MEA of PEMFCs.

operating temperature < 100°C), (2) HT-PEMFCs (H_2/O_2 [or air] fuel, operating temperature ~ 120°C–200°C), (3) direct methanol fuel cells (DMFCs; methanol–water mixture/O_2 [or air] fuel, operating temperature < 90°C), (4) AMFCs (H_2/O_2 [or air] fuel, operating temperature < 100°C), and (5) direct methanol AMFCs (DMAMFCs; methanol–water mixture/O_2 [or air] fuel, operating temperature < 90°C). The membrane electrode assemblies (MEAs) of all the aforementioned PEMFCs have similar structure. The MEA of PEMFC (either LT-PEMFC or HT-PEMFC) consists of a PEM located at its center with the anode and cathode catalyst layers (CLs) located next on both surfaces of the PEM and two gas diffusion layers (GDLs) located on the outer surfaces of CLs (Figure 1.7) [23,24]. The CL consists of carbon-supported Pt (or metal) catalyst particles and polyelectrolyte binder, and the GDL consists of a porous carbon paper (or carbon cloth) treated with a hydrophobic microporous layer. The catalyst particles are the sites where the electrochemical reaction takes place, and the GDLs are the sites where the input fuel (i.e., H_2 gas in the PEMFCs and AMFCs and methanol/water mixture in the DMFCs and DMAMFCs) and oxidant gas O_2 (or air) diffuse into the anode and cathode CLs, respectively [23,24].

1.3.2.1 Main Functions and Required Properties of Polymer Electrolyte Membranes in the Polymer Electrolyte Membrane Fuel Cells

The PEM acts as a barrier for the anode fuel and the cathode oxidant gases and also an electrolyte for transporting proton H^+ ion from anode to cathode in the PEMFCs and DMFCs or an electrolyte for transporting hydroxide OH^- ion from cathode to anode in the AMFCs [23,24]. The PEM in the PEMFCs can be described in two parts. The first part consists of the polymeric membrane material, which includes the polymer backbone, the side chains, fillers, and supporting materials. The second part is the proton carrier, including the water, H_3PO_4, and other ionic mediums (e.g., ionic liquids) [25–27]. There is an optimum amount of proton carrier in the PEM of an MEA for maximum performance of the fuel cell. Increasing the proton carrier content weakens the membrane mechanical property; however, decreasing the proton carrier content reduces the membrane proton conductivity. The PEM must retain an optimum quantity of proton carrier either under operating or nonoperating conditions. The loss of the proton carrier results in a decrease in the proton conductivity and the degradation of the membranes. Since the PEM plays both the roles of the barrier for the H_2 fuel and O_2 (or air) gases and the electrolyte for transporting H^+ ion in the MEAs, the basic required properties of a PEM for fuel cell applications are (1) good mechanical strength and toughness, (2) high thermal stability and chemical stability, (3) good barrier for anode H_2 and cathode O_2 (or air) gases, (4) good proton conductance, and (5) low electron conductance.

In an MEA (Figure 1.7), the length of the proton transfer pathway depends on the thickness (L) of the PEM. Thus, the resistance (R) of a PEM depends on both the ionic conductivity (i.e., σ) and L and is proportional to L/σ. The thicker membrane possesses a longer proton transfer pathway and a larger R and thus worsens fuel cell performance. However, a thinner membrane increases the possibility for the fuel and oxidant gas to cross over the membrane, leading to a lower fuel cell output power. Thus, there is an optimum L for the PEM for maximum fuel cell performance. A PEM possessing suitable mechanical properties, that is, tensile strength, tensile strain, and toughness, has a higher tolerance to the impact force from the high-flow-rate fuel gas, and thus its thickness can be reduced. The development of a PEM with high mechanical strength, high ionic conductivity, and lower thickness is an important issue for a high-performance PEMFC.

While lots of PEMs have been developed, the sulfonated hydrocarbon polymer electrolyte (SHCPE) and perfluorocarbon sulfonated ionomer (PFSI) membranes are most widely investigated for LT-PEMFCs and DMFCs, and the PBI and its modified polymer membranes doped with H_3PO_4 are most widely applied to HT-PEMFCs. In the following sections, we discuss the morphologies and conducting mechanisms of the SHCPEs, PFSIs, and PBI doped with H_3PO_4 (PBI/H_3PO_4) membranes.

1.3.2.2 Sulfonated Polymer Electrolyte Membranes

Polyanions possessing $-SO_3H$ functional group are more often used as proton-conducting PEMs than the other polyanions possessing $-COOH$ and $-PO_3H_2$ anionic groups, because of the higher degree of dissociation and higher proton conductivity of the $-SO_3H$ functional group. Sulfonated styrene–divinylbenzene copolymers were first used as the PEM of PEMFCs in the Space Program in the early 1960s. Although poor durability and fuel cell performance are reported, several sulfonated hydrocarbon polyanions, such as sulfonated polystyrene (SPS), sulfonated polysulfones (SPSUs), sulfonated poly(ether sulfone)s, sulfonated polyimides (SPIs), and sulfonated poly(ether ether ketone)s (SPEEKs), are still widely studied, due to their ease of synthesis and cheap cost. Compared to SHCPEs, the PFSI PEMs, for example, Nafion (a product of DuPont Co.), exhibit better operation and performance under normal conditions and have been found to be the technology standard material for PEMFCs.

1.3.2.2.1 Conducting Mechanism and Factors Controlling the Properties of Sulfonated Polymer Electrolyte Membranes

The sulfonated polymers are generally amphiphilic; they are composed of hydrophobic and hydrophilic subunits. Hydrophobic domains are responsible for providing suitable mechanical properties to the PEMs, which prevent from dissolving in proton carrier (i.e., water) or other solvents such as methanol and from penetrations of H_2 and O_2 (or air) gases when applied in fuel cell operations. Hydrophilic domains are responsible for retaining water and help the PEMs in proton transport and water transport pathways [28,29]. Proton conductivity takes place via diffusion using water as proton carriers in the sulfonated PEMs. In these PEMs, H_2O molecules act as vehicles and form H_3O^+ ions for carrying H^+ ions to diffuse across the membrane via the ionic $-SO_3^{3-}$ group aggregated regions. Figure 1.8 illustrates the proton conductivity diffusion (or vehicle) mechanism of sulfonated PEMs.

The properties of sulfonated PEMs are controlled by (1) the chemical structures of the polymer repeat unit; (2) the polymer molecular weight and molecular weight distribution; (3) the microstructure of the polymers (including the cis- and trans-orientations; the random, alternating, and block structures; and the linear, branched, and cross-linked structures); (4) the degree of sulfonation of the polymers; (5) the morphology of the solid PEMs, that is, the hydrophobic and hydrophilic domain

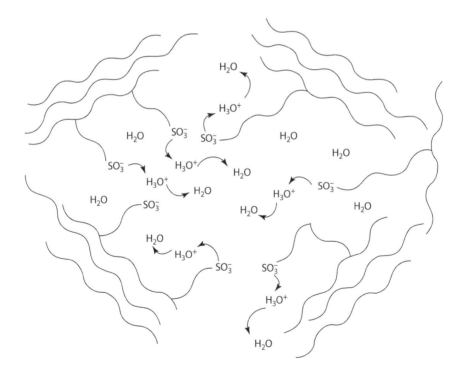

FIGURE 1.8 Diffusion (or vehicle) mechanism of proton conduction in sulfonated PEMs containing water as proton carrier. (Modified with permission from Haubold, H.G. et al., *Electrochim. Acta*, 46, 1559, 2001.)

phases and the amorphous and crystalline domain phases; and (6) the secondary bonding forces among the polymer chains, including the dipole–dipole, hydrogen-bonding, and van der Waals interactions [30].

In PEMs, the polymer chemical structure, molecular weight, and molecular weight distribution are the most important factors controlling the mechanical properties (e.g., tensile strength, flexibility, and toughness) and thermal properties (e.g., T_g of amorphous phase and T_m of crystalline phase) [31]. Usually, PEMs with polymer M_w larger than 5×10^4 g mol^{-1} and T_g higher than $T_{cell} + 20°C$ (where T_{cell} is the fuel cell operation temperature) are suitable for PEMFC applications. Modification of the PEMs with proper cross-linked structures may improve their mechanical strength and reduce the possibility of dissolving in water and methanol when applying to the PEMFCs and DMFCs. However, higher degree of PEM cross-link density may result in larger brittleness (or lower toughness) and lower free volume, which lead to reduce both the water retention and proton conductivity. Modification of the polymer backbone with long grafted side chains results in the increase in the flexibility and free volume of PEMs. The larger free volume causes the PEM to have a larger water retention and higher proton conductivity. However, the larger free volume may also lead the PEMs to be dissolved easily in water and methanol when applying to PEMFC and DMFC applications. Therefore, optimizations of the cross-linked and the grafted side chain structures are necessary for PEMs with a high fuel cell performance.

The most important factors influencing the water uptake and proton conductivity of PEMs are the degree of sulfonation of the polymer (or the sulfonic acid equivalent weight [EW]) and the hydrophilic aggregate domain morphology of the solid PEMs. Increasing the sulfonic acid functional group content (or decreasing the sulfonic acid EW) and the hydrophilic aggregate domain size results in increase in the swollen water content and thus the proton conductivity. However, increasing the sulfonic acid functional group content also weakens mechanical strength of the PEMs, which makes easier for them to be dissolved in water and easier for H_2, O_2, and air gases to penetrate across the membrane. Thus, optimization of the degree of sulfonation (or sulfonic acid EW) and hydrophilic domain morphology is necessary to design a PEM possessing high proton conductivity with suitable mechanical properties.

The morphology of a PEM depends not only on the chemical structure and molecular weight of the polymer but also on the PEM preparation procedures. The method of PEM fabrications (such as extrusion, solution casting, and hot pressing), the solvents used for dissolving polymers, the additives (such as fillers, plasticizers, and reinforcing materials) for physical and chemical property modifications, the solvents and solvent evaporation procedures for solution-cast membranes, and the thermal annealing temperature and annealing processes are all important factors controlling the PEM morphology.

In spite of many PEMs developed and reported in literature, Nafion (a trade name by DuPont Co., which is a perfluorosulfonic acid ionomer [PFSI]) is considered the benchmark of PEMs [23,32], and it has the largest number of researchers devoted for its study. In the following sections, we discuss the morphology of Nafion PEM (Sections 1.3.2.2.2 and 1.3.2.2.3) and sulfonated hydrocarbon PEM (Section 1.3.2.2.4).

1.3.2.2.2 Morphology of Nafion Membrane

Nafion PEM has excellent ionic conductivity, cation selectivity, and chemical, thermal, and mechanical stability, and it is one of the most successful PEMs applied to LT-PEMFCs and DMFCs [23,32]. Its chemical structure consists of a hydrophobic perfluorocarbon backbone and hydrophilic vinyl ether side chains terminated with sulfonic acid groups (Figure 1.9) [33,34] with a molecular weight of $M_w = \sim 2 \times 10^5$ g mol^{-1} [35,36]. The hydrophobic perfluorocarbon backbone and the hydrophilic sulfonated vinyl ether side chains have solubility parameters of $\delta_1 = 9.7$ (cal cm^{-3})$^{1/2}$ and $\delta_2 = 17.3$ (cal cm^{-3})$^{1/2}$, respectively, and they are incompatible with each other [37]. Phase separation in heterogeneous nanoscale structures between the hydrophobic fluorocarbon backbones and the hydrophilic sulfonated ionic vinyl ether pendants occurs in the polymers. The nanostructure of the ionic aggregates is strongly correlated with the proton transport properties of the membranes, and the perfluorocarbon backbone aggregated structure is correlated with the mechanical strength of the membranes. The unique chemical structure and the membrane morphology result in the useful properties of the Nafion PEMs, such as chemical and thermal stability, solubility, mechanical properties, and ionic conductivity [38].

The commercial Nafion membranes of DuPont Co. are fabricated by extrusion. Its morphology has been investigated using small-angle x-ray scattering (SAXS) and small-angle neutron scattering (SANS) [39–52], transmission electron microscopy (TEM) [44–46,53–55], and atomic force microscopy (AFM) [47,55]. SAXS and SANS spectra of Nafion exhibit a broad maximum ionomer peak in the medium angular range and an upturn of scattered intensity at very low angles (Figure 1.10 [50]). Based on the information of the observed scattering profiles, three structure models, that is, ionic cluster structure, layer structure, and intrinsic fibrillar structure models, have been proposed. The ionic cluster structure model (Figure 1.11 [41]) proposes that the hydrophilic vinyl ether sulfonate side chain groups form spherical ionic clusters of sizes in the order of nanometer range, which are embedded in the hydrophobic fluorocarbon backbone phase with ionic channels connecting among the ionic clusters [39–41]. These hydrophilic ionic clusters are the regions for swelling water and proton transport [33,34,40–48]. The layered structure model (Figure 1.12 [49]) proposes the perfluorocarbon backbones form thin crystalline layered structures (i.e., the circle rods in Figure 1.12, which are aligned vertically into the

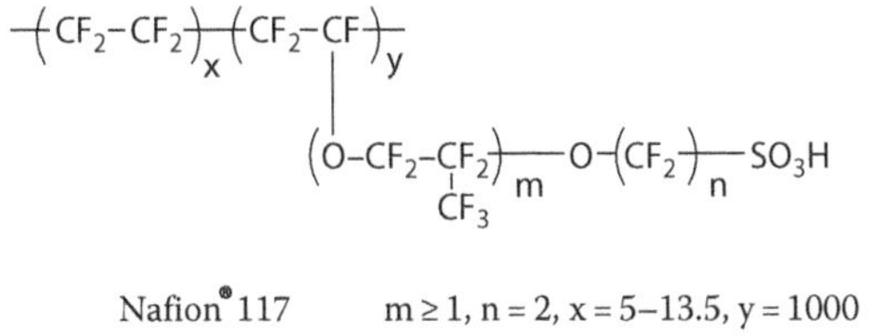

$$-\left(CF_2-CF_2\right)_x-\left(CF_2-CF\right)_y$$

$$\left(O-CF_2-CF_2\right)_m - O-\left(CF_2\right)_n - SO_3H$$
$$\qquad\qquad\; \overset{|}{CF_3}$$

Nafion® 117 $m \geq 1, n = 2, x = 5\text{–}13.5, y = 1000$

FIGURE 1.9 Chemical structure of Nafion. (From Grot, W.G., Nafion as a separator in electrolytic cells, *Nafion Perfluorinated Membranes Product Bulletin*, DuPont Co., Wilmington, DE, 1986; Grot, W.G., Process for making articles coated with a liquid composition of perfluorinated ion exchange resin, US Patent 4453991, 1986.)

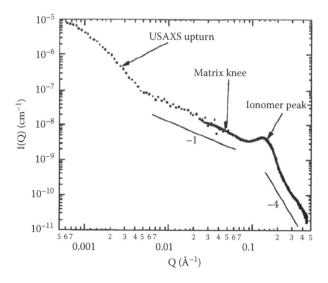

FIGURE 1.10 SAXS and SANS profile of Nafion membrane. (Reproduced with permission from Rubatat, L. et al., *Macromolecules*, 35, 4050, 2002.)

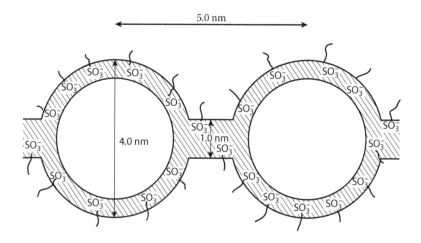

FIGURE 1.11 Ionic cluster–network model for Nafion membranes. (Reproduced with permission from Hsu, W.Y. and Gierke, T.D., *J. Membr. Sci.*, 13, 307, 1983.)

paper plane surface) with their surfaces being covered with hydrophilic vinyl ether sulfonate ionic side chains (i.e., the curved lines) [49]. There are defects in the ionic cluster structure and layer structure models. The ionic cluster model does not agree with the two scattering behaviors at the scattering vector $q = \sim 0.05$ A^{-1} of the crystalline and $q = \sim 0.15$ A^{-1} of the ionic cluster regions: (1) the ionic cluster scattering peak position depends linearly on the swollen water volume fraction, φ_w, in the membrane [50,51]; and (2) the scattering Porod behavior exhibits that the total ionic cluster

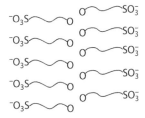

FIGURE 1.12 Layered structure model for Nafion membranes. The Nafion perfluorocarbon backbones (O) are in the direction vertical to the paper plane, and the sulfonated side chains (~) are on the plane surface of the paper and are vertical in direction to the perfluorocarbon backbones. (Reproduced with permission from Starkweather, H.W., *Macromolecules*, 15, 320, 1982.)

surface area in the membranes is independent of the φ_w [41,42]. The flaws of the layer structure model are that (1) the model does not account the scattering intensity upturn behavior at very low scattering angles and (2) it lacks the agreement between the swollen water induced-shift dependency of the ionic region and the crystalline region scattering peaks.

The intrinsic fibrillar structure model (Figure 1.13) proposes that the Nafion membrane is composed of assembly of bundles of fibrils [50,51]. The fibrils are aggregates of elongated and collapsed polymeric chains with widths of 3–5 nm and degrees of orientation of mesoscopic scale. The fibrils are surrounded with negative charges on their surfaces. This model accounts (1) the linear relationship of the shift of ionomer scattering peak with φ_w, while $\varphi_w < 40\%$, and (2) an upturn of the scattering profile

500 Å

FIGURE 1.13 Fibrillar structure model for Nafion membranes. (Reproduced with permission from Rubatat, L. et al., *Macromolecules*, 37, 7772, 2004.)

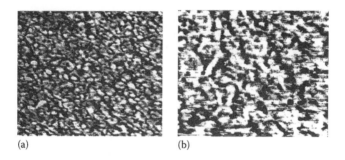

(a)   (b)

FIGURE 1.14 TM-AFM low-energy phase images of Nafion-117-K$^+$ membranes after exposure to (a) ambient humidity and (b) liquid water at room temperature. (Reproduced with permission from McLean, R.S. et al., *Macromolecules*, 33, 6541, 2000.)

at very low scattering angles (scattering vectors $q < 0.01$ A^{-1}) for membranes possess low φ_w, in which the bundles of fibrils are packed together face to face.

TEM and AFM were also used for Nafion membrane morphology characterizations. TEM images of Nafion membranes without swelling with water showed the existence of ionic clusters with roughly spherical shape and sizes of 3–10 nm [53,54]. McLean et al. [55] using taping mode AFM (TM-AFM) showed remarkable images of ionic clusters in Nafion membranes. These ionic clusters exhibited strong dependence on the content of swelling water. Figure 1.14 illustrates the low-energy phase images of Nafion-117-K$^+$ after exposure to ambient humidity at room temperature (Figure 14a) and to water (Figure 14b). The bright regions in the images are ionic cluster domains. Uniform distribution of ionic clusters with sizes of ~4–10 nm in the membrane when exposed to ambient humidity is shown in Figure 14a, and larger ionic channel-like structures with sizes of ~7–15 nm in the narrowest dimension in the membrane after exposure to water can be seen in Figure 14b. These studies demonstrate that the sulfonate ionic groups aggregate in the perfluorinated polymer matrix that allows water swelling and ionic transport in the polymer matrix. Porat et al. performed TEM studies on thin Nafion films recast from ethanol/water mixture solutions [54]. They demonstrated that poly(tetrafluoro ethylene) (PTFE)-like single-crystal structures distributed randomly in the recast film. The average distance between the randomly distributed PTFE-like crystal structures was around several microns. Lin et al. [56] using TEM micrograph demonstrated fibril-like structures, which were similar to the intrinsic fibrillar structure proposed by Rubatat et al. [50,51], in the Nafion ionomer membranes casted from N,N′-dimethyl acetamide solutions after treatment with an electric field while the membranes were annealed at 120°C. All of the preceding studies demonstrated that the sulfonate ionic groups aggregate in the perfluorinated polymer matrix that allows for water swelling and ionic transport in the polymer matrix.

1.3.2.2.3 Nafion Ionomer Solution-Cast Membranes

An important step in the modification of Nafion membranes is the solution-cast process, in which the membrane is formed after evaporating the solvent from an ionomer solution-cast film. One of the important Nafion modifications is the reinforcement of Nafion ionomer resin with high-mechanical-strength porous fiber films such as porous poly(tetra fluoro ethylene) (PTFE) fiber film, electrospun poly(vinylidene fluoride) (PVdF) nanofiber film, electrospun poly(vinylidene fluoride-co-hexafluoropropylene) (PVdF-co-HFP) nanofiber film, mineral nanofiber, and cross-linked electrospun poly(vinyl alcohol) (PVA) films. In the fiber-reinforced Nafion membranes, the high-mechanical-strength fiber film acts a supporting material, and it is responsible for the mechanical strength of the membrane, while the Nafion resin is responsible for the ionic conductivity of the membrane. The other important Nafion modifications are blending Nafion with hydrocarbon polymers (such as PBI, PVA, polyaniline, and sulfonated poly(ether ketone) [SPEK]) and inorganic nanomaterials (such as silicate, silica–PWA, $ZrHPO_4$, and ZrO). All these modifications need Nafion ionomer solutions for impregnating with porous fiber films or blending with hydrocarbon polymers and inorganic nanoparticles (Inor-NPs). The structures and morphologies of the modified Nafion membranes are strongly influenced by the Nafion ionomer nanostructures in casting solutions and the membrane casting and annealing processes. The solvents and the microstructures of Nafion molecules in the solutions are thus important for casting membrane modifications.

1.3.2.2.3.1　Nafion Ionomer Structures in Dilute Solutions The $\delta_1 = 9.7$ (cal $cm^{-3})^{1/2}$ of perfluorocarbon backbones, $\delta_2 = 17.3$ (cal $cm^{-3})^{1/2}$ of sulfonated vinyl ether side chains, and the ionic property of side chain sulfonic acid groups lead Nafion molecular morphology in solutions to strongly depend on the δ and dielectric constant (i.e., ε) of the solvents. The Nafion solution properties and Nafion microstructures in dilute alcohol/water mixture, DMAc, and N,N'-dimethyl formamide (DMF) solutions have been studied using viscometer, SANS, SAXS, electron spin resonance, dynamic light scattering (DLS), and TEM by several researchers [57–69]. These results exhibited that Nafion polymers in dilute alcohol/water mixture solutions form cylinder structures via hydrophobic perfluorocarbon backbone aggregations with ionic side chains surrounding on the surfaces of the cylinders (Figure 1.15) [60,61]. The sizes of the hydrophobic aggregated cylinders increased with the increasing solvent ε and also with the decreasing of the solvent compatibility with the Nafion perfluorocarbon backbone. The DLS studies revealed aggregates of two different sizes in Nafion in the alcohol/water mixture. The smaller aggregates were attributed to the primary cylinder-like hydrophobic perfluorocarbon backbone aggregates with the ionic side chains surrounding their surfaces. The larger aggregates were attributed to the secondary electrostatic interactions of the primary cylinder aggregates through the side chain ionic pairs (Figure 1.16) [64,65]. Ma et al. [66] and Ngo et al. [67,68], using freeze-drying technique and TEM, investigated the Nafion nanostructures in the dilute solutions ([Nafion] = 0.6 mg cm^{-3}). They found that Nafion molecules formed primary coil-like structures in the solutions as the ε and δ values of the solvent were smaller than 16 and close to $\delta_1 = 9.7$ (cal $cm^{-3})^{1/2}$ (which is the δ of Nafion perfluorocarbon backbone), respectively, and rodlike structures as the ε and δ values of the solvents were around ~39–44 and ~17–18 (cal $cm^{-3})^{1/2}$, respectively. The primary rodlike structures might aggregate via ionic interactions and form larger

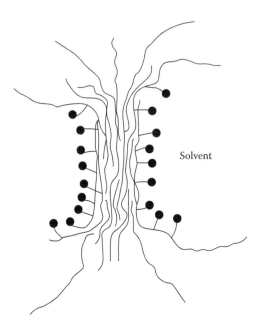

FIGURE 1.15 Nafion molecular aggregated structure in alcohol/water solution. (Reproduced with permission from Szajdzinska-Pietek, E. and Schlick, S., *Langmuir*, 10, 2188, 1994.)

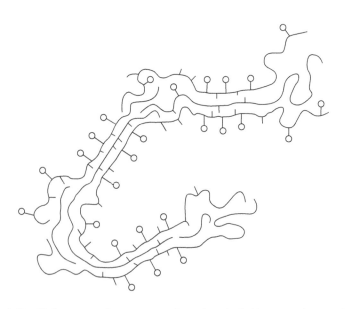

FIGURE 1.16 Nafion secondary aggregations in alcohol/water solvents. (~~~~) Perfluorocarbon backbone. (—o) Sulfonated side chain. (Reproduced with permission from Lee, S.J. et al., *Polymer*, 45, 2853, 2004.)

secondary aggregated particles, when the ε and δ values of the solvents were larger than 44 and 18 (cal cm^{-3})$^{1/2}$, respectively.

1.3.2.2.3.2 *Morphology of Nafion Ionomer Solution-Cast Membranes*   Earlier, several researchers [69–74] reported that the solution-cast Nafion membranes did not have the same chemical and physical properties as the commercial Nafion membranes. Especially the solution-cast membranes prepared by evaporating the solvent at room temperature were brittle and soluble in water and organic solvents. Subsequently, Zook and Leddy [75], de Almeida and Kawano [76], Laporta et al. [77], and Siroma et al. [78] reported the importance of the annealing process for preparing the solution-cast membranes possessing good chemical and physical stability. They obtained quasi-insoluble cast membranes from Nafion commercial ionomer solution by annealing the cast membranes at 130°C–160°C to effect polymer molecular rearrangements forming an insoluble and continuous fluorocarbon backbone crystalline phase. Martin et al. [69], Moore et al. [70,71], and Silva et al. [79] showed that the solubility of the solution-cast Nafion membranes depends on the casting solvents and the solvent evaporating temperature. The membranes casted from high-boiling-point (T_b) solvents, that is, DMF (T_b = 153°C), dimethyl sulfoxide (T_b = 189°C), and ethylene glycol (T_b = 198°C), by evaporating the solvents from the casting solutions at temperatures 10°C–20°C below the solvent T_b exhibited low solubility in alcohol–water mixtures and had the desirable physical, mechanical, and chemical properties equivalent to commercial Nafion membranes. Thomas et al. [80] investigated the influence of annealing on the cast membranes of the Nafion and quaternary ammonium bromide salt mixtures. They showed that the membrane selectivity against anions was increased by annealing. Using SAXS and differential scanning calorimetric (DSC) and thermal gravimetric analysis, Ma et al. reported that the morphology, water uptake, proton conductivity, and methanol permeability of the solution-cast Nafion membranes, which were annealed at 125°C–150°C (around 20°C–45°C above Nafion T_g) after evaporating the solvents from the solution-cast films, strongly depend on the Nafion molecular structures in the cast solutions and thus depend on the δ and ε of the cast solvents [66]. Table 1.1 summarizes the water uptake, proton conductivity, and methanol permeability of solution-cast Nafion membranes using solvents having different δ's and ε's.

1.3.2.2.4 *Morphology of Sulfonated Hydrocarbon Polyelectrolyte Membranes*
The morphologies of the aliphatic and aromatic SHCPEs have also been reported in literature [81–88]. Essafi et al., using SAXS and SANS, studied the relation of the morphology of SPIs with the polyelectrolyte ionic content, ionic group distribution, and polymer chain flexibility [88]. The scattering profiles demonstrated the presence of ionic scattering peak for flexible SPIs, indicating the presence of large ionic domains. However, the ionic scattering peak was not observed for rigid SPIs despite of the large content of ionic groups presented in the rigid SPIs, suggesting a low degree of hydrophilic and hydrophobic phase separation in the rigid sulfonated polymers. Wang et al. [81] and Harrison et al. [82], using DSC and AFM, investigated the morphology of SPSUs (which were polymerized from disodium 3,3′-disulfonate-4,4′-dichlorodiphenyl sulfone with various biphenol monomers). They showed

TABLE 1.1

Water Uptake, Proton Conductivity (σ), and Methanol Permeability (P) of Nafion Membranes Prepared by Solution Casting from Solvents Possessing Various Solubility Parameters (δ's) and Dielectric Constants (ε's) and Annealed at 150°C for 90 min

Solvent	Thickness[b] (10^{-2} cm)	δ[b] (cal cm^{-3})$^{1/2}$	ε[b]	Water Uptake[a,b] (wt%)	σ (at 70°C, 95% RH)[a,b] (10^{-3} S cm^{-1})	P(at 70°C)[a,b] (10^{-6} cm^2 s^{-1})
Methanol–water (4/1 g/g)	1.74	16.3	40.3	5.99	17.6	4.44
Ethanol–water (4/1 g/g)	1.75	15.0	32.2	5.76	8.11	4.11
Isopropanol–water (4/1 g/g)	1.74	14.4	28.5	5.26	7.01	3.40
NMF	1.71	16.1	182.4	11.38	1.36	0.923
DMF	1.73	12.2	36.7	10.44	1.81	1.20
DMAc	1.72	10.8	37.8	8.94	4.34	1.60

[a] The membranes were stored in distilled water at 25°C for 7 days before water uptake, σ, and P measurements.

[b] Data obtained with permission from Tables 1, 2-a, 8-a, and 9 of [66].

that the sizes of ionic hydrophilic domains and their connectivity varied with the degree of sulfonation of the polymer. The isolated ionic clusters and the continuous hydrophilic ionic domains were observed in the SPSU membranes when the SPSU contained less than 40 mol% and more than 60 mol%, respectively, degree of sulfonation. Kim et al. synthesized a series of SPS-b-PEB-bPS copolymers and investigated the morphology and properties of the solution-casting membranes [85–87]. The proton conductivity, methanol permeability, and the membrane morphology strongly depend on the degree of sulfonation of the SPS-b-PEB-bPS copolymers and the casting solvents (i.e., mixture of methanol [MeOH] and tetrahydrofuran [THF]). Both the proton conductivity and methanol permeability increased with increasing the degree of sulfonation and the proton conductivity.

In summary, the aforementioned reports demonstrated that the backbones of the aliphatic and aromatic SHCPEs are less hydrophobic and the sulfonic acid groups of the aliphatic and aromatic SHCPEs are less acidic and less polar than those of the PFSIs [88–90]. The aromatic SHCPEs exhibit a lesser degree of phase separation between the hydrophilic and hydrophobic domain phase than the PFSIs, due to the more rigid aromatic hydrocarbon backbones and the shorter sulfonated side chains of the aromatic SHCPEs [88]. The morphology difference between Nafion-117 (swollen with 40 vol% of water) and the SPEK (65% degree of sulfonation and swollen with 41 vol% of water) membranes was demonstrated by Kreuer [91]. Kreuer, comparing to Nafion-117, demonstrated that the SAXS of sulfonated poly(ether ether ketone ketone) exhibited a broader ionomer peak at higher scattering angles

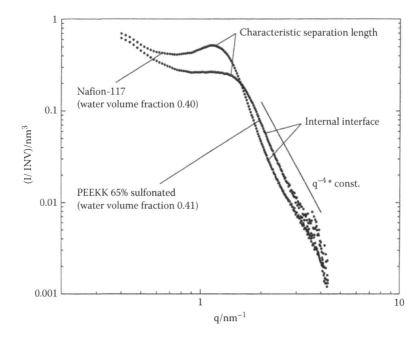

FIGURE 1.17 SAXS spectra of hydrated Nafion-117 (swollen with 40 vol% of water) and a hydrated SPEK (65% degree of sulfonation and swollen with 41 vol% of water). (Reproduced with permission from Kreuer, K.D., *J. Membr. Sci.*, 185, 29, 2001.)

and higher intensity in the Porod regime (Figure 1.17), suggesting smaller separation length with wider distribution and larger internal interface between the hydrophilic and hydrophobic domains in the SPEK than in Nafion-117 [91]. Based on the SAXS results, Kreuer proposed an illustration of the morphology of Nafion and SPEK as shown in Figure 1.18. Compared to Nafion, the water-swollen hydrophilic ionic channels in the SPEK are narrower, of less hydrophilic–hydrophobic separation, and more branched and possess more dead-end *pockets*. The less acidic and less polar sulfonic acid groups and the less degree of phase separation between the hydrophilic and hydrophobic domains of the aliphatic and aromatic SHCPEs than the PFSIs lead the SHCPE membranes to have lower proton conductivity than the PFSI membranes.

1.3.2.2.5 Modifications of Nafion Membranes

Nafion is currently the most widely used PEM in PEMFC applications and has been the technology standard materials of the PEM and the electrolyte binder of the CL of PEMFCs. Lots of researchers devoted efforts to enhance the mechanical strength to reduce membrane thickness, improve the moisture retention capability to enhance conductivity, and reduce the methanol crossover for DMFCs. Two of the major modifications are (1) fiber-reinforced Nafion composite membranes and (2) Inor-NP hybridizing Nafion composite membranes, which are discussed in the following two sections.

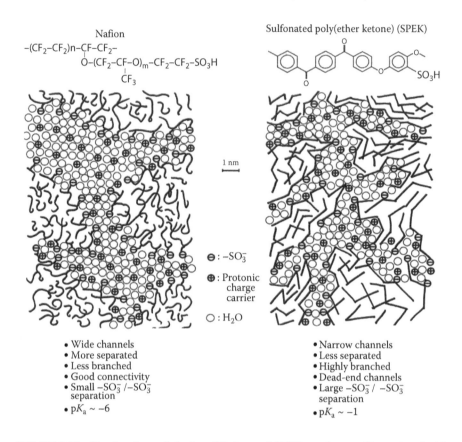

Nafion

$-(CF_2-CF_2)n-CF-CF_2-$
$\quad\quad O-(CF_2-CF-O)_m-CF_2-CF_2-SO_3H$
$\quad\quad\quad\quad CF_3$

Sulfonated poly(ether ketone) (SPEK)

1 nm

$\ominus : -SO_3^-$

$\oplus :$ Protonic charge carrier

$\bigcirc : H_2O$

• Wide channels
• More separated
• Less branched
• Good connectivity
• Small $-SO_3^-/-SO_3^-$ separation
• $pK_a \sim -6$

• Narrow channels
• Less separated
• Highly branched
• Dead-end channels
• Large $-SO_3^-/-SO_3^-$ separation
• $pK_a \sim -1$

FIGURE 1.18 Simulated morphologies of Nafion and SPEK membranes. (Reproduced with permission from Kreuer, K.D., *J. Membr. Sci.*, 185, 29, 2001.)

1.3.2.2.5.1 Fiber-Reinforced Nafion Composite Membranes   The ultrathin fiber-reinforced Nafion composite membrane (Gore-Select membrane) was first fabricated using porous PTFE thin film (thickness $L = \sim10-20$ μm; Figure 1.19a) of high mechanical strength as a supporting material for impregnating with Nafion ionomer solution [92–96]. Since then, lots of paper regarding the properties, modifications, and applications of Nafion/PTFE composite membranes to PEMFCs and DMFCs were reported [97–118]. Although the room-temperature conductivity σ of the Nafion/PTFE membrane is lower than that of Nafion membranes, its area resistance ($R = L/\sigma$) is significantly lower due to its thinness (see Table 1.2). After initial reports of Nafion/PTFE composite membranes, the porous films, such as poly(ethylene terephthalate) (PETE) [119], polycarbonate [120], and polyethylene (PE) [121,122], and the electrospun polymer nanofiber films, such as PVdF [123,124], PVdF-co-HFP [125], and PVA nanofiber films (Figure 1.19b) [126–131], were also used as supporting films for Nafion/fiber-reinforced composite PEMs. In Table 1.2, we summarize several literature report properties of fiber-reinforced Nafion PEMs. Figure 1.20a and b shows the SEM micrographs of the plane surfaces of Nafion/PTFE and

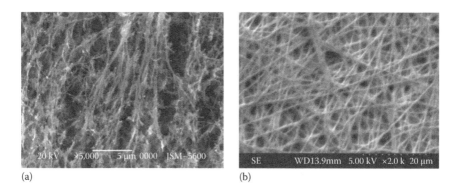

(a)   (b)

FIGURE 1.19 SEM micrograph of (a) porous PTFE thin film. (Reproduced with permission from Lin, H.L. et al., *J. Membr. Sci.*, 237, 1, 2004.) (b) Electrospun PVAc nanofiber film cross-linked with glutaraldehyde. (Reproduced with permission from Lin, H.L. et al., *J. Membr. Sci.*, 365, 114, 2010.)

Nafion/PVA composite membranes, respectively. These SEM micrographs demonstrate that the Nafion ionomers were homogeneously impregnated into the nano-pores of the support films and no void was observed in the micrographs. Because of the high mechanical strength of the porous support films, the fiber-reinforced Nafion composite membranes have higher mechanical strength than the neat-Nafion membrane (Table 1.2). Thus, the thickness of the Nafion/fiber composite membranes can be reduced to make it thinner than the neat-Nafion membranes, for example, Nafion-117 (thickness ~175 μm), Nafion-115 (thickness ~125 μm), and Nafion-212 (thickness ~50 μm). These thinner composite PEMs contain much less of the expensive Nafion resin than the thicker neat-Nafion membranes; thus, the cost of the PEMs can be reduced. Besides lower cost, other advantages of the composite PEMs include good mechanical strength in both swollen and unswollen states, lower thickness compared to the widely used neat Nafion-112 and Nafion-117 membranes, and easier property modifications (e.g., methanol permeability reduction for DMFCs and moisture content enhancement at temperatures above 100°C for HT-PEMFCs). In these thin composite PEMs, the porous supporting film (its chemical structures do not consist of any ionic functional groups) provides the membrane with mechanical strength, and the Nafion (its chemical structures consist of sulfonic acid functional groups) provides the membrane proton–conducting paths.

Methanol crossover is the main issue in lowering the performance of DMFCs. In order to reduce the methanol crossover through the PEMs, instead of using the thinner membranes (i.e., Nafion-112 and Nafion-212), most of the researchers have used the thicker Nafion-117 membrane in DMFCs. Several researchers have made efforts on reducing the methanol crossover through the membranes by blending the Nafion PEMs with low methanol compatible polymers, such as PBI [132–137], PVA [138–141], polyaniline [142], and SPEK [143], or by blending Nafion PEM with Inor-NPs, which is discussed in the next section (Section 1.3.2.2.5.2). The thicknesses of these modified Nafion membranes were similar to (or higher than) that of the neat-Nafion membrane (i.e., Nafion-117). Though the methanol crossover from these

TABLE 1.2
Tensile Stress, Elongation, Conductivity (σ), and Methanol Permeability (P_{MeOH}) of Fiber-Reinforced Nafion Composite Membranes

Membrane	Thickness L (μm)	Tensile Stress (MPa)	Elongation (%)	σ (10^{-2} S cm^{-1})	L/σ (Ω cm^2)	P_{MeOH} (10^{-6} cm^2 s^{-1})	Reference
1. Nafion/PTFE							Tang et al. [108]
Nafion/PTFE	14	34.1	—	6.2 (60°C, 100% RH)	0.023	—	
Nafion-211	25	21.3	—	9.1 (60°C, 100% RH)	0.028	—	
2. Nafion/PTFE							Si et al. [190]
Nafion/PTFE	18	—	—	0.35 (120°C, 31% RH)	0.52	—	
		—	—	0.69 (105°C, 51% RH)	0.26	—	
				1.8 (80°C, 100% RH)	0.10	—	
Nafion/PTFE/Zr(HPO$_4$)$_2$ (Nafion/Zr(HPO$_4$)$_2$ = 9/1 g/g)							
	18	—	—	0.75 (120°C, 31% RH)	0.24	—	
	18	—	—	1.5 (105°C, 51% RH)	0.12	—	
	18	—	—	2.0 (80°C, 100% RH)	0.09	—	
Nafion-1135	88	—	—	1.2 (120°C, 31% RH)	0.76	—	
	88	—	—	2.4 (105°C, 51% RH)	0.36	—	
	88	—	—	7.3 (80°C, 100% RH)	0.12	—	

(Continued)

TABLE 1.2 (*Continued*)
Tensile Stress, Elongation, Conductivity (σ), and Methanol Permeability (P_{MeOH}) of Fiber-Reinforced Nafion Composite Membranes

Membrane	Thickness L (μm)	Tensile Stress (MPa)	Elongation (%)	σ (10^{-2} S cm^{-1})	L/σ (Ω cm^2)	P_{MeOH} (10^{-6} cm^2 s^{-1})	Reference
3. Nafion/PVdF							
Nafion/PVdF	105	—	—	0.16 (60°C, 100% RH)	6.77	—	Choi et al. [123]
Nafion-115	125	—	—	3.2 (60°C, 100% RH)	0.39	—	
4. Nafion/PE							
Nafion/PE	26	—	—	0.6 (60°C, 100% RH)	0.43	0.95 (25°C)	Adjemian et al. [149]
Nafion-117	185	—	—	6.3 (60°C, 100% RH)	0.29	1.50 (25°C)	
5. Nafion/cross-linked electrospun PVA fiber							
Nafion/PVA	19	34.8	5.7	5.5 (80°C, 95% RH)	0.035	—	Molla et al. [128]
Nafion/PVA	25	—	—	5.0 (80°C, 95% RH)	0.050	—	
Nafion/PVA	31	—	—	4.0 (80°C, 95% RH)	0.078	—	
Nafion/PVA	39	37.5	10.2	3.0 (80°C, 95% RH)	0.13	—	
Nafion-212	50	—	—	10.0 (80°C, 95% RH)	0.050	—	
Cast Nafion	37	24.8	41.6	—	—	—	
6. Nafion/cross-linked electrospun PVA fiber							
Nafion/PVA	50	—	—	1.07 (70°C, 95% RH)	0.46	4.86 (70°C)	Lin et al. [126]
Nafion-117	175	—	—	1.92 (70°C, 95% RH)	0.91	3.99 (70°C)	

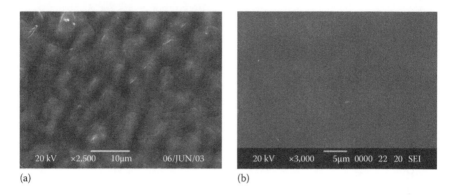

(a)   (b)

FIGURE 1.20 SEM micrographs of the plane surfaces of (a) Nafion/PTFE. (Reproduced with permission from Lin, H.L. et al., *J. Membr. Sci.*, 237, 1, 2004.) (b) Nafion/PVA composites membranes. (Reproduced with permission from Lin, H.L. et al., *J. Membr. Sci.*, 365, 114, 2010.)

modified membranes is reduced (which causes an improvement in DMFC performance), the proton transfer resistance R of these membranes is increased (which causes reduction in DMFC performance). The advantage of applying the thin nanofiber-reinforced composite membranes to the DMFCs is that the methanol crossover can be reduced without increasing the membrane resistance (i.e., $R = L/\sigma$). The DMFC performances of these thin composite membranes were found being better than the thicker Nafion-117 PEM. Table 1.2 summarizes mechanical properties, conductivity, and methanol permeability of fiber-reinforced Nafion composite membranes reported in literature.

1.3.2.2.5.2 Inorganic Nanoparticle Hybridizing Nafion Composite Membranes Nafion membranes work successfully for LT-PEMFCs (temp < 90°C); however, they suffer from problems of (1) low proton conductivity when operating the PEMFCs at temperatures above 90°C [23,144,145] due to the low moisture content of the membrane and (2) large methanol crossover when applying to DMFCs [23,144–146]. Recent reports have shown that hybridizing Inor-NPs, such as zirconium oxide [147,148], silicone oxide [149–164], silicate compound [165–173], and zirconium hydrogen phosphate (ZrHPO$_4$) [174–182], can help Nafion membranes in retaining moisture at temperatures above 90°C and thus improves PEMFC performance when operating at 90°C–130°C (a medium–high-temperature [MHT] regime) [149,150,174,176–179,182,183]. Besides applications in MHT-PEMFCs, hybridizing Inor-NPs in the Nafion membranes can reduce *methanol* crossover and improves DMFC performances [169,179–182,184,185].

In the Inor-NP/polyelectrolyte hybrid membranes, the polyelectrolyte serves as matrix component and Inor-NPs are dispersed in the organic polymers [144]. The advantage of incorporating hygroscopic Inor-NPs (SiO$_2$, TiO$_2$, etc.) into PEMs is that the dispersed hydrophilic Inor-NPs improve both the water management and proton conductivity of the PEM by improving the self-humidification of the PEM near the anode side, by back-diffusing the water produced at the cathode to anode [146], and/or by reducing the electroosmotic drag from anode to cathode. In case, the incorporated Inor-NPs are also solid proton conductors (e.g., zirconium phosphate and heteropolyacid) and thus increase the membrane conductivity [176,177]. The mechanical properties of the PEM are also enhanced when the inorganic–polymer component interaction is favored. The disadvantage of hybridizing inorganic solid particles in the PEMs is the nonhomogeneity of the dispersed solid particles. The inorganic solid particles are likely to agglomerate in the membranes and thus increase the lower limit of the hybrid membrane's thickness. The ionic conducting pathway for most of the Inor-NPs is across their surface rather than through their bulk, which tends to increase with decreasing the inorganic particle sizes. The particle size can be reduced by using in situ solgel hybrid method to form Inor-NPs within the hybrid membranes [148,152,153,155,164,166], which is discussed in the following section.

Two processes, *directing casting* and *in situ solgel*, have been reported in the Nafion/Inor-NP hybrid membrane preparations. In *in situ solgel* process (Figure 1.21), the Nafion membrane is used as an interactive template to swell the precursor solutions (i.e., inorganic alkoxide alcohol aqueous solutions and organoalkoxysilane alcohol aqueous solutions) to direct the condensation polymerization of the inorganic

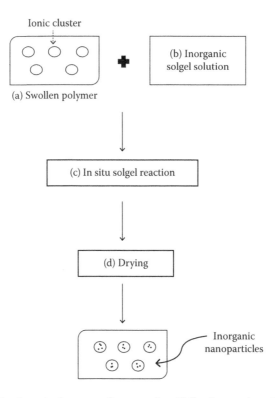

FIGURE 1.21 In situ solgel process for preparing Nafion/inorganic oxide nanoparticles hybrid composite membranes. (Reproduced with permission from Deng, Q. et al., *J. Appl. Polym. Sci.*, 68, 747, 1998.)

(a) Solgel solution—hydrolysis

$(R_1)_y Si(OR_2)_x + H_2O \dashrightarrow (R_1)_y Si(OH)_x + xR_2OH$ $(x + y = 4)$

(b) In situ solgel reaction—condensation

$(R_1)_y Si(OH)_x + (R_1)_y Si(OH)_x \dashrightarrow (R_1)_y Si-O-Si(OH)_x$

$\dashrightarrow$ Poly(organo silicate)

FIGURE 1.22 Chemical reaction of alkyl-siloxane solgel reaction. (Reproduced with permission from Deng, Q. et al., *J. Appl. Polym. Sci.*, 68, 747, 1998.)

alkoxides and organoalkoxysilanes to form inorganic oxide nanoparticles inside the hydrophilic ionic cluster regions in the Nafion membrane [152–156,165,166]. The Nafion membranes used for swelling the precursor solutions can be either the commercial Nafion membranes (Nafion-117, Nafion-115, Nafion-112, Nafion-212, etc.) or the solution-cast membranes annealed at temperatures above Nafion T_g (~105°C) for 1–3 h. Phase separation between the hydrophobic perfluorocarbon backbone aggregated domain and the hydrophilic ionic aggregated domain is well developed in these membranes (Figure 1.21a). After impregnating the Nafion membrane into the precursor inorganic alkoxide or organoalkoxysilane alcohol aqueous solutions (Figures 1.21b and 1.22a), the inorganic alkoxide or organoalkoxysilane molecules migrate into the hydrophilic ionic aggregated domains to proceed condensation (Figures 1.21c and 1.22b) and form Inor-NPs (Figure 21d) [152]. Thus, the Inor-NP sizes and distributions are controlled by the sizes and morphology of the ionic domains in the Nafion membranes.

In the *direct casting* process, the Inor-NPs (SiO_2, TiO_2, ZrO_2, ZrH_2PO_4, etc.) are directly mixed into a Nafion solution, and the mixture solution is cast onto a glass (or Teflon) plate. After evaporating the solvent, the composite hybrid membrane is obtained by annealing the cast hybrid film at temperature above T_g (~105°C) of Nafion for 1–3 h [157,158,164,186–189]. The morphologies of these hybrid membranes are controlled by (1) δ and ε of the casting solvents, (2) the interaction of the Inor-NPs with the ionic sulfonate groups and thus by the concentration of Inor-NP in the casting solution, and (3) the annealing temperature and time of the casting membrane. Compared to the Inor-NPs in the membranes fabricated by using the *in situ solgel* process, the inorganic particles are less homogeneous both in size and distributions in the *direct casting* membranes.

Modification of Nafion/porous PTFE composite membranes with Inor-NPs using *in situ solgel* and *direct casting* processes has also been reported in literature [105,106,109–113,190,191]. There are reports dealing with the methanol crossover, proton conductivity, and DMFC test results of Nafion/Inor-NP and Nafion/PTFE/ Inor-NP composite membranes at temperatures around 50°C–80°C and comparing these results with those of benchmark Nafion-117 membrane [111,147,157,163,164, 168,171,188]. Most of the reports demonstrated that the Nafion/Inor-NP and Nafion/ PTFE/Inor-NP composite membranes have lower methanol crossover and better DMFC performance than Nafion-117 and Nafion/PTFE, respectively. The Nafion/ InNP and Nafion/PTFE/InNP composite membranes also exhibit higher water uptake, higher proton conductance, and better fuel cell performance than the neat Nafion-112 and Nafion/PTFE, respectively, at low humidity (30%–60% RH) and high temperature (110°C–130°C) [149,175–179,182,183,189,191]. Table 1.3 lists the

TABLE 1.3

Conductivity (σ), Water Uptake, and Methanol Permeability (P_{MeOH}) of Inor-NP Hybridizing Nafion PEMs

Membrane	Thickness L (μm)	σ (10^{-2} S cm^{-1})	L/σ (Ω cm^2)	Water Uptake (%)	P_{MeOH} (10^{-7} cm^2 s^{-1})	Membrane Preparation Process and Reference
1. Nafion/SiO$_2$ (12.4% SiO$_2$)	125	7.0	0.018	14.0	—	Solgel; Jung et al. [163]
Nafion/SiO$_2$ (10% SiO$_2$)	125	8.0	0.16	13.0	—	
Nafion/SiO$_2$ (6.4% SiO$_2$)	125	7.5	0.17	11.0	—	
Nafion-115	125	7.4	0.17	4.7	—	
2. Nafion/SiO$_2$ (13.0% SiO$_2$)	125	0.52 (80°C, 40% RH)	2.4	4.0 (80°C, 40% RH)	—	Solgel; Rodgers et al. [192]
Nafion/SiO$_2$ (9.9% SiO$_2$)	125	0.62 (80°C, 40% RH)	2.0	2.6 (80°C, 40% RH)	—	
Nafion/SiO$_2$ (5.9% SiO$_2$)	125	0.43 (80°C, 40% RH)	2.9	2.1 (80°C, 40% RH)	—	
Nafion-115	125	0.78 (80°C, 40% RH)	1.6	1.2 (80°C, 40% RH)	—	
3. Nafion/SiO$_2$ (7% SiO$_2$)	125	0.30 (25°C, 30% RH)	4.2	40 (25°C, 100% RH)	—	Solgel; Amjadi et al. [193]
Nafion/SiO$_2$ (5% SiO$_2$)	125	0.43 (25°C, 30% RH)	2.9	38 (25°C, 100% RH)	—	
Nafion/SiO$_2$ (2% SiO$_2$)	125	0.71 (25°C, 30% RH)	1.8	35 (25°C, 100% RH)	—	
Nafion-117	125	1.21 (25°C, 30% RH)	1.0	31 (25°C, 100% RH)	—	
4. Nafion/ZrHPO$_4$ (67/23 g/g)	125	11.0 (80°C, 100% RH)	0.11	—	—	In situ solgel; Costamgana et al. [175]
Nafion-115	125	12.9 (80°C, 100% RH)	0.097	—	—	

(Continued)

TABLE 1.3 (Continued)
Conductivity (σ), Water Uptake, and Methanol Permeability (P_{MeOH}) of Inor-NP Hybridizing Nafion PEMs

Membrane	Thickness L (μm)	σ (10^{-2} S cm^{-1})	L/σ (Ω cm^2)	Water Uptake (%)	P_{MeOH} (10^{-7} cm^2 s^{-1})	Membrane Preparation Process and Reference
5. Nafion/SiO$_2$/PWA (97/2.1/0.9 g/g/g) (PWA = H$_3$PW$_{12}$O$_4$)						
	100	24.5 (135°C, 100% RH)	0.041	0.15 (130°C, 30% RH)	—	Cast; Arico et al. [194]
		14.2 (80°C, 100% RH)	0.070	1.03 (60°C, 30% RH)	—	
Nafion/SiO$_2$ (97/3 g/g)	100	17.9 (135°C, 100% RH)	0.056	0.13 (130°C, 30% RH)	—	Cast; Arisco et al. [194]
		12.5 (80°C, 100% RH)	0.080	1.03 (60°C, 30% RH)	—	
Nafion/ZrO$_2$ (97/3 g/g)	100	12.3 (135°C, 100% RH)	0.081	—	—	Cast; Arisco et al. [194]
		9.6 (135°C, 100% RH)	0.10	—		
Nafion/Al$_2$O$_3$ (97/3 g/g)	100	8.8 (135°C, 100% RH)	0.11	0.11 (130°C, 30% RH)	—	Cast; Arisco et al. [194]
		7.9 (80°C, 100% RH)	0.13	1.03 (60°C, 30% RH)	—	
Recast Nafion		6.6 (135°C, 100% RH)	0.15	0.10 (130°C, 30% RH)	—	Cast; Arisco et al. [194]
		5.4 (80°C, 100% RH)	0.19	1.03 (60°C, 30% RH)	—	
6. Nafion/STA (STA 2.0 mM; STA = SiW$_{12}$O$_{40}^{4-}$)						
	100	2.05 (25°C, 100% RH)	0.49	60 (25°C)	—	Cast; Tazi and Savadogo [186]
Nafion/STA/thiophene (TH) (STA 2.0 mM; TH 6 vol%)	120	9.50 (25°C, 100% RH)	0.126	—	—	Cast; Tazi and Savadogo [186]
Nafion-117	175	1.23 (25°C, 100% RH)	0.142	27 (25°C)	—	

(Continued)

TABLE 1.3 (Continued)

Conductivity (σ), Water Uptake, and Methanol Permeability (P_{MeOH}) of Inor-NP Hybridizing Nafion PEMs

Membrane	Thickness L (μm)	σ (10^{-2} S cm^{-1})	L/σ (Ω cm^2)	Water Uptake (%)	P_{MeOH} (10^{-7} cm^2 s^{-1})	Membrane Preparation Process and Reference
7. Nafion/SiO$_2$/PWA (Nafion/SiO$_2$/PWA = 1/74/7 g/μg/μg) (PWA = H$_3$PW$_{12}$O$_4$)						
	175	2.2 (25°C, 100% RH)	0.80	—	3.0	In situ solgel; Xu et al. [184]
Nafion/SiO$_2$/PWA (Nafion/SiO$_2$/PWA = 1/292/41 g/μg/μg)						
	175	2.4 (25°C, 100% RH)	0.73	—	4.7	
Nafion/SiO$_2$ (Nafion/SiO$_2$ = 1/74 g/μg)						
	175	0.5 (25°C, 100% RH)	3.5	—	3.5	In situ solgel; Xu et al. [184]
Nafion/SiO$_2$ (Nafion/SiO$_2$ = 1/292 g/μg)						
	175	0.002 (25°C, 100% RH)	875	—	8.6	
Nafion-117	175	1.5 (25°C, 100% RH)	1.2	—	4.6	
8. Nafion/P(PSA)-SiO$_2$ (96/4 g/g)	—	4.1 (80°C, 100% RH)	—	9.66	8.45	Cast; Tay et al. [195]
Pristine Nafion	—	2.7 (80°C, 100% RH)	—	22.5	31.1	
9. Nafion/SiO$_2$—O—SO$_3$H (a silica modified with sulfonic acid group; particle size ~3 nm)						
	175	21.3 (100°C, 100% RH)	0.082	52.7		In situ solgel; Kumar et al. [196]
		19.5 (80°C, 100% RH)	0.090			
Nafion/SiO$_2$ (particle size ~3 nm)						
	175	14.2 (100°C, 100% RH)	—	45.8	—	In situ solgel; Kumar et al. [196]
		13.5 (80°C, 100% RH)	—			
Nafion-117	175	9.6 (100°C, 100% RH)	0.182	37.6	—	
		11.7 (80°C, 100% RH)	0.150			

(Continued)

TABLE 1.3 (Continued)

Conductivity (σ), Water Uptake, and Methanol Permeability (P_{MeOH}) of Inor-NP Hybridizing Nafion PEMs

Membrane	Thickness L (μm)	σ (10^{-2} S cm^{-1})	L/σ (Ω cm^2)	Water Uptake (%)	P_{MeOH} (10^{-7} cm^2 s^{-1})	Membrane Preparation Process and Reference
10. Nafion/TEOS (Nafion/SiO$_2$)	88	8.9 (80°C, 100% RH)	0.099	18.5	—	Cast; Joseph et al. [197]
		1.4 (80°C, 50% RH)	0.63	—	—	
Nafion/TEOS-PETES (Nafion/–SiO$_2$–O–ethyl–PO$_3$H$_2$)						
	91	15.2 (80°C, 100% RH)	0.60	24.9	—	
		4.9 (80°C, 50% RH)	0.19	—	—	
Recast Nafion	68	8.0 (80°C, 100% RH)	0.085	14.9	—	
		0.8 (80°C, 50% RH)	0.85	—	—	
11. Nafion/ZrO$_2$ (Nafion/ZrO$_2$ = 95/5 g/g)						In situ solgel; Thampan et al. [198]
	50	1.85 (120°C, 40% RH)	0.27	8.0 (120°C, 40% RH)	—	
		0.17 (120°C, 10% RH)	2.9	6.0 (120°C, 10% RH)	—	
		1.45 (90°C, 40% RH)	0.34	—	—	
		0.15 (90°C, 40% RH)	3.33	—	—	
Nafion-112	50	1.80 (120°C, 40% RH)	0.28	5.5 (120°C, 40% RH)	—	
		0.15 (120°C, 10% RH)	3.3	3.0 (120°C, 10% RH)	—	
		1.22 (90°C, 40% RH)	0.41	—	—	
		0.09 (90°C, 10% RH)	5.6	—	—	

(Continued)

TABLE 1.3 (Continued)
Conductivity (σ), Water Uptake, and Methanol Permeability (P_{MeOH}) of Inor-NP Hybridizing Nafion PEMs

Membrane	Thickness L (μm)	σ (10^{-2} S cm^{-1})	L/σ (Ω cm^2)	Water Uptake (%)	P_{MeOH} (10^{-7} cm^2 s^{-1})	Membrane Preparation Process and Reference
Nafion/ZrO$_2$ (Nafion/ZrO$_2$ = 95/5 g/g)						
	50	1.50 (120°C, 40% RH)	0.33	7.0 (120°C, 40% RH)	—	Cast; Thampan et al. [198]
		0.17 (120°C, 10% RH)	2.94	5.0 (120°C, 10% RH)	—	
		1.30 (90°C, 40% RH)	0.38	—	—	
		0.07 (90°C, 10% RH)	7.14	—	—	
Recast Nafion	50	1.70 (120°C, 40% RH)	0.29	—	—	
		0.09 (120°C, 10% RH)	5.55	—	—	
		1.19 (90°C, 40% RH)	0.42	—	—	
		0.09 (90°C, 10% RH)	5.6	—	—	
12. Nafion/SiO$_2$–SZr (SiO$_2$-supported sulfonated Zr) (Nafion/SiO$_2$–SZr = 97/3 g/g)						
	50	7.0 (60°C, 100% RH)	0.071	54.8	—	Cast; Bi et al. [199]
		6.6 (60°C, 0% RH)	0.076	—	—	
Nafion/SiO$_2$ (Nafion/SiO$_2$–SZr = 97/3 g/g)						
	50	5.5 (60°C, 100% RH)	0.091	50.5	—	Cast; Bi et al. [199]
		3.8 (60°C, 0% RH)	0.13	—	—	
Recast Nafion	50	6.6 (60°C, 100% RH)	0.076	30.2	—	
		2.6 (60°C, 0% RH)	0.19	—	—	

(Continued)

TABLE 1.3 (Continued)
Conductivity (σ), Water Uptake, and Methanol Permeability (P_{MeOH}) of Inor-NP Hybridizing Nafion PEMs

Membrane	Thickness L (μm)	σ (10^{-2} S cm^{-1})	L/σ (Ω cm^2)	Water Uptake (%)	P_{MeOH} (10^{-7} cm^2 s^{-1})	Membrane Preparation Process and Reference
13. Nafion/ZrHPO$_4$ (NF/ZrP)						Cast; Arbizzani et al. [200]
NF/ZrP = 94/6	200	0.43 (80°C, 100% RH)	4.7	—	—	
NF/ZrP = 96/4	210	0.52 (80°C, 100% RH)	4.0	—	0.55 (80°C)	
NF/ZrP = 98/2	195	3.2 (80°C, 100% RH)	0.61	—	0.74 (80°C)	
Recast Nafion	200	5.3 (80°C, 100% RH)	0.38	—	1.0 (80°C)	
14. Nafion/Cs-PWA (85/15 g/g) (PWA = phosphotungstic acid)						Cast; Amirinejad et al. [201]
	70	1.19 (120°C, 50% RH)	0.59	48 (25°C)	—	
		1.40 (80°C, 100% RH)	0.50	—	—	
		0.80 (80°C, 47% RH)	0.88	—	—	
Recast Nafion	80	0.35 (80°C, 100% RH)	2.3	42 (25°C)	—	
		0.21 (80°C, 47% RH)	3.8	—	—	
15. Nafion/CsPMo (10/1 g/g) (Cs$_{2.5}$H$_{0.5}$ PMo$_{12}$O$_{40}$)						Cast; Amirinejad et al. [202]
	67	6.7 (120°C, 0% RH)	0.1	76 (25°C)	—	
		5.7 (25°C, 90% RH)	0.12	—	—	
		2.9 (25°C, 30% RH)	0.23	—	—	
Nafion/CsPW (10/1 g/g) (Cs$_{2.5}$H$_{0.5}$ PW$_{12}$O$_{40}$)						Cast; Amirinejad et al. [202]
	65	6.0 (120°C, 0% RH)	0.11	62 (25°C)	—	
		5.2 (25°C, 90% RH)	0.13	—	—	
		2.5 (25°C, 30% RH)	0.26	—	—	
Recast Nafion	62	0.2 (120°C, 0% RH)	3.1	—	33 (25°C)	
		3.6 (25°C, 90% RH)	0.17	—	—	
		1.8 (25°C, 30% RH)	0.34	—	—	

proton conductivity, water uptake, and methanol permeability data of Nafion/Inor-NP composite membranes summarized from literature [163,175,184,186,192–202].

1.3.2.3 PBI Doped with H_3PO_4 Membranes

In comparison to LT-PEMFCs (operating temperature < 100°C), HT-PEMFCs (operating temperature ~ 120°C –200°C) are advantageous due to their faster anode and cathode electrochemical reaction kinetics, greater tolerance of Pt catalyst to CO impurities in hydrogen fuel, and simple water and thermal management [203,204]. PBI, which is a hydrocarbon polymer possessing high mechanical strength, good chemical and thermal stability, and nonflammable property at high temperatures [205], was first reported in 1995 to dope with phosphoric acid for fabricating PBI/H_3PO_4 Lewis acid–base complex membranes for HT-PEMFC applications [206]. Since then, the phosphoric acid–doped PBI, poly(2,5-benzimidazole) (AB-PBI), pyridine PBI (PyPBI), and their modifications have been extensively investigated (Figure 1.23) [207–210]. In the PBI/H_3PO_4 membranes, the PBI imidazole –NH and –N=C– groups, which are the Lewis bases, provide electron donating sites for coordinating with the Lewis acid, HO–PO$_3$H$_2$. The presence of H_3PO_4 in the membranes helps proton conductance. Currently, these PBI/H_3PO_4 series membranes have shown to be more successful than other membranes in HT-PEMFCs [207].

1.3.2.3.1 Preparation of PBI Doped with H_3PO_4 Membranes

The conventional process for preparing the PBI/H_3PO_4 (or AB-PBI/H_3PO_4, PyPBI/H_3PO_4, and other H_3PO_4–doped imidazole polymers) complex membranes is doping the neat PBI (or other imidazole polymers) membranes directly into a H_3PO_4 aqueous solution. The PBI, AB-PBI, and PyPBI polymers are polymerized from the 3,3'-diamino benzidine (DABZ) with isophthalic acid (IPA) monomers,

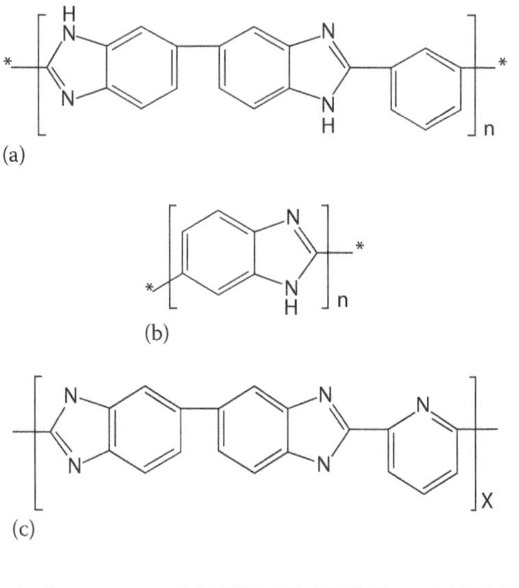

(a)

(b)

(c)

FIGURE 1.23 Chemical structures of (a) PBI, (b) AB-PBI, and (c) PyPBI.

3,4-diaminobenzoic acid monomer, and DABZ with pyridine 2,6-dicarboxylic acid monomers, respectively, using polyphosphoric acid (PPA) as the polymerization solvent [211,212]. The polymerization proceeds under a N_2 atmosphere at 200°C–220°C. After polymerization, PBI powder is isolated from the polymerization product mixture using an alkali aqueous solution and then rinsed with deionized water. The PBI powder is dissolved in DMAc solvent, and the resulting solution cast is processed to obtain neat PBI membranes. The membranes are then doped with a H_3PO_4 aqueous solution to obtain PBI/H_3PO_4 complex membranes.

Another process for fabricating PBI/H_3PO_4 complex membranes is the *PBI/PPA in situ solgel transition* process, which was developed by Xiao et al. [213]. In this process, the polymerization is carried out in a similar way as that of the conventional process, which is described in the previous paragraph. After finishing the polymerization, the PBI/PPA mixture is cast onto a glass plate at 200°C–220°C and the PPA is hydrolyzed in situ to phosphoric acid. These membranes have higher H_3PO_4 doping level than the membranes fabricated using conventional process. The membranes also show good mechanical properties, excellent conductivities, and long-term stability when operating at temperatures over 150°C. The solgel transition is induced by the hydrolysis of the PPA solvent. PPA (a good solvent for PBI) is converted to H_3PO_4 (a poor solvent of PBI) via a hydrolysis reaction following the absorption of moisture during the membrane postcasting process. The in situ PPA hydrolysis process was confirmed with a ^{31}P spectroscopy study [213].

1.3.2.3.2   Conducting Mechanism of PBI Doped with H_3PO_4 Membranes

H_3PO_4 molecules act as proton carriers in the PBI/H_3PO_4 series membranes. Li et al. suggested a useful H_3PO_4 doping level of ~300–750 wt%, taking into consideration both the factors conductivity and mechanical strength [214]. Proton conductance takes place in these membranes via hoping mechanism [26]. Figure 1.24 illustrates the H^+ ions transporting along the PBI imidazole >N– and –NH groups and the three –OH and one –P=O groups of H_3PO_4 molecules. The main difference in the proton carrier functions between the H_2O molecules in PFSI (e.g., Nafion) and SHCPE (e.g., SPEEK and SPSU) PEMs and H_3PO_4 molecules in PBI/H_3PO_4 membranes is their different proton-carrying mechanisms. The H_2O molecules behave like vehicles carrying H^+ ions (i.e., H_3O^+) and diffusing via the ionic aggregated channels in the PFSI and SHCPE membranes (Section 1.3.2.2.1, Figure 1.8). However, the H_3PO_4 molecules behave like islands cooperating with imidazole –NH and –N=C groups to provide binding sites for H^+ ions to hop in the PBI/H_3PO_4 complex membranes. No water molecules are required to conduct the protons, and thus the PBI/H_3PO_4 membranes are suitable for low-humidity HT-PEMFCs.

1.3.2.3.3   Modifications of PBI Doped with H_3PO_4 Membranes

The rigid aromatic backbone chemical structure and the strong intermolecular H-bond interactions between the –NH and –N=C groups of PBI polymers (Figure 1.25) [215] make the PBI membranes brittle, which restricts the H_3PO_4 doping level and limits the applications of PBI membranes to HT-PEMFCs. Several efforts have been exerted in modifying PBI membranes to improve their proton conductivity and mechanical properties for HT-PEMFC applications, which have been

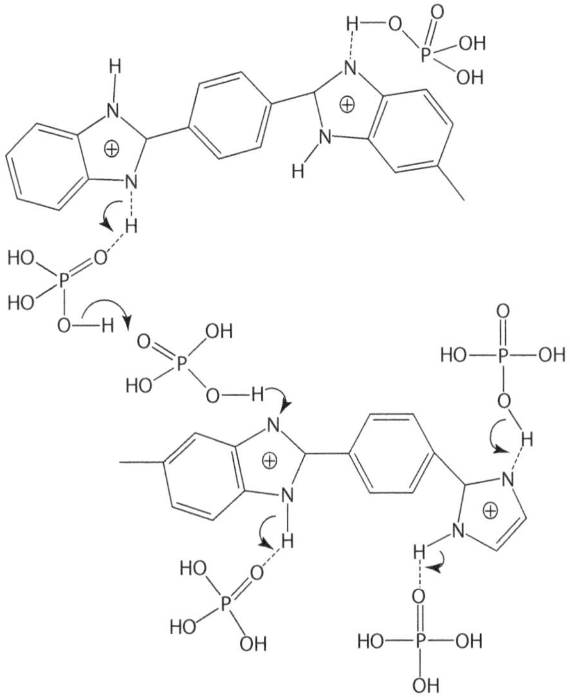

FIGURE 1.24 Proton hopping mechanism in the H_3PO_4 doped PBI membranes, in which H_3PO_4 molecules act as proton carrier. (Modified with permission from Ma, Y.L. et al., *J. Electrochem. Soc.*, 151, A8, 2004.)

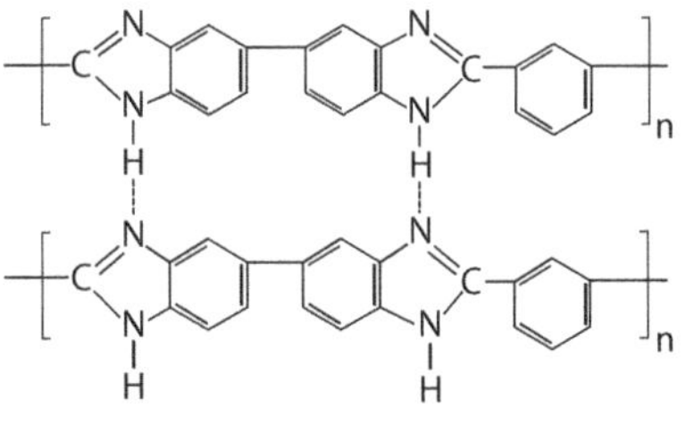

FIGURE 1.25 Interpolymer –NH and –N=C hydrogen bond interactions among PBI polymers. (Reproduced with permission from Lin, H.L. et al., *J. Power Sourc.*, 181, 228, 2008.)

reviewed by Li et al. [207], Rikakawa and Sanui [208], and Mader et al. [209] and Chandan et al. [210]. These modifications include (1) introducing sulfonated or phosphonated groups on PBI either by substituting $-OSO_2H$ or $-OPO_3H_2$ groups directly on the backbone of PBI or by grafting $-R-OSO_2H$ or $-R-OPO_3H_2$ (where R is an alkyl or an aryl group) on the –NH groups of PBI, (2) hybridizing the PBI membrane with Inor-NPs, (3) introducing interpolymer cross-links among the PBI polymer chains, and (4) reinforcing the membrane with high-mechanical-strength porous supporting films.

1.3.2.3.3.1   Chemical Modifications of PBI with Acidic Functional Groups   The chemical modifications of PBI with acidic functional groups include postsulfonation of PBI [216–219] and direct polymerization of sulfonated monomers to form sulfonated PBIs [220–224], grafting of PBI with alkyl sulfonic and aryl sulfonic substituents [208,225–235], grafting of PBI with alkyl phosphoric and aryl phosphoric substituents [235,236], grafting of PBI with imidazole substituents [237,238], and grafting of PBI with pyridine substituents [239]. The introducing of sulfonated groups; grafting of sulfonated, phosphonated, imidazole, and pyridine side chains onto the PBI backbones; and blending of these side chain grafted polymers into the PBIs reduce the PBI interpolymer hydrogen bonding (Figure 1.25) and provide more imidazole –NH and –N=C– Lewis base coordinating sites for the H_3PO_4 molecules, thus increasing the membrane phosphoric acid doping level, which leads to higher proton conductivity. The grafting of alkyl- and aryl-sulfonated or phosphonated side chains also increases the free volume of the PBI membranes, which helps the membrane to retain more excess free H_3PO_4 molecules and improves proton conductivity. The disadvantages of these sulfonate- and phosphonate-modified PBIs are (1) lower thermal stability due to the desulfonation and dephosphonation by heating [208,225] and (2) high dissolution in phosphoric acid aqueous solutions when the degree of substituted sulfonic or phosphoric acid groups is larger than 20 mol% [223,233,240,241]. Thus, optimization of the degree of sulfonic or phosphoric acid substitution is necessary for a highest fuel cell performance.

The blending of PBI with sulfonated or phosphonated polymers, such as SPSU [240] and SPI [242], has also been investigated. The Lewis acid–base interactions between the $-SO_3H$ (or $-OPO_3H_2$) groups bonded on the polymers with the imidazole >NH and >N=C groups bonded on the PBI (i.e., $-OS(=O)_2OH...N=C$, or $-OP(O_2H)OH...N=C$) [243] lead to interpolymer cross-linking and provide suitable mechanical strength to PBI membranes. The membranes exhibit moderate swelling due to the interpolymer Lewis acid–base cross-linking and suitable mechanical strengths. Compared to neat PBI membranes, the sulfonated PBI membranes display higher proton conductivity, higher solubility, and higher degree of cross-linking due to the Lewis acid–base interactions, but lower thermal decomposition temperature due to the desulfonation by heating [208,214,225]. The alkyl sulfonic and aryl sulfonic grafted PBIs exhibit more thermal stable and more flexibility in mechanical properties. Their electrochemical and mechanical properties can be controlled by the content of substituent and the lengths of alkyl and aryl side chains [207,208]. It had been shown that the alkyl sulfonic grafted PBIs have water uptake capabilities and proton conductivities similar to sulfonated PBIs without sacrificing desirable properties such as thermal stability, mechanical strength, solubility in organic solvents, and chemical resistance [208,225]. Table 1.4 summarizes the phosphoric acid doping level and conductivity of the modified PBI membranes compiled from earlier studies.

1.3.2.3.3.2   PBI/Inorganic Nanoparticle Hybrid Membranes   Compared to Nafion membrane, smaller hydrophilic and hydrophobic phase separations occur

TABLE 1.4
Conductivity (σ) of Modified PBI Membranes with Various H_3PO_4 Doping Levels (PA_{dop})

Membrane	PA_{dop} (wt%)	σ (10^{-2} S cm^{-1})	Temp °C	RH %	Reference
1. PBI	200	1.6	140	5	Ma et al. [26]
PBI	200	4.2	180	5	Ma et al. [26]
2. PBI–m–Bz–SO$_3$H (x%)—PBI grafted with –CH$_2$–C$_6$H$_4$–SO$_3$H (x% = degree of grafting)					
PBI–m–Bz–SO$_3$H (50%)	0	0.13	40	100	Glipa et al. [228]
PBI–m–Bz–SO$_3$H (75%)	0	1.3	40	100	Glipa et al. [228]
PBI	0	0.01	40	100	Glipa et al. [228]
3. PBI/SPSU-x; blend of PBI and SPSU (PBI/SPSU = 3/1 by wt), x = degree of PSU sulfonation was x mol%					Hasiotis et al. [240]
PBI/SPSU-36	159	7.5	160	80	
PBI/SPSU-70	159	10.0	160	80	
4. PBI/PBI-BS-64; blend of PBI and PBI grafted with butylsulfonate (PBI-BS) (PBI/PBI-BS = 6/4 by wt)					Lin et al. [234]
PBI/PBI-BS-64	220	3.1	160	0	

(Continued)

TABLE 1.4 (Continued)

Conductivity (σ) of Modified PBI Membranes with Various H_3PO_4 Doping Levels (PA_{dop})

Membrane	PA_{dop} (wt%)	σ (10^{-2} S cm^{-1})	Temp °C	RH %	Reference
5. bSPI (80/20)—polyimide grafted with $-C_3H_6-SO_3H$ (80% degree of sulfonation); oPBI = ether-PBI; PA = phosphoric acid)					
bSPI (80/20)	0	$<10^{-2}$	120	30	Iizuka et al. [242]
bSPI (80/20)/oPBI	0	$<10^{-2}$	120	30	
PA-bSPI (80/20)/oPBI	70	4.5	120	30	
6. PBI-BI (x%) benzimidazole grafted with PBI (degree of grafting x%)					Yang et al. [237]
PBI-BI (16.5%)	22.5 mol/mol	29.5	180	0	
PBI-BI (5.3%)	13.1 mol/mol	15.2	180	0	
PBI	10.7 mol/mol	8.1	180	0	
7. pPBI-BI = benzimidazole grafted with para-PBI; mPBI = meta-PBI					
pPBI-BI	81	16	180	0	Kim et al. [238]
mPBI	65	6.0	180	0	
8. OH-PyPBI = PBI containing hydroxyl pyridine monomer					
OH-PyPBI	8.6	10.2	180	0	Yang et al. [239]
mPBI	7.6	7.5	180	0	
9. PBI-EP-12 = PBI grafted with ethyl phosphate; degree grafting was 12 mol%					
PBI-EP-12	255	7.0	140	0	Sukurmar et al. [241]

Temp and RH stand for temperature and relative humidity for measurements.

in PBI membrane. Most of researchers prepared PBI/Inor-NP hybrid membranes using casting process rather than in situ process. Inor-NP/PBI hybrid membranes have been prepared using zirconium phosphate [244], phosphotungstic acid [244], silicotungstic acid (STA) [244], phosphotungstic acid on SiO_2 support [245] titanium oxide [246], maleimide-modified silica nanoparticle [247], tetraethoxy silane (TEOS) [248], diethyl phosphate ethyl-triethoxysilane (DEPE-TEOS) [249], silicon oxide [137], mesoporous silica [250,251], sulfonated silica [252], and sulfonated graphite oxide [253]. After doping with H_3PO_4, these hybrid membranes exhibit higher phosphoric acid doping level and thus higher proton conductivity than the neat PBI membranes, due to the Inor-NPs that help the membranes to retain H_3PO_4. The hybridization of the Inor-NP also improves the membrane mechanical strength. However, a large content of Inor-NPs in the hybrid membranes causes brittleness in the membranes and reduces their long-time operating durability. Thus, an optimum content of Inor-NP is necessary for a hybrid membrane to have a highest fuel cell performance. Table 1.5 summarizes the Inor-NP content, H_3PO_4 doping level, and conductivity of several Inor-NP/PBI hybrid membranes compiled from previous studies.

1.3.2.3.3.3 Cross-Linked PBI Membranes The mechanical properties of the PBI membranes can be improved by introducing a proper degree of interpolymer cross-linking in the membranes. Two chemical modification processes for preparing cross-linked PBI membranes have been reported. One is the ionic cross-linking using the Lewis acid–base complex polymer concept [243] by hybridizing PBI with the sulfonated polymers or phosphonated polymers (mentioned in the previous Section 1.3.2.3.3.1). The other is the covalent cross-linking in that PBI membranes are prepared by blending PBI with cross-linkers that possess reacting functional groups such as epoxide resins [254–257], dibromopropane [256], para-xylene [258,259], silane–epoxide resin [260], divinyl sulfone [261], and dichloromethyl phosphoric acid [262] and perform covalent cross-linking reaction between the PBI imidazole groups and the cross-linkers. Compared to ionic cross-link PBIs, the covalent cross-link PBIs exhibit better thermal stabilities, higher mechanical strength, lower free volume for retaining H_3PO_4 due to the restricted motion of polymer chains, and lower proton conductivity. However, the higher mechanical strength of the covalent cross-link PBI membranes allows them to have a lower thickness than the non-cross-link PBI and ionic cross-link membranes when they are used for PEMFC applications without lowering the membrane's H_2/O_2 (or air) gas barrier property. The lower thickness of the membrane causes a lower proton transport resistance ($R \sim L/\sigma$) across the membrane. Table 1.6 summarizes the mechanical properties, H_3PO_4 doping level, and conductivity of several cross-linked PBI membranes compiled from earlier studies.

1.3.2.3.3.4 Reinforcement of PBI Membranes with High-Mechanical-Strength Supporting Films Similar to fiber-reinforced PFSI proton exchange membranes (PEMs) (Section 1.3.2.2.5.1), porous PTFE as supporting films for PBI [263,264] and quaternized polysulfone (QNPSU; poly(SU–$R_1R_2R_3$)–N^+) [265,266] composite membranes for HT-PEMFCs have also been reported. Lin et al., because of poor compatibility of PTFE with PBI, prepared PBI/PTFE composite membranes using

TABLE 1.5

Mechanical Properties (Undoped Membranes), H₃PO₄ Doping Levels (PA_{dop}), and Conductivities (σ) of Inor-NP

Membrane	Thickness L (μm)	Tensile Stress (MPa) Undoped	Elongation (%) Undoped	PA_{dop} wt% [mol PA]/[mol Repeat Unit]	σ (10^{-2} S cm^{-1})	Temp °C	RH %	Preparation Process and Reference
1. PBI/Zr(HPO₄)₂ blend								
PBI	—	—	—	5.6 mol/mol	5.7	170	10	Cast; He et al. [244]
PBI/Zr(HPO₄)₂ (85/15)	—	—	—	5.6 mol/mol	7.4	170	10	
PBI/Zr(HPO₄)₂ (80/20)	—	—	—	5.6 mol/mol	6.7	170	10	
2. PBI/PWA blend; (PWA = H₃PW₁₂O₄₀)								
PBI	—	—	—	4.4 mol/mol	5.2	170	5	Cast; He et al. [244]
PBI/PWA (7/3)	—	—	—	4.4 mol/mol	4.1	170	5	
3. PBI/SiWA blend (SiWA = H₄SiW₁₂O₄₀)								
PBI	—	—	—	4.4 mol/mol	5.2	170	5	Cast; He et al. [244]
PBI/SiWA (7/3)	—	—	—	5.1 mol/mol	3.6	170	5	
4. PBI/[SiWA/SiO₂] blend; (SiWA = H₄SiW₁₂O₄₀; SiWA/SiO₂ = SiWA on SiO₂ support; SiWA/SiO₂ = 45/55 g/g)								
PBI/[SiWA/SiO₂] (70/30 g/g)	—	—	—	0	0.08	160	100	Cast; Staiti [245]
PBI/[SiWA/SiO₂] (50/50 g/g)	—	—	—	0	0.10	160	100	
PBI/[SiWA/SiO₂] (30/70 g/g)	—	—	—	0	0.20	160	100	
5. PBI/TiO₂ blend								
PBI	—	—	—	11.3 mol/mol	5.7	150	0	Cast; Lobato et al. [246]
PBI	—	—	—	11.3 mol/mol	4.8	175	0	
PBI/TiO₂ (98/2 g/g)	—	—	—	15.3 mol/mol	12.7	150	0	
PBI/TiO₂ (98/2 g/g)	—	—	—	15.3 mol/mol	11.7	175	0	

(Continued)

TABLE 1.5 (Continued)
Mechanical Properties (Undoped Membranes), H_3PO_4 Doping Levels (PA_{dop}), and Conductivities (σ) of Inor-NP

Membrane	Thickness L (μm)	Tensile Stress (MPa) Undoped	Elongation (%) Undoped	PA_{dop} wt% [mol PAI/mol Repeat Unit]	σ (10^{-2} S cm^{-1})	Temp °C	RH %	Preparation Process and Reference
6. PBI/SNP-PBI blend (SNP-PBI = silica nanoparticle modified with N-(p-carboxyphenyl)maleimide and PBI on particle surface)								
PBI	—	78	30.0	416 wt%	3.9	160	0	Cast; Suryani et al. [247]
PBI/SNP-PBI (95/5 g/g)	—	85	10.0	419 wt%	2.0	160	0	
PBI/SNP-PBI (90/10 g/g)	—	99	17.3	385 wt%	5.0	160	0	
PBI/SNP-PBI (85/15 g/g)	—	90	5.2	381 wt%	4.7	160	0	
Tensile tests of undoped membranes were performed at 25°C, an ambient environment.								
7. PBI/TEOS blend (TEOS)								
PBI	—	90	27	—	—	—	—	Cast solgel; Ossiander et al. [248]
PBI/TEOS (4/10 g/g)	—	136	53	—	—	—	—	
PBI/TEOS (8/10 g/g)	—	127	51	—	—	—	—	
PBI/TEOS (12/10 g/g)	—	115	41	—	—	—	—	
Tensile tests of undoped membranes were performed at room temperature.								
8. PBI/DEPE-TEOS blend; (DEPE-TEOS)								
PBI	70	6.6	5.5	6.3 mol/mol	2.0	160	0	Cast solgel; Lin et al. [249]
PBI/DEPE-TEOS (75/25)	77	7.0	7.1	6.9 mol/mol	3.3	160	0	
PBI/DEPE-TEOS (50/50)	72	7.2	6.4	7.3 mol/mol	3.7	160	0	
PBI/DEPE-TEOS (25/75)	74	7.4	2.8	8.6 mol/mol	4.3	160	0	
Tensile tests of undoped membranes were performed at 25°C and ~45% RH.								

Modified PBI membranes. Temp and RH stand for temperature and relative humidity for measurements.

TABLE 1.6
Mechanical Properties (Undoped Membranes), H₃PO₄ Doping Levels (PA_{dop}), and Conductivities (σ) of Cross-Linked PBI Membranes

Membrane	Tensile Stress (MPa) Undoped	Elongation (%) Undoped	PA_{dop} (wt%) mol H₃PO₄/mol Repeat Unit	σ (10^{-2} S cm⁻¹)	Temp °C	RH %	Reference
1. Epoxide cross-linked PBI							
(1.1) PBI	77.8	8.0	5.97 mol/mol	1.4	160	0	Wang et al. [254]
Epx-C-PBI-5%	97.7	26.4	4.99	1.3	160	0	
Epx-C-PBI-10%	101.6	32.7	4.97	0.84	160	0	
Epx-C-PBI-15%	104.2	8.9	4.86	0.70	160	0	
Epx = 1,3-bis(2,3-epoxypropoxy)-2,2-dimethylpropane (NGDE); 5%, 10%, and 15% are the degrees of cross-linking							
(1.2) pPBI	43.7	31.7	291 wt%	3.8	160	0	Wang et al. [255]
Epx-C-pPBI-10%	79.1	21.5	406	3.1	160	0	
Epx-C-pPBI-15%	79.5	18.6	324	—	—	—	
pPBI = porous PBI prepared using 50 wt% porogen							
Epx = 4,4′-diglycidyl (3,3′,5′-tetramethylbiphenyl); 10% and 15% are the degrees of cross-linking							
(1.3) Epx-C-A-PBI (100% –NH₂)	72	7.0	Brittle	—	—	—	Xu et al. [256]
Epx-C-A-PBI (66% –NH₂)	70	45	160	—	—	—	
Epx-C-A-PBI (50% –NH₂)	100	56	210	2.5	160	0	
A-PBI = amino pendent PBI (PBI synthesized from 3,3′-diaminobenzidine, 5-aminoisophthalic acid, and IPA)							
Epx = ethylene glycol diglycidyl ether cross-linker							

(*Continued*)

TABLE 1.6 (Continued)

Mechanical Properties (Undoped Membranes), H₃PO₄ Doping Levels (PA_{dop}), and Conductivities (σ) of Cross-Linked PBI Membranes

Membrane	Tensile Stress (MPa) Undoped	Elongation (%) Undoped	PA_{dop} (wt%) mol H₃PO₄/mol Repeat Unit	σ (10^{-2} S cm⁻¹)	Temp °C	RH %	Reference
2. 1,3-dibromopropane cross-linked A-PBI							
Dbp-C-A-PBI (100% –NH₂)	55	3.3	120	—	—	—	Xu et al. [256]
Dbp-C-A-PBI (66% –NH₂)	58	13	240	7.0	160	0	
Dbp-C-A-PBI (50% –NH₂)	74	22	290	2.5	160	0	

A-PBI = amino pendent PBI (PBI synthesized from 3,3'-diaminobenzidine, 5-aminoisophthalic acid, and IPA)

Dbp = 1,3-dibromopropane cross-linker

Membrane	Tensile Stress (MPa) Undoped	Elongation (%) Undoped	PA_{dop} (wt%) mol H₃PO₄/mol Repeat Unit	σ (10^{-2} S cm⁻¹)	Temp °C	RH %	Reference
3. Para-xylene cross-linked PBI							
(3.1) Px-C-PBI-13			283	4.8	150	3	Li et al. [258]
Px-C-PBI-13			283	6.3	180	3	

Px = para-xylene dibromide cross-linker; PBI-13 = PBI containing 13 mol% of cross-link

Membrane	Tensile Stress (MPa) Undoped	Elongation (%) Undoped	PA_{dop} (wt%) mol H₃PO₄/mol Repeat Unit	σ (10^{-2} S cm⁻¹)	Temp °C	RH %	Reference
(3.2) Para-xylene cross-linked pPBI							
Pristine PBI	73	—	4.7 mol/mol	0.80	160	0	Shen et al. [259]
Px-C-pPBI-30	35	—	7.2	1.2	160	0	
Px-C-pPBI-45	36	—	12.5	2.0	160	0	
Px-C-pPBI-60	50	—	13.8	3.0	160	0	

Px = para-xylene dichloride cross-linker; pPBI = porous PBI with 36–38% porosity.

Px-C-pPBI-30, Px-C-pPBI-45, and Px-C-pPBI-60 are cross-linked porous PBIs containing 30, 45, and 60 mol% of cross-link, respectively.

(Continued)

TABLE 1.6 (Continued)
Mechanical Properties (Undoped Membranes), H$_3$PO$_4$ Doping Levels (PA_{dop}), and Conductivities (σ) of Cross-Linked PBI Membranes

Membrane	Tensile Stress (MPa) Undoped	Elongation (%) Undoped	PA_{dop} (wt%) mol H$_3$PO$_4$/mol Repeat Unit	σ (10^{-2} S cm^{-1})	Temp °C	RH %	Reference
4. Silane–epoxy cross-linked PBI							
PBI	132	11.4	4.6 mol/mol	1.3	160	0	Solution cast; Wang et al. [260]
SiE-C-PBI-1	182	26.5	7.3	5.6	160	0	
SiE-C-PBI-3	174	22.8	8.0	6.2	160	0	
SiE-C-PBI-5	172	17.0	7.2	4.4	160	0	
SiE = γ-(2,3-epoxypropoxy) propyltrimethoxysilane cross-linker							
SiE-C-PBI-1, SiE-C-PBI-3, and SiE-C-PBI-5 are cross-linked PBIs containing 1%, 3%, and 5% of cross-link, respectively							
5. Divinylsulfone cross-linked PBI							
PBI	122 (180°C)	42	~10 mol/mol	10.2	—	—	Aili et al. [261]
DVS-C-PBI-7	103 (180°C)	27	~10	9.4	—	—	
DVS = divinyl sulfone cross-linker							
DVS-C-PBI-7 = divinyl sulfone cross-linked PBI containing 7% degree of cross-linking							
6. Dichloromethyl phosphonic acid cross-linked PBI							
PBI	82	38	6 mol/mol	4.2	180	4	Noye et al. [262]
DCMP-C-PBI-20	15	124	8	5.9	—	—	

Tensile tests that were performed at 20°C

DCMP = dichloromethyl phosphonic acid cross-linker; DCMP-C-PBI-20 = 20% DCMP cross-linked PBI

Temp and RH stand for temperature and relative humidity for σ measurements.

Nafion as a coupling agent [263,264]. The chemical structure of Nafion consists of a perfluorocarbon main chain, which is compatible with the PTFE supporting thin film, and ether sulfonic acid side chains, which have $-SO_3H$ groups for interaction with PBI imidazole groups via the Lewis acid–base interaction. A good fuel cell performance for the PBI/PTFE and QNPSU/PTFE composite membranes doped with H_3PO_4 has also been demonstrated when applied to HT-PEMFCs (temp = 150°C–170°C) with low-humidity H_2/O_2 fuel gases. Owing to the high mechanical strength of the reinforced supporting material, the thickness of the composite membranes can be reduced to ~35–40 μm, and thus the proton transport resistance across the composite membrane can be reduced and enhance the fuel cell performance [263–266].

1.4 POLYMER ELECTROLYTE MEMBRANE APPLICATIONS IN ELECTROCHEMICAL DEVICES

The theoretical energy density for a lithium secondary battery is around 2–3 times higher than for Ni-Cd or Pd-acid cells. Polymer electrolyte-lithium batteries have advantages of low cost and high density. They are widely used as power sources of electronic devices including small portable electronics (e.g., video camera, headphone stereo, portable telephone, portable CD player, notebook computer, and personal communication equipment), electric vehicle, and start–light–ignition. The essential requirements for these applications are low cost, high energy density both in volume and weight, safety, robustness, reliability, long cycle life, simplified system, and manufacturability.

The PEMFCs generate power from a watt to hundred kilowatts. They can be used in applications of local power generation including LT-PEMFC in transportation, LT-PEMFC and HT-PEMFC in stationary power generator, LT-PEMFC and HT-PEMFC in backup power, and DMFC and LT-PEMFC in portable power generator. The main advantages for using the PEMFCs as power generators are the high efficiency and low emissions. The main disadvantages are the cost of fuel cells and the cost and availability of hydrogen. These disadvantages are also the obstacles for the commercialization of PEMFCs.

The transportation applications of PEMFCs include automobiles, buses, scooters, bicycles, and utility vehicles (e.g., forklifts, material-handling industrial vehicles, golf cars, lawn maintenance vehicles, and airport people movers). These applications have already been demonstrated. Besides the high cost of fuel cells, lack of hydrogen infrastructure (hydrogen production and distribution) is also a big obstacle for the commercialization of fuel cell automobiles and buses. However, several world's leading car companies, for example, Mercedes Benz, Honda, Toyota, Hyundai, and General Motors, have made big progress on both the fuel cell–powered automobile technology and the cost reduction. There are already thousands of hydrogen refueling stations in the United States, Canada, Japan, and Western Europe. The fuel cell–powered utility vehicles have advantages of zero emission, low noise, suitability for indoor usage, and large extended range.

The PEMFC stationary power systems also have advantages of low noise and low emission and are suitable for indoor stationary power applications. The stationary fuel cell–powered system has another advantage for people to install distributed power generation systems. The PEMFC is also suitable for backup-power (or an uninterruptable power supply) applications. It is a device with a power level of around 1–100 kW and provides instantaneous, uninterruptable power for no more than 30–60 min. The essential requirement for the backup powers is the ability to start instantly upon power outage within a few milliseconds. The key features for the application of small fuel cells to portable power systems are small size, light weight, long running time, and simplified system. The critical issues for the portable fuel cell power systems are the fuel and the fuel storage. Because of bulkiness and weight of the hydrogen gas storage tank, most portable fuel cells use methanol as fuel either directly (so-called DMFCs) or via microreformers.

1.5   CHAPTER SUMMARY

Two main classes of PEMs for Li batteries and fuel cell applications are discussed in this chapter. One is the Lewis acid–base coordinated PEMs and the other is sulfonated PEMs. The two main Lewis acid–base complex PEMs are PEO–Li salt and PBI/H_3PO_4 for Li batteries and HT-PEMFCs, respectively. The main sulfonated PEMs for LT-PEMFCs are perfluorosulfonated PEMs (e.g., DuPont Nafion PEMs) and sulfonated hydrocarbon PEMs (e.g., SPEEK, SPSU, and SPI). The conductivity of the Lewis acid–base complex PEMs results from the hopping of the Lewis acid cations among the vacant Lewis basic electron donating sites inside the PEMs. Because of low T_g of PEO ($T_g \sim -60°C$), the PEO segmental relaxation motion helps the motion of Li^+ ions and enhances the conductance of PEO–Li-salt PEMs. However, the cation hopping is the major contribution to the conductivity in the PBI/H_3PO_4 PEMs. Because of the much higher T_g of PBI ($T_g \sim 400°C$) than the HT-PEMFC operating temperature (120°C–200°C) and the rigidity of the PBI backbones, no local segmental relaxation motion of PBI occurs in the PBI/H_3PO_4 PEMs. No additional low-molecular-weight solvent is needed to enhance the conductivity of the Lewis acid–base complex PEMs. However, the sulfonated PEMs are usually swollen with low-molecular-weight water molecules, which act as proton carriers (i.e., vehicles) and combine with H^+ ions to form H_3O^+ ions. The transportation of the vehicles, that is, the diffusion of H_3O^+ ions in the hydrophilic ionic aggregated regions, results in the conductivity of the sulfonated PEMs. Nafion, a sulfonated PEM, is currently the most widely used PEM for LT-PEMFCs. It needs water for proton conductivity and suffers with the problems of high methanol crossover when working in DMFCs and low proton conductivity when working in a low-humidity environment and at temperatures above 90°C. Most of the modifications of Nafion PEMs are concentrated on (1) enhancing the mechanical strength and reducing the membrane thickness and thus reducing the resistance of the membranes, (2) improving the water retention ability to enhance the proton conductivity for working at low-humidity environment and at temperatures above 90°C, and (3) reducing methanol crossover without losing proton conductivity for DMFC applications.

REFERENCES

1. Gary F.M., *Solid Polymer Electrolytes—Fundamentals and Technological Applications*, VCH Publishers, New York, 1991.
2. Gary F.M., *Polymer Electrolytes*, The Royal Society of Chemistry, Cambridge, U.K., 1997.
3. Eisenberg, A. and Rinaldo, M. (1990) Polyelectrolytes and ionomers, *Polym. Bull.*, 24, 671.
4. Dautzenberg, H., Jaeger, W., Kotz, J., Philipp, B., Seidel, Ch., and Stscherbina, D., *Polyelectrolytes: Formation, Characterization and Application*, Hanser Publisher, Munich, Germany, 1994.
5. Tanford, C., *Physical Chemistry of Macromolecules*, John Wiley & Sons, Inc., New York, 1961, Chapter 7.
6. Hara, M., *Polyelectrolytes: Science and Technology*, Marcel Dekker, Inc., New York, 1993.
7. Fenton, D.E., Parker, J.M., and Wright, P.V. (1973) Complexes of alkali metal ions with poly(ethylene oxide), *Polymer*, 14, 589–589.
8. Meyer, W.H. (1998) Polymer electrolytes for lithium-ion batteries. *Adv. Mater.*, 10, 439–448.
9. Appetecchi, G.B. and Passerini, S. (2002) Poly(ethylene oxide)-LiN($SO_2CF_2CF_3$)$_2$ polymer electrolytes, *J. Electrochem. Soc.*, 149, A891–A897.
10. Villano, P., Carewska, M., Appetecchi, G.B., and Passerini, S. (2002) PEO-LiN($SO_2CF_2CF_3$)$_2$ polymer electrolytes, *J. Electrochem. Soc.*, 149, A1282–A1285.
11. Appetecchi, G.B., Zane, D., and Scrosati, B. (2004) PEO-based electrolyte membranes based on $LiBC_4O_8$ salt, *J. Electrochem. Soc.*, 151, A1369–A1374.
12. Fan, L., Wang, X., Long, F., and Wang, X. (2008) Enhanced ionic conductivities in composite polymer electrolytes by using succinonitrile as a plasticizer, *Solid State Ionics*, 179, 1772–1775.
13. Chen-Yang, Y.W., Wang, Y., Chen, Y., Li, Y., Chen, H., and Chiu, H. (2008) Influence of silica aerogel on the properties of polyethylene oxide-based nanocomposite polymer electrolytes for lithium battery, *J. Power Sourc.*, 182, 340–348.
14. Shen, C., Wang, J., Tang, Z., Wang, H., Lian, H., Zhang, J., and Cao, C. (2009) Physicochemical properties of poly(ethylene oxide)-based composite polymer electrolytes with a silane-modified mesoporous silica SBA-15, *Electrochim. Acta*, 54, 3490–3494.
15. An, S., Jeong, I., Won, M., Jeong, E., and Shim, Y. (2009) Effect of additives in PEO/PAA/PMAA composite solid polymer electrolytes on the ionic conductivity and Li ion battery performance, *J. Appl. Electrochem.*, 39, 1573–1578.
16. Marzantowicz, M., Dygas, J.R., Krok, F., Tomaszewska, A., Florjańczyk, Z., Zygadło-Monikowska, E., and Lapienis, G. (2009) Star-branched poly(ethylene oxide) LiN(CF_3SO_2)$_2$: A promising polymer electrolyte, *J. Power Sourc.*, 194, 51–57.
17. Ndeugueu, J.L., Ikeda, M., and Aniya, M. (2010) Correlation between the temperature range of cooperativity and the fragility index in ion conducting polymers, *Solid State Ionics*, 181, 16–19.
18. Kim, G.T., Appetecchi, G., Carewska, M., Joost, M., Balducci, A., Winter, M., and Passerini, S. (2010) UV cross-linked, lithium-conducting ternary polymer electrolytes containing ionic liquids, *J. Power Sourc.*, 195, 6130–6137.
19. Yang, M. and Hou, J. (2012) Membranes in lithium ion batteries, *Membranes*, 2, 367–383.
20. Orinakova, R., Fedorkova, A., and Orinak, A. (2013) Effect of Ppy/PEG conducting polymer film on electrochemical performance of $LiFePO_4$ cathode material for Li-ion batteries, *Chem. Paper*, 67, 860–875.
21. Miessler, G.L. and Tarr, D.A., *Inorganic Chemistry*, 3rd edn., Prentice Hall, Upper Saddle River, NJ, 2004, Chapter 6.

22. Van Krevelen, D.W. and Hoftyzer, P.J., *Properties of Polymers*, Elsevier Scientific Publishing Co., Amsterdam, the Netherlands, 1976.
23. Barbir, F., *PEM Fuel Cells: Theory and Practice*, Elsevier Academic Press, Burlington, MA, 2005.
24. Larminie, J. and Dicks, A., *Fuel Cell Systems Explained*, John Wiley & Sons, Ltd., Chichester, U.K., 2000.
25. Eikerling, M., Koryshev, A.A., and Spohr, E., Proton-conducting polymer electrolyte membranes: Water and structure in charge, in *Fuel Cell Handbook-I*, Ed. Scherer, G.G., Advances in Polymer Science, Vol. 215, Springer-Verlag, Berlin, Germany, 2008, pp. 15–54.
26. Ma, Y.L., Wainright, J.S., Litt, M.H., and Savinell, R.F. (2004) Conductivity of PBI membranes for high-temperature polymer electrolyte fuel cells, *J. Electrochem. Soc.*, 151(1), A8–A16.
27. Yi, S.Z., Zhang, F.F., Li, W., Huang, C., Zhang, H.N., and Pan, M. (2011) Anhydrous elevated-temperature polymer electrolyte membranes based on ionic liquids, *J. Membr. Sci.*, 366, 349–355.
28. Yoshitake, M. and Watakabe, A., Perfluorinated ionic polymers for PEFCs (including supported PFSA), in *Fuel Cell Handbook-I*, Ed. Scherer, G.G., Advances in Polymer Science, Vol. 215, Springer-Verlag, Berlin, Germany, 2008, pp. 127–155.
29. Haubold, H.G., Vad, Th., and Jungbluth, J.P. (2001) Nano structure of Nafion: A SAXS study, *Electrochim. Acta*, 46, 1559–1563.
30. Yang, Y., Siu, A., Peckham, T.J., and Holdcroft, S., Structure and morphological features of acid-bearing polymers for PEM fuel cells, in *Fuel Cell Handbook-I*, Ed. Scherer, G.G., Advances in Polymer Science, Vol. 215, Springer-Verlag, Berlin, Germany, 2008, pp. 55–126.
31. Sperling, L.H., *Introduction to Physical Polymer Science*, 4th edn., Wiley-Interscience, Hoboken, NJ, 2006.
32. Peighambardoust, S.J., Rowshanzamir, S., and Amjadi, M. (2010) Review of the proton exchange membranes for fuel cell applications, *Int. J. Hydrogen Energ.*, 35, 9349–9384.
33. Grot, W.G., Nafion as a separator in electrolytic cells, *Nafion Perfluorinated Membranes Product Bulletin*, DuPont Co., Wilmington, DE, 1986.
34. Grot, W.G., Process for making articles coated with a liquid composition of perfluorinated ion exchange resin, US Patent 4453991, 1986.
35. Lousenberg, R.D. (2005) Molar mass distributions and viscosity behavior of perfluorinated sulfonic acid polyelectrolyte aqueous dispersions, *J. Polym. Sci. Part B Poly. Phys.*, 43, 421–428.
36. Liu, W.H., Yu, T.L., and Lin, H.L. (2007) Static light scattering and transmission microscopy study of dilute Nafion solutions, *e-Polymers*, no. 109, 1–8.
37. Yeo, R. (1980) Dual cohesive energy densities of perfluorosulfonic acid (Nafion) membrane, *Polymer*, 21, 432–435.
38. Mauritz, K.A. and Moore, B. (2004) State of understanding of Nafion, *Chem. Rev.*, 104, 4535–4585.
39. Roche, E.J., Pineri, M., Duplessix, R., and Levelut, A.M. (1981) Small-angle scattering studies of NAFION membranes, *J. Polym. Sci. Polym. Phys. Ed.*, 19, 1–11
40. Gierke, T.D., Munn, G.E., and Wilson, F.C. (1981) Morphology in Nafion perfluorinated membrane products, as determined by wide- and small-angle X-ray studies, *J. Polym. Sci. Polym. Phys. Ed.*, 19, 1687–1704.
41. Hsu, W.Y. and Gierke, T.D. (1983) Ion transport and clustering in Nafion perfluorinated membranes, *J. Membr. Sci.*, 13, 307–326.
42. Fujimura, M., Hashimoto, T., and Kawai, H. (1981) Small-angle x-ray scattering study of perfluorinated ionomer membranes. 1. Origin of two scattering maxima, *Macromolecules*, 14, 1309–1315.

43. Fujimura, M., Hashimoto, T., and Kawai, H. (1982) Small-angle x-ray scattering study of perfluorinated ionomer membranes. 2. Models for ionic scattering maximum, *Macromolecules*, 15, 136–144.

44. Kumar, S. and Pineri, M. (1986) Interpretation of small angle X-ray and neutron scattering data for perfluorosulfonated ionomer membranes, *J. Polym. Sci. Polym. Phys. Ed.*, 24, 1767–1782.

45. Halim, J., Scherer, G.G., and Stamm, M. (1994) Characterization of recast Nafion films by small- and wide-angle X-ray scattering, *Macromol. Chem. Phys.*, 195, 3783–3788.

46. Elliot, J.A., Hanna, S., Elliot, A.M.S., and Cooley, G.E. (2000) Interpretation of the small-angle X-ray scattering from swollen and oriented perfluorinated ionomer membranes, *Macromolecules*, 33, 4161–4171.

47. Gebel, G. and Lambard, J. (1997) Small-angle scattering study of water-swollen perfluorinated ionomer membranes, *Macromolecules*, 30, 7914–7920.

48. Gebel, G. (2000) Structural evolution of water swollen perfluorosulfonated ionomers from dry membrane to solution, *Polymer*, 41, 5829–5838.

49. Starkweather, H.W. (1982) Crystallinity in perfluorosulfonic acid ionomers and related polymers, *Macromolecules*, 15, 320–3.

50. Rubatat, L., Rollet, A.L., Gebel, G., and Diat, O. (2002) Evidence of elongated polymeric aggregates in Nafion, *Macromolecules*, 35, 4050–4055.

51. Rubatat, L., Gebel, G., and Diat, O. (2004) Fibrillar structure of nafion: Matching fourier and real space studies of corresponding films and solutions, *Macromolecules*, 37, 7772–7783.

52. Gebel, G. and Diat, O. (2005) Neutron and X-ray scattering: Suitable tools for studying ionomer membranes, *Fuel Cells*, 5, 261–276.

53. Xue, T., Trent, J.S., and Osseo-Asare, K. (1989) Characterization of Nafion membranes by transmission electron microscopy, *J. Membr. Sci.*, 45, 261–271.

54. Porat, Z., Fryer, J.R., Huxham, M., and Rubinstein, I. (1995) Electron microscopy investigation of the microstructure of Nafion films, *J. Phys. Chem.*, 99, 4667–4671.

55. McLean, R.S., Doyle, M., and Sauer, B.B. (2000) High-resolution imaging of ionic domains and crystal morphology in ionomers using AFM techniques, *Macromolecules*, 33, 6541–6550.

56. Lin, H.L., Yu, T.L., and Han, F.H. (2006) A method for improving ionic conductivity of Nafion membranes and its application to PEMFC, *J. Polym. Res.*, 13, 379–385.

57. Aldebert, P., Gebel, G., Loppinet, B., and Nakamura, N. (1995) Polyelectrolyte effect in perfluorosulfonated ionomer solutions, *Polymer*, 36, 431–434.

58. Aldebert, P., Dreyfus, B., and Pineri, M. (1986) Small-angle neutron scattering of perfluorosulfonated ionomers in solution, *Macromolecules*, 19, 2651–2653.

59. Loppinet, B., Gebel, G., and Williams, C.E. (1997) Small-angle scattering study of perfluorosulfonated ionomer solutions, *J. Phys. Chem. B*, 101, 1884–1892.

60. Szajdzinska-Pietek, E. and Schlick, S. (1994) Self-assembling of perfluorinated polymeric surfactants in water. Electron spin resonance spectra of nitroxide spin probes in Nafion solutions and swollen membranes, *Langmuir*, 10, 1101–1109.

61. Szajdzinska-Pietek, E. and Schlick, S. (1994) Self-assembling of perfluorinated polymeric surfactants in nonaqueous solvents. Electron spin resonance spectra of nitroxide spin probes in Nafion solutions and swollen membranes, *Langmuir*, 10, 2188–2196.

62. Li, H. and Schlick, S. (1995) Effect of solvents on phase separation in perfluorinated ionomers, from electron spin resonance of VO_2^+ in swollen membranes and solutions, *Polymer*, 36, 1141–1146.

63. Cirkel, P.A., Okada, T., and Kinugasa, S. (1999) Equilibrium aggregation in perfluorinated ionomer solutions, *Macromolecules*, 32, 531–533.

64. Jiang, S., Xia, K.Q., and Xu, G. (2001) Effect of additives on self-assembling behavior of Nafion in aqueous media, *Macromolecules*, 34, 7783–7788.

65. Lee, S.J., Yu, T.L., Lin, H.L., Liu, W.H., and Lai, C.L. (2004) Solution properties of Nafion in methanol/water mixture solvents, *Polymer*, 45, 2853–2862.

66. Ma, C.H., Yu, T.L., Lin, H.L., Huang, Y.T., Chen, Y.L., Jeng, U.S., Lai, Y.H., and Sun, Y.S. (2009) Morphology and properties of Nafion membranes prepared by solutions casting, *Polymer*, 50, 1764–1777.

67. Ngo, T.T., Yu, T.L., and Lin, H.L. (2013) Influence of the composition of isopropyl alcohol/water mixture solvents in catalyst ink solutions on proton exchange membrane fuel cell performance, *J. Power Sourc.*, 225, 293–303.

68. Ngo, T.T., Yu, T.L., and Lin, H.L. (2013) Nafion-based membrane electrode assemblies prepared from catalyst inks containing alcohol/water solvent mixtures, *J. Power Sourc.*, 238, 1–10.

69. Martin, C.R., Rhoades, T.A., and Ferguson, J.A. (1982) Dissolution of perfluorinated ion containing polymers, *Anal. Chem.*, 54, 1639–1641.

70. Moore, R.B. and Martin, C.R. (1986) Procedures for preparing solution-cast perfluorosulfonate ionomer films and membranes, *Anal. Chem.*, 58, 2569–2570.

71. Moore, R.B. and Martin, C.R. (1988) Chemical and morphological properties of solution cast perfluorosulfonate ionomers, *Macromolecules*, 21, 1334–1339.

72. Weber, J., Janda, P., Kavan, L., and Jegorov, A. (1986) Electrochemical, IR and XPS study of Nafion films prepared from hexamethylphosphortriamide solution, J. *Electroanal. Chem.*, 199, 81–92.

73. Weber, J., Janda, P., Kavan, L., and Jegorov, A. (1986) Study of Nafion films on electrodes prepared from dimethyl acetamide solution, *J. Electroanal. Chem.*, 200, 379–381.

74. Tsatsas, A.T. and Risen, J.R.W.M. (1993) Studies on solution cast perfluorocarbonsulfonic acid ionomers, *J. Polym. Sci. B Polym. Phys.*, 31, 1223–1227.

75. Zook, L.A. and Leddy, J. (1996) Density and solubility of Nafion: Recast, annealed, and commercial films, *Anal. Chem.*, 68, 3793–3796.

76. de Almeida, S.H. and Kawano, Y. (1999) Thermal behavior of Nafion membranes, *J. Therm. Anal. Calorim.*, 58, 569–577.

77. Laporta, M., Pegorato, M., and Zanderighi, L. (2000) Recast Nafion-117 thin film from water solution, *Macromol. Mater. Eng.*, 282, 22–29.

78. Siroma, Z., Fujiwara, N., Ioroi, T., Yamazaki, S., Yasuda, K., and Miyazaki, Y. (2004) Dissolution of Nafion membrane and recast Nafion film in mixtures of methanol and water, *J. Power Sourc.*, 126, 41–45.

79. Silva, R.F., de Francesco, M., and Pozio, A. (2004) Solution-cast Nafion ionomer membranes: Preparation and characterization, *Electrochim. Acta*, 49, 3211–3219.

80. Thomas, T.J., Ponnusamy, K.E., Chang, N.M., Galmore, K., and Minteer, S.D. (2003) Effects of annealing on mixture-cast membranes of Nafion and quaternary ammonium bromide salts, *J. Membr. Sci.*, 212, 55–66.

81. Wang, F., Hickner, M., Kim, Y.S., Zawodzinski, T.A., and McGrath, J.E. (2002) Direct polymerization of sulfonated poly(arylene ether sulfone) random (statistical) copolymers: Candidates for new proton exchange membranes, *J. Membr. Sci.*, 197, 231–242.

82. Harrison, W.L., Wang, F., Mecham, J.B., Bhanu, V.A., Hill, M., Kim, Y.S., and McGrath, J.E. (2003) Influence of the bisphenol structure on the direct synthesis of sulfonated poly(arylene ether) copolymers, *J. Polym. Sci. A Polym. Chem.*, 41, 2264–2276.

83. Xing, D. and Kerres, J. (2006) Improved performance of sulfonated polyarylene ethers for proton exchange membrane fuel cells, *Polymer Adv. Tech.*, 17, 591–597.

84. Bae, B., Miyatake, K., and Watanabe, M. (2008) Sulfonated poly(arylene ether sulfone) ionomers containing fluorenyl groups for fuel cell applications, *J. Membr. Sci.*, 310, 110–118.

85. Kim, J., Kim, B., and Jung, B. (2002) Proton conductivities and methanol permeabilities of membranes made from partially sulfonated polystyrene-block-poly(ethylene-ran-butylene)-block-polystyrene copolymers, *J. Membr. Sci.*, 207, 129–137.
86. Kim, B., Kim, J., and Jung, B. (2005) Morphology and transport properties and methanol through partially sulfonated copolymers, *J. Membr. Sci.*, 250, 175–182.
87. Kim, J., Kim, B., Jung, B., Kang, Y.S., Ha, H.Y., Oh, I.H., and Ihn, K.J. (2002) Effect of casting solvent on morphology and physical properties of partially sulfonated polystyrene-block-poly(ethylene-ran-butylene)-block-polystyrene copolymers, *Macromol. Rapid Commun.*, 23, 753–756.
88. Essafi, W., Gebel, G., and Mercier, R. (2004) Sulfonated polyimide ionomers: A structural study, *Macromolecules*, 37, 1431–1440.
89. Marestin, C., Gebel, G., Diat, O., and Mercier, R., Sulfonated polyimides, in *Fuel Cell Handbook-II*, Ed. Scherer, G.G., Advances in Polymer Science, Vol. 216, Springer-Verlag, Berlin, Germany, 2008, pp. 185–258.
90. Ma, C., Zhang, L., Mukerjee, S., Ofer, D., and Nair, N. (2003) An investigation of proton conduction in select PEM's and reaction layer interfaces-designed for elevated temperature operation, *J. Membr. Sci.*, 219, 123–136.
91. Kreuer, K.D. (2001) On the development of proton conducting polymer membranes for hydrogen and methanol fuel cells, *J. Membr. Sci.*, 185, 29–39.
92. Ukihashi, H., Asawa, T., and Gunjima, T., Cation exchange membrane of fluorinated polymer containing polytetrafluoroethylene fibrils for electrolysis and preparation thereof, US Patent 4218542, 1980.
93. Penner, R.M. and Martin, C.R. (1985) Ion-transporting composite membranes. I. Nafion-impregnated Gore-Tex membranes, *J. Electrochem. Soc.*, 132, 514–515.
94. Liu, C. and Martin, C.R. (1990) Ion-transporting composite membranes. II. Ion transport mechanism in Nafion-impregnated Gore-Tex membranes, *J. Electrochem. Soc.*, 137, 510–517.
95. Bahar, B, Hobson, A.R., and Kolde, J.A., Ultra-thin film integral composite membrane, US Patent 5547551, 1996.
96. Nouel, K.M. and Fedkiew, P.S. (1998) Nafion® based composite polymer electrolyte membrane, *Electrochim. Acta*, 43, 2381–2387.
97. Shim, J., Ha, H.Y., Hong, S.A., and Oh, U.W. (2002) Characteristics of the Nafion ionomer-impregnated composite membrane for polymer electrolyte fuel cells, *J. Power Sourc.*, 109, 412–417.
98. Liu, F., Yi, B., Xing, D., Yu, J., and Zang, H. (2003) Nafion/PTFE composite membranes for fuel cell applications, *J. Membr. Sci.*, 212, 213–223.
99. Liu, F., Yi, B., Xing, D., Yu, J., Hou, Z., and Fu, Y. (2003) Development of novel self-humidifying composite membranes fuel cells, *J. Power Sourc.*, 124, 81–89.
100. Lin, H.L., Yu, T.L., Shen, K.S., and Huang, L.N. (2004) Effect of Triton-X on the preparation of Nafion/PTFE composite membranes, *J. Membr. Sci.*, 237, 1–7.
101. Yu, T.L., Lin, H.L., Shen, K.S., Chang, Y.C., Jung, B., and Haung, J.C. (2004) Nafion/PTFE composite membranes for fuel cell applications, *J. Polym. Res.*, 11, 217–224.
102. Ahn, S.Y., Lee, Y.C., Ha, H.Y., Hong, S.A., and Oh, I.H. (2004) Properties of the reinforced composite membranes formed by melt soluble ion conducting polymer resins for PEMFCs, *Electrochim. Acta*, 50, 571–573.
103. Xu, F., Mu, S., and Pan, M. (2011) Mineral nanofibre reinforced composite polymer electrolyte membranes with enhanced water retention capability in PEM fuel cells, *J. Membr. Sci.*, 377, 134–140.
104. Lin, H.L., Yu, T.L., Huang, L.N., Chen, L.C., Shen, K.S., and Chung, B. (2005) Nafion/PTFE composite membranes for direct methanol fuel cell applications, *J. Power Sourc.*, 150, 11–19.

105. Huang, L.N., Chen, L.C., Yu, T.L., and Lin, H.L. (2006) Nafion/PTFE/silicate composite membranes for direct methanol fuel cells, *J. Power Sourc.*, 161, 1096–1105.

106. Wang, L., Xing, D.M., Liu, Y.H., Cai, Y.H., Shao, Z.G., Zhai, Y.F., Zhong, H.X., Yi, B.L., and Zhang, H.M. (2006) Pt/SiO$_2$ catalyst as an addition to Nafion/PTFE self-humidifying composite membrane, *J. Power Sourc.*, 161, 61–67.

107. Zhang, Y., Zhang, H., Zhu, X., Gang, L., Bi, C., and Liang, Y. (2007) Fabrication and characterization of a PTFE-reinforced integral composite membrane for self-humidified PEMFC, *J. Power Sourc.*, 165, 786–792.

108. Tang, H., Pan, M., Jiang, S.P., Wang, X., and Ruan, Y. (2007) Fabrication and characterization of PFSI/ePTFE composite proton exchange membranes of polymer electrolyte fuel cells, *Electrochim. Acta*, 52, 5304–5311.

109. Lin, H.L. and Chang, T.J. (2008) Preparation of Nafion/PTFE/Zr(HPO$_4$)$_2$ composite membranes by direct impregnation, *J. Membr. Sci.*, 325, 880–886.

110. Chen, L.C., Yu, T.L., Lin, H.L., and Yeh, S.H. (2008) Nafion/PTFE and zirconium phosphate modified Nafion/PTFE composite membranes for direct methanol fuel cells, *J. Membr. Sci.*, 307, 10–20.

111. Lin, H.L., Yeh, S.H., Yu, T.L., and Chen, L.C. (2009) Silicate and zirconium phosphate modified Nafion/PTFE composite membranes for high temperature PEMFC, *J. Polym. Res.*, 16, 519–527.

112. Jung, G.B., Weng, F.B., Su, A., Wang, J.S., Yu, T.L., Lin, H.L., Yang, T.F., and Chan, S.H. (2008) Nafion/PTFE/silicate membranes for high-temperature proton exchange membrane fuel cells, *Int. J. Hydrogen Energ.*, 33, 2413–2417.

113. Weng, F.B., Wang, J.S., and Yu, T.L. (2008) The study of PTFE/Nafion/silicate membranes operating at low relative humidity and elevated temperature, *J. Chin. Inst. Chem. Eng.*, 39, 429–433.

114. Jao, T.C., Ke, S.T., Chi, P.H., Jung, B., and Chan, S.H. (2010) Degradation on a PTFE/Nafion membrane electrode assembly with accelerating degradation technique, *Int. J. Hydrogen Energ.*, 35, 6941–6949.

115. Pattabiraman, K. and Ramya, K. (2011) Phosphotungstic acid modified expanded PTFE based Nafion composite, *J. New Mat. Electrochem. Sys.*, 14, 73–78.

116. Jung, G.B., Weng, F.B., Peng, C.C., and Jao, T.C. (2011) The development of PTFE/Nafion/TEOS membranes for application in moderate and high temperature proton exchange membrane fuel cells, *Int. J. Hydrogen Energ.*, 36, 6045–6050.

117. Yang, L., Li, H., Ai, F., Chen, X., Tang, J., Zhu, Y., Wang, C., Yuan, W.Z., Zhang, Y., and Zhang, Y. (2013) A new method to prepare high performance fluorinated sulfonic acid ionomer/porous expanded polytetrafluoroethylene composite membranes based on perfluorinated sulfonyl fluoride polymer solution, *J. Power Sourc.*, 243, 392–396.

118. Xing, D., He, G., Hou, Z., Ming, P., and Song, S. (2013) Properties and morphology of Nafion/polytetrafluoroethylene composite membrane fabricated by a solution-spray process, *Int. J. Hydrogen Energ.*, 38, 8400–8408.

119. Shim, J.H., Koo, I.G., and Lee, W.M. (2005) Nafion-impregnated polyethyleneterephthalate film used as the electrolyte for direct methanol fuel cells, *Electrochim. Acta*, 50, 2385–2391.

120. Kim, K.H., Ahn, S.Y., Oh, I.H., Ha, H.Y., Hong, S.A., Kim, M.S., Lee, Y., and Lee, Y.C. (2004) Characteristics of the Nafion-impregnated polycarbonate composite membranes for PEMFCs, *Electrochim. Acta*, 50, 577–581.

121. Hakan Yildrim, M., Stamatialis, D., and Wessling, M. (2008) Dimensionally stable Nafion-polyethylene composite membranes for direct methanol fuel cell applications, *J. Membr. Sci.*, 321, 364–372.

122. Cho, M.S., Son, H.D., Nam, J.D., Suh, S.J., and Lee, Y. (2006) Proton conducting membrane using multi-layer acid-base complex formation on porous PE film, *J. Membr. Sci.*, 284, 155–160.

123. Choi, S.W., Fu, Y.Z., Ahn, Y.R., Jo, S.M., and Manthiram, A. (2008) Nafion-impregnated electrospun polyvinylidene fluoride composite membrane for direct methanol fuel cells, *J. Power Sourc.*, 180, 167–171.

124. Jang, W.G., Hou, J., and Byun, H.S. (2011) Preparation and characterization of PVdF nanofiber ion exchange membrane for the PEMFC application, *Desalin. Water Treat.*, 34, 315–320.

125. Wei, X. and Yates, M.Z. (2010) Control of Nafion/poly(vinylidene fluoride-co-hexafluoropropylene) composite membrane microstructure to improve performance in direct methanol fuel cells, *J. Electrochem. Soc.*, 157, B522–528.

126. Lin, H.L., Wang, S.H., Chiu, C.K., Yu, T.L., Chen, L.C., Huang, C.C., Chen, T.H., and Lin, J.M. (2010) Preparation of Nafion/poly(vinyl alcohol) electro-spun fiber composite membranes for direct methanol fuel cells, *J. Membr. Sci.*, 365, 114–122.

127. Molla, S., Compan, V., Lafuente, S.L., and Prats, J. (2011) On the methanol permeability through pristine nafion and PVA membranes measured by different techniques. A comparison of methodologies, *Fuel Cells*, 11, 897–906.

128. Molla, S., Compan, V., Gimenez, E., Blazquez, A., and Urdanpilleta, I. (2011) Novel ultrathin composite membranes of Nafion/PVA for PEMFCs, *Int. J. Hydrogen Energ.*, 36, 11025–11033.

129. Molla, S. and Compan, V. (2011) Polyvinyl alcohol nanofiber reinforced nafion membranes for fuel cell applications, *J. Membr. Sci.*, 372, 191–200.

130. Molla, S. and Compan, V. (2011) Performance of composite Nafion/PVA membranes for direct methanol fuel cells, *J. Power Sourc.*, 196, 2699–2708.

131. Lin, H.L. and Wang, S.H. (2014) Nafion/poly(vinyl alcohol) nano-fiber composite and Nafion/poly(vinyl alcohol)blend membranes for direct methanol fuel cells, *J. Membr. Sci.*, 452, 253–262.

132. Hu, J., Luo, J., Wagner, P., Conrad, O., and Agert, C. (2009) Anhydrous proton conducting membranes based on electro-deficient nanoparticles/PBI-OO/PFSA blend composites for high temperature PEMFC, *Electrochem. Comm.*, 11, 2324–2327.

133. Hu, J., Luo, J., Wagner, P., Agert, C., and Conrad, O. (2011) Thermal behaviours and single cell performance of PBI-OO/PFSA blend membranes composited with Lewis acid nanoparticles for intermediate temperature DMFC, *Fuel Cells*, 11, 756–763.

134. Ajmad, H., Kamarudin, S.K., Hasran, U.A., and Daud, W.R.W. (2011) A novel hybrid Nafion-PBI-ZP membrane for direct methanol fuel cells, *Int. J. Hydrogen Energ.*, 36, 14668–14677.

135. Aili, D., Hansen, M.K., Pan, C., Li, Q., Christensen, E., Jensen, J.O., and Bjerrum, N.J. (2011) Phosphoric acid doped membranes based on Nafion, PBI and their blends—Membrane preparation, characterization and steam electrolysis testing, *Int. J. Hydrogen Energ.*, 36, 6985–6993.

136. Ainla, A. and Brandell, D. (2007) Nafion/polybenzimidazole (PBI) composite membranes for DMFC applications, *Solid State Ionics*, 178, 581–585.

137. Wang, I., Advani, S.G., and Prasad, A.K. (2013) PBI/Nafion/SiO$_2$ hybrid membrane for high-temperature low-humidity fuel cell applications, *Electrochim. Acta*, 105, 530–534.

138. Shao, Z.G., Wang, X., and Hsing, I.M. (2002) Composite Nafion/poly(vinyl alcohol) membranes for direct methanol fuel cell, *J. Membr. Sci.*, 210, 147–153.

139. Xu, W., Liu, C., Xue, X., Su, Y., Lv, Y., Xing, W., and Lu, T. (2004) New proton exchange membranes based on poly(vinyl alcohol) for DMFCs, *Solid State Ionics*, 171, 121–127.

140. DeLuca, N.W. and Elabd, Y.A. (2006) Nafion/poly(vinyl alcohol) blends: Effect of composition and annealing temperature on transport properties, *J. Membr. Sci.*, 282, 217–224.

141. DeLuca, N.W. and Elabd, Y.A. (2006) Direct methanol fuel cell performance of Nafion/poly(vinyl alcohol) blend membranes, *J. Power Sourc.*, 163, 386–391.

142. Wang, C.H., Chen, C.C., Hsu, H.C., Du, H.Y., Chen, L.C., Shih, H.C., Stejskal, J., Chen, K.H., Chen, C.P., and Huang, J.Y. (2009) Low methanol-permeable polyaniline/nafion composite membrane for direct methanol fuel cells, *J. Power Sourc.*, 190, 279–284.

143. Tsai, J.C., Cheng, H.P., Kuo, J.F., Huang, Y.H., and Chen, C.Y. (2009) Blended Nafion/ s-PEEK direct methanol fuel membranes for reduced methanol permeability, *J. Power Sourc.*, 189, 958–965.

144. Jones, D.J. and Roziere, J., Inorganic/organic composite membranes, in *Handbook of Fuel Cells*, Eds. Vielstich, W., Lamm, A., and Gasteiger, H.A., Vol. 3, John Wiley & Sons, 2003, Chapter 35.

145. Lin, J.C., Kunz, H.R., and Fenton, J.M., Membrane/electrode additives for low humidification operation, in *Handbook of Fuel Cells*, Eds. Vielstich, W., Lamm, A., and Gasteiger, H.A., Vol. 3, John Wiley & Sons, 2003, Chapter 36.

146. Watanabe, M., Uchida, H., and Emori, E. (1998) Polymer electrolyte membranes incorporated with nanometer-size particles of Pt and/or metal-oxides: Experimental analysis of the self-humidification and suppression of gas-crossover in fuel cells, *J. Phys. Chem. B*, 102, 3129–3137.

147. Silva, V.S., Ruffmann, B., Silva, H., Silva, V.B., Mendes, A., Maderra, L.M., and Nunes, S. (2006) Zirconium oxide hybrid membranes for direct methanol fuel cells—Evaluation of transport properties, *J. Membr. Sci.*, 284, 137–144.

148. Apichatachutapan, W., Moore, R.B., and Mauritz, K.A. (1996) Asymmetric Nafion/(zirconium oxide) hybrid membranes via in situ sol-gel chemistry, *J. Appl. Polym. Sci.*, 62, 417–426.

149. Adjemian, K.T., Lee, S.J., Srinvasan, S., Benziger, J., and Bocarsly, A.B. (2002) Silicon oxide nafion composite membranes for proton-exchange membrane fuel cell operation at 80–140°C, *J. Electrochem. Soc.*, 149, A256–A261.

150. Adjemian, K.T., Srinvasan, S., Benziger, J., and Bocarsly, A.B. (2002) Investigation of PEMFC operation above 100°C employing perfluorosulfonic acid silicon oxide composite membranes, *J. Power Sourc.*, 109, 343–356.

151. Linag, Z.X., Zhao, T.S., and Prabhuram, J. (2006) Diphenyl silicate incorporated Nafion membranes for reduction of methanol crossover in direct methanol fuel cells, *J. Membr. Sci.*, 283, 219–224.

152. Deng, Q., Moore, R.B., and Mauritz, K.A. (1998) Nafion®/(SiO₂, ORMOSIL, and dimethylsiloxane) hybrids via in situ sol-gel reactions: Characterization of fundamental properties, *J. Appl. Polym. Sci.*, 68, 747–763.

153. Shao, P.L., Moore, R.B., and Mauritz, K.A. (1996) Perfluorosulfonate ionomer]/[SiO₂-TiO₂] nanocomposites via polymer-in situ sol-gel chemistry: Sequential alkoxide procedure, *J. Polym. Sci. B Polym. Phys. Ed.*, 34, 873–882.

154. Deng, Q., Cable, K.M., Moore, R.B., and Mauritz, K.A. (1996) Small angle X-ray scattering of Nafion/[silicon oxide] and Nafion/ormosil nanocomposites, *J. Polym. Sci. B Polym. Phys.*, 34, 1917–1923.

155. Gummaraju, R.V., Moore, R.B., and Mauritz, K.A. (1996) Asymmetric [Nafion]/[silicon oxide] hybrid membranes via the in situ sol-gel reaction for tetraethoxysilane, *J. Polym. Sci. B Polym. Phys. Ed.*, 34, 2382–2392.

156. Deng, Q., Wilkie, C.A., Moore, R.B., and Mauritz, K.A. (1998) TGA-Fti.r. investigation of the thermal degradation of Nafion and Nafion/[silicon oxide]-based nanocomposites, *Polymer*, 39, 5961–5972.

157. Antonucci, P.L., Arico, A.S., Creti, P., Ramunni, E., and Antonucci, V. (1999) Investigation of direct methanol fuel cell based on a composite Nafion-silica electrolyte for high temperature operation, *Solid State Ionics*, 125, 431–437.

158. Watanabe, M., Uchida, H., Seki, Y., Emori, M., and Stonehart, P. (1996) Self-humidifying polymer electrolyte membranes for fuel cells, *J. Electrochem. Soc.*, 143, 3847–3852.

159. Mauritz, K.A. and Warren, R.M. (1989) Microstructural evolution of a silicon oxide phase in a perfluorosulfonic acid ionomer by an in situ sol-gel reaction. 1. Infrared spectroscopic studies, *Macromolecules*, 22, 1730–1734.
160. Mauritz, K.A., Stefanithis, I.D., Davis, S.V., Scheetz, R.W., Pope, R.K., Wilkes, G.L., and Haung, H.H. (1995) Microstructural evolution of a silicon oxide phase in a perfluorosulfonic acid ionomer by an in situ sol–gel reaction, *J. Appl. Polym. Sci.*, 55, 181–190.
161. Shao, P.L., Mauritz, K.A., and Moore, R.B. (1985) Perfluorosulfonate ionomer]/[mixed inorganic oxide] nanocomposites via polymer in situ sol-gel chemistry, *Chem. Mater.*, 7, 192–200.
162. Roberson, M.A.F. and Mauritz, K.A. (1998) Infrared investigation of the silicon oxide phase in [perfluoro-carboxylate/sulfonate (bilayer)]/[silicon oxide] nanocomposite membranes, *J. Polym. Sci. B Polym. Phys.*, 36, 595–606.
163. Jung, D.H., Cho, S.Y., Peck, D.H., Shin, D.R., and Kim, J.S. (2002) Performance evaluation of a Nafion/silicon oxide hybrid membranes for direct methanol fuel cell, *J. Power Sourc.*, 106, 173–177.
164. Mauritz, K.A. (1998) Organic-inorganic hybrid materials: Perfluorinated ionomers as sol-gel polymerization templates for inorganic alkoxides, *Mater. Sci. Eng. C*, 6, 121–133.
165. Mauritz, K.A. and Payne, J.T. (2000) Perfluorosulfonate ionomer]/silicate hybrid membrane via base-catalyzed in situ sol-gel processes for tetraethylorthosilicate, *J. Membr. Sci.*, 168, 39–51.
166. Young, S.K. and Mauritz, K.A. (2002) Nafion/(organically modified silicate) nanocomposites via polymer in situ sol-gel reactions: Mechanical tensile properties, *J. Polym. Sci. B Polym. Phys.*, 40, 2237–2247.
167. Greso, A.J., Moore, R.B., Cable, K.M., Jarrett, W.L., and Mauritz, K.A. (1997) Chemical modification of a Nafion sulfonyl fluoride precursor via in situ sol-gel reactions, *Polymer*, 38, 1345–1356.
168. Staiti, P., Arico, A.S., Baglio, V., Lufrano, F., Passalacqua, E., and Antonucci, V. (2001) Hybrid Nafion–silica membranes doped with heteropolyacids for application in direct methanol fuel cells, *Solid State Ionics*, 145, 101–107.
169. Lu, G.Q., Wang, C.Y., Yen, T.J., and Zhang, X. (2004) Development and characterization of a silicon-based micro direct methanol fuel cell, *Electrochem. Acta*, 49, 821.
170. Liang, Z.X., Zhao, T.S., and Prabhuram, J. (2006) Diphenyl silicate incorporated Nafion membranes for reduction of methanol crossover in direct methanol fuel cells, *J. Membr. Sci.*, 283, 219–224.
171. Deng, Q., Moore, R.B., and Mauritz, K.A. (1995) Novel nafion/ORMOSIL hybrids via in situ sol-gel reactions. 1. Probe of ORMOSIL phase nanostructures by infrared spectroscopy, *Chem. Mater.*, 7, 2259–2268.
172. Young, S.K. and Mauritz, K.A. (2001) Dynamic mechanical analyses of Nafion/organically modified silicate nanocomposites, *J. Polym. Sci. B Polym. Phys.*, 39, 1282–1295.
173. Thomassin, J.M., Pagnoulle, C., Bizzari, D., Caldarella, G., Germain, A., and Jerome, R. (2004) Nafion-layered silicate nanocomposite membrane for fuel cell application, *e-Polymers*, no. 018.
174. Yang, C., Srinivasan, S., Arisco, A.S., and Creti, P. (2001) Composite Nafion/zirconium phosphate membranes for direct methanol fuel cell operation at high temperature, *Electrochem. Solid State Lett.*, 4, A31–34.
175. Costamagna, P., Yang, C., Bocarsly, A.B., and Srinivasan, S. (2002) Nafion 115/zirconium phosphate composite membranes for operation of PEMFCs above 100°C, *Electrochim. Acta*, 47, 1023–1033.
176. Yang, C., Srinivasan, S., Bocarsly, A.B., Tulyani, S., and Benziger, J.B. (2004) A comparison of physical properties and fuel cell performance of Nafion and zirconium phosphate/Nafion composite membranes, *J. Membr. Sci.*, 237, 145–161.

177. Xie, Z., Navessin, T., Shi, Z., Chow, R., and Holdcroft, S. (2006) Gas diffusion electrodes containing ZHP/Nafion for PEMFC operation at 120°C, *J. Electroanal. Chem.*, 596, 38–46.
178. Zhai, Y., Zhang, H., Hu, J., and Yi, B. (2006) Preparation and characterization of sulfated zirconia (SO_4^{2-}/ZrO_2)/Nafion composite membranes for PEMFC operation at high temperature low humidity, *J. Membr. Sci.*, 280, 148–155.
179. Chalkova, E., Fedkin, M.V., Komarneni, S., and Lvova, S.N. (2007) Nafion/zirconium phosphate composite membranes for PEMFC operating at up to 120°C and down to 13% RH, *J. Electrochem. Soc.*, 154, B288–295.
180. Bauer, F. and Willert-Porada, M. (2005) Characterization of zirconium and titanium phosphates and direct methanol fuel cell (DMFC) performance of functionally graded Nafion(R) composite membranes prepared out of them, *J. Power Sourc.*, 145, 101–107.
181. Bauer, F. and Willert-Porada, M. (2006) Comparison between Nafion® and a Nafion® zirconium phosphate nano-composite in fuel cell applications, *Fuel Cells*, 6, 261–269.
182. Song, Y., Fenton, J.M., Kunz, H.R., Bonville, L.J., and Williams, M.V. (2005) High-performance PEMFCs at elevated temperatures using Nafion-112 membranes, *J. Electrochem. Soc.*, 152, A539–A544.
183. Kim, J.D., Mori, T., and Honma, I. (2006) Organic–inorganic hybrid membranes for a PEMFC operation at intermediate temperatures, *J. Electrochem. Soc.*, 153, A508–A514.
184. Xu, W., Lu, T., Liu, C., and Xing, W. (2005) Low methanol permeable composite Nafion/silica/PWA membranes for low temperature direct methanol fuel cells, *Electrochem. Acta*, 50, 3280–3285.
185. Jung, D.H., Cho, S.Y., Peck, D.H., Shin, D.R., and Kim, J.S. (2003) Preparation and performance of a Nafion/montmorillonite nano-composite membrane for direct methanol fuel cell, *J. Power Sourc.*, 118, 205–211.
186. Tazi, B. and Savadogo, O. (2000) Parameters of PEM fuel-cells based on new membranes fabricated from Nafion, silicotungstic acid and thiophene, *Electrochim. Acta*, 45, 4329–4339.
187. Adjemian, K.T., Dominey, R., Krishnan, L., Ota, H., Majsztrisk, P., Zhang, T., Mann, J. et al. (2006) Function and characterization of metal oxide–Nafion composite membranes for elevated-temperature H_2/O_2 PEM fuel cells, *Chem. Mater.*, 18, 2238–2248.
188. Satterfield, M.B., Majsztrik, P.W., Ota, H., Benziger, J.B., and Bocarsly, A.B. (2006) Mechanical properties of Nafion and titania/Nafion composite membranes for polymer electrolyte membrane fuel cells, *J. Polym. Sci. B Polym. Phys.*, 44, 2327–2345.
189. Adjemian, K.T., Srinvasan, S., Benziger, J., and Bocarsly, A.B. (2002) Investigation of PEMFC operation above 100°C employing perfluorosulfonic acid silicon oxide composite membranes, *J. Power Sourc.*, 109, 356–364.
190. Si, Y., Kunz, H.R., and Fenton, J.M. (2004) Nafion-Teflon-$Zr(HPO_4)_2$ composite membranes for high-temperature PEMFCs, *J. Electrochem. Soc.*, 51, A623–A631.
191. Jiang, R., Kunz, H.R., and Fenton, J.M. (2006) Influence of temperature and relative humidity on performance and CO tolerance of PEM fuel cells with Nafion-Teflon- $Zr(HPO_4)_2$ high temperature composite membranes, *Electrochim. Acta*, 51, 5596–5510.
192. Rodgers, M.P., Shi, Z., and Holdcroft, S. (2008) Transport properties of composite membranes containing silicon dioxide and Nafion, *J. Membr. Sci.*, 325, 346–356.
193. Amjadi, M., Rowshanzamir, S., Peighambardoust, S.J., and Sedghi, S. (2012) Preparation, characterization and cell performance of durable nafion/SiO_2 hybrid membrane for high-temperature fuel cells, *J. Power Sourc.*, 210, 350–357.
194. Arisco, A.S., Baglio, V., Di Blasi, A., and Antonucci, V. (2008) FTIR spectroscopic investigation of inorganic fillers for composite DMFC membranes, *Electrochem. Commun.*, 5, 862–866.

195. Tay, S.W., Zhang, X., Liu, Z., Hong, L., and Chan, S.H. (2007) Composite Nafion membrane embedded with hybrid nanofillers for promoting direct methanol fuel cell performance, *J. Membr. Sci.*, 321, 139–145.

196. Kumar, G.G., Kim, A.R., Nahm, K.S., and Elizabeth, R. (2009) Nafion membranes modified with silica sulfuric acid for the elevated temperature and lower humidity operation of PEMFC, *Int. J. Hydrogen Energ.*, 34, 9788–9794.

197. Joseph, J., Tseng, C.Y., and Hwang, B.J. (2011) Phosphoric acid-grafted mesostructures silica/Nafion hybrid membranes for fuel cell applications, *J. Power Sourc.*, 196, 7363–7371.

198. Thampan, T.M., Jalani, N.H., Choi, P., and Datta, R. (2005) Systematic approach to design higher temperature composite PEMs, *J. Electrochem. Soc.*, 152(2), A316–A325.

199. Bi, C., Zhang, H., Zhang, Y., Zhu, X., Ma, Y., Dai, H., and Xiao, S. (2008) Fabrication and investigation of SO_2 supported sulfated zirconia/Nafion self humidifying membrane for proton exchange membrane fuel cell applications, *J. Power Sourc.*, 184, 197–203.

200. Arbizzani, C., Donnadio, A., Pica, M., Sganappa, M.,Varzi, A., Casciola, M., and Mastragostino, M. (2010) Methanol permeability and performance of Nafion-zirconium phosphate composite membranes in active and passive direct methanol fuel cells, *J. Power Sourc.*, 195, 7751–7756.

201. Amirinejad, M., Madaeni, S.S., Navarra, M.A, Rafiee, E., and Scrosati, B. (2011) Preparation and characterization of phosphotungstic acid-derived salt/Nafion nanocomposite membranes for proton exchange membrane fuel cells, *J. Power Sourc.*, 196, 988–998.

202. Amirinejad, M., Madaeni, S.S., Rafiee, E., and Amirinejad, S. (2011) Cesium hydrogen salt of heteropolyacids/Nafion nanocomposite membranes for proton exchange membrane fuel cells, *J. Membr. Sci.*, 377, 89–98.

203. Li, Q., He, R.H., Jensen, O.J., and Bjerrum, N.J. (2003) Approaches and recent development of polymer electrolyte membranes for fuel cells operating above 100°C, *Chem. Mater.*, 15, 4896–915.

204. Hickner, M.A., Ghassemi, H., Kim, Y.S., Eins, B.R., and McGrath, J.E. (2004) Alternative polymer systems for proton exchange membranes (PEMs), *Chem. Rev.*, 104, 4587–612.

205. Vogel, H. and Marvel, C.S. (1996) Polybenzimidazoles, new thermally stable polymers, *J. Polym. Sci. Polym. Chem.*, 34, 1125–53.

206. Wainright, J.S., Wang, T., Weng, D., Savinell, R.F., and Litt, M. (1995) Acid-doped polybenzimidazoles: A new polymer electrolyte, *J. Electrochem. Soc.*, 142, L121.

207. Li, Q., Jensen, J.O., Savinell, R.F., and Bjerrum, N.J. (2009) High temperature proton exchange membranes based on polybenzimidazoles for fuel cells, *Prog. Polym. Sci.*, 34, 449–477.

208. Rikakawa, M. and Sanui, K. (2000) Proton conducting polymer electrolyte membranes based on hydrocarbon polymers, *Prog. Polym. Sci.*, 25, 1463–1502.

209. Mader, J., Xiao, L., Schmidt, T.J., and Benicwicz, B.C., Polybenzimidazole/acid complex as high temperature membranes, in *Fuel Cell Handbook-II*, Ed. Scherer, G.G., Advances in Polymer Science, Vol. 216, Springer-Verlag, Berlin, Germany, 2008, pp. 63–124.

210. Chandan, A., Hattenberger, M., EI-kharouf, A., Du, S., Dhir, A., Self, V., Pollet, B.G., Ingram, A., and Bujalski, W. (2013) High temperature (HT) polymer electrolyte membrane fuel cells (PEMFC)—A review, *J. Power Sourc.*, 231, 264–278.

211. Ueda, M., Sato, M., and Mochizuki, A. (1985) Poly(benzimidazole) synthesis by direct reaction of diacids and diamines, *Macromolecules*, 18, 2723–2726.

212. Iwakura, Y., Uno, K., and Imai, Y. (1964) Polyphenylenebenzimidazoles, *J. Polym. Sci. A*, 2, 2605–2615.

213. Xiao, L., Zhang, H., Scanlon, E., Ramanthan, L.S., Choe, E.W., Rogers, D., Apple, T., and Benicewicz, B.C. (2005) High-temperature polybenzimidazole fuel cell membranes via sol-gel process, *Chem. Mater.*, 17, 5328–5333.

214. Li, Q., Hjuler, H.A., and Bjerrum, N.J. (2001) Phosphoric acid doped polybenzimidazole membranes: Physiochemical characterization and fuel cell applications, *J. Appl. Electrochem.*, 31, 773–779.

215. Lin, H.L., Chen, Y.C., Li, C.C., Cheng, C.P., and Yu, T.L. (2008) Preparation of PBI/PTFE composite membranes from PBI in *N,N*-dimethyl acetamide solutions with various concentrations of LiCl, *J. Power Sourc.*, 181, 228–236.

216. Juan, A., Aslvador, B., and Pedro G.R. (2002) Proton-conducting polymers based on benzimidazoles and sulfonated benzimidazoles, *J. Polym. Sci. Part A Polym Chem.*, 40, 3703–3710.

217. Staiti, P., Lufrano, F., Arico, A.S., Passalacqua, E., and Antonucci, V. (2001) Sulfonated PBI membranes-preparation and physico-chemical characterization, *J. Membr. Sci.*, 188, 71–78.

218. Peron, J., Ruiz, E., Jones, D.J., and Roziere, A. (2008) Solution sulfonation of a novel PBI. A proton electrolyte for fuel cell application, *J. Membr. Sci.*, 314, 247–256.

219. Li, Q., Rudbeck, H.C., Chromik, A., Jensen, J.O., Pan, C., Steenberg, T., Calverley, M., and Kerres, J. (2010) Properties, degradation and high temperature fuel cell test of different types of PBI and PBI blend membranes, *J. Membr. Sci.*, 347, 260–270.

220. Mader, J.A. and Benicewicz, B.C. (2010) Sulfonated PBI for high temperature PEM fuel cells, *Macromolecules*, 43, 6706–6715.

221. Uno, K., Niume, K., Iwata, Y., Toada, F., and Iwakura, Y. (1997) Synthesis of PBI with sulfonic acid groups, *J. Polym. Sci. A Polym. Chem.*, 15, 1309–1318.

222. Li, S., Xu, H., Guo, X., Fang, J., Fang, L., and Yin, J. (2011) Synthesis and properties of novel sulfonated PBI from disodium 4,6-bis(4-carboxyphenoxy) benzene-1,3-disulfonate, *J. Power Sourc.*, 196, 3039–3047.

223. Sukurmar, P.R., Wu, W., Markova, D., Unsal, O., Klapper, M, and Mullen, K. (2007) Functionalized poly(benzimidazole) as membrane materials for fuel cells, *Macromol. Chem. Phys.*, 208, 2258–2267.

224. Villa, D.C., Angioni, S., Quartarone, E., Righetti, P.P., and Mustarelli, P. (2013) New sulfonated PBIs for PEMFC application, *Fuel Cells*, 13, 98–103.

225. Bae, J.M., Honma, I., Murata, M., Yamamoto, T., Rikukawa, M., and Ogata, N. (2002) Properties of selected sulfonated polymers as proton-conducting electrolytes for polymer electrolyte fuel cells, *Solid State Ionics*, 147, 189–194.

226. Gieselman, M.B. and Reynolds, J.R. (1992) Water soluble PBI-based polyelectrolytes, *Macromolecules*, 25, 4832–4834.

227. Gieselman, M.B. and Reynolds, J.R. (1993) Aramid and imidazole based polyelectrolytes: Physical properties and ternary phase behavior with PBI in methylsulfonic acid, *Macromolecules*, 26, 5633–5642.

228. Glipa, X., Haddad, M.E., Jones, D.J., and Roziere, J. (1997) Synthesis and characterization of sulfonated PBI: A highly conducting proton exchange polymer, *Solid State Ionics*, 97, 323–331.

229. Jones D.J. and Roziere, J. (2001) Recent development in the functionalisation of PBI and polyetherketone for fuel cell applications, *J. Membr. Sci.*, 185, 41–58.

230. Roziere, J., Jones, D.J., Marrony, M., Glipa, X., and Mula, B. (2001) On the doping of sulfonated PBI with strong bases, *Solid State Ionics*, 145, 61–68.

231. Kawahara, M., Rikukawa, M., and Sanui, K. (2000) Relationship between absorbed water and proton conductivity in sulfopropylated poly(benzimidazole), *Polym. Adv. Technol.*, 11, 1–5.

232. Jannasch, P. (2005) Membrane materials by chemical grafting of aromatic main-chain polymers, *Fuel Cells*, 5, 248–260.

233. Tan, N., Xiao, G., Yan, D., and Sun, G. (2010) Preparation and properties of PBIs with sulfophenylsulfonyl pendant groups for proton exchange membranes, *J. Membr. Sci.*, 353, 51–59.

234. Lin, H.L., Hu, C.R., Lai, S.W., and Yu, T.L. (2012) Polybenzimidazole and butylsulfonate grafted polybenzimidazole blends for proton exchange membrane fuel cells, *J. Membr. Sci.*, 389, 399–406.

235. Maity, S., Sannigrahi, A., Ghosh, S., and Jana, T. (2013) N-alkyl polybenzimidazole: Effect of alkyl chain length, *Eur. Polymer J.*, 49, 2280–2292.

236. Sinigersky, V., Budurova, D., Penchev, H., Ublekov, F., and Radev, I. (2013) Polybenzimidazole-graft-polyvinylphosphonic acid—Proton conducting fuel cell membranes, *J. Appl. Polym. Sci.*, 129, 1223–1231.

237. Yang, J., Aili, D., Li, Q., Xu, Y., Liu, P., Che, Q., Jensen, J.O., Bjerrum, N.J., and He, R. (2013) Benzimidazole grafted polybenzimidazoles for proton exchange membrane fuel cells, *Polym. Chem.*, 4, 4768–4775.

238. Kim, S.K., Kim, T.H., Jung, J.W., and Lee, J.C. (2009) Polybenzimidazole containing benzimidazole side groups for high temperature fuel cell applications, *Polymer*, 50, 3495–3502.

239. Yang, J., Xu, Y., Zhou, L., Che, Q., He, R., and Li, Q. (2013) Hydroxyl pyridine containing polybenzimidazole membranes for proton exchange membrane fuel cells, *J. Membr. Sci.*, 446, 318–325.

240. Hasiotis, C., Li, Q., Deimede, V., Kallitsis, J.K., Kontoyannis, C.G., and Bjerrum, N.J. (2001) Development and characterization of acid-doped polybenzimidazole/sulfonated polysulfone blend polymer electrolytes for fuel cells, *J. Electrochem. Soc.*, 148, A513–A519.

241. Sukurmar, P.R., Wu, W, Markova, D., Unsal, O., Klapper, M., and Mullen, K. (2007) Functionalized poly(benzimidazole) as membrane materials for fuel cells, *Macromol. Chem. Phys.*, 208, 2258–2267.

242. Iizuka, Y., Tanaka, M., and Kawakami, H. (2012) Preparation and proton conductivity of phosphoric acid-doped blend membranes composed of sulfonated block copolyimides and polybenzimidazole, *Polym. Int.*, 62, 703–708.

243. Kerres, J., Ullrich, A., Meier, F., and Haring, T. (1999) Synthesis and characterization of novel acid–base polymer blends for application in membrane fuel cells, *Solid State Ionics*, 125, 243–249.

244. He, R., Li, Q., Xiao, G., and Bjerrum, N.J. (2003) Proton conductivity of phosphoric acid doped polybenzimidazole and its composites with inorganic proton conductors, *J. Membr. Sci.*, 226, 169–184.

245. Staiti, P. (2001) Proton conductive membranes based on silicotungstic acid/silica and polybenzimidazole, *Mater. Lett.*, 47, 241–246.

246. Lobato, J., Canizares, P., Rodrigo, M.A., Ubeda, D., and Pinar, F.J. (2011) A novel titanium PBI-based composite membrane for high temperature PEMFCs, *J. Membr. Sci.*, 369, 105–111.

247. Suryani, C.M., Chang, Y.N., Lai, J.Y., and Liu, Y.L. (2012) Polybenzimidazole (PBI)-functionalized silica nanoparticles modified PBI nanocomposite membranes for proton exchange membrane fuel cells, *J. Membr. Sci.*, 403–404, 1–7.

248. Ossiander, T., Heinzl, C., Gleich, S., Schonberger, F., Volk, P., Welsch, M., and Scheu, C. (2014) Influence of the size and shape of silica nanoparticles on the properties and degradation of a PBI-based high temperature polymer electrolyte membrane, *J. Membr. Sci.*, 454, 12–19.

249. Lin, H.L., Tang, T.H., Hu, C.R., and Yu, T.L. (2012) Poly(benzimidazole)/silica-ethylphosphoric acid hybrid membranes for proton exchange membrane fuel cells, *J. Power Sourc.*, 201, 72– 80.

250. Zeng, J., Jin, B., Shen, P.K., He, B., Lamb, K., De Marco, R., and Jiang, S.P. (2013) Stack performance of phosphotungstic acid functionalized mesoporous silica (HPW-mesosilica) nanocomposite high temperature proton exchange membrane fuel cells, *Int. J. Hydrogen Energ.*, 38, 12830–12837.

251. Choi, S., Coronas, J., Lai, Z., Yust, D., Onorato, F., and Tsapatsis, M. (2008) Fabrication and gas separation properties of polybenzimidazole/nanoporous silicates hybrid membranes, *J. Membr. Sci.*, 316, 145–152.

252. Suryani, C.M. and Liu, Y.L. (2009) Preparation and properties of nanocomposite membranes of polybenzimidazole/sulfonated silica nano-articles for proton exchange membranes, *J. Membr. Sci.*, 332, 121–128.

253. Xu, C., Cao,Y., Kumar, R., Wu, X., Wang, X., and Scott, K. (2011) A polybenzimidazole/sulfonated graphite oxide composite membrane for high temperature polymer membrane fuel cells, *J. Mater. Chem.*, 21, 11359–11364.

254. Wang, S., Zhang, G., Han, M., Li, H., Zhang, Y., Ni, J., Ma, W. et al. (2011) Novel epoxy-based crosslinked polybenzimidazole for high temperature proton exchange membrane fuel cells, *Int. J. Hydrogen Energ.*, 36, 8412–8421.

255. Wang, S., Zhao, C., Ma, W., Zhang, G., Liu, Z., Ni, J., Li, M., Zhang, N., and Na, H. (2012) Preparation and properties of epoxy-cross-linked porous polybenzimidazole for high temperature proton exchange membrane fuel cells, *J. Membr. Sci.*, 411–412, 54–63.

256. Xu, N., Guo, X., Fang, J., Xu, H., and Yin, J. (2009) Synthesis of novel polybenzimidazoles with pendant amino groups and the formation of their crosslinked membranes for medium temperature fuel cell applications, *J. Appl. Polym. Sci.*, 47, 6992–7002.

257. Lin, H.L., Chou, Y.C., Yu, T.L., and Lai, S.W. (2012) Poly(benzimidazole)-epoxide crosslink membranes for high temperature proton exchange membrane fuel cells, *Int. J. Hydrogen Energ.*, 37, 383–392.

258. Li, Q., Pan, C., Jensen, J.O., Noye, P., and Bjerrum, N.J. (2007) Cross-linked polybenzimidazole membranes for fuel cells, *Chem. Mater.*, 19, 350–352.

259. Shen, C.H., Jheng, L.C., Hsu, S.L.C., and Wang, J.T.W. (2011) Phosphoric acid-doped cross-linked porous polybenzimidazole membranes for proton exchange membrane fuel cells, *J. Mater. Chem.*, 21, 15660–15665.

260. Wang, S., Zhao, C., Ma, W., Zhang, N., Zhang, Y., Zhang, G., Liu, Z., and Na, H. (2013) Silane-cross-linked polybenzimidazole with improved conductivity for high temperature proton exchange membrane fuel cells, *J. Mater. Chem. A*, 1, 621–629.

261. Aili, D., Li, Q., Christensen, E., Jensen, J.O., and Bjerrum, N.J. (2011) Crosslinking of polybenzimidazole membranes by divinylsulfone post-treatment for high temperature proton exchange membrane fuel cell applications, *Polym. Int.*, 60, 1201–1207.

262. Noye, P., Li, Q., Pan, C., and Bjerrum, N.J. (2008) Cross-linked polybenzimidazole membranes for high temperature proton exchange membrane fuel cells with dichloromethyl phosphinic acid as a crosslinker, *Polymers Adv. Tech.*, 19, 1270–1275.

263. Lin, H.L., Hsieh, Y.S., Chiu, C.W., Yu, T.L., and Chen, L.C. (2009) Durability and stability test of proton exchange membrane fuel cells prepared from PBI/PTFE composite membrane, *J. Power Sourc.*, 193, 170–174.

264. Lin, H.L., Huang, J.R., Chen, Y.T., Su, P.H., Yu, T.L., and Chan, S.H. (2012) Polybenzimidazole/poly(tetrafluoro ethylene) composite membranes for high temperature proton exchange membrane fuel cells, *J. Polym. Res.*, 19, 9875–8.

265. Li, M., Scott, K., and Wu, X. (2009) A poly($R_1R_2R_3$)-N^+/H_3PO_4 composite membrane for phosphoric acid polymer electrolyte membrane fuel cells, *J. Power Sourc.*, 194, 811–814.

266. Li, M. and Scott, K. (2011) A polytetrafluoroethylene/quaternized polysulfone membrane for high temperature polymer electrolyte membrane fuel cells, *J. Power Sourc.*, 196, 1894–1898.

2 Perfluorinated Polymer Electrolyte Membranes

Hong Li and Yongming Zhang

CONTENTS

2.1 INTRODUCTION

Perfluorinated polymers exhibit outstanding properties such as chemical resistance, high-temperature stability, and long lifetime.[1] Perfluorinated ionomers (PFSIs) that consist of a linear fluorocarbon polymer and a small percentage of pending acid side groups are particularly interesting and important materials for the preparation of polymer electrolyte membranes (PEMs). Due to perfluorinated backbone, PFSI membranes are relatively chemically and thermally stable and have good mechanical properties. Moreover, PFSI membranes have good proton conductivity. PFSI membrane is an important and most widely used PEM in chlor-alkali industry and PEM fuel cells (PEMFCs).

Up to date, there are a number of excellent review articles on the perfluorinated electrolyte membrane. The early book (1982) by Eisenberg and Yeager, *Perfluorinated Ionomer Membranes*, provides an overview of early work on PFSI membranes and useful and classical information about the membranes.[2] Later in 1996, Heitner-Wirguin reported a comprehensive review on perfluorinated membranes including structure, properties, and applications.[3] In 2003, Doyle and Rajendran provided a comprehensive review on synthesis, characterization, properties, and morphology of perfluorinated membranes.[4] In 2004, Mauritz and Moore provided a state-of-the-art review of structure investigation on Nafion® membranes.[5] Most recently, Devanathan summarized progress in understanding perfluorinated membrane.[6] Zhang and Shen reported recent development of perfluorosulfonic acid (PFSA) ionomer membranes for fuel cells.[7] In this chapter, we summarize the basic concept and fundamental studies on perfluorinated sulfonic ionomer membranes including ionomer synthesis, membrane preparation and characterization, membrane structure models, and membrane conduction mechanism. Their application in the electrochemical devices for energy storage and conversion is also described briefly. We gather some experimental data into the chapter so that researchers can use them as a resource for studies. The discussion of PFSI with other acid groups such as carboxylic acid, phosphonic acid, and sulfonamide is not included here.

2.2 IONOMER SYNTHESIS AND MEMBRANE PREPARATION

2.2.1 Typical Functional Monomer for Perfluorosulfonic Acid Ionomer

PFSA ionomers were first developed and produced by the Dupont Company.[8] These materials are synthesized by copolymerization of perfluorovinyl ether sulfonyl fluoride (PFVESF) monomer with tetrafluoroethylene (TFE). The typical functional monomer structures for PFSA ionomers are shown in Table 2.1. They can be generally divided into two types: the long-chain PFVESF with two ether oxygens and the short-chain PFVESF with only one ether oxygen.

The long-chain PFVESF monomer used in most commercial systems (Nafion by the Dupont Company, Flemion® by the Asahi Glass Company, Aciplex® by the Asahi Chemical Company, Dyfion® by the Dongyue Company) was synthesized through reaction of sulfur trioxide (SO_3) with TFE to create the cyclic sultone, which can rearrange to form rearranged sultone with a sulfonyl fluoride end group and an acid fluoride. The latter reacts with two equivalents of hexafluoropropylene oxide to produce sulfonyl fluoride adducts and after pyrolysis in the presence of sodium carbonate (Na_2CO_3) leads to the perfluorosulfonyl fluoride ethyl propyl vinyl ether (PSEPVE), as shown in Figure 2.1.[9] Sultone is key reactant in this process. It should be noted that the process of sultone synthesis involves dangerous factors such as TFE under conditions of high pressure and temperature and the explosive mixture of SO_3 and sultone. Figure 2.2 shows the ^{19}F nuclear magnetic resonance (NMR) spectrum of PSEPVE monomer: perfluoro[2-(2-fluorosulfonylethoxy) propyl vinyl ether] in $CDCl_3$. All of the peaks of the fluorine atoms were labeled.

The short-chain PFVESF monomers resemble the long-chain perfluorosulfonic acid monomers except that the side chain of the former is shorter and contains one ether oxygen. The synthesis of the short-chain monomer is substantially more challenging than the long-chain monomer. The DuPont, Dow® Chemical, and Solvay Solexis Companies produce $CF_2=CFOCF_2CF_2SO_2F$ monomer by different process (Figures 2.1, 2.3, and 2.4).[10] The Asahi Glass Company developed $CF_2=CFOCF_2CF_2CF_2SO_2F$ (Figure 2.5).[11] And 3M Company synthesized

---

TABLE 2.1

Typical Functional Monomer Structures for PFSA Ionomers

Long-chain PFVESF monomer	$CF_2=CFOCF_2CF(CF_3)OCF_2CF_2SO_2F$
Short-chain PFVESF monomers	$CF_2=CFO(CF_2)_nSO_2F$, $n=2$–4

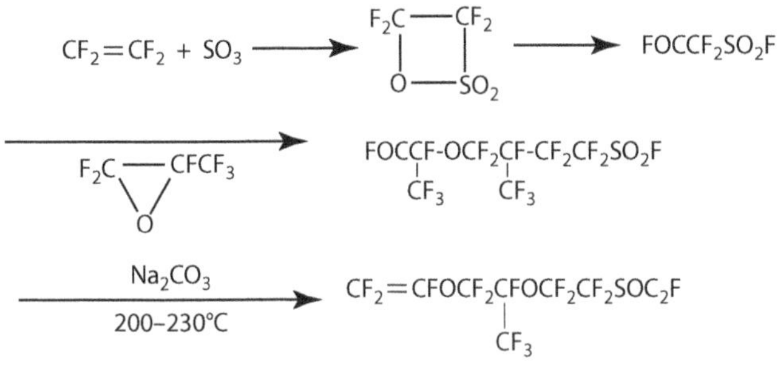

FIGURE 2.1 Synthesis process of long-side-chain PFVESF according to the DuPont Company.

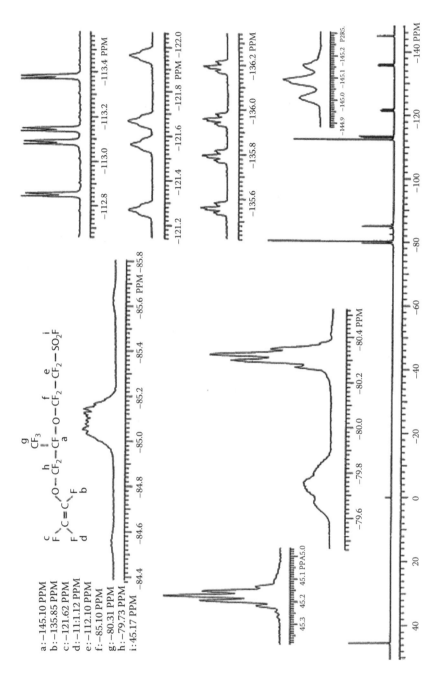

FIGURE 2.2 ^{19}F NMR spectrum of perfluoro[2-(2-fluorosulfonylethoxy) propyl vinyl ether] in CDCl$_3$.

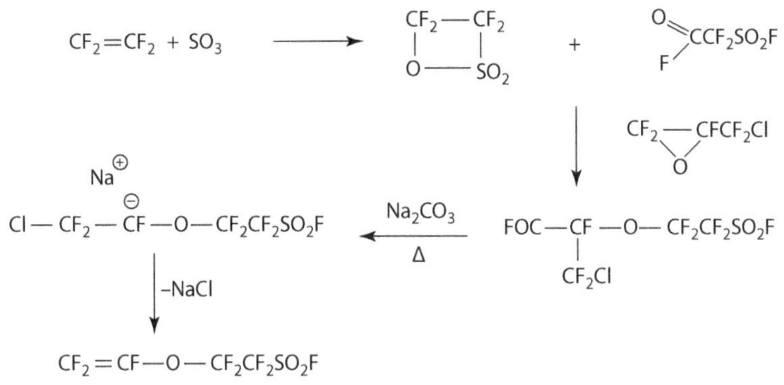

$$CF_2{=}CF_2 + SO_3 \longrightarrow \underset{O \rule[0.5ex]{1.5em}{0.4pt} SO_2}{\overset{CF_2 \rule[0.5ex]{1.5em}{0.4pt} CF_2}{|\qquad|}} \quad + \quad \overset{O}{\underset{F}{\diagdown}}{\!}CCF_2SO_2F$$

$$\downarrow \quad \overset{CF_2 \rule[0.5ex]{1em}{0.4pt} CFCF_2Cl}{\diagdown\; O\; \diagup}$$

$$\underset{|}{\overset{Na^{\oplus}}{Cl-CF_2-\underset{|}{\overset{\ominus}{CF}}-O-CF_2CF_2SO_2F}} \xleftarrow[\Delta]{Na_2CO_3} FOC-\underset{|}{\overset{}{CF}}-O-CF_2CF_2SO_2F$$

$$\quad\quad CF_2Cl \qquad\qquad\qquad\qquad CF_2Cl$$

$$\downarrow\; -NaCl$$

$$CF_2{=}CF-O-CF_2CF_2SO_2F$$

FIGURE 2.3 Synthesis route of the short-side-chain PFVESF monomer by the Dow Chemical Company.

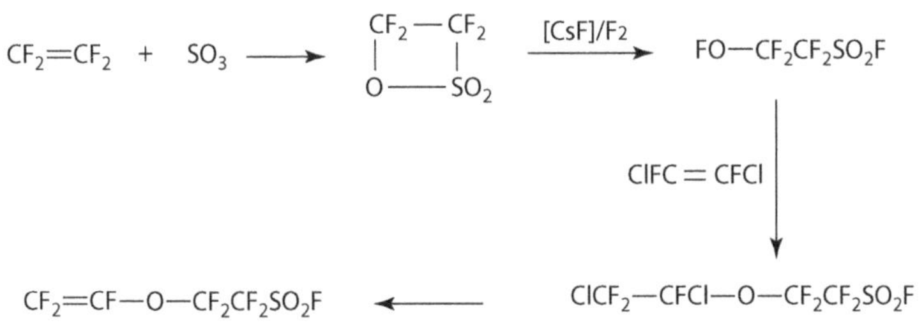

$$CF_2{=}CF_2 + SO_3 \longrightarrow \underset{O \rule[0.5ex]{1.5em}{0.4pt} SO_2}{\overset{CF_2 \rule[0.5ex]{1.5em}{0.4pt} CF_2}{|\qquad|}} \xrightarrow{[CsF]/F2} FO-CF_2CF_2SO_2F$$

$$\downarrow\; ClFC{=}CFCl$$

$$CF_2{=}CF-O-CF_2CF_2SO_2F \longleftarrow ClCF_2-CFCl-O-CF_2CF_2SO_2F$$

FIGURE 2.4 Synthesis route of the short-side-chain PFVESF monomer by Solvay Solexis.

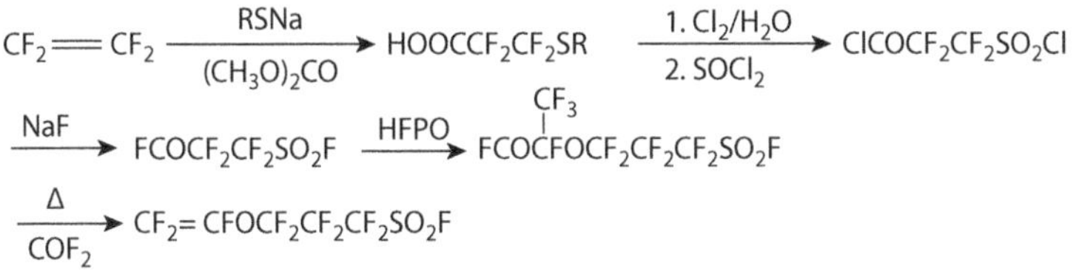

$$CF_2{=}CF_2 \xrightarrow[(CH_3O)_2CO]{RSNa} HOOCCF_2CF_2SR \xrightarrow[2.\ SOCl_2]{1.\ Cl_2/H_2O} ClCOCF_2CF_2SO_2Cl$$

$$\xrightarrow{NaF} FCOCF_2CF_2SO_2F \xrightarrow{HFPO} FCO\underset{|}{\overset{CF_3}{C}}FOCF_2CF_2CF_2SO_2F$$

$$\xrightarrow[COF_2]{\Delta} CF_2{=}CFOCF_2CF_2CF_2SO_2F$$

FIGURE 2.5 Synthesis route of the short-side-chain PFVESF monomer developed by the Asahi Glass Company.

$CF_2{=}CFOCF_2CF_2CF_2SO_2F$ (Figure 2.6). The synthesis of the short-chain PFVESF monomer involves numerous steps, and present yields are not satisfying. However, the short-side-chain PFSA (SSC-PFSA) shows significant improvement in performance compared to the long-side-chain PFSA (LSC-PFSA) and the best balance of properties for fuel cell application in terms of proton conductivity, dimensional stability, mechanical strength, and beginning of life fuel cell performance.[10c]

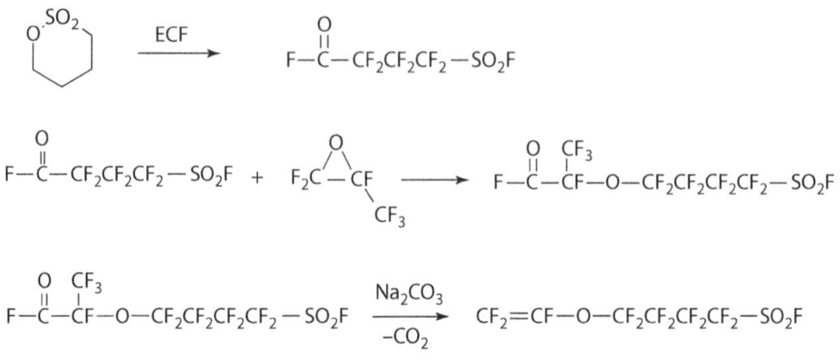

FIGURE 2.6 Synthesis of the short-side-chain PFVESF monomer used in the 3M ionomers (ECF, electrochemical fluorination).

2.2.2 POLYMER SYNTHESIS METHODS AND PROCESSES

2.2.2.1 Polymer Synthesis

PFSA ionomers are generated by three steps. The first one involves copolymerization of PFVESF with TFE, resulting in sulfonyl fluoride precursor (PFSF). Then, the precursor was converted to the sodium sulfonate form using a hydrolysis process with NaOH solution. Conversion of Na⁺ form ionomer to the H⁺ form is accomplished using a fresh solution of nitric acid (HNO_3) or sulfuric acid (H_2SO_4). The chemical structure of the resulting ionomer is given in Table 2.2, and some commercial PFSA membranes used for PEMFC are listed in Table 2.2.[12] Among them, Nafion membranes are the representative based on LSC-PFSA ionomers, and Aquivion™ (Hyflon® Ion is its former trademark) membranes available from Solvay Solexis are the current representative based on SSC-PFSA ionomers.

The production of the PFSA ionomer is by free radical copolymerization of a PFVESF with TFE in a chlorofluorocarbons (such as 1,1,2-trichloro-1,2,2-rifluoroethane) or other perfluorinated organic solvents (such as perfluoro (methylcyclohexane), perfluorodimethylcyclobutane, and perfluoroctan).[13] The reactivity ratios of TFE(r_{TFE}) and PFVESF(r_{PFVESF}) are 8 and 0.08, respectively.[14] Thus, it can be seen the reactivity of TFE is much higher than that of PFVESF. So, to obtain copolymer with an appreciable content of PFVESF units, considerable excess of PFVESF monomer is required. And unreacted PFVESF monomer can be reused after separation. It is favorable to carry out polymerization at as low temperature as possible because some side reactions such as β-scission reaction of PFVESF macroradicals and transfer reaction of chain to the solvent grow sharply with increasing temperature.[13a] These side reactions reduce the copolymer molecular weight and broaden the copolymer molecular dispersity. Perfluorocarbon peroxides and azo compounds such as aco-*bis*-isobutyronitrile are used as initiators of radical copolymerization of PFVESF and TFE. Perfluoroperoxides are preferred as an initiator because they permit the polymerization temperature to be reduced to 45°C and the end group of the resulting copolymer is perfluorinated.

TABLE 2.2
Commercial PFSA Membranes by Producer

$$—(CF_2—CF_2)_x—(CF_2—CF)_y—$$
$$(O—CF_2—CF)_m—O—(CF_2)_n—SO_3H$$
$$CF_3$$

Structure Parameter	Trade Name and Type	EW (g/mol)	Thickness (μm)
$m=1; x=5–13.5$	DuPont		
$n=2; y=1$	Nafion N120	1200	260
	Nafion N117	1100	175
	Nafion N115	1100	125
	Nafion N112	1100	80
	Nafion NR211	990–1050	25
	Nafion NR212	990–1050	50
$m=0, 1; n=1–5$	Asahi Glass		
	Flemion-T	1000	120
	Flemion-S	1000	80
	Flemion-R	1000	50
$m=0; n=2–5;$	Asahi Chemicals		
$x=1.5–14$	Aciplex-S	1000–1200	25–100
$m=0; n=2; x=3.6–10$	Dow Chemical		
	Dow	800	125
$m=0; n=2; x=5–10$	Aquivion		
	Aquivion E79	790	30, 50
	Aquivion E87	870	30, 50
$m=0, n=4$	3M		
	3M ionomer		

Source: Li, Q. et al., *Chem. Mater.*, 15(26), 4896, 2003.

Aqueous emulsion and suspension polymerizations are also employed in polymer synthesis. The copolymerization of PFVESF with TFE is carried out in the presence of a water-soluble initiator and emulsifying agent. The most common perfluoroalkyl surfactant is ammonium perfluorooctanoate.[10a] Because of recent environmental concerns with regard to perfluorooctanoic acid and salts, several different approaches have attempted to reduce or eliminate the use of perfluoroalkyl surfactants in the polymerization of fluorine-containing monomers. Partially fluorinated surfactants, siloxane surfactants,[15] and fluoroether surfactants[16] have been reported in fluoropolymer synthesis. On the other hand, aqueous initiators usually lead to incorporation of unstable carboxylic acid and acid fluoride end groups resulting in polymer instability. Besides, the PFSF in an aqueous media can hydrolyze to form acid form ionomer, which cannot be melting processed.

New polymerization method for the PFSA ionomer such as supercritical carbon dioxide (SC-CO_2) polymerization is applied in our lab.[17] CO_2 is environmentally benign, nonflammable, and readily available. Moreover, it can offer some benefit for the safety profile of the polymerization process. When TFE is stored and handled in the presence of CO_2, it forms an pazeotrope that is less likely to deflagrate from oxygen initiation.[18] The reactivity ratios for TFE (r_{TFE}) and PFVESF (r_{PFVESF}) are calculated to be 7.85, 0.079, and 7.92, 0.087 according to Fineman–Ross and Kelen–Tudos methods, respectively, which are close to the literature data for conventional solution copolymerization of TFE and PFVESF in Freon-113. The PFSA ionomers synthesized in SC-CO_2 have comparable ion-exchange capacity (IEC) value and thermal stability to those of the fluorinated polymer prepared in emulsion. This new polymerization method may find application for preparation of the PFSA ionomers in the future.

2.2.3 MEMBRANE FORMATION

2.2.3.1 Membrane Preparation Method

There are two practical methods to prepare the PFSA ionomer membrane. One is melt extrusion and the other is solution cast. The former is suitable for continuous production and no solvent needed, and thus it will not cause harm to the environment. The PFSF displays a single α-relaxation at about 0°C, so the PFSF can be easily melt processed.[19] Extrusion at 180°C–220°C is usually used for the PFSF. The temperature required for processing the acid or salt form ionomer is substantially increased and may reach the decomposition point of the ionomer. As a result, the acid or salt form ionomer is no longer melt processable.

The general procedure to prepare solution-cast PFSA membrane is as follows. First, since the PFSF cannot dissolve in common solvents, it should be converted to Na^+ form ionomer by hydrolysis with aqueous NaOH. Afterward, the Na^+ form ionomer is immersed into H_2SO_4 or HNO_3 solution to convert to H^+ form ionomer. Then, the H^+ form ionomer dissolves in solvent. The reported solvents include dimethylformamide (DMF), dimethyl sulfoxide (DMSO), ethylene glycol (EG), and mixture of water with alcohols. Finally, the membranes were fabricated by solution cast of the obtained H^+ form ionomer solution.

Solution-cast process has great effect on the property of the membrane. Casting the PFSA ionomer solution at room temperatures produces brittle membranes soluble in water and organic solvents. We studied the morphology of the room casting membrane. The scanning electron microscopy (SEM) image shows that the membrane is assembled loosely by aggregates with the size of 60–150 nm. While when the membrane is fabricated at higher temperature, the aggregates can fuse into each other well to form dense membranes. Increasing casting temperature increases membrane insolubility, membrane density, and tensile stress at the break of the membranes.[20]

Recently, we develop a new method to prepare PFSA membranes from their precursor in hexafluoropropene trimer (HFPT) solvent. Dissolution of the PFSF has been achieved by swelling it in HFPT (2 wt%) and heating for 2 h at 250°C under pressure. Then, the PFSF membranes were fabricated at 80°C or 120°C. After hydrolysis, these membranes are converted into PFSA membranes. Compared with those of traditional solution-casted membranes, the obtained membranes exhibit higher dimensional stability, proton conductivity, and chemical stability and remarkably decreased methanol crossover.[21]

2.3 MEMBRANE CHARACTERIZATION

2.3.1 SPECTRAL STUDIES

2.3.1.1 Infrared Spectra

Infrared spectroscopy has been applied as useful method to study PFSA membranes. According to the chemical structure of the PFSA, there are three structural components: the fluorocarbon main chain, the ether-linker fluorinated side chains, and the ionic end groups $-SO_3^-M^+$. Besides, water is a very important component in the membranes. All these four components have available assignments on the Fourier transform infrared (FT-IR) spectrum of the membranes. Figure 2.7 shows the FT-IR spectrum of the Na$^+$-type perfluorosulfonated ionomer membrane. And Table 2.3 lists the positions of the main absorption band, together with the best available assignments to vibrational mode of the structural components. The most intense absorptions in the spectrum are from the fluorocarbon main chain. Their assignment can be made by comparison with polytetrafluoroethylene (PTFE) and with PFSA membranes literature data.[22] The band at 1156 cm^{-1} is attributed to symmetric C–F stretching, while the band at 1231 cm^{-1} is due to asymmetric C–F stretching. The ether linkages C–O–C in the side chains give rise to the band at 982 cm^{-1}, which is assigned to C–O–C symmetric stretch. The band at 1062 cm^{-1} is due to $-SO_3^-$ symmetric stretching. Although the ionomer sample is vacuum dried before FT-IR measurement, the spectrum contains distinct broadband in the region between 3200 and 3700 cm^{-1} due to the fundamental stretching of water, while the band at 1667 is attributed to the bending of water.

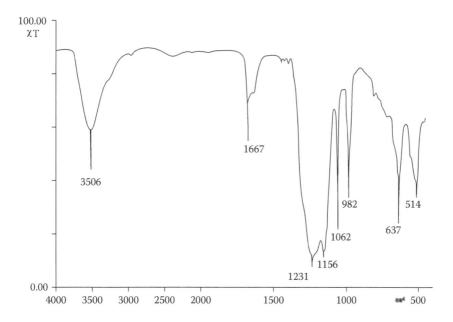

FIGURE 2.7 Infrared transmittance spectrum of Na$^+$ perfluorosulfonic ionomer.

TABLE 2.3

Infrared Absorption Bands of PFSA–Na⁺ Ionomer and Raman Spectroscopy Peaks of PFVESF Precursor (870 g/mol)

FT-IR		Raman	
Band Position (cm⁻¹)	Assignment	Band Position (cm⁻¹)	Assignment
3506	$\nu_3(H_2O)$	292	$\tau(CF_2)$
1667	$\nu_2(H_2O)$	385	$\delta(CF_2)$
1231	$\nu_{as}(CF_2)$	575	$\delta(CF_2)$
1156	$\nu_s(CF_2)$	732	$\nu_s(CF_2)$
1062	$\nu_s(-SO_3^-)$,	798	$\nu(C-S)$
982	$\nu_s(C-O-C)$	989	$\nu_s(C-O-C)$
637	$\omega(CF_2)$	1217	$\nu_{as}(CF2)$
514	$\delta(CF_2)$	1297	$\nu(C-C)$
3506	$\nu_3(H_2O)$	1378	$\nu(C-C)$
1667	$\nu_2(H_2O)$	1469	$\nu_s(SO_2)$
1231	$\nu_{as}(CF_2)$		

It needs to note that the bands of water and $-SO_3^-$ depend on the hydration degree of the membranes.[22a,e,f,23] Gruger studied the FT-IR spectra features in the 4000–1500 cm⁻¹ region change with the hydration level of the membrane.[22f] Meanwhile, spectra of PFSA membranes with different counterions differ somewhat. For dry membranes with Li⁺, Na⁺, K⁺, and Rb⁺ counterions, the band due to symmetric stretching vibration of $-SO_3^-$ groups $[\nu_s(SO_3^-)]$ red shifted with the decrease of the size of counterions. While for hydrated membranes, no red shift occurs.[24] Accordingly, FT-IR was applied to study on the state of water, the microstructure of the membranes, and water response to structural changes in the membranes.

Perusich performed detailed FT-IR study of the sulfonyl fluoride, potassium salt, and sulfonic acid forms of Nafion and established the FT-IR method to determine the equivalent weight (EW) of PFSI.[25] For thin films (<1.1 mil), the EW of PFSI can be obtained by calibrating the C–F/C–O–C absorbance band ratio. While for thick films (5–25 mil), the EW of PFSI can be obtained by calibrating the C–F/–SO₂F absorbance band ratio.

Attenuated total reflection (ATR) is a useful method to study the surface structural of PFSA membrane, and it can circumvent the film thickness problem. Liang found that $\nu_{as}(CF_2)$, $\nu_s(CF_2)$, and $\nu_s(SO)$ of ATR spectra are red shifted by 20.4, 10.7, and 4.8 cm⁻¹ compared with that of transmission IR, which indicates that chemical environment is different for membrane surface and bulk.[22b]

2.3.1.2 Raman Spectroscopy

Figure 2.8 shows the Raman spectrum of the long-side-chain PFVESF (LSC-PFSF) (870 g/mol), which was obtained by excitation with 632.8 nm radiation from a He–Ne gas laser and blue diode laser with 473 nm operated at about 17 mW (~12 mW on the sample) on LabRAM ARMIS IR.[26]

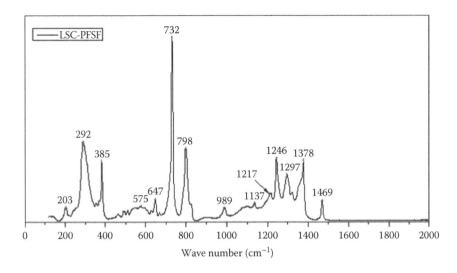

FIGURE 2.8 Raman spectrum of the long-chain PFVESF (870 g/mol). (From Gao, H., Stabilisation of perfluorosulfonic acid membrane by cross-linking and inorganic–organic composite formation. Application in medium temperature proton exchange membrane fuel cells, PhD dissertation, University of Montpellier II, Shanghai Jiao Tong University, Shanghai, China, 2010.)

And Table 2.3 lists the maxima of the lines observed in the Raman spectrum and their assignments. These peak positions correlated well with results obtained by Gruger.[22f] As for PFSA ionomer, the peaks due to SO_3^- should appeared at 1060 cm^{-1} $\nu_s(SO_3^-)$ and 1216 cm^{-1} $\nu_{as}(SO_3^-)$.

2.3.1.3 NMR

NMR is a powerful technique for the study of the molecular structure and morphology of polymeric materials. For PFSA, ^{19}F NMR and ^{13}C NMR are used to study the chemical structure and morphology.[27] Chen and Schmidt-Rohr reported the ^{19}F and ^{13}C assignments of Nafion 117 by 2D^{13}C–^{19}F heteronuclear correlation and ^{19}F-exchange NMR experiments.[27b] The backbone and side group CF sites are assigned to the signals at −138 and −144 ppm, respectively. The CF_3 and OCF_2 signals are near −80.4 and −79.8 ppm, respectively. The SCF_2 group is at −117 ppm. Takasaki et al. clearly determined all ^{19}F and ^{13}C assignments of Nafion (EW = 1100) in lower alcohol and water solution.[27a] The results are shown in Table 2.4. Meanwhile, on the basis of the NMR assignments and the integrated values, the presence of aggregates in a PFSA ionomer solution is suggested.

Compared to PFSA ionomer, a new peak due to $-SO_2F$ appears at 44 ppm in ^{19}F NMR spectrum of the PFSF.[26]

^1H NMR is a powerful technique to pursue both quantity and motion of water involved in a Nafion membrane. Zhao et al. reports analytical results of equilibrium swelling water using ^1H NMR.[23] There are two types of absorbing water. Type 1 is the strongly adsorbing water on the sulfonic acid groups to form a water shell, while type

TABLE 2.4
^{13}C NMR and ^{19}F NMR Assignment of Nafion (EW = 1100)

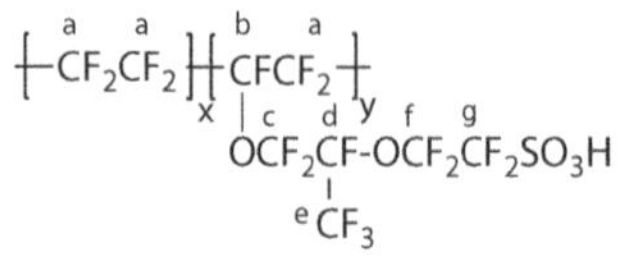

$$\left[CF_2CF_2\right]_x \left[CFCF_2\right]_y$$
$$\underset{c}{|}\ \underset{d}{O}CF_2\underset{}{CF}\text{-}OCF_2CF_2SO_3H$$
$$\underset{e}{|}\ CF_3$$

(a, a, b, a labels above; c, d, f, g; e label)

	^{13}C NMR				^{19}F NMR	
	Chemical Shift (ppm)	Multiplicity	$^1J_{CF}$, $^2J_{CF}$ (Hz)	Line Width (Hz)	Chemical Shift (ppm)	Line Width (Hz)
a	112.5	t	290	97	−123.2	780
b	109.2	d	265	92	−140	317
c	118.0	t–d		75	−80.3, −80.1	Not clear
d	104.0	d–s	267, 37	27	−146.1	153
e	119.4	q–d	287, 32	30	−80.3, −80.1	Not clear
f	119.0	t–t	291, 34	51	−80.3, −80.1	Not clear
g	113.5	t–t	290, 37	22	−118.9	134

Source: Takasaki, M. et al., *Macromolecules*, 38(14), 6031, 2005.

2 is the weakly adsorbing water molecules on the water shell. Wakai et al. employed ^{1}H NMR spectroscopy to reveal the hydration process of a Nafion membrane.[28] The results suggest that the hydration process comprises three distinct steps. The first hydration is a very strong adsorption of water probably on the hydroxyl group of the sulfonic acid group, and the second one is a relatively weak adsorption on another site of the sulfonic acid group. The third hydration is readily assigned to excess bulk (liquid-like) water.

2.3.2 Morphological Studies

PFSA ionomer is known to have microphase-separated structure with hydrophobic regions that provide excellent structural integrity and hydrophilic domains that facilitate proton transport. The morphology of PFSA is known to change with hydration level, process condition, and thermal history. Many experimental techniques including SEM, transmission electron spectroscopy (TEM), atom force microscopy (AFM), small-angle x-ray scattering (SAXS), and x-ray diffraction (XRD) are applied to detect the morphology of the PFSA membranes.

2.3.2.1 Scanning Electron Microscopy

SEM has been applied to study on the surface morphology and cross-sectional morphology of the PFSA membranes and their composite membranes.[20,29] Luan and coworkers carried out SEM study on surface morphology of PFSA ionomer film prepared from DMF solution at different temperatures. The film prepared at low temperature is assembled loosely by sphere-like globules of about 20 nm, and some of the globules assembled to form aggregates about 60–150 nm, while the surface of the membrane prepared at high temperature is flat and dense.[20] Yang and coworkers compared the cross-sectional morphologies of ePTFE-reinforced composite PFSA membranes and found that the ePTFE substrate is well impregnated with the PFSA in PFSF precursor solution made composite membrane, whereas discontinuous layers are observed in its PFSA solution made counterpart.[29b] Su et al. investigated the microstructure of pristine PFSA membrane and modified PFSA membrane with polysiloxane by SEM and confirmed no observable microphase separation occurred in modified PFSA membrane.[29c]

SEM equipped with an energy-dispersive spectroscopy (EDS) provides image of the materials as well as chemical analysis. Bi et al. investigated the cross-sectional morphology of the Nafion/SiO_2-supported sulfated zirconia composite membrane and obtained the silicon and zirconium element distribution in the polymer matrix by SEM-EDS.[30] Chalkova et al. carried out surface and cross-sectional morphology study of Nafion/TiO_2 composite membranes by SEM-EDS.[31]

2.3.2.2 Transmission Electron Spectroscopy

TEM can provide a direct visualization of the size, shapes, and distribution of ionic clusters, crystallites, and the continuous perfluorinated matrix. Many TEM studies have been performed for study of ionic cluster structure of PFSA membrane and have generally found evidence of small 3–10 nm ionic clusters approximately in spherical shape.[4] Due to poor image contrast in PFSA membrane, dye-staining and ultrathin sectioning techniques are applied in sample preparation. On the other hand, the image of PFSA membrane shows strong effect on electron beam–induced damage.[32]

Ceynowa performed a TEM study of microtomed Nafion 124 membranes. To increase the electron density contrast, H^+ form membranes were converted to the Pb^{2+} form using 1.0 M $Pb(NO_3)_2$ for 60 h. Then, the sample was dehydrated using ethanol solutions, rehydrated in 1, 2-epoxypropane, and embedded into epoxy resin. Ultrathin sections were cut to 60–80 nm size. TEM images of the film indicate ion cluster of 3–6 nm in diameter uniformly distributed in the polymer matrix.[33]

Xue et al. carried out TEM investigations of Nafion 117 morphology.[34] The solution-cast films of thicknesses in the range 50–100 nm were exposed to RuO_4 vapor, which reacts with water to form RuO_2 microparticles. TEM images showed three phases in micrographs and spherical clusters in the diameter range of 2.5–5 nm were surrounded by interfaces and uniformly distributed in polymer matrix.

TEM investigation of thin sections (60 nm) of Nafion (1100 EW) in the cesium ion form was performed by Fujimura et al.[35] The heavy Cs^+ ion associated with the sulfonate groups provides electron density contrast. TEM image of the sample shows isolated dark particles of a few nanometers in diameter dispersed in the matrix.

Wang et al. studied the micromorphology of Nafion in different dilute solutions by TEM.[36] Thin Nafion membranes were stained by mercury nitrate solutions. The size of the ionic domains and the morphology of the proton transport channels in the Nafion membrane varied with the solvent.

However, Yakovlev et al. pointed out that the images of Nafion show strong effects of electron beam–induced damage.[32] At elevated dose rates, the membrane is damaged even before the first image is collected. The low level of contrast in low-dose images is a good representation of Nafion structure. In contrast to the claims that ion exchange enhances the contrast in Nafion, they observed no effect on contrast in undamaged material. They suggest that using highly a conductive substrate is one of the most promising methods to reduce damage.

2.3.2.3   Atom Force Microscopy

AFM is one of the foremost tools for imaging, measuring, and manipulating matter at the nanoscale. For PFSA membrane, AFM has been applied to both measurements of surface textural features as well as the direct examination of ionic clustering.

Lehmani et al. carried out an early tapping mode (TM) AFM study of Nafion 117 membranes.[37] The results show *supernodular aggregates* of spherical domains having diameters of ~45 nm. This domain size is greater than the structure size of an x-ray study. James et al. performed TM AFM studies of Nafion 115-H$^+$.[38] The results show clusters having a range of size from 5 to 30 nm, significantly larger than 4 nm structures proposed from an x-ray study. The authors attributed it to formation of cluster aggregates. McLean et al. used AFM methods to resolve the surface and near-surface morphology of the ionic domains in Nafion membranes.[39] A new tapping AFM method where low oscillation amplitudes were used was applied to resolve ionic domains down to about 0–5 nm below the surface. The low amplitude experiments revealed ionic domains of a size of 4–5 nm for typical ambient humidity conditioned Nafion 117 (K$^+$). Upon exposure to water, the ionic features became enlarged with a size of about 4–10 nm in diameter. In other operating modes, the fluoropolymer crystal or aggregate domains were imaged using tapping AFM. The dimensions of fluorocarbon crystalline domains are on the order of 10 nm. McLean et al. also studied the near-surface region of Nafion membranes using very light TM. The results verified that there are essentially no ionic species at the outermost few angstroms of the surface and the concentration of ionic species is substantially greater just under the surface. Affoune et al. investigated the surface morphology modification of the Nafion 117 membrane by AFM.[40] They found Nafion topography considerably changes when samples absorb water.

Conductive AFM was used to characterize Nafion membrane. O'hayre and coworker carried out conductive AFM of Nafion 117 membrane. By AFM impedance technique, local activity of the electrolyte membrane is visualized. Hydrophilic domains with a size on the order of several hundred nanometers were resolved.[41] Aleksandrova et al. introduce an electrochemical AFM method to measure the ion current through the Nafion membrane and to visualize ionically conductive channels on the membrane surface.[42] From simultaneously recorded current images, ionically conductive region was identified, and the conductive region has the ends of channel structures with a diameter of 9–25 nm in width. O'Dea et al. investigated the surface morphology of Nafion at hydrated and dehydrated conditions using AFM.[43] They observed that Nafion's surface morphology evolves from a low density of isolated hydrophilic domains at dehydrated conditions to a network of elongated, segmented, and wormlike domains that span several micrometers at hydrated conditions.

2.3.2.4   SAXS

SAXS is mainly one of the important experimental techniques for studying on the ion cluster and crystal structure of the PFSA ionomer. Almost all the morphological models of PFSA membrane are from simulating results of SAXS data.[44–48] The scattering characteristics of Nafion's nanostructure are the scattering maximum around the scattering vector $q = 1$–2 nm^{-1}, which arises from the periodic ionic clusters (which yields the domain- or d-spacing) and a broader peak observed at lower q value (0.4–0.8 nm^{-1}), named the matrix knee, which is associated with the intercrystalline spacing in the polymer matrix.[49] q can be obtained from the following equation:

$$q = \frac{4\pi \sin(\theta)}{\lambda} \tag{2.1}$$

where θ and λ are the scattering angles and the wavelength of the x-rays, respectively.

The effects of EW, cation form, temperature, and water content on the morphology of the PFSA ionomer and morphology change brought about by swelling with water/methanol mixture are investigated by SAXS technique.[50] Table 2.5 lists SAXS spacing of H$^+$ form PFSA membrane as function of EW. The results indicate that the interlamellar spacing (D) increases with increasing EW, which corresponds to increasing crystallinity and perfection of the crystals with EW. SAXS spacing of various PFSA membranes with different cations under dry state was also determined by Hashimoto et al.[50a] For a given EW and anions, the ionic spacing for Na$^+$ form membranes is greater than that for H$^+$ membranes.

Orfino and Holdcroft performed SAXS studies of Nafion 117 acid form samples that were in the dry and wet states. One of the conclusions was that for the wet membranes, the cluster's number density is determined to be three times less than

---

TABLE 2.5

Physical Properties of PFSA Membrane as a Function of EW Determined by SAXS and XRD

	Water Uptake		Crystalline Phase		SAXS Spacing (nm)[c] Lamellar (D)	SAXS Spacing (nm)[c] Ionic (S)	
EW	(wt%)	N[a]	FWHM[b] (deg.)	X_c (wt%)	Dry	Dry	Wet
1100	32	19.6	1.6	12	18.3	2.6	5.5
1150	23	14.7	—	—	—	—	—
1200	20	13.3	1.4	19	18.7	2.7	5.2
1400	10	7.8	1.0	20	22.2	2.8	4.4
1500	8	6.7	0.9	22	24.9	2.8	4.4

Source: Hashimoto, T. et al., Structure of sulfonated and carboxylated perfluorinated ionomer membranes, in: *Perfluorinated Ionomer Membranes*, American Chemical Society, Vol. 180, 1982, pp. 217–248.

[a] Number of water molecules per one –SO$_3$H group.

[b] FWHM of decomposed profile due to crystals in the wide-angle x-ray diffraction profile.

[c] Spacing estimated from the Bragg equation.

air-dried Nafion, but the radius of the ionic cores is larger. The average channel length is in the range of 0.3–0.88 nm.[51]

Grazing-incidence small-angle scattering (GISAXS), which combines the accessible length scales of small-angle scattering (SAS, SAXS or SANS [small-angle neutron scattering]) and the surface sensitivity of grazing-incidence diffraction (GID), is a powerful tool to study nanostructured surfaces and thin film. Modestino et al. carried out in situ GISAXS measurements on the nanostructure of thin-film Nafion ionomers.[41] The results demonstrate water uptake has the strong effects on internal morphology of Nafion thin films.

2.3.2.5   X-Ray Diffraction

XRD is used to determine the crystallinity of the PFSI membranes. According to the reports,[20,50d,52] Nafion membrane samples display a broad diffraction feature at scattering angle $2\theta = 8°–25°$, which can be deconvoluted into two peaks. The broad scattering peak at $2\theta \approx 15.5°$ and the sharp peak at $2\theta \approx 17.5°$ are assigned to diffraction maxima associated with amorphous and crystalline regions of the membranes, respectively. The crystallinity (X_c) can be calculated with the following equation from the deconvoluted peaks centered at $17.5°$ and $15.5°$:

$$Xc_{cr} = \frac{\int_0^\infty I_{cr}(s)s^2 ds}{\int_0^\infty \left[I_{cr}(s) + I_{am}(s)\right]s^2 ds} \qquad (2.2)$$

where
  X_c is the calculated crystallinity
  I_{cr} and I_{am} are the relative intensity of the deconvoluted peaks, centered at $17.5°$
      and $15.5°$, respectively
  s is the magnitude of the reciprocal-lattice vector, given by

$$s = \frac{2\sin(\theta)}{\lambda} \qquad (2.3)$$

where
  2θ is the diffraction angle (°)
  λ is the x-ray wavelength

Hashimoto estimated X_c, the lattice spacing of the crystalline peak and full width at half maximum (FWHM) of the crystalline peaks of the PFSA membranes with different EWs based on the XRD measurements.[50a] The results are listed in Table 2.5. From Table 2.5, X_c increases with increasing EW of the membranes due to a decreasing number of noncrystallizable units. FWHM and the lattice spacing decrease slightly with increasing EW, indicating that the perfection of the crystals also becomes high with EW.

Grazing-incidence x-ray diffraction (GIXRD) is a method of characterizing the ordered and amorphous nanostructure of the surface of materials. Tang et al.

reported the determination of the crystallization properties of the solution-cast PFSI membrane surface by a combination of GIXRD and GISAXS.[53] Crystallinity and crystallite size at the membrane surface as a function of the penetration depth were calculated from GIXRD patterns of the membrane. A crystallite-rich skin with 4 nm thickness is observed on the PFSA membrane surface. And the skin possesses a crystallinity that is 40% higher than the bulk.

2.3.3 PHYSICAL PROPERTIES

2.3.3.1 Solubility

The solubility parameter (δ) of a material is a fundamental property of a material and is a measure of the intermolecular forces in a given substance. Yeo determined the solubility parameter of Nafion membrane experimentally by the swelling technique.[54] The results indicated that Nafion exhibits two solubility parameter values δ_{21} and δ_{22}, which have been tentatively ascribed to the fluorocarbon backbone and the ion clusters of the membranes, respectively. For Nafion 1100, δ_{21} is equal to ~10.0 and δ_{22} is equal to ~16.8. While for Nafion 1200, δ_{21} and δ_{22} are ~9.6 and ~17.3, respectively.

The solubility property of PFSA membranes with no cross-linking depends on their crystalline domains to inhibit the dissolution. PFSAs with EW of less than 1000 are soluble in most solvents, primarily due to the absence of crystallinity.[55] However, PFSAs with EW of more than 1000 usually are insoluble in all solvents at temperature lower than 180°C. The degree of solubility increases with temperatures. At temperatures higher than 180°C, crystalline domains of these polymers melt down and the polymers are dissolved.

Moore and Martin first reported that cast Nafion films at high temperatures from high-boiling solvents (e.g., DMSO, EG, and DMF) were insoluble.[52] They introduced percent solubility W_{sol} that is calculated from the ratio of the final weight ($M_{residue}$) of Nafion film divided by the initial weight (M_{mem}), multiplied by 100%. The final weight of Nafion film is obtained by weighting the film that is prepared from the solution:

$$W_{sol} = \frac{M_{residue}}{M_{mem}} \times 100\% \qquad (2.4)$$

Zook and Leddy reported commercial PFSA membranes (Nafion 117, EW = 1100), and recast films (from Nafion suspension, 5% w/v in alcohol/water, EW = 1100, heating the films in an oven) heated at 140°C for as little as 10 min were insoluble in a sonicated ethanol/water mixture, while unheated recast films were 22%–100% soluble under the same conditions.[56]

Siroma et al. examined degrees of dissolution of Nafion 117 membranes and recast films made from Nafion solution in methanol/water mixtures.[57] At 80°C, more than 30% of the membrane was dissolved in mixed solvents with methanol concentrations of higher than 80%. Even at 35°C, solvents containing methanol higher than 60 mol% dissolved the membrane. Heat treatment of the recast film improved the resistance to dissolution. The recast films treated at 160°C for constant period (1 min) were not dissolved by any of the solvents at room temperature.

2.3.3.2 Thermal Stability

Perfluoropolymers exhibit high thermal stability. Gao reported that for the sulfonyl fluoride precursor PFSF (870 g/mol), the 1% weight loss temperature is 385°C.[26] The incorporation of ions into their structure reduces the thermal stability. The thermal decomposition proceeded in three stages: loss of water; loss of sulfonic groups, denoting a breaking of the C–S bonds; and the destruction of the perfluorinated matrix. There is no mass loss up to 340°C. The 5% weight loss temperature is 368°C. The first stage (340°C–400°C) is associated with a desulfonation process, and the second stage (400°C–600°C) is related to the side chain and PTFE backbone decompositions.

Feldheim et al. reported a strong dependence of the thermal stability of Nafion on the nature of the counterion.[60] Nafion films show improved thermal stability as the size of the counter cation decreases with the exceptional of Li^+ form film. The thermal stabilities of the alkali metal–exchanged Nafion films follow this trend shown:

$$Na^+ > K^+ > Rb^+ > Cs^+ \sim Li^+$$

Iwai and Yamanishi investigated the effect of exchanged ions on the thermal stability of Nafion N117CS membranes.[59] They confirmed that the Na-exchange Nafion N117CS membrane possessed better thermal stability than K^+ and Li^+ form Nafion membranes. They also found the thermal stability of an aluminum (III)-exchange Nafion membrane is significantly lower, which agrees with the result early reported by Sun et al.[60]

2.3.3.3 Mechanical Properties

The mechanical properties of PFSA membrane are important to its practical use and lifetimes. Iwai and Yamanishi measured the tensile strength and percent elongation at the break of the cation-exchange Nafion membranes at room temperature based on the American Society for Testing and Materials standard tests (ASTM D 1822L).[59] The results are shown in Table 2.6. Humidity and temperature have effect on the mechanical properties of PFSA membrane. Young's modulus and the proportional limit stress of the PFSA membrane decrease as humidity and temperature increase. At the same time, higher temperature leads to lower break stress and higher break strain. However, humidity has little effect on the break stress and break strain.[61]

In an early study, Yeo and Eisenberg performed dynamic mechanical studies on H^+ form Nafion membrane (EW = 1365).[62] They note three mechanical peaks, labeled α, β, and γ. The α-peak occurs at about 110°C. This relaxation was initially considered as the glass transition of the fluorocarbon matrix. The β-peak was seen at about 20°C. With increasing water content, the β-peak migrates to lower temperature, which suggested that β-peak associated with molecular motions within the ionic domains. The γ-peaks at about −100°C were attributed to short-range molecular motions in the TFE phase. Later, Kyu and Eisenberg discussed dynamic mechanical relaxations for the same Nafion membrane.[63] They found the α-peak is very

TABLE 2.6

Tensile Strength of Various Cation-Exchange Nafion N117CS Membranes

Form	Tensile Strength (MPa)		Elongation at Break		Water Uptake (%)
	Average	Standard Deviation	Average	Standard Deviation	
H+ form	41.67	1.26	106.31	3.85	10.26
Li+ form	48.05	4.65	70.61	5.46	10.65
Na+ form	45.16	1.69	62.61	3.57	7.89
K+ form	37.14	2.10	72.12	5.23	7.53
Mg+ form	30.14	1.96	77.00	7.32	16.67
Ca+ form	33.98	1.31	85.32	4.57	14.71
Al+ form	33.11	1.42	80.68	11.53	18.88

Source: Iwai, Y. et al., *Polym. Degrad. Stab.*, 94(4), 679, 2009.

sensitive to ion type, degree of neutralization, and water content, so they reassigned the α-peak as the glass transition of the polar regions and assigned the β-peak as the glass transition of the Nafion matrix.

The effect of counterion type and size on the dynamic mechanical properties of Nafion has been studied by Cable et al.[19] The PFSF displays a single α-peak near 0°C. However, once the copolymer is converted to the Na+ form, the α-peak relaxation shifts to a temperature near 250°C. For H+ form Nafion, the α- and β-relaxation occurred at temperatures about 100°C.

2.3.4 Ion-Exchange Capacity, Water Uptake, and Swelling Ratio

2.3.4.1 Ion-Exchange Capacity

IEC is defined by the total amount of ions bound to the ion-exchange matrix. It is characterized as the number of counterion equivalents in a specified amount of matrix.

IEC is basic and important physical property for ionomers since many key properties of ionomers such as ionic conductivity, water uptake, and degree of swelling depend directly on IEC.

IEC value of the PFSA membranes can be obtained by titration, analysis of atomic sulfur, and FT-IR. Luan et al. determined the IEC of PFSA ionomer from the titration of the released amount of proton of the preweighed dry acid–type samples in 2 M sodium chloride with 0.1 M sodium hydroxide by using a phenolphthalein indicator, and it was recorded as an average value for each sample in units of milliequivalents NaOH per gram of the dry acid form membrane (mequiv./g).[20]

For Nafion, EW, a measure of the ionic concentration within ionomer, is more often used than IEC. EW is the number of grams of dry Nafion per mole of sulfonic acid groups when the material is in the acid form (g/equiv.). The EW of Nafion membrane can be expressed using the following equation. The lower the

EW, the lower the TFE concentration and the higher the perfluorinated vinyl ether concentration:

$$EW = 100m + 444 \tag{2.5}$$

where

> m is the number of TFE groups on average per perfluorinated vinyl ether monomer
> 100 and 444 are the molecular weights of TFE and perfluorinated vinyl ether monomer, respectively

EW is equal to 1000/IEC. Like IEC, EW is conveniently measured either by FT-IR techniques or by analysis of atomic sulfur. FT-IR methods for EW measurements are described in the literature.[25]

2.3.4.2   Water Uptake

Water uptake of PFSA membranes has been a frequent topic of study over the years due to its relationship to membrane properties such as ionic conductivity.

Water uptake (W_{H_2O}) is expressed in terms of weight percent calculated based on the weight of the wet sample:

$$W_{H_2O} = \frac{\left(W_{wet} - W_{dry}\right)}{W_{dry}} \times 100\% \tag{2.6}$$

where W_{dry} and W_{wet} are the weights of dry and corresponding water-boiled samples, respectively. The relationship between the water content (λ) in terms of the number of water molecules per sulfonic acid site and the water uptake is illustrated by

$$\lambda = \frac{EW \times W_{H_2O}}{M_{H_2O}} \tag{2.7}$$

where M_{H_2O} is the molecular weight of water.

The previous history and pretreatment of the membranes have a marked impact on water uptake of the membrane due to morphology change in ionic cluster and crystallinity. The water uptake of the membrane is strongly dependent on the preceding drying method used.[64] The membrane dried at room temperature takes up roughly twice as much water as the membrane dried at elevated temperatures. Luan et al. reported that swelling and water uptake of the solution-cast PFSA membranes decreased with increasing casting temperature.[65]

Broka and Ekdunge investigated water uptake from the vapor phase by Nafion 117 membrane and recast film.[66] The results show water vapor uptake by both Nafion 117 membrane and recast film decreased with increasing temperature. The lower water uptake at higher temperatures has also been reported by other researchers. Broka and Ekdunge also found PFSA membrane water uptake from the liquid water is higher than those for water vapor.[66] This phenomenon is known as Schroeder's paradox. Zawodzinski et al. explored the isopiestic sorption curve for Nafion 117-H⁺.

They found water uptake from the vapor phase is notably lower with that from liquid water, with 14 waters per sulfonate absorbed from the vapor phase and 22 from the liquid phase. They considered the difficulty in condensing vapor within the pores of the membrane as a possible reason for lower water uptake from the vapor phase.[64]

Hinatsu et al. performed an extensive study on the water uptake of PFSI membranes.[67] Water uptake from liquid water increased with decreasing EW of the ionomer and increased with temperature for absorption. Water content from the vapor phase decreased with temperature down to 10 waters per sulfonate site at 100% RH at 80°C.

2.3.4.3 Swell Ratio

There are three kinds of swell ratio: length swell ratio, area swell ratio, and volume swell ratio.

The length swell ratio can be calculated by

$$SR = \frac{\left(L_{wet} - L_{dry}\right)}{L_{dry}} \times 100\% \qquad (2.8)$$

where L_{wet} and L_{dry} are the lengths of wet and dry membranes, respectively.

The area swell ratio (SR) can be calculated by

$$SR = \frac{\left(S_{wet} - S_{dry}\right)}{S_{dry}} \times 100\% \qquad (2.9)$$

where S_{wet} and S_{dry} are the surface areas of wet and dry membranes, respectively.

The volume swell ratio is illustrated by

$$SR = \frac{\left(V_{wet} - V_{dry}\right)}{V_{dry}} \times 100\% \qquad (2.10)$$

where V_{wet} and V_{dry} are the volumes of wet and dry membranes, respectively.

Bauer et al. performed humidity scans of the Nafion 117-H$^+$ membrane swelling isotherms.[68] The results indicated that the length swell ratio increased with the humidity and the increase in length is more pronounced at high humidity and at temperature more than 75°C. A similar behavior for the sorption isotherm of Nafion at room temperature has been described by Laporta et al.[22a] and Zawodzinski et al.[64]

Besides water, PFSI membranes swell in various solvents including alcohols, glycol, and amines. Yeo performed comprehensive swelling studies of Nafion in K$^+$ form in organic solvent.[54] The results showed that there were two distinct swelling envelopes corresponding to dual cohesive energy densities of the membrane: one is ascribed to the organic part of the membrane, whereas the other is attributed to the ion clusters in the membrane. Solvent uptake of PFSI membrane also depended on membrane pretreatment and the solubility parameter of the solvent.[55] Meanwhile, the counterion influences swelling of the membranes drastically, and the solvent uptake decreases in the following sequence: H$^+$ > Li$^+$ > Na$^+$ > K$^+$.

2.3.5   PROTON CONDUCTIVITY

Proton conductivity is a crucial criterion for evaluating PEM conducting property and fuel cell performance, and it is theoretically as well as practically important. The high proton conductivity of PFSA membranes makes them an important class of proton conductors.

Proton conductivity of PFSA membranes is most frequently measured from the alternating current (AC) impedance spectroscopic technique. The impedance (Z) is defined as the ratio of the voltage to the current at a given frequency, and it is represented as a complex quantity that consists of a real part (resistance, Z′) and an imaginary part (reactance, Z″) with a phase angle θ as described in the following equation:

$$Z = Z' + jZ'' \tag{2.11}$$

Since the approximated value at which reactive impedance Z″ in the impedance measurement (Equation 2.11) is so small as to be neglected, the resistance of the membranes is estimated and recorded as the impedance (Z) at the extrapolated intercept on the real axis of the impedance curve within a reasonable frequency range. Proton conductivity (σ) of PEM can be calculated from

$$\sigma = \frac{L}{RA} \tag{2.12}$$

where
  L corresponds to the electrode separation: the gap between electrodes for in-plane measurements and thickness of the film for through-plane measurements
  A denotes the membrane cross-sectional area in case of the in-plane setup and the area of the electrodes in case of the through-plane setup
  R is the measured resistance of the membrane

Difference in membrane pretreatments and other experimental conditions such as cell geometry, data analysis, humidification, and temperature can cause variability in proton conductivity data. Zawodzinski et al. studied the proton conductivity of Nafion 117 membrane as functions of membrane water content.[64] The results show the conductivity decreases roughly linearly with decreasing water content. Sone et al. reported the dependence of conductivity of Nafion 117 membrane on both relative humidity and temperature.[69] The conductivity of Nafion 117 without heat treatment was 7.8×10^{-2} S/cm at ambient temperature and 100% RH (vapor). When water content (water per acid site) increased from 2 to 4, the proton conductivity increases exponentially, and above 4, this relationship is linear. The conductivity of the membrane decreases with the increase of temperature from 20°C to 45°C due to loss of water. In contrast, from 45°C to 80°C, the conductivity increases with temperature since the water content remains rather constant, and the activation energy was lower than 2 kJ/mol. Kopitzke et al. carried out a study of the temperature dependence of the ionic conductivity of Nafion 117-H⁺ membranes.[70]

The conductivity of the membrane was measured under liquid H_2O condition. The conductivity of the membrane was 9.2×10^{-2} S/cm at 20°C with a water content of 11.5 H_2O per sulfonic acid site. The activation energy was 7.82 kJ/mol over the full temperature range.

To improve proton transport of PFSA membranes at high temperatures and in order to operate PFSA membranes at temperatures above 100°C, inorganic compounds such as SiO_2[71] and TiO_2[72] were employed as additives to retain water in the Nafion membranes for an acceptable proton conductivity. Proton conductors such as zirconium phosphate,[73] heteropolyacid,[74] and heterocycle compounds[75] including imidazole,[73a,76] benzimidazole,[77] triazole,[77a,78] and polyfunctional phosphonic acid[79] were also added in the Nafion membrane. In addition, ionic liquids were applied to fabricate composite Nafion membranes due to their anhydrous high conductivity and good thermal stability.[80] Besides, Nafion/Nafion-functionalized multiwalled carbon nanotube composite membrane exhibits a remarkable improvement in proton conductivity compared to the pristine Nafion membrane.[81]

Another hot research topic about the PFSA proton conductivity is about the direction of the conductivity. There are two kinds of proton conductivity: tangential/in-plane conductivity and normal/through-plane conductivity. Commonly, proton conductivity is measured along the plane of the membrane, since the measurement of the conductivity in the in-plane direction is much easier to carry out with higher stability, reproducibility, and accuracy. However, in practice, the membrane requires proton conduction perpendicular to the membrane. So, through-plane conductivity of the membrane may have significant effect on the performance of fuel cells.

There are two different viewpoints about whether Nafion membrane is isotropic or anisotropic. Anantaraman and Gardner reported that the tangential conductivity is almost 3.6 times higher than the normal conductivity using coaxial probe method; thus, the specific conductivity of Nafion 117 is anisotropic.[82] Ma et al. measured proton conductivity of Nafion 117 membranes in the direction of thickness by means of two- to four-probe methods.[83] Heavy conductivity anisotropy with $\sigma_{in-plane}/\sigma_{thickness} = 2.5$–$5$ over the thickness and in-plane directions for Nafion 117 membranes pretreated by hot-pressing at a high temperature of 150°C and pressures more than 600 kg/cm^2 was found. For membranes untreated and/or treated at 150°C without pressing, no big fading in proton conductivity was measured. Pozio and coworkers reported conductivity measurements on hot-pressed carbon paper/Nafion membrane (Nafion 112, 115, 117)/carbon paper samples immersed in water and pressure between 1.37 and 1.47 kg/cm^2.[84] The results showed that normal and tangential direction conductivity measurements gave the same results and the membrane is isotropic. Soboleva et al. investigated proton conductivity of Nafion membranes (Nafion 112, 115, 1135, 117, and 211) in the X, Y, Z direction using electrochemical impedance spectroscopy with two-probe electrochemical cell.[85] The results demonstrated a slight anisotropy with the in-plane conductivity over the through-plane conductivity ($\sigma_{in-plane}/\sigma_{through-plane} = 1.0$–$1.4$) with the exception of Nafion 211. Yun et al. studied the effect of pressure on through-plane proton conductivity of Nafion membranes.[86] These results demonstrated that the through-plane conductivity decreased with increasing pressure.

2.3.6 CHEMICAL STABILITY

Chemical stability of PFSA membranes is one of most crucial issues affecting their commercial applicability in fuel cells. Despite their relatively good chemical resistance, PFSA membranes suffer from chemical degradation during fuel cell operation. Peroxyl radical (OOH·) and hydroxyl radical (OH·) are responsible for PFSA membrane degradation.[87] The radicals attack on the remaining H-containing terminal bonds on polymer end group and side chains.[88] The radicals are generated by the decomposition of hydrogen peroxide. Formation of hydrogen peroxide can take place at the cathode and anode.[89] Cipollini proposed that compared to hydroxyl radical, peroxyl radical is less reactive and will attack only the ionomer end group, not the C–S or C–O–C.[90]

During the polymerization of PFSA ionomer, carboxylic acid (–COOH) end group is unavoidably generated due to the hydrolysis of persulfate initiators. Chemical degradation of PFSA membranes starts from the hydrogen abstraction of –COOH group by OH·, forming perfluorocarbon radicals, water, and carbon dioxide (Equation 2.13).[91] The perfluorocarbon radical may then combine with a hydroxyl radical, producing an alcohol intermediate that rearranges to yield hydrogen fluoride and an acid fluoride (Equation 2.14). Finally, the acid fluoride hydrolyzes into HF and carboxylate end groups for the next degradation cycle (Equation 2.15).[87]

$$R_f\text{-}CF_2COOH + OH· \rightarrow R_f\text{-}CF_2· + CO_2 + H_2O \qquad (2.13)$$

$$R_f\text{-}CF_2· + OH· \rightarrow R_f\text{-}CF_2OH \rightarrow R_f\text{-}COF + HF \qquad (2.14)$$

$$R_f\text{-}CF_2· + H_2O \rightarrow R_f\text{-}COOH + HF \qquad (2.15)$$

In addition to the radical attack of the end group, hydroxyl radical OH· will attack the C–S and C–O–C bond in the copolymer side chain.[90] Loss of a great number of sulfonate groups or whole side chains would affect the proton conductivity of the membranes. Ghassemzadeh et al. studied chemical degradation of PFSA after fuel cell in situ tests by solid-state NMR spectroscopy. The NMR spectra prove that degradation mostly takes place within the polymer side chains.[92]

Besides end group and side-chain degradation, Coms proposed a main chain degradation mechanism.[89a] H· radical that is generated in the reaction of OH· radical with H_2 can attack any secondary or tertiary C–F bonds on the main chain in PFSAs, resulting in main chain scission. On the other hand, defects in the main chain such as *residual* C–H bonds or C=C bonds, which may form in the polymer during the manufacturing process, are sufficient to initiate radical reactions and eventually degrade PFSA.

The catalyst Pt has a strong impact on membrane degradation. Ghassemzadeh reported that Nafion is stable when it is exposed to H_2 and O_2 without Pt.[93] Nafion membrane degrades much rapidly with only a change in catalyst and the other component of the fuel cell remaining the same. Similarly, membrane degradation studies at open-circuit conditions have demonstrated that PFSA membrane degradation is insignificant in the absence of H_2, O_2, or catalyst.[94] Mittal et al. studied Nafion membrane degradation mechanisms in PEMFCs.[95] The results showed that membrane degradation occurs because molecular H_2 and O_2 react on the surface of the Pt catalyst to form the membrane-degrading species.

PEMFC contamination can also adversely impact membrane performance and life. The common impurities in fuel cell include anions, cations, CO_X, H_2S, NH_3, SO_x, NO_x, and volatile organic compounds.[87,96] Cationic impurities, including alkaline metals and ammonium, can infiltrate the membrane, considerably reducing performance.[96] Metal ions, such as Fe^{2+} and Cu^{2+}, can catalyze the radical formation reactions, strongly accelerating the chemical degradation of membranes in a PEMFC.[97] Stainless steel is unsuitable as a material for end plates in PEMFCs.[98]

Ex situ accelerated methods were applied to study the degradation of PFSA membranes. Ex situ accelerated chemical degradation experimentation of PFSA most commonly employs Fenton's testing. Fenton's reagents include hydrogen peroxide with Fe^{2+} ions in order to produce hydroxyl radicals as follows:

$$H_2O_2 + Fe^{2+} \rightarrow Fe^{3+} + OH^{\cdot} + OH^{-} \tag{2.16}$$

Ghassemzadeh and Holdcroft reported molecular-level quantification of chemical degradation of PFSA ionomer membranes in the presence of hydroxyl radicals generated using Fenton's reagent.[99] The results show the backbone is stable and that degradation occurs solely on the side chain, with the most significant attack occurring toward the end of the side chain. Besides accelerated degradation of PFSA, aging phenomena have also been shown to occur in Nafion membranes under long storage time at high temperature and relative humidity (80°C and 80% RH, respectively).[100] The aging mechanism consists in sulfonic acid group condensation resulting in anhydride formation between Nafion side chains. This degradation leads to significant evolutions of Nafion properties such as IEC, ionic conductivity, water uptake, and mechanical properties.

To mitigate the membrane degradation, several strategies are adopted: (1) development of *chemically stabilized* grades of PFSA, for example, Nafion 211, for which the concentration of terminal carboxylic acid groups was decreased to negligible levels; (2) use of a perfluorinated initiator instead of persulfate initiators, resulting in the form of perfluorinated end groups[17,101]; (3) composition with PTFE; (4) introduction of peroxide-decomposition catalysts such as MnO_2 to suppress H_2O_2 formation and eliminate existing H_2O_2[102]; (5) utilization of inorganic OH radical scavengers such as cesium (III) or CeO_2[103]; and (6) incorporation of organic radical scavenger such as terephthalic acid.[104]

2.4 MEMBRANE STRUCTURE MODEL AND CONDUCTION MECHANISM

2.4.1 MEMBRANE STRUCTURE MODEL

Mauritz and Moore provided a state-of-the-art review of structure investigation on Nafion membranes.[5] Devanathan reported recent progress in understanding Nafion.[6]

Perfluorosulfonated membranes have a microscopic phase-separated structure with hydrophobic regions and hydrophilic domain. Hydrophobic regions provide the mechanical support and hydrophilic ionic domains provide proton transport channel. Many morphological models for PFSA have been developed based on SAXS and wide-angle x-ray scattering (WAXS) experiments of the membranes. However, because of the random chemical structure of the PFSA copolymer, morphological variation with water content and complexity of coorganized crystalline and ionic domains, limited characteristic detail proved by the SAXS and WAXS experiments, the structure of the PFSI has been still subject of debate.[5] Here, a brief description of seven membrane structure models is provided.

> *Cluster–network model*: Gierke et al. proposed the cluster–network model
> (also referred to as the inverse micelle model).[50b,d,105] This model has been
> the most widely referenced model in the history of PFSA membrane.
> This cluster–network model suggests that the solvent and ion-exchange
> site phase separates from the fluorocarbon matrix into inverted micellar
> structures that are connected by short narrow channels. The ionic cluster
> dimension obtained from SAXS scan of PFSA is 3–5 nm. One nanometer
> channels connect the clusters. The intensity of the reflection associated with
> ionic cluster increases with decreasing EW. The Bragg spacing associated
> with ionic cluster increases with increasing water content and decreasing
> EW. Recently, the nanoscale distribution of distinctly different water mol-
> ecules in Nafion has been chemically imaged using high-spatial-resolution
> AFM-IR spectroscopy.[106] The results have revealed ionically bound water
> molecules that were clustered in domains, while the hydrated free bulk-like
> water molecules were found within transportable channels.
>
> *Core–shell model*: Fujimura et al. reported the results of detailed SAXS and
> WAXS experiments on Nafion (in H^+, Na^+, and Cs^+ forms).[35,49] Two scat-
> tering maxima at $s \sim 0.07$ and 0.3 nm^{-1} (the s is defined as $s = (2 \sin \theta)/\lambda$)
> were obtained and attributed to morphological features associated with
> crystalline and ionic domains, respectively. The intensity of reflection asso-
> ciated with crystalline increased with increasing EW. The origin of scat-
> tering maximum at $s = 0.07$ nm^{-1} was due to an average spacing between
> crystalline lamellar platelets. With increasing water content, the cluster
> size increases. Moreover, the microscopic degree of swelling determined
> by SAXS was found to be much greater than the macroscopic degree of
> swelling. They also observed the microscopic draw ratio is slightly less
> than the macroscopic draw ratio. From these observations, an intraparticle
> core–shell model was proposed (as shown in Figure 2.9). This model sug-
> gests that an ion-rich core is surrounded by an ion-poor shell that is com-
> posed mainly of perfluorocarbon chains. The short-range order distance
> within the core–shell particle gives rise to the ionic scattering maximum.
> The core–shell particles are dispersed in the matrix of the intermediate
> ionic phase. This model does not give a clear view of long-range structure.
>
> *Local-order model*: Based on the observation from SANS experiments
> on Nafion samples and excellent agreement between the theoretical and

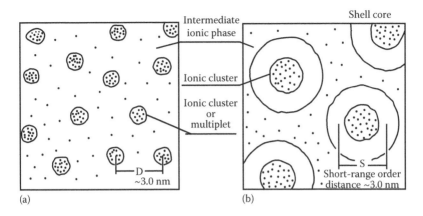

FIGURE 2.9 Core-shell model for Nafion membrane. (From Fujimura, M. et al., *Macromolecules*, 15(1), 136, 1982.)

experimental SANS curves, Dreyfus et al. proposed a local-order model to define the spatial distribution of spherically shaped ionic clusters in Nafion.[107] The model describes the distribution of spherical, hydrated clusters in a locally ordered structure with four first neighbors located at a well-defined distance embedded in a matrix of completely disordered clusters. From the model, the specific surface area per charge site on the cluster was found to be relatively constant 73A^2. The number of charge sites per cluster was found to increase from ~25 to 45 with increasing hydration. However, the number of charge sites per cluster was calculated from the radius of the scattering particle that was unrealistically large.

Lamellar model: Using the SAXS data of Gierke et al. over a limited range of water contents, Litt indicated that the *d* spacings are proportional to the volume of absorbed water.[108] Based on this observation, a lamellar model was proposed. In this model, hydrophilic layers are separated by thin, lamellar PTFE-like crystallites. This model provides a simple explanation for the swelling behavior of Nafion, but it ignored the crystalline interlamellar long spacing. Martin and Moore studied SSC-PFSAs and found that the interlamellar spacing varied with water content in a manner significantly different from that of the intercluster spacings.[109] This observation demonstrated that the lamellar model oversimplifies the morphology of Nafion.

Discrete sandwich: Haubold et al. proposed a variation of the lamellar model, discrete sandwich.[44] The rectangle with a sandwich structure was chosen as the structure element. The sandwich consists of shell (the outer portion) and core (inner liquid portion). The former consists of the side chains and the latter consists of water/methanol molecules. These structural elements were proposed to be juxtaposed in a linear fashion and random distributed inside the membrane. The core thickness *c*, total shell thickness *s*, and their lateral dimensions *a* and *b* are calculated. *a* and *b* is between 15 and 45 A, and the total thickness (*c* + *s*) is about 60 A.

Polymer-bundle model: Gebel investigated the structural evolution of PFSA membranes from dry materials to highly swollen membranes and solutions. On the basis of a semiquantitative analysis of scattering data of Nafion, Gebel proposed a conceptual description for the swelling and dissolution process.[110] For the dry membrane, isolated, spherical ionic clusters with diameter of roughly 15A embedded in the polymer matrix. With the absorption of water, the cluster swells to hold pools of water. When the water volume fraction increased to 0.3–0.5, structural reorganization occurs, and spherical ionic domains connected with cylinders of water dispersed in the polymer matrix. With increasing water volume fraction, a connected network of polymer rods forms. Finally, at water volume fraction between 0.5 and 0.9, this connected rodlike network swells. This model provides a plausible mechanism for evolution in structure from low water content to high water content. However, there is no experiment evidence to prove phase inversion point at water volume fraction of 0.5.

Parallel water-channel (inverted micelle cylinder) model: Schmidt-Rohr and Chen quantitatively simulated previously published SAXS data of hydrated Nafion and found that none of the models discussed earlier agree with the SAXS data. They proposed a model featuring long parallel water channels in cylindrical inverted micelles (Figure 2.10).[46] At 20 vol% water, the water channels have diameters of between 1.8 and 3.5 nm, with an average of 2.4 nm. Nafion crystallites are elongated and parallel to the water channels, with cross sections of about $(5 \text{ nm})^2$. The model with wide parallel water channel can easily explain important features of Nafion, including

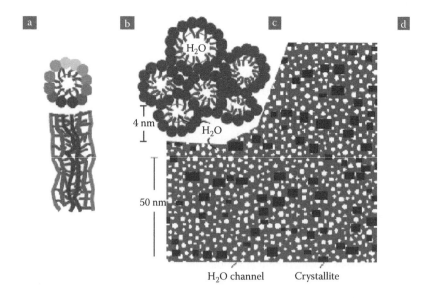

FIGURE 2.10 Parallel water-channel (inverted micelle cylinder) model for Nafion membrane. (From Schmidt-Rohr, K. and Chen, Q., *Nat. Mater.*, 7(1), 75, 2008.)

fast diffusion of water and protons through Nafion and its persistence at low temperatures.

Bicontinuous network of ionic clusters: Elliott and co-workers demonstrated a unified morphological description of PFSAs based on both statistical (MaxEnt) and thermodynamic (DPD) descriptions, which broadly favours a bicontinuous network of ionic clusters embedded in a matrix of fluorocarbon chains. Elliott, 2011[49] The existence of a continuous network of water-filled channels explains the high water diffusion coefficient of water in Nafion.

2.4.2 Membrane Conduction Mechanism

Due to the importance of proton transport on fuel cell performance of PFSA membranes, it is desirable to understand proton transport in membranes. Kreuer et al. provided a state-of-the-art review of proton transport mechanisms both in homogenous media and heterogeneous systems.[111] Transport of protons in PFSA membranes occurs through water-swollen, hydrophilic channels that form as a result of nanophase separation of hydrophilic and hydrophobic segments of the ionomer. Proton conduction in such heterogeneous systems can occur through the Grotthuss mechanism (structure diffuse), vehicular mechanism, and surface mechanism (Figure 2.11).[112]

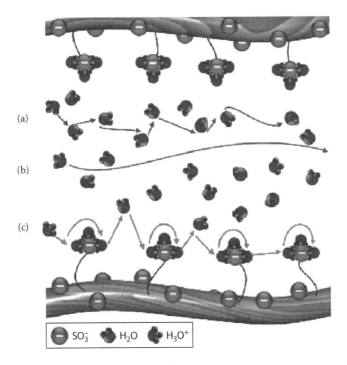

FIGURE 2.11 Schematic illustration of different modes of proton conduction in a solid polymer electrolyte: where) Grotthuss, vehicular, and (C) surface mechanisms. (From Peckham, T.J. and Holdcroft, S., *Adv. Mater.*, 22(42), 4667, 2010.)

At high water contents (i.e., Nafion with more than 14 water molecules per sulfonic acid group, $\lambda > 14$), the majority of excess protons are located in the central part of the hydrated hydrophilic nanochannels, and the mobility of protons is fast. In this region, the water is bulk like, and proton transport is similar to those in pure water and moves mainly via structure diffusion (the Grotthuss mechanism). The main proton carriers are $H_5O_2^+$ and $H_9O_4^+$, and the transport occurs via the passing of protons from one water molecule to the next by the forming and breaking of hydrogen bonds. With decreasing water content, the concentration of excess protons is increasing, which restrains the proton transfer. Consequently, proton mobility decreases and proton moves mainly via the vehicular mechanism and Grotthuss mechanism. At very low water contents (Nafion with less than six water molecules per sulfonic acid group, $\lambda < 6$), surface transport becomes increasingly important. In this mechanism, proton transports between $-SO_3^-$ groups located on the walls of hydrophilic channels.[113] Due to the strong electrostatic attraction of $-SO_3^-$ groups, the proton mobility through the surface is considerably smaller than that of the bulk.

Choi et al. proposed a pore transport model to describe proton diffusion within Nafion.[113b] The diffusion coefficients are predicted. The surface diffusion coefficient is 1.01×10^{-7} cm²/s at room temperature; the vehicular diffusion coefficient is 1.71×10^{-5} cm²/s and the Grotthuss diffusion coefficient is 7×10^{-5} cm²/s. The Grotthuss diffusion is the fastest proton transport mechanism within Nafion. The surface diffusion coefficient is much lower than the other two diffusion coefficients. The surface diffusion does not contribute significantly to the overall conductivity of protons except at low water levels.

Choe et al. investigated the nature of proton dynamics in Nafion.[114] The results show the hydrogen bond network of water is completely different in very low water content ($\lambda = 4.25$) and at mediate water content ($\lambda = 12.75$). It is discontinuous in the former and continuous in the latter. The disconnected hydrogen bonds around the sulfonic groups in $\lambda = 4.25$ hinder the proton transfer.

Paddison and Paul studied the diffusion of protons in fully hydrated Nafion ($\lambda = 22.5$) with nonequilibrium statistical mechanical transport model.[115] The model calculated a diffusion coefficient for a proton moving along the pore center of 1.92×10^{-5} cm²/s. Within 4A of the pore center proton transfer via the Grotthuss mechanism, while within 8A of the wall of the pore, the transport of the proton is predominantly vehicular in nature.

2.5 MEMBRANE APPLICATIONS IN ELECTROCHEMICAL DEVICES FOR ENERGY STORAGE AND CONVERSION

PFSA membranes have excellent chemical inertness and mechanical integrity in a corrosive and oxidative environment, and their superior properties allowed for broad application in electrochemical devices and other fields such as superacid catalysis, gas drying or humidification, sensors, and metal-ion recovery. Here, we refer their important applications in electrochemical devices for energy storage and conversion including PEMFC, chlor-alkali production, water electrolysis, vanadium redox flow batteries, lithium-ion batteries (LIBs), and solar cells.

2.5.1 APPLICATIONS FOR PEM IN FUEL CELLS

PFSA membrane found its application in hydrogen/oxygen fuel cells that served as power supplies for the satellites developed and launched in the 1960s.[116] The PFSA membrane is used as the PEM (separator/electrolyte) in fuel cells (as shown in Figure 2.12). A fuel cell is an electrochemical device that directly converts the chemical energy of a fuel (hydrogen or methanol) into electrical energy. Fuel cells have higher efficiency compared to conventional power-generated technology, and they offer the possibility of zero emission at the point of use. PEM served as (1) a separator to prevent mixing of reactants, (2) a proton conductor from anode to cathode, (3) an electrical insulator to drive electrons through an external path to the cathode, and (4) a structural framework to support the electrocatalysts (in the case of catalyst-coated membrane).[7]

The membrane must be resistant to the harsh oxidative environment at the anode as well as the reducing environment at the cathode. In addition, to achieve high efficiency, the membrane must possess the following desirable properties: high proton conductivity and zero electronic conductivity, adequate mechanical strength and stability, chemical and electrochemical stability under operating conditions, extremely low fuel or oxygen crossover to maximize coulombic efficiency, and production costs compatible with intended application.[117]

PFSA membrane is by far the most studied proton electrolyte for PEMFC. There are three advantages to the use of PFSA membranes in PEMFCs. First, due to PTFE-based backbone, PFSA membranes are relatively strong and stable in both oxidative and reductive environments. In fact, durability of 60,000 h has been reported.[118] Second, the proton conductivity achieved in PFSA membrane can be as high as 0.13 S/cm at 75°C and 100% relative humidity.[68] A cell resistance is as low as 0.05 Ω cm^2 for a 100 μm thick membrane with voltage loss of only 50 mV at 1 A/cm^2.[119] Third, PFSA has relatively good mechanical properties. For Nafion

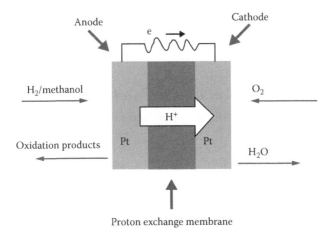

FIGURE 2.12 Schematic of hydrogen and methanol fueled polymer electrolyte membrane fuel cell (PEMFC) with a proton-conducting membrane.

(N115, N117, N110) under room temperature and 50% RH, its maximum tensile strength is 30–40 MPa and its elongation at break is more than 200%.[120] So, the PFSA membranes are the most commonly used PEMs and have served as benchmarks.

Water plays a vital role for proton transportation in PFSA, and the conductivity of PFSA decreases seriously at the temperature lower than 0°C and higher than 100°C. So, PFSA is unsuitable for application at high temperature (>80°C) and low humidity (below ~80% RH). Besides this shortcoming, PFSA suffers from other drawbacks, such as poor mechanical and chemical stability at elevated temperature, insufficient resistance to methanol crossover, and high cost. To overcome these disadvantages, many efforts have been devoted to conventional PFSI modification. Several different approaches for the modification of PFSI membranes have been explored, including (1) physical or chemical treatment, (2) chemical crosslinking, (3) reinforcement by porous support materials, and (4) addition of organic or inorganic compounds. Zhang and Shen give us a very detailed review on recent development of PFSI membranes for fuel cells.[7] Subianto et al. also summarized and commented the many approaches developed for improving the mechanical properties of PFSA membranes.[120]

Other strategy to modify the PFSA is to shorten the side-chain length. At given EW in a random copolymer of PFSA, SSC-PFSA includes a higher level of TFE in the backbone. So, the SSC-PFSA ionomer has the higher crystallinity and the higher glass transition temperature (T_g) compared to the long-side-chain one.[10c,109] The higher crystallinity of the backbone improves the membrane mechanical properties and lowers its tendency to swell in water. SSC-PFSA ionomers include the Dow, Hyflon/Aquivion, and 3M ionomers (listed in Table 2.2). The thermal stability of the SSC-PFSA polymer (Aquivion) is very high. Thermogravimetric analysis shows weight loss of 1% at temperatures as high as 420°C.[10c] Table 2.7 lists some of the properties of SSC-PFSA membranes as well as Nafion membrane evaluated at the Asahi Kasei.[121] Both the 3M- and Dow-type ionomer had higher T_g than the Nafion. And the 3M ionomer displayed higher conductivity and greater thermal stability in TGA experiments both in air and argon. SSC-PFSA membranes showed improved fuel cell performance in operating fuel cells.[10c,122] In the mid-1980s, Ballard Power

---

TABLE 2.7

Some Properties of PFSA Membranes Measured at Asahi Kasei

PFSA	Nafion (EW 950)	Dow Ionomer (EW 740)	3M Ionomer (EW 800)
Conductivity (mS/cm) (80°C, 100% RH)	100	130	140
T_g (°C)	123	148	144
Decomposition temperature (air) (°C)	319	319	362
Decomposition temperature (Ar) (°C)	317	312	393

Source: Hamrock, S.J. and Yandrasits, M.A., *J. Macromol. Sci., Polym. Rev.*, 46(3), 219, 2006.

Systems showed significant improvements in fuel cell performance using SSC-PFSA membranes obtained from the Dow Chemical Company.[123] Merlo et al. reported that the fuel cell performance of Hyflon Ion–based MEAs at medium (75°C) and high (120°C) temperatures demonstrated extremely good power output.[122a] Stassi et al. reported that a maximum power density of about 870 mW/cm^2 was obtained with Aquivion at 130 (3 bar, abs.; 100% RH). This value was significantly better compared to the power density of 620 mW/cm^2 measured with Nafion membrane.[122c]

2.5.2 APPLICATIONS FOR CHLOR-ALKALI ELECTROLYSIS

Chlor-alkali industry is one of the largest electrochemical operations in the world, the main product being chlorine (Cl_2) and sodium hydroxide (NaOH) generated at the same time by the electrolysis of sodium chloride (NaCl) solutions. The main technologies applied for chlor-alkali production are mercury, diaphragm, and membrane cell electrolysis. Because of environmental problems and operating cost disadvantages associated with mercury and diaphragm cell, modern production methods are using membrane cell.

In membrane cell process, purified saturated brine is passed into the anode chamber of the cell where the Cl$^-$ ions are oxidized at the anode, losing electrons to become Cl_2:

$$2Cl^- \rightarrow Cl_2 + 2e^-$$

At the cathode, positive H$^+$ ions pulled from water molecules are reduced by the electrons provided by the electrolytic current, to H$_2$, releasing OH$^-$ ions into the solution:

$$2H_2O + 2e^- \rightarrow H_2 + 2OH^-$$

In addition to keeping Cl_2 and H$_2$ gases separated, the ion-exchanging membrane at the center of the cell prevents the unfavorable back migration of OH$^-$ ions from cathode chamber, while allowing the Na$^+$ ions to pass to the cathode chamber where they react with the OH$^-$ ions to produce caustic soda (NaOH). The overall reaction for the electrolysis of brine is as follows:

$$2NaCl + 2H_2O \rightarrow Cl_2 + H_2 + 2NaOH$$

Nafion membranes were found to have good ion-exchange properties as well as high resistance to the harsh environment of the chlor-alkali cell. Nafion membranes were first used as permselective membrane separator in a commercial chlor-alkali plant in 1975. The first membranes, made from the Nafion, were practical only at low caustic concentrations, since the caustic efficiency decreased significantly at high caustic strengths. To overcome this problem, asymmetric membranes having sulfonic acid groups on the anode side and converted groups on the cathode side were developed. Later, a perfluorocarboxylate membrane, Flemion, which exhibited better resistance to caustic back migration, was produced. The low electrical resistivity of persulfonate-based membranes and the low caustic back-migration characteristics of perfluorocarboxylate membranes were coupled by converting the sulfonic acid groups of Nafion to carboxylate groups on the cathode side to realize the beneficial properties of both the membrane types. Today's membranes comprise a perfluorosulfonate polymer layer, a PTFE reinforcing fabric, and a perfluorocarboxylate polymer.

2.5.3   APPLICATIONS FOR WATER ELECTROLYSIS

Hydrogen is one of the most promising energy carriers because it can be used as fuel in almost every application where fossil fuels are being, but without harmful emissions. PEM for water electrolysis as a hydrogen production device has high efficiency and current density capability and enables safe differential pressure operation. PEM electrolyzers can be operated at high current densities over 1.5 A/cm^2, whereas the current density of alkaline electrolyzer is limited under 0.5 A/cm^2. The voltage efficiency of alkaline electrolysis at 0.2–0.5 A/cm^2 is equivalent to that of PEM electrolysis at 1.5–2.5 A/cm^2.[124] PEM electrolysis also produces very pure hydrogen, with none of the typical catalyst poisons that may be found in hydrogen produced from reforming. PFSA membrane is a commonly used PEM due to its excellent chemical stability, thermal stability, mechanical strength, and proton conductivity.[125] The one disadvantage of PEM electrolysis is the acidic environment within the membrane, which limits the catalysts to noble metal. Reducing catalyst loading and developing nonnoble catalysts are current and long-term research directions on PEM electrolysis.

2.5.4   APPLICATIONS FOR VANADIUM REDOX FLOW BATTERIES

The vanadium redox flow battery (VRB) proposed by Skyllas-Kazacos and coworkers has attracted many attentions due to its long cycle life, flexible design, fast response time, deep-discharge capability, and low cost in energy storage.[126] The VRB employs V^{2+}/V^{3+} and VO^{2+}/VO_2^+ redox couples in the negative and positive half-cell electrolytes, respectively, which are separated by an ion-exchange membrane. The half-cell electrode reactions are as follows[127]:

$$\text{For positive electrode, } VO^{2+} + H_2O - e^- \rightarrow VO_2^+ + 2H^+$$

$$\text{For negative electrode, } V^{3+} + e^- \rightarrow V^{2+}$$

Ion-exchange membrane is one of the key materials for VRB. The function of the membrane is to prevent cross mixing of the positive and negative electrolytes, while still allowing the transport of ions, such as proton in the case of VRB-Nafion, to complete the circuit during the passage of current. The ideal membrane should possess low permeability of vanadium ions, high proton conductivity, good chemical stability, and low cost. PFSA polymers such as Nafion are the most commonly used PEM for research in VRB owing to their high proton conductivity and good chemical and thermal stability in VO_2^+ solution. Very high voltage efficiency (around 90%) was shown with Nafion due to its good proton conductivity.[128] However, Nafion membrane suffers from the crossover of vanadium ions, which results in decreases in energy efficiency.[129] There has been extensive research on the modification of Nafion-based membrane to reduce vanadium ion crossover. Li et al. reported a review on the development of modified Nafion membranes for VRBs.[128a] Hybrid membranes such as SiO_2/Nafion,[129a] TiO_2/Nafion,[130] zirconium phosphates/Nafion,[131] and composite membranes such as cationic charged polyethylenimine layer Nafion,[132] Nafion with pyrrole and polypyrrole coatings,[133] Nafion/polycation poly(diallyldimethylammonium chloride)/polyanion poly(sodium styrene sulfonate) (PSS)[134] and Nafion/PVDF blend membranes,[135] PTFE/Nafion,[136] and Nafion-g-P(*N,N*-dimethylaminoethyl methacrylate) membranes[137] have been prepared. These membranes show lower crossover of vanadium ions.

2.5.5 APPLICATION FOR LITHIUM BATTERIES

LIBs are the most popular rechargeable batteries for portable electronics and are also growing in popularity for military, electric vehicle, and aerospace applications. Most present commercial LIBs use liquid electrolytes. Such batteries using liquid electrolytes show poor thermal stability and require relatively stringent safety precautions. The use of lithiated ion-exchange membranes swollen with nonaqueous liquid solvents as electrolytes may be a promising approach for the development of future polymer electrolytes for LIB applications. However, early research demonstrated that the lithiated Nafion membranes in nonaqueous media, especially in some organic carbonate-based solvents (such as propylene carbonate [PC] or ethylene carbonate [EC]) commonly used in LIBs, had relatively poor ionic conductivity ($<10^{-4}$ S/cm).[138] Navarrini et al. synthesized a lithiated short-side-chain perfluorinated sulfonic ionomer (EW = 750 g/equiv.) and prepared a successive membrane by using the melt-press process.[139] This new polymer electrolyte swelled in PC showed ionic conductivity of 6×10^{-4} S/cm at room temperature. Cai et al. prepared PC swollen lithiated perfluorinated sulfonic ion-exchange membranes.[140] The ionic conductivity of new lithiated perfluorinated sulfonic acid membranes swollen in PC was 4.63×10^{-4} S/cm at room temperature. Later, Liu et al. prepared lithiated perfluorinated sulfonic ion-exchange membranes swollen with mixed EC and PC solvents, and the ionic conductivity of the membranes (EW = 847.5 g/mmol) improved to reach the values $>10^{-3}$ S/cm at room temperature.[141]

The rechargeable Li–air battery has attracted intensive attention since it is able to produce very high specific energies, 3505 W h/kg based on the reaction $2Li + O_2 \leftrightarrow Li_2O_2$ in nonaqueous media.[142] Cheng and Scott prepared a Li metal–compatible and Li metal–stable Li-Nafion binder and membrane, which is used to construct a novel rechargeable Li–O_2/air solid battery.[143] The resulted battery achieved a high capacity over 1000 mA h/(g solids) with good rate capability and capacity retention.

Due to high theoretical gravimetric capacity of 1675 mA h/g and theoretical energy density of 2600 W h/kg,[144] lithium–sulfur (Li–S) batteries are also considered as one of the most probable candidates to meet the ever-increasing demand for high-energy storage system. Tang et al. prepared Nafion-coated sulfur–carbon-coated electrodes for Li–S batteries.[145] The results demonstrated that the Nafion layer is quite effective in reducing shuttle effect and enhancing the stability and the reversibility of the electrode. Bauer et al. used Nafion-coated poly-propylene separator as cation-selective membrane for Li–S batteries to suppress polysulfide diffusion enhancing the charge efficiency of the cells especially at low charge/discharge rates.[146]

2.5.6 Applications for Solar Cells

PFSI has already been used to tune the properties of poly(3,4-ethylenedioxythio-phene)/poly(styrene sulfonate) (PEDOT/PSS) in polymer solar cell.[147] Lee et al. reported single spin coating of poly(3,4-ethylenedioxythiophene)/poly(styrene sul-fonate) composition with PFSI forms a hole-injection layer with a gradient work function by self-organization of the PFSI, which resulted in a remarkably improved device lifetime and efficiency.[147a]

According to reports, PFSI can be used as effective dispersant and support for catalyst in dye-sensitized solar cells (DSSCs) due to its intrinsic chemical struc-ture (hydrophilic side groups and hydrophobic backbone) and excellent chemical and photochemical stability. DSSCs are a type of photovoltaic device based on the charge transfer process between dye molecules and an oxide. DSSCs are promis-ing next-generation alternatives to conventional silicon-based photovoltaic devices owing to their low manufacturing cost and potentially high conversion efficiency.[148] To improve the performance of DSSCs, Sun et al. fabricated graphene–TiO_2 compos-ite photoanodes.[149] To overcome graphene aggregation, graphene was dispersed by Nafion to form highly stable Nafion-functionalized graphene dispersion. The DSSC incorporating 0.5 wt% graphene in the TiO_2 photoanode demonstrates a power con-version efficiency of 4.28%, which is 59% higher than that without graphene. Yeh et al. prepared a composite film of graphene/Nafion, which was used as the catalytic film on the counter electrode of a DSSC.[150] Nafion is demonstrated to be an excellent dispersant for inhibiting the aggregation of graphene. A solar-to-electricity conver-sion efficiency (g) of 8.19% was achieved.

In addition, PFSA also has been used in water-splitting dye-sensitized photo-electrochemical cells, where visible light drives water splitting to produce H_2 and O_2. Park reported that visible-light-induced production of H_2 on Nafion-coated TiO_2 using simple Ru(bpy)$_3^{2+}$ as a sensitizer is more efficient than that of Ru(dcbpy)$_3$–TiO_2.[151] Brimblecombe et al. described a method for inducing the $[Mn_4O_4]^{7+}$ cubane core supported by phosphinate ligands to undertake catalytic water oxidation by transfer into Nafion.[152] They proposed that Nafion provides an oxidatively inert sup-port and hydrophobic environment that red-shifts the electronic absorptions of the cluster, while allowing it to interact with water molecules and providing proton con-duction sites essential for continuous turnover.

2.6 SUMMARY AND OUTLOOK

This chapter is a survey of the fundamental studies on perfluorinated sulfonic iono-mer membranes including ionomer synthesis, membrane characterization, membrane conduction mechanism, and structure models based on recent as well as historical literature. Applications in the electrochemical devices for energy storage and conversion are also described in brief.

Since perfluorinated PEM, Nafion, was developed by the Dupont Company in the 1960s, it has an over 50 years of history. The tremendous fundamental and applied research results built up over these decades on the membrane, which contribute to much deeper understanding of the membrane and the wider application of the membrane. In the foreseeable future, perfluorinated PEM continues to be the most widely studied and employed membrane for PEMFCs due to its excellent oxidative stability and superior proton conductivity. However, there are some drawbacks for perfluorinated PEM such as poor mechanical and chemical stability and poor performance at elevated temperature, insufficient resistance to methanol crossover, and high cost. Both chemical and physical modifications on the membrane are the current and future hot points of researcher works. At the same time, a better understanding of the membrane nanostructures and their relationship to the proton transport mechanism is needed to enhance the performance of the membrane.

REFERENCES

1. Scheirs, J., *Modern Fluoropolymers*, John Wiley & Sons, Ltd., Chichester: 1997.
2. Eisenberg, A., Yeager, H. L., Preface. In *Perfluorinated Ionomer Membranes*, American Chemical Society: 1982; Vol. 180, p. ix.
3. Heitner-Wirguin, C., Recent advances in perfluorinated ionomer membranes: Structure, properties and applications. *J. Membr. Sci.* 1996, *120* (1), 1–33.
4. Doyle, M., Rajendran, G., Perfluorinated membranes. In *Handbook of Fuel Cells*, W. Vielstich, A. Lamm, H. A. Gasteiger, H. Yokokawa (eds.), John Wiley & Sons, Ltd: Chichester, U.K., 2010.
5. Mauritz, K. A., Moore, R. B., State of understanding of Nafion. *Chem. Rev.* 2004, *104* (10), 4535–4585.
6. Devanathan, R., Recent developments in proton exchange membranes for fuel cells. *Energy Environ. Sci.* 2008, *1* (1), 101–119.
7. Zhang, H., Shen, P. K., Recent development of polymer electrolyte membranes for fuel cells. *Chem. Rev.* 2012, *112* (5), 2780–2832.
8. (a) Connolly, D. J., Gresham, W. F., Sulfo derivatives of perfluorovinyl ether monomers. USP 3282875, 1966; (b) Gibbs, H. H., Vienna, V. W., Griffin, R. N., Fluorocarbon sulfonyl fluorides, USP 3041317, 1962.
9. Vaughan, D. J., "Nafion", An Electrochemical Traffic Controller, *DuPont Innovation* 1973, *4*, 10.
10. (a) Carl, W. P., Ezzell, B. R., Low equivalent weight sulfonic fluoropolymers. USP 4940525, 1990; (b) Arcella, V., Ghielmi, A., Tommasi, G., High performance perfluoropolymer films for membranes. *Ann. N. Y. Acad. Sci.* 2003, *984*, 226–244; (c) Arcella, V., Troglia, C., Ghielmi, A., Hyflon ion membranes for fuel cells. *Ind. Eng. Chem. Res.* 2005, *44* (20), 7646–7651.
11. Yamabe, M., Miyake, H., Arai, K., Fluoropolymer cation exchange membranes. Japanese Patent 5,228,588, 1977.
12. Li, Q., He, R., Jensen, J. O., Bjerrum, N. J., Approaches and recent development of polymer electrolyte membranes for fuel cells operating above 100°C. *Chem. Mater.* 2003, *15* (26), 4896–4915.

13. (a) Kirsh, Y. E., Smirnov, S. A., Popkov, Y. M., Timashev, S. F., Perfluorinated carbon-chain copolymers with functional groups and cation exchange membranes based on them: synthesis, structure and properties, *Russ. Chem. Rev.* 1990, *59*, 560; (b) Connolly, D. J., Gresham, W. F., Fluorocarbon vinyl ether polymers. US3282875A, 1964.

14. Seko, M., Ogawa, S., Kimoto, K., Perfluorocarboxylic acid membrane and membrane chlor-alkali process developed by Asahi Chemical Industry. In *Perfluorinated Ionomer, Membranes*, A. Eisenberg, H. L. Yeager (eds.), American Chemical Society: Washington, DC, Vol. 180, pp. 365–410, 1982.

15. Sayed, Y. A., Mehdi, D., Lotfi, H., Roice, A. W., Polymerization of halogen-containing monomers using siloxane surfactant, US6841616, 2003.

16. Larichev, R., Wilmington, B., Aqueous polymerization of fluorinated monomer using hydrogen containing branched fluoroether surfactant EP 2367862, 2011.

17. Xu, A., Zhao, J., Yuan, W. Z., Li, H., Zhang, H., Wang, L., Zhang, Y., Tetrafluoroethylene copolymers with sulfonyl fluoride pendants: Syntheses in supercritical carbon dioxide, polymerization behaviors, and properties. *Macromol. Chem. Phys.* 2011, *212* (14), 1497–1509.

18. Van, B. D. J., Shiflett, M. B., Yokozeki, A., Tetrafluoroethylene is made safe (non-explosive) to handle by forming a liquid solution of tetrafluoroethylene and carbon dioxide, one in the other, in a pressurized container US 5345013, 1994.

19. Cable, K. M., Tailored morphology–property relationships in perfluorosulfonate iono-mer. PhD dissertation. University of Southern Mississippi, Hattiesburg, MS, 1996.

20. Luan, Y., Zhang, H., Zhang, Y., Li, L., Li, H., Liu, Y., Study on structural evolution of perfluorosulfonic ionomer from concentrated DMF-based solution to membranes. *J. Membr. Sci.* 2008, *319* (1–2), 91–101.

21. Yang, L., Tang, J., Li, L., Ai, F., Chen, X., Yuan, W. Z., Zhang, Y., Properties of precursor solution cast PFSI membranes with various ion exchange capacities and annealing temperatures. *RSC Adv.* 2013, *3* (20), 7289–7295.

22. (a) Laporta, M., Pegoraro, M., Zanderighi, L., Perfluorosulfonated membrane (Nafion): FT-IR study of the state of water with increasing humidity. *PCCP* 1999, *1* (19), 4619–4628; (b) Liang, Z., Chen, W., Liu, J., Wang, S., Zhou, Z., Li, W., Sun, G., Xin, Q., FT-IR study of the microstructure of Nafion® membrane. *J. Membr. Sci.* 2004, *233* (1–2), 39–44; (c) Wang, Y., Kawano, Y., Aubuchon, S. R., Palmer, R. A., TGA and time-dependent FTIR study of dehydrating Nafion–Na membrane. *Macromolecules* 2003, *36* (4), 1138–1146; (d) Basnayake, R., Peterson, G. R., Casadonte, D. J., Korzeniewski, C., Hydration and interfacial water in Nafion membrane probed by transmission infrared spectroscopy. *J. Phys. Chem. B* 2006, *110* (47), 23938–23943; (e) Ludvigsson, M., Lindgren, J., Tegenfeldt, J., FTIR study of water in cast Nafion films. *Electrochim. Acta* 2000, *45* (14), 2267–2271; (f) Gruger, A., Régis, A., Schmatko, T., Colomban, P., Nanostructure of Nafion® membranes at different states of hydration: An IR and Raman study. *Vib. Spectrosc.* 2001, *26* (2), 215–225.

23. Zhao, Q., Majsztrik, P., Benziger, J., Diffusion and interfacial transport of water in Nafion. *J. Phys. Chem. B* 2011, *115* (12), 2717–2727.

24. Lowry, S. R., Mauritz, K. A., An investigation of ionic hydration effects in perfluorosul-fonate ionomers by Fourier transform infrared spectroscopy. *J. Am. Chem. Soc.* 1980, *102* (14), 4665–4667.

25. Perusich, S. A., FTIR equivalent weight determination of perfluorosulfonate polymers. *J. Appl. Polym. Sci.* 2011, *120* (1), 165–183.

26. Gao, H., Stabilisation of perfluorosulfonic acid membrane by cross-linking and inorganic–organic composite formation. Application in medium temperature proton exchange membrane fuel cells. PhD dissertation. University of Montpellier II, Shanghai Jiao Tong University, Shanghai, China, 2010.

27. (a) Takasaki, M., Kimura, K., Kawaguchi, K., Abe, A., Katagiri, G., Structural analysis of a perfluorosulfonate ionomer in solution by 19F and 13C NMR. *Macromolecules* 2005, *38* (14), 6031–6037; (b) Chen, Q., Schmidt-Rohr, K., 19F and 13C NMR signal assignment and analysis in a perfluorinated ionomer (Nafion) by two-dimensional solid-state NMR. *Macromolecules* 2004, *37* (16), 5995–6003; (c) Page, K. A., Cable, K. M., Moore, R. B., Molecular origins of the thermal transitions and dynamic mechanical relaxations in perfluorosulfonate ionomers. *Macromolecules* 2005, *38* (15), 6472–6484.

28. Wakai, C., Shimoaka, T., Hasegawa, T., Analysis of the hydration process and rotational dynamics of water in a Nafion membrane studied by 1H NMR spectroscopy. *Anal. Chem.* 2013, *85* (15), 7581–7587.

29. (a) Liu, F., Yi, B., Xing, D., Yu, J., Zhang, H., Nafion/PTFE composite membranes for fuel cell applications. *J. Membr. Sci.* 2003, *212* (1–2), 213–223; (b) Yang, L., Li, H., Ai, F., Chen, X., Tang, J., Zhu, Y., Wang, C., Yuan, W. Z., Zhang, Y., A new method to prepare high performance perfluorinated sulfonic acid ionomer/porous expanded polytetrafluoroethylene composite membranes based on perfluorinated sulfonyl fluoride polymer solution. *J. Power Sources* 2013, *243* (0), 392–395; (c) Su, L., Pei, S., Li, L., Li, H., Zhang, Y., Yu, W., Zhou, C., Preparation of polysiloxane/perfluorosulfonic acid nanocomposite membranes in supercritical carbon dioxide system for direct methanol fuel cell. *Int. J. Hydrogen Energy* 2009, *34* (16), 6892–6901; (d) Laporta, M., Pegoraro, M., Zanderighi, L., Recast Nafion-117 thin film from water solution. *Macromol. Mater. Eng.* 2000, *282* (1), 22–29; (e) Adjemian, K. T., Lee, S. J., Srinivasan, S., Benziger, J., Bocarsly, A. B., Silicon oxide Nafion composite membranes for proton-exchange membrane fuel cell operation at 80–140°C. *J. Electrochem. Soc.* 2002, *149* (3), A256–A261; (f) Zhang, H., Huang, H., Shen, P. K., Methanol-blocking Nafion composite membranes fabricated by layer-by-layer self-assembly for direct methanol fuel cells. *Int. J. Hydrogen Energy* 2012, *37* (8), 6875–6879.

30. Bi, C., Zhang, H., Zhang, Y., Zhu, X., Ma, Y., Dai, H., Xiao, S., Fabrication and investigation of SiO_2 supported sulfated zirconia/Nafion® self-humidifying membrane for proton exchange membrane fuel cell applications. *J. Power Sources* 2008, *184* (1), 197–203.

31. Chalkova, E., Fedkin, M. V., Wesolowski, D. J., Lvov, S. N., Effect of TiO_2 surface properties on performance of Nafion-based composite membranes in high temperature and low relative humidity PEM fuel cells. *J. Electrochem. Soc.* 2005, *152* (9), A1742–A1747.

32. Yakovlev, S., Balsara, N., Downing, K., Insights on the study of Nafion nanoscale morphology by transmission electron microscopy. *Membranes* 2013, *3* (4), 424–439.

33. Ceynowa, J., Electron microscopy investigation of ion exchange membranes. *Polymer* 1978, *19* (1), 73–76.

34. Xue, T., Trent, J. S., Osseo-Asare, K., Characterization of Nafion® membranes by transmission electron microscopy. *J. Membr. Sci.* 1989, *45* (3), 261–271.

35. Fujimura, M., Hashimoto, T., Kawai, H., Small-angle x-ray scattering study of perfluorinated ionomer membranes. 2. Models for ionic scattering maximum. *Macromolecules* 1982, *15* (1), 136–144.

36. Wang, Z., Tang, H., Li, J., Zeng, Y., Chen, L., Pan, M., Insight into the structural construction of a perfluorosulfonic acid membrane derived from a polymeric dispersion. *J. Power Sources* 2014, *256* (0), 383–393.

37. Lehmani, A., Durand-Vidal, S., Turq, P., Surface morphology of Nafion 117 membrane by tapping mode atomic force microscope. *J. Appl. Polym. Sci.* 1998, *68* (3), 503–508.

38. James, P. J., McMaster, T. J., Newton, J. M., Miles, M. J., In situ rehydration of perfluorosulphonate ion-exchange membrane studied by AFM. *Polymer* 2000, *41* (11), 4223–4231.

39. McLean, R. S., Doyle, M., Sauer, B. B., High-resolution imaging of ionic domains and crystal morphology in ionomers using AFM techniques. *Macromolecules* 2000, *33* (17), 6541–6550.

40. Affoune, A. M., Yamada, A., Umeda, M., Surface observation of solvent-impregnated Nafion membrane with atomic force microscopy. *Langmuir* 2004, *20* (17), 6965–6968.

41. O'Hayre, R., Lee, M.,Prinz, F. B., Ionic and electronic impedance imaging using atomic force microscopy. *J. Appl. Phys.* 2004, *95* (12), 8382–8392.

42. Aleksandrova, E., Hiesgen, R., Andreas Friedrich, K., Roduner, E., Electrochemical atomic force microscopy study of proton conductivity in a Nafion membrane. *PCCP* 2007, *9* (21), 2735–2743.

43. O'Dea, J. R., Economou, N. J., Buratto, S. K., Surface morphology of Nafion at hydrated and dehydrated conditions. *Macromolecules* 2013, *46* (6), 2267–2274.

44. Haubold, H. G., Vad, T., Jungbluth, H., Hiller, P., Nano structure of NAFION: A SAXS study. *Electrochim. Acta* 2001, *46* (10–11), 1559–1563.

45. Rubatat, L., Rollet, A. L., Gebel, G., Diat, O., Evidence of elongated polymeric aggregates in Nafion. *Macromolecules* 2002, *35* (10), 4050–4055.

46. Kreuer, K. D., Paddison, S. J., Spohr, E., Schuster, M., Transport in proton conductors for fuel-cell applications: Simulations, elementary reactions, and phenomenology. *Chem. Rev.* 2004, *104* (10), 4637–4678.

47. Elliott, J. A., Wu, D., Paddison, S. J., Moore, R. B., A unified morphological description of Nafion membranes from SAXS and mesoscale simulations. *Soft Matter* 2011, *7* (15), 6820–6827.

48. Kusoglu, A., Modestino, M. A., Hexemer, A., Segalman, R. A., Weber, A. Z., Subsecond morphological changes in Nafion during water uptake detected by small-angle x-ray scattering. *ACS Macro Lett.* 2011, *1* (1), 33–36.

49. Fujimura, M., Hashimoto, T., Kawai, H., Small-angle x-ray scattering study of perfluorinated ionomer membranes. 1. Origin of two scattering maxima. *Macromolecules* 1981, *14* (5), 1309–1315.

50. (a) Hashimoto, T., Fujimura, M., Kawai, H., Structure of sulfonated and carboxylated perfluorinated ionomer membranes. In *Perfluorinated Ionomer Membranes*, A. Eisenberg1, H. L. Yeager (eds.), American Chemical Society: Washington, DC, Vol. 180, pp. 217–248, 1982; (b) Gierke, T. D., Munn, G. E., Wilson, F. C., Morphology of perfluorosulfonated membrane products. In *Perfluorinated Ionomer Membranes*, A. Eisenberg1, H. L. Yeager, American Chemical Society: Washington, DC, Vol. 180, pp. 195–216, 1982; (c) Roche, E. J., Pineri, M., Duplessix, R., Levelut, A. M., Small-angle scattering studies of nafion membranes. *J. Polym. Sci. Pol. Phys.* 1981, *19* (1), 1–11; (d) Gierke, T. D., Munn, G. E., Wilson, F. C., The morphology in Nafion perfluorinated membrane products, as determined by wide- and small-angle x-ray studies. *J. Polym. Sci. Pol. Phys.* 1981, *19* (11), 1687–1704.

51. Orfino, F. P., Holdcroft, S., The morphology of Nafion: Are ion clusters bridged by channels or single ionic sites? *J. New Mat. Electrochem. Syst.* 2000, *3*, 287–292.

52. Moore, R. B., Martin, C. R., Chemical and morphological properties of solution-cast perfluorosulfonate ionomers. *Macromolecules* 1988, *21* (5), 1334–1339.

53. Tang, J., Yuan, W., Zhang, J., Li, H., Zhang, Y., Evidence for a crystallite-rich skin on perfluorosulfonate ionomer membranes. *RSC Adv.* 2013, *3* (23), 8947–8952.

54. Yeo, R. S., Dual cohesive energy densities of perfluorosulphonic acid (Nafion) membrane. *Polymer* 1980, *21* (4), 432–435.

55. Yeo, R. S., Cheng, C. H., Swelling studies of perfluorinated ionomer membranes. *J. Appl. Polym. Sci.* 1986, *32* (7), 5733–5741.

56. Zook, L. A., Leddy, J., Density and solubility of Nafion: Recast, annealed, and commercial films. *Anal. Chem.* 1996, *68*, 3793–3796.

57. Siroma, Z., Fujiwara, N., Ioroi, T., Yamazaki, S., Yasuda, K., Miyazaki, Y., Dissolution of Nafion® membrane and recast Nafion® film in mixtures of methanol and water. *J. Power Sources* 2004, *126* (1–2), 41–45.

58. Feldheim, D. L., Lawson, D. R., Martin, C. R., Influence of the sulfonate countercation on the thermal stability of nafion perfluorosulfonate membranes. *J. Polym. Sci., Part B: Polym. Phys.* 1993, *31* (8), 953–957.

59. Iwai, Y., Yamanishi, T., Thermal stability of ion-exchange Nafion N117CS membranes. *Polym. Degrad. Stab.* 2009, *94* (4), 679–687.

60. Sun, L., Thrasher, J. S., Studies of the thermal behavior of Nafion® membranes treated with aluminum(III). *Polym. Degrad. Stab.* 2005, *89* (1), 43–49.

61. Tang, Y., Karlsson, A. M., Santare, M. H., Gilbert, M., Cleghorn, S., Johnson, W. B., An experimental investigation of humidity and temperature effects on the mechanical properties of perfluorosulfonic acid membrane. *Mater. Sci. Eng., A* 2006, *425* (1–2), 297–304.

62. Yeo, S. C., Eisenberg, A., Physical properties and supermolecular structure of perfluorinated ion-containing (nafion) polymers. *J. Appl. Polym. Sci.* 1977, *21* (4), 875–898.

63. Kyu, T., Eisenberg, A. D. I., Mechanical relaxations in perfluorosulfonate ionomer membranes. In *Perfluorinated Ionomer Membranes*, Adi Eisenberg, Howard L. Yeager American Chemical Society: Washington, 1982; Vol. 180, pp. 79–110.

64. Zawodzinski, T. A., Derouin, C., Radzinski, S., Sherman, R. J., Smith, V. T., Springer, T. E., Gottesfeld, S., Water uptake by and transport through Nafion® 117 membranes. *J. Electrochem. Soc.* 1993, *140* (4), 1041–1047.

65. Luan, Y., Zhang, Y., Zhang, H., Li, L., Li, H., Liu, Y., Annealing effect of perfluorosulfonated ionomer membranes on proton conductivity and methanol permeability. *J. Appl. Polym. Sci.* 2008, *107* (1), 396–402.

66. Broka, K., Ekdunge, P., Oxygen and hydrogen permeation properties and water uptake of Nafion® 117 membrane and recast film for PEM fuel cell. *J. Appl. Electrochem.* 1997, *27* (2), 117–123.

67. Hinatsu, J. T., Mizuhata, M., Takenaka, H., Water uptake of perfluorosulfonic acid membranes from liquid water and water vapor. *J. Electrochem. Soc.* 1994, *141* (6), 1493–1498.

68. Bauer, F., Denneler, S., Willert-Porada, M., Influence of temperature and humidity on the mechanical properties of Nafion® 117 polymer electrolyte membrane. *J. Polym. Sci., Part B: Polym. Phys.* 2005, *43* (7), 786–795.

69. Sone, Y., Ekdunge, P., Simonsson, D., Proton conductivity of Nafion 117 as measured by a four-electrode AC impedance method. *J. Electrochem. Soc.* 1996, *143* (4), 1254–1259.

70. Kopitzke, R. W., Linkous, C. A., Anderson, H. R., Nelson, G. L., Conductivity and water uptake of aromatic-based proton exchange membrane electrolytes. *J. Electrochem. Soc.* 2000, *147* (5), 1677–1681.

71. (a) Deng, Q., Moore, R. B., Mauritz, K. A., Nafion®/(SiO$_2$, ORMOSIL, and dimethylsiloxane) hybrids via in situ sol–gel reactions: Characterization of fundamental properties. *J. Appl. Polym. Sci.* 1998, *68* (5), 747–763; (b) Tang, H., Wan, Z., Pan, M., Jiang, S. P., Self-assembled Nafion–silica nanoparticles for elevated-high temperature polymer electrolyte membrane fuel cells. *Electrochem. Commun.* 2007, *9* (8), 2003–2008; (c) Shao, Z.-G., Joghee, P., Hsing, I. M., Preparation and characterization of hybrid Nafion–silica membrane doped with phosphotungstic acid for high temperature operation of proton exchange membrane fuel cells. *J. Membr. Sci.* 2004, *229* (1–2), 43–51; (d) Antonucci, V., Di Blasi, A., Baglio, V., Ornelas, R., Matteucci, F., Ledesma-Garcia, J., Arriaga, L. G., Aricò, A. S., High temperature operation of a composite membrane-based solid polymer electrolyte water electrolyser. *Electrochim. Acta* 2008, *53* (24), 7350–7356; (e) Tang, H. L., Pan, M., Synthesis and characterization of a self-assembled Nafion/silica nanocomposite membrane for polymer electrolyte membrane fuel cells. *J. Phys. Chem. C* 2008, *112* (30), 11556–11568.

72. (a) Shao, Z.-G., Xu, H., Li, M., Hsing, I. M., Hybrid Nafion–inorganic oxides membrane doped with heteropolyacids for high temperature operation of proton exchange membrane fuel cell. *Solid State Ionics* 2006, *177* (7–8), 779–785; (b) Chalkova, E., Pague, M. B., Fedkin, M. V., Wesolowski, D. J., Lvov, S. N., Nafion/TiO_2 proton conductive composite membranes for PEMFCs operating at elevated temperature and reduced relative humidity. *J. Electrochem. Soc.* 2005, *152* (6), A1035–A1040.

73. (a) Yang, C., Costamagna, P., Srinivasan, S., Benziger, J., Bocarsly, A. B., Approaches and technical challenges to high temperature operation of proton exchange membrane fuel cells. *J. Power Sources* 2001, *103* (1), 1–9; (b) Costamagna, P., Yang, C., Bocarsly, A. B., Srinivasan, S., Nafion® 115/zirconium phosphate composite membranes for operation of PEMFCs above 100°C. *Electrochim. Acta* 2002, *47* (7), 1023–1033; (c) Alberti, G., Casciola, M., Capitani, D., Donnadio, A., Narducci, R., Pica, M., Sganappa, M., Novel Nafion–zirconium phosphate nanocomposite membranes with enhanced stability of proton conductivity at medium temperature and high relative humidity. *Electrochim. Acta* 2007, *52* (28), 8125–8132; (d) Bauer, F., Willert-Porada, M., Characterisation of zirconium and titanium phosphates and direct methanol fuel cell (DMFC) performance of functionally graded Nafion(R) composite membranes prepared out of them. *J. Power Sources* 2005, *145* (2), 101–107.

74. (a) Staiti, P., Aricò, A. S., Baglio, V., Lufrano, F., Passalacqua, E., Antonucci, V., Hybrid Nafion–silica membranes doped with heteropolyacids for application in direct methanol fuel cells. *Solid State Ionics* 2001, *145* (1–4), 101–107; (b) Ramani, V., Kunz, H. R., Fenton, J. M., Stabilized heteropolyacid/Nafion® composite membranes for elevated temperature/low relative humidity PEFC operation. *Electrochim. Acta* 2005, *50* (5), 1181–1187.

75. Çelik, S. Ü., Bozkurt, A., Hosseini, S. S., Alternatives toward proton conductive anhydrous membranes for fuel cells: Heterocyclic protogenic solvents comprising polymer electrolytes. *Prog. Polym. Sci.* 2012, *37* (9), 1265–1291.

76. Fu, Y.-Z., Manthiram, A., Nafion–imidazole–H_3PO_4 composite membranes for proton exchange membrane fuel cells. *J. Electrochem. Soc.* 2007, *154* (1), B8–B12.

77. (a) Kim, J.-D., Mori, T., Hayashi, S., Honma, I., Anhydrous proton-conducting properties of Nafion–1,2,4-triazole and Nafion–benzimidazole membranes for polymer electrolyte fuel cells. *J. Electrochem. Soc.* 2007, *154* (4), A290–A294; (b) Fu, Y., Manthiram, A., Guiver, M. D., Blend membranes based on sulfonated poly(ether ether ketone) and polysulfone bearing benzimidazole side groups for proton exchange membrane fuel cells. *Electrochem. Commun.* 2006, *8* (8), 1386–1390.

78. Sen, U., Bozkurt, A., Ata, A., Nafion/poly(1-vinyl-1,2,4-triazole) blends as proton conducting membranes for polymer electrolyte membrane fuel cells. *J. Power Sources* 2010, *195* (23), 7720–7726.

79. (a) Montoneri, E., Boffa, V., Bottigliengo, S., Casciola, M., Sganappa, M., Marigo, A., Speranza, G., Minati, L., Torrengo, S., Alberti, G., Bertinetti, L., A new polyfunctional acid material for solid state proton conductivity in dry environment: Nafion doped with difluoromethandiphosphonic acid. *Solid State Ionics* 2010, *181* (13–14), 578–585; (b) Traer, J. W., Montoneri, E., Samoson, A., Past, J., Tuherm, T., Goward, G. R., Unraveling the complex hydrogen bonding of a dual-functionality proton conductor using ultrafast magic angle spinning NMR. *Chem. Mater.* 2006, *18* (20), 4747–4754.

80. (a) Doyle, M., Choi, S. K., Proulx, G., High-temperature proton conducting membranes based on perfluorinated ionomer membrane-ionic liquid composites. *J. Electrochem. Soc.* 2000, *147* (1), 34–37; (b) Schmidt, C., Glück, T., Schmidt-Naake, G., Modification of Nafion membranes by impregnation with ionic liquids. *Chem. Eng. Technol.* 2008, *31* (1), 13–22; (c) Yang, J., Che, Q., Zhou, L., He, R., Savinell, R. F., Studies of a high temperature proton exchange membrane based on incorporating an ionic liquid cation 1-butyl-3-methylimidazolium into a Nafion matrix. *Electrochim. Acta* 2011, *56* (17), 5940–5946.

81. (a) Kannan, R., Kakade, B. A., Pillai, V. K., Polymer electrolyte fuel cells using Nafion-based composite membranes with functionalized carbon nanotubes. *Angew. Chem. Int. Ed.* 2008, *47* (14), 2653–2656; (b) Thomassin, J.-M., Kollar, J., Caldarella, G., Germain, A., Jérôme, R., Detrembleur, C., Beneficial effect of carbon nanotubes on the performances of Nafion membranes in fuel cell applications. *J. Membr. Sci.* 2007, *303* (1–2), 252–257; (c) Liu, Y.-L., Su, Y.-H., Chang, C.-M., Suryani, Wang, D.-M., Lai, J.-Y., Preparation and applications of Nafion-functionalized multiwalled carbon nanotubes for proton exchange membrane fuel cells. *J. Mater. Chem.* 2010, *20* (21), 4409–4416.

82. Anantaraman, A. V., Gardner, C. L., Studies on ion-exchange membranes. Part 1. Effect of humidity on the conductivity of Nafion®. *J. Electroanal. Chem.* 1996, *414* (2), 115–120.

83. Ma, S., Siroma, Z., Tanaka, H., Anisotropic conductivity over in-plane and thickness directions in Nafion-117. *J. Electrochem. Soc.* 2006, *153* (12), A2274–A2281.

84. Silva, R. F., De Francesco, M., Pozio, A., Tangential and normal conductivities of Nafion® membranes used in polymer electrolyte fuel cells. *J. Power Sources* 2004, *134* (1), 18–26.

85. Soboleva, T., Xie, Z., Shi, Z., Tsang, E., Navessin, T., Holdcroft, S., Investigation of the through-plane impedance technique for evaluation of anisotropy of proton conducting polymer membranes. *J. Electroanal. Chem.* 2008, *622* (2), 145–152.

86. Yun, S.-H., Shin, S.-H., Lee, J.-Y., Seo, S.-J., Oh, S.-H., Choi, Y.-W., Moon, S.-H., Effect of pressure on through-plane proton conductivity of polymer electrolyte membranes. *J. Membr. Sci.* 2012, *417–418* (0), 210–216.

87. Rodgers, M. P., Bonville, L. J., Kunz, H. R., Slattery, D. K., Fenton, J. M., Fuel cell perfluorinated sulfonic acid membrane degradation correlating accelerated stress testing and lifetime. *Chem. Rev.* 2012, *112* (11), 6075–6103.

88. Pianca, M., Barchiesi, E., Esposto, G., Radice, S., End groups in fluoropolymers. *J. Fluorine Chem.* 1999, *95* (1–2), 71–84.

89. (a) Coms, F. D., The chemistry of fuel cell membrane chemical degradation. *ECS Trans.* 2008, *16* (2), 235–255; (b) Endoh, E., Hommura, S., Terazono, S., Widjaja, H., Anzai, J., Degradation mechanism of the PFSA membrane and influence of deposited Pt in the membrane. *ECS Trans.* 2007, *11* (1), 1083–1091; (c) Panchenko, A., Dilger, H., Kerres, J., Hein, M., Ullrich, A., Kaz, T., Roduner, E., In-situ spin trap electron paramagnetic resonance study of fuel cell processes. *PCCP* 2004, *6* (11), 2891–2894.

90. Cipollini, N. E., Chemical aspects of membrane degradation. *Meeting Abstracts* 2007, *MA2007-02* (9), 500.

91. Borup, R., Meyers, J., Pivovar, B., Kim, Y. S., Mukundan, R., Garland, N., Myers, D. et al., Scientific aspects of polymer electrolyte fuel cell durability and degradation. *Chem. Rev.* 2007, *107* (10), 3904–3951.

92. Ghassemzadeh, L., Marrony, M., Barrera, R., Kreuer, K. D., Maier, J., Muller, K., Chemical degradation of proton conducting perfluorosulfonic acid ionomer membranes studied by solid-state nuclear magnetic resonance spectroscopy. *J. Power Sources* 2009, *186* (2), 334–338.

93. Ghassemzadeh, L., Kreuer, K.-D., Maier, J., Müller, K., Chemical degradation of Nafion membranes under mimic fuel cell conditions as investigated by solid-state NMR spectroscopy. *J. Phys. Chem. C* 2010, *114* (34), 14635–14645.

94. Mittal, V. O., Kunz, H. R., Fenton, J. M., Effect of catalyst properties on membrane degradation rate and the underlying degradation mechanism in PEMFCs. *J. Electrochem. Soc.* 2006, *153* (9), A1755–A1759.

95. Mittal, V. O., Kunz, H. R., Fenton, J. M., Membrane degradation mechanisms in PEMFCs. *J. Electrochem. Soc.* 2007, *154* (7), B652–B656.

96. Cheng, X., Shi, Z., Glass, N., Zhang, L., Zhang, J., Song, D., Liu, Z.-S., Wang, H., Shen, J., A review of PEM hydrogen fuel cell contamination: Impacts, mechanisms, and mitigation. *J. Power Sources* 2007, *165* (2), 739–756.

97. Inaba, M., Kinumoto, T., Kiriake, M., Umebayashi, R., Tasaka, A., Ogumi, Z., Gas crossover and membrane degradation in polymer electrolyte fuel cells. *Electrochim. Acta* 2006, *51* (26), 5746–5753.

98. Kundu, S., Simon, L. C., Fowler, M. W., Comparison of two accelerated Nafion™ degradation experiments. *Polym. Degrad. Stab.* 2008, *93* (1), 214–224.

99. Ghassemzadeh, L., Holdcroft, S., Quantifying the structural changes of perfluorosulfonated acid ionomer upon reaction with hydroxyl radicals. *J. Am. Chem. Soc.* 2013, *135* (22), 8181–8184.

100. (a) Naudy, S., Collette, F., Thominette, F., Gebel, G., Espuche, E., Influence of hygrothermal aging on the gas and water transport properties of Nafion® membranes. *J. Membr. Sci.* 2014, *451* (0), 293–304; (b) Collette, F. M., Thominette, F., Mendil-Jakani, H., Gebel, G., Structure and transport properties of solution-cast Nafion® membranes subjected to hygrothermal aging. *J. Membr. Sci.* 2013, *435* (0), 242–252; (c) Collette, F. M., Lorentz, C., Gebel, G., Thominette, F., Hygrothermal aging of Nafion®. *J. Membr. Sci.* 2009, *330* (1–2), 21–29.

101. Bunyard, W. C., Kadla, J. F., DeSimone, J. M., Viscosity effects on the thermal decomposition of bis(perfluoro-2-N-propoxypropionyl) peroxide in dense carbon dioxide and fluorinated solvents. *J. Am. Chem. Soc.* 2001, *123* (30), 7199–7206.

102. Trogadas, P., Ramani, V., Pt/C/MnO$_2$ hybrid electrocatalysts for degradation mitigation in polymer electrolyte fuel cells. *J. Power Sources* 2007, *174* (1), 159–163.

103. (a) Zhao, D., Yi, B. L., Zhang, H. M., Yu, H. M., Wang, L., Ma, Y. W., Xing, D. M., Cesium substituted 12-tungstophosphoric (CsxH3–xPW12O40) loaded on ceria-degradation mitigation in polymer electrolyte membranes. *J. Power Sources* 2009, *190* (2), 301–306; (b) Danilczuk, M., Perkowski, A. J., Schlick, S., Ranking the stability of perfluorinated membranes used in fuel cells to attack by hydroxyl radicals and the effect of Ce(III): A competitive kinetics approach based on spin trapping ESR. *Macromolecules* 2010, *43* (7), 3352–3358; (c) Wang, Z., Tang, H., Zhang, H., Lei, M., Chen, R., Xiao, P., Pan, M., Synthesis of Nafion/CeO$_2$ hybrid for chemically durable proton exchange membrane of fuel cell. *J. Membr. Sci.* 2012, *421–422* (0), 201–210; (d) Prabhakaran, V., Arges, C. G., Ramani, V., Investigation of polymer electrolyte membrane chemical degradation and degradation mitigation using in situ fluorescence spectroscopy. *Proc. Natl. Acad. Sci. U. S. A.* 2012, *109* (4), 1029–1034.

104. Zhu, Y., Pei, S., Tang, J., Li, H., Wang, L., Yuan, W. Z., Zhang, Y., Enhanced chemical durability of perfluorosulfonic acid membranes through incorporation of terephthalic acid as radical scavenger. *J. Membr. Sci.* 2013, *432* (0), 66–72.

105. Hsu, W. Y., Gierke, T. D., Elastic theory for ionic clustering in perfluorinated ionomers. *Macromolecules* 1982, *15* (1), 101–105.

106. Awatani, T., Midorikawa, H., Kojima, N., Ye, J., Marcott, C., Morphology of water transport channels and hydrophobic clusters in Nafion from high spatial resolution AFM-IR spectroscopy and imaging. *Electrochem. Commun.* 2013, *30* (0), 5–8.

107. Dreyfus, B., Gebel, G., Aldebert, P., Pineri, M., Escoubes, M., Thomas, M., Distribution of the 'micelles' in hydrated perfluorinated ionomer membranes from SANS experiments, *J. Phys. (Paris)* 1990, *51*, 1341.

108. Litt, M. H., Reevaluation of NAFION morphology. *Polym. Prepr.* 2001, *38*, 80–81.

109. Moore, R. B., Martin, C. R., Morphology and chemical properties of the Dow perfluorosulfonate ionomers. *Macromolecules* 1989, *22* (9), 3594–3599.

110. Gebel, G., Structural evolution of water swollen perfluorosulfonated ionomers from dry membrane to solution. *Polymer* 2000, *41* (15), 5829–5838.

111. Kreuer, K. D., Paddison, S. J., Spohr, E., Schuster, M., Transport in proton conductors for fuel-cell applications: Simulations, elementary reactions, and phenomenology. *Chem. Rev.* 2004, *104* (10), 4637–4678.

112. Peckham, T. J., Holdcroft, S., Structure–morphology–property relationships of nonperfluorinated proton-conducting membranes. *Adv. Mater.* 2010, *22* (42), 4667–4690.

113. (a) Eikerling, M., Kornyshev, A. A., Proton transfer in a single pore of a polymer electrolyte membrane. *J. Electroanal. Chem.* 2001, *502* (1–2), 1–14; (b) Choi, P., Jalani, N. H., Datta, R., Thermodynamics and proton transport in Nafion: II. Proton diffusion mechanisms and conductivity. *J. Electrochem. Soc.* 2005, *152* (3), E123–E130.

114. Choe, Y. K., Tsuchida, E., Ikeshoji, T., Yamakawa, S., Hyodo, S., Nature of proton dynamics in a polymer electrolyte membrane, nafion: A first-principles molecular dynamics study. *PCCP* 2009, *11* (20), 3892–3899.

115. Paddison, S. J., Paul, R., The nature of proton transport in fully hydrated Nafion (R). *PCCP* 2002, *4* (7), 1158–1163.

116. Banerjee, S., Curtin, D. E., Nafion® perfluorinated membranes in fuel cells. *J. Fluorine Chem.* 2004, *125* (8), 1211–1216.

117. Smitha, B., Sridhar, S., Khan, A. A., Solid polymer electrolyte membranes for fuel cell applications—A review. *J. Membr. Sci.* 2005, *259* (1–2), 10–26.

118. Rozière, J., Jones, D. J., Non-fluorinated polymer materials for proton exchange membrane fuel cells. *Annu. Rev. Mater. Res.* 2003, *33* (1), 503–555.

119. Gottesfeld, S., Zawodzinski, T. A., Polymer electrolyte fuel cells. In *Advances in Electrochemical Science and Engineering*, R.C. Alkire, D.M. Kolb (eds.), Wiley-VCH Verlag GmbH: London, U.K., pp. 195–301, 2008.

120. Subianto, S., Pica, M., Casciola, M., Cojocaru, P., Merlo, L., Hards, G., Jones, D. J., Physical and chemical modification routes leading to improved mechanical properties of perfluorosulfonic acid membranes for PEM fuel cells. *J. Power Sources* 2013, *233* (0), 216–230.

121. Hamrock, S. J., Yandrasits, M. A., Proton exchange membranes for fuel cell applications. *J. Macromol. Sci., Polym. Rev.* 2006, *46* (3), 219–244.

122. (a) Merlo, L., Ghielmi, A., Cirillo, L., Gebert, M., Arcella, V., Membrane electrode assemblies based on HYFLON® ion for an evolving fuel cell technology. *Sep. Sci. Technol.* 2007, *42* (13), 2891–2908; (b) Tu, Z., Zhang, H., Luo, Z., Liu, J., Wan, Z., Pan, M., Evaluation of 5 kW proton exchange membrane fuel cell stack operated at 95°C under ambient pressure. *J. Power Sources* 2013, *222* (0), 277–281; (c) Stassi, A., Gatto, I., Passalacqua, E., Antonucci, V., Arico, A. S., Merlo, L., Oldani, C., Pagano, E., Performance comparison of long and short-side chain perfluorosulfonic membranes for high temperature polymer electrolyte membrane fuel cell operation. *J. Power Sources* 2011, *196* (21), 8925–8930; (d) Li, J., Pan, M., Tang, H., Understanding short-side-chain perfluorinated sulfonic acid and its application for high temperature polymer electrolyte membrane fuel cells. *RSC Adv.* 2014, *4* (8), 3944–3965.

123. Prater, K., The renaissance of the solid polymer fuel cell. *J. Power Sources* 1990, *29* (1–2), 239–250.

124. Ayers, K. E., Anderson, E. B., Capuano, C., Carter, B., Dalton, L., Hanlon, G., Manco, J., Niedzwiecki, M., Research advances towards low cost, high efficiency PEM electrolysis. *ECS Trans.* 2010, *33* (1), 3–15.

125. (a) Millet, P., Durand, R., Pineri, M., Preparation of new solid polymer electrolyte composites for water electrolysis. *Int. J. Hydrogen Energy* 1990, *15* (4), 245–253; (b) Millet, P., Water electrolysis using EME technology: Electric potential distribution inside a Nafion membrane during electrolysis. *Electrochim. Acta* 1994, *39* (17), 2501–2506.

126. Sum, E., Rychcik, M., Skyllas-Kazacos, M., Investigation of the V(V)/V(IV) system for use in the positive half-cell of a redox battery. *J. Power Sources* 1985, *16* (2), 85–95.

127. Ponce de León, C., Frías-Ferrer, A., González-García, J., Szánto, D. A., Walsh, F. C., Redox flow cells for energy conversion. *J. Power Sources* 2006, *160* (1), 716–732.

128. (a) Li, X., Zhang, H., Mai, Z., Zhang, H., Vankelecom, I., Ion exchange membranes for vanadium redox flow battery (VRB) applications. *Energy Environ. Sci.* 2011, *4* (4), 1147–1160; (b) Lawton, J. S., Jones, A., Zawodzinski, T., Concentration dependence of VO^{2+} crossover of Nafion for vanadium redox flow batteries. *J. Electrochem. Soc.* 2013, *160* (4), A697–A702.

129. (a) Xi, J., Wu, Z., Qiu, X., Chen, L., Nafion/SiO_2 hybrid membrane for vanadium redox flow battery. *J. Power Sources* 2007, *166* (2), 531–536; (b) Hwang, G.-J., Ohya, H., Crosslinking of anion exchange membrane by accelerated electron radiation as a separator for the all-vanadium redox flow battery. *J. Membr. Sci.* 1997, *132* (1), 55–61; (c) Luo, X., Lu, Z., Xi, J., Wu, Z., Zhu, W., Chen, L., Qiu, X., Influences of permeation of vanadium ions through PVDF-g-PSSA membranes on performances of vanadium redox flow batteries. *J. Phys. Chem. B* 2005, *109* (43), 20310–20314; (d) Lawton, J. S., Aaron, D. S., Tang, Z., Zawodzinski, T. A., Qualitative behavior of vanadium ions in Nafion membranes using electron spin resonance. *J. Membr. Sci.* 2013, *428* (0), 38–45.

130. Teng, X., Zhao, Y., Xi, J., Wu, Z., Qiu, X., Chen, L., Nafion/organic silica modified TiO_2 composite membrane for vanadium redox flow battery via in situ sol–gel reactions. *J. Membr. Sci.* 2009, *341* (1–2), 149–154.

131. Sang, S., Wu, Q., Huang, K., Preparation of zirconium phosphate (ZrP)/Nafion1135 composite membrane and H^+/VO^{2+} transfer property investigation. *J. Membr. Sci.* 2007, *305* (1–2), 118–124.

132. Luo, Q., Zhang, H., Chen, J., Qian, P., Zhai, Y., Modification of Nafion membrane using interfacial polymerization for vanadium redox flow battery applications. *J. Membr. Sci.* 2008, *311* (1–2), 98–103.

133. Schwenzer, B., Kim, S., Vijayakumar, M., Yang, Z., Liu, J., Correlation of structural differences between Nafion/polyaniline and Nafion/polypyrrole composite membranes and observed transport properties. *J. Membr. Sci.* 2011, *372* (1–2), 11–19.

134. Xi, J., Wu, Z., Teng, X., Zhao, Y., Chen, L., Qiu, X., Self-assembled polyelectrolyte multilayer modified Nafion membrane with suppressed vanadium ion crossover for vanadium redox flow batteries. *J. Mater. Chem.* 2008, *18* (11), 1232–1238.

135. Mai, Z., Zhang, H., Li, X., Xiao, S., Zhang, H., Nafion/polyvinylidene fluoride blend membranes with improved ion selectivity for vanadium redox flow battery application. *J. Power Sources* 2011, *196* (13), 5737–5741.

136. (a) Teng, X., Dai, J., Su, J., Zhu, Y., Liu, H., Song, Z., A high performance polytetrafluoroethylene/Nafion composite membrane for vanadium redox flow battery application. *J. Power Sources* 2013, *240* (0), 131–139; (b) Teng, X., Sun, C., Dai, J., Liu, H., Su, J., Li, F., Solution casting Nafion/polytetrafluoroethylene membrane for vanadium redox flow battery application. *Electrochim. Acta* 2013, *88* (0), 725–734.

137. Ma, J., Wang, S., Peng, J., Yuan, J., Yu, C., Li, J., Ju, X., Zhai, M., Covalently incorporating a cationic charged layer onto Nafion membrane by radiation-induced graft copolymerization to reduce vanadium ion crossover. *Eur. Polym. J.* 2013, *49* (7), 1832–1840.

138. (a) Doyle, M., Lewittes, M. E., Roelofs, M. G., Perusich, S. A., Lowrey, R. E., Relationship between ionic conductivity of perfluorinated ionomeric membranes and nonaqueous solvent properties. *J. Membr. Sci.* 2001, *184* (2), 257–273; (b) Liang, H. Y., Qiu, X. P., Zhang, S. C., Zhu, W. T., Chen, L. Q., Study of lithiated Nafion ionomer for lithium batteries. *J. Appl. Electrochem.* 2004, *34* (12), 1211–1214.

139. Navarrini, W., Scrosati, B., Panero, S., Ghielmi, A., Sanguineti, A., Geniram, G., Lithiated short side chain perfluorinated sulfonic ionomeric membranes: Water content and conductivity. *J. Power Sources* 2008, *178* (2), 783–788.

140. Cai, Z., Liu, Y., Liu, S., Li, L., Zhang, Y., High performance of lithium-ion polymer battery based on non-aqueous lithiated perfluorinated sulfonic ion-exchange membranes. *Energy Environ. Sci.* 2012, *5* (2), 5690–5693.

141. Liu, Y., Cai, Z., Tan, L., Li, L., Ion exchange membranes as electrolyte for high performance Li-ion batteries. *Energy Environ. Sci.* 2012, *5* (10), 9007–9013.

142. Bruce, P. G., Freunberger, S. A., Hardwick, L. J., Tarascon, J. M., LigO 2 and LigS batteries with high energy storage. *Nat. Mater.* 2012, *11* (1), 19–29.

143. Cheng, H., Scott, K., Improving performance of rechargeable Li-air batteries from using Li-Nafion® binder. *Electrochim. Acta* 2014, *116* (0), 51–58.

144. Fu, Y., Manthiram, A., Enhanced cyclability of lithium-sulfur batteries by a polymer acid-doped polypyrrole mixed ionic-electronic conductor. *Chem. Mater.* 2012, *24* (15), 3081–3087.

145. Tang, Q., Shan, Z., Wang, L., Qin, X., Zhu, K., Tian, J., Liu, X., Nafion coated sulfur–carbon electrode for high performance lithium–sulfur batteries. *J. Power Sources* 2014, *246* (0), 253–259.

146. Bauer, I., Thieme, S., Brückner, J., Althues, H., Kaskel, S., Reduced polysulfide shuttle in lithium–sulfur batteries using Nafion-based separators. *J. Power Sources* 2014, *251* (0), 417–422.

147. (a) Lee, T. W., Chung, Y., Kwon, O., Park, J. J., Self-organized gradient hole injection to improve the performance of polymer electroluminescent devices. *Adv. Funct. Mater.* 2007, *17* (3), 390–396; (b) Moet, D. J. D., de Bruyn, P., Blom, P. W. M., High work function transparent middle electrode for organic tandem solar cells. *Appl. Phys. Lett.* 2010, *96* (15), 153504–153504(3).

148. O'Regan, B., Gratzel, M., A low-cost, high-efficiency solar cell based on dye-sensitized colloidal TiO_2 films. *Nature* 1991, *353* (6346), 737–740.

149. Sun, S., Gao, L., Liu, Y., Enhanced dye-sensitized solar cell using graphene-TiO_2 photoanode prepared by heterogeneous coagulation. *Appl. Phys. Lett.* 2010, *96* (8), 083113–083113(3).

150. Yeh, M.-H., Sun, C.-L., Su, J.-S., Lin, L.-Y., Lee, C.-P., Chen, C.-Y., Wu, C.-G., Vittal, R., Ho, K.-C., A low-cost counter electrode of ITO glass coated with a graphene/Nafion® composite film for use in dye-sensitized solar cells. *Carbon* 2012, *50* (11), 4192–4202.

151. (a) Park, H., Choi, W., Visible-light-sensitized production of hydrogen using perfluorosulfonate polymer-coated TiO_2 nanoparticles: An alternative approach to sensitizer anchoring. *Langmuir* 2006, *22* (6), 2906–2911; (b) Park, J., Yi, J., Tachikawa, T., Majima, T., Choi, W., Guanidinium-enhanced production of hydrogen on Nafion-coated dye/TiO_2 under visible light. *J. Phys. Chem. Lett.* 2010, *1* (9), 1351–1355.

152. Brimblecombe, R., Kolling, D. R. J., Bond, A. M., Dismukes, G. C., Swiegers, G. F., Spiccia, L., Sustained water oxidation by $[Mn_4O_4]^{7+}$ core complexes inspired by oxygenic photosynthesis. *Inorg. Chem.* 2009, *48* (15), 7269–7279.

3 Sulfonated Polyimide Membranes

Xinbing Chen

CONTENTS

3.1 INTRODUCTION

It is well known that polyimides (PIs) are a class of thermally stable polymers that are related to their stiff aromatic backbones. In the past decades, PIs have been extensively studied owing to both scientific and industrial interests. In membrane-based technologies, PI membranes have been used for a long time for a wide variety of applications due to their excellent thermal stability, high mechanical strength and modulus, good film forming ability, and superior chemical resistance [1]. These merits are just what are required for the polymer electrolyte membrane (PEM) materials employed in fuel cell systems.

To date, a larger number of sulfonated PIs (SPIs) have been developed as potential PEM materials for fuel cell applications. Although in early studies five-membered-ring PIs have been investigated as PEMs [2–4], sulfonated six-membered-ring (naphthalenic) PIs have been identified to be promising candidates for fuel cells. Due to their lower ring strain as well as lower electron affinity of the dianhydride moieties, six-membered-ring PIs have superior chemical and thermal stability compared to the more common five-membered-ring PIs [5].

Previous studies on SPIs mainly include investigation of the effect of distribution of ionic groups along the polymer backbone and variations in polymer structure on the membrane properties [6–11], swelling behaviors in water and gas transport mechanisms [12], proton conductivity and hydrolytic stability [13], and their use in membrane electrode assemblies [14]. Generally, proton conductivity of a membrane is determined by its ion-exchange capacity (IEC), and higher IEC tends to give higher conductivity. However, proton conductivity is also closely related to membrane morphology. At the same IEC level, random co-PI membranes usually possess lower conductivity but better hydrolytic and oxidative stability than the corresponding sequenced/block co-PI membranes [11]. Further studies on structure–property relationships in SPIs reveal that hydrolytic stability of membranes is increased by introducing flexible linkages into polymer backbone, and it has also been reported that the higher the basicity of the diamine moieties constituting the backbone, the greater the hydrolytic stability [13]. Sulfonated naphthalenic PI membranes with promising fuel cell performance and stability for more than 5000 h in single fuel cell operation at 80°C under low-humidity conditions have been reported [15].

This chapter presents an overview of the synthesis, membrane characterization, membrane stability, chemical and electrochemical properties, and fuel cell applications of PEM based on SPIs that have been made during the past decades.

3.2   MEMBRANE SYNTHESIS

3.2.1   Materials Requirement and Selection

As one of the core components of the PEM fuel cells (PEMFCs), the PEMs must meet the following requirements to fulfill practical applications: (a) chemical and electrochemical stability, (b) high mechanical strength, (c) low permeability to reactant species, (d) high proton conductivity but zero electronic conductivity, and (e) low production cost. Since the chemical structure of SPI membranes has great impact on their performances, it is essential to properly design the molecular structures of the dianhydride and the diamine monomers and to select appropriate polymerization technique.

It is known that sulfonated polymers, in which the sulfonic acid groups are attached to the polymer main chain, sometimes show huge water uptake at high IECs, resulting in unfavorable swelling of the membranes and a dramatic loss in their mechanical properties. To overcome this problem, polymer membranes with controlled morphologies that exhibit ionic transport channels formed by phase-separated hydrophilic and hydrophobic domain structures have been developed via block, graft, random–graft, and block–graft copolymerization techniques [16–18]. Furthermore, polymers with highly branched, globular macromolecular structures, such as dendritic and hyperbranched polymers, have also attracted increasing attention due to their unique properties in comparison with the linear analogues [19–22]. A series of sulfonated star-hyperbranched PIs for the proton exchange membrane have been prepared via surface modification of the dendritic and hyperbranched polymers. The star-hyperbranched PIs composed of a hydrophobic hyperbranched polymer for polymer stability and a hydrophilic SPI as the proton transport site have been synthesized, and their membranes have been prepared without any cross-linking [23].

3.2.2 POLYMER SYNTHESIS METHODS AND PROCESSES

The most common method for preparing five-membered-ring aromatic PIs involves two-step reactions between aromatic tetracarboxylic acid dianhydrides and diamines. The first step is the condensation polymerization of a dianhydride and a diamine in an aprotic solvent such as N,N-dimethylacetamide (DMAc), N,N-dimethylformamide (DMF), and 1-methyl-2-pyrrolidon (NMP) at room temperature yielding a processable poly(amic) acid precursor. In general, the formation of poly(amic) acid is completed within 24 h or less, depending on monomer reactivity. The second step is the cyclodehydration reaction (imidization) that is accomplished by thermal (heating to elevated temperatures), chemical (incorporating a chemical dehydration agent), or azeotropic solution imidization. This is a representative and the most practical way to synthesize high-performance PIs.

SPIs can be prepared by postsulfonation of common non-SPIs or by direct condensation polymerization from sulfonated monomers. Postsulfonation has been systematically studied with a number of aromatic polymers such as poly(ether ether ketone) and poly(4-phenoxybenzoyl-1,4-phenylene) [24]. It is reported that sulfonation with chlorosulfonic or fuming sulfuric acid is sometimes accompanied by degradation of the polymers and the solubility of polymers changes, while the degree of sulfonation increases. The reaction rate and the degree of sulfonation of polymers in sulfuric acid can be controlled by varying the reaction time and the acid concentration [25]. The first case of the post-SPI (PSPI) membranes was based on the PI derived from 4,4′-(hexafluoroisopropylidene)diphthalic anhydride (6-FDA) and 4,4′-oxydianiline (ODA) [26]. Postsulfonated membranes based on 3,3′,4,4′-benzophenonetetracarboxylic dianhydride (BTDA) and ODA seem to be unsuitable for PEMFC application due to their low mechanical stability. In the case of postsulfonation of the PIs that were prepared using pyromellitic dianhydride as a comonomer, the PSPI with high mechanical stability was obtained, and their properties were also investigated [27]. To avoid excess swelling or even dissolution of the membranes in water, the degrees of sulfonation must be controlled at a moderate level via careful controlling of postsulfonation conditions.

Direct polymerization from sulfonated monomers is the most widely used method for the synthesis of the SPIs. The synthesis is usually performed by polycondensation reaction of sulfonated diamine monomers with tetracarboxylic dianhydride monomers in m-cresol in the presence of triethyl amine (Et_3N) and benzoic acid at around 180°C [4,7,10,11,13,28–34]. Sulfonated diamines are inner salts that are formed due to the interaction of the sulfonic acid groups and the basic amino groups, and the function of the Et_3N is to liberate the protonated amino groups of sulfonated diamines for polymerization with the 1,4,5,8-naphthalene tetracarboxylic dianhydride (NTDA), while benzoic acid functioned as a catalyst. Nonsulfonated diamines are often used as comonomers to control the degree of sulfonation of the polymers. This method was first reported by Mercier and coworkers [11], and thereafter a large number of SPI copolymers have been synthesized from various sulfonated diamines, nonsulfonated diamines, and tetracarboxylic dianhydrides by many research groups. Examples [12,27,33,35–76] of the dianhydride monomers including both the commercially available monomers and the lab-synthesized ones are shown in Figure 3.1.

A few sulfonated dianhydride monomers (Figure 3.2) have also been reported and employed for synthesis of the SPIs [70,77]:

It is noted that the commercially available sulfonated diamine monomers are quite few and the SPI membranes derived from those monomers showed rather poor water stability owing to easy hydrolysis of the imide rings. To improve the water stability of SPI membranes, a great number of sulfonated diamine monomers have been synthesized either by direct sulfonation of the corresponding nonsulfonated diamines using fuming sulfuric acid as the sulfonating reagent or by multiple-step

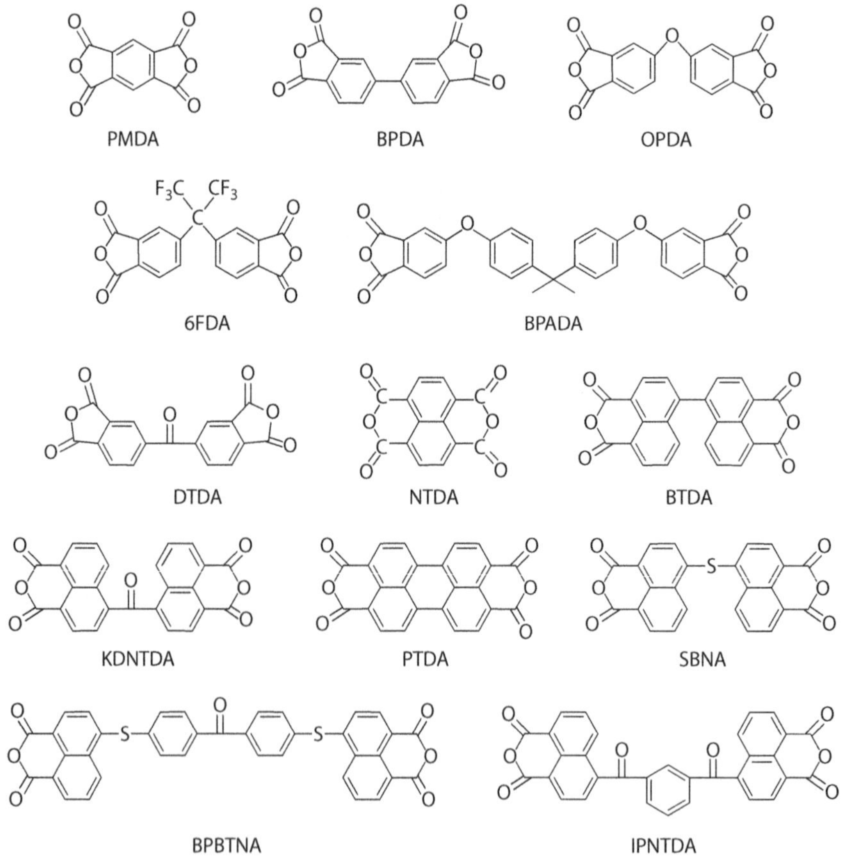

FIGURE 3.1   Chemical structures of various dianhydride monomers for preparation of SPIs.

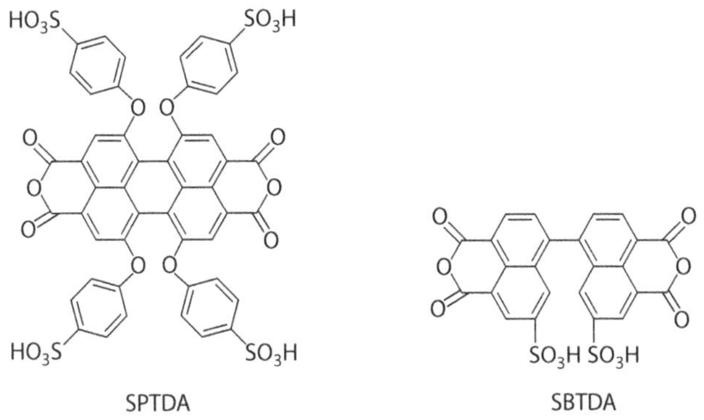

FIGURE 3.2 Chemical structures of sulfonated dianhydride monomers for preparation of SPIs.

reactions. The sulfonated diamine monomers can be classified into two types: main-chain type and side-chain type. The main-chain types of sulfonated diamines refer to those of which sulfonic acid groups are directly attached to the backbones without a spacer, while the side-chain types of sulfonated diamines refer to those of which sulfonic acid groups are located at the end of side chains [9,10,13,30,35,59]. Examples of main-chain type of sulfonated diamines are shown in Figure 3.3 [33,36,37,39–41,43,45–47,50,53,54,57,60,62,63,66–70,73,76,78–82]:

Examples of side-chain type of sulfonated diamines are shown in Figure 3.4 [15,35,42,48,52,65–67,79,83,84]:

Sulfonated co-PIs are generally synthesized by condensation copolymerization of dianhydrides, sulfonated diamines, and nonsulfonated diamines. The copolymerizations can be conducted in three manners: random, sequential, and multiblock copolymerizations. A random copolymer is usually synthesized by reaction of a dianhydride monomer with a solution mixture of a sulfonated diamine and a nonsulfonated diamine in *m*-cresol in the presence of Et₃N and benzoic acid at 180°C. The IEC can be controlled by regulating the molar ratio between the sulfonated diamine and the nonsulfonated diamine. Sequential copolymerization is typically performed by one-pot two-step reactions. First, a sulfonated diamine monomer is polymerized with excess amount of a dianhydride monomer to yield an anhydride-terminated SPI oligomer. Second, the resulting SPI oligomer is further reacted with stoichiometric amount of a nonsulfonated diamine monomer to give the sequenced sulfonated co-PI. In the first step reaction, when excess amount of a sulfonated diamine monomer is used to react with a dianhydride monomer, an amine-terminated SPI oligomer is formed. Accordingly, in the second step reaction, a stoichiometric amount of a dianhydride monomer is added to undergo further polymerization with the SPI oligomer. By regulating the molar ratio between the sulfonated diamine and the dianhydride, the sequence length can be controlled. The synthesis of multiblock co-PIs is usually performed by two-pot three-step reaction procedures. In one pot, an anhydride-terminated SPI oligomer (hydrophilic block) is synthesized by polymerization of excess amount of a dianhydride with a sulfonated diamine, while in

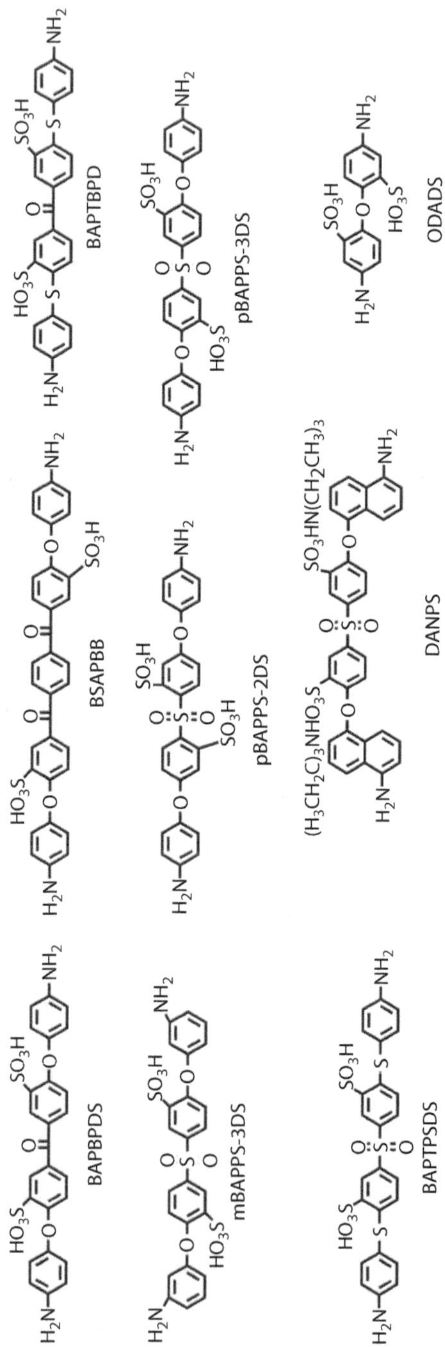

FIGURE 3.3 Chemical structures of various main-chain type of sulfonated diamine monomers.

(Continued)

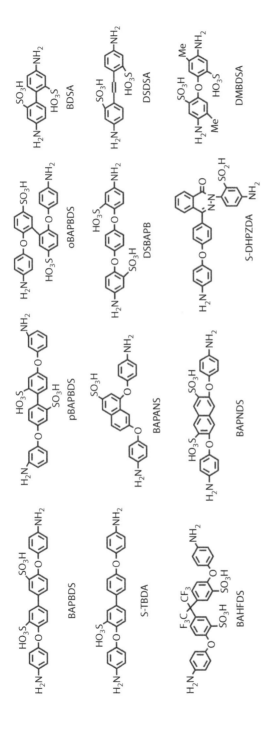

FIGURE 3.3 (Continued) Chemical structures of various main-chain type of sulfonated diamine monomers.

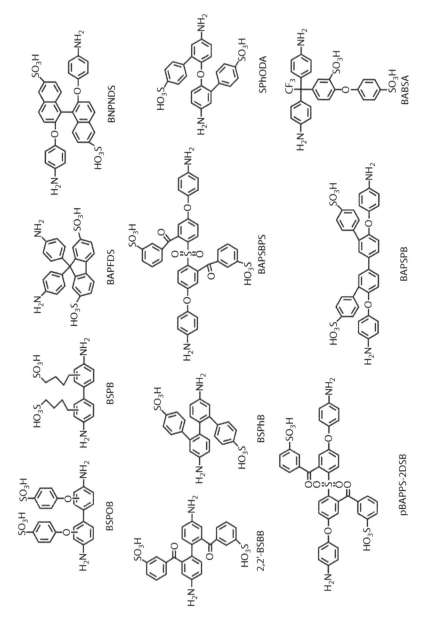

FIGURE 3.4   Chemical structures of various side-chain type of sulfonated diamine monomers.

another pot, an amine-terminated PI oligomer (hydrophobic block) is synthesized via the polymerization of excess amount of a nonsulfonated diamine and a nonsulfonated dianhydride. Transferring the oligomer solution of one of the pots to another pot and continuing further polymerization yield the desired sulfonated multiblock co-PI. Of course, the multiblock copolymerization can also be achieved from the amine-terminated SPI oligomer and the anhydride-terminated PI oligomer. By regulating the molar ratio between the dianhydride and the sulfonated diamine or nonsulfonated diamine, the hydrophilic block length and the hydrophobic block length can be controlled.

3.2.3 MEMBRANE FORMATION

In lab, the SPI membranes are generally fabricated by solution cast method. A polymer solution with a concentration ranging from a few percent to 20% (wt/v) may be employed depending on the viscosity of the polymer solution that ensures feasibility for casting. The solvent in which a SPI is well soluble is generally selected to prepare the polymer solution. For the six-membered-ring SPIs, m-cresol is, in most cases, the best solvent, and therefore it is the most frequently used solvent for casting membranes. For the five-membered-ring SPIs, some aprotic solvents such as DMAc or NMP and dimethylsulfoxide (DMSO) are usually employed because they are good solvents for the polymers. The SPI solutions are needed to be filtered and degased prior to casting onto clean glass plates or petri dishes. Then the glass plates or petri dishes are placed in an oven and strictly kept horizontal. Upon heating (110°C–120°C for the solutions in m-cresol, 70°C–80°C for the solutions in DMAc, NMP, or DMSO), the solvents are gradually evaporated and finally solid membranes are formed. The membranes can be peeled from the glass plates or petri dishes by immersion in deionized water for a while or by storage in air for a long period of time (a few hours to 1 day). To completely remove the residual solvents, the membranes are needed to soak in hot methanol or acetone for a period of time (a few hours to 1 day). Because the synthesized SPIs are in their triethylammonium salt form, proton exchange treatment is necessary to convert the membranes to their proton form. The proton exchange treatment is usually performed by soaking the SPI membranes in diluted (0.1–1.0 M) hydrochloric acid or sulfuric acid at room temperature for a couple of days. Then the membranes are taken out, thoroughly washed with deionized water till the rinsed water becomes neutral and finally dried at 120°C–180°C for a period of time (a few hours to 1 day) in vacuum. It should be noted that to avoid any possible wrinkle formation owing to shrinking during drying process, the membranes are usually fixed on stainless steel frames prior to vacuum drying.

3.3 MEMBRANE CHARACTERIZATIONS

3.3.1 SPECTRAL STUDIES

Nuclear magnetic resonance spectroscopy and Fourier transform infrared (FT-IR) spectroscopy are two common techniques for structural characterizations of organic compounds and polymers, and they have been widely used to characterize the

chemical structures of the SPIs synthesized by postsulfonation or by direct polycondensation of sulfonated/nonsulfonated diamines and dianhydrides.

FT-IR is a powerful technique to characterize the presence of functional groups.
The absorption bands around 1100 and 1030 cm^{-1} are assigned to the asymmetric stretch and symmetric stretch of sulfonate groups. For the five-membered-ring
SPIs, the strong absorption bands assigned to the carbonyl groups of imide rings
typically appear around 1780 cm^{-1} (C=O asymmetric stretch) and 1720 cm^{-1} (C=O
symmetric stretch), while the strong absorption band assigned to C–N stretch usually
appears around 1380 cm^{-1}. Another useful absorption band assigned to C=O bending typically appears around 725 cm^{-1}. For the six-membered-ring SPIs, the strong
absorption bands assigned to the carbonyl groups of imide rings typically appear at
1710 cm^{-1} (C=O asymmetric stretch) and 1670 cm^{-1} (C=O symmetric stretch), which
are significantly lower than those of the five-membered-ring SPIs. The absorption
band assigned to C–N stretch also shifts to a lower wavenumber, ~1350 cm^{-1}.

3.3.2 MORPHOLOGICAL STUDIES

For ionomers, it is generally agreed that the ion pairs form multiplets that aggregate in
larger entities called clusters [85,86], which are embedded in the hydrophobic polymer matrix, resulting in a nanophase separation [87]. Hydrophobic domains impart
mechanical stability to the membranes, whereas hydrated, hydrophilic domains
serve to transport protons and water. The morphology of PEMs has chiefly been
investigated through small-angle x-ray scattering (SAXS), small-angle neutron scattering (SANS), scanned electron microscopy, transmission electron microscopy, and
atomic force microscopy. The SAXS or SANS spectra of ionomers usually exhibit
a small-angle scattering maximum in intensity called the ionomer peak. Several
models [88,89] were proposed considering that the ionomer peak originates from
either the shape of the clusters or their spatial distribution. In the presence of water,
the clusters swell, inducing a shift of the ionomer peak position toward lower angles.

The aim of studying the structure of ionomers is to determine the relationships between the microstructure and the macroscopic properties such as the
mechanical properties or the ionic conductivity. The five-membered-ring (phthalic)
SPIs displayed an intense ionomer peak at very low angles by SANS, while the six-
membered-ring (naphthalenic) SPIs exhibit a strong scattering at long angles, but no
peak was observed. The study using micro-SAXS experiments has shown that the
structure of naphthalenic SPIs is highly anisotropic, and the results were interpreted
as originating from a multiscale foliated structure that induces anisotropic transport
properties. Contrary to naphthalenic SPIs, the swelling properties of phthalic SPIs
are isotropic along the three directions. These results can be attributed to differences
in the polymer chain flexibility [90].

The morphology of a series of SPI ionomers can be investigated by SAXS and
SANS as a function of different parameters such as the ionic content, the ionic groups
distribution, and the flexibility of the polymer chain. Both SAXS and SANS data
exhibit the presence of an ionomer peak, which indicates the presence of large ionic
domains for most of the polymers. The dimension and the distribution of the ionic
aggregates depend significantly on the ionic group's distribution along the polymer

and on structural evolution at nanoscopic scale accompany the swelling process. It was found that the ionomer peak was not observed for very rigid polymers despite the presence of large ionic domains [76].

3.3.3 PHYSICAL PROPERTIES

Solubility in common organic solvents is an important concern to the SPIs because the membranes are usually fabricated by solution cast method as foregoing described. Generally, conventional non-SPIs exhibited poor solubility in common organic solvents owing to their highly stiff backbones and strong intermolecular and intramolecular charge transfer. The solubility of PIs is mainly determined by their chemical structures of the dianhydride and the diamine moieties, and in some cases, imidization method (thermal or chemical) and additives may also affect the solubility. The structural features include the bridging and bulky groups, the position of substituents, and the sequence of diamine units along the polymer chains [9,11,13,91]. Typically, the ether (–O–) group between two phenylene rings can make a different intrasegmental configuration to offer better solubility, and m-amino-substituted diamine leads to a greater improvement in the solubility than the p-amino-substituted diamine. Similar structure–solubility relationship has also been observed with the SPIs. Moreover, the sulfonic acid group itself is a bulky group, which is favorable for improving the polymer solubility.

Most SPIs in their triethylammonium salt form are well soluble in m-cresol (acidic solvent) but less soluble or even completely insoluble in aprotic solvents such as DMAc or NMP. By proper molecular design of diamines and/or dianhydrides, the solubility of the SPIs can be improved. For example, it has been reported that the SPIs derived from a flexible aromatic–aliphatic diamine and m-amino-substituted diamine are well soluble in NMP, which is a less toxic solvent than m-cresol [74]. Replacing NTDA with other dianhydride monomers such as 4,4'-ketone dinaphthalene 1,1',8,8'-tetracarboxylic dianhydride (KDNTDA) and isophthatic dinaphthalene 1,1',8,8'-tetracarboxylic dianhydride (IPNTDA) also gives the SPIs with improved solubility.

Viscosities provide information on the size (molecular weight) of polymer molecules in solution, and for Newtonian fluid, the relationship between viscosity and molecular weight of a polymer can be interpreted by Mark–Houwink equation. It is well known that gel permeation chromatography (GPC) is the most widely used technique to measure the molecular weight of polymers. Tetrahydrofuran (THF) is the most frequently used eluent for GPC measurement, and sometimes high boiling point solvents such as N,N-dimethylforamide (DMF) and NMP may also be used. Because almost all the SPIs are insoluble in THF and only a few SPIs are soluble in DMF or NMP, the GPC method is seldom used for measurement of their molecular weights. In contrast, viscosity measurement is facile, and in principle, any kind of solvent can be used as long as it can dissolve the SPIs without chemical reactions. As a result, viscosity measurement has been widely used as a tool to characterize the molecular weight of the SPIs. The viscosity measurement is usually performed with an Ubbelohde viscometer, thermostatically controlled in a water bath.

It is well known that fuel cells are preferentially operated at elevated temperatures (>100°C) because of the advantages of high CO tolerance of electrodes, easy water management, and fast electrochemical kinetics, and therefore thermal stability is an important property of the SPI membranes, which is usually evaluated by thermogravimetric analysis (TGA). Aromatic SPI membranes are usually thermally stable up to ~300°C, and two-stage weight loss owing to polymer degradation is often observed from TGA curves. Typically, the first stage weight loss due to decomposition of sulfonic acid groups starts at around 250°C–300°C, while the second stage weight loss due to decomposition of polymer main chains occurs at around 500°C–600°C [10,92,93]. From the first stage weight loss, the degree of sulfonation can be roughly estimated. It should be noted that the cations also affect the thermal stability of the sulfonate groups and the SPIs in their proton form and triethylammonium form generally exhibit similar decomposition temperatures, whereas the SPIs in their metallic (e.g., sodium) salt form exhibit much higher thermal stability (typically, >400°C).

Mechanical properties of PEMs are one of the key factors affecting fuel cell durability. The evaluation of membrane mechanical properties is usually performed by tensile test. High tensile stress and large elongation at break suggest good mechanical properties of the membranes. Mechanical properties are determined by many factors such as polymer molecular weight, the chemical structure of polymer main chain, polymer morphology, and test conditions (temperatures, relative humidity [RH], etc.). Among them, polymer molecular weight is the most important factor affecting membrane mechanical properties, and the higher the molecular weight, the higher the mechanical strength. For high-molecular-weight SPI membranes, the tensile stress values are usually above 80 MPa, which are much higher than that (~25 MPa) of Nafion®. Besides tensile strength and elongation at break values, dimensional stability upon swelling, resistance to crack formation, and propagation also have to be considered. Creep of sulfonated polymer membranes is likely to occur because the water-swollen membrane is plasticized and the membrane is under a constant compaction force in the fuel cells [94]. This may lead to membrane thinning and, eventually, puncturing and pinhole formation. An effect especially pertaining to swelling of the polymer upon water sorption is a fatigue-type phenomenon, where the membrane electrode assembly is subjected to dry–wet cycles, leading to periodic stress buildup relaxation in the membrane and, ultimately, to crack formation. This has been observed to be a membrane failure mode [95].

3.3.4 Ion-Exchange Capacity, Water Uptake, and Swelling Ratio

IEC of a polymer is defined as the milliequivalent ionic (e.g., proton) groups per gram of polymer, and its unit is mequiv g^{-1} (also noted as meq g^{-1} in some literature). The IEC is a physical quantity to characterize the averaged ionic density of an ionic-conducting polymer. Sometimes, this mass-based IEC may be misleading due to the fact that different polymer backbone chemistries lead to different masses in relation to the acid functionality. Therefore, a volume-based IEC (unit in mequiv cm^{-3}) may be more meaningful to quote, but that does not seem to be customary. Generally, the membranes with higher IECs tend to have higher proton conductivities, and therefore high IEC is essential to ensure high proton conductivity. However, it should be noted that high IEC values often cause side effects such as unacceptable dimensional changes and the loss of the mechanical integrity, especially at high humidity and high temperature [96,97].

Sulfonated polymer membranes are highly hydrophilic and thus exhibit the feature of sorption of considerable amount of water molecules. The water uptake is mainly determined by the IEC value of a membrane, and the higher the IEC, the larger the water uptake. The water uptake of sulfonated polymer membranes is known to have a profound effect on membrane conductivity and mechanical properties [98]. Water molecules dissociate acid functionality and facilitate proton transport. However, excessively high levels of water uptake can result in membrane fragility and large dimensional change, which leads to the loss of mechanical properties.

One can imagine that primarily the absorbed water molecules will be accommodated in the free volume of the polymer membranes. As the free volume is completely filled, further sorption of water molecules will cause dimensional expansion (membrane swelling), which is usually described in terms of swelling ratio. A swelling ratio refers to the ratio of the swollen volume or dimension to the nonswollen volume or dimension of a membrane. Membrane swelling may be isotropic or anisotropic. Isotropic swelling means the swelling ratios of both in-plane direction and through-plane direction are identical, while anisotropic swelling means significant difference between the swelling ratios of in-plane direction and through-plane direction. In the case of anisotropic swelling, the SPI membranes generally exhibit larger through-plane swelling ratio (Δt) than the in-plane swelling ratio (Δl), and the ratio of Δt to Δl, which is denoted as $\Delta_{t/l}$, is often used to depict the extent of swelling anisotropy of the PEMs. The swelling behaviors of the SPI membranes are closely related to their chemical structures. It is found that the swelling behaviors of the NTDA-based SPIs strongly depend on the chemical structures of the sulfonated diamine moieties [52,66,83,84,99–104]. The SPIs derived from sulfonated benzidine-type diamines such as 2,2'-benzidinedisulfonic acid (BDSA), 2,2'- or 3,3'-bis(sulfopropoxy) benzidine (BSPB), 2,2'- or 3,3'-bis(sulfophenoxy) benzidine (BSPOB), and 2,2'-*bis*(4-sulfophenyl) benzidine usually exhibit strong anisotropic swelling with the $\Delta_{t/l}$ values typically in the range of 6–13. In contrast, the SPIs derived from the sulfonated diamines containing flexible ether bond(s) in their main chains such as the 4,4'-bis(4-aminophenoxy)biphenyl-3,3'-disulfonic acid (BAPBDS), 4,4'-oxydianiline-2,2'-disulfonicacid (ODADS), and 4,4'-bis (4-aminophenoxy)-3,3'-bis(4-sulfophenyl) biphenyl (BAPSPB) show only moderate anisotropic membrane swelling with the $\Delta_{t/l}$ values typically in the range of 2–4 or even isotropic swelling. In addition, the block co-PI membranes and the sequenced co-PI membranes generally exhibit strong anisotropic swelling behavior, whereas the random co-PI membranes generally show isotropic swelling or moderate/weak swelling. The anisotropic membrane swelling is considered to be related to the polymer chain alignment in-plane direction of membrane resulting from the highly stiff rodlike polymer backbone. The SPI backbones derived from the NTDA and the benzidine-based diamines with pendant sulfonic acid groups seem to cause the better alignment of polymer chain in-plane direction, whereas the flexible ether bonds in the main chain tend to disturb the polymer chain alignment. This is a reason that the BSPOB- and BSPB-based SPIs exhibited much larger anisotropic membrane swelling ratio than the BAPBDS- and BAPSPB-based SPIs [83,100–103]. It is noted that the dianhydride moieties with the significantly different rigidity hardly affected the anisotropy degree of membrane swelling of the BAPBDS-based SPIs.

3.3.5 Proton Conductivity

Proton conductivity is one of the most important performances of a PEM since it determines the fuel cell performance. High proton conductivity ensures low internal resistance of fuel cells and thus high power density output. Proton conductivity is affected by many factors such as IEC, RH, temperature, and polymer morphology. Higher IEC tends to give higher proton conductivity because more protons are available for transport. However, too high IEC often causes excess swelling of membranes. As a result, it is essential to control IEC at a reasonable level. Water also plays important role in proton conduction. First, water is essential to dissociation of sulfonic acid groups into sulfonate anions and protons. Second, water molecules act as a carrier for proton transport. Without the aid of water molecules, protons and sulfonate anions are tightly bonded together leading to extremely low proton conductivity. This is why proton conductivity is strongly dependent on RH and at low RH the conductivity becomes extremely low. At a given RH, proton conductivity increases with increasing temperature. However, it should be noted that at the temperatures above the boiling point of water (100°C), RH decreases with increasing temperature, and therefore pressurization of the system is often needed to ensure high RH. Polymer morphology is also an important factor influencing proton conductivity. It is well known that in Nafion there are a lot of well-defined ionic channels that can facilitate proton transport. For sulfonated hydrocarbon polymer membranes including SPIs, such ionic channels are seldom observed. However, by block copolymerization technique and by using side-chain-type sulfonated diamine monomers, the SPIs with perfect nanophase separation can be obtained, which has been proved to be favorable for enhancing proton conductivity at low RH. In addition, the acidity of sulfonic acid groups should also be a factor affecting proton conductivity. The strong acidity of $-(CF_2)_2-SO_3H$ pendant groups is considered to positively contribute to the high proton conductivity of Nafion.

The measurement of proton conductivity is usually performed by a four-probe or two-probe AC impedance method. Because the measurement of in-plane conductivity is much easier to operate than the measurement of through-plane conductivity especially at low RHs (extremely long time is needed to achieve water vapor sorption equilibrium because both surfaces of the sample membrane are covered by electrodes), the in-plane conductivity has been widely used to characterize the proton conductivity of a PEM [105]. However, in some cases, proton conductivity exhibits strong anisotropy, that is, in-plane conductivity is significantly different from through-plane conductivity. Since only the through-plane conductivity contributes to fuel cell performance, it is very important to develop a facile and effective method for measurement of through-plane proton conductivity. In liquid water, the sorption kinetics is relatively fast and the sorption equilibrium can be achieved in a short period of time. As a result, the through-plane proton conductivity in liquid water can be relatively easily measured.

A big challenge in the field of PEMFCs is to develop PEMs with high proton conductivity at elevated temperatures (>100°C) and low RHs because of the merits of fast electrochemical kinetics, high tolerance of catalyst to CO poison, and easy management of water when the fuel cells are operated at elevated temperatures.

Many methods such as incorporation of hydrophilic nanoparticles or mesoporous materials into SPI membranes, block copolymerization, and structural modification of sulfonated diamine monomers have been attempted to enhance the proton conductivity of SPI membranes at low RHs. Among them, block copolymerization seems to be a promising method. Another strategy is to develop anhydrous proton-conducting membranes such as ionic liquids, which is a hot topic of current researches.

3.3.6 CHEMICAL STABILITY (HYDROLYTIC STABILITY AND RADICAL OXIDATIVE STABILITY)

Maintaining the chemical and mechanical integrity of the membrane over the anticipated lifetime is a key requirement, and it deserves attention already in the early stage of membrane development. PEMs undergo chemical degradation through polymer chain scission and loss of functional groups or constituents (blocks, side chains, blend component), caused by HO and HOO radicals, which are formed in situ through interaction of H_2 and O_2 with the noble metal catalyst on both anode and cathode sides [106]. SPI membranes with phthalic imide rings suffer from hydrolytic decomposition in harsh fuel cell conditions [5,91]. When sulfonated five-membered-ring (phthalic) PI membranes are used for proton exchange membranes in fuel cells, they degrade quickly and become brittle and fragile. This is because under strong acid conditions, the hydrolytic rate of the phthalimide structure is very fast leading to chain scission [91]. Unlike the five-membered-ring SPIs, the six-membered-ring (naphthalenic) imide structures do not quickly degrade and are much more stable to hydrolysis. The hydrolytic stability is usually evaluated by measuring the changes in mechanical properties, weight loss, IEC, and proton conductivity after immersion in liquid water at elevated temperatures (e.g., 80°C–130°C) for a period of time.

The stability of SPI membranes in water has been one of the major research subjects because it is one of the important factors affecting membrane performance [107–109]. In general, the imide rings of PIs have the tendency to hydrolyze due to the attack of water molecules on the carbonyl carbon of imide rings. The hydrolytic stability of the imide rings is strongly dependent on the basicity of the sulfonated diamine moiety. Introducing the diamine moieties with high basicity has been proved to be highly effective for improving the water stability of PIs because the high basicity of the sulfonated diamine moieties causes deactivation of the carbonyl carbon of imido rings via resonance [110,111]. On the basis of the same principle, replacing the NTDA with low-electron-affinity dianhydride monomers such as the BTDA, 3,4,9,10-perylene tetracarboxylic dianhydride (PTDA), and benzophenone-4,4'-bis(4-thio-1,8-naphthalic anhydride) (BPBTNA) (Figure 3.1) is also very effective to improve the hydrolytic stability of SPI membranes.

During the fuel cell operation, peroxy radicals are formed and they can bring about a severe degradation of the membrane material [112–114]. The degradation leads first to the embrittlement and then to the development of cracks and total destruction of the membrane. For practical applications, membrane lifetime of at least several thousand hours is necessary. Since the long-term longevitytesting may only be performed with the membranes at the advanced stage of development, the membrane stability toward oxidation is generally examined by observing the dissolving behavior or weight loss in Fenton's reagent (3 ppm Fe^{2+} in 3% H_2O_2) at elevated temperatures (e.g., 80°C), which is one of the standard tests for oxidative stability.

3.4 MEMBRANE CONDUCTION MECHANISM AND THEORETICAL MODELING

Many researches on proton transport behaviors through SPI membranes have been reported [9–11,13,115]. Proton transport occurs either via hopping of protons from one water molecule to the next (the Grotthuss mechanism) or via the net transport of H_3O^+ or other aggregates of water and H^+ (vehicle mechanism). Generally, it is believed that protons can be transported along with hydrogen-bonded ionic channels and cationic mixtures such as H_3O^+, $H_5O_2^+$, and $H_9O_4^+$ in the water medium [116–118]. In a fully hydrated state, sulfonated polymers may dissociate immobile sulfonic acid groups and mobile protons in aqueous solution. Then, the free protons move through a localized ionic network within fully water-swollen sulfonated polymer membranes. Accordingly, the proper water content level should be maintained in sulfonated polymer membranes in order to guarantee high proton conductivity. Simultaneously, the water state such as free water, frozen-bound water, and non-frozen-bound water should be deemed to improve proton conduction across the sulfonated polymer membranes.

3.5 MEMBRANE APPLICATIONS IN ELECTROCHEMICAL DEVICES FOR ENERGY STORAGE AND CONVERSION

3.5.1 APPLICATIONS FOR PEM FOR FUEL CELLS

Fuel cells are efficient devices to generate electric power via electrochemical reaction of fuels with oxygen and therefore have attracted attention as a clean powering system. Proton exchange membrane fuel cells (also known as PEMFC) are a promising energy converter to reduce CO_2 emission, which is expected as a cogeneration system and automobile power source. PEMFCs convert chemical energy of the fuel (hydrogen, methanol, or formic acid) into electrical energy with high efficiency (~60%) and minimal environmental pollution.

A proton-conducting membrane with high conductivity and good membrane stability is one of the most important factors in relation to high PEMFC performance. To improve fuel cell performance, it is desirable to operate PEMFCs at higher temperatures (>90°C). Faster oxygen reduction reactions at the cathode, enhanced CO tolerance of the catalyst at the anode, and higher efficiency of heat recovery are the major benefits of operating PEMFCs at higher temperatures. However, a main problem prohibiting the practical use of SPIs is the water stability of their membranes, which is related to the mechanical stability of highly hydrated membranes and thus the fuel cell lifetime. Fortunately, some recent works have been achieved in this field, and good results for fuel cell performance have been reported [44,119,120].

3.5.2 APPLICATIONS FOR VANADIUM REDOX FLOW BATTERIES

The vanadium redox flow battery (VRB) has attracted considerable attention owing to its long cycle life, flexible design, fast response time, deep discharge capability, low cost, and low pollution. Therefore, it has a wide range of applications such as remote area power systems, emergency backup usage, and uninterruptable power sources. The proton-conducting membrane is one of the key components for VRB, which effectively prevents the crossover of catholyte and anolyte and provides a conducting pathway to complete the circuit during the passage of current. An ideal proton-conducting membrane for VRB should possess high proton conductivity, low vanadium ion permeability, excellent chemical stability, good mechanical strength, and low price. In recent years, a series of SPI-based composite membranes have been investigated; the results indicate that the composite membranes with excellent properties are promising proton-conducting membranes for VRB [121,122].

3.6 CHALLENGES AND PERSPECTIVES

Although SPI membranes have shown great promise as high-performance PEMs for fuel cell applications, there are still some challenges/problems to be solved. First, hydrolytic stability is one of the major concerns associated with SPI membranes. Although greatly improved hydrolytic stability has been achieved by many methods such as branching, cross-linking, semi-interpenetrating, introducing flexible and aliphatic linkages, or using highly basic sulfonated diamines in SPI structures, it is still desirable to further improve the hydrolytic stability of the SPI membranes to meet the requirements for long-term durability of fuel cells. Second, like many other sulfonated hydrocarbon polymer membranes, SPI membranes generally exhibit poor radical oxidative stability that needs to be largely improved. Recent researches have demonstrated that the incorporation of benzimidazole groups into SPI backbone is an effective way to enhance the radical oxidative stability [28]. Third, almost all the sulfonated polymer membranes exhibit rather low proton conductivity at low RH. For this reason, fuel cells have to operate under strong humidification conditions making the fuel cell systems very complex. It is a big challenge to develop the SPI membranes with high proton conductivity at low RH. Block copolymerization seems to be one of the most promising methods to enhance the proton conductivity at low RH. Therefore, design and synthesis of new monomers and the relevant SPI membranes with high performance are current research activities in this field.

REFERENCES

1. Vandezande, P., Gevers, L. E. M., Vankelecom, I. F. J., Solvent resistant nanofiltration: Separating on a molecular level, *Chem. Soc. Rev.*, 2008, 37, 365–405.
2. Gunduz, N., McGrath, J. E., Synthesis and characterization of sulfonated polyimides, *Polym. Prepr. (Am. Chem. Soc., Div. Polym. Chem.)*, 2000, 41, 182–183.
3. Shobha, H. K., Sankarapandian, M., Glass, T. E., McGrath, J. E., Sulfonated aromatic diamines as precursors for polyimides for proton exchange membranes, *Polym. Prepr. (Am. Chem. Soc., Div. Polym. Chem.)*, 2000, 41, 1298–1299.

4. Woo, Y., Oh, S. Y., Kang, Y. S., Jung, B., Synthesis and characterization of sulfonated polyimide membranes for direct methanol fuel cell, *J. Membr. Sci.*, 2003, 220, 31–45.

5. Rusanov, A. L., Novel bis(naphthalic anhydrides) and their polyheteroarylenes with improved processability, *Adv. Polym. Sci.*, 1994, 111, 115–175.

6. Zhang, Y., Litt, M., Savinell, R. F., Wainright, J. S., Molecular design considerations in the synthesis of high conductivity PEMs for fuel cells, *Polym. Prepr. (Am. Chem. Soc., Div. Polym. Chem.)*, 1999, 40, 480–481.

7. Cornet, N., Diat, O., Gebel, G., Jousse, F., Marsacq, D., Mercier, R., Pineri, M., Sulfonated polyimide membranes: A new type of ion-conducting membrane for electrochemical applications, *J. New Mater. Electrochem. Syst.*, 2000, 3, 33–42.

8. Cornet, N., Beaudoing, G., Gebel, G., Influence of the structure of sulfonated polyimide membranes on transport properties, *Sep. Purif. Technol.*, 2001, 22–23, 681–687.

9. Cuo, X., Fang, J., Watari, T., Tanaka, K., Kita, H., Okamoto, K., Novel sulfonated polyimides as polyelectrolytes for fuel cell application. 2. Synthesis and proton conductivity of polyimides from 9,9-*bis*(4-aminophenyl)fluorine-2,7-disulfonic acid, *Macromolecules*, 2002, 35, 6707–6713.

10. Fang, J., Guo, X., Haroda, S., Watari, T., Tanaka, K., Kita, H., Okamoto, K., Novel sulfonated polyimides as polyelectrolytes for fuel cell application. 1. Synthesis, proton conductivity, and water stability of polyimides from 4,4′-diaminodiphenyl ether-2,2′-disulfonic acid, *Macromolecules*, 2002, 35, 9022–9028.

11. Genies, C., Mercier, R., Sillion, B., Cornet, N., Gebel, G., Pineri, M., Soluble sulfonated naphthalenic polyimides as materials for proton exchange membranes, *Polymer*, 2001, 42, 359–373.

12. Piroux, F., Espuch, E., Mercier, R., Pineri, M., Gebel, G., Gas transport mechanism in sulfonated polyimides: Consequences on gas selectivity, *J. Membr. Sci.*, 2002, 209, 241–253.

13. Watari, T., Fang, J., Tanaka, K., Kita, H., Okamoto, K., Hirano, T., Synthesis, water stability and proton conductivity of novel sulfonated polyimides from 4,4′-*bis*(4-aminophenoxy)biphenyl-3,3′-disulfonic acid, *J. Membr. Sci.*, 2004, 230, 111–120.

14. Besse, S., Capron, P., Diat, O., Gebel, G., Jousse, F., Marsacq, D., Pineri, M., Marestin, C., Mercier, R., Sulfonated polyimides for fuel cell electrode membrane assemblies (EMA), *J. New Mater. Electrochem. Syst.*, 2002, 5, 109–112.

15. Kabasawa, A., Saito, J., Yano, H., Miyatake, K., Uchida, H., Watanabe, M., Durability of a novel sulfonated polyimide membrane in polymer electrolyte fuel cell operation, *Electrochim. Acta*, 2009, 54, 1076–1082.

16. Nakano, T., Nagaoka, S., Kawakami, H., Proton conductivity of sulfonated long-chain-block copolyimide films, *Kobunshi Ronbunshu*, 2006, 63, 200–204.

17. Asano, N., Miyatake, K., Watanabe, M., Sulfonated block polyimide copolymers as a proton-conductive membrane, *J. Polym. Sci. A Polym. Chem.*, 2006, 44, 2744–2748.

18. Nakano, T., Nagaoka, S., Kawakami, H., Preparation of novel sulfonated block copolyimides for proton conductivity membranes, *Polym. Adv. Technol.*, 2005, 16, 753–757.

19. Deffieux, A., Schappacher, M., Hirao, A., Watanabe, T., Synthesis and AFM structural imaging of dendrimer-like star-branched polystyrenes, *J. Am. Chem. Soc.*, 2008, 130, 5670–5672.

20. Chojnowski, J., Fortuniak, W., Scibiorek, M., Rozga-Wijas, K., Grzelka, A., Mizerska, U., 3-Chloropropyl functionalized dendrigraft polysiloxanes and dendritic polyelectrolytes, *Macromolecules*, 2007, 40, 9339–9347.

21. Pérignon, N., Marty, J. D., Mingotaud, A. F., Dumont, M., Rico-Lattes, I., Mingotaud, C., Hyperbranched polymers analogous to PAMAM dendrimers for the formation and stabilization of gold nanoparticles, *Macromolecules*, 2007, 40, 3034–3041.

22. Tobita, H., Random degradation of branched polymers. 2. Multiple branches, *Macromolecules*, 1996, 9, 3010–3021.

23. Suda, T., Yamazaki, K., Kawakami, H., Syntheses of sulfonated star-hyperbranched polyimides and their proton exchange membrane properties, *J. Power Sources*, 2010, 195, 4641–4646.

24. Rikukawa, M., Sanui, K., Proton-conducting polymer electrolyte membranes based on hydrocarbon polymers, *Prog. Polym. Sci.*, 2000, 25, 1463–1502.

25. Bishop, M. T., Karasz, F. E., Russo, P. S., Langley, K. H., Solubility and properties of a poly(aryl ether ketone) in strong acids, *Macromolecules*, 1985, 18, 86–93.

26. Wang, J. L., Lee, M. H., Yang, J., Synthesis and characterization of sulphonated polymer derived from 6FDA-ODA polyimide, *Polym. Bull.*, 2005, 55, 357–365.

27. Deligo, H., Vatansever, S., Oksuzomer, F., Koc, S. N., Ozgumus, S., Gurkaynak, M. A., Preparation and characterization of sulfonated polyimide ionomers via post-sulfonation method for fuel cell applications, *Polym. Adv. Technol.*, 2008, 19, 1126–1132.

28. Li, W., Guo, X., Fang, J., Synthesis and properties of sulfonated polyimide-polybenzimidazole copolymers as proton exchange membranes, *J. Mater. Sci.*, 2014, 49, 2745–2753.

29. Einsla, B. R., Hong, Y. T., Kim, Y. S., Wang, F., Gunduz, N., McGrath, J. E., Sulfonated naphthalene dianhydride based polyimide copolymers for proton-exchange-membrane fuel cells. I. Monomer and copolymer synthesis, *J. Polym. Sci. A Polym. Chem.*, 2004, 42, 862–874.

30. Guo, X., Fang, J., Tanaka, K., Kita, H., Okamoto, K. I., Synthesis and properties of novel sulfonated polyimides from 2,2'-*bis*(4-aminophenoxy)biphenyl-5,5'-disulfonic acid, *J. Polym. Sci. A Polym. Chem.*, 2004, 42, 1432–1440.

31. Mistri, E. A., Mohanty, A. K., Banerjee, S., Komber, H., Voit, B., Naphthalene dianhydride based semifluorinated sulfonated copoly(ether imide)s: Synthesis, characterization and proton exchange properties, *J. Membr. Sci.*, 2013, 441, 168–177.

32. Meier-Haack, J., Teager, A., Vogel, C., Schlenstedt, K., Lenk, W., Lehmann, D., Membranes from sulfonated block copolymers for use in fuel cells, *Sep. Purif. Technol.*, 2005, 41, 207–220.

33. Sundar, S., Jang, W., Lee, C., Shul, Y., Han, H., Crosslinked sulfonated polyimide networks as polymer electrolyte membranes in fuel cells, *J. Polym. Sci. B Polym. Phys.*, 2005, 43, 2370–2379.

34. Zou, L., Anthamatten, M., Synthesis and characterization of polyimide-polysiloxane segmented copolymers for fuel cell applications, *J. Polym. Sci. A Polym. Chem.*, 2007, 45, 3747–3758.

35. Chen, X., Chen, P., Okamoto, K., Synthesis and properties of novel side-chain-type sulfonated polyimides, *Polym. Bull.*, 2009, 63, 1–14.

36. Song, Y., Jin, Y., Liang, Q., Li, K., Zhang, Y., Hu, W., Jiang, Z., Liu, B., Novel sulfonated polyimides containing multiple cyano groups for polymer electrolyte membranes, *J. Power Sources*, 2013, 238, 236–244.

37. Kasahara, A., Takahashi, A., Oyama, T., Photosensitive sulfonated polyimides utilizing alkaline-developable negative-tone reaction development patterning, *J. Photopolym. Sci. Technol.*, 2011, 24, 269–272.

38. Wei, H., Chen, G., Cao, L., Zhang, Q., Yan, Q., Fang, X., Enhanced hydrolytic stability of sulfonated polyimide ionomers using *bis*(naphthalic anhydrides) with low electron affinity, *J. Mater. Chem. A*, 2013, 35, 10412–10421.

39. Shang, Y., Xie, X., Jin, H., Guo, J., Wang, Y., Feng, S., Wang, S., Xu, J., Synthesis and characterization of novel sulfonated naphthalenic polyimides as proton conductive membrane for DMFC applications, *Eur. polym. J.*, 2006, 42, 2987–2993.

40. Li, N., Cui, Z., Zhang, S., Xing, W., Sulfonated polyimides bearing benzimidazole groups for proton exchange membranes, *Polymer*, 2007, 48, 7255–7263.

41. Alvarez-Gallego, Y., Ruffmann, B., Silva, V., Silva, H., Lozano, A. E., Campa, J. G., Nunes, S. P., Abajo, J., Sulfonated polynaphthalimides with benzimidazole pendant groups, *Polymer*, 2008, 49, 3875–3883.

42. Sun, F., Wang, T., Yang, S., Fan, L., Synthesis and characterization of sulfonated polyimides bearing sulfonated aromatic pendant group for DMFC applications, *Polymer*, 2010, 51, 3887–3898.

43. Yamazaki, K., Tang, Y., Kawakami, H., Evaluating oxygen diffusion, surface exchange and oxygen semi-permeation in $Ln_2NiO_4+\sigma$ membranes (Ln = La, Pr and Nd) *J. Membr. Sci.*, 2010, 362, 234–240.

44. Akbarian-Feizi, L., Mehdipour-Ataei, S., Yeganeh, H., Survey of sulfonated polyimide membrane as a good candidate for Nafion substitution in fuel cell, *Int. J. Hydrogen Energy*, 2010, 35, 9385–9397.

45. Kim, Y. K., Park, H. B., Lee, Y. M., Carbon molecular sieve membranes derived from metal-substituted sulfonated polyimide and their gas separation properties, *J. Membr. Sci.*, 2003, 226, 145–158.

46. Islam, M. N., Zhou, W., Honda, T., Tanaka, K., Kita, H., Okamoto, K. I., Preparation and gas separation performance of flexible pyrolytic membranes by low-temperature pyrolysis of sulfonated polyimides. *J. Membr. Sci.*, 2005, 261, 17–26.

47. Ye, X., Bai, H., Ho, W. S., Synthesis and characterization of new sulfonated polyimides as proton-exchange membranes for fuel cells, *J. Membr. Sci.*, 2006, 279, 570–577.

48. Li, N., Cui, Z., Zhang, S., Xing, W., Synthesis and characterization of rigid-rod sulfonated polyimides bearing sulfobenzoyl side groups as proton exchange membranes, *J. Membr. Sci.*, 2007, 295, 148–158.

49. Zhai, F., Guo, X., Fang, J., Xu, H., Synthesis and properties of novel sulfonated polyimide membranes for direct methanol fuel cell application, *J. Membr. Sci.*, 2007, 296, 102–109.

50. Bai, H., Ho, W. S., New poly(ethylene oxide) soft segment-containing sulfonated polyimide copolymers for high temperature proton-exchange membrane fuel cells, *J. Membr. Sci.*, 2008, 313, 75–85.

51. Chen, X., Yin, Y., Chen, P., Kita, H., Okamoto, K. I., Synthesis and properties of novel sulfonated polyimides derived from naphthalenic dianhydride for fuel cell application, *J. Membr. Sci.*, 2008, 313, 106–119.

52. Hu, Z., Yin, Y., Okamoto, K., Moriyama, Y., Morikawa, A., Synthesis and characterization of sulfonated polyimides from 2,2'-*bis*(4-sulfophenyl)-4,4'-oxydianiline as polymer electrolyte membranes for fuel cell applications, *J. Membr. Sci.*, 2009, 329, 146–152.

53. Liu, C., Li, L., Liu, Z., Guo, M., Jing, L., Liu, B., Jiang, Z., Matsumoto, T., Guiver, M. D., Sulfonated naphthalenic polyimides containing ether and ketone linkages as polymer electrolyte membranes, *J. Membr. Sci.*, 2011, 366, 73–81.

54. Mistri, E. A., Mohanty, A. K., Banerjee, S., Synthesis and characterization of new fluorinated poly(ether imide) copolymers with controlled degree of sulfonation for proton exchange membranes, *J. Membr. Sci.*, 2012, 411, 117–129.

55. Blázquez, J. A., Iruin, J. J., Eceolaza, S., Marestin, C., Mercier, R., Mecerreyes, D., Miguel, O., Vela, A., Marcilla, R., Solvent and acidification method effects in the performance of new sulfonated copolyimides membranes in PEM-fuel cells, *J. Power Sources*, 2005, 151, 63–68.

56. Fang, J., Guo, X., Xu, H., Okamoto, K. I., Sulfonated polyimides: Synthesis, proton conductivity and water stability, *J. Power Sources*, 2006, 159, 4–11.

57. Liu, H., Lee, M. H., Lee, J., Synthesis of new sulfonated polyimide and its photo-crosslinking for polymer electrolyte membrane fuel cells, *Macromol. Res.*, 2009, 17, 725–728.

58. Kim, H., Litt, M. H., Nam, S. Y., Shin, E., Synthesis and characterization of sulfonated polyimide polymer electrolyte membranes, *Macromol. Res.*, 2003, 11, 458–466.

59. Yin, Y., Fang, J., Watari, T., Tanaka, K., Kita, H., Okamoto, K. I., Synthesis and properties of highly sulfonated proton conducting polyimides from *bis*(3-sulfopropoxy)benzidine diamines, *J. Mater. Chem.*, 2004, 14, 1062–1070.

60. Guo, X., Zhai, F., Fang, J., Laguna, M. F., Lopez-Gonzalez, M., Riande, E., Permselectivity and conductivity of membranes based on sulfonated naphthalenic copolyimides, *J. Phys. Chem. B*, 2007, 111, 13694–13702.

61. Miyatake, K., Asano, N., Tombe, T., Watanabe, M., Effect of cross-linking on polyimide ionomer membranes, *Electrochemistry*, 2007, 75, 122–125.

62. Yin, Y., Hayashi, S., Yamada, O., Kita, H., Okamoto, K. I., Branched/crosslinked sulfonated polyimide membranes for polymer electrolyte fuel cells, *Macromol. Rapid Commun.*, 2005, 26, 696–700.

63. Rabiee, A., Mehdipour-Ataei, S., Banihashemi, A., Yeganeh, H., Preparation of new membranes based on sulfonated aromatic copolyimides, *Polym. Adv. Tech.*, 2008, 19, 361–370.

64. Yasuda, T., Li, Y., Miyatake, K., Hirai, M., Nanasawa, M., Watanabe, M., Synthesis and properties of polyimides bearing acid groups on long pendant aliphatic chains, *J. Polym. Sci. A Polym. Chem.*, 2006, 44, 3995–4005.

65. Li, Y., Jin, R., Wang, Z., Cui, Z., Xing, W., Gao, L., Synthesis and properties of novel sulfonated polyimides containing binaphthyl groups as proton-exchange membranes for fuel cells, *J. Polym. Sci. A Polym. Chem.*, 2007, 45, 222–231.

66. Chen, S., Yin, Y., Kita, H., Okamoto, K. I., Synthesis and properties of sulfonated polyimides from homologous sulfonated diamines bearing *bis*(aminophenoxyphenyl) sulfone, *J. Polym. Sci. A Polym. Chem.*, 2007, 45, 2797–2811.

67. Chen, X., Chen, K., Chen, P., Higa, M., Okamoto, K. I., Effects of tetracarboxylic dianhydrides on the properties of sulfonated polyimides, *J. Polym. Sci. A Polym. Chem.*, 2010, 48, 905–915.

68. Lei, R., Kang, C., Huang, Y., Qiu, X. P., Ji, X. L., Xing, W., Gao, L., Sulfonated polyimides containing pyridine groups as proton exchange membrane materials, *Chin. J. Polym. Sci.*, 2011, 29, 532–539.

69. Liu, H., Kim, J. H., Li, X., Lee, M. H., Synthesis and characterization of photosensitive poly(imide sulfonates) for fuel cell application, *J. Wuhan Univ. Technol.-Mater. Sci. Ed.*, 2013, 28, 635–642.

70. Zhang, F., Li, N., Zhang, S., Li, S., Ionomers based on multisulfonated perylene dianhydride: Synthesis and properties of water resistant sulfonated polyimides, *J. Power Sources*, 2010, 195, 2159–2165.

71. Sugioka, T., Hay, A. S., Synthesis of novel poly(thioether -naphthalimide)s that utilize hydrazine as the diamine, *J. Polym. Sci. A Polym. Chem.*, 2001, 39, 1040–1050.

72. Gunduz, N., Virginia Polytechnic Institute and State University, 2001, Doctoral dissertation, Synthesis and characterization of sulfonated polyimides as proton exchange membranes for fuel cells.

73. Watari, T., Wang, H., Kuwahara, K., Tanaka, K., Kita, H., Okamoto, K. I., Water vapor sorption and diffusion properties of sulfonated polyimide membranes, *J. Membr. Sci.*, 2003, 219, 137–147.

74. Lee, C. H., Park, H. B., Chung, Y. S., Lee, Y. M., Freeman, B. D., Water sorption, proton conduction, and methanol permeation properties of sulfonated polyimide membranes cross-linked with *N,N-bis*(2-hydroxyethyl)-2-aminoethane sulfonic acid (BES), *Macromolecules*, 2006, 39, 755–764.

75. Qiu, Z., Wu, S., Li, Z., Zhang, S., Xing, W., Liu, C., Sulfonated poly(arylene-co-naphthalimide)s synthesized by copolymerization of primarily sulfonated monomer and fluorinated naphthalimide dichlorides as novel polymers for proton exchange membranes, *Macromolecules*, 2006, 39, 6425–6432.

76. Essafi, W., Gebel, G., Mercier, R., Sulfonated polyimide ionomers: A structural study, *Macromolecules*, 2004, 37, 1431–1440.

77. Li, N., Cui, Z., Zhang, S., Li, S., Synthesis and properties of novel polyimides from sulfonated binaphthalene dianhydride for proton exchange membranes, *J. Polym. Sci. A Polym. Chem.*, 2008, 46, 2820–2832.

78. Genies, C., Mercier, R., Sillion, B., Petiaud, R., Cornet, N., Gebel, G., Pineri, M., Stability study of sulfonated phthalic and naphthalenic polyimide structures in aqueous medium, *Polymer*, 2001, 42, 5097–5105.

79. Li, Y., Jin, R., Cui, Z., Wang, Z., Xing, W., Qiu, X., Ji, X., Gao, L., Synthesis and characterization of novel sulfonated polyimides from 1,4-*bis*(4-aminophenoxy)-naphthyl-2,7-disulfonic acid, *Polymer*, 2007, 48, 2280–2287.

80. Wei, H., Fang, X., Novel aromatic polyimide ionomers for proton exchange membranes: Enhancing the hydrolytic stability, *Polymer*, 2011, 52, 2735–2739.

81. Jung, E. Y., Chae, B., Kwon, S. J., Kim, H. C., Lee, S. W., Synthesis and characterization of sulfonated copolyimides via thermal imidization for polymer electrolyte membrane application, *Solid State Ionics*, 2012, 216, 95–99.

82. Chen, G., Zhang, X., Wang, J., Zhang, S., Synthesis and characterization of soluble poly(amide-imide)s bearing triethylamine sulfonate groups as gas dehumidification membrane material, *J. Appl. Polym. Sci.*, 2007, 106, 3179–3184.

83. Chen, K., Chen, X., Yaguchi, K., Endo, N., Higa, M., Okamoto, K. I., Synthesis and properties of novel sulfonated polyimides bearing sulfophenyl pendant groups for fuel cell application, *Polymer*, 2009, 50, 510–518.

84. Chen, S., Yin, Y., Tanaka, K., Kita, H., Okamoto, K. I., Synthesis and properties of novel side-chain-sulfonated polyimides from *bis*[4-(4-aminophenoxy)-2-(3-sulfobenzoyl)] phenyl sulfone, *Polymer*, 2006, 47, 2660–2669.

85. Eisenberg, A., Clustering of ions in organic polymers. a theoretical approach, *Macromolecules*, 1970, 3, 147–154.

86. Dreyfus, B., Model for the clustering of multiplets in ionomers, *Macromolecules*, 1985, 18, 284–292.

87. Longworth, R., Vaughan, D., Physical structure of ionomers. V. Internal friction and dielectric relaxation of ethylene copolymers and ionomers, *Polym. Prepr. (Am. Chem. Soc., Div. Polym. Chem.)*, 1968, 9, 525.

88. Macknight, W. J., Taggart, W. P., Stein, R. S., Model for the structure of ionomers, *J. Polym. Sci. Polym. Symp.*, 1974, 45, 113–128.

89. Gierke, T. D., Munn, G. E., Wilson, F. C., The morphology in Nafion perfluorinated membrane products, as determined by wide- and small-angle x-ray studies, *J. Polym. Sci. Polym. Phys. Ed.*, 1981, 19, 1687–1704.

90. Blachot, J. F., Diat, O., Putaux, J. L., Rollet, A. L., Rubatat, L., Vallois, C., Muller, M., Gebel, G., Anisotropy of structure and transport properties in sulfonated polyimide membranes, *J. Membr. Sci.*, 2003, 214, 31–42.

91. Zhang, F., Li, N., Zhang, S., Preparation and characterization of sulfonated poly(arylene-co-naphthalimide)s for use as proton exchange membranes, *J. Appl. Polym. Sci.*, 2010, 118, 3187–3196.

92. Tamai, S., Kuroki, T., Shibuya, A., Yamaguchi, A., Synthesis and characterization of thermally stable semicrystalline polyimide based on 3,4′-oxydianiline and 3,3′,4,4′-biphenyltetra-carboxylic dianhydride, *Polymer*, 2001, 42, 2373–2378.

93. Wang, F., Hicker, M., Kim, Y. S., Zawodzinski, T. A., McGrath, J. E., Direct polymerization of sulfonated poly(arylene ether sulfone) random (statistical) copolymers: Candidates for new proton exchange membranes, *J. Membr. Sci.*, 2002, 197, 231–242.

94. Makharia, R., Kocha, S. S., Yu, P. T., *208th Meeting of the Electrochemical Society*, Los Angeles, CA, October 16–21, 2005, Abstract #1165.

95. Gasteiger, H. A., *International Conference on Solid State Ionics (SSI-15)*, Baden-Baden, Germany, July 17–22, 2005, Oral Contribution #72.

96. Liu, B., Robertson, G., Kim, D., Guiver, M. D., Hu, W., Jiang, Z., Aromatic poly(ether ketone)s with pendant sulfonic acid phenyl groups prepared by a mild sulfonation method for proton exchange membranes, *Macromolecules*, 2007, 40, 1934–1944.

97. Liu, B., Kim, Y., Hu, W., Robertson, G. P., Pivovar, B. S., Guiver, M. D., Homopolymer-like sulfonated phenyl- and diphenyl-poly(arylene ether ketone)s for fuel cell applications, *J. Power Sources*, 2008, 185, 899–903.

98. Kopitzke, R. W., Linkous, C. A., Anderson, H. R., Nelson, G. L., Conductivity and water uptake of aromatic-based proton exchange membrane electrolytes, *J. Electrochem. Soc.*, 2000, 147, 1677–1681.

99. Yin, Y., Yamada, O., Tanaka, K., Okamoto, K. I., On the development of naphthalene-based sulfonated polyimide membranes for fuel cell applications, *Polym. J.*, 2006, 38, 197–219.

100. Hu, Z., Yin, Y., Yaguchi, K., Endo, N., Higa, M., Okamoto, K. I., Synthesis and properties of sulfonated multiblock copolynaphthalimides, *Polymer*, 2009, 50, 2933–2943.

101. Fang, J., Guo, X., Litt, M., Synthesis and properties of novel sulfonated polyimides for fuel cell application, *Trans. Mater. Res. Soc. Jpn.*, 2004, 29, 2541–2546.

102. Suto, Y., Yin, Y., Kita, H., Okamoto, K. I., Synthesis and properties of sulfonated polyimides from sulfophenoxy benzidines, *J. Photopolym. Sci. Technol.*, 2006, 19, 273–274.

103. Sutou, Y., Yin, Y., Kita, H., Chen, S., Kita, H., Okamoto, K. I., Wang, H., Kawasato, H., Synthesis and properties of sulfonated polyimides derived from *bis*(sulfophenoxy) benzidines, *J. Polym. Sci. A Polym. Chem.*, 2009, 47, 1463–1477.

104. Hu, Z., Yin, Y., Kita, H., Okamoto, K. I., Suto, Y., Wang, H., Kawasato, H., Synthesis and properties of novel sulfonated polyimides bearing sulfophenyl pendant groups for polymer electrolyte fuel cell application, *Polymer*, 2007, 48, 1962–1971.

105. Zawodzinski, T. A., Neeman, M., Sillerud, L. O., Gottesfeld, S., Determination of water diffusion coefficients in perfluorosulfonate ionomeric membranes, *J. Phys. Chem.*, 1991, 95, 6040–6044.

106. Mittal, V., Kunz, H. R., Fenton, J. M., Factors accelerating membrane degradation rate and the underlying degradation mechanism in PEMFC, *ECS Trans.*, 2006, 1, 275–282.

107. Miyatake, K., Zhou, H., Uchida, H., Watanabe, M., Highly proton conductive polyimide electrolytes containing fluorenyl groups, *Chem. Commun.*, 2003, 3, 368–369.

108. Einsla, B. R., Kim, Y. S., Hickner, M. A., Hong, Y. T., Hill, M. L., Pivovar, B. S., McGrath, J. E., Sulfonated naphthalene dianhydride based polyimide copolymers for proton-exchange-membrane fuel cells. II. Membrane properties and fuel cell performance, *J. Membr. Sci.*, 2005, 255, 141–148.

109. Yasuda, T., Miyatake, K., Hirai, M., Nanasawa, M., Watanabe, M., Synthesis and properties of polyimide ionomers containing sulfoalkoxy and fluorenyl groups, *J. Polym. Sci. A Polym. Chem.*, 2005, 43, 4439–4445.

110. Asano, N., Miyatake, K., Watanabe, M., Hydrolytically stable polyimide ionomer for fuel cell applications, *Chem. Mater.*, 2004, 16, 2841–2843.

111. Asano, N., Aoki, M., Suzuki, S., Miyatake, K., Uchida, H., Watanabe, M., Aliphatic/aromatic polyimide ionomers as a proton conductive membrane for fuel cell applications, *J. Am. Chem. Soc.*, 2006, 128, 1762–1769.

112. Steck, A. E., Stone, C., Development of BAM membrane for fuel cell applications, *Proceedings of the Second International Symposium on New Mater. Fuel Cell Modern Battery Systems II*, Montreal, Quebec, Canada, 1997, pp. 792–807.

113. Mattsson, B., Ericson, H., Torell, L. M., Sundholm, F., Degradation of a fuel cell membrane, as revealed by micro-Raman spectroscopy, *Electrochim. Acta*, 2000, 45, 1405–1408.

114. Hübner, G., Roduner, E., EPR investigation of HO·radical initiated degradation reactions of sulfonated aromatics as model compounds for fuel cell proton conducting membranes, *J. Mater. Chem.*, 1999, 9, 409–418.

115. Yin, Y., Fang, J., Cui, Y., Tanaka, K., Kita, H., Okamoto, K. I., Synthesis, proton conductivity and methanol permeability of a novel sulfonated polyimide from 3-(2′,4′-diaminophenoxy)propane sulfonic acid, *Polymer*, 2003, 44, 4509–4518.

116. Eikerling, M., Kornyshev, A. A., Kuznetsov, A. M., Ulstrup, J., Walbran, S., Mechanisms of proton conductance in polymer electrolyte membranes, *J. Phys. Chem. B*, 2001, 105, 3646–3662.

117. Kornshev, A. A., Kuznetsov, A. M., Spohr, E., Ulstrup, J., Kinetics of proton transport in water, *J. Phys. Chem. B*, 2003, 107, 3351–3366.

118. Spohr, E., Commer, P., Kornyshev, A. A., Enhancing proton mobility in polymer electrolyte membranes: Lessons from molecular dynamics simulations, *J. Phys. Chem. B*, 2002, 106, 10560–10569.

119. Endo, N., Matsuda, K., Yaguchi, K., Hu, Z., Chen, K., Higa, M., Okamoto, K., Cross-linked sulfonated polyimide membranes for polymer electrolyte fuel cells, *J. Electrochem. Soc.*, 2009, 156, B628–B633.

120. Devanathan, R., Recent developments in proton exchange membranes for fuel cells, *Energy Environ. Sci.*, 2008, 1, 101–119.

121. Yue, M., Zhang, Y., Wang, L., Sulfonated polyimide/chitosan composite membrane for vanadium redox flow battery: Influence of the infiltration time with chitosan solution, *Solid State Ionics*, 2012, 217, 6–12.

122. Li, J., Zhang, Y., Wang, L., Preparation and characterization of sulfonated polyimide/TiO_2 composite membrane for vanadium redox flow battery, *J. Solid State Electrochem.*, 2014, 18, 729–737.

4 Sulfonated Poly(Ether Sulfone) Membranes

Tomoya Higashihara and Mitsuru Ueda

CONTENTS

ABBREVIATIONS

6F-BPA	4,4′-(Hexafluoroisopropylidene)diphenol
AFM	Atomic force microscope
BNSH	Sulfonated binaphthyl-based poly(arylene ether sulfone)
BP	Biphenyl
BPS	Biphenyl-based poly(arylene ether sulfone)
BPSH	Randomly sulfonated biphenyl-based poly(arylene ether sulfone)
CM	1-Ethynyl-2,4-difluorobenzene
DCDPS	4,4′-Dichlorodiphenylsulfone
DFBP	Decafluorobiphenyl
DMA	Dynamic mechanical analysis
DMAc	Dimethylacetamide
DMF	Dimethylformamide
DMFC	Direct methanol fuel cell
DMSO	Dimethyl sulfoxide
DS	Degree of sulfonation
ESPSN	Sulfonated poly(phenylene sulfide nitrile)
EW	Equivalent weight
FTIR	Fourier transform infrared spectrometry
HFB	Hexafluorobenzene
HSP	Sulfonated poly(arylene ether sulfone)
IEC	Ion exchange capacity
MB	Multiblock sulfonated-fluorinated poly(arylene ether sulfone)
MEA	Membrane electrode assembly
MP	$M(HPO_4)_2 \cdot nH_2O$ (M = Sn, Zr, Ti)
MW	Molecular weight
NMP	N-Methylpyrrolidone
NMR	Nuclear magnetic resonance
NS	Nanosheet
PEEKK	Poly(ether ether ketone ketone)
PEFC	Polymer electrolyte fuel cells
PEM	Proton exchange membrane
PEMFC	Proton exchange membrane fuel cell
PES	Poly(ether sulfone)
PES-Br	Poly(oxy-4,4′-(3,3′-dibromobiphenylene)-oxy-4,4′-diphenylsulfone)
PES-PSA	Perfluorosulfonated poly(arylene ether sulfone)
PPMA	Phosphorus pentoxide-methanesulfonic acid
PPO	Poly(propylene oxide)

PSA-K	Potassium 1,1,2,2-tetrafluoro-2-(1,1,2,2-tetrafluoro-2-iodoethoxy)ethane sulfonate
RH	Relative humidity
SAXS	Small-angle x-ray scattering
SDCDPS	3,3′-Disulfonate-4,4′-dichlorodiphenylsulfone
SFM	Scanning force microscope
SHQ	Potassium 2,5-dihydroxybenzenesulfonate
s-MWNT	Sulfonated multiwalled carbon nanotubes
SnP	$Sn(HPO_4)_2 \cdot nH_2O$
SPES	Sulfonated poly(ether sulfone)
SPESK	Sulfonated poly(arylene ether sulfone ketone)
SPSU	Sulfonated poly(sulfone)
STEM	Scanning transmission electron microscope
TEM	Transmission electron microscope
TGA	Thermogravimetric analysis
WAXS	Wide-angle x-ray scattering
WU	Water uptake
XESPSN	Cross-linked sulfonated poly(phenylene sulfide nitrile)
XRD	X-ray diffraction

4.1 INTRODUCTION

Ever-increasing energy consumption is inescapable for progress in human society. Most of the energy we use today is provided by the combustion of fossil fuels.
However, fossil fuel combustion has resulted in air pollution and global warming,
which affects all living things on earth. In addition, the depletion of fossil fuel reserves
is a great concern. These health and environmental concerns call for new technologies for energy conversion and power generation that are more efficient and environmentally friendly. Fuel cells are one of the promising future power sources due
to their advantages such as high efficiency, high energy density, quiet operation, and
environmental friendliness. Among them, the polymer electrolyte fuel cells (PEFCs)
are considered to be the most promising power source for portable and automotive
applications in which proton exchange membranes (PEMs) play an important role
in PEFCs, which are responsible for the proton transfer from the anode to the cathode and the entire fuel cell performance. Currently, perfluorinated polymers, such as
Nafion® or Flemion®, are the state-of-the-art materials because of their good physical
and chemical stabilities along with a high proton conductivity under a wide range of
relative humidity (RH) at moderate operating temperatures (Figure 4.1) [1,2].

However, high operating temperatures, which are required for actual operations,
reduce their properties. In addition, their shortcomings, such as high cost and high
methanol permeation, limit their application. To remedy these problems, aromatic
polymers are regarded as potential candidates for PEM applications because of their
availability, processability, low cost, and wide variety of chemical compositions. In
addition, the high mechanical, thermal, chemical, and electrochemical properties of
aromatic polymers, which promise a high adaptability in a fuel cell environment,

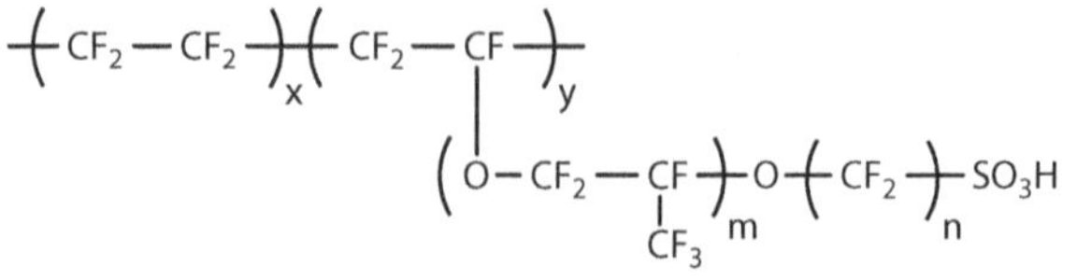

$x = 6-10, y = 1000$

Nafion : $m = 1$, $n = 2$

Flemion : $m = 1$, $n = 2$

FIGURE 4.1 Chemical structures of perfluorosulfonic acid membranes.

drive development as alternative PEM materials to the perfluorosulfonic acid membranes. A number of sulfonated aromatic polymers, such as poly(phenylene)s, poly(ether ether ketone)s, poly(ether sulfone)s (PESs), poly(arylene ether)s, and polyimides, have been developed as alternate candidates. Generally, membranes based on these polymers can achieve a high proton conductivity by increasing their ion exchange capacities (IECs) but result in significant water uptake (WU) and a dramatic loss of mechanical properties [3–7].

Among them, the sulfonated poly(ether sulfone) (SPES) is the most studied aromatic polymer for PEMs, which has excellent properties due to the combination of sulfone and ether groups. The former affords rigidity and polarity, and the latter provides flexibility to the membranes. Thus, the focus of this chapter will be on this class of membranes and the recent advances in understanding their use in fuel cells.

4.2 MEMBRANE SYNTHESIS

4.2.1 REQUIREMENTS FOR PEMs

The key materials for PEFCs are catalysts and PEMs because they dominate the fuel cell performance. In general, PEMs are based on polymer electrolytes, which consist of two phases on a nanoscale level, one a structural (polymer) domain and the other a functional (water) domain with acid groups. The chemical and mechanical properties of the membrane can be attributed to the polymer domain, while the transport properties could be attributed to the water domain.

PEMs typically require a high performance including (1) high proton conductivity even at a low RH and high temperature, (2) low electronic conductivity, (3) low permeability to fuel and oxidant, (4) low water transport through diffusion and electroosmosis, (5) oxidative and hydrolytic stability, (6) high dimensional stability, (7) excellent physical and chemical durabilities in both the dry and hydrated states, (8) low cost, (9) capability for fabrication into membrane electrode assemblies (MEAs), and (10) high electrochemical fuel cell performance over a long time period (>5000 h) [4,5].

4.2.2 POLYMER SYNTHETIC METHODS

Aromatic polymers, such as PESs, poly(ether ether ketone)s, and polyimides, are used as polymer matrices for PEMs due to the high performance described earlier. Sulfonated PEMs are generally prepared by two methods: the postsulfonation of aromatic polymers usually leading to a random functionalization along the polymer main chains and direct copolymerization of the sulfonated monomers to afford random copolymers. Both methods are discussed in the following.

4.2.2.1 Postsulfonation of Aromatic Polymers

The most common and accessible way to obtain sulfonated aromatic polymers is postsulfonation, which is usually achieved by an electrophilic substitution reaction (Scheme 4.1). In general, sulfonation reagents, such as concentrated sulfuric acid, fuming sulfuric acid, chlorosulfonic acid, and sulfur trioxide, are used for the introduction of sulfonic acid groups to the aromatic polymers. While postsulfonation is an accessible method, the reaction is often restricted due to the difficult reaction control (the degree and position of sulfonation) and side reactions such as cross-linking and degradation of the polymer chains. Therefore, a postmodification reaction significantly depends on the reaction time, temperature, and type of sulfonation reagent.

One of the most commonly used polymers for aromatic PEMs is PES due to its high thermal, mechanical, and chemical resistances. Generally, the sulfonation of PES is performed using sulfonation reagents. When concentrated sulfuric acid is used as the sulfonation reagent in an organic solvent-free reaction, polymer degradation equally occurs, leading to a decrease in the product recovery yield [8].

Since PES can be dissolved in chlorinated solvents, chlorosulfonic acid is used for the sulfonation. However, it is reported that chlorosulfonic acid often induces chain cleavages during the sulfonation of PES [9]. In addition, SPES is precipitated from the chlorinated solvents during the sulfonation, and control of the sulfonation degree is difficult [10,11]. Thus, trimethylsilyl chlorosulfonate is used as an alternative sulfonation reagent [12–15]. Due to the formation of a trimethylsilyl ester, SPES remains soluble in the chlorinated solvents. Furthermore, trimethylsilyl chlorosulfonate can suppress the chain breakings of PES because it is milder than chlorosulfonic acid. Genova-Dimitrova et al. studied the sulfonation of the bisphenol-A-based PES using chlorosulfonic acid and trimethylsilyl chlorosulfonate [13]. The authors found that a strong sulfonating agent, chlorosulfonic acid, yielded a heterogeneous reaction that must be prevented by adding several drops of dimethylformamide (DMF), finally leading to chain cleavage, indicated by the viscometric measurements. On the other hand, the reaction mixture remained perfectly homogeneous

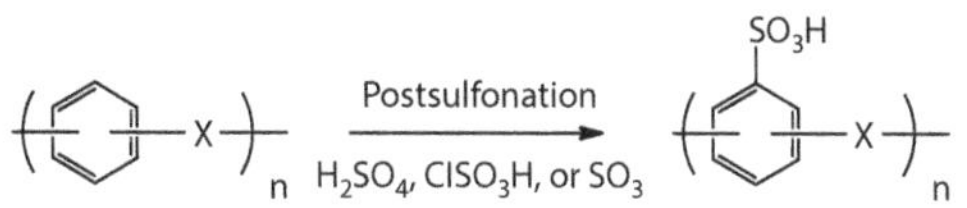

SCHEME 4.1   Postsulfonation of aromatic polymers.

with $(CH_3)_3SiSO_3Cl$, which did not affect the polymer backbone. One drawback of $(CH_3)_3SiSO_3Cl$ is that the reaction rate significantly decreased after a degree of sulfonation (DS) of 0.74 (theoretical sulfonation degree = 2, which means two sulfonic groups per one monomer unit). Furthermore, the real sulfonation degree is limited to about 0.85 even when a threefold excess of the sulfonating agent was employed [13]. For the bisphenol-A-based systems, not more than one sulfonic acid group can be introduced into the repeating unit (Figure 4.2) [16].

On the other hand, PES containing fluorenyl groups can be achieved with up to two sulfonic acid groups per one monomer repeating unit using chlorosulfonic acid (Scheme 4.2) [17].

An alternative route has also been reported, that is, a metalation route to SPEMs, which consisted of three steps, that is, metalation of aromatic polymers, sulfination by SO_2 gas, and oxidation as shown in Scheme 4.3 [18]. Postsulfonation usually occurs at the *ortho* position of the benzene rings activated by electron-donating groups such as ether linkages; in contrast, sulfonic acid groups are introduced to the electron-poor benzene rings by metalation as shown in Scheme 4.3. Therefore, the resulting sulfonated aromatic polymers are more stable and show a higher acidity. Similarly, PES has been polysulfophenylated by lithiation and a reaction with 2-sulfobenzoic acid cyclic anhydride (Scheme 4.4) [19].

4.2.2.2 Direct Copolymerization of Sulfonated Monomers

Although postsulfonation is a facile method for preparing sulfonated aromatic polymers, there are several shortcomings, typified by the lack of control over the degree and location of the sulfonation. Therefore, the direct copolymerization of sulfonated

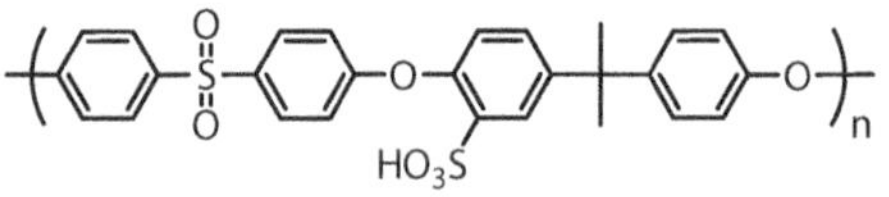

FIGURE 4.2   Sulfonated structure of PES.

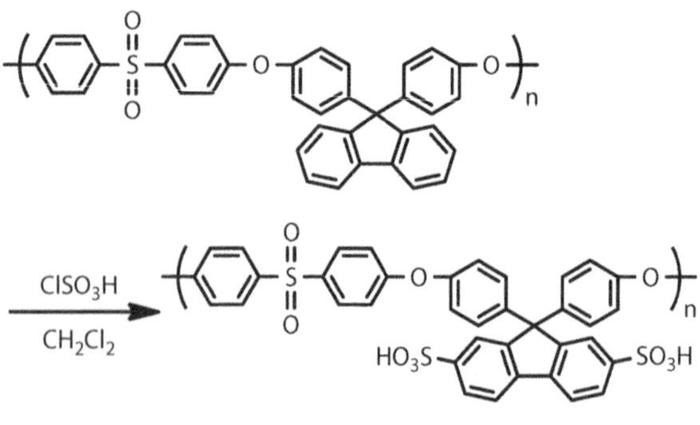

SCHEME 4.2   Sulfonation of PES containing fluorenyl groups.

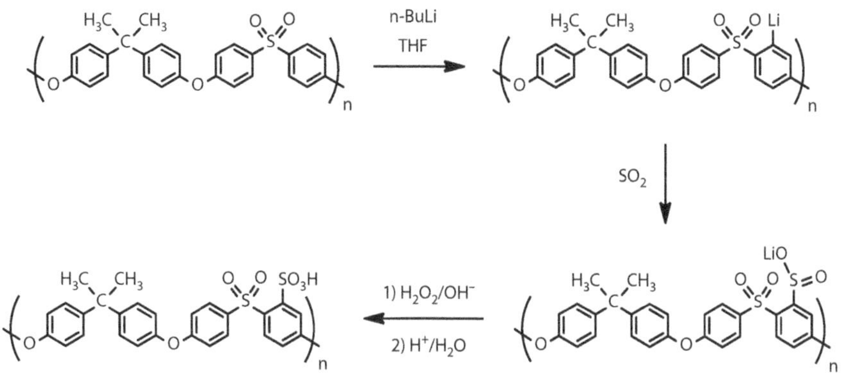

SCHEME 4.3 Sulfonation of PES by lithiation, sulfination, and oxidation.

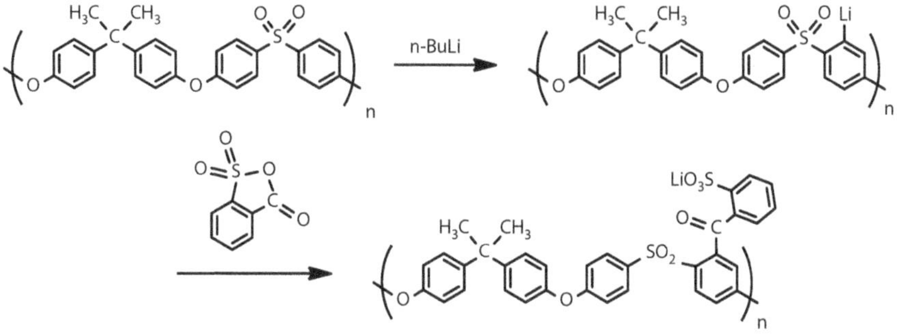

SCHEME 4.4 Sulfophenylation of PES by lithiation and reaction with 2-sulfobenzoic acid cyclic anhydride.

monomers is attractive as an alternative method for the preparation of sulfonated aromatic polymers. The use of sulfonated monomers is advantageous for controlling the amount of sulfonic acid groups and forming sulfonated polymers with defined structures. In addition, this method can be applied to the synthesis of multiblock copolymers containing sulfonated blocks.

In contrast to the postsulfonation reactions of aromatic polymers, small molecules allow sulfonation on deactivated sites, such as aromatic rings with electron-withdrawing groups, since strong sulfonation reagents, such as fuming sulfuric acid, can be applied. One of the most common sulfonated monomers (Figure 4.3) is 3,3′-disulfonated 4,4′-dichlorodiphenyl sulfone (Figure 4.3a) [20]. This monomer can be used for the general nucleophilic aromatic substitution polymerization and provides SPESs [21]. A number of SPES copolymers have been reported, and a typical example is shown in Scheme 4.5 [22]. The DS and the chemical composition can be easily controlled by changing the ratio of the sulfonated monomers and types of monomers, respectively [23,24].

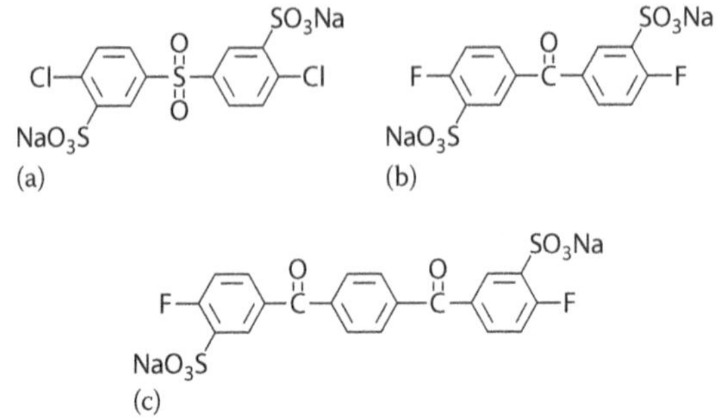

FIGURE 4.3 Structure of (a) 3,3′-disulfonated 4,4′-dichlorodiphenyl sulfone, (b) 3,3′-disulfonated 4,4′-difluorodiphenyl ketone, and (c) 1,4-*bis*(3-sodium sulfonated-4-fluorobenzoyl)benzene.

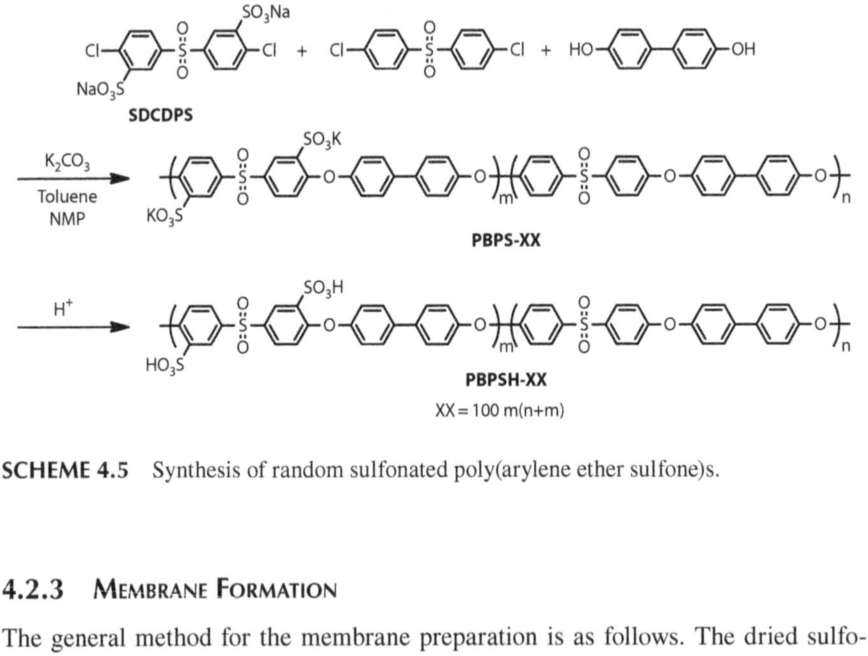

SCHEME 4.5 Synthesis of random sulfonated poly(arylene ether sulfone)s.

4.2.3 MEMBRANE FORMATION

The general method for the membrane preparation is as follows. The dried sulfonated polymers in the sodium salt form are readily dissolved to make 5–10 wt% solutions in dipolar aprotic solvents, such as dimethyl sulfoxide (DMSO), DMF, *N,N*-dimethylacetamide (DMAc), and *N*-methylpyrrolidone (NMP), at 60°C. The solutions are filtered, cast onto glass plates using a doctor blade, dried at 60°C for 12 h, and then treated in a vacuum at 100°C for 10 h. The as-cast membranes are immersed in a 2.0 M H_2SO_4 solution for 24 h at room temperature and then thoroughly washed with deionized water. Tough, ductile ionomer membranes were prepared with a controlled thickness in the range of 30–70 μm, depending on the casting solution concentration.

4.3 MEMBRANE CHARACTERIZATION

4.3.1 SPECTRAL STUDIES (IR, NMR)

The structures of the PEMs were confirmed by Fourier transform infrared (FTIR), ^{1}H-, and ^{13}C NMR spectroscopy. A typical example from Wang et al. [21] is as follows. Aromatic poly(arylene ether sulfone)s (BPSH-XXs) containing up to two pendant sulfonate groups per one repeat unit were prepared by the potassium carbonate–mediated direct aromatic nucleophilic substitution polycondensation of disodium 3,3′-disulfonate-4,4′-dichlorodiphenylsulfone (SDCDPS), 4,4′-dichlorodiphenylsulfone (DCDPS), and 4,4′-biphenol as shown in Scheme 4.5.

FTIR spectroscopy was used to analyze the characteristic bands corresponding to the sulfonate groups in the polymers. The introduction of the sodium sulfonate groups was confirmed by the FTIR spectra (Figure 4.4) in which strong characteristic peaks at 1030 and 1098 cm^{-1} assigned to symmetric and asymmetric stretchings of the sodium sulfonated groups are observed for all the sulfonated copolymers from PBPS-10 to PBPS-60, and the densities of these two characteristic peaks increased with the higher SDCDPS content. The symmetric stretching of the sulfonate groups at 1030 cm^{-1} can be compared to internal standards, such as the in-chain diphenyl ether absorption. Quantitative studies of the FTIR spectra are provided in Figure 4.5. A linear relationship exists between the DS and the density ratio of the sulfonate stretching (1030 cm^{-1}) to the diphenyl ether bands (1006 cm^{-1}). This result proves that the sulfonate groups are indeed quantitatively introduced to the copolymers as expected.

Two series of multiblock copolymers based on poly(arylene ether sulfone)s were synthesized by a coupling reaction between the phenoxide-terminated fully disulfonated poly(arylene ether sulfone) (BPSH-100) and decafluorobiphenyl (DFBP) or hexafluorobenzene (HFB) end-capped unsulfonated poly(arylene ether sulfone)

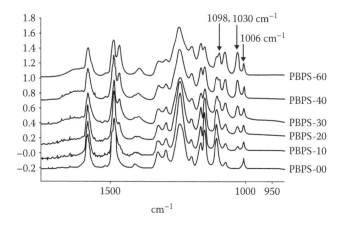

FIGURE 4.4 Influence of the DS on the FTIR of sulfonated poly(arylene ether sulfone) copolymer. (Reprinted from Wang, F. et al., *J. Membr. Sci.*, 197(1–2), 231, 2002. With permission.)

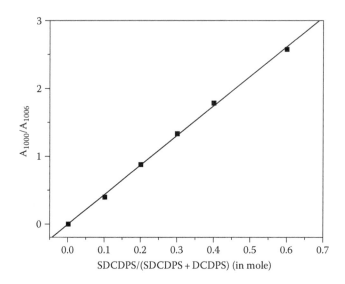

FIGURE 4.5 FTIR correlation of sulfonate groups and internal standard in polymer chain. (Reprinted from Wang, F. et al., *J. Membr. Sci.*, 197(1–2), 231, 2002. With permission.)

(BPS-0) as hydrophilic and hydrophobic blocks, respectively [25]. Disulfonated poly(arylene ether sulfone) hydrophilic oligomers (BPSH-100) and unsulfonated poly(arylene ether sulfone) hydrophobic oligomers (BPS-0) with a phenoxide telechelic functionality were synthesized via a step-growth polymerization (Scheme 4.6).

The ^{1}H NMR spectra of both BPSH-100 and BPS-0 are presented in Figure 4.6. Four small peaks at 6.80, 7.05, 7.40, and 6.80 ppm in Figure 4.6a are assigned to the protons on the BP moieties, which are located at the end of the oligomers. On the other hand, the peaks at 7.10, 7.65 (Figure 4.6a), and 7.10, 7.90 ppm (Figure 4.6b) are assigned to the protons on the BP moieties in the middle of the main chains of BPSH-100 and BPS-0, respectively. By comparing the integrations of the end group BP and the main chain BP, the number-average molecular weights of the oligomers were determined.

The BPS-0 hydrophobic blocks were then end-capped with HFB or DFBP to produce the fluorine-terminated end group functionality (Scheme 4.7). The ^{1}H NMR spectra of the end-capped oligomers are shown in Figure 4.7 in which the phenoxide bearing BP peaks at 6.80, 7.40, and 7.60 ppm disappear, indicating that all of the phenoxide groups react with the end-capping reagents.

Finally, two series of multiblock copolymers were synthesized by a coupling reaction between the phenoxide-terminated BPSH-100 oligomers and DFBP or the HFB end-capped BPS-0 oligomers (Scheme 4.8). The structure of the resulting multiblock copolymers was confirmed by the ^{1}H and ^{13}C NMR spectra. Figure 4.8 shows the ^{1}H NMR spectra of the multiblock copolymers. The phenoxide end group peaks completely disappeared from the hydrophilic oligomer confirming that the coupling reaction was successful. In the ^{13}C NMR spectra of the BPSH35 random copolymer,

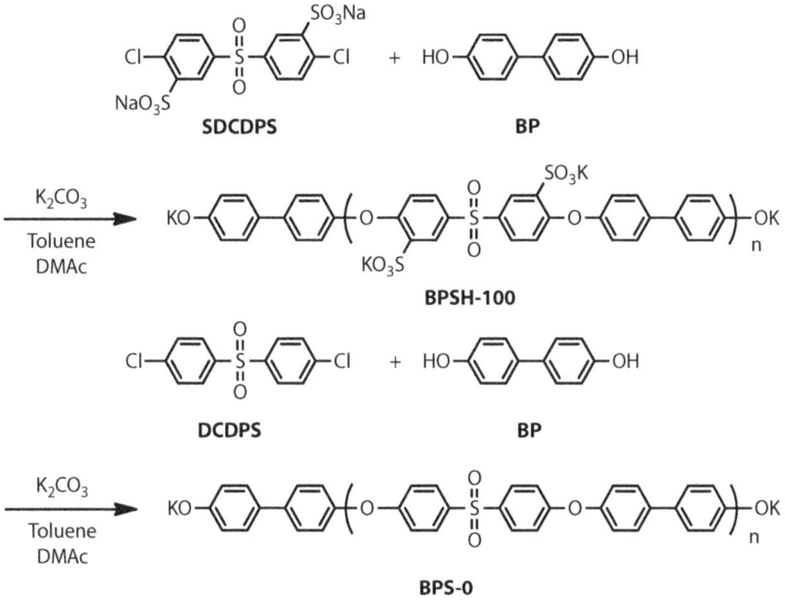

SCHEME 4.6 Synthetic routes for a fully disulfonated hydrophilic oligomer with phenoxide telechelic functionality (BPSH-100) and unsulfonated hydrophobic oligomer with phenoxide telechelic functionality (BPS-0).

each peak from the random copolymer shows a multiplet, suggesting a random sequence of the sulfonated moieties on the main chain. On the other hand, the multiblock copolymer shows sharp narrow peaks, explaining the ordered sequences in the copolymer. These results clearly show the prevention of randomization by the ether–ether chain exchange reaction.

4.3.2 Morphological Studies (AFM, TEM, X-Ray)

The microphase separation is induced by the segregation of the hydrophobic and hydrophilic domains in SPES in order to minimize the contact energy between them. The surface morphology of the microphase separation has been significantly investigated by atomic force microscopy (AFM) or scanning force microscopy (SFM). Typical examples of the AFM tapping phase images of SPES (BPS-00, BPSH-20, BPSH-40, BPSH-60, see Section 4.3.1) and Nafion 117 are shown in Figure 4.9 [21]. For the BPS-00 without sulfonate groups, a featureless phase morphology was observed. In contrast, for BPSH-20, BPSH-40, and BPSH-60, interpenetrating cluster-like structures with a diameter of 10–25 nm were clearly observed. The dark phase was assigned to a softer region of the hydrophilic sulfonic acid groups containing water. The domain size and degree of connection increase with the increasing DS from 0% to 60%. This observation indicated that the large channels of an ionic rich phase are significantly related to the DS, that is, the IEC values.

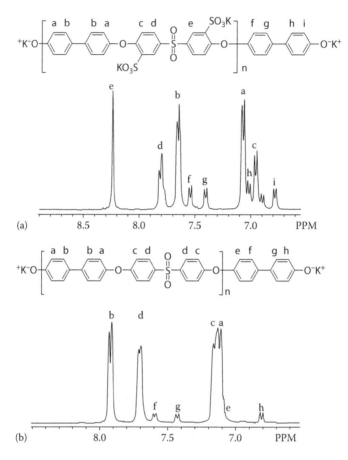

FIGURE 4.6 ¹H NMR spectra of (a) phenoxide-terminated BPSH-100 hydrophilic oligomer and (b) phenoxide-terminated BPS-0 hydrophobic oligomer. (Reprinted from Lee, H.S. et al., *Polymer*, 49(3), 715, 2008. With permission.)

Transmission electron microscopy (TEM) and scanning TEM (STEM) are also promising tools for identifying the morphology of not only the surface but also the morphology of the entire bulk films. Especially, for fuel cell applications, the connection of proton channels across the membrane (out-of-plane direction) is quite important; therefore, the cross-sectional view of the membrane has been widely studied. In a typical experiment [26], the sulfonated multiblock copoly(ether sulfone) membrane was stained with lead by ion exchange of the sulfonic acid groups by immersing them in a large excess of a $PbNO_3$ aqueous solution for 3 days, then rinsed with water. The stained membrane was embedded in epoxy resin and sectioned to provide a 70 nm thick membrane. The cross-sectional TEM image of the membrane is shown in Figure 4.10. Since the membranes were stained with Pb ions, the dark and bright regions are assigned to the hydrophilic and hydrophobic domains, respectively. Indeed, the hydrophilic domains (5–10 nm width) are connected to each other, and such connected hydrophilic domains in the

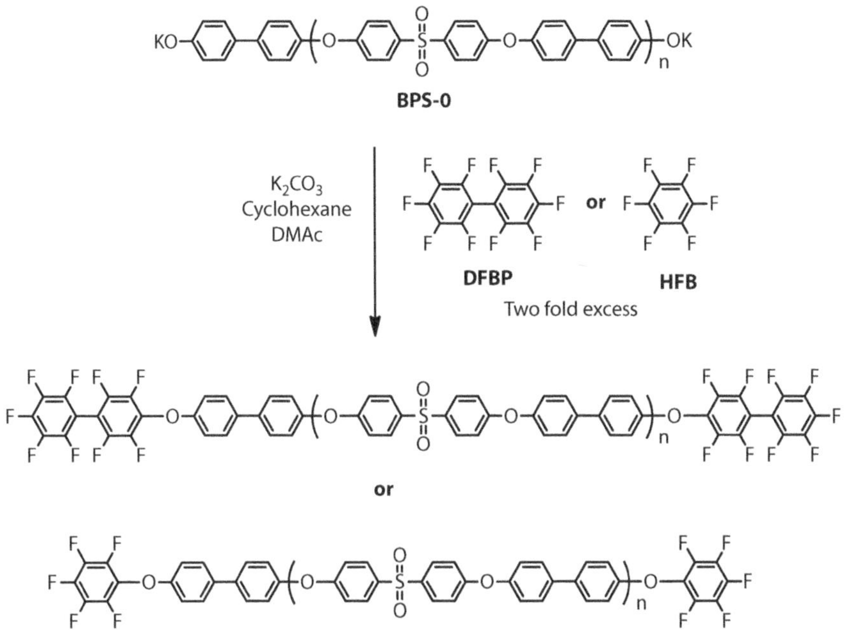

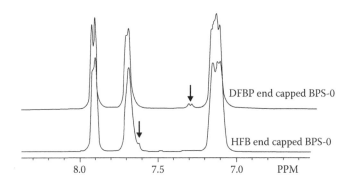

SCHEME 4.7   End capping of BPS-0 hydrophobic oligomer with DFBP or HFB.

FIGURE 4.7   ¹H NMR spectra of BPSH and HFB end-capped BPS-0 hydrophobic oligo-mers. Black arrows indicate the peaks from the end-capped BP moieties. (Reprinted from Lee, H.S. et al., *Polymer*, 49(3), 715, 2008. With permission.)

membrane thickness direction presumably function as the ionic channels in the out-of-plane direction.

On the other hand, the investigation of the x-ray diffraction (XRD) including wide-angle x-ray scattering (WAXS) and small-angle x-ray scattering (SAXS) has also been utilized to clarify the morphology of the membranes. For instance, the morphological difference between Nafion 117 and sulfonated poly(ether ether ketone ketone) (PEEKK) membranes, which bear similar main chains to the SPES,

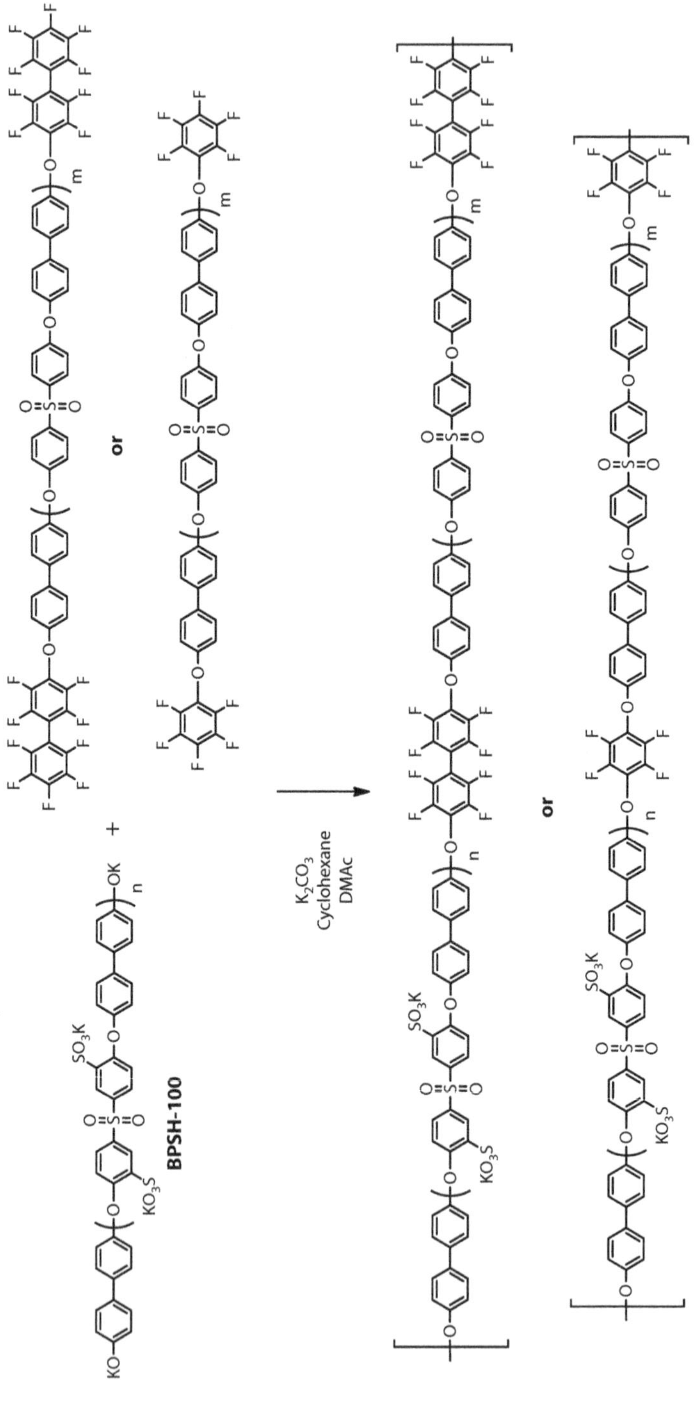

SCHEME 4.8 Synthesis of segmented sulfonated multiblock copolymers (BPSH–BPS) with different linkage groups.

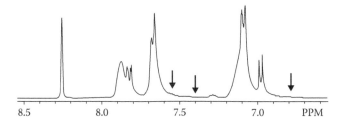

FIGURE 4.8 ¹H NMR spectrum of BPSH5–BPS5 with DFBP linkage. Black arrows indicate the disappearance of the end groups on the hydrophilic blocks after the coupling reaction with fluorine-terminated hydrophobic blocks. (Reprinted from Lee, H.S. et al., *Polymer*, 49(3), 715, 2008. With permission.)

can be revealed based on the results of SAXS experiments [27]. As shown in Figure 4.11, the hydrated PEEKK membrane possesses an ionomer peak at a higher scattering angle with a broader shape compared to Nafion 117. Also, the scattering intensity at the high scattering angles is greater for PEEKK than Nafion 117. This indicates that the mean distance of the phase separation between the hydrophobic and hydrophilic domains for PEEKK is smaller than that of Nafion 117 along with a wider distribution and a larger interface area. The morphological difference is illustrated in Figure 4.12.

4.3.3 Physical Properties (Solubility, Viscosity, Thermal Stability, and Mechanical Properties)

The thermal stability of the polymers was investigated by thermogravimetric analysis (TGA). The sulfonated PEMs are preheated at 150°C for 30 min in the TGA furnace to remove the moisture. The TGA measurements are normally run from 50°C to 700°C at a heating rate of 10°C/min under nitrogen or air. As a typical example, the TGA traces for the sulfonated multiblock PES are shown in Figure 4.13 [28]. The loss of the sulfonic acid groups starts above 350°C in the acid form of SPES, while thermal degradation of the main chains occurs above 500°C. On the other hand, the degradation of the salt form initiates at around 500°C.

The initial weight loss at around 100°C is caused by evaporation of water molecules bound in the hydrophilic sulfonated polysulfone (SPSU) membrane (Figure 4.14) [29]. The secondary weight loss between 200°C and 400°C is mainly due to the thermally activated decomposition of the sulfonic acid groups in the polymer chains, which is confirmed by the evolution of SO and SO_2 gases detected in the mass spectra. Above 550°C, it was found that the production of characteristic pyrolyzate with a hydroxyl end group such as phenol as well as benzene occurs.

Good mechanical properties of the membranes are one of the necessary requirements for their effective use in proton exchange membrane fuel cell (PEMFC) applications. The membranes require a tensile stress at a maximum load over 20 MPa and elongation at break over 20% in a low humidity with the general trend of a lower maximum stress for the higher IEC.

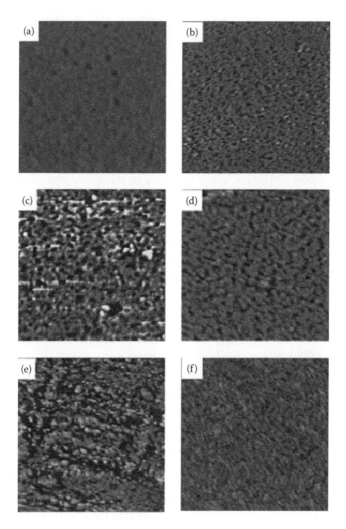

FIGURE 4.9 AFM tapping phase image for sulfonated poly(arylene ether sulfone) copolymers and Nafion 117: (a) BPS-00; (b) BPSH-20; (c) BPSH-40; (d) BPSH-50; (e) BPSH-60; (f) Nafion 117 (acid form). (Reprinted from Wang, F. et al., *J. Membr. Sci.*, 197(1–2), 231,2002. With permission.)

4.3.4 ION EXCHANGE CAPACITY, WATER UPTAKE, AND SWELLING RATIO

4.3.4.1 Ion Exchange Capacity

The IEC, the DS, and equivalent weight (EW) are all related measures of the relative concentration of the acid groups within a polymer. The EW is defined as the grams of polymer per equivalent of the sulfonated moiety. The IEC is inversely related to the EW as seen in Equation 4.1. Examples of the IEC values for Nafion and BPSH-40 are shown in Figure 4.15.

Sulfonated multiblock copoly(ether sulfone)

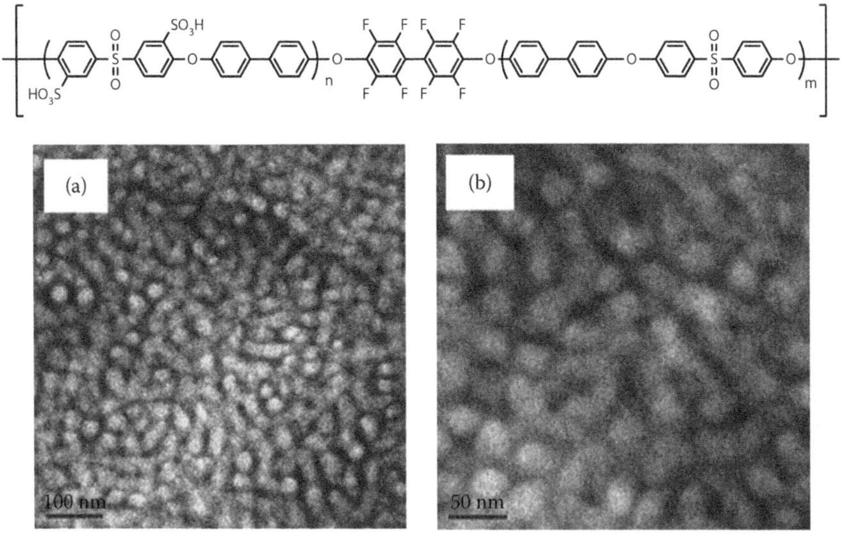

FIGURE 4.10 TEM images of sulfonated multiblock copoly(ether sulfone) membrane: (a) low resolution and (b) high resolution. (Reprinted from Nakabayashi, K. et al., *J. Polym. Sci. A: Polym. Chem.*, 46(22), 7332, 2008. With permission.)

$$IEC = \frac{1000}{EW} = \frac{mmol\ of\ sulfonic\ acid}{g\ of\ polymer} \quad (4.1)$$

The IEC values of the membranes can be determined by titration. A sample membrane in the proton form is immersed in a 2 M NaCl aqueous solution for 48 h to exchange the H^+ ions with Na^+ ions. The released protons are titrated by a 0.01 M NaOH solution using phenolphthalein as the indicator (see Equation 4.2):

$$IEC = \frac{consumed\ NaOH\ (mL) \times molarity\ of\ NaOH}{weight\ of\ dried\ membrane} \ (mequiv.\ g^{-1}) \quad (4.2)$$

Figure 4.16 shows the relationship between the proton conductivity and IEC values for BPSH. The proton conductivity generally increases with the increasing IEC values.

4.3.4.2 Water Uptake

The WU of the membranes is a key factor for sustaining the proton conductivity because water molecules play an important role as proton transportation carriers in membranes. As protons migrate with the water as hydronium ions in the vehicle mechanism or via exchange of hydrogen bonding with water molecules in the

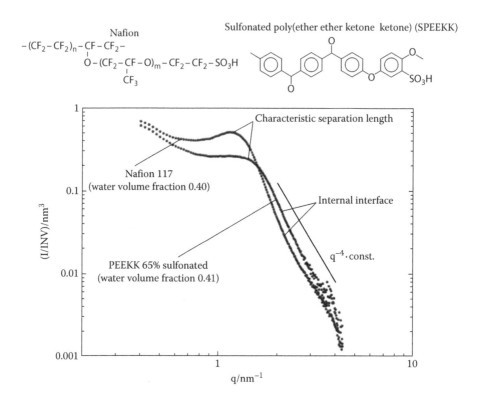

FIGURE 4.11 SAXS spectra of hydrated Nafion and a hydrated sulfonated poly(ether ketone). (Reprinted after modification from Kreuer, K.D., *J. Membr. Sci.*, 185(1), 29, 2001. With permission.)

Grotthuss mechanism, the proton conductivity is directly related to the WU, which is calculated by

$$\text{Water uptake (WU)}(\%) = \frac{\text{mass(wet)} - \text{mass(dry)}}{\text{mass(dry)}} \times 100 \qquad (4.3)$$

where mass (wet) and mass (dry) refer to the mass of the wet membrane and the mass of the dry membrane, respectively.

The hydration number (λ), number of water molecules absorbed per sulfonic acid, can be calculated from the mass WU and the ion content of the dry copolymer as

$$\lambda = \frac{\text{mass(wet)} - \text{mass(dry)}}{\text{mass(dry)} \times \text{MW(water)} \times \text{IEC}} \qquad (4.4)$$

where MW (water) is the molecular weight of water (18.01 g/mol).

The WU of SPESs increases with the increasing IEC values. However, membranes with an excess WU dramatically swell, forming a hydrogel that would not be useful as a PEM (see Figure 4.17 [21]).

Nafion   PEEKK

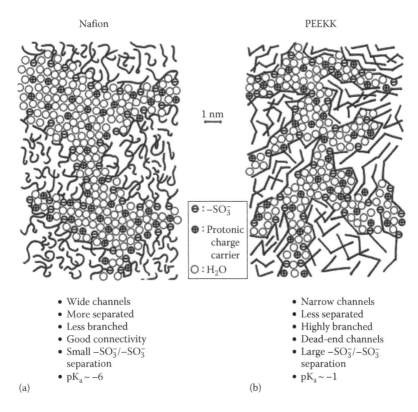

1 nm

⊖ : –SO₃⁻

⊕ : Protonic
      charge
      carrier

○ : H₂O

(a)
- Wide channels
- More separated
- Less branched
- Good connectivity
- Small –SO_3^-/–SO_3^-
  separation
- $pK_a \sim -6$

(b)
- Narrow channels
- Less separated
- Highly branched
- Dead-end channels
- Large –SO_3^-/–SO_3^-
  separation
- $pK_a \sim -1$

FIGURE 4.12 Schematic representation of the microstructures of (a) Nafion and (b) a sulfonated poly(ether ketone). (Reprinted after modification from Kreuer, K.D., *J. Membr. Sci.*, 185(1), 29, 2001. With permission.)

4.3.4.3 Swelling Ratio

Dimensional swelling should also be considered in conjunction with the WU, because excessive swelling significantly decreases the dimensional stability and mechanical properties of the polymer films.

4.3.5 PROTON CONDUCTIVITY

Proton conductivity is one of the key properties for predicting the PEM suitability, since a high conductivity is necessary for their effective utilization in fuel cell devices [21]. Conductivity measurements are performed on the acid form of the membranes using the special cells. This cell geometry is chosen to ensure that the membrane resistance dominates the response of the system. An impedance spectrum is recorded from 10 MHz to 10 Hz using an impedance/gain-phase analyzer. The resistance of the film is taken at a frequency that produces the minimum imaginary response. All the impedance measurements are performed at various temperatures and various RHs.

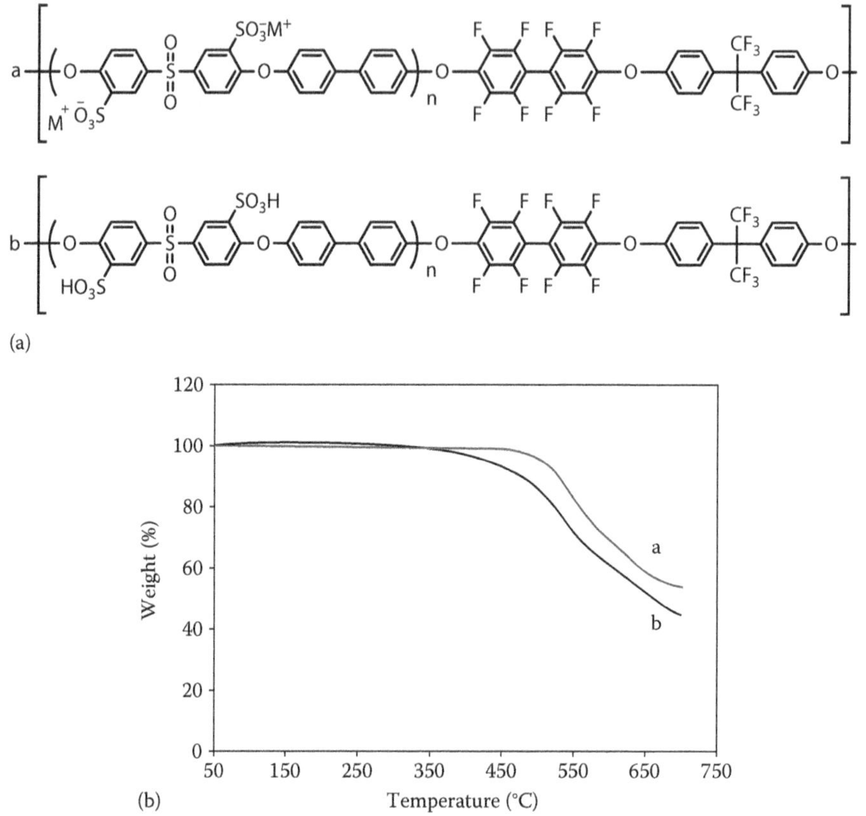

(a)

(b)

FIGURE 4.13 TGA traces of sulfonated–fluorinated poly(arylene ether sulfone)s in the salt form (a) and acid form (b) (under N_2). (Reprinted after modification from Ghassemi, H. et al., *Polymer*, 47(11), 4132, 2006. With permission.)

4.3.6 CHEMICAL STABILITY

The chemical stability of the PESs has been well reviewed elsewhere [30]. One of the drawbacks of sulfonated aromatic polymers is their weak chemical stability. Although the mechanism by which polymers decompose under fuel cell operating conditions has not been clarified, it is likely due to radical attacks (HO· or HOO·) on the polymers [31]. Oxidation and hydrolysis can easily occur on hydrophilic sites around the sulfonic acid groups to induce the decomposition of the polymer backbones and elimination of the sulfonic acid groups (Scheme 4.9).

4.3.6.1 Hydrolytic Stability

It is well known that sulfonated aromates undergo desulfonation when subjected to high temperatures and high water activities:

$$Ar\text{–}SO_3H + H_2O \rightarrow Ar\text{–}H + H_2SO_4 \text{ (Ar: aromatic ring)}$$

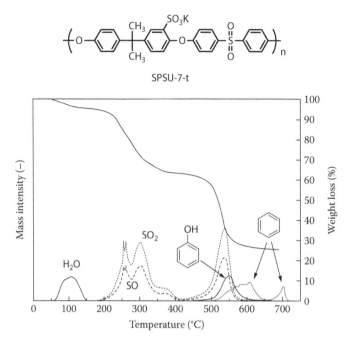

SPSU-7-t

FIGURE 4.14 Thermogravimetry–mass spectroscopy (TG–MS) curve of SPSU-7-t membrane. (Reprinted after modification from Park, H.B. et al., *J. Membr. Sci.*, 247(1–2), 103, 2005. With permission.)

e.g. Nafion 117 : IEC = 0.91

BPSH-40 : IEC = 1.71

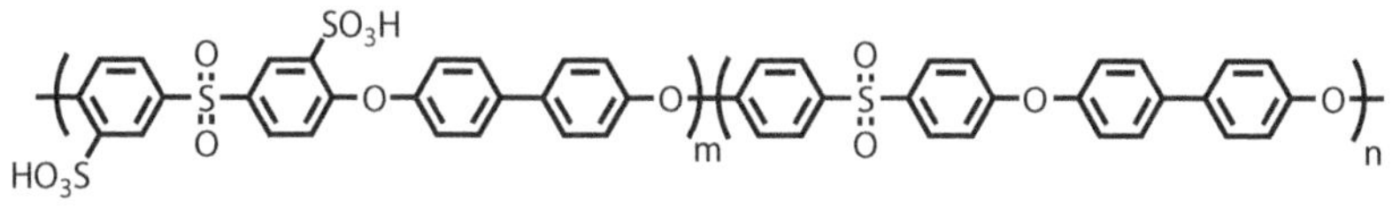

(m / n = 4 : 6)

FIGURE 4.15 IEC values for Nafion 117 and BPSH-40.

The hydrolytic stability of aromatic sulfonic acid groups depends on the electronic structure of the aromatic ring; while electron-donating groups generally facilitate desulfonation, electron-withdrawing groups impede desulfonation. The hydrolytic stability of the polymers has been determined by cycling TGA measurements between $T = 105°C$ (110) and $180°C$ in an all-water-vapor atmosphere $(p(H_2O) = 1 \text{ atm} = 10^5 \text{ Pa})$. Reversible weight changes due to water desorption and

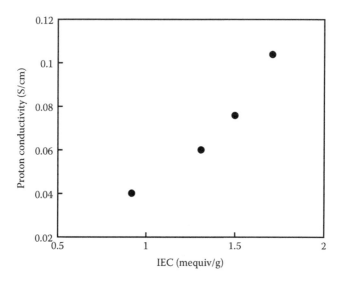

FIGURE 4.16 Proton conductivity of BPSH in water at 30°C.

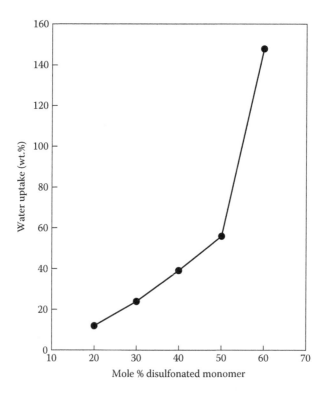

FIGURE 4.17 Influence of the DS on the WU of sulfonated poly(arylene ether sulfone) copolymers. (Reprinted from Wang, F. et al., *J. Membr. Sci.*, 197(1–2), 231, 2002. With permission.)

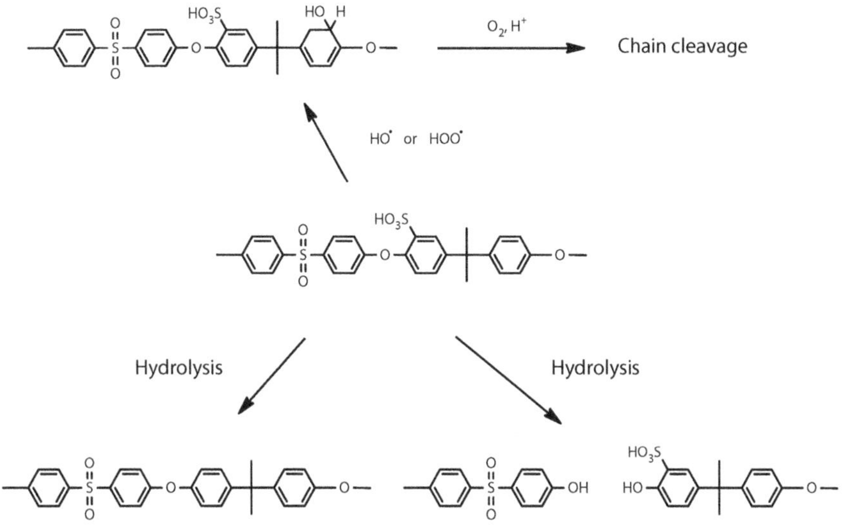

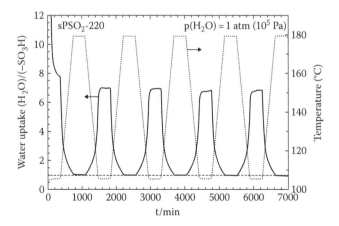

SCHEME 4.9 Possible decomposition mechanism of poly(arylene ether sulfone).

absorption during the heating and cooling runs indicate a high hydrolytic stability in the scanned temperature range. Any desulfonation reaction is expected to show up as some irreversibility in the weight changes. As shown in Figure 4.18, the sulfonated poly(phenylene sulfone), sPSO$_2$-220 (IEC of 4.3–4.5), is totally stable under these high-temperature and high-water-activity conditions [32].

FIGURE 4.18 Reversible water desorption/absorption for oxidized polymer sPSO$_2$-220 in the temperature range $T=105°C–180°C$ at a water pressure of $p(H_2O)=1$ atm (10^5 Pa) as obtained by TGA (heating and cooling rate 0.2°C/min). (Reprinted from Schuster, M. et al., *Macromolecules*, 42(8), 3129, 2009. With permission.)

4.3.6.2 Oxidative Stability

It is well known that the poor oxidative stability of the membranes may cause failure during long-term fuel cell operation. Hydroxy or hydroperoxy radicals attack the SPESs at the carbon atoms ortho to the ether linkages due to high electron density and initiate decomposition. The prepared membranes undergo the oxidative stability test by immersing the samples in Fenton's reagent (3% H_2O_2, 20 ppm $FeSO_4$) at room temperature or (3% H_2O_2, 2 ppm $FeSO_4$) at 80°C for 1 h. The resistance to oxidation of the membranes is evaluated by observing their dissolving behavior. This method is regarded as one of the standard tests to gauge the relative oxidative stability and to simulate accelerated fuel cell operating conditions.

However, there is no strong correlation between the oxidative stability evaluated by Fenton's reagent and the durability of membranes during PEMFC operation. The following several reasons are proposed: (1) PEM degradation in fuel cells consists of a complex combination of different degradation processes that are strongly influenced by the membrane, fabrication, and operating conditions. (2) The excessive peroxide content in the accelerated Fenton's test is unrealistically high under normal fuel cell operating conditions. (3) The concentration of $H_2O_2^+$ in the MEAs partly depends on the gas permeability of the membranes.

4.4 MEMBRANE CONDUCTING MECHANISM

4.4.1 PROTON CONDUCTION MECHANISM

The proton conduction in PEMs is explained by two mechanisms, that is, the vehicle and Grotthuss mechanisms (Figure 4.19). For the sulfonated PEMs, the vehicle mechanism, in which protons are transferred by the migration of hydronium ions (H_3O^+), is dominant at high temperatures or a low water content (under low humidity conditions). On the other hand, the Grotthuss mechanism, which involves very rapid proton hopping between neighboring sites involving the $H_9O_4^+$ and $H_5O_2^+$ cluster ions, is dominant at low temperatures or a high water content (under high humidity conditions). For the phosphoric-acid- or heterocycle-containing materials

Vehicle mechanism

H_3O^+

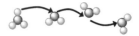

Migration of hydronium ions

(a)

Grotthuss mechanism

Proton hopping though the hydrogen-bonding network

(b)

FIGURE 4.19 Schematic illustration of (a) vehicle mechanism and (b) Grotthuss mechanism.

and phosphonated PEMs, the Grotthuss mechanism is dominant because they can form hydrogen-bonding networks through which protons or protonic defect diffuses.

4.5 MEMBRANE APPLICATIONS OF PEMs FOR FUEL CELLS

4.5.1 PROBLEMS OF AROMATIC PEMs

As described in the previous sections, aromatic PEMs have many merits and the possibility to become alternative materials to the perfluorosulfonic acid membranes. However, several problems that must be solved for actual use still remain. The major problem of aromatic PEMs is their low proton conductivity under low humidity conditions. To reduce the requirement for humidification in the fuel cell system, the performance of PEMs under low humidity conditions is quite important.

Perfluorosulfonic acid membranes, such as Nafion, exhibit a high proton conductivity over a wide range of RHs at moderate temperatures, while conventional aromatic PEMs show significant decreases in proton conductivity under low humidity conditions. The difference between the perfluorosulfonic acid membranes and aromatic PEMs can be explained by their morphology. Since perfluorosulfonic acid membranes have an extremely hydrophobic perfluorinated backbone and extremely hydrophilic sulfonic acid functional groups, hydrophilic/hydrophobic phase separation occurs on a nanoscale, especially in the presence of water. On the other hand, aromatic PEMs have a less pronounced hydrophilic/hydrophobic phase-separated structure due to their less acidic sulfonic acid groups and lower flexibility of the polymer backbone. Kreuer compared Nafion and PEEKK and came to the following conclusion [27]. Nafion has wider proton transport channels, fewer channel dead ends, shorter interanionic distances, and stronger acid groups than PEEKK, which allow high proton conduction (Figure 4.12).

The low proton conductivity of aromatic PEMs, especially under low humidity conditions, results from a less pronounced phase-separated structure and lower connectivity of the proton transport channels. Therefore, the formation of a defined phase-separated structure and well-connected proton transport channels would be the key factors in improving the proton conductivity of the aromatic PEMs.

4.5.2 RECENT DEVELOPMENTS OF AROMATIC SPESs

Many research studies have focused on improving the nanophase-separated structures between the hydrophilic and hydrophobic units to increase the proton conductivity of the aromatic ionomers under low RH. In this section, recent approaches to improve the membrane properties, especially the proton conductivity, which is usually the first characteristic considered when evaluating membranes for fuel cells, and morphology will be discussed as follows: (1) multiblock SPES copolymers, (2) locally and densely SPES, (3) SPES with high IEC values and high free volume, (4) SPES with pendant perfluoroalkyl sulfonic acids, (5) cross-linked SPES, and (6) thermally annealed SPESs.

4.5.2.1 Multiblock Copolymers

The low conductivity of the sulfonated aromatic PEMs at low RH has been attributed to the fact that sulfonated aromatic polymers have fewer connected water domains as well as more phase mixing of the hydrophobic and hydrophilic domains [33]. Block copolymers, consisting of covalently bonded chemically dissimilar sequences, exhibit highly periodic microphase-separated structures [34]. The characteristic lengths of the structures are determined by the molecular size and are in the 10–100 nm range. Therefore, many research studies have focused on the development of sulfonated multiblock copoly(ether sulfone)s for improving the nanophase-separated structures and proton conductivity under a low RH.

In 2006, McGrath and coworkers reported multiblock sulfonated–fluorinated poly(arylene ether sulfone)s (MB) for PEMs by the nucleophilic aromatic substitution of the dialkali metal salt of bisphenol-terminated poly(arylene ether sulfone) and decafluorobiphenyl-terminated poly(arylene ether) [28] (Scheme 4.10).

The polymer films had IEC values of 0.95–2.29 meq./g and showed WU values ranging from 40% to more than 400%. These membranes showed an excellent proton conductivity up to 0.32 S/cm in water (IEC = 2.29). The relationship between the RH and proton conductivity for the MB-150 (IEC = 1.50) and Nafion 112 is shown in Figure 4.20.

The MB-150 exhibited a higher proton conductivity than the Nafion 112 at low RH. The tapping mode AFM images of the membranes (IEC = 1.50) revealed a well-defined phase separation and a distinct morphological architecture compared to the random copolymers (Figure 4.21). They concluded that the existence of well-connected hydrophilic domains was most likely the reason for the high proton conductivity of the multiblock copolymers.

The same groups also studied the proton conductivity as a function of the RH for the block BisAF-BPSH(x:y)K (x:y = 3:3, 5:5, and 8:8) series with different block lengths (Figure 4.22) [35]. The block BisAF-BPSH (8:8)K membrane showed the highest proton conductivity among the block BisAF-BPSH (x:y)K copolymers over a wide range of RH values, but the lowest WU and IEC values. This behavior was explained by increasing the extent of the nanophase separation and the connectivity between the hydrophilic domains with an increase in the block length.

Another synthetic approach has been reported by Nakabayashi et al. [36]. As described earlier, multiblock sulfonated–fluorinated poly(arylene ether sulfone)s were prepared by the bisphenol-terminated poly(arylene ether sulfone) and deca-fluorobiphenyl-terminated poly(arylene ether), which was obtained by an excess amount of DFBP and bisphenol-terminated poly(arylene ether sulfone). However, considering the high reactivity of DFBP, multiblock copolymers can be obtained by connecting hydroxy-terminated oligomers with a small amount of DFBP. That is, DFB functions as a chain extender. The synthetic procedure is depicted in Scheme 4.11.

The multiblock copoly(arylene ether)s (BCP1 and BCP2) (IEC.1.71–2.15 meq./g) were synthesized by the polycondensation of three compounds, that is, the hydrophilic dialkali metal salt of bisphenol-terminated poly(arylene ether sulfone), hydrophobic bisphenol-terminated poly(arylene ether sulfone), and the chain extender DFBP.

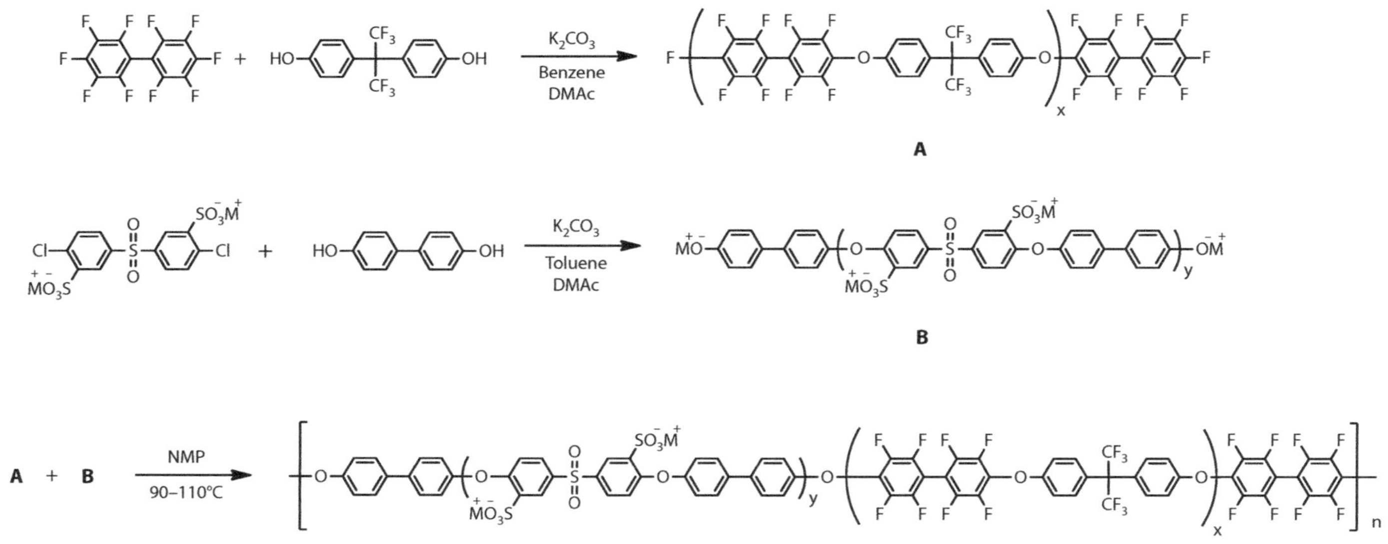

SCHEME 4.10 Synthesis of multiblock copoly(arylene ether sulfone) (MB-150).

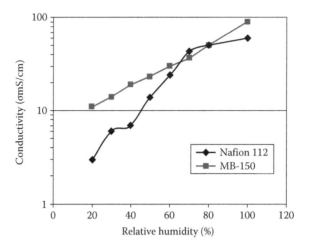

FIGURE 4.20 Influence of RH on proton conductivity of multiblock copoly(arylene ether sulfone) at 30°C. (Reprinted from Ghassemi, H. et al., *Polymer*, 47(11), 4132, 2006. With permission.)

Multiblock copoly(ether sulfone)s with high molecular weights (up to $M_n = 122,000$) were successfully obtained under moderate reaction conditions (120°C for 10–18 h). Under these conditions, the ether–ether interchange reaction was not observed by the ^{13}C NMR spectroscopy. All polymers gave tough and flexible films and demonstrated excellent oxidative and dimensional stability and good WU properties under low RH (Table 4.1).

The membranes maintained a comparatively high proton conductivity of 6.0×10^{-3} S/cm at 50% RH. Furthermore, all membranes showed a higher proton conductivity than the sulfonated random block copolymer (Figure 4.23).

AFM observations of the multiblock copolymer membrane supported the formation of a clear hydrophilic/hydrophobic phase-separated structure, which simultaneously had a good proton conductivity and dimensional stability. It should be mentioned here that the obtained copolymers are not complete alternating multiblock copolymers in which each hydrophilic/hydrophobic segment is alternately connected, but are random multiblock copolymers. However, the aforementioned data indicated that the polymer structure of the obtained multiblock copolymers with the chain extender system was highly controlled [36].

Furthermore, several sulfonated multiblock copoly(ether sulfone)s were prepared by changing the hydrophilic and hydrophobic block lengths in the presence of DFBP as a chain extender, and the influence of each length on the membrane properties was studied. The resulting multiblock copolymers with high molecular weights ($M_n >$ 50,000, $M_w > 150,000$) were successfully obtained under mild conditions (120°C for 12 h) and produced tough, flexible, and transparent membranes by the NMP solution casting method. All membranes (IEC = 2.0 meq./g) demonstrated excellent oxidative and dimensional stabilities and good WU properties under low RH. The membranes possessing the longest block lengths demonstrated the relatively high

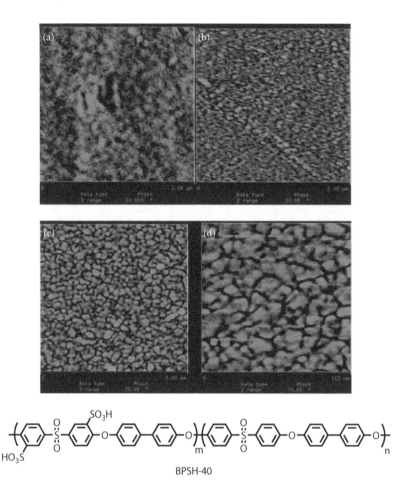

FIGURE 4.21 AFM tapping phase image for Nafion 112 (a), BPSH-40 (b), and multiblock MB-150 (c and d). (Reprinted from Ghassemi, H. et al., *Polymer*, 47(11), 4132, 2006. With permission.)

proton conductivity of 7.0×10^{-3} S/cm at 80°C and 50% RH. From the AFM images, the hydrophilic domains of the membranes are larger and better connected to each other, and that induces good proton conduction. In addition, TEM observations also confirmed the formation of the cross-sectional phase separation of the membranes (see Figure 4.10) [26].

It is interesting to compare the properties of the random multiblock copolymers and alternating multiblock copolymers in which the A and B segments are alternatively connected (see Figure 4.24), because the latter is expected to produce a highly phase-separated morphology as compared to the former [37].

Two sulfonated block copoly(ether sulfone)s with the same chemical composition were prepared by a chain extender method and an end-capping method, respectively, and investigated for their WU and proton conductivity. The results indicated that

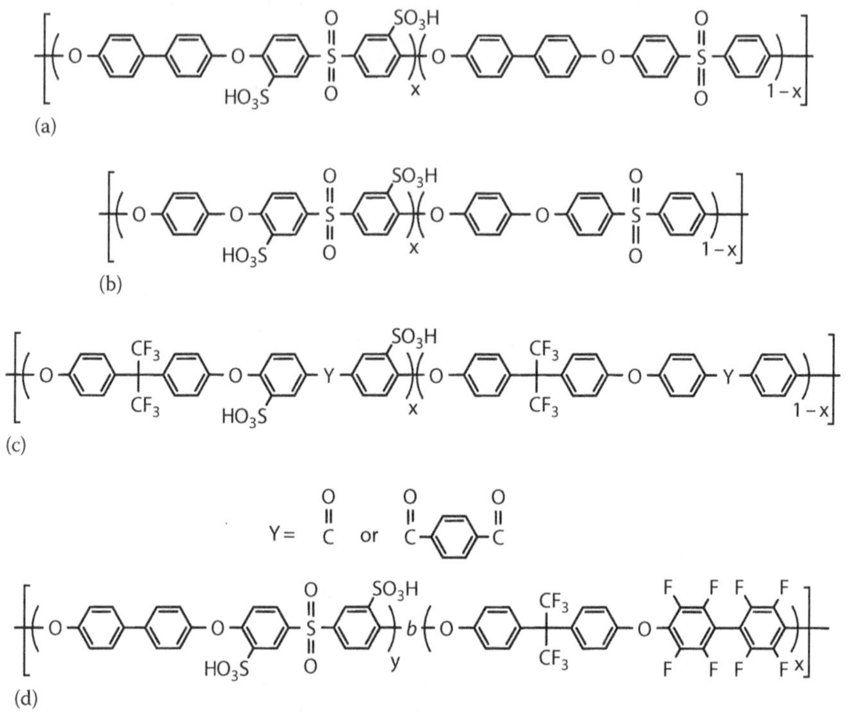

(a)

(b)

(c)

(d)

FIGURE 4.22 Copolymer chemical structures of (a) BPSH-xx, (b) HQSH-xx, (c) B-ketone-xx and PB-diketone-*xx*, and (d) block BisAF-BPSH(*x:y*)K.

alternating multiblock copolymers showed a better performance (8.6×10^{-3} S/cm, 50% RH) than the random multiblock copolymers (6.1×10^{-3} S/cm, 50% RH) having similar IEC values (~2 meq./g) probably due to better proton paths induced by the controlled structure.

On the other hand, three sulfonated aromatic polymers with different sequences, that is, alternating, random, and multiblock sequences (Figure 4.25), were studied in order to better understand the relationship between the molecular structure, morphology, and properties of the PEMs as a function of the RH [38].

Table 4.2 shows the IEC and WU for each polymer. The three membranes have similar WU and IEC values. The proton conductivity of these PEMs is shown in Figure 4.26.

The conductivity of Nafion is higher than those of the sulfonated aromatic PEMs (4.2 mS/cm at 50% RH). The relative slopes of the log (conductivity) versus RH plots for each polymer are of interest and can illustrate the effect of humidity on each membrane's conductivity. The slope indicates the dependency of proton conductivity on the RH. Both Nafion and the multiblock copolymer exhibit similar slopes (~0.026), and their slopes are lower than those of other random and alternating polymers, meaning that their conductivity is less dependent on the RH than those of the random and alternating polymers. To obtain more information on the difference in

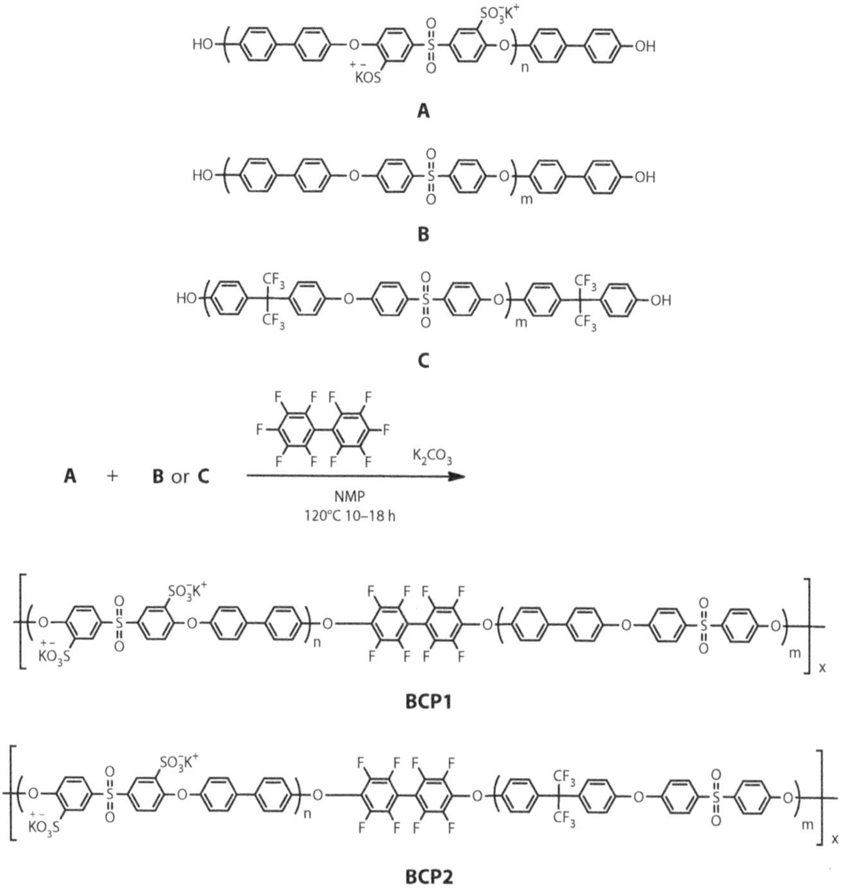

SCHEME 4.11 Synthesis of multiblock copoly(arylene ether sulfone)s (BCP1 and BCP2).

the proton conductivities for these PEMs, a morphological study of these membranes was carried out by tapping mode SFM in the range of 8%–75% RH. These results are shown in Figure 4.27.

The alternating polymer shows large hydrophobic and hydrophilic domains with a poor continuity between the hydrophilic domains. The random copolymer exhibits a disordered morphology with some connectivity between the hydrophilic regions, but no well-defined ionic pathways for the proton or water transport. On the other hand, the multiblock copolymer shows a well-defined *fingerprint-type* structure with continuous hydrophilic and hydrophobic pathways. The authors concluded that this microstructure was probably responsible for the fast proton transport even at a low humidity, as well as the faster water transport than those of the alternating and random polymers.

The fuel cell performances and polarization characteristics of the single cells using each of the PEMs were also evaluated at different gas inlet humidities

TABLE 4.1

IEC Value, WU, Dimensional Change, and Oxidative Stability of BCP1 and BCP2

Run	IEC (meq./g) Theory	IEC (meq./g) ^{1}H NMR	IEC (meq./g) Titration^a	WU wt%	λ^b	Δl	Δt	Oxidative Stability^c wt%
BCP1	1.90	1.77	1.71	45	15.2	0.10	0.12	100
BCP1	2.10	2.08	2.17	64	17.3	0.06	0.10	98
BCP2	1.74	1.71	1.76	39	12.8	0.05	0.09	99
BCP2	2.10	2.03	2.15	53	13.5	0.03	0.14	97

^{a} Titration with 0.02 M NaOH aq.
^{b} The number of water molecules per sulfonic acid group.
^{c} Residue after treatment with Fenton's reagent (3% H_2O_2 aqueous solution containing 2 ppm $FeSO_4$).

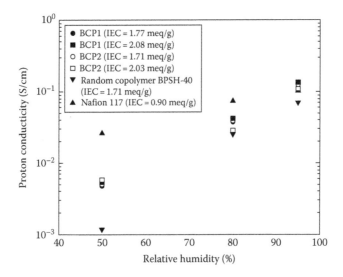

FIGURE 4.23   Humidity dependence of proton conductivity of BCP1, BCP2, BPSH-40, and Nafion 117 membranes at 80°C.

(70% and 40% RH) (Figure 4.28). The fuel cell performance of a single cell using this multiblock copolymer at 40% RH is comparable to that of Nafion, which performs much better than the alternating and random polymers. These results clearly show that the formation of well-defined channels in the PEMs is extremely important for achieving a high proton conduction and fuel cell performance.

Li et al. reported the performance of comb-shaped copoly(arylene ether sulfone)s [39]. Their chemical structures and simplified illustration are shown in Figure 4.29, and the properties of the comb-shaped polymer membranes are summarized in Table 4.3.

Random multiblock copolymer

Alternating multiblock copolymer

■ Hydrophilic segment
■ Hydrophobic segment

FIGURE 4.24 Images of random multiblock and alternating multiblock copolymer.

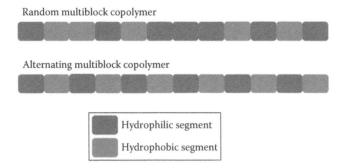

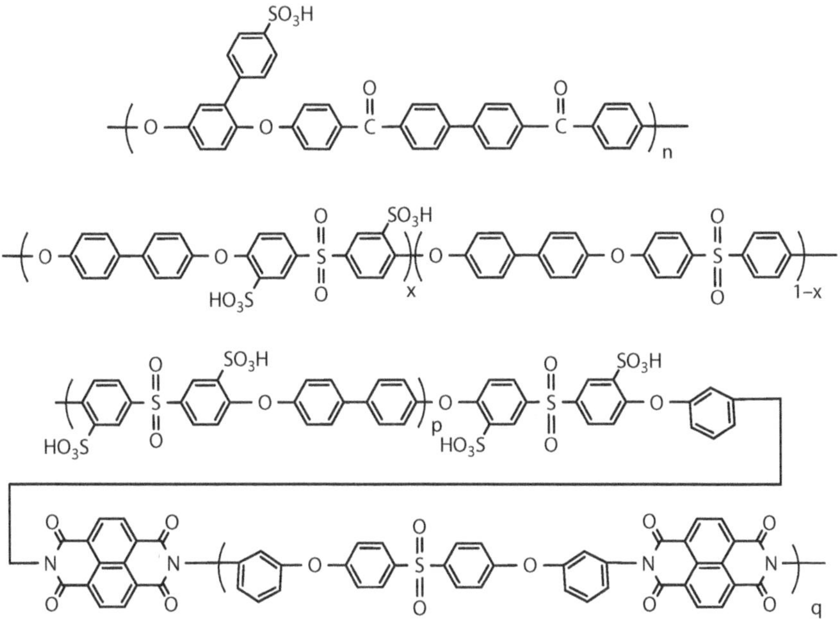

FIGURE 4.25 Structure of sulfonated aromatic PEMs used (a, Ph–PEEKDK alternating polymer; b, BPSH-35 random copolymer; c, BPSH-15–PI-15 multiblock copolymer).

Figure 4.30 shows TEM images of the lead-stained comb-shaped copolymer membranes; the dark areas correspond to the hydrophilic poly(propylene oxide) (PPO) graft chains. The TEM images clearly show phase-separated wormlike domains.

A mesoscale simulation was employed to investigate the phase morphology of the comb-shaped polymer membranes. In the periodic 3D images, all the membranes display well-connected hydrophilic nanochannels, which become thicker and more interconnected as the graft chain length increased.

All the membranes, even with low IEC values, exhibit a high proton conductivity over the 30%–90% RH range (Figure 4.31a) at 90°C. These conductivities are higher

TABLE 4.2

Intrinsic Viscosity and IEC Data for the PEMs

PEM	$[\eta]$ (dL/g)	IEC (meq./cm³) Theory	IEC (meq./cm³) Expel[a]	WU (wt%)	IEC$_{v(wet)}$ (meq./cm³)
Alternating (Ph–PEEKDK)	3.35[b]	1.60	1.60	32	1.41
Random (BPSH-55)	0.80[c]	1.53	1.50	40	1.31
Multiblock (BPSH-15–PI-15)	0.67[c]	1.51	1.55	51	1.21
Nafion 212			0.95–1.01[d]	19	1.45

Source: Data from Einsla, M.L. et al., *Chem. Mater.*, 20(17), 5636, 2008.

[a] Titration with standard NaOH aq.
[b] Measured at 30°C in DMAc (no salt).
[c] Measured at 25°C in NMP with 0.05 M LiBr.
[d] From the DuPont material data sheet.

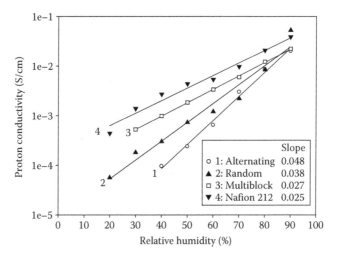

FIGURE 4.26 Proton conductivity as a function of RH at 80°C (Nafion is shown for comparison). (Reprinted from Einsla, M.L. et al., *Chem. Mater.*, 20(17), 5636, 2008. With permission.)

than or similar to those of Nafion 212. Moreover, the proton conductivity at 30% RH and 90°C over 24 h is nearly constant with values of about 10^{-2} S/cm (Figure 4.31b). These membranes will be suitable PEM materials for automotive fuel cells, taking into consideration the requirements for membrane durability even under harsh operating conditions such as high temperature and low RH.

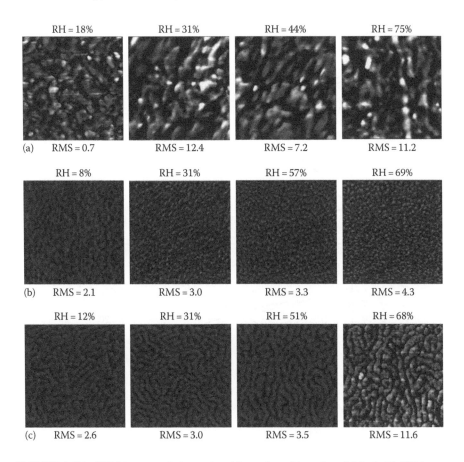

FIGURE 4.27 SFM images of alternating (a), random (b), and multiblock (c) PEMs as a function of RH. Image size is 1 μm, and phase range is 40° for all images. (Reprinted from Einsla, M.L. et al., *Chem. Mater.*, 20(17), 5636, 2008. With permission.)

4.5.3 LOCALLY AND DENSELY SULFONATED POLYMERS

The proton conducting properties of the PEMs basically rely on the IEC and WU values, especially for the sulfonated aromatic polymers. Although highly sulfonated polymers show a high proton conductivity, their high WU results in a dramatic loss of their mechanical properties. As a strategy to improve the proton conductivity and mechanical properties, a multiblock copolymer system has been extensively studied [25,26,28,36,37]. By introducing hydrophilic and hydrophobic units in a single molecule, well-defined phase separation, which allows effective proton conduction, is formed in these systems as described previously. The key point of the systems is to develop a high contrast in the polarity between hydrophilic and hydrophobic units. In addition, the well-segregated hydrophobic networks apart from hydrophilic domains tend to survive a radial attack derived from water molecules and improve the mechanical properties in the multiblock copolymer system.

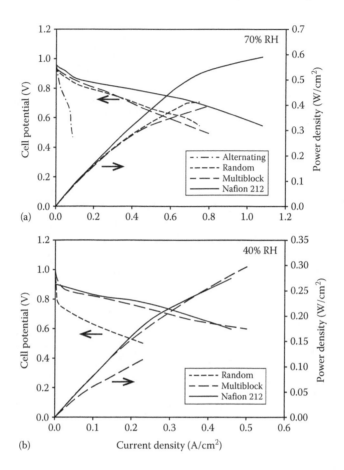

FIGURE 4.28 Hydrogen–air fuel cell performance of the alternating, random, and multiblock PEMs at 100°C with gas inlet humidification at 70% (a) and 40% (b) (Nafion is shown for comparison). (Reprinted Einsla, M.L. et al., *Chem. Mater.*, 20(17), 5636, 2008. With permission.)

Locally and densely sulfonated homopolymers, in which the concentrated sulfonic acid units in a membrane allow for the creation of hydrophilic domains (making the matrix hydrophobic), have been investigated as another approach to efficient PEMs.

Hay et al. prepared poly(arylene ether)s with highly sulfonated units only on the end groups in order to enhance the difference between the hydrophilic and hydrophobic units (Figure 4.32) [40,41]. 1-(4-Hydroxyphenyl)-2,3,4,5,6-pentaphenylbenzene was introduced as the end group bearing a number of pendant phenyl rings to provide a number of postsulfonation sites. The membrane with IEC = 1.1 meq./g showed a significantly phase-separated morphology and the proton conductivity of 0.029 S/cm at 80°C under 80% RH. Thus, locally and densely sulfonated polymers formed hydrophilic/hydrophobic phase-separated structures because of the high contrast in polarity between them.

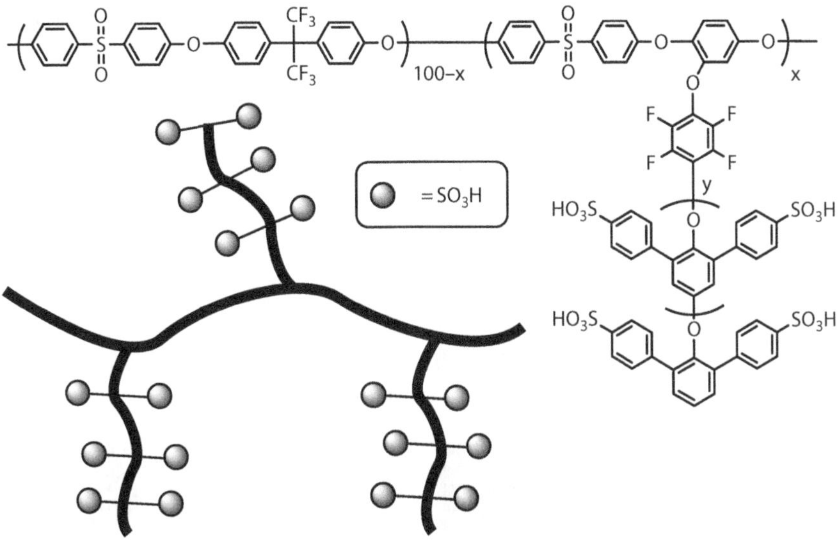

FIGURE 4.29 Structure and simplified illustration of fully aromatic comb-shaped copolymers.

TABLE 4.3

Properties of Comb-Shaped Copolymer Membranes

Run	IEC (meq./g) Titration[a]	IEC$_v$ (meq./g)	DS[b] (%)	WU[c] (wt%)	σ[c]
X5-Y6	0.92	1.24	100	28.2	15.2
X5-Y9	1.28	1.77	100	52.3	17.3
X3-Y14	1.26	1.74	100	60.5	12.8
X5-Y14	1.72	2.46	100	75.6	13.5
Nafion 112	0.90	1.78		19.2	

Source: Data from Li, N. et al., *Angew. Chem. Int. Ed.*, 50(39), 9158, 2011.

[a] Measured by acid–base titration.

[b] DS = degree of sulfonation.

[c] Measured in water at room temperature.

Ueda and coworkers applied the idea of hydrophilic/hydrophobic phase-separated structures to random copolymer systems by introducing densely sulfonated hydrophilic units because the use of a linear homopolymer system was simple and direct [42]. Locally and densely SPESs (eight sulfonic acid groups/monomer) were successfully prepared by the nucleophilic substitution of 4,4′-dichlorodiphenylsulfone with 1,2,4,5-tetrakis([1,1′-biphenyl]-2-oxy)-3,6-*bis*(4-hydroxyphenoxy)benzene and

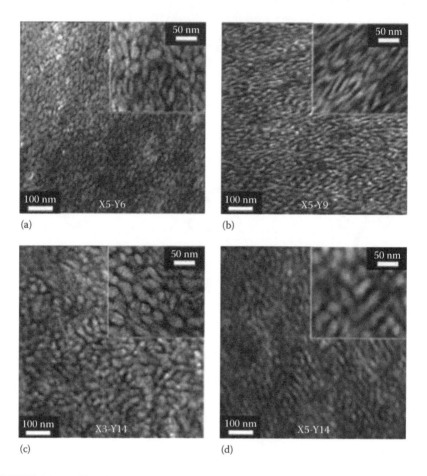

FIGURE 4.30　TEM phase images of comb-shaped copolymer membranes with IEC values in the range 0.92–1.72 meq./g., (a) X5-Y6, (b) X5-Y9, (c) X3-Y14, and (d) X5-Y14. (Reprinted from Li, N. et al., *Angew. Chem. Int. Ed.*, 50(39), 9158, 2011. With permission.)

2,2-*bis*(4-hydroxyphenyl)hexafluoropropane, followed by sulfonation using chlorosulfonic acid (Scheme 4.12).

The proton conductivities of the SPES membrane were measured at 80°C in the range of 30%–95% RH. The results are shown in Figure 4.33, and those of BPSH-40 and Nafion 117 are compared. The proton conductivity of the membrane (IEC = 2.03) is maintained at the same level as that of Nafion 117 over the measured range (30%–95% RH). This result suggests that SPES membranes have well-defined phase-separated structures because of their high proton conductivity under low RH conditions. In fact, AFM and TEM observations of the SPES membranes supported the clear phase-separated structures between the hydrophilic and hydrophobic units.

To heighten the contrast in the polarity between the hydrophilic and hydrophobic units, more densely sulfonated units (10 sulfonic acid groups/monomer) than

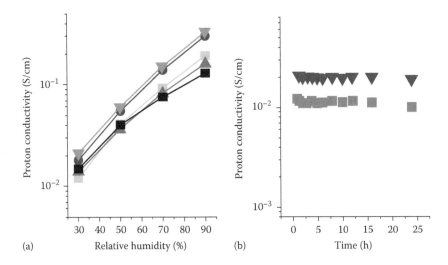

(a)   Relative humidity (%)   (b)   Time (h)

FIGURE 4.31   (a) Proton conductivity of 4(*X*5-*Y*6) (■), 4(*X*5-*Y*9) (•), 4(*X*3-*Y*14) (▲), 4(*X*5-*Y*14) (▼), and Nafion 112 (■) membranes at 90°C as a function of RH; (b) proton conductivity of 4(*X*5-*Y*6) (■) and 4(*X*5-*Y*14) (▼) membranes as a function of test time at 30% RH and 90°C. (From Li, N. et al., *Angew. Chem. Int. Ed.*, 50(39), 9158, 2011.)

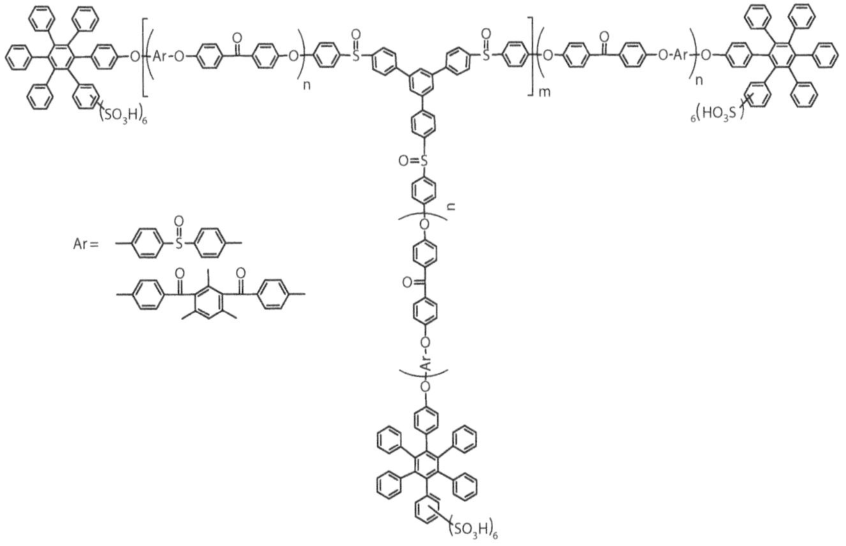

FIGURE 4.32   Chemical structure of end-functionalized poly(sulfide ketone)s with six sulfonic acid moieties at α,ω-chain ends.

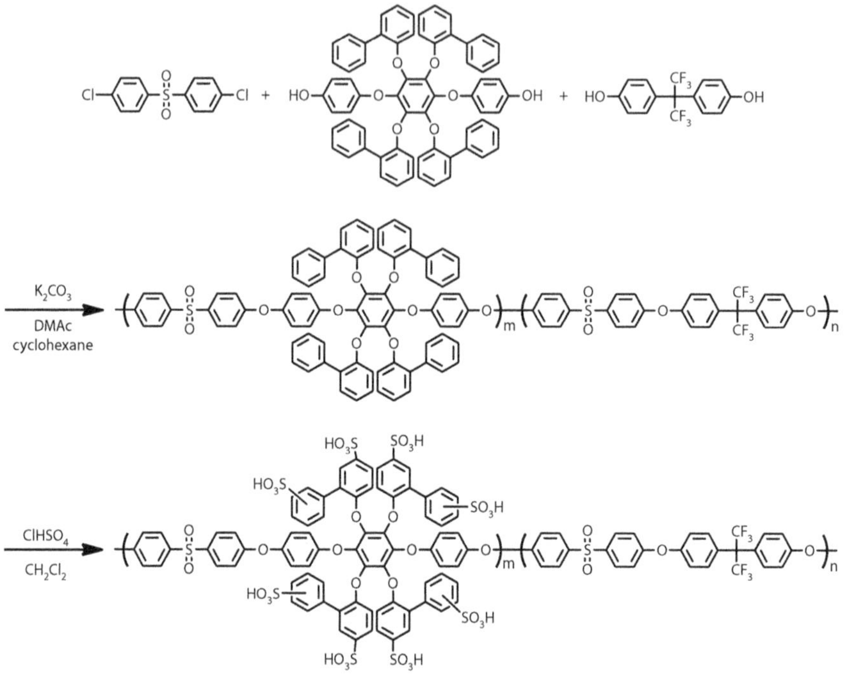

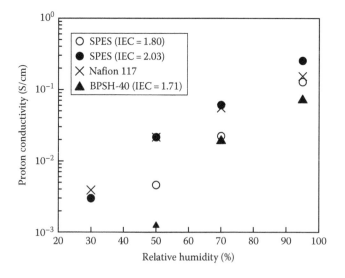

SCHEME 4.12 Synthesis of PES bearing eight sulfonic acid moieties distributed within the chains.

FIGURE 4.33 RH dependence of WU of locally and densely sulfonated PES, BPSH-40, and Nafion 117 at 80°C. (Reprinted from Matsumoto, K. et al., *Macromolecules*, 42(4), 1161, 2009. With permission.)

those of the SPESs (8 sulfonic acid groups/monomer) were prepared [43]. Each hydrophilic unit had 10 sulfonic acid groups, and the significant difference in polarity was expected to form well-defined phase-separated structures and connected proton paths. PESs were prepared by the nucleophilic substitution of *bis*(4-fluorophenyl) sulfone with 1,2,4,5-tetrakis ([1,1′-biphenyl]-2-oxy)-3,6-*bis*(4-hydroxyphenoxy)benzene and *bis*(4-hydroxyphenyl) sulfide, followed by oxidation using *m*-chloroperoxybenzoic acid (Scheme 4.13).

The proton conductivity of the SPES membrane (10 sulfonic acid groups/ monomer) with an IEC value of 2.38 meq./g (4.0×10^{-3} S/cm) surpassed that (3.0×10^{-3} S/cm) of the SPES (8 sulfonic acid groups/monomer) and was comparable to that of Nafion 117 at 30% RH. Furthermore, the AFM observation clearly supported the formation of well-connected proton paths. These results clearly indicated that the high contrast in polarity between the hydrophilic and hydrophobic units

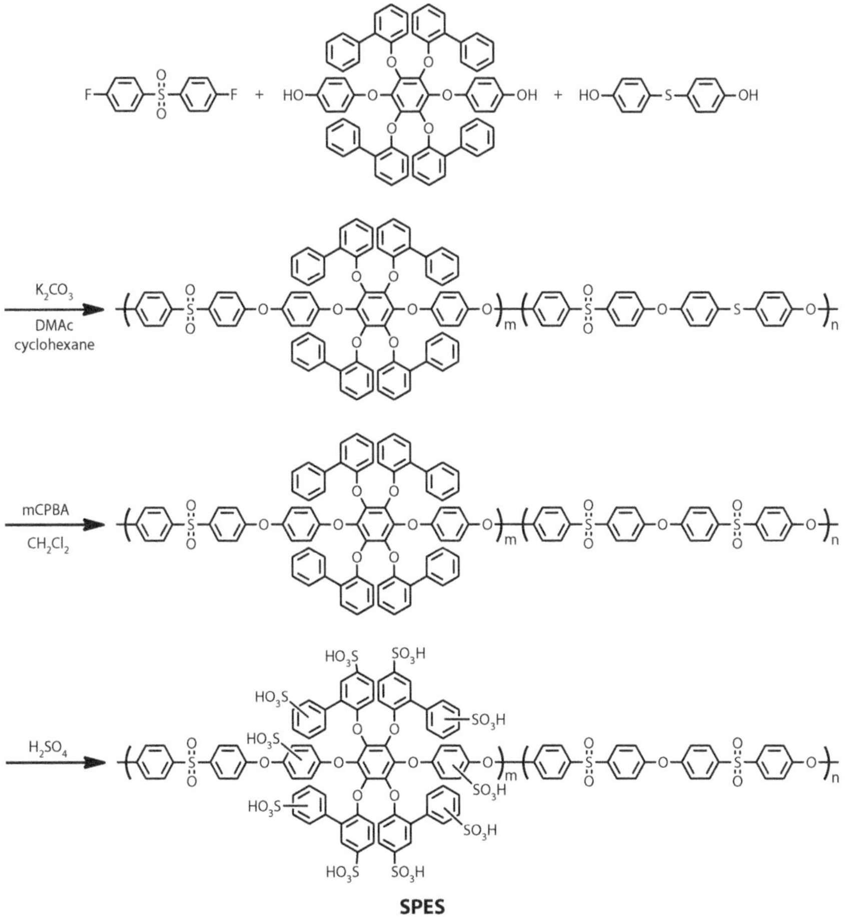

SPES

SCHEME 4.13 Synthesis of PES bearing 10 sulfonic acid moieties distributed within the chains.

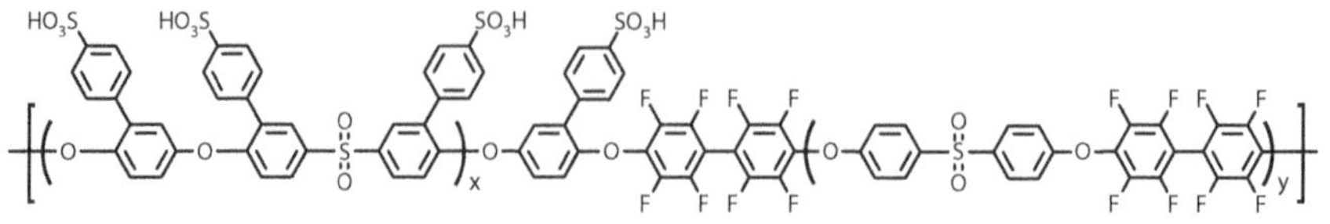

FIGURE 4.34 Chemical structure of the segmented sulfonated copoly(ether sulfone).

enabled the formation of effective proton paths even in a random copolymer system as well as in an amphiphilic block copolymer system.

Guiver et al. reported the synthesis of segmented copoly(arylene ether sulfone) membranes having densely sulfonated pendent phenyl blocks (Figure 4.34) [44]. These polymers were synthesized by the coupling reaction of the phenoxide-terminated oligomers with *bis*(4-hydroxyphenyl) sulfone and DFBP, followed by postpolymerization sulfonation of the blocks containing the pendent phenyl substituents.

The resulting polymers produced transparent, flexible, and tough membranes by solution casting. Morphological observations by TEM and AFM showed that the high local concentration and regular sequence of the pendent sulfonic acid groups within the hydrophilic blocks enhanced the nanophase separation between the hydrophobic and hydrophilic blocks (Figure 4.35).

The ionomer membrane with $X20Y20$ (X and Y refer to the number of hydrophilic and hydrophobic repeat units, respectively) and IEC (1.82 meq./g) had a proton conductivity of 3.6×10^{-2} S/cm at 80°C and 50% RH, which was comparable to that of the Nafion membrane (4.0×10^{-2} S/cm).

Quite recently, a series of pendant-type locally and densely sulfonated PESs were successfully obtained by a nucleophilic substitution reaction, followed by postsulfonation using concentrated sulfuric acid (Figure 4.36) [45]. The polymers with IEC values ranging from 0.98 to 1.66 meq./g afforded considerable proton conductivity and water absorption. The SPES with an IEC value of 1.66 meq./g showed a proton conductivity (0.131 S/cm) equal to Nafion 117 at 100°C in fully hydrated state. Their excellent performance is attributed to the internal structure of the polymers, which formed a distinct phase separation between the hydrophilic and hydrophobic moieties observed by SAXS.

4.5.4 SPES with High IEC Values and High Free Volume

The simplest and most promising approach among all the methodologies accessible to improve the proton conductivity is to increase the IEC values of the PEMs. By increasing the IEC values, the proton channels should become wider and better connected. Schuster et al. reported the synthesis of a 100% sulfonated poly(phenylene sulfone) with an extremely high IEC value (IEC = 4.5 meq./g) and its application to PEMs (Scheme 4.14) [32].

The preparation followed a two-step process involving a polycondensation reaction of the sulfonated difluorodiphenyl sulfone with sodium sulfide, yielding sulfonated poly(phenylene sulfide sulfone), and subsequent oxidation of the targeted poly(phenylene sulfone). Molecular weights of up to 61,000 g/mol were obtained,

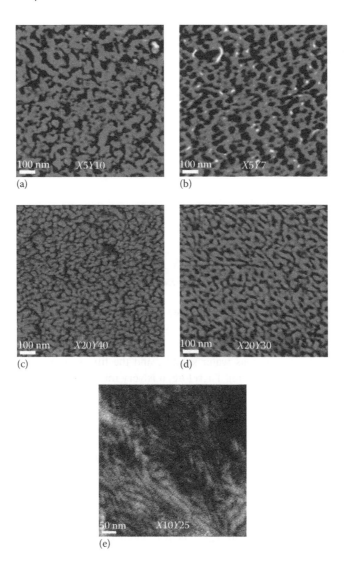

FIGURE 4.35 AFM image of the segmented sulfonated copoly(ether sulfone) membranes (a–d) and TEM image of 3(X10Y25, IEC = 1.20 meq./g) membrane (e). (Reprinted from Li, N. et al., *Macromolecules*, 44(12), 4901, 2011. With permission.)

corresponding to intrinsic viscosities of up to 0.73 dL/g. The obtained polymers exhibited a very high density (1.75 g/cm^3 in the dry state), no glass transition or melting temperature, and very high thermo-oxidative and hydrothermal stabilities. The latter is attributed to the specific molecular structure consisting of very electron-poor aromatic rings. As a consequence of the high IEC and the high water content in terms of the mass or volume fraction at a high temperature (110°C–160°C) and low RH (50%–15%), the proton conductivity exceeded that of Nafion by a factor of 5–7.

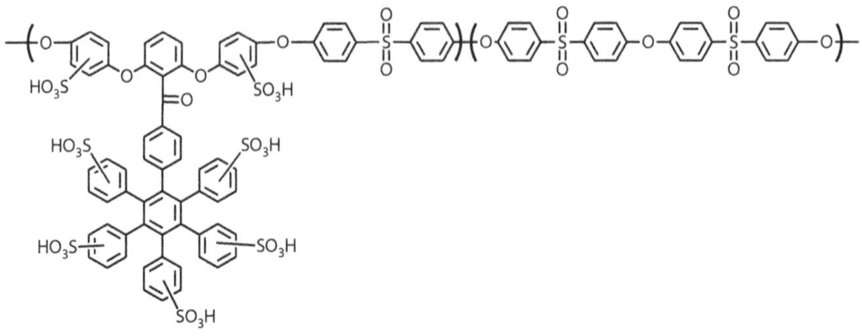

FIGURE 4.36 Chemical structure of the pendant-type locally and densely sulfonated PES.

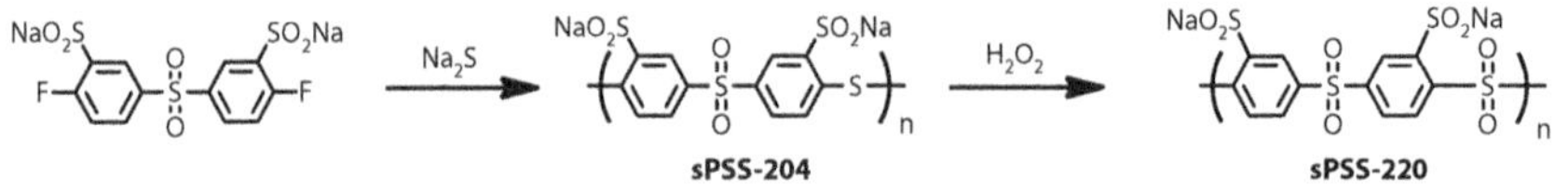

SCHEME 4.14 Synthesis of 100% sulfonated poly(phenylene sulfone) with extremely high IEC values (IEC = 4.5 meq./g).

However, the membrane was water soluble, and the film-forming properties were rather poor (brittle in dry state). Therefore, it is very important to form insoluble and dimensionally stable membranes while maintaining a high IEC value level.

Although membranes with high IEC values obviously display high proton conductivities, they are usually soluble or swell too much in water. To overcome this trade-off between the proton conductivity and water resistance, both high IEC values and the introduction of a highly hydrophobic structure seem to be a simple and effective way because these factors simultaneously contribute to the high proton conductivity and high water resistance, respectively. The cardo monomer having a bulky structure is a good candidate for providing a high hydrophobicity to PEMs due to its high free volume. Watanabe and coworkers reported the synthesis of sulfonated poly(arylene ether) (IEC up to 3.26 meq./g) ionomers containing fluorenyl groups that generated a large free volume [46]. The chemical structures of these ionomers are shown in Figure 4.37.

Poly(arylene ether sulfone) with an extremely high IEC value (IEC = 3.26 meq./g) was insoluble in water and formed a tough, flexible, and transparent membrane. As shown in Figure 4.38, by increasing the IEC, the proton conductivity remarkably increases at a low RH. The membrane (IEC = 3.26 meq./g) shows a proton conductivity of 5.6×10^{-3} S/cm at 80°C and 20% RH, which is comparable to that of the Nafion membrane. To investigate the hydrophilic/hydrophobic microphase separation and the proton transport pathway of the membranes, STEM observations were carried out (Figure 4.39). The membranes with the IEC values of 1.63–2.51 meq./g exhibited spherical ionic clusters of relatively uniform size. These ionic clusters were not connected to each other, but were rather isolated. The connectivity of these ionic clusters was significantly improved in the PES-d (IEC = 3.26 meq./g)

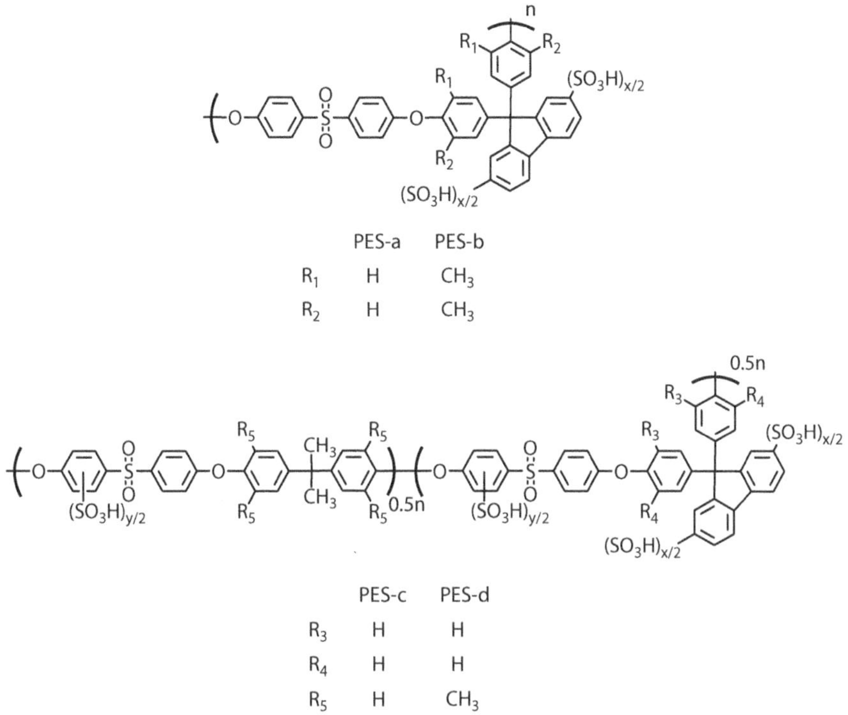

FIGURE 4.37 Chemical structure of the sulfonated poly(arylene ether) ionomers containing fluorenyl groups.

membrane (Figure 4.39c) and was similar to that of the Nafion 112 membrane (Figure 4.39d).

Furthermore, the durability of the membrane is shown in Figure 4.40. Although there is a decline in the proton conductivity with time, after 10,000 h, the proton conductivities are still at acceptable levels for fuel cell operation. The membranes retained their strength, flexibility, and high molecular weight after the 10,000 h.

Ueda et al. focused their attention on a binaphthyl structure, which has a high hydrophobicity and a large free volume derived from its twisted structure, and was expected to show a low WU even after sulfonation. The monomer, 4,4'-di (1-naphthoxy)diphenyl sulfone, was successfully prepared by a nucleophilic aromatic substitution reaction between 4,4'-dichlorodiphenyl sulfone and 1-naphthol. A highly SPES with highly hydrophobic binaphthyl units (IEC = 3.19 meq./g, BNSH-100(OX)) was obtained by the oxidative polymerization of 4,4'-di(1-naphthoxy) diphenyl sulfone with ferric chloride as an oxidant, followed by postsulfonation into the binaphthyl units (Scheme 4.15).

The BNSH-100(OX) membrane showed a higher proton conductivity than that of Nafion 117 over the wide range of 30%–95% RH (Figure 4.41). A high mechanical strength and good dimensional stability were also observed regardless of its high IEC value (3.19 meq./g). Consequently, BNSH-100(OX) overcame the aforementioned

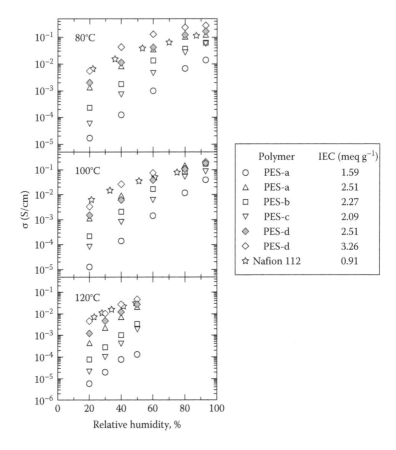

Polymer	IEC (meq g^{-1})
○ PES-a	1.59
△ PES-a	2.51
□ PES-b	2.27
▽ PES-c	2.09
◆ PES-d	2.51
◇ PES-d	3.26
☆ Nafion 112	0.91

FIGURE 4.38 Humidity dependence of the proton conductivity of PES membranes and Nafion 112 at 80°C, 100°C, and 120°C. (Reprinted after modification from Miyatake, K. et al., *J. Am. Chem. Soc.*, 129(13), 3879, 2007. With permission.)

trade-off and demonstrated a high performance for hydrocarbon PEMs [47]. They also reported the facile synthesis of a highly SPES with binaphthyl units (BNSH-100) by the nucleophilic aromatic polymerization of disodium 3,3′-disulfo-4,4′-difluorophenyl sulfone with 1,1′-binaphthyl-4,4′-diol [48]. BNSH-100 has two sulfonic acid groups in a diphenyl sulfone unit, and the binaphthyl unit can behave as the hydrophobic part (Scheme 4.16).

Regardless of its high IEC value, the BNSH-100 membrane exhibited a high mechanical strength in the dry state and a controlled dimensional stability in both its length and thickness in the hydrated state (80°C and 95% RH). In particular, the BNSH-100 membrane maintained a relatively high WU even at a low RH (11.6 wt% at 30% RH). The proton conductivity of the BNSH-100 membrane was higher than that of Nafion 117 in the range of 30%–95% RH. Even at 30% RH, the BNSH-100 membrane demonstrated an excellent proton conductivity of 6.3×10^{-3} S/cm, which was 1.5 times higher than that of Nafion 117 under the same conditions.

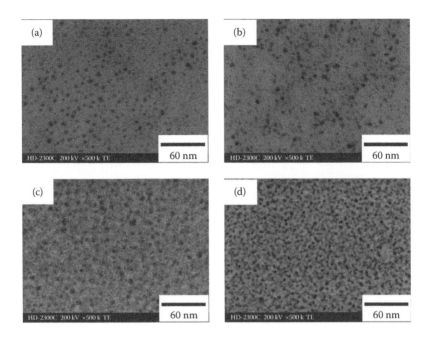

FIGURE 4.39 STEM images of (a) PES-a (IEC = 1.59 meq./g), (b) PES-a (IEC = 2.51 meq./g), (c) PES-d (IEC = 3.26 meq./g), and (d) Nafion 112 membranes. (Reprinted after modification from Miyatake, K. et al., *J. Am. Chem. Soc.*, 129(13), 3879, 2007. With permission.)

AFM observations of the BNSH-100 membrane supported the formation of the surface hydrophilic/hydrophobic phase-separated structure derived from the high IEC value and the highly hydrophobic binaphthyl units (Figure 4.42).

Watanabe and coworkers prepared sulfonated poly(arylene ether sulfone ketone) (SPESK) multiblock membranes (Figure 4.43) [49]. The properties and stability of the SPESK block copolymers are summarized in Table 4.4.

The IECs of the SPESK block copolymers range from 1.07 to 1.62 meq./g. The oxidative stability of the membranes is not good probably due to the oxidative degradation of the phenylene carbon atoms *ortho* to the ether bonds by attack of the hydroxyl radicals. On the other hand, the hydrolytic stability of these membranes is excellent as no degradation is observed. STEM and SAXS measurements supported the formation of the clear phase separation of the hydrophobic and hydrophilic blocks. Figure 4.44 shows the dependence of the WU and in-plane proton conductivity on the RH at 80°C and 110°C. The *X*30*Y*8 SPESK (IEC = 1.62 meq./g) shows a high WU even at 20% RH and a proton conductivity similar to that of Nafion over a wide range of RH values at 80°C.

4.5.5 SPES with Pendant Perfluoroalkyl Sulfonic Acids

Another approach to improve the performance of the aromatic PEMs as well as high IEC values is the introduction of perfluoroalkyl sulfonic acids in place of the normal sulfonic acids. Compared to the acidity of the sulfonated aromatic polymers

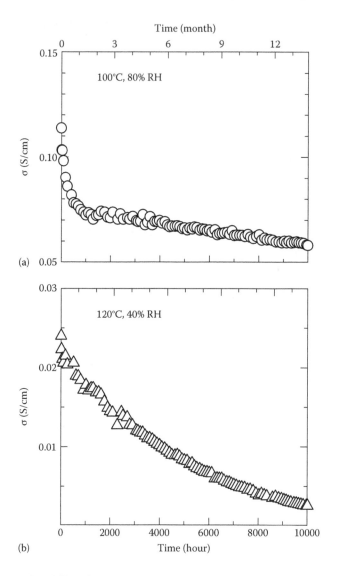

FIGURE 4.40 Durability of the proton conductivity of the PES-d (IEC = 3.26 meq./g) membrane (a) at 100°C and 80% RH and (b) at 120°C and 40% RH. (Reprinted from Miyatake, K. et al., *J. Am. Chem. Soc.*, 129(13), 3879, 2007. With permission.)

(pKa ~ −1), the higher acidity of the sulfonated perfluoropolymers (pKa ~ −6) contributed to the more effective proton concentration and proton mobility, thus leading to the higher proton conductivity.

Yoshimura et al. reported the synthesis of polymers having poly(arylene ether sulfone) in the main chain and –CF$_2$CF$_2$OCF$_2$CF$_2$SO$_3$H in the side chain (PES-PSA). This polymer was synthesized by dehalogen coupling of poly(oxy-4,4′-(3,3′-dibromobiphenylene)-oxy-4,4′-diphenylsulfone) (PES-Br) and potassium

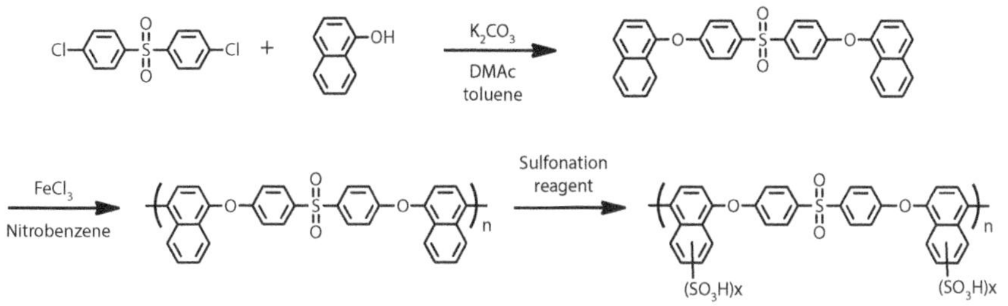

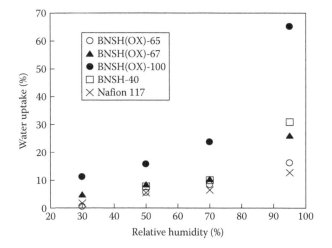

SCHEME 4.15 Synthesis of SPES with highly hydrophobic binaphthyl units (BNSH(OX)).

FIGURE 4.41 RH dependence of WU of BNSH, BPSH-40, and Nafion 117 membranes at 80°C. (Reprinted after modification from Matsumoto, K. et al., *J. Polym. Sci. A: Polym. Chem.*, 47(21), 5827, 2009. With permission.)

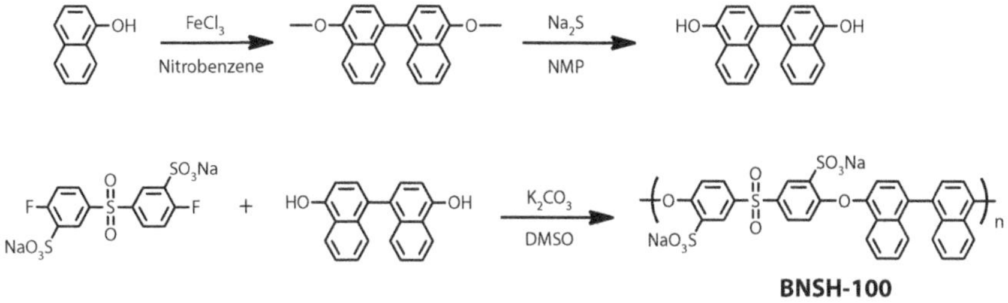

SCHEME 4.16 Synthesis of SPES with highly hydrophobic binaphthyl units (BNSH).

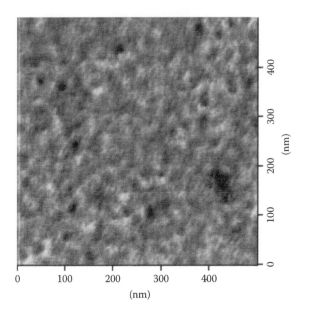

FIGURE 4.42 AFM tapping mode phase image of BNSH-100 membrane: Scan sizes are 500×500 nm². (Reprinted from Nakagawa, T. et al., *J. Mater. Chem.*, 20, 6662, 2010. With permission.)

1,1,2,2-tetrafluoro-2-(1,1,2,2-tetrafluoro-2-iodoethoxy)ethanesulfonate   (PSA-K) using copper metal, followed by treatment with aqueous HCl (Scheme 4.17) [50].

The proton conductivity of PES-PSA with the IEC value of 1.58 mmol/g was 0.12 S/cm at 80°C and 90% RH. PES-PSA with an IEC of 1.34 mmol/g was used as the membrane for the PEMFC, and the maximum power output at 80°C was 805 mW/cm² when fully humidified hydrogen and air were provided. A dynamic mechanical analysis (DMA) measurement and a tensile test revealed that the PES-PSA had a higher α-relaxation temperature than Nafion and higher flexibility than SPES.

Poly(arylene ether sulfone)s containing perfluorinated sulfonic acid groups were prepared by the polycondensation of 2,7-dibromo (or 2,7-diiodo)-9,9-*bis*(4-hydroxyphenyl)fluorene with *bis*(4-fluorophenyl)sulfone, followed by perfluorosulfonation via the Ullmann reaction (Scheme 4.18).

Despite its higher IEC values, the proton conductivity of the membranes was lower than that of the Nafion membrane. The authors concluded that acidity was a crucial factor to determine the proton conducting properties. However, other structural factors, such as the hydrophobicity and flexibility of the main chain, would have to be optimized for further improving the proton conductivity of the aromatic ionomer membranes [51]. On the other hand, Ueda and coworkers reported a novel PES containing binaphthyl units with pendant perfluoroalkyl sulfonic acids (BNSH–PSA) for PEM [52]. The BNSH–PSA (IEC = 1.91 meq./g) was prepared by the aromatic nucleophilic substitution reaction of 1,1′-binaphthyl-4,4′-diol and 4,4′-dichlorodiphenylsulfone, followed by bromination with bromine, and then the Ullmann coupling reaction with PSA-K (Scheme 4.19).

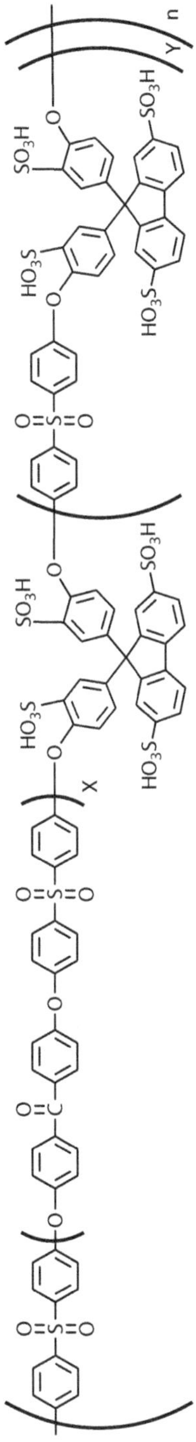

FIGURE 4.43 Chemical structure of SPESK multiblock membrane.

TABLE 4.4
Properties and Stability of SPESK Block Copolymers

Run	IEC (meq./g) Titration[a]	DS[b] (%)	Molecular Weight (kDa)[c] (MWD)[d]		Residual Weight after Stability Test	
			Before Sulfonation	After Sulfonation	Oxidative Test[e]	Hydrolytic Test[f]
X60Y8	1.07	100	150 (1.9)	192 (2.4)	49	100
X60Y12	1.41	100	170 (2.3)	360 (3.0)	45	100
X30Y8	1.62	100	125 (1.7)	252 (3.2)	40	100
Nafion 112	0.91	—	—	—	—	

Source: Data from Bae, B. et al., *Angew. Chem. Int. Ed.*, 49(2), 317, 2010.

a   The IEC values were determined by titration.
b   The DS was determined by NMR spectroscopy.
c   Weight-averaged molecular weight.
d   Molecular-weight distribution.
e   The polymer was heated in the Fenton's reagent (FeSO$_4$ (2ppm) in 3% H$_2$O$_2$) at 80°C for 1 h.
f   In an accelerated hydrolytic test, the polymer was heated in water at 140°C for 24 h.

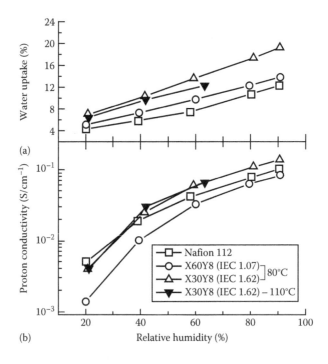

FIGURE 4.44 (a) WU and (b) in-plane proton conductivity of Nafion 112 and block SPESKs as a function of RH at 80°C and 110°C. The IEC values are given in meq./g. (Reprinted from Bae, B. et al., *Angew. Chem. Int. Ed.*, 49(2), 317, 2010. With permission.)

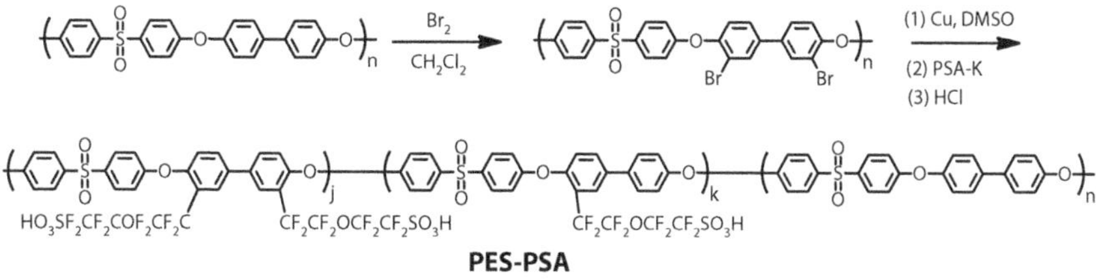

SCHEME 4.17 Synthesis of polymers having poly(arylene ether sulfone) in the main chain and $-CF_2CF_2OCF_2CF_2SO_3H$ in the side chain (PES-PSA).

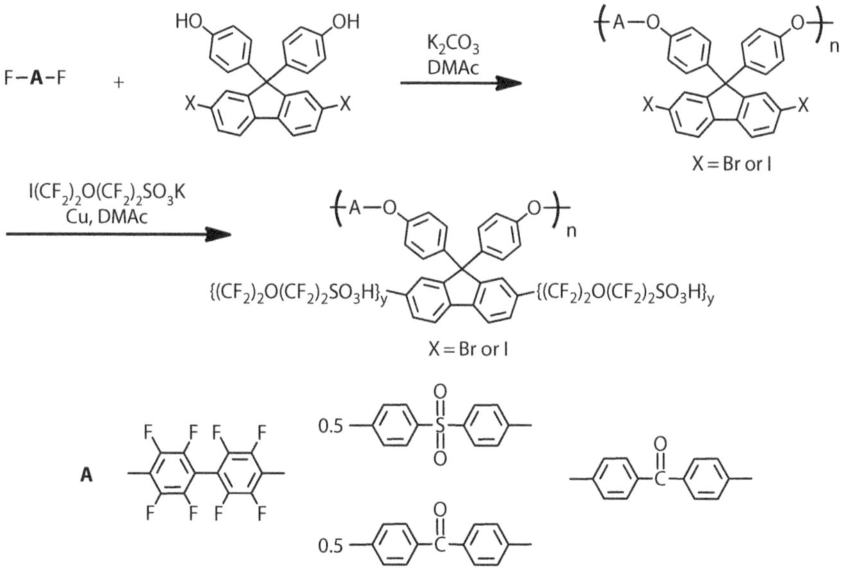

SCHEME 4.18 Synthesis of poly(arylene ether sulfone)s containing perfluorinated sulfonic acid groups.

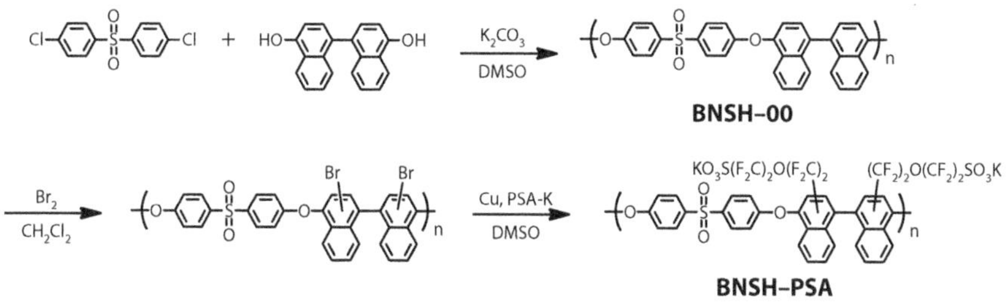

SCHEME 4.19 Synthesis of PES containing binaphthyl units with pendant perfluoroalkyl sulfonic acids (BNSH–PSA).

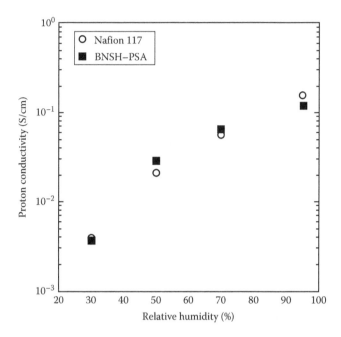

FIGURE 4.45 Humidity dependence of proton conductivity of BNSH–PSA and Nafion 117 membranes at 80°C. (Reprinted from Nakagawa, T. et al., *J. Polym. Sci. A: Polym. Chem.*, 49(14), 2997, 2011. With permission.)

The BNSH–PSA membrane with highly acidic perfluoroalkyl sulfonic acids prepared by solution casting showed a high oxidative and dimensional stability. It also produced a high proton conductivity comparable to the Nafion 117 membrane in the range of 30%–95% RH, regardless of the moderate IEC value (1.91 meq./g) (Figure 4.45). Furthermore, AFM observation supported the formation of the phase-separated structure in which the hydrophilic domains were well dispersed and connected to each other.

4.5.6 CROSS-LINKED SPES

To realize both a high proton conductivity and high durability of membranes, the combination of a multiblock polymer structure and high IEC values for high-performance PEM materials was introduced. Even multiblock copolymers, however, sometimes show a high degree of water swelling or partially dissolve in the water for IEC >3.00 meq./g. To overcome this problem, cross-linking PEM materials with high IEC values has been developed. Ueda et al. reported PEM materials based on the cross-linked highly sulfonated multiblock copoly(ether sulfone)s (IEC = 2.99–3.40 meq./g) [53]. Cross-linked membranes were successfully prepared by the chemical reaction of the sulfonic acid groups in the polymers and 1,4-diphenoxybenzene in the presence of phosphorus pentoxide/methanesulfonic acid (1:10 by weight, PPMA) (Scheme 4.20).

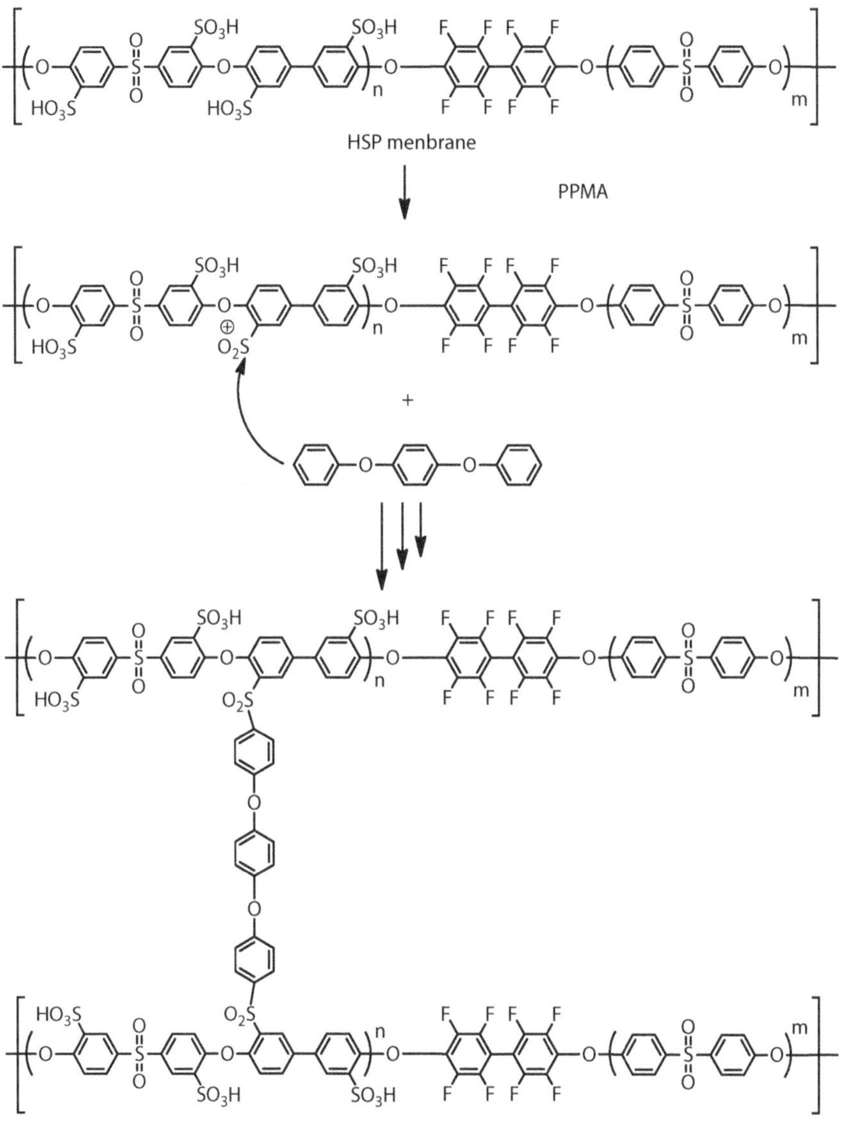

SCHEME 4.20 Preparation of PEM materials based on cross-linked highly sulfonated multiblock copoly(ether sulfone)s.

The resulting cross-linked membranes show a proton conductivity significantly higher than that of Nafion 117 in the wide range of 30%–95% RH (Figure 4.46). In particular, the proton conductivity of the membrane with the IEC value of 3.40 meq./g reached 1.0×10^{-2} S/cm at 80°C and 30% RH, which is 2.4 times higher than that of Nafion 117 under the same conditions. The high WU of the HSPx3 membrane over

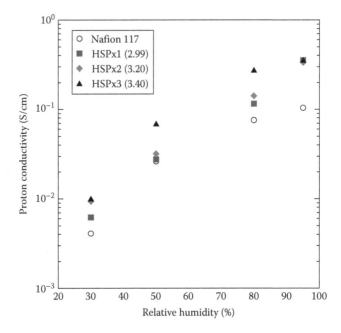

FIGURE 4.46 Humidity dependence of proton conductivity of HSPx1–3 and Nafion 117 membranes at 80°C. (Reprinted from Nakabayashi, K. et al., *Macromolecules*, 43(13), 5756, 2010. With permission.)

the entire range of RHs led to such an excellent proton conductivity. The HSPx1–3 membranes showed a good WU at a low RH, and in particular, the HSPx2 and HSPx3 membranes maintained the high WU of 14.9 and 17.6 wt%, respectively, even at 30% RH (Figure 4.47). These WU values, respectively, correspond to 2.59 and 2.88 water molecules per one sulfonic acid group (λ), whereas the λ value of Nafion 117 was 2.99 under the same conditions.

Consequently, these results demonstrated that the strategy to combine the cross-linked multiblock polymer structure with the high IEC value is promising for producing high-performance PEM materials, which have both a high proton conductivity and high water resistance.

Lee and coworkers investigated the cross-linking density effect of fluorinated aromatic polyethers on the transport properties. The copolymers were synthesized using potassium 2,5-dihydroxybenzenesulfonate (SHQ), 4,4′-(hexafluoroisopropylidene)diphenol (6F-BPA), DFBP, and 1-ethynyl-2,4-difluorobenzene (CM) as a cross-linking moiety by polycondensation (Scheme 4.21) [54].

Cross-linked membranes were produced by the trimerization of the ethynyl moiety, which formed a benzene ring by thermal curing. Figure 4.48 presents the WU and swelling ratios of the cross-linked membranes as a function of the molar ratio of the sulfonated monomer. As expected, the WU and swelling ratios of the membranes increase with the DS because the presence of sulfonic acid groups in the polymer main chain increases the ionic nature of the sulfonated copolymers.

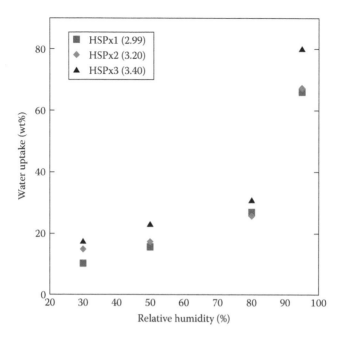

FIGURE 4.47 Humidity dependence on WU of HSPx1–3 membranes at 80°C. (Reprinted from Nakabayashi, K. et al., *Macromolecules*, 43(13), 5756, 2010. With permission.)

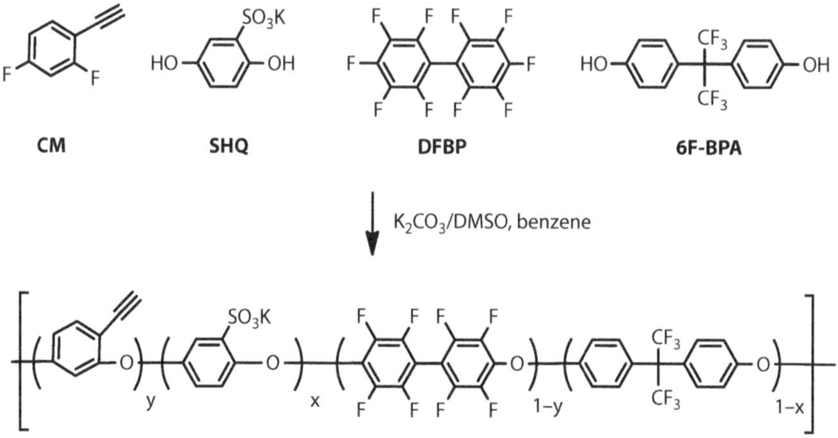

SCHEME 4.21 Synthesis of cross-linkable copolymers (SFPE*x*-CM*y*).

Figure 4.49 shows the proton conductivity and methanol permeability through the cross-linked membranes as a function of the CM feed ratio. The proton conductivity and methanol permeability of the cross-linked membranes decrease with increasing cross-linking density because cross-linking reduces the size of the ionic channels, resulting in the suppressed transport of the protons and methanol. Nevertheless,

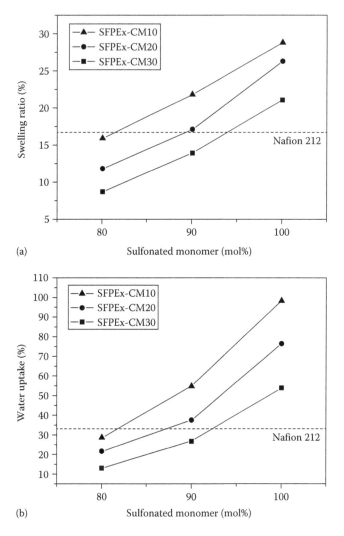

FIGURE 4.48 Swelling ratio (a) and WU (b) versus the molar ratio of sulfonated mono-mer. In the SFPEx-CM10, SFPEx-CM20, and SFPEx-CM30, x is the molar ratio of sul-fonated monomer, that is, x=80, 90, and 100 mol%. (Reprinted from Jeong, M.H. et al., *Macromolecules*, 42(5), 1652, 2009. With permission.)

the proton conductivity and methanol permeability of the membranes still had reasonable values.

A similar approach has been reported by Lee and coworkers. Sulfonated poly(phenylene sulfide nitrile) (ESPSN) random copolymers were successfully con-verted to cross-linked polymers by the thermal trimerization of the ethynyl moiety at 250°C for 4 h (Scheme 4.22) [55]. The XESPSN50 and 60 membranes exhibited suitable WU and λ values and an excellent oxidative stability.

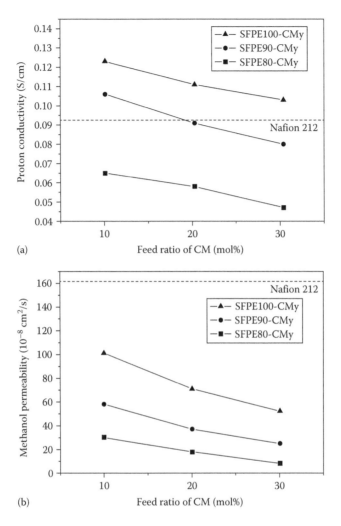

FIGURE 4.49 Influence of the CM feed ratios on proton conductivity (a) and methanol permeability (b). In the SFPE100-CM*y*, SFPE90-CM*y*, and SFPE80-CM*y*, *y* is the feed ratio of CM, that is, *y* = 10, 20, and 30 mol%. Reprinted from Jeong, M.H. et al., *Macromolecules*, 42(5), 1652, 2009. With permission.)

The proton conductivity of XESPSN60 at 30% RH was 3.2 and 6.2 times higher than those of ESPSN60 and Nafion 212, respectively. This improved proton conduction was explained by the formation of an ordered nanophase morphology due to the high local density of the sulfonic acid groups within the hydrophilic domain during cross-linking.

XESPSN60 demonstrated the highest PEFC performance among the tested samples. At 0.6 V, its current density and maximum power density were around 160% and 150% higher than those of ESPSN60 and Nafion 212, respectively.

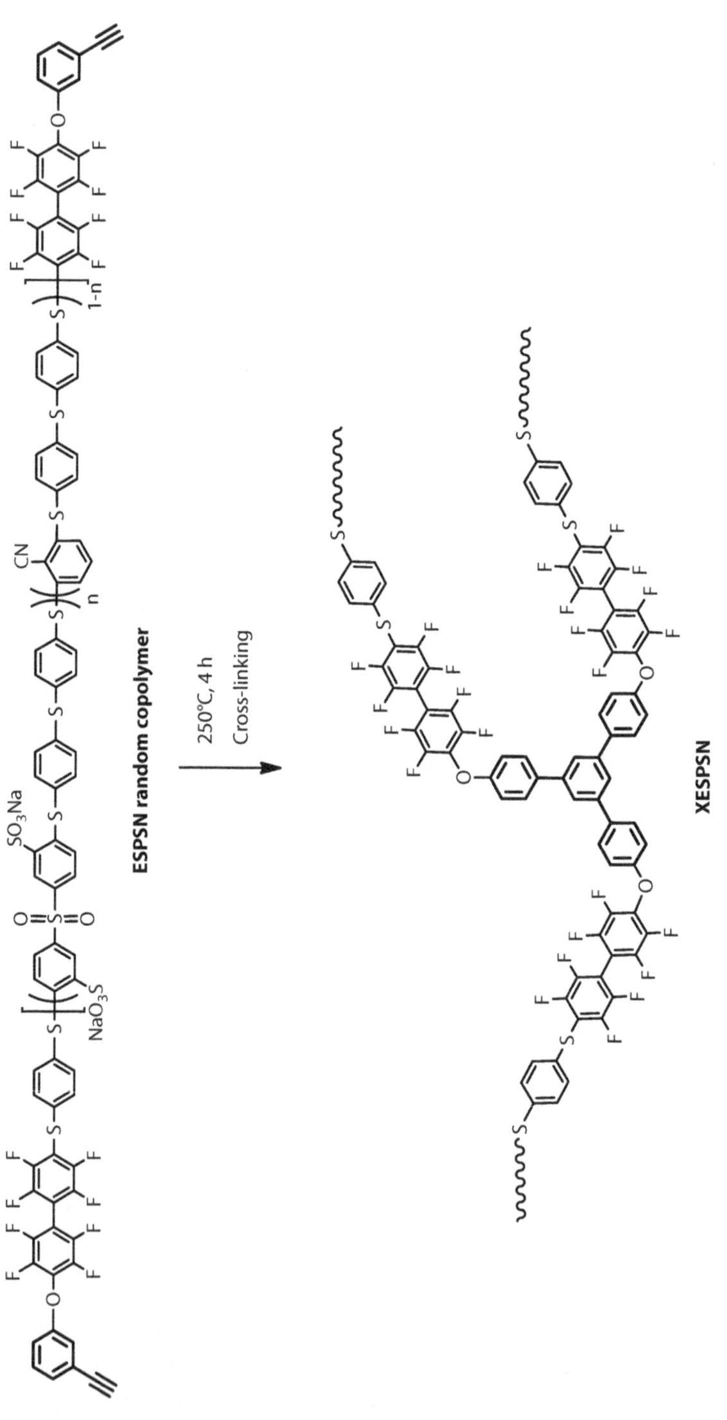

SCHEME 4.22   Synthesis of cross-linked copolymers (XESPSN).

4.5.7 COMPOSITE-TYPE SPESS

At high temperature (>100°C) and low RH (<50%), the proton conducting channels lose water, reducing the cluster size and degree of interconnection. As an
attempt to overcome these problems, most researchers focus on modifying the polymer membranes with inorganic particles, which act as a water trapping site. Silica
nanoparticle–based composite SPESs, which were prepared by the ball milling technique, were reported to avoid the trade-off between water swelling and mechanical
properties and to retain water at >100°C [56]. While the thermal stability was unaffected by the silica load, Young's modulus and tensile strength increased with the
silica contents.

The IEC values decreased with the increasing silica content as expected. The WU
of membranes was between 34% and 38%, independent of the IEC, which provided
the hydrophilic nature of the silica. The proton conductivity had a maximum at a
5 wt% silica content that can be explained as follows: silica improves the hydration and thus conductivity, but with decreasing IEC values, the membrane resistance
increases. Furthermore, in the fuel cell (H_2/air, 120°C, RH <50%), it showed a current density of 173 mA/cm^2 at 0.7 V, which was 3.4 times higher than for a pristine
membrane.

Kim et al. studied the effect of sulfonated multiwalled carbon nanotubes
(s-MWNTs)/SPES composite membranes on the membrane properties, including
the proton conductivity, WU, mechanical properties, and methanol for direct methanol fuel cells (DMFCs) [57]. The s-MWNTs were prepared by the oxidation of
MWNTs with H_2SO_4 and HNO_3, followed by treatment of MWNTs–COOH with
$SOCl_2$, and finally the reaction of MWNTs–COCl with aminomethanesulfonic acid
(Scheme 4.23).

The IEC values (from 0.8 up to 1.7 meq./g) and the WU (from 4% to 7.2%
at 80°C) of the SPES membranes increased with the increasing s-MWNT weight

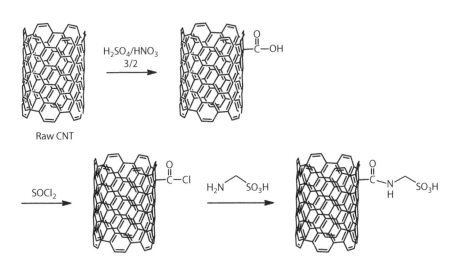

SCHEME 4.23 Sulfonation of s-MWNTs.

ratio (from 0 to 5 wt%) in the SPES membranes. The tensile strength of the SPES/ s-MWNT membranes increased by increasing the contents of s-MWNT (from 0% with 61 MPa to 5% with 65.1 MPa) due to the strong interaction between the carbonyl groups of s-MWNT and sulfonic acids of SPES. Furthermore, the composite membranes exhibited proton conductivities ranging from 3.9×10^{-4} (0% of s-MWNT) to 7.1×10^{-3} S/cm (5 wt% of s-MWNT) at 80°C, as well as a low methanol permeability ranging from 4.4×10^{-10} to 9.1×10^{-10} cm²/s at 25°C (Nafion, 2.9×10^{-6} cm²/s).

Layered metal phosphate hydrates, $M(HPO_4)_2 \cdot nH_2O$ (M = Sn, Zr, Ti) (MP), are layered compounds with water molecules in the interlayer that show high proton conductivities [58,59]. Among these hydrates, $Sn(HPO_4)_2 \cdot nH_2O$ (SnP) shows a high and stable proton conductivity above 100°C [58].

Based on these results, Sugata and coworkers investigated the properties of the SPES/SnP-NS membranes through formation of hydrogen-bonded networks with SnP [60]. The WU of the membranes decreased with increasing SnP contents (from 43 wt% with 0 vol% SnP to 20 wt% with 40 vol% SnP). The T_g and T_s of the membranes increased with the increasing SnP content due to the presence of a network of hydrogen bonds at the SPES–SnP interface. Although the 10 vol%

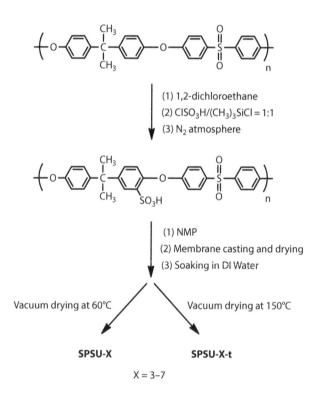

SCHEME 4.24 Preparation procedure of SPSU membranes.

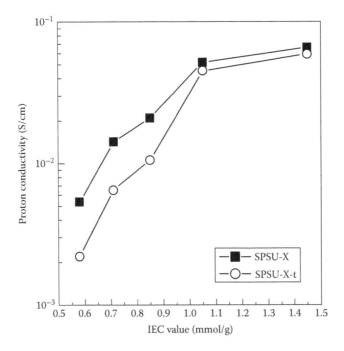

FIGURE 4.50 Proton conductivity of the SPSU-X and the SPSU-X-t membranes as a function of IEC value. (Reprinted from Park, H.B. et al., *J. Membr. Sci.*, 247(1–2), 103, 2005. With permission.)

SnP-NS membrane showed a high conductivity of 5.9×10^{-2} S/cm at 150°C under saturated water vapor pressure, at low RH, the proton conductivity remained constant even with the increasing SnP content due to the very small amount of water in the SPES–SnP interfacial layer. It is probable that the hydrogen-bonded network no longer effectively functions as a proton conducting pathway.

4.5.8 THERMALLY ANNEALED SPESs

Thermal annealing is a well-known process to stabilize the glassy polymer membrane, which induces a densification of the polymer matrix and leads to a restricted chain mobility. Thus, the membrane plasticization caused by the absorbing of gas and liquid molecules is hindered [61].

The heat treatment of SPSU membranes (Scheme 4.24) at a higher temperature led to an increased T_g and simultaneously a decreased WU, proton conductivity, and methanol permeability (Figures 4.50 and 4.51) [29]. Annealing below the T_g accelerated the equilibrium process and physical aging of the SPSU ionomer membranes subsequently inducing a more compact chain packing structure, which eventually affected the proton and methanol transport through these ionomer membranes.

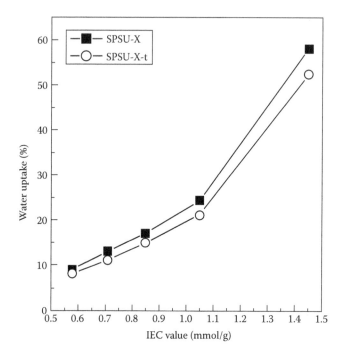

FIGURE 4.51 WU (%) in the SPSU-X and the SPSU-X-t membranes as a function of IEC value. Reprinted from Park, H.B. et al., *J. Membr. Sci.*, 247(1–2), 103, 2005. With permission.)

4.6   CHAPTER SUMMARY

In this chapter, the synthesis, characterization, and PEFC application of SPES membranes are reviewed. In order to develop efficient and practical PEM materials to replace currently used perfluorosulfonic acid membranes, the synthesis of new SPESs and their performance as PEMs has been widely investigated by several research groups as introduced in this chapter. The promising candidates for the breakthrough in PEM application are categorized into the following five materials: locally and densely sulfonated polymers (Section 4.5.3), SPES with high IEC values and high free volumes (Section 4.5.4), SPES with pendant perfluoroalkyl sulfonic acids (Section 4.5.5), cross-linked SPES (Section 4.5.6), and composite-type SPESs (Section 4.5.7). All methods target generation of clear and well-connected proton channels while maintaining the water stability for a high proton conduction even at a low RH and high operating temperature. By further pursuing new and/or optimal polymer structures, extraordinarily useful materials possessing well-balanced properties are likely to emerge for fuel cell applications in the near future.

ACKNOWLEDGMENT

This work was supported in part by Industrial Technology Research Grant Program in 2011 (#11B02008c) from the New Energy and Industrial Technology Development

Organization (NEDO) of Japan. Partial support from Toyota Motor Corporation is also gratefully acknowledged.

REFERENCES

1. Savadogo, O. 1998. Emerging membranes for electrochemical systems. I. Solid polymer electrolyte membranes for fuel cell systems. *J. New Mater. Electrochem. Syst.* 1:47–66.
2. Rikukawa, M. and Sanui, K. 2000. Proton-conducting polymer electrolyte membranes based on hydrocarbon polymers. *Prog. Polym. Sci.* 25:1463–1502.
3. Mauritz, K. A. and Moore, R. B. 2004. State of understanding of Nafion. *Chem. Rev.* 104:4535–4585.
4. Hickner, M. A., Ghassemi, H., Kim, Y. S., Einsla, B. R., and McGrath, J. E. 2004. Alternative polymer systems for proton exchange membranes (PEMs). *Chem. Rev.* 104:4587–4612.
5. Park, C. H., Lee, C. H., Guiver, M. D., and Lee, Y. M. 2011. Sulfonated hydrocarbon membranes for medium-temperature and low-humidity proton exchange membrane fuel cells (PEMFCs). *Prog. Polym. Sci.* 36:1443–1498.
6. Hickner, M. A. and Pivovar, B. S. 2005. The chemical and structural nature of proton exchange membrane fuel cell properties. *Fuel Cells* 5:213–229.
7. Higashihara, T., Matsumoto, K., and Ueda, M. 2009. Sulfonated aromatic hydrocarbon polymers as proton exchange membranes for fuel cells. *Polymer* 50:5341–5357.
8. Blanco, J. F., Ngyuyen, Q. T., and Schaetzel, P. 2002. Sulfonation of polysulfones: Suitability of the sulfonated materials for asymmetric membrane preparation. *J. Appl. Polym. Sci.* 84:2461–2473.
9. Baradie, B., Poinsignon, C., Sanchez, J. Y., Piffard, Y., Vitter, G., Bestaoui, N., Foscallo, D., Denoyelle, A., Delabouglise, D., and Vaujany, M. 1998. Thermostable ionomeric filled membrane for H_2/O_2 fuel cell. *J. Power Source* 74:8–16.
10. Manea, C. and Mulder, M. 2002. Characterization of polymer blends of polyethersulfone/sulfonated polysulfone and polyethersulfone/sulfonated polyetheretherketone for direct methanol fuel cell applications. *J. Membr. Sci.* 206:443–453.
11. Chen, M. H., Chiao, T. C., and Tseng, T. W. 1996. Preparation of sulfonated polysulfone/polysulfone and aminated polysulfone/polysulfone blend membranes. *J. Appl. Polym. Sci.* 61:1205–1209.
12. Dyck, A., Fritsch, D., and Nunes, S. P. 2002. Proton-conductive membranes of sulfonated polyphenylsulfone. *J. Appl. Polym. Sci.* 86:2820–2827.
13. Genova-Dimitrova, P., Baradie, B., Foscallo, D., Poinsignon, C., and Sanchez, J. Y. 2001. Ionomeric membranes for proton exchange membrane fuel cell (PEMFC): Sulfonated polysulfone associated with phosphatoantimonic acid. *J. Membr. Sci.* 185:59–71.
14. Lufrano, F., Gatto, I., Staiti, P., Antonucci, V., and Passalacqua, E. 2001. Sulfonated polysulfone ionomer membranes for fuel cells. *Solid State Ionics* 145:47–51.
15. Lufrano, F., Squadrito, G., Patti, A., and Passalcqua, E. 2000. Sulfonated polysulfone ionomer membranes for fuel cells. *J. Appl. Polym. Sci.* 77:1250–1256.
16. Quentin, J. P. 1973. Sulfonated polyarylether sulfones. U.S. Patent, 3,709,841.
17. Chikashige, Y., Chikyu, Y., Miyatake, K., and Watanabe, M. 2005. Poly(arylene ether) ionomers containing sulfofluorenyl groups for fuel cell applications. *Macromolecules* 38:7121–7126.
18. Kerres, J., Cui, W., and Reichle, S. 1996. New sulfonated engineering polymers via the metalation route. 1. Sulfonated poly(ethersu1fone) PSU Udel via metalation-sulfination-oxidation. *J. Polym. Sci. A: Polym. Chem.* 34:2421–2438.
19. Lafitte, B., Karlsson, L. E., and Jannasch, P. 2002. Sulfophenylation of polysulfones for proton-conducting fuel cell membranes. *Macromol. Rapid Commun.* 23:896–900.

20. Ueda, M., Toyota, H., Ochi, T., Sugiyama, J., Yonetake, K., Masuko, T., and Teramoto, T. 1993. Synthesis and characterization of aromatic poly(ether sulfone)s containing pendant sodium sulfonate groups. *J. Polym. Sci. A: Polym. Chem.* 31:853–858.

21. Wang, F., Hickner, M., Kim, Y. S., Zawodzinski, T. A., and McGrath, J. E. 2002. Direct polymerization of sulfonated poly(arylene ether sulfone) random (statistical) copolymers: Candidates for new proton exchange membranes. *J. Membr. Sci.* 197:231–242.

22. Wang, F., Hickner, M., Ji, Q., Harrison, W., Mecham, J., Zawodzinski, T. A., and McGrath, J. E. 2001. Synthesis of highly sulfonated poly(arylene ether sulfone) random (statistical) copolymers via direct copolymerization. *Macromol. Symp.* 175:387–395.

23. Harrison, W., Hickner, M., Kim, Y. S., and McGrath, J. E. 2005. Monomer building blocks: Synthesis, characterization, and performance—A topical review. *Fuel Cells* 5:201–212.

24. Bae, B., Miyatake, K., and Watanabe, M. 2008. Sulfonated poly(arylene ether sulfone) ionomers containing fluorenyl groups for fuel cell applications. *J. Membr. Sci.* 310:110–118.

25. Lee, H. S., Roy, A., Lane, O., Dunn, S., and McGrath, J. E. 2008. Hydrophilic–hydrophobic multiblock copolymers based on poly(arylene ether sulfone) via low-temperature coupling reactions for proton exchange membrane fuel cells. *Polymer* 49:715–723.

26. Nakabayashi, K., Matsumoto, K., and Ueda, M. 2008. Influence of adjusted hydrophilic–hydrophobic lengths in sulfonated multiblock copoly(ether sulfone) membranes for fuel cell application. *J. Polym. Sci. A: Polym. Chem.* 46:7332–7341.

27. Kreuer, K. D. 2001. On the development of proton conducting polymer membranes for hydrogen and methanol fuel cells. *J. Membr. Sci.* 185:29–39.

28. Ghassemi, H., McGrath, J. E., and Zawodzinski Jr., T. A. 2006. Multiblock sulfonated–fluorinated poly(arylene ether)s for a proton exchange membrane fuel cell. *Polymer* 47(11):4132–4139.

29. Park, H. B., Shin, H. S., Lee, Y. M., and Rhim, J. W. 2005. Annealing effect of sulfonated polysulfone ionomer membranes on proton conductivity and methanol transport. *J. Membr. Sci.* 247:103–110.

30. Borup, R., Meyers, J., Pivovar, B., Kim, Y. S., Mukundan, R., Garland, N., Myers, D. et al. 2007. Scientific aspects of polymer electrolyte fuel cell durability and degradation. *Chem. Rev.* 107:3904–3951.

31. Hübner, G. and Roduner, E. 1999. EPR investigation of HO/radical initiated degradation reactions of sulfonated aromatics as model compounds for fuel cell proton conducting membranes. *J. Mater. Chem.* 9:409–418.

32. Schuster, M., de Araujo, C. C., Atanasov, V., Andersen, H. T., Kreuer, K. D., and Maier, J. 2009. Highly sulfonated poly(phenylene sulfone): Preparation and stability issues. *Macromolecules* 42:3129–3137.

33. Genies, C., Mercier, R., Sillion, B., Cornet, N., Gebel, G., and Pineri, M. 2001. Soluble sulfonated naphthalenic polyimides as materials for proton exchange membranes. *Polymer* 42:359–373.

34. Khandpur, A. K., Foerster, S., Bates, F. S., Hamley, I. W., Ryan, A. J., Bras, W., Almdal, K., and Mortensen, K. 1995. Polyisoprene-polystyrene diblock copolymer phase diagram near the order-disorder transition. *Macromolecules* 28:8796–8806.

35. Roy, A., Hickner, M. A., Yu, X., Li, Y., Glass, T. E., and McGrath, J. E. 2006. Influence of chemical composition and sequence length on the transport properties of proton exchange membranes. *J. Polym. Sci. B: Polym. Phys.* 44:2226–2239.

36. Nakabayashi, K., Matsumoto, K., and Ueda, M. 2008. Synthesis and properties of sulfonated multiblock copoly(ether sulfone)s by a chain extender. *J. Polym. Sci. A: Polym. Chem.* 46:3947–3957.

37. Nakabayashi, K., Matsumoto, K., Higashihara, T., and Ueda, M. 2009. Influence of polymer structure in sulfonated block copoly(ether sulfone) membranes for fuel cell application. *Polym. J.* 41:332–337.
38. Einsla, M. L., Kim, Y. S., Hawley, M., Lee, H. S., McGrath, J. E., Liu, B., Guiver, M. D., and Pivovar, B. S. 2008. Toward improved conductivity of sulfonated aromatic proton exchange membranes at low relative humidity. *Chem. Mater.* 20:5636–5642.
39. Li, N., Wang, C., Lee, S. Y., Park, C. H., Lee, Y. M., and Guiver, M. D. 2011. Enhancement of proton transport by nanochannels in comb-shaped copoly(arylene ether sulfone)s. *Angew. Chem. Int. Ed.* 50:9158–9161.
40. Matsumura, S., Hlil, A. R., Lepiller, C., Gaudet, J., Guay, D., and Hay, A. S. 2008. Ionomers for proton exchange membrane fuel cells with sulfonic acid groups on the end groups: Novel linear aromatic poly(sulfide–ketone)s. *Macromolecules* 41:277–280.
41. Matsumura, S., Hlil, A. R., Lepiller, C., Gaudet, J., Guay, D., Shi, Z., Holdcroft, S., and Hay, A. S. 2008. Ionomers for proton exchange membrane fuel cells with sulfonic acid groups on the end groups: Novel branched poly(ether–ketone)s. *Macromolecules* 41:281–284.
42. Matsumoto, K., Higashihara, T., and Ueda, M. 2009. Locally and densely sulfonated poly(ether sulfone)s as proton exchange membrane. *Macromolecules* 42:1161–1166.
43. Matsumoto, K., Higashihara, T., and Ueda, M. 2009. Locally sulfonated poly(ether sulfone)s with highly sulfonated units as proton exchange membrane. *J. Polym. Sci. A: Polym. Chem.* 47:3444–3453.
44. Li, N., Hwang, D. S., Lee, S. Y., Liu, Y. L., Lee, Y. M., and Guiver, M. D. 2011. Densely sulfophenylated segmented copoly(arylene ether sulfone) proton exchange membranes. *Macromolecules* 44:4901–4910.
45. Feng, S., Shen, K., Wang, Y., Pang, J., and Jiang, Z. 2013. Concentrated sulfonated poly(ether sulfone)s as proton exchange membranes. *J. Power Sources* 224:42–49.
46. Miyatake, K., Chikashige, Y., Higuchi, E., and Watanabe, M. 2007. Tuned polymer electrolyte membranes based on aromatic polyethers for fuel cell applications. *J. Am. Chem. Soc.* 129:3879–3887.
47. Matsumoto, K., Nakagawa, T., Higashihara, T., and Ueda, M. 2009. Sulfonated poly(ether sulfone)s with binaphthyl units as proton exchange membranes for fuel cell application. *J. Polym. Sci. A: Polym. Chem.* 47:5827–5834.
48. Nakagawa, T., Nakabayashi, K., Higashihara, T., and Ueda, M. 2010. A high performance polymer electrolyte membrane based on sulfonated poly(ether sulfone) with binaphthyl units. *J. Mater. Chem.* 20:6662–6667.
49. Bae, B., Yoda, T., Miyatake, K., Uchida, H., and Watanabe, M. 2010. Proton-conductive aromatic ionomers containing highly sulfonated blocks for high-temperature-operable fuel cells. *Angew. Chem. Int. Ed.* 49:317–320.
50. Yoshimura, K. and Iwasaki, K. 2009. Aromatic polymer with pendant perfluoroalkyl sulfonic acid for fuel cell applications. *Macromolecules* 42:9302–9306.
51. Miyatake, K., Shimura, T., Mikamiac, T., and Watanabe, M. 2009. Aromatic ionomers with superacid groups. *Chem. Commun.* 42:6403–6405.
52. Nakagawa, T., Nakabayashi, K., Higashihara, T., and Ueda, M. 2011. Polymer electrolyte membrane based on poly(ether sulfone) containing binaphthyl units with pendant perfluoroalkyl sulfonic acids. *J. Polym. Sci. A: Polym. Chem.* 49:2997–3003.
53. Nakabayashi, K., Higashihara, T., and Ueda, M. 2010. Polymer electrolyte membranes based on cross-linked highly sulfonated multiblock copoly(ether sulfone)s. *Macromolecules* 43:5756–5761.
54. Jeong, M. H., Lee, K. S., and Lee, J. S. 2009. Cross-linking density effect of fluorinated aromatic polyethers on transport properties. *Macromolecules* 42:1652–1658.

55. Lee, S. Y., Kang, N. R., Shin, D. W., Lee, C. H., Lee, K. S., Guiver, M. D., Li, N., and Lee, Y. M. 2012. Morphological transformation during cross-linking of a highly sulfonated poly(phenylene sulfide nitrile) random copolymer. *Energy Environ. Sci.* 5:9795–9802.

56. Krishnan, N. N., Henkensmeier, D., Jang, J. H., Kim, H. J., Rebbin, V., Oh, I. H., Hong, S. A., Nam, S. W., and Lim, T. H. 2011. Sulfonated poly(ether sulfone)-based silica nanocomposite membranes for high temperature polymer electrolyte fuel cell applications. *Int. J. Hydrogen Energ.* 36:7152–7161.

57. Yun, S., Heo, Y., Im, H., and Kim, J. 2012. Sulfonated multiwalled carbon nanotube/ sulfonated poly(ether sulfone) composite membrane with low methanol permeability for direct methanol fuel cells. *J. Appl. Polym. Sci.* 126:E513–E521.

58. Kawakami, Y. and Miyayama, M. 2006. Proton conducting properties of layered metal phosphate hydrates. *Key Eng. Mater.* 320:267–270.

59. Alberti, G., Casciola, M., Costantino, U., Levi, G., and Ricciardi, G. 1978. On the mechanism of diffusion and ionic transport in crystalline insoluble acid salts of tetravalent metals—I. Electrical conductance of zirconium bis(monohydrogen ortho-phosphate) monohydrate with a layered structure. *J. Inorg. Nucl. Chem.* 40:533–537.

60. Sugata, S., Suzuki, S., Miyayama, M., Traversa, E., and Licoccia, S. 2012. Effects of tin phosphate nanosheet addition on proton-conducting properties of sulfonated poly(ether sulfone) membranes. *Solid State Ion.* 228:8–13.

61. Ismail, A. F. and Lorna, W. 2003. Suppression of plasticization in polysulfone membranes for gas separations by heat-treatment technique. *Sep. Purif. Technol.* 30:37–46.

5 Sulfonated Poly(Ether Ketone) Membranes

Chengji Zhao, Gang Zhang, and Hui Na

CONTENTS

5.1 INTRODUCTION

Aromatic poly(ether ketone)s (PEKs) are a class of high-performance engineering thermoplastics, which have good solvent resistance, good mechanical properties, and high thermal oxidative stability.[1,2] Various types of commercial PEKs are listed in Table 5.1. This class of advanced materials is currently receiving considerable attention for potential applications in aerospace, automobile, electronics, and other

TABLE 5.1

Introduction of Commercial PEKs

Product	Structure	Tg (°C)	Tm (°C)
ICI(Vitrex) PEEK		143	334
ICI(Vitrex) PEK		162	373
ICI(Vitrex) HTX		205	386
Dupont (Aretone) PEKK		156	343
BASF(Ultra) PEK		173	371
Hoechst (Hostatec) PEEKK		162	373

high-tech areas.[3,4] In fact, it is difficult for PEKs to be used as thin films and coating materials because of their poor solubility and processability. Functionalized PEKs with pendant groups have been successfully prepared to address this problem.[5-7]

Sulfonic acid groups have been also attached onto PEK backbones to be used as polymer electrolyte membranes (PEMs).[8] Sulfonated poly(ether ketone) (SPEK) not only maintains the physical and mechanical properties of PEK itself but also provides the hydrophilicity and ion-exchange ability for electrochemical applications, such as fuel cell,[9-12] vanadium redox flow battery (VRB),[13,14] water electrolysis,[15,16] and supercapacitor.[17,18] An ideal PEM used in electrochemical devices should possess the following characteristics: (1) high ionic conductivity; (2) minimal water/methanol transport; (3) high resistance to dehydration; (4) low gas permeability; (5) low swelling; (6) high mechanical strength; (7) high resistance to oxidation, reduction, and hydrolysis; and (8) low cost.[19] Compared to the state-of-the-art membrane material Nafion®, SPEK shows better thermal and mechanical stabilities, lower fuel crossover, comparable ionic conductivity, and, particularly, lower cost. Therefore, SPEK has been intensively developed as a potential alternative PEM to the commercial Nafion.

SPEK has been earlier prepared via modification of the parent polymer, in which sulfonic acid groups were grafted on the polymer chain by the sulfonation agents, including concentrated sulfuric acid, complex sulfuric trioxide, chlorosulfuric acid, and methanesulfuric acid.[20,21] An attractive alternative approach is the direct copolymerization of sulfonated monomers. The direct synthesis of SPEK from the sulfonated monomer has been proven more advantageous than that of postsulfonation.[22] Some of the advantages are listed as follows: (1) Compared with the postsulfonated PEK, the concentration as well as the positions of the sulfonate groups (e.g., *meta-*, *para-*, and *ortho-*) within the directly synthesized SPEKs can be readily controlled. This allows one to control the degree of sulfonation (DS) easily. (2) The direct-sulfonation method avoids the cross-linking and other side reactions, which may result in better thermal stability and mechanical properties. Several sulfonated monomers have been developed, and various types of SPEKs with controlled DS were then prepared in the past decades. Generally, these polymers show suitable conductivities only at high ion-exchange capacities (IECs). Unfortunately, too high loading of acidic groups may lead to undesirable large swelling and thus resulting in a dramatic loss of mechanical properties. Furthermore, there is a significant reduction in the proton conductivity when the temperature is higher than 100°C due to the evaporation of molecular water. The renewed interest has led to many studies directed toward decreasing water swelling and improving proton conductivity of SPEKs at lower IEC or under higher temperature/low humidity conditions.[23,24]

It has been reported that the morphology plays an important role in determining the PEM's macroscopic properties, especially the proton conductivity.[25] The model discussed by Kreuer on the basis of small-angle x-ray scattering (SAXS) data for conventional SPEK shows a less pronounced phase separation than that of Nafion, that is, morphology with narrower, less connected hydrophilic channels than those of Nafion.[26] One promising way to achieve more distinct phase separation between hydrophilic and hydrophobic domains of SPEK copolymers is precisely controlling

the polymer architectures, including block copolymers, high-free volume copolymers, grafted/branched copolymers, and highly sulfonated monomer-based copolymers.[27] They have demonstrated that these copolymers show much improved proton
conductivity over the random copolymers, especially at low humidity. Therefore,
this chapter introduces the most recent work in the synthesis of SPEK with different polymer architectures and their applications in electrochemistry devices, paying
special attention on the development of SPEK membranes for proton exchange membrane fuel cells (PEMFCs) and direct methanol fuel cells (DMFCs).

5.2 MEMBRANE SYNTHESIS

5.2.1 MATERIALS REQUIREMENT AND SELECTION

There are two major methods for the synthesis of SPEKs: postsulfonation and the
direct copolymerization of sulfonated monomers. For the postsulfonation method,
the commercially available PEK families, including poly(ether ether ketone) (PEEK),
poly(ether ether ketone ketone) (PEEKK), poly(ether ether ketone) with cardo
groups (PEEK-WC), and poly(arylene ether ketone) (PAEK), are needed for modification, which are mainly provided by Victrex company in the market nowadays.
Due to the semicrystalline nature of certain PEKs and their corresponding poor
solubility in common organic solvents, sulfonation procedures were first achieved
in concentrated sulfuric acid. Other sulfonation agents were also employed in the
sulfonation process, such as chlorosulfonic acid, fuming sulfuric acid, chlorosulfonic acid, acetyl sulfate, sulfur trioxide complexes, and trimethylsilyl chlorosulfonate ($(CH_3)_3SiSO_3Cl$). The selection of sulfonation agents is dependent on the level of
reactivity required and the need for exact control of the DS. For the direct-sulfonation
method, a presulfonated monomer is required for the polymerization. Several sulfonated dihalo monomers that have been used for SPEK synthesis for PEMs are shown
in Figure 5.1.[9–11,22,28] The synthesis of sodium 5,5′-carbonylbis(2-fluorobenzene sulfonate) was firstly reported by Wang et al. in 1998.[29] The Na research group reported
the use of 1,4-bi(3-sodium sulfonate-4-fluorobenzoyl)benzene and 1,3-*bis*(3-sodium
sulfonate-4-fluorobenzoyl)benzene as the sulfonated monomers to prepare sulfonated poly(ether ether ketone ketone)s (SPEEKKs).[10,28] Several groups then prepared
a series of disulfonated PAEK copolymers via aromatic nucleophilic substitution
from these sulfonated monomers, nonsulfonated dihalo monomers, and bisphenol
monomers.[30,31] In this way, DS is easily controlled, and at the same time, it avoids the
polymer chain degradation during the sulfonation process. The detailed procedures
of these two methods will be described in Section 5.2.2.

5.2.2 POLYMER SYNTHESIS METHODS AND PROCESSES

5.2.2.1 Postsulfonation Method

PEEK derived from hydroquinone and 4,4′-difluorobenzophenone is a kind of semicrystalline aromatic polymer with high performance. Due to its crystal structure,
insolubility in polar solvents makes postsulfonation relatively difficult and heterogeneous. After sulfonation on the phenyl rings, the crystallinity decreased and

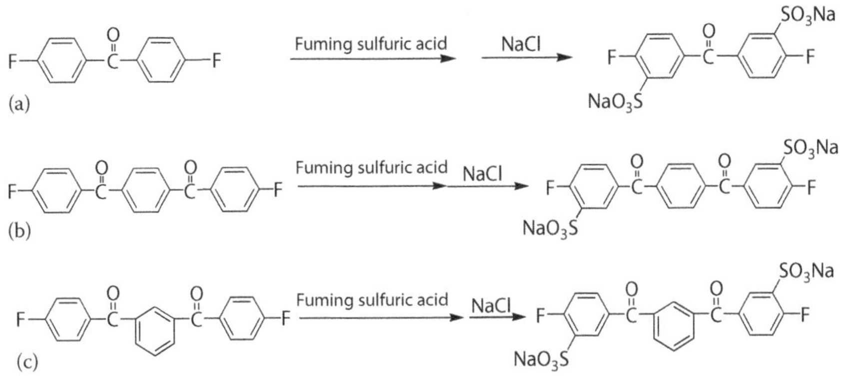

FIGURE 5.1 The preparation of (a) sodium 5,5'-carbonylbis(2-fluorobenzene sulfonate), (b) 1,4-bi(3-sodium sulfonate-4-fluorobenzoyl) benzene, and (c) 1,3-*bis*(3-sodium sulfonate-4-fluorobenzoyl)benzene.

solubility increased. Sulfonation of PEEK with chlorosulfonic acid or fuming sulfuric acid usually produces benzenesulfonic acid in the process, which might cause the degradation of the polymer.[32,33] Therefore, a milder concentrated sulfuric acid (95%–98%) is typically used (Figure 5.2). The DS could be controlled by changing the reaction time, temperature, and the acid concentration to provide the host polymers with the DS in the range of 30%–100%.[34,35] However, the heterogeneity of the reaction in sulfuric acid causes difficulty in the reproducibility in subsequent experiments or batches. The sulfonation of PEEK preferentially takes place at the aromatic ring flanked by two ether links, due to the higher electron density of the ring.[36–38] Since the electron density of the other two aromatic rings in the repeat unit is relatively low due to the electron-attracting nature of the neighboring ketone group, no more than one sulfonic acid group per repeat unit could be achieved.

Chlorosulfonic acid or fuming sulfuric acid was also used to be sulfonation agents for the host polymer with special structures. Hamciuc et al. reported the sulfonation and chlorosulfonation of a PEK polymer containing hexafluoroisopropylidene groups by using 100% chlorosulfonic acid in 2 h.[39] The introduction of the hexafluoro moiety greatly decreased the chemical degradation experienced by the bisphenol A–based PEK or parent PEEK. Meng and coworkers synthesized several PAEKs with hindered and bulky groups, such as fluorine, biphenyl, and

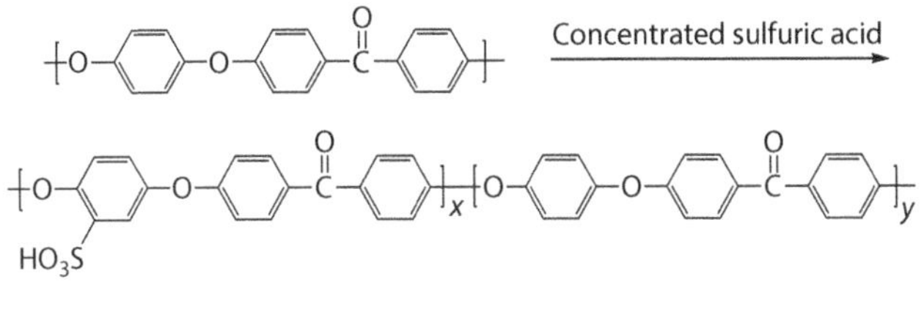

FIGURE 5.2 Synthesis of postsulfonated PEEK.

tetraphenylmethane moieties.[40] These polymers were then sulfonated with chlorosulfonic acid in methylene chloride at room temperature for a few hours (Figure
5.3a). The DS can be controlled by simply varying the molar amount of chlorosulfonic acid used. Due to the rigid structure, these membranes exhibited excellent hydrolytic and oxidative stabilities along with good mechanical properties.
Poly(phthalazinone ether ketone) (PPEK) is a new high-performance polymer with
excellent chemical and oxidative resistance, mechanical strength, high thermal stability, and very high glass transition temperatures. Gao et al. reported the sulfonation reactions of PPEK with dilute fuming sulfuric acid as both the solvent and
sulfonation agent (Figure 5.3b).[41] They initially attempted to carry out the sulfonation in concentrated sulfuric acids; however, only very low DS was obtained even
after a long reaction time and high temperature. Due to the low nucleophilicity of
poly(phthalazinone)s, fuming sulfuric acid was chosen as a strong sulfonation agent
for the preparation of high DS derivatives. It is interesting to observe that the viscosities of SPPEK increased with increasing DS, indicating that the polymer chain
was not degraded during sulfonation.

Since sulfonation is an electrophilic reaction, the position of the attached sulfonic
acid groups depends on the substituent present on the phenyl ring. Many researchers proposed a method to control the sulfonation sites by adjusting the molecular
structures of the host polymers. Li et al. prepared a series of PAEK copolymers containing two types of segments: a sulfonatable segment with phenyl pendant groups
and a nonsulfonatable one with (3-trifluoromethyl) phenyl pendant groups (Figure
5.4a).[42] No sulfonation reaction was observed for the pendant ring containing trifluoromethylated phenyl, while the phenyl groups are readily sulfonated under rapid and
mild conditions using concentrated sulfuric acid. Thus, the DS was controlled by the
composition of sulfonatable/nonsulfonatable segments. Jeong et al. also reported the

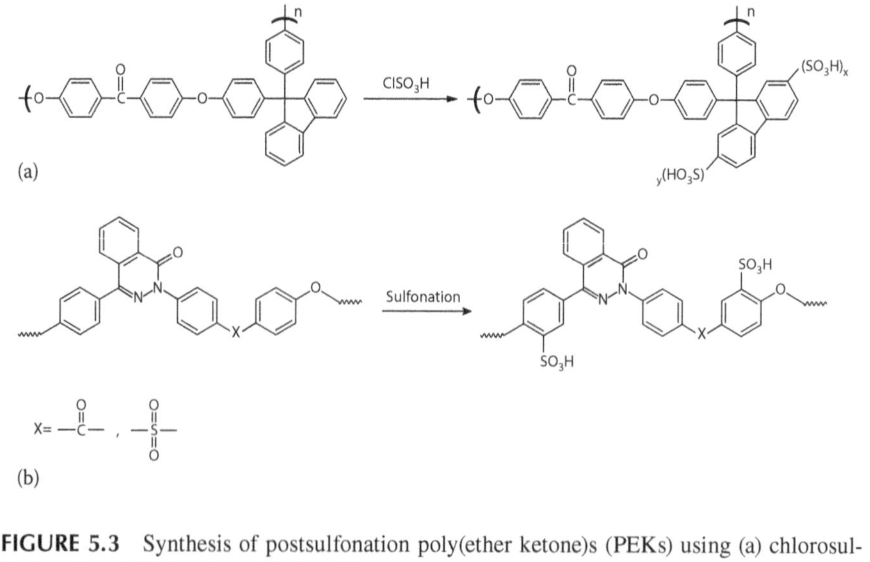

FIGURE 5.3 Synthesis of postsulfonation poly(ether ketone)s (PEKs) using (a) chlorosulfonic acid and (b) fuming sulfuric acid.

selective and quantitative sulfonation of PAEKs containing pendant phenyl rings by chlorosulfonic acid (Figure 5.4b).[35] The nonsulfonatable segments were derived from the polymerization of 4,4'-(hexafluoroisopropylidene) diphenol. It was discovered that only one sulfonic acid group can be selectively introduced without degradation per each repeating unit of polymer due to requirements for controlling reaction conditions such as reaction time and temperature. The Guiver group incorporated a variety of side substituents into PEEK, including phenyl, methylated phenyl, trifluoromethylated phenyl, and phenoxyphenyl groups (Figure 5.4c).[43] PEEKs with selected pendant groups, such as phenyl and 4-methylphenyl, were much more easily sulfonated than the other pendant groups under mild reaction conditions. This provided the opportunity to achieve SPEEKs with a controlled sulfonated position and DS.

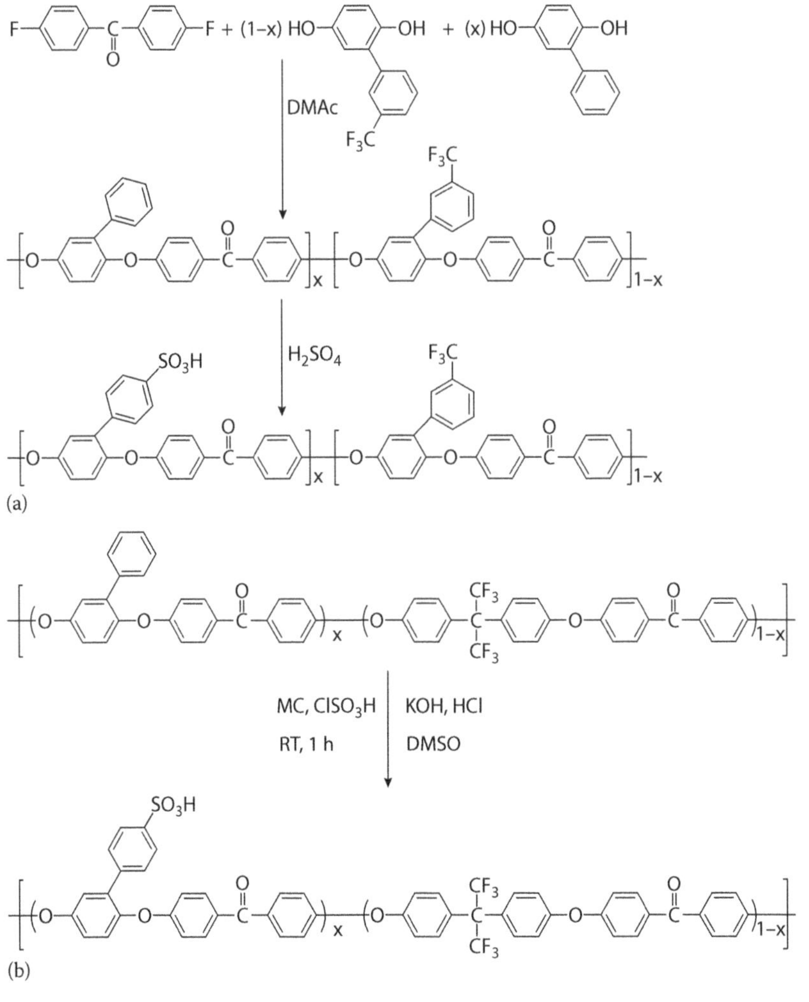

FIGURE 5.4 Selective sulfonation of PEK polymers (a) containing pendant phenyl rings by concentrated sulfuric acid or (b) chlorosulfonic acid. *(Continued)*

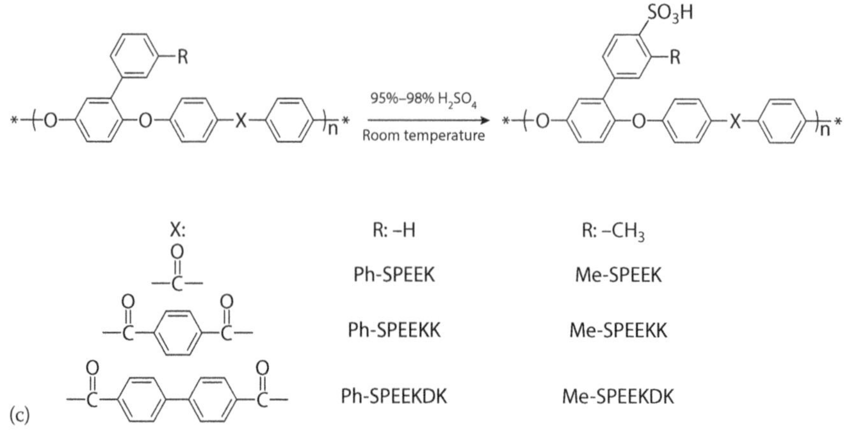

FIGURE 5.4 (Continued) Selective sulfonation of PEK polymers (c) containing pendant 4-methylphenyl rings.

Besides these homopolymers and random copolymers, postsulfonation could be applied to PEK copolymers with special polymer architectures to achieve a more distinct hydrophilic–hydrophobic microphase separation structure. Highly phase-separated polymers can form well-defined nanosized channels for proton conducting, thus resulting in less dependence of humidity and temperature. In addition, an overall promoted performance such as reduced water uptake and better dimensional stability can be achieved compared with random copolymer. Ueda and coworkers have designed star-shaped block copolymers as shown in Figure 5.5. Star-shaped sulfonated block polymers were synthesized via a Friedel–Crafts reaction.[44] Subsequently, concentrated sulfuric acid was applied to treat the polymer. Sulfonated sites were attached directly to the backbone, and all other hydrophobic blocks were placed at the end of each chain. The resulting block copolymer had good solubility in common polar aprotic solvents. After recasting the polymer, the membranes were tough and flexible. TEM test showed the star-shaped block polymers have more clearly hydrophilic–hydrophobic phase separation.

The grafting of sulfonation groups onto polymer side chains is another efficient way to form nanochannel in PEMs. Currently, several research groups are focusing their work on the development of side-chain-type sulfonated hydrocarbon-based polymers by grafting. Na and coworkers prepared a new monomer 1,5-*bis*(4-fluorobenzoyl)-2,6-dimethoxynaphthalene (DMNF) and corresponding naphthalene-based PAEK copolymers containing methoxy groups (Figure 5.6a).[45] The methoxy groups were converted to reactive hydroxyl groups, which reacted with 1,4-butane sultone to obtain sulfonated polymers bearing aliphatic sulfonic acid groups. The copolymers possessed high molecular weights, revealed by their high viscosities and formation of tough and flexible membranes. TEM images revealed that sulfonate groups aggregated into hydrophilic clusters formed a continuous network at a DS of 0.8 (SNPAEK-80). The same group then developed another two types of bisphenol monomers containing one or two methoxy groups (Figure 5.6b,c).[46,47] According

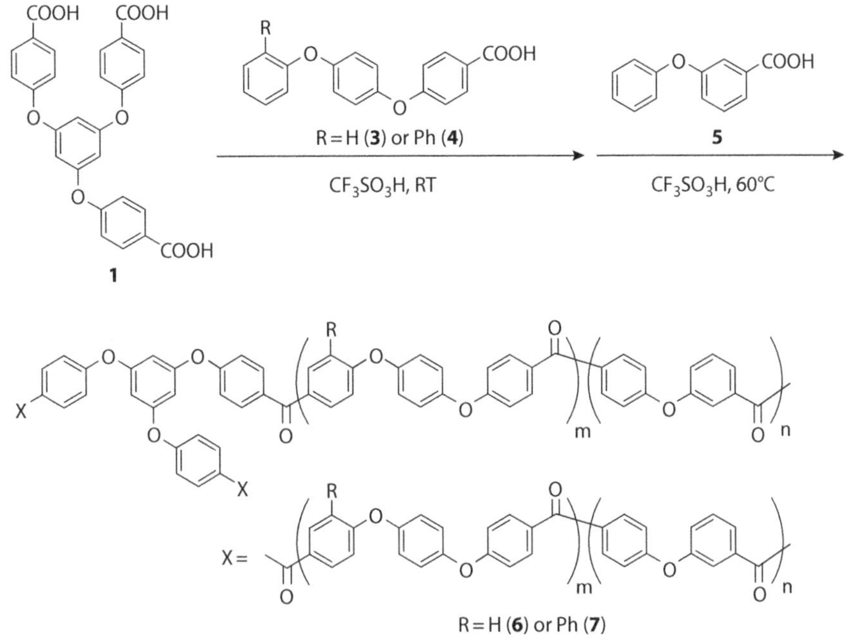

FIGURE 5.5 Synthetic route of star-shaped sulfonated block copoly(ether ketone)s.

to the similar procedure, SPAEKs bearing pendant aliphatic sulfonic acid groups were also prepared as proton-conductive materials. Compared with main-chain-type SPAEK membranes, these side-chain-type SPAEK membranes exhibited relatively low water uptake, low swelling ratio, but advantageous proton conductivity.

Compared with the aliphatic sulfonic acid groups, perfluoroalkylsulfonic acids were much stronger acids with pKa value of ~−6 and thus tend to be better proton conductors. Grafting of superacidic perfluoroalkylsulfonic acids onto the polymer side chains was proposed to increase the ionic conductivity of PEMs at a wide humidity range. Miyatake and coworkers synthesized a series of iodo-containing poly(arylene ether)s by the polymerization of 2,7-diiodo-9,9-*bis*(4-hydroxyphenyl) fluorene with difluorinated compounds, such as *bis*(4-fluorophenyl)ketone, under nucleophilic substitution conditions (Figure 5.7).[48] The iodo groups on the fluorenyl groups were converted to perfluorosulfonic acid groups via the Cu-catalyzed Ullmann coupling reaction. The superacid-modified ionomers showed thermal and gas permeation properties similar to those of the conventional sulfonated aromatic ionomers. In contrast, PEKs containing superacid groups showed well-developed, interconnected ionic clusters and thus high proton conductivity compared to the conventional aromatic ionomer membranes.

5.2.2.2 Direct Copolymerization of Sulfonated Monomers

A novel approach alternative to postsulfonation has been developed by several research groups—the copolymerization of sulfonated monomers. Direct copolymerization offers several advantages over the postsulfonation reactions, including ease

of control over the DS by varying the molar ratio of sulfonated monomer to nonsulfonated monomers, degradation-free preparation, and the lack of cross-linking reactions.[49] Furthermore, postsulfonation technique places only one sulfonic acid moiety
per repeat unit on the activated phenyl ring, and direct sulfonation of disulfonated
monomers can easily and reproducibly place up to two sulfonic acid moieties per
repeat unit on the deactivated phenyl ring. This direct-sulfonation method produces
polymer architectures with sulfonic acid moieties that are less susceptible to hydrolyzation because of the difficulty in generating the required carbocation intermediate
and thus resulting in a thermally more stable polymer chain.

Wang et al. initially made a significant progress in preparing wholly aromatic
SPAEKs by direct copolymerization of sodium 5,5′-carbonylbis(2-fluorobenzene
sulfonate) and several bisphenol monomers.[29] The measured thermal stability was
high for these polymers in air and nitrogen. However, no proton conductivity values
were reported since only the salt form was examined and it was not the interest of
the researchers. Gil et al. used tetramethylbiphenol to polymerize more thermally
stable SPEEK via direct copolymerization as competitive candidates for PEMFCs.[22]
Compared with Nafion membrane, these directly synthesized SPEEK membranes
show comparable water swelling, proton conductivity, and thermal stability, but
improved IEC and significantly reduced methanol crossover. The properties of the
direct-synthesized SPEEK membranes are greatly improved than those of the postsulfonated PEEK membranes.

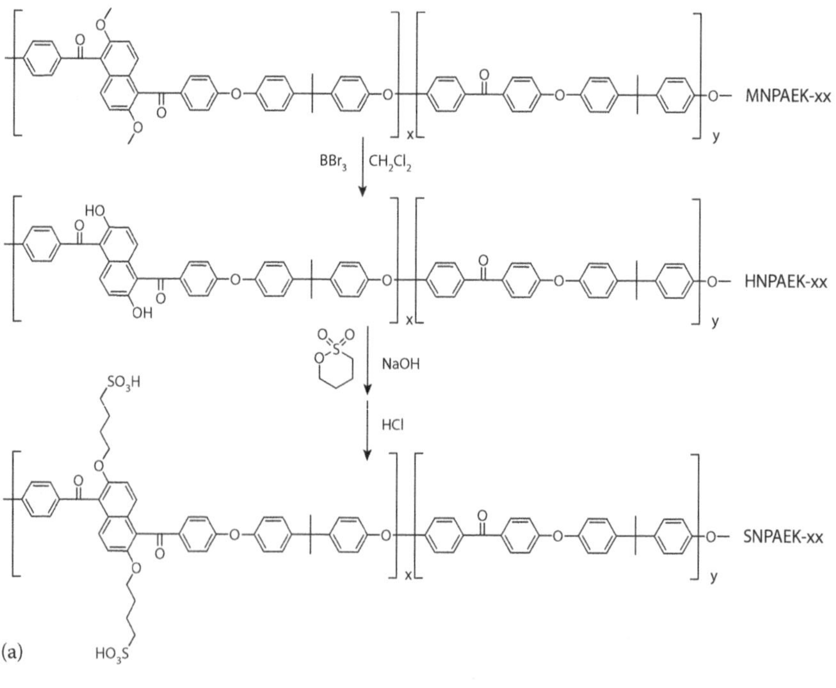

FIGURE 5.6   The synthetic route for grafting of sulfonation groups onto PAEK side chains.
(Continued)

Na and coworkers reported the preparation of another two difluoro monomers, 1,4-bi(3-sodium sulfonate-4-fluorobenzoyl) benzene and 1,3-*bis*(3-sodium sulfonate-4-fluorobenzoyl)benzene.[10,28] A large number of SPAEKs with different structures were then prepared via aromatic nucleophilic substitution from these sulfonated monomers, nonsulfonated dihalo monomers, and bisphenol monomers in the presence of potassium carbonate in polar solvents.[9–11,28–31] Various types of these sulfonated polymers are illustrated in Figure 5.8. Compared to the postsulfonation method, the DS of SPAEKs by direct polymerization can be controlled very easily and precisely, which enables us to finely tune the properties of membranes and maximizes the overall performance of membranes for applications in PEMFCs and DMFCs.

It has been proved by many research groups that SPAEKs are promising for PEM applications, owing to their good thermal stability and appropriate conductivity at a high DS. However, a number of investigations have demonstrated that water uptake increased with the DS increasing.[49,50] Too much water uptake resulted in the loss of mechanical strength and higher methanol permeability. Cross-linking, including covalent cross-linking and ionic cross-linking, is thought to be an effective approach to reduce water swelling and methanol crossover, as well as an effective

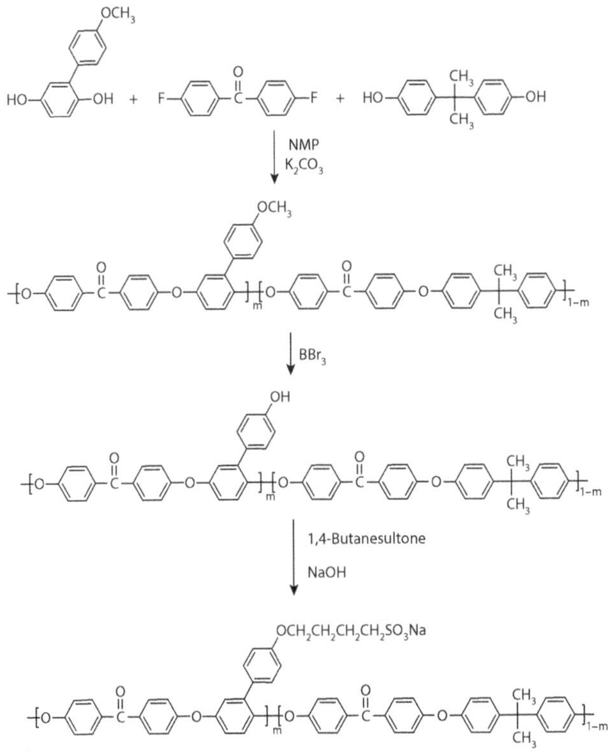

(b)

FIGURE 5.6 (Continued) The synthetic route for grafting of sulfonation groups onto PAEK side chains.   (*Continued*)

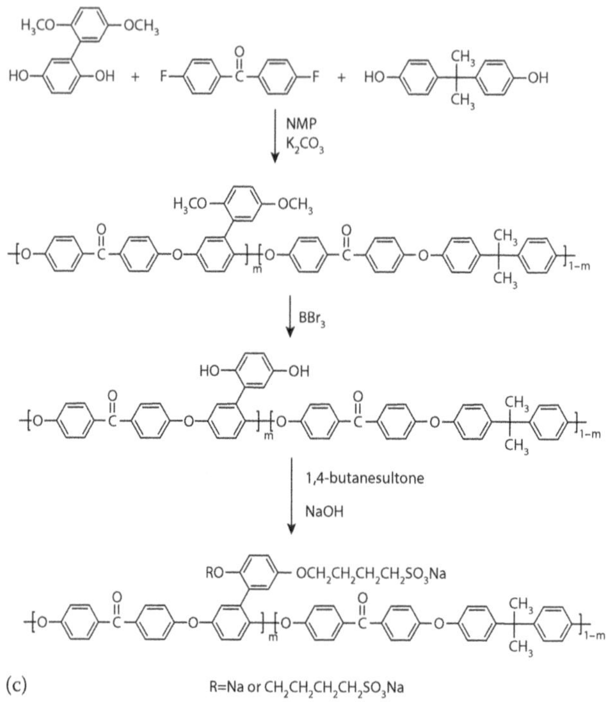

FIGURE 5.6 (Continued) The synthetic route for grafting of sulfonation groups onto PAEK side chains.

method to enhance mechanical strength. Zhong et al. prepared a series of cross-linkable SPEEKs containing propenyl groups by the direct copolymerization of sulfonated monomers and diallyl bisphenol A (Figure 5.9).[51–53] The photochemical cross-linking of the SPEEK membranes was carried out by dissolving benzophenone and triethylamine photoinitiator system in the membrane casting solution and then exposing the resulting membranes after solvent evaporation to UV light. The membrane performance can be controlled by adjusting the photoirradiation time. The experimental results showed that the cross-linked SPEEK (XSPEEK) membranes with photoirradiation of 10 min had the optimum performance for PEMs. Compared with the non-XSPEEK membranes, the XSPEEK membranes with photoirradiation of 10 min markedly improved thermal stabilities and mechanical properties as well as hydrolytic and oxidative stabilities and greatly reduced water uptake and methanol crossover with only slight sacrifice in proton conductivity. They were particularly promising as PEMs for DMFC applications.

Na and coworkers developed a self-cross-linked SPAEK bearing carboxylic acid groups.[54] Cross-linked membranes were obtained by the Friedel–Crafts reaction at 160°C between carboxylic acid groups and nucleophilic phenyl rings and the cross-linked mechanism is shown in Figure 5.10. The sulfonic acid groups in the copolymer act as a benign solid catalyst, serving as not only proton transport facilitators but also as acid sources for the Friedel–Crafts reaction with the aid of vacuum for

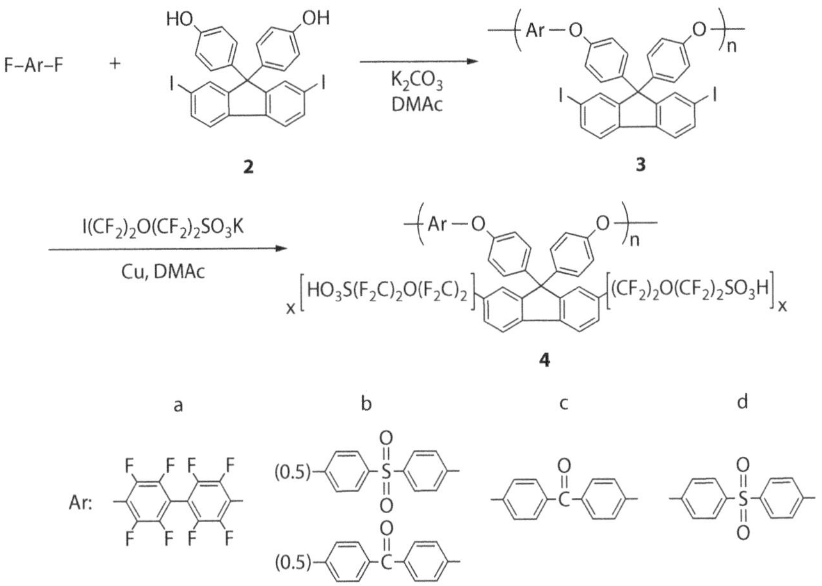

FIGURE 5.7 Synthesis of poly(arylene ether)s containing superacid groups derived from different difluorinated compounds (a–d).

dehydration. This cross-linked method is very simple; sulfonic acid groups are not involved in the cross-linking reaction, while other cross-linkers are not introduced into the system, so that the cross-linked membrane can achieve reasonable proton conductivity without decreasing IEC. They also reported the synthesis and characterization of a series of novel SPAEKs containing simultaneously sulfonic groups and benzimidazole groups.[55–57] The introduction of basic benzimidazole group is expected to induce a strong interchain interaction between the basic amino groups and the sulfonic acid groups, which is also called *ionic cross-linking*. The achieved results suggested that ionic cross-linked membranes possessed improved thermal and oxidative stabilities, lower water uptake, and improved methanol-resistant properties.

Side-chain-type sulfonated polymers are getting much attention due to their distinctly separated nanophases. Pang et al. synthesized a new sulfonated monomer with sulfonic acid groups on flexible aliphatic chains and prepared the novel sulfonated poly(arylene ether)s with pendant sulfoalkyl groups by a direct copolymerization method.[58] The sulfonated monomer was prepared by an anhydrous aluminum chloride-catalyzed Friedel–Crafts acylation of 1-bromo-3-phenylpropane with 2,6-difluorobenzoyl chloride and subsequently sulfonated with Na_2SO_3 (Figure 5.11). The generated membranes exhibit high proton conductivities and a very low swelling ratio at high temperature.

Besides the use of dihalo sulfonated monomers, the Guiver group introduced several commercial sulfonated bisphenol monomers to prepare SPEK polymers.[59] For example, sodium 6,7-dihydroxy-2-naphthalenesulfonate (DHNS) was purchased from Rintech, Inc. and recrystallized before use. The sulfonated poly(arylene ether ether ketone ketone) copolymer containing pendant sulfonic acid groups was

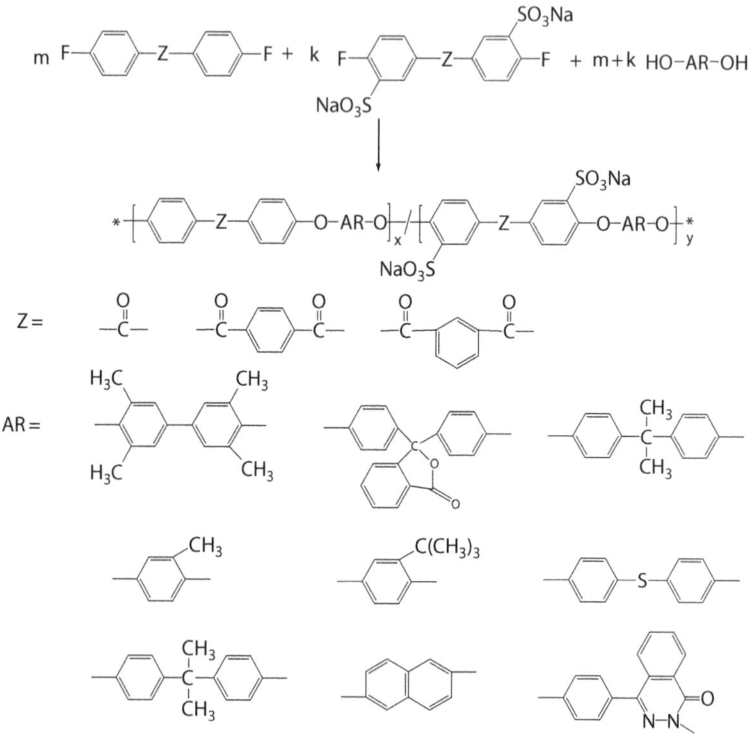

FIGURE 5.8 The structures of the SPEK families achieved by directly polymerization of sulfonated monomers.

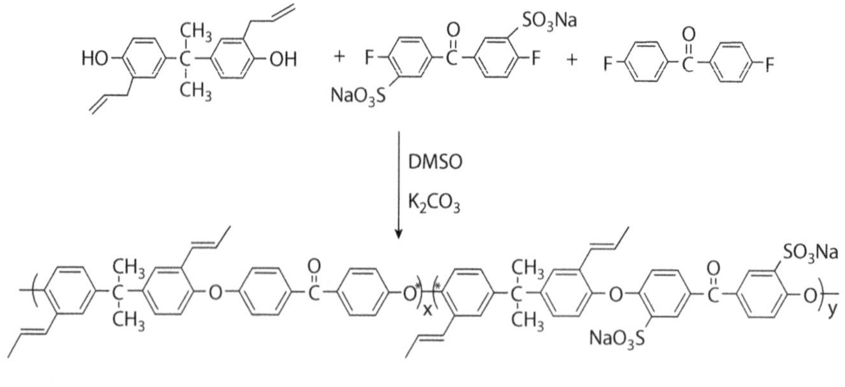

FIGURE 5.9 The synthesis of a UV-cross-linkable SPEEK.

synthesized from DHNS and other commercially available monomers (Figure 5.12a). Wang et al. described the synthesis and characterization of SPAEKs prepared from sulfonated hydroquinone, 4,4′-difluorobenzophenone, and (co)bisphenols (Figure 5.12b).[60] Sulfonated hydroquinone was synthesized by sulfonation of hydroquinone with concentrated sulfuric acid at room temperature, followed by neutralization.

5.2.3 MEMBRANE FORMATION

Organic polar solvents, such as N-methyl-2-pyrrolidinon (NMP), N,N-dimethylformamide (DMF), and dimethyl sulfoxide (DMSO), were used to form SPEK polymer solutions.[61,62] The concentration of polymer was set to 20 wt.% in the casting solution. At room temperature, this viscous solution was cast on a glass plate with a casting knife. The membrane was dried in a convection oven; the hot air was streaming tangential from the side. The temperature of the oven was programmed to remain at 40°C for 4 h, subsequently increased to 60°C and maintained at this

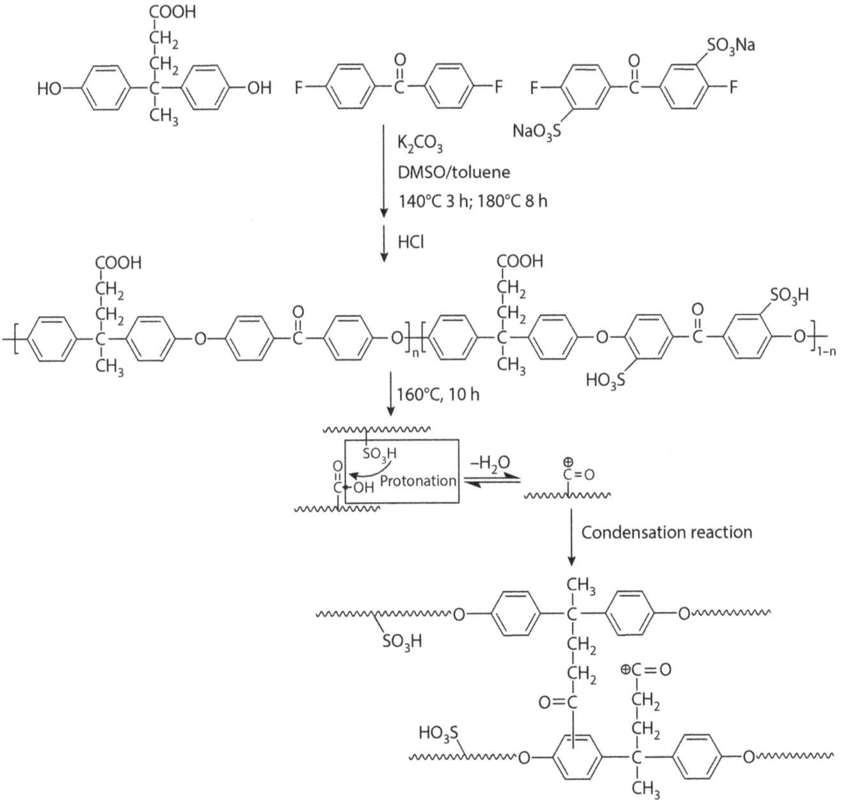

(a)

FIGURE 5.10   The schematic representation of the fabrication of (a) covalent cross-linked.

(*Continued*)

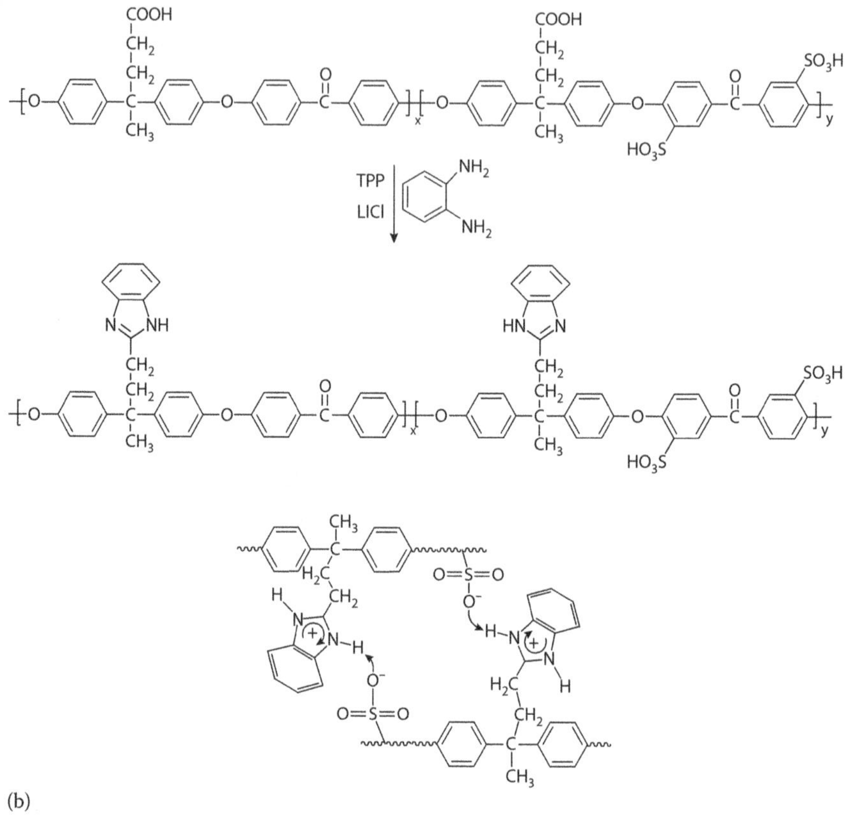

(b)

FIGURE 5.10 (Continued) The schematic representation of the fabrication of (b) ionic cross-linked sulfonated PAEK membranes.

temperature for 2 h, and finally the temperature was increased to 90°C and kept constant for another 4 h. After this treatment, the membranes can be handled, but still contain solvent. Therefore, the membrane was placed in a vacuum oven for more than 24 h at 100°C. The resulting membranes were removed from the glass plate and were acidified in 1 M HCl solution overnight, and the membranes in acidic form were then rinsed with deionized water to remove any excess acid.

5.3   MEMBRANE CHARACTERIZATION

The properties of the obtained SPEK membranes are needed to be investigated to determine whether the membranes satisfied the requirement of the special application. A wide range of analytic techniques could be employed to characterize membranes. Not only the structure of membranes but also the other parameters need to be measured, like water uptake and proton conductivity. This part offers an overview of various characterizations and detailed experimental conditions, and analysis can be found in Section 5.3.1 and 5.3.2.

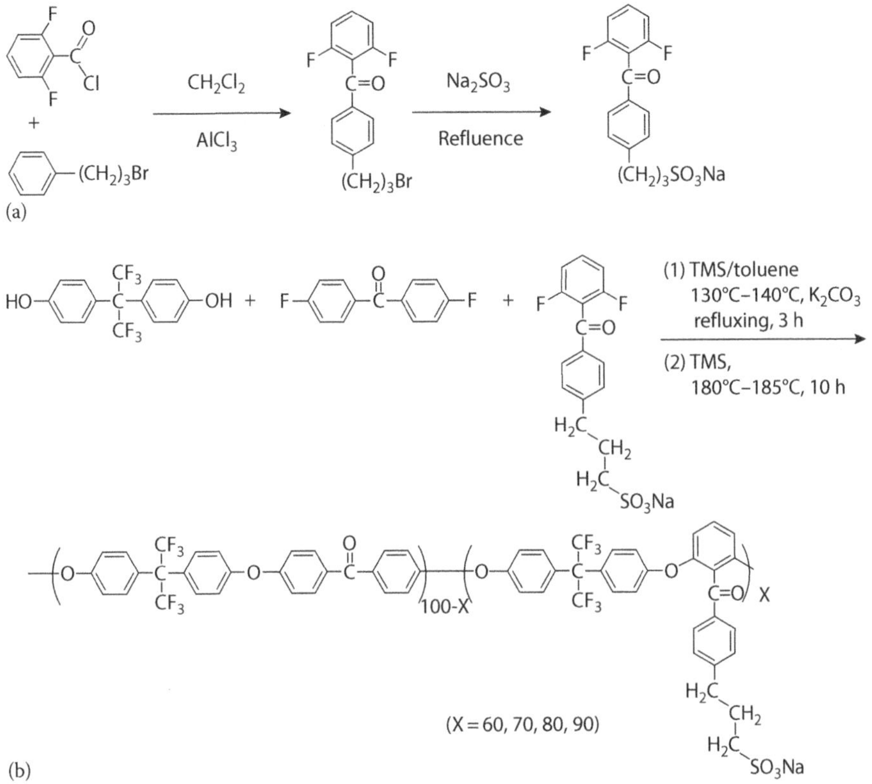

FIGURE 5.11 Synthesis of (a) sulfonated monomer and (b) sulfonated poly(arylene ether)s with pendant sulfoalkyl groups by direct copolymerization.

5.3.1 SPECTRAL STUDIES (IR, NMR, RAMAN, ETC.)

5.3.1.1 Fourier Transform Infrared

Fourier transform infrared (FTIR) spectroscopy was used to identify the presence of sulfonic acid groups in polymer samples. It was also used to observe the interactions between SPEEK polymer and solvent in the SPEEK membrane. Also, FTIR spectroscopy has been widely used to characterize functional groups in polymers and organic compounds. Membrane based on SPEK, which usually introduces new functional groups, can be readily monitored by FTIR. FTIR spectra were recorded by using either powder samples inside a diamond cell or KBr pellets composed of 50 mg of IR spectroscopic grade KBr and 1 mg polymer sample.[38]

The comparative FTIR spectra of PEEKK and SPEEKK are shown in Figure 5.13. The new absorptions at 1247, 1082, and 1027 cm^{-1} can be assigned to asymmetric and symmetric O=S=O stretching vibrations of sulfonic acid groups.[50] The absorption band at 690 cm^{-1} can be assigned to the S–O stretching of sulfonic acid groups. There are no related stretchings with sulfonic acid groups in nonsulfonated

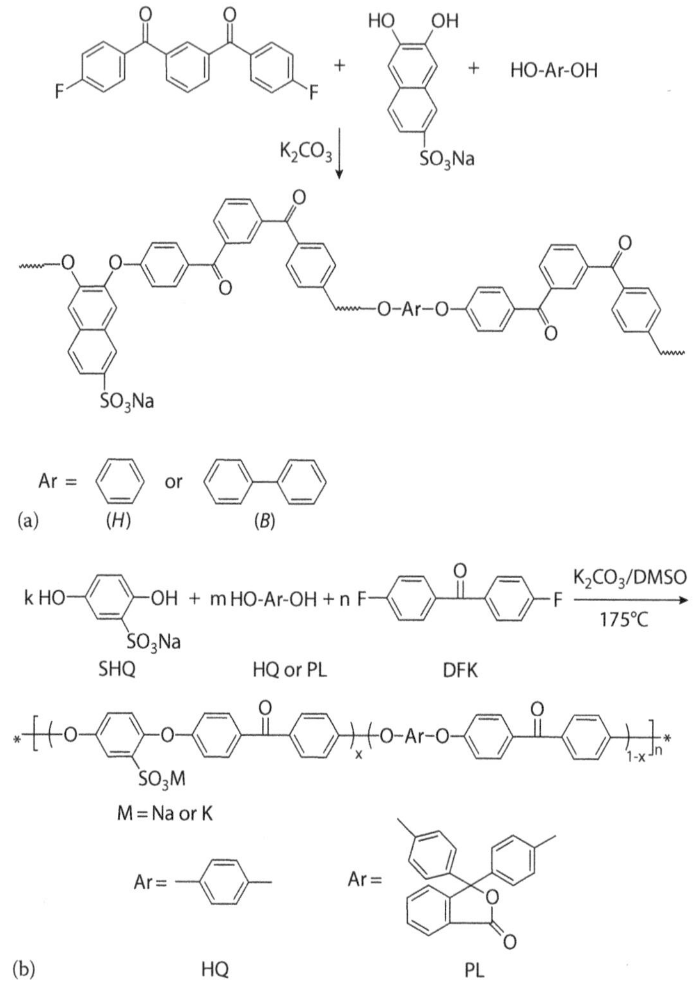

FIGURE 5.12 Synthesis of SPEKs by direct copolymerization of (a) sodium 6,7-dihydroxy-2-naphthalenesulfonate and (b) sulfonated hydroquinone.

PEEKK sample, which suggests the successful introduction of sulfonic acid groups in SPEEKK.

5.3.1.2 Nuclear Magnetic Resonance

The structure of SPEKs was usually determined by nuclear magnetic resonance (NMR) analysis.[38,63] For each analysis, ~3 wt.% polymer solution was prepared in deuterated dimethyl sulfoxide (DMSO-d_6) for ^{1}HNMR and ~15 wt.% for ^{13}CNMR. The chemical shift of tetramethylsilane was used as the internal standard reference. The typical ^{1}HNMR spectrum and its chemical shift assignment for SPEEK obtained by the postsulfonation method are also shown in Figure 5.14.

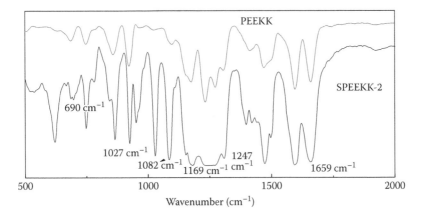

FIGURE 5.13   Comparative FTIR of PEEK and SPEEKK.

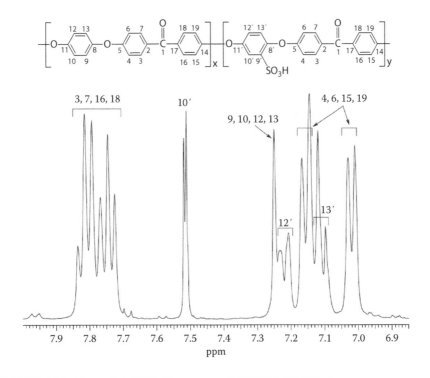

FIGURE 5.14   The typical ¹H NMR spectrum of SPEEK in DMSO-d_6.

5.3.2 Morphological Studies (SEM, TEM, AFM, Etc.)

5.3.2.1 Scanning Electron Microscopy

The morphologies of SPEK membranes are usually investigated using scanning electron microscopy (SEM). Specimens for the SEM were prepared by freezing the dry membrane samples in liquid nitrogen up to 10 min and breaking them to produce a cross section. Fresh cross-sectional cryogenic fractures of the membranes were vacuum sputtered with a thin layer of gold by using an ion sputtering before viewing on the SEM with a potential of 10 kV under magnifications ranging from 500× to 2000×.

5.3.2.2 Transmission Electron Microscopy

For transmission electron microscopy (TEM) observations, the membranes were stained with lead or silver ions by ion exchange of the sulfonic acid groups in 0.5 M lead acetate or silver nitrate aqueous solution, rinsed with deionized water, and dried in vacuum oven for 12 h. The stained membranes were embedded in epoxy resin, sectioned to 90 nm thickness with Leica Ultracut UCT microtome, and placed on copper grids for TEM.

It has been known that silver or lead ions can chelate with the sulfonate groups, resulting in silver or lead clusters within the SPEEK matrix.[22,49,50] Figure 5.15 shows the TEM images of the silver-chelated SPEEK, indicating the formation of silver particles dispersing within the SPEEK matrix. No definite conclusion can be made about the density of particles due to the different thickness of the TEM samples; however, the silver clusters, respectively, show approximate sizes of 5, 8, and 15 nm for the polymers with the DS of 0.44, 0.76, and 1.1. This indicates that a higher DS results in a larger hydrophilic domain and thus a larger proton transport channel.

5.3.2.3 Atomic Force Microscopy

Atomic force microscopy (AFM) has an important advantage in the study of surfaces in quantifying both surface morphology and surface interactions in a single instrument. Tapping-mode AFM images were obtained at ambient conditions. Both height and phase images were recorded simultaneously using the retrace signal. Etched Si tips with a resonance frequency of approximately 250–300 kHz and a

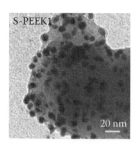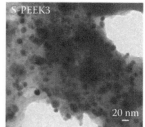

FIGURE 5.15 TEM micrographs of the SPEEK polymers with different DS.

FIGURE 5.16 AFM images of SPEEKK with different DS: (a, b, c) topography of SPEEKK; (d, e, f) the phase images of SPEEKK.

spring constant of about 2 N m^{-1} were used; the scan rate was in the range from 1.0 to 2.0 Hz. Each scan line contains 256 pixels, and a whole image is imposed of 256 scan lines. The membranes were obtained by spin coating a polymer solution in 0.05 g mL^{-1} DMF onto freshly cleaned silicon wafer at 3000 rpm for 50 s at room temperature.

Na and coworkers used AFM to investigate the phase structure of SPEEKKs with different DS.[64] The topography and phase images of SPEEKK membranes are shown in Figure 5.16. For the phase images of SPEEKK membranes, the dark regions were assigned to the softer regions, which represent the hydrophilic sulfonic acid groups. And the light regions were assigned to the hydrophobic polymer backbone. The size and continuity of these two regions play an important role on the transport properties of membranes. Larger and more continuous hydrophilic regions were found with DS of SPEEKK increasing from 0.8 to 1.2. The phase-separated structures may be beneficial to proton transport channel formation, which further lead to the increase of proton conductivity.

5.3.2.4 Small-Angle X-Ray Scattering

SAXS measurements were performed on a Philips PW-1710 x-ray diffractometer equipped with a Kratky small-angle x-ray camera, a step-scanning device, and a scintillation counter for recording the scattering intensity.[64] All measurements were made with a Cu Kα radiation source operating at 40 kV and 30 mA. The wavelength (λ) is 0.1542 nm. Hydrated membranes were analyzed in a solid sample holder at 25°C. The scattering vector q (nm⁻¹) was calculated according to the following equation:

$$q = \frac{4\pi \sin\theta}{\lambda} \qquad (5.1)$$

where q is the scattering angle. The characteristic separation length d (nm), that is, the Bragg spacing, was calculated using the following equation:

$$d = \frac{2\pi}{q} \qquad (5.2)$$

SAXS data for SPEEKK with different DS are shown in Figure 5.17. The maximum at large angle is the Bragg peak, which is derived from the ionic clusters. The scattering maximum is clearly related to the ionic groups of SPEEKK. With the content of sulfonic acid groups increasing, the scattering maximum shifted to lower scattering vectors. SPEEKK polymers with DS of 0.78, 0.97, and 1.23 showed the scattering maximum at 1.615, 1.542, and 1.368 nm⁻¹, respectively. The *center to center distances* calculated from *interparticle model* were 3.88, 4.07, and 4.59 nm, respectively. The result indicated that SPEEKK membranes exhibited the larger size of ionic clusters or more clearly phase-separated structures with the DS increasing.

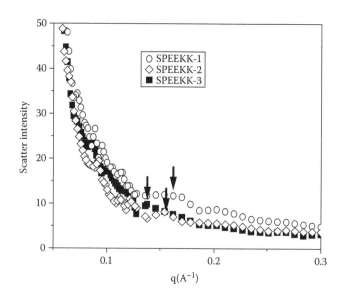

FIGURE 5.17 SAXS curves of the SPEEKK polymers with different DS.

5.3.3 PHYSICAL PROPERTIES (SOLUBILITY, VISCOSITY, THERMAL STABILITY, MECHANICAL PROPERTIES, ETC.)

5.3.3.1 Solubility

The solubility of SPEKs allows casting from organic solution and offers a more convenient and less expensive process. Sulfonation of the host polymer enhanced the hydrophilicity and solubility of the membranes. Most of the SPEEK polymers could be soluble in DMF, DMSO, and other polar aprotic solvents.

5.3.3.2 Viscosity

Intrinsic viscosities of polymers are among the most significant characteristic of SPEKs, and solution viscometer is one of the simplest methods to study the interactions and properties of polymer. By measuring the solution viscosity, we should be able to get an idea about molecular weight. The viscosities of SPEK polymers were usually measured in DMF at $25°C \pm 1°C$ with polymer concentration of 5 mg dL^{-1}. The viscosities of SPEK polymers indicate high molecular weight of resulting polymers.

5.3.3.3 Thermal Stability

The thermal stability of SPEK was usually studied by heating samples in a thermogravimetric analysis (TGA). TGA was used to determine weight loss. According to the procedure proposed by most of researchers, the samples were preheated under N_2 from room temperature to 100°C at 10°C min^{-1} to remove moisture, then cooled to 90°C, and then reheated from that temperature to 800°C at 10°C min^{-1}.

5.3.3.4 Mechanical Properties

The mechanical properties are usually studied by tensile testing using a universal tensile test machine at room temperature at a constant crosshead speed of 2 mm min^{-1}. The uniaxial tensile measurements were conducted on the membrane samples of 4 mm width and 15 mm length.[65] Prior to the measurements, the membrane samples were stabilized at ambient temperature and humidity, which is around 35% relative humidity.

5.3.4 ION-EXCHANGE CAPACITY, WATER UPTAKE, AND SWELLING RATIO

IEC is usually defined as the mmol of attached SO_3^{-1} sites per gram of polymer. The IEC values play a definitive role in determining the proton conductivity of membranes. IEC values were usually obtained by a titration method. Firstly, SPEK membranes in H^+ form were immersed in NaCl solutions for 24 h to exchange all the H^+ ions by Na^+ ions. Then, the H^+ ions in solution were then titrated with NaOH aqueous solution using a phenolphthalein indicator. IEC values of SPEK membranes could be calculated from the titration data via the following equation:

$$IEC \ (mequiv. \ g^{-1}) = \frac{consumed \ NaOH \times molarity \ NaOH}{weight \ dried \ membrane} \tag{5.3}$$

Water uptake and swelling ratio of membranes play very important roles in the properties of SPEK membranes. First, the proton conduction in SPEK membranes is largely influenced by the water content in the membranes, since proton conduction requires a significant amount of water to coordinate with a proton as it moves through the membrane. The proton conductivity usually increases as the water uptake increases. Ionic membranes are not good proton conductors in the dry state. However, they become conductive and functioning when hydrated. Meanwhile, the ionized sulfonic acid groups provide counterions to the protons for transport in the PEMs. So it is very important to select membranes with appropriate levels of sulfonic acid groups and water uptake. The sulfonic acid groups form hydrophilic hydrated domains contributing to the proton conductivity, and the nonsulfonated backbone forms a well-networked hydrophobic domain providing mechanical strength. Most of SPEKs exhibit hydrophobic/hydrophilic nanophase separation in the presence of water. The water uptake and swelling ratio were measured by the difference in the weight and length between the dry and corresponding water swollen membranes, respectively. The membranes were dried at 100°C for 48 h, then their weight and length were measured in the dry state. Then the membranes were soaked in distilled water at different temperatures for 48 h. After swelling completely, the samples were taken out and then wiped to quickly remove the excess water that adhered to the surface with a tissue paper. The weight and length in the wet state were measured. The water uptake of SPEK membrane was calculated from the following formula:

$$\text{Water uptake (WU)} = \frac{W_{wet} - W_{dry}}{W_{dry}} \times 100\% \qquad (5.4)$$

where W_{wet} and W_{dry} are the weights of the dry and corresponding water swollen membranes, respectively.

Also, the swelling ratio was calculated with the following equation:

$$\text{Swelling ratio (\%)} = \frac{l_w - l_d}{l_d} \times 100 \qquad (5.5)$$

where l_d and l_w are the lengths of the dry and corresponding water swollen membranes, respectively.

5.3.5 PROTON CONDUCTIVITY

Proton conductivity is a key performance property for PEMs. The lower the conductivity is, the higher the resistive losses are in the cell during operation. It can be measured by a four-probe ac impedance method from 1 Hz to 1 MHz using a potentiostat/galvanostat/frequency response analyzer. The membranes were fixed in a Teflon cell made of two outer gold wires to feed current to the sample and two inner gold wires to measure the voltage drops (Figure 5.18). The cell was placed in a thermocontrolled chamber in liquid water for measurement. Conductivity measurements under fully hydrated conditions were carried out with the cell immersed in liquid water. All samples were equilibrated in water for at least 24 h before

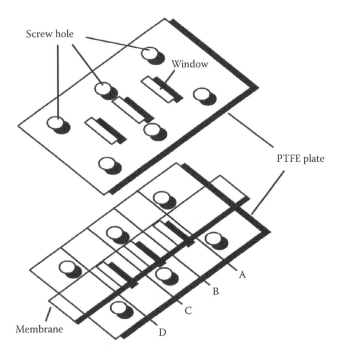

FIGURE 5.18 The test cell for in-plane proton conductivity measurements.

conductivity measurements. At a given temperature, the samples were equilibrated for at least 30 min before any measurement. Repeated measurements were taken at that given temperature with a 10 min interval until no more change in conductivity was observed. The proton conductivity of the membrane was calculated from the observed sample resistance from the equation

$$\sigma = \frac{L}{RS} \tag{5.6}$$

where
  σ is the proton conductivity (in S cm^{-1})
  L is the distance between the electrodes used to measure the potential
  R is the impedance of the membrane (in ohm), which was measured at the frequency that produced the minimum imaginary response
  S is the membrane section area (in cm^2)

To further test the influence of humidity of environment on conductivity of polymer membranes, membranes were placed in several closed chambers with different relative humidity levels generated by the different saturated salt solutions. These humidity chambers were made of glass having a height of 19 cm and diameter of 6 cm. The five different standard saturated aqueous salt solutions of MgCl$_2$ (~33% RH), Mg(NO$_3$)$_2$ (~53% RH), NaNO$_2$ (~65% RH), NaCl (~75% RH), and K$_2$SO$_4$ (~97% RH) were used to act as humidity sources. The saturated salt solutions were placed in the chambers for three days to ensure that the air in the bottle reached equilibrium state.

5.3.6 CHEMICAL STABILITY

Extensive studies have been conducted on the chemical degradation of PEMs.[66–70] It is known that typical chemical degradation of a PEM originates from chemical attack by hydrogen peroxide radicals, resulting in breakage of the membrane's backbone and side-chain groups and subsequent loss of mechanical strength and proton conductivity, thus leading to an increase in resistance and declining cell performance. It is possible to simulate the degradation of PEMs using an accelerated test medium such as Fenton's reagent (3 wt.% hydrogen peroxide aqueous solution containing 4 ppm Fe^{2+}).[66,67] The ferrous ion catalyzes the decomposition of peroxide to peroxy and/or hydroxyl radicals that chemically attack the SPEK polymers.

The oxidative stability of SPEK membranes was investigated by immersing the membranes into Fenton's reagent at 80°C. The membranes were evaluated by visual observation and degradation of the membranes by weight loss after 90 min was recorded. The time that the membrane took to break into pieces after shaking was also recorded to evaluate the oxidative stability.

5.4 MEMBRANE CONDUCTION MECHANISM AND THEORETICAL MODELING

The proton conductivity plays an important role in determining fuel cell performance. Therefore, it is very necessary to investigate the membrane conduction mechanism. For the pristine SPEK membrane, the proton conduction is dependent on water. The hydrophobic domain (polymer backbone) provides the morphological stability and prevents the membrane from dissolving in water. The sulfonic acid functional groups aggregate to form hydrophilic domains that are hydrated in the presence of water. And the connected hydrophilic domain is responsible for the transport of protons and water.

5.4.1 MEMBRANE CONDUCTION MECHANISM MODELING

It is a very complex process for protons to conduct through the sulfonated membranes either Nafion or SPEK membranes. At present, proton conduction mechanism is still filled with controversies. There are two widely accepted proton conduction mechanisms: vehicle mechanism and Grotthuss mechanism. The vehicle mechanism could be described as that an excess proton can travel on top of a host molecule through the solvent. The Grotthuss mechanism is described as that proton jumps between water molecules and is followed by a local molecular rearrangement.[71] This rearrangement (most often a rotation), in turn, allows the next jump. In most cases, these two proton conduction mechanisms are not entirely separate from one another and occur simultaneously to some degree. There always are many factors determining the proton conduction mechanism, for example, the length of sulfonate groups' chain. It has also been found that the proton conduction mechanism is strongly related to water concentration. For example, Ochi et al. have analyzed the diffusion coefficients using NMR and electrical conductivity measurements.[72] The results showed that the dominant mode of proton conduction at low RH is via the vehicle mechanism, while for PEMs in a high humidity environment, protons are rapidly exchanged between hydrated proton exchange sites via the Grotthuss mechanism.

5.5 MEMBRANE APPLICATIONS IN ELECTROCHEMICAL DEVICES FOR ENERGY STORAGE AND CONVERSION

5.5.1 APPLICATIONS FOR PEM FUEL CELLS

In the last few years, PEMFC has gained more and more attention since it represents an alternative technology for power production in the field of both portable and stationary applications due to its ability to convert the chemical energy of a fuel (hydrogen, methanol, and ethanol, respectively) directly into electrical energy with relatively high efficiency as well as its adaptable size and relatively low operating temperature. Today, Nafion is recognized as the most used proton exchange membrane for PEMFC applications. It is based on a perfluorinated polymer, which shows good thermal stability and high proton conductivity as main benefits. On the contrary, Nafion is an expensive material and suffers high fuel crossover (particularly, methanol crossover in DMFC applications) besides the proton conductivity loss above 100°C. Therefore, in the last decades, many scientists paid special attention on the development of new materials based on nonfluorinated polymers as an alternative to Nafion.

Several studies pointed out SPEK membranes seem to constitute a real alternative to commercial Nafion. In some cases, it shows comparable (or superior) proton conductivity performances to Nafion as well as superiority in terms of thermochemical properties, lower fuel crossover, and, particularly, lower costs. Researchers have made various attempts to get high single-cell performance using SPEK as proton exchange membranes. Here, we give a few examples about the application of SPEK for PEMFCs and DMFCs in the open literature, including new types of SPEK membranes, hybrid/blend membranes, modification of SPEKs, and incorporation of inorganic materials into SPEK-based composited membranes. The conditions of MEA and single-cell performance test are also described in this chapter and summarized in Table 5.2.

Liu et al. synthesized fully aromatic poly(ether ketone)s based on the monomer p-biphenyl-hydroquinone.[73] Then they prepared site-selective sulfonated biphenylated poly(ether ether ketone)s (BiPh-SPEEKDKs) by mild and rapid sulfonation reactions. MEA fabrication and fuel cell operating conditions were conducted according to Los Alamos National Laboratory protocols. The cell was preconditioned under H_2/air fuel cell operation at 0.7 V for 2 h. Pt–Ru black was used for anode and Pt black for cathode and 5% commercially available Nafion dispersion. The catalyst ink was painted onto the membrane until the catalyst loading reached 8 mg cm^{-2} for anode and 6 mg cm^{-2} for cathode. Active area is 5 cm^2. The cell was held at 80°C; 1 and 2 M aqueous methanol solution was fed to the anode with a flow rate of 1.8 mL min^{-1}; 90°C humidified air was fed at 500 sccm without back pressure. The current density of BiPh-SPEEKDK at 0.5 V at 80°C in 1 and 2 M methanol reached

TABLE 5.2

Conditions of MEA and Single-Cell Performance Test of SPEK Membranes

Structure and Modified Method	MEA Conditions	Cathode		Anode		Temp. (°C)	Cell Performances
		Cat.	Fuel	Cat.	Fuel		
Sulfonated biphenylated poly(aryl ether ketone)s		Pt black 6 mg cm⁻²	Humidified air 500 sccm	Pt–Ru 8 mg cm⁻²	2 M methanol 1.8 mL min⁻¹	80	Current density at 0.5 V: 117 mA cm⁻² OCV, 0.588 V Max power density: 26 mW cm⁻²
Nafion/aminated SPEEK		Pt/Ru 0.6 mg cm⁻²	Dry O₂ 0.1 L min⁻¹	Pt/Ru 1.2 mg cm⁻²	2 M methanol 2 mL min⁻¹		
Grafted porous PTFE/partially fluorinated SPEEK composite membrane	Hot-pressing 160°C 1 MPa 2 min	Pt 0.7 mg cm⁻²	Humidified O₂ 0.2 MPa	Pt 0.3 mg cm⁻²	Humidified H₂ 0.2 MPa	80	The Ohmic polarization zone not significantly vary during 40 h test
SPEEK-sulfonated GO composite membranes	Hot-pressed 125°C	Pt/C 0.38 mg cm⁻²	Dry O₂ 90 mL min⁻¹	Pt/C 0.38 mg cm⁻²	30% RH H₂ 0.1 L min⁻¹	80	OCV, 0.95 V 378 mW cm⁻²
Imidazole microcapsules/SPEEK	60 kg cm⁻² 3 min	0.38 mg cm⁻²	Humidified O₂	0.38 mg cm⁻²	Humidified H₂	60	Max power density: 104 mW cm⁻²
SPEEK/SDBS-GO composite membrane	Hot-pressing 140°C 50 kg cm⁻² 3 min	Pt/C 2.5 mg cm⁻²	Humidified O₂ 0.2 L min⁻¹	Pt–Ru/C 2.5 mg cm⁻²	1 M methanol 1 mL cm⁻²	65	Max power density: 62 mW cm⁻²
Sulfonated cross-linker Cross-linked PAEK	Hot-pressing 120°C 6.9 MPa 3 min	Pt 1.3 mg cm⁻²	O₂ 0.2 L min⁻¹	Pt/Ru 1.3 mg cm⁻²	Methanol 2 M 5 mL min⁻¹	80	Max power density: 74 mW cm⁻²
SPEEK membrane embedded by dopamine-modified nanotubes	Hot-pressing 140°C 4.0 MPa 3 min	Pt 0.25 mg cm⁻²	Humidified O₂ 0.2 L min⁻¹	Pt 0.1 mg cm⁻²	Humidified H₂ 0.15 L min⁻¹	85	OCV, 0.95 V Max power density: 110.7 mW cm⁻²
Functionalized graphene oxide nanocomposite SPEEK membrane		Pt/C 0.2 mg Pt cm⁻²	O₂ 0.5 L min⁻¹	Pt/C 0.1 mg Pt cm⁻²	H₂ 0.2 L min⁻¹	120	Max power density: 150 mW cm⁻²
Pendant SPAEK membranes cross-linked with proton-conducting reagent	Hot-pressing 100°C 1000 psi 1 min	Pt 1.0 mg cm⁻²	O₂ 0.2 L min⁻¹	Pt/Ru 1.0 mg cm⁻²	2 M methanol 5 mL min⁻¹	80	Max power density: 71 mW cm⁻²

195 and 117 mA cm^{-2}, respectively (Figure 5.19). The DMFC performance of the MEA based on BiPh-SPEEKDK was comparable to or even better than Nafion® 112 and 117. The improved performance of BiPh-SPEEKDK could be explained by a combination of lower high-frequency resistance (95 mΩ cm^2 at 2 M) and lower methanol crossover limiting current (328 mA cm^{-2} at 2 M). Under the same test conditions, Nafion 112 had higher resistance (106 mΩ cm^2 at 2 M) and higher methanol crossover limiting current (527 mA cm^{-2} at 2 M).

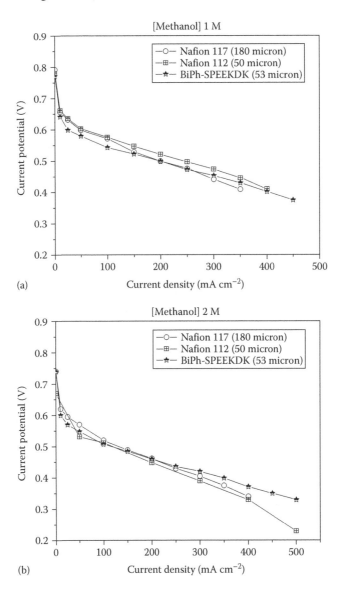

FIGURE 5.19 DMFC performances of BiPh-SPEEKDK and Nafion membranes: (a) methanol feed concentration at 1 M and (b) methanol feed concentration at 2 M.

Lin and Tsai synthesized a series of novel main-chain-type and side-chain-type sulfonated poly(ether ether ketone)s (MS-SPEEKs) with different side-chain ratios by reacting the sulfonic acid groups of pristine SPEEKs with 2-aminoethanesulfonic acid to improve the nanophase-separated morphology of the material.[74] In the cell performance test, the anode and cathode catalysts were applied to carbon paper by brushing. The anode and cathode consisted of commercial Pt–Ru and Pt with a loading of 3 mg cm^{-2}. The anode was supplied with 2 M methanol at a flow rate of 2 mL min^{-1}, and the cathode was supplied with O_2 at a rate of 100 mL min^{-1}. Single-cell performance was evaluated using a DMFC unit with a cross-sectional area of 4 cm^2. The cell temperature was kept at 60°C and 70°C. Single cells prepared with MS-SPEEK membranes at 60°C exhibited high open-circuit voltages (OCVs) of 0.77–0.81 V. At 70°C, the OCV was 0.75–0.82 V, which were all higher than that of Nafion®, 117 (0.65 V). The highest power density obtained in the experiment was 94–142 mW cm^{-2} at 60°C–70°C, which was greater than that of Nafion 117.

The same group also prepared blend membranes consisting of Nafion and aminated SPEEK, which combined the sulfonic groups and amino groups in the same polymeric chain.[75] In DMFC applications, the anode and cathode consisted of commercial 20 wt.% Pt/Ru (1:1) in Vulcan carbon (E-TEK) with a Pt loading of 1.2 and 0.6 mg cm^{-2}, respectively. Methanol (2 M) was supplied to the anode by a micropump at 2 mL min^{-1}, while the cathode was supplied with dry oxygen at a rate of 100 mL min^{-1}. The results showed that the single cells of all of the blend membranes have exhibited higher OCV (0.579–0.588 V) than that of Nafion® 115 (OCV 0.571 V). The blend membrane with a content of 1% aminated SPEEK had the highest power density (26 mW cm^{-2}), compared with that of the other blended membranes and Nafion 115 (22 mW cm^{-2}).

Bi et al. fabricated a low-cost and mechanically stabilized composite membrane based on grafted porous polytetrafluoroethylene (grPTFE) substrate and partially fluorinated sulfonated poly(arylene ether ketone) (FSPAEK) polymer for PEMFC applications.[76] The MEA was fabricated by hot-pressing the anode, the membrane, and the cathode at 160°C and 1 MPa for 2 min. The Pt loadings in the anode and the cathode were 0.3 and 0.7 mg cm^{-2}, respectively. Nafion solution was only used as an electrode bonding material. The Nafion loading in both the anode and the cathode was 0.4 mg cm^{-2}. The active area of MEAs was 5 cm^2. PEMFC single cell was operated at 80°C and 0.2 MPa with fully humidified H_2/O_2. The single cell was operated at 1000 mA cm^{-2} during the day (10 h) and shut down during the night. The results showed that the OCV of FSPAEK/grPTFE composite membrane did not significantly drop, which showed H_2/O_2 crossover of FSPAEK/grPTFE composite membrane does not remarkably change.

Zhou et al. synthesized a kind of SPAEK containing free carboxylic group.[77] Then the free carboxylic group was activated with N-hydroxysuccinimide and reacted with disulfonic acid–benzidine diamines, a functional cross-linking reagent containing both sulfonic acid group and amine group, to prepare a series of cross-linked SPAEK (CSPAEK) membranes. The condition of the single cell test was as follows: The electrodes were loaded with 1.3 mg cm^{-2} of the catalyst using the Pt/Ru alloy as the anode and Pt as the cathode. A solution of the 5 wt.% Nafion ionomer was overlaid on the catalyst layer as the binder to the membrane. The MEA was prepared through hot-pressing at 120°C and 6.9 MPa for 3 min. A 2 M methanol solution was fed to the anode at a flow rate of 5 mL min⁻¹, and oxygen was fed to the cathode at a flow rate of 200 mL min⁻¹. The single cell with an active electrode area of 6.25 cm² was tested at 80°C. The results showed that all the CSPAEK membranes had exhibited higher power density at a constant current density than Nafion 117 except CSPAEK-10. The CSPAEK-20% cell showed the highest cell performance with a power density of 74 mW cm⁻², and that of Nafion 117 was 65 mW cm⁻² under the same test conditions.

Kim et al. also synthesized PAEK bearing a pendant sulfonic acid group by reacting the activated carboxylic group with 2-aminoethanesulfonic acid or with the cross-linker disulfonic acid–benzidine diamines.[78] The MEAs were fabricated for the cell performance test. CSPAEK membrane was inserted between the two electrodes, Pt/Ru (1/1) anode and Pt cathode, at the amount of 1.0 mg cm⁻² each. Nafion solution (5 wt.%) was used to bind the membrane and catalyst. MEA was prepared under application of high pressure (1000 psi) for 1 min at 100°C. The cell test was conducted at 80°C. Methanol (2 M) was introduced at the anode side at 5 mL min⁻¹ and the oxygen to the cathode side at 200 mL min⁻¹. The cell area was 6.25 cm². The maximum power density of Nafion 117 was 66.5 mW cm⁻², whereas the 20% cross-linked membrane showed better performance, 71 mW cm⁻². All the membranes' OCVs equal to that of Nafion 117.

Li et al. prepared a kind of acid–base PEM membrane by introducing polypyrrole (Ppy) into sulfonated poly(ether ether ketone) (SPEEK) membranes by polymerization in SPEEK solutions, thus improving their methanol resistance.[79] Cell performance was evaluated by using a DMFC unit cell, and the catalysts used at the anode and the cathode were Pt–Ru/C and Pt/C, respectively. The catalyst loading was 2.5 mg cm⁻² at both the anode and the cathode. A 2 M methanol solution was fed to the anode at a flow rate of 82 mL min⁻¹, while the cathode was fed with dry O₂ at a rate of 2000 mL min⁻¹ via a flow meter. The cell temperature was kept at 70°C. The cell performance of the SPEEK membrane was a little better than its composite membranes with a power of 115 mW. The highest power of the composite membranes was about 95 mW. But after the fuel cell performance experiment, SPEEK membranes show much more serious swelling than the composite membranes, despite the fact that they have a little better performance.

Inspired by the bioadhesion principle, Zhang et al. synthesized dopamine-modified halloysite nanotubes (DHNTs) bearing –NH₂ and –NH– groups by directly immersing natural halloysite nanotubes into dopamine aqueous solution under mild conditions.[80] DHNTs were then embedded into SPEEK matrix to prepare hybrid membranes. The MEAs were prepared by hot-pressing method at 140°C and 4.0 MPa for 3 min with an active area of 4.0 cm² (2.0 cm × 2.0 cm). The Pt catalyst loadings were 0.10 and 0.25 mg cm⁻² for anode and cathode, respectively. The single cell was operated at 65°C and 85°C with humidity H₂/O₂, and the humidification temperatures were 65°C and 85°C, respectively. The flux rates of H₂ and O₂ were 150 and 200 mL min⁻¹, respectively. Prior to the measurement, the single cell was kept under the operation conditions for 4 h. For SPEEK control membrane, the OCV, maximum current density, and power density were 0.89 V, 263.2 mA cm⁻², and 52.6 mV cm⁻² at 65°C, respectively, while those of DHNT-filled hybrid membranes were 0.92 V,

368.3 mA cm^{-2}, and 79.8 mV cm^{-2}, which were much higher. As the temperature increased from 65°C to 85°C, the maximum current densities of SPEEK/DHNT-15 increased to 400.5 mA cm^{-2}. Meanwhile, the maximum power densities increased to 110.7 mV cm^{-2}.

Meenakshi et al. prepared a series of mixed matrix PEM membranes by blending SPEEK with methane sulfonic acid (MSA) and zeolite 4A for DMFC.[81] Sixty weight percentage of Pt–Ru (1:1 atomic ratio) and forty weight percentage of Pt catalyst supported on Vulcan XC-72R carbon mixed with binder were coated on the GDL to constitute the catalyst layer on the anode and the cathode, respectively. The catalyst loading was 2 mg cm^{-2}. The active area for the DMFC was 4 cm^2. MEAs were attained by sandwiching the respective membrane between the two electrodes followed by hot-pressing at 100°C for 3 min at a pressure of 1.96 MPa. The DMFC cells were tested at 70°C with 2 M methanol at a flow rate of 2 mL min^{-1} at the anode side and oxygen at a flow rate of 300 mL min^{-1} through the cathode side at atmospheric pressure. The peak power density of 130 mW cm^{-2} was achieved for MEAs comprising SPEEK–MSA (20 wt.%) blend, which was higher in comparison with the other percentages of MSA, pristine SPEEK, and Nafion 117. Performance was further enhanced, that is, peak power density of 159 mW cm^{-2} was observed for the mixed matrix membrane comprising zeolite 4A (4 wt.%)–SPEEK–MSA (20 wt.%).

Kumar et al. (2014) prepared SPEEK/sulfonated graphene oxide composite membranes for PEMFCs. Sulfonated graphene oxide nanosheets were incorporated into the SPEEK structure to interconnect the proton transport channels.[82] The MEA was hot-pressed at 125°C and a pressure of 5.88 MPa for 3 min. Both the anode and the cathode consist of 20% Pt/C with a Pt loading of 0.38 mg cm^{-2}. The active electrode area was 1 cm^2. H$_2$ and O$_2$ were fed to each side of the cell at a flow rate of 100 and 70 mL min^{-1}, respectively. The cell was conditioned at 0.3 V for 1 h before polarization studies. The OCV of both MEAs was greater than 0.95 V, indicating negligible gas crossover. SGO/SPEEK membrane showed a maximum power density of 378 mW cm^{-2} at 80°C and 30% RH, which is much higher than that of SPEEK (250 mW cm^{-2}).

5.5.2 Applications for Vanadium Redox Flow Batteries

Renewable energies like solar and wind power are receiving great attention under the driving forces of fossil energy shortage and environmental load. However, the intermittent and random nature of these renewable energies seriously affects the final quality of the output electricity as well as their stability in the grid. Electric energy storage becomes a valid solution to solve these problems. Combined with renewable energies, energy storage can increase the quality and the stability of photovoltaic and wind-generated electricity. As one kind of energy techniques, VRB has emerged as a promising alternative to existing large-scale power conversion systems due to its perfect combination of high energy efficiency, high safety, and reliability. VRB composes of two tanks filled with active species of vanadium ions in different valance states, two pumps, and a battery cell. The conversion between electricity and chemical energy is realized by reduction and oxidation of vanadium ions. Nafion is still the most commonly used membranes in VRB due to its superior proton conductivity and chemical stability. However, the low ion selectivity and extremely high cost have hindered the application in commercialization of VRB. Due to the tunable ion conductivity and low cost, SPEK polymers have been investigated in the VRB field.

Zhang and coworker prepared a Nafion/SPEEK layered composite membrane. The N/S membrane showed lower permeability of vanadium ions accompanied by higher area resistance compared with Nafion membrane.[83] The VRB single cell employing N/S membrane exhibited lower voltage efficiency and higher coulombic efficiency compared with that employing Nafion membrane. Although the overall energy efficiency of VRB single cell with N/S membrane was a little lower, its good chemical stability and low cost made it a promising membrane used in VRB system. After then, they used the direct copolymerized sulfonated poly(tetramethydiphenylether ether ketone) membranes with varied DS values for applications in VRB.[84] The introduction of rigid biphenyl groups was aimed to enhance the chemical and mechanical stabilities of membranes under VRB medium. The characterized physiochemical properties showed that SPEEK membranes possessed much higher selectivity in VRB system than Nafion 115. It was probably owing to less connected ionic cluster region in SPEEK membrane, which reduced the permeation rate of vanadium ions. In VRB single cell test, the SPEEK membranes exhibited comparative, even superior performances compared with Nafion 115 (Figure 5.20). In the 80-cycle charge–discharge test, the SPEEK40 membrane exhibited stable performance and its internal structure remained dense.

Recently, they fabricated the polytetrafluoroethylene (PTFE)-reinforced SPEEK composite membranes to improve their chemical stability and ion selectivity in VRB applications.[13] Compared to pristine SPEEK membranes, the composite membranes exhibited the lower swelling ratio and higher mechanical stability due to the PTFE reinforcement. As expected, the SPEEK/PTFE membranes showed much higher coulombic efficiency and energy efficiency than those of SPEEK membrane. In addition, the composite membranes exhibited much better stability than pristine SPEEK under VRB operating conditions.

Yan and coworkers fabricated SPEEK/polypropylene/perfluorosulfonic acid layered composite membrane and SPES/SPEEK membrane for VRB.[85] These membranes served as a shield preventing the oxidation degradation of the composite membrane by VO^{2+} ions in the positive half-cell electrolyte of the VRB, while the SPEEK layer blocked the vanadium ion permeation. The composite membranes displayed much lower vanadium ion permeability than that of Nafion® 212. In a VRB single cell, these composite membranes showed higher coulombic efficiency and energy efficiency and a lower self-discharge rate than those of Nafion 212.

In summary, there are urgent needs in developing high-quality and low-cost membranes that are believed to be vital to achieve cost-effective VRB systems. SPEK and its composite membranes showed much lower vanadium ion permeability and comparable performance of VRB single cell. However, these membranes suffered from low chemical stability in VRB medium, which are strongly acidic, oxidizing, and with high electric potential. Enhancing the oxidation stability of the membrane for VRB application is still the biggest challenge.

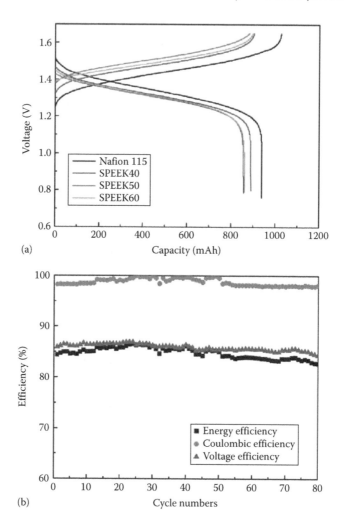

FIGURE 5.20 (a) Charge and discharge curves of SPEEK and Nafion 115 membranes in VRB operation (50 mA cm^{-2}); (b) charge–discharge cycling performance (60 mA cm^{-2}) of the cell with SPEEK40.

5.5.3 APPLICATIONS FOR WATER ELECTROLYSIS

Hydrogen is an efficient energy carrier and considered to be the best way to store the energy derived from the renewable and intermittent power sources. In particular, with the capacity of localized renewable energy sources surpassing one billion watts, a suitable system should prepare hydrogen with the utilization of those obtained energy. Solid polymer electrolyte (SPE) water electrolysis technique for hydrogen production has many advantages, such as environment friendly, high efficiency, low energy consumption, and good performance stability and security. The expensive price of SPE membrane electrode assembly is one of the most important barriers, which restrict its industrialization. And one part of the existing SPE membrane electrode assemblies is the costly perfluorosulfonic acid membrane. In order to reduce the cost of the SPE, SPEK polymers have been attracting more and more attention.

Kang and coworkers applied the composite PEMs consisting of covalently XSPEEK with tungstophosphoric acid (TPA) to the SPE water electrolysis applications.[86] Covalently cross-linking and blending treatment was proposed to achieve high proton conductivity without sacrificing mechanical stability in swollen state. The results exhibited that the membrane properties, including proton conductivity, tensile strength, swelling, dimensional stability, and antioxidative stability, were enhanced by cross-linking. Moreover, addition of TPA significantly improved the electrochemical and mechanical properties of the membrane. After assembling in MEA, their electrochemical properties and single-cell performance were investigated with regard to application in water electrolysis. The cell voltage during water electrolysis was 1.78 V at 1 A cm^{-2} at 80°C with a platinum loading of 1.28 mg cm^{-2}. The electrochemical surface area and roughness factor of the MEA prepared were 25.11 m^2 g^{-1} and 321.4 cm^2 Pt cm^{-2}, respectively.

The same group then blended SPEEK with sulfonated polysulfone and TPA to serve as an SPE for water electrolysis.[87] The blending was also aimed to avoid water swelling at elevated temperatures and maintain sufficient mechanical strength. Electrochemical and mechanical properties were improved with the addition of TPA agent, and the blend membrane exhibited good antioxidative stability. The cell voltages of MEA at 80°C can reach up to 1.83 and 1.90 V at 1 A cm^{-2} with platinum loadings of 1.12 and 1.01 mg cm^{-2}, respectively.

Wei et al. applied SPEEK/poly(ether sulfone) blend membranes to be used as an SPE for water electrolysis.[16] MEA was fabricated by incorporating Ir anode layer and Pt cathode layer with the membrane by a decal method. An electrolytic current of 1655 mA cm^{-2} was obtained at 2 V and 80°C with the SPEEK-based MEA and under suitable fabrication conditions. Although the result was still not as good as a Nafion-based MEA, it also suggested that SPEK membrane could be an alternative to Nafion in SPE water electrolysis for hydrogen production.

In summary, MEA fabricated based on SPEK was much cheaper than that based on Nafion for achieving a comparable performance in water electrolysis. Therefore, it would enable a more economically competitive hydrogen production by SPE water electrolysis. To accomplish this goal, however, more detailed work is needed, including MEA fabrication improvement, MEA active area scale-up, and stability experiments.

5.5.4 Applications for Supercapacitors

Supercapacitors are promising devices for delivering high power density. Growing demands in digital communications, electrical vehicle, and other devices that require electrical energy at a high power levels in short pulses have prompted the development of supercapacitors. Samui and coworkers reported the use of SPEEK membrane for the applications in supercapacitors.[17] Figure 5.21 shows the structure

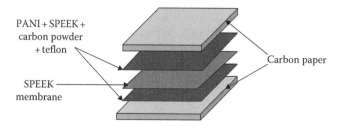

PANI + SPEEK +
carbon powder
+ teflon

Carbon paper

SPEEK
membrane

FIGURE 5.21   Construction of a unit cell assembly of all-solid supercapacitors.

of an all-solid capacitor using SPEEK and polyaniline (PANI). The unit cell consisted of two electrodes separated by a SPEEK membrane with a thickness of
50 µm. The composite electrode was made from PANI, SPEEK, electronically conducting carbon black, and PTFE. SPEEK employed here acted as both separator
and electrolyte. Compared to the conventional supercapacitor using liquid electrolytes, all-solid capacitor using SPEEK as the solid electrolyte is easily handled and
reliable without electrolyte leakage. The performance of the supercapacitor was
studied using cyclic voltammetry (CV), charge–discharge testing, and impedance
analysis. The capacitance measured from CV was 27 F g^{-1} of active polymer material. Charge and discharge curves showed a similar capacitance value, which was
relatively low due to the low concentration of sulfonic acid groups in the electrode
matrix.

Recently, they attempted to use an XSPEEK as the solid electrolyte and separator
membrane in a PANI-based supercapacitor.[18] The specific capacitance of PANI with
PANI/XSPEEK weight ratio of 1:0.5 in the electrode was 480 F g^{-1} and found to be
the highest reported for an all-solid supercapacitor based on PANI and the proton-
conducting SPE. The study demonstrated that a SPEEK-based electrolyte may be a
promising material for supercapacitors.

5.6   TYPICAL EXAMPLE ANALYSIS FROM MATERIAL SELECTION, SYNTHESIS, CHARACTERIZATION, AND APPLICATIONS

5.6.1   POSTSULFONATED POLY(ETHER ETHER KETONE)

Typically, 20 g of PEEK was dissolved in 800 mL of concentrated (95%–98%) sulfuric acid and vigorously stirred at 60°C under flowing argon. The duration of the
reaction was varied from 3 to 8 h to obtain the desired DS. To stop sulfonation
reaction, the polymer solution was decanted into a large excess of ice cold water
under continuous mechanical stirring. The suspension containing the SPEEK was
then centrifuged, and residual sulfuric acid was removed by dialyzing the recovered
polymer in deionized water using dialyzing tubing polymer until reaching the pH of
water. The resulting suspension was centrifuged again, and the polymer was dried
at 80°C in oven. The DS of SPEEK was controlled by the reaction time in the range
of 50%–80%. The postsulfonated PEEK membranes show a water uptake of 120%
at a DS of 80%. The proton conductivity is 0.008 S cm^{-1} at 25°C and 0.035 S cm^{-1}
at 80°C, respectively.

5.6.2 SULFONATED POLY(ETHER ETHER KETONE KETONE) PREPARED BY DIRECT COPOLYMERIZATION OF SULFONATED MONOMERS

5.6.2.1 Monomer

The synthesis of sulfonated monomer (1,4-bi(3-sodium sulfonate-4-fluorobenzoyl) benzene) was accorded to a procedure described as follows: 1,4-bi(4-fluorobenzoyl) benzene was first sulfonated with fuming sulfuric acid, followed by neutralizing with NaOH and precipitating with NaCl. The crude product was recrystallized with the mixture of methanol and water. The yield of this monomer was 81%. IR (KBr, cm^{-1}), 1656 (C=O), 1211, 1093, 621 (Ar-SO$_3$Na). ^{1}H-NMR (500 MHz, DMSO, ppm), 8.12–8.15 (dd, 2.3 Hz, 4.5 Hz), 7.87 (s) 7.81–7.84 (m, 2.5, 2.0, 5.3 Hz), 7.34–7.38 (dd, 5.3 Hz, 5.3 Hz). ^{13}C-NMR (500 Hz, DMSO, ppm), 194.74, 163.34, 161.29, 141.01, 133.98, 133.91, 132.81, 131.66, 130.29, 117.53.

5.6.2.2 Polymer Synthesis

As shown in Figure 5.22, SPEEKKs were synthesized via the nucleophilic aromatic substitution reactions of the following monomers: 1,4-bi(4-fluorobenzoyl) benzene (monomer k), 1,4-bi(3-sodium sulfonate-4-fluorobenzoyl)benzene (monomer m), and 4,4'-dihydroxydiphenyl propane(monomer l). The DS was controlled by adjusting the ratio of the monomer m to the monomer k. In this reaction, the monomer l was fixed at 100 mmol, which is equal to the total amount of the monomer m and the monomer k. After mixing the monomers in DMSO/toluene system, the mixture was stirred at 140°C for 4 h and then raised to 170°C for 6 h. The reaction mixture was cooled to room temperature and poured into acetone. The inorganic salts were removed by washing with boiling water. IR (KBr, cm^{-1}) 1247, 1078, 699 (Ar-SO$_3$Na), 1160 (–O–), 1656 (C=O), 2969 (–CH$_3$).

5.6.2.3 Preparation of Membranes

SPEEKK membranes were cast onto glass plates from their DMF solutions (5%–10%). The SPEEKK membranes (in sodium salt form) were transformed to their acid forms by soaking in 1.0 M HCl solution for 24 h. After that, the resulting membranes

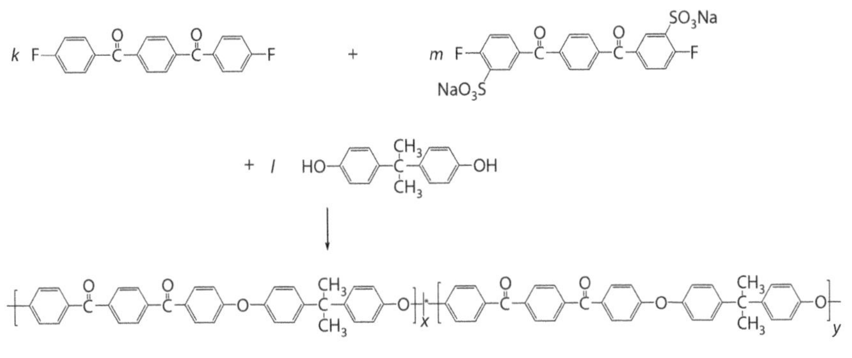

FIGURE 5.22 The synthesis of SPEEKK polymers.

(in acid form) were immersed and washed with deionized water. The thickness of all membrane samples was in the range of 50 and 150 μm.

5.6.2.4 Characterization and Applications

Tapping-mode phase images of the SPEEKK membranes were recorded under ambient conditions on a 1 μm × 1 μm size scale in order to investigate the morphology of SPEEKKs. In the phase images, dark regions were assigned to a softer region, which represented the hydrophilic sulfonic acid groups. The domain sizes of clusters increased with the DS increasing, thus providing more or larger proton transport channels. SPEEKK membranes showed relatively better mechanical properties than that of commercial Nafion membrane. Compared with Nafion, SPEEKK membranes show relatively lower water uptake, higher thermal stability, and comparable proton conductivity. The methanol permeability of the SPEEKK membranes was about one order of magnitude lower than that of Nafion, making these SPEEKK membranes good alternative to reduce problems associated with high methanol crossover in DMFCs.

5.6.3 NAPHTHALENE-BASED POLY(ARYLENE ETHER KETONE) COPOLYMERS CONTAINING SULFOBUTYL PENDANT GROUPS

5.6.3.1 Monomer

Anhydrous ferric chloride (1.65 g) was added in small portions to a cold solution (0°C–5°C) with a mixture of 2,6-dimethoxylnaphthalene (9.4 g) and 4-fluorobenzoyl chloride (17.4 g) dissolved in chloroform. And the reaction was then kept at this temperature for 24 h. The resulting mixture was poured into diluted hydrochloric acid. The product was removed by decantation and the brown solid was precipitated in methanol. The crude product was then recrystallized twice from DMF/water mixture (10:1, v/v).

Yield, 82%. Mp (DSC), 262°C. ^1H NMR (500 MHz, DMSO-d_6, δ, ppm), 3.79 (m, 6 H), 7.12 (t, $J = 8.56$, 8.56 Hz, 4 H), 7.27 (d, $J = 8.77$ Hz, 2 H), 7.63 (d, $J = 9.27$ Hz, 2 H), 7.89 (dd, $J = 5.51$, 8.66 Hz, 4 H). ^{13}C NMR (125 MHz, CDCl$_3$, δ, ppm), 57.04, 115.39, 116.33, 123.34, 127.41, 127.70, 132.77, 134.74, 153.17, 167.59, 196.23. Anal. Calcd for $C_{26}H_{18}F_2O_4$: C, 72.22; H, 4.20. Found: C, 72.25; H, 4.28.

5.6.3.1.1 Copolymerization of Naphthalene-Based Poly(Arylene Ether Ketone) Copolymers Containing Methoxy Groups (MNPAEK-xx)

A typical synthetic procedure, illustrated by the preparation of MNPAEK-80 copolymers (DMNF/DFB = 80/20), is described as follows: In a 250 mL three-necked flask equipped with a mechanical stirrer, a Dean-Stark trap, and a nitrogen inlet, DMNF (6.912 g, 0.016 mol), DFB (0.872 g, 0.004 mol), bisphenol A (4.56 g, 0.02 mol), K$_2$CO$_3$ (3.036 g, 0.022 mol), NMP (52 mL), and toluene (20 mL) were placed. The mixture was heated at 140°C for about 3 h to remove the water by azeotropic distillation with toluene and then was slowly raised to 180°C and maintained at that temperature for 6 h. The high-viscosity mixture was coagulated into a large excess of deionized water with vigorous stirring. The resulting fibrous copolymer was washed thoroughly with water several times and dried under vacuum at 100°C for 24 h. The copolymer was denoted MNPAEK-80, where *xx* (80) refers to the DMNF molar content.

^{1}H NMR (500 MHz, CDCl$_3$, δ, ppm), 1.69 (–CH$_3$), 3.78 (–OCH$_3$), 6.86–7.04 (ArH *ortho* to ether), 7.20–7.33 (ArH and naphthalene β-H), 7.6 (naphthalene α-H), 7.75–7.86 (ArH *ortho* to carbonyl). ^{13}C NMR (125 MHz, CDCl$_3$, δ, ppm), 31.41, 42.81, 57.14, 115.35, 117.53, 120.26, 123.82, 127.44, 127.54, 128.75, 132.46, 132.80, 147.38, 152.93, 153.48, 163.10, 196.35.

5.6.3.1.2 Demethylation of MNPAEK to Naphthalene-Based Poly(Arylene Ether Ketone) Copolymers Bearing Hydroxyl Groups

The methoxy groups in MNPAEK copolymers were converted into hydroxyl functionalities according to a modified procedure. Typically, 1 g MNPAEK-80 (2.59 mmol methoxy groups) was dissolved in 10 mL dried CH$_2$Cl$_2$. The solution was cooled down to 0°C (ice bath) and a threefold excess (0.7 mL 7.78 mmol) BBr$_3$ dissolved in 5 mL CH$_2$Cl$_2$ was added dropwise. The reaction mixture was stirred at room temperature under nitrogen for 12 h. Then the temperature was increased to reflux and the reaction proceeded at these conditions for another 6 h. The mixture was poured into 100 mL ice water to hydrolyze the BBr$_3$ and the boron complexes and then washed with methanol and deionized water. The resulting copolymers (HNPAEK-*xx*) dried under vacuum at 100°C for 24 h.

^{1}H NMR (500 MHz, DMSO-d_6, δ, ppm), 1.62 (–CH$_3$), 6.95–7.05 (ArH *ortho* to ether), 7.13 (naphthalene α-H), 7.21–7.33 (ArH and naphthalene β-H), 7.65–7.79 (ArH *ortho* to carbonyl), 9.83(–OH). ^{13}C NMR (125 MHz, DMSO-d_6, δ, ppm), 31.41, 42.75, 117.93, 118.03, 120.44, 121.01, 126.67, 126.76, 129.18, 132.67, 133.17, 147.41, 150.67, 153.50, 162.61, 196.51.

5.6.3.1.3 Sulfobutylation of HNPAEK-xx to Sulfonated Naphthalene-Based Poly(Arylene Ether Ketone) Copolymers(SNPAEK-xx)

The sulfobutylation experiment was conducted in such a way that HNPAEK-*xx* was dissolved in dry NMP or DMSO and the hydroxyl groups were deprotonated with NaOH first and then reacted with excessive 1,4-butane sultone. General description: 1 g HNPAEK-80 (2.59 mmol hydroxyl groups) was dissolved in 20 mL DMSO in a three-necked flask. 0.2 g NaOH (5 mmol) was then added to the solution and the reaction mixture was stirred for 6 h at room temperature. 0.39 g 1,4-butane sultone (2.85 mmol) was then added dropwise to the solution, and the reaction was heated to 160°C, stirred at that temperature for 12 h. The resulting sulfonated copolymers were precipitated in 250 mL acetone and washed by boiling water several times. The obtained polymer was immersed in a large excess of HCl (5 wt.%) solution over 24 h to make the salt form convert into the acid form. The product (SNPAEK-80) was finally dried in a vacuum at 80°C.

^{1}H NMR (500 MHz, DMSO-d_6, δ, ppm), 1.38–1.55 (–CH$_2$CH$_2$–), 1.66 (–CH$_3$), 2.38 (–CH$_2$SO$_3$H), 3.99 (–OCH$_2$–), 4.20–5.00 (–SO$_3$H), 6.88–7.12 (ArH *ortho* to ether), 7.26–7.34 (ArH and naphthalene β-H), 7.51 (naphthalene α-H), 7.70–7.80 (ArH *ortho* to carbonyl). ^{13}C NMR (125 MHz, CDCl$_3$, δ, ppm), 21.77, 28.54, 31.44,

42.78, 51.53, 69.53, 117.53, 118.03, 120.42, 123.84, 127.09, 127.31, 129.21, 132.51, 133.40, 147.44, 152.52, 153.48, 162.76, 195.92.

5.6.3.2 Preparation of Membranes

SNPAEK-xx membranes were cast onto glass plates from their DMAc solutions (10 wt.%) and dried at 60°C for 24 h. The membranes were then peeled off from the substrates. To remove any excess of the solvent, the membranes were dried under vacuum at 80°C for 48 h.

5.6.3.3 Characterization and Applications

The temperature of glassy transition of sodium-form SNPAEK-xx increased from 206°C to 301°C when sulfobutylation content increased from 40% to 80%. They showed good mechanical properties with tensile strength, elongation at break, and Young's modulus in the ranges of 43–49 MPa, 15.34%–26.75%, and 1.41–1.72 GPa in dry state, respectively. The SNPAEK-xx membranes showed anisotropic membrane swelling in water with larger swelling in thickness than in plane, which was helpful to improve the dimensional stability. SNPAEK-70 and SNPAEK-80 exhibited high proton conductivity of 0.114 and 0.179 S cm^{-1} at 80°C, respectively. The methanol permeability values of SNPAEK-70 and SNPAEK-80 were 2.57×10^{-7} and 4.49×10^{-7} cm^2 s^{-1}, respectively. TEM images revealed that sulfonic acid groups aggregated into hydrophilic clusters formed a continuous network at 80% (SNPAEK-80). The SNPAEK-xx membranes exhibited low methanol permeability, lower water swelling, and high proton conductivity and could be promising materials as alternative to Nafion membrane for DMFC applications.

5.7 CHALLENGES AND PERSPECTIVES

SPEKs have been regarded as one of the most promising alternative to the traditional perfluorosulfonic polymers such as Nafion. However, in the foreseeable future, the target of replacing perfluorosulfonic polymers is still challenging. Firstly, there is a lack of the long-term stability data under real operation conditions. Actually, SPEKs appeared to degrade fast than perfluorosulfonic polymers in the electrochemical or chemical environment, for example, the free radicals formed in a fuel cell and strongly acidic and oxidizing VRB mediums. The electron-donating ether linkages on the SPEK main chain are thought to be unstable under harsh conditions. Furthermore, the durability data reported in the open literatures were obtained by accelerated Fenton's regent tests under respective laboratory. It is difficult to compare the long performance and durability of SPEK membranes with varied structures and compositions. As a consequence, the durability data based on an appropriate test protocol are needed to be measured and compared in future studies. Secondly, the proton conductivity of SPEK membranes at higher temperature and lower humidity is still lower than DOE target (proton conductivity > 0.1 S cm^{-1} at 80°C and 50% RH). The control of morphology and thus improvement of proton transport in SPEK membranes have been under intense investigation. This is achieved by synthesis of block SPEKs, grafting sulfonic acid groups onto the aromatic and/or aliphatic side chain, introduction of superacids, etc. These SPEK membranes usually possess a microphase-separated structure. A detailed fundamental work in understanding the relationships between the structure and morphology is needed. This is the key to develop novel SPEK copolymers with long-range, well-ordered morphologies to provide water and ionic transport channels. Thirdly, to achieve a high power density in fuel cell and water electrolysis, more detailed work is needed, including MEA fabrication improvement, MEA active area scale-up, and membrane/catalyst interface optimization.

5.8 CHAPTER SUMMARY

A number of SPEKs with different structures have been prepared by postsulfonation, chemical grafting, and direct copolymerization of sulfonated monomers. To reduce water swelling and methanol permeability of SPEK membranes at a high DS, polymers containing covalent or ionic cross-linking groups and polymer with the desired polymer architectures were also developed. They have been reviewed as SPEs in electrochemical devices, such as fuel cells, vanadium flow batteries, water electrolysis, and supercapacitors, due to their availability, processability, excellent thermal and chemical stabilities, good mechanical properties, and low cost. To realize SPEK replacement of Nafion for electrochemical devices, a more thorough study on the morphology and durability of these membranes in operation conditions should be carried out.

REFERENCES

1. Dhara, M.G., Banerjee, S., Fluorinated high-performance polymers: Poly (arylene ether)s and aromatic polyimides containing trifluoromethyl groups, *Prog. Polym. Sci.*, 2010, 35, 1022–1077.
2. Patel, P., Hull, T.R., McCabe, R.W., Flath, D., Grasmeder, J., Percy, M., Mechanism of thermal decomposition of poly(ether ether ketone) (PEEK) from a review of decomposition studies, *Polym. Degrad. Stabil.*, 2010, 95, 709–718.
3. Shekar, R.I., Kotresh, T.M., Rao, P.M.D., Kumar, K., Properties of high modulus PEEK yarns for aerospace applications, *J. Appl. Polym. Sci.*, 2009, 112, 2497–2510.
4. Goyal, R.K., Rokade, K.A., Kapadia, A.S., Selukar, B.S., Garnaik, B., PEEK/SiO$_2$ composites with high thermal stability for electronic applications, *Electron. Mater. Lett.*, 2013, 9, 95–100.
5. Ma, W.J., Zhao, C.J., Lin, H.D., Zhang, G., Na, H., Poly (aryl ether ketone)s with bromomethyl groups: Synthesis and quaternary amination, *J. Appl. Polym. Sci.*, 2011, 120, 3477–3483.
6. Liu, B.J., Hu, W., Chen, C.H., Jiang, Z.H., Zhang, W.J., Wu, Z.W., Matsumoto, T., Soluble aromatic poly(ether ketone)s with a pendant 3,5-ditrifluoromethylphenyl group, *Polymer*, 2004, 45, 3241–3247.
7. Wang, F., Roovers, J., Functionalization of poly(aryl ether ether ketone) (PEEK): Synthesis and properties of aldehyde and carboxylic acid substituted PEEK, *Macromolecules*, 1993, 26, 5295–5302.
8. Kreuer, K.D., Ion conducting membranes for fuel cells and other electrochemical devices, *Chem. Mater.*, 2014, 26, 361–380.
9. Li, X.F., Na, H., Lu, H., Novel sulfonated poly(ether ether ketone ketone) derived from bisphenol S, *J. Appl. Polym. Sci.*, 2004, 94, 1569–1574.

10. Li, X.F., Zhao, C.J., Lu, H., Wang, Z., Na, H., Direct synthesis of sulfonated poly(ether ether ketone ketone)s (SPEEKKs) proton exchange membranes for fuel cell application, *Polymer*, 2005, 46, 5820–5827.

11. Zhao, C.J., Li, X.F., Lin, H.D., Shao, K., Na, H., Sulfonated poly(arylene ether ketone)s prepared by direct copolymerization as proton exchange membranes: Synthesis and comparative investigation on transport properties, *J. Appl. Polym. Sci.*, 2008, 108, 671–680.

12. Li, H.T., Cui, Z.M., Zhao, C.J., Wu, J., Fu, T.Z., Zhang, Y., Shao, K., Zhang, H.Q., Na, H., Xing, W., Synthesis and property of a novel sulfonated poly(ether ether ketone) with high selectivity for direct methanol fuel cell applications, *J. Membr. Sci.*, 2009, 343, 164–170.

13. Wei, W.P., Zhang, H.M., Li, X.F., Mai, Z.S., Zhang, H.Z., Poly(tetrafluoroethylene) reinforced sulfonated poly(ether ether ketone) membranes for vanadium redox flow battery application, *J. Power Sources*, 2012, 208, 421–425.

14. Ling, X., Jia, C.K., Liu, J.G., Yan, C.W., Preparation and characterization of sulfonated poly(ether sulfone)/sulfonated poly(ether ether ketone) blend membrane for vanadium redox flow battery, *J. Membr. Sci.*, 2012, 415, 306–312.

15. Song, M.A., Ha, S.I., Park, D.Y., Ryu, C.H., Kang, A.S., Moon, S.B., Chung, J.H., Development and characterization of covalently cross-linked SPEEK/Cs-TPA/CeO$_2$ composite membrane and membrane electrode assembly for water electrolysis, *Int. J. Hydrogen Energy*, 2013, 38, 10502–10510.

16. Wei, G., Xu, L., Huang, C., Wang, Y., SPE water electrolysis with SPEEK/PES blend membrane, *Int. J. Hydrogen Energy*, 2010, 35, 7778–7783.

17. Sivaraman, P., Hande, V.R., Mishra, V.S., Rao, C., Samui, A.B., All-solid supercapacitor based on polyaniline and sulfonated poly(ether ether ketone), *J. Power Sources*, 2003, 124, 351–354.

18. Sivaraman, P., Kushwaha, R.K., Shashidhara, K., Hande, V.R., Thakur, A.P., Samui, A.B., Khandpekar, M.M., All solid supercapacitor based on polyaniline and crosslinked sulfonated poly[ether ether ketone], *Electrochim. Acta*, 2010, 55, 2451–2456.

19. Rikukawa, M., Sanui, K., Proton-conducting polymer electrolyte membranes based on hydrocarbon polymers, *Prog. Polym. Sci.*, 2000, 25, 1463–1502.

20. Daoust, D., Devaux, J., Godard, P., Mechanism and kinetics of poly(ether ether ketone) (PEEK) sulfonation in concentrated sulfuric acid at room temperature. Part 1. Qualitative comparison between polymer and monomer model compound sulfonation, *Polym. Int.*, 2001, 50, 917–924.

21. Takamuku, S., Akizuki, K., Abe, M., Kanesaka, H., Synthesis and characterization of postsulfonated poly(arylene ether sulfone) diblock copolymers for proton exchange membranes, *J. Polym. Sci., Part A: Polym. Chem.*, 2009, 47, 700–712.

22. Gil, M., Ji, X.L., Li, X.F., Na, H., Hampsey, J.E., Lu, Y.F., Direct synthesis of sulfonated aromatic poly(ether ether ketone) proton exchange membranes for fuel cell applications, *J. Membr. Sci.*, 2004, 234, 75–81.

23. Zhang, H., Shen, P.K., Advances in the high performance polymer electrolyte membranes for fuel cells, *Chem. Soc. Rev.*, 2012, 41, 2382–2394.

24. Zhang, H., Shen, P.K., Recent development of polymer electrolyte membranes for fuel cells, *Chem. Rev.*, 2012, 112, 2780–2832.

25. Zhao, C.J., Lin, H.D., Li, X.F., Na, H., Morphological investigations of block sulfonated poly(arylene ether ketone) copolymers as potential proton exchange membranes, *Polym. Adv. Technol.*, 2011, 22, 2173–2181.

26. Kreuer, K.D., On the development of proton conducting polymer membranes for hydrogen and methanol fuel cells, *J. Membr. Sci.*, 2001, 185, 29–39.

27. Li, N.W., Guiver, M.D., Ion transport by nanochannels in ion-containing aromatic copolymers, *Macromolecules*, 2014, 47, 2175–2198.

28. Zhang, G., Fu, T., Shao, K., Li, X.F., Zhao, C.J., Na, H., Zhang, H., Novel sulfonated poly(ether ether ketone ketone)s for direct methanol fuel cells usage: Synthesis, water uptake, methanol diffusion coefficient and proton conductivity, *J. Power Sources*, 2009, 189, 875–881.

29. Wang, F., Chen, T., Xu, J., Sodium sulfonate-functionalized poly(ether ether ketone)s, *Macromol. Chem. Phys.*, 1998, 199, 1421–1426.

30. Xing, P., Robertson, G.P., Guiver, M.D., Mikhailenko, S.D., Kaliaguine, S., Sulfonated poly(aryl ether ketone)s containing the hexafluoroisopropylidene diphenyl moiety prepared by direct copolymerization, as proton exchange membranes for fuel cell application, *Macromolecules*, 2004, 37, 7960–7967.

31. Hickner, M.A., Ghassemi, H., Kim, Y.S., Einsla, B.R., McGrath, J.E., Alternative polymer systems for proton exchange membranes (PEMs), *Chem. Rev.*, 2004, 104, 4587–4612.

32. Lakshmi, R.T.S.M., Meier-Haack, J., Schlenstedt, Komber, K.H., Choudhary, V., Varma, I.K., Synthesis, characterisation and membrane properties of sulphonated poly(aryl ether sulphone) copolymers, *React. Funct. Polym.*, 2006, 66, 634–644.

33. Unveren, E.E., Erdogan, T., Çelebi, S.S., Inan, T.Y., Role of post-sulfonation of poly(ether ether sulfone) in proton conductivity and chemical stability of its proton exchange membranes for fuel cell, *Int. J. Hydrogen Energy*, 2010, 35, 3736–3744.

34. Sakaguchi, Y., Kaji, A., Kitamura, K., Takase, S., Omote, K., Asako, Y., Kimura, K., Polymer electrolyte membranes derived from novel fluorine-containing poly(arylene ether ketone)s by controlled post-sulfonation, *Polymer*, 2012, 53, 4388–4398.

35. Jeong, M.H., Lee, K.S., Hong, Y.T., Lee, J.S., Selective and quantitative sulfonation of poly(arylene ether ketone)s containing pendant phenyl rings by chlorosulfonic acid, *J. Membr. Sci.*, 2008, 314, 212–220.

36. Bishop, M.T., Karasz, F.E., Russo, P.S., Langley, K.H., Solubility and properties of a poly(aryl ether ketone) in strong acids, *Macromolecules*, 1985, 18, 86–93.

37. Alberti, G., Casciola, M., Massinelli, L., Bauer, B., Polymeric proton conducting membranes for medium temperature fuel cells (110–160°C), *J. Membr. Sci.*, 2001, 185, 73–81.

38. Xing, P., Robertson, G.P., Guiver, M.D., Mikhailenko, S.D., Wang, K., Kaliaguine, S., Synthesis and characterization of sulfonated poly(ether ether ketone) for proton exchange membranes, *J. Membr. Sci.*, 2004, 229, 95–106.

39. Hamciuc, C., Bruma, M., Klapper, M., Sulfonated poly(ether-ketone)s containing hexafluoroisopropylidene groups, *J. Macro. Sci.: Pure Appl. Chem., A*, 2001, 38, 659–671.

40. Shang, X., Tian, S., Kong, L., Meng, Y., Synthesis and characterization of sulfonated fluorene-containing poly(arylene ether ketone) for proton exchange membrane, *J. Membr. Sci.*, 2005, 266, 94–101.

41. Gao, Y., Robertson, G.P., Guiver, M.D., Jian, X., Mikhailenko, S.D., Wang, K., Kaliaguine, S., Sulfonation of poly(phthalazinones) with fuming sulfuric acid mixtures for proton exchange membrane materials, *J. Membr. Sci.*, 2003, 227, 39–50.

42. Li, Z.L., Liu, X.C., Chao, D.M., Zhang, W.J., Controllable sulfonation of aromatic poly(arylene ether ketone)s containing different pendant phenyl rings, *J. Power Sources*, 2009, 193, 477–482.

43. Liu, B.J., Robertson, G.P., Kim, D.S., Guiver, M.D., Hu, W., Jiang, Z.H., Aromatic poly(ether ketone)s with pendant sulfonic acid phenyl groups prepared by a mild sulfonation method for proton exchange membranes, *Macromolecules*, 2007, 40, 1934–1944.

44. Matsumoto, K., Higashihara, T., Ueda, M., Star-shaped sulfonated block copoly(ether ketone)s as proton exchange membranes, *Macromolecules*, 2008, 41, 7560–7565.

45. Shao, K., Zhu, J., Zhao, C.J., Li, X.F., Cui, Z.M., Zhang, Y., Li, H.T., Xu, D., Zhang, G., Fu, T.Z., Wu, J., Na, H., Xing, W., Naphthalene-based poly(arylene ether ketone) copolymers containing sulfobutyl pendant groups for proton exchange membranes, *J. Polym. Sci. Part A: Polym. Chem.*, 2009, 47, 5772–5783.

46. Zhang, Y., Wan, Y., Zhao, C.J., Shao, K., Zhang, G., Li, H.T., Lin, H.D., Na, H., Novel side-chain-type sulfonated poly(arylene ether ketone) with pendant sulfoalkyl groups for direct methanol fuel cells, *Polymer*, 2009, 50, 4471–4478.

47. Zhang, Y., Zhang, G., Wan, Y., Zhao, C.J., Shao, K., Li, H.T., Han, M.M., Zhu, J., Xu, S., Liu, Z.G., Na, H., Synthesis and characterization of poly(arylene ether ketone)s bearing pendant sulfonic acid groups for proton exchange membrane materials, *J. Polym. Sci. Part A: Polym. Chem.*, 2010, 48, 5824–5832.

48. Mikami, T., Miyatake, K., Watanabe, M., Poly(arylene ether)s containing superacid groups as proton exchange membranes, *Appl. Mater. Interfaces*, 2010, 2, 1714–1721.

49. Harrison, W.L., Wang, F., Mecham, J.B., Bhanu, V.A., Hill, M., Kim, Y.S., McGrath, J.E., Influence of the bisphenol structure on the direct synthesis of sulfonated poly(arylene ether) copolymers. *J. Polym. Sci. Part A: Polym. Chem.*, 2003, 41, 2264–2276; Li, X.F., Wang, Z., Lu, H., Zhao, C.J., Na, H., Zhao, C., Electrochemical properties of sulfonated PEEK used for ion exchange membranes, *J. Membr. Sci.*, 2005, 254, 147–155.

50. Li, X.F., Liu, C.P., Lu, H., Zhao, C.J., Wang, Z., Xing, W., Na, H., Preparation and characterization of sulfonated poly(ether ether ketone ketone) proton exchange membranes for fuel cell application, *J. Membr. Sci.*, 2005, 255, 149–155.

51. Zhong, S.L., Fu, T.Z., Dou, Z.Y., Zhao, C.J., Na, H., Preparation and evaluation of a proton exchange membrane based on crosslinkable sulfonated poly(ether ether ketone)s, *J. Power Sources*, 2006, 162, 51–57.

53. Zhong, S.L., Cui, X.J., Cai, H.L., Fu, T.Z., Zhao, C.J., Na, H., Crosslinked sulfonated poly(ether ether ketone) proton exchange membranes for direct methanol fuel cell applications, *J. Power Sources*, 2007, 164, 65–72.

54. Zhong, S.L., Liu, C.G., Na, H., Preparation and properties of UV irradiation-induced crosslinked sulfonated poly(ether ether ketone) proton exchange membranes, *J. Membr. Sci.*, 2009, 326, 400–407.

54. Zhang, Y., Wan, Y., Zhang, G., Shao, K., Zhao, C.J., Li, H.T., Na, H., Preparation and properties of novel cross-linked sulfonated poly(arylene ether ketone) for direct methanol fuel cell application, *J. Membr. Sci.*, 2010, 348, 353–359.

55. Zhang, Y., Shao, K., Zhao, C.J., Zhang, G., Li, H.T., Fu, T.Z., Na, H., Novel sulfonated poly(ether ether ketone) with pendant benzimidazole groups as a proton exchange membrane for direct methanol fuel cells, *J. Power Sources*, 2009, 194, 175–181.

56. Lin, H.D., Zhao, C.J., Cui, Z.M., Ma, W.J., Fu, T.Z., Na, H., Xing, W., Novel sulfonated poly(arylene ether ketone) copolymers bearing carboxylic or benzimidazole pendant groups for proton exchange membranes, *J. Power Sources*, 2009, 193, 507–514.

57. Li, H.T., Wu, J., Zhao, C.J., Zhang, G., Zhang, Y., Shao, K., Xu, D., Lin, H.D., Han, M.M., Na, H., Proton-conducting membranes based on benzimidazole containing sulfonated poly(ether ether ketone) compared with their carboxyl acid form, *Int. J. Hydrogen Energy*, 2009, 34, 8622–8629.

58. Pang, J., Zhang, H., Li, X., Ren, D., Jiang, Z., Low water swelling and high proton conducting sulfonated poly(arylene ether) with pendant sulfoalkyl groups for proton exchange membranes, *Macro. Rapid Commun.*, 2007, 28, 2332–2338.

59. Gao, Y., Robertson, G.P., Guiver, M.D., Mikhailenko, S.D., Li, X., Kaliaguine, S., Synthesis of poly(arylene ether ether ketone ketone) copolymers containing pendant sulfonic acid groups bonded to naphthalene as proton exchange membrane materials, *Macromolecules*, 2004, 37, 6748–6754.

60. Wang, F., Chen, T., Xu, J., Liu, T., Jiang, H., Qi, Y., Liu, S., Li, X., Synthesis and characterization of poly(arylene ether ketone)(co)polymers containing sulfonate groups, *Polymer*, 2006, 47, 4148–4153.

61. Robertson, G.P., Mikhailenko, S.D., Wang, K., Xing, P., Guiver, M.D., Kaliaguine, S., Casting solvent interactions with sulfonated poly(ether ether ketone) during proton exchange membrane fabrication, *J. Membr. Sci.*, 2003, 219, 113–121.

62. Kaliaguine, S., Mikhailenko, S.D., Wang, K.P., Xing, P., Robertson, G., Guiver, M., Properties of SPEEK based PEMs for fuel cell application, *Catal. Today*, 2003, 82, 213–222.

63. Xing, P., Robertson, G.P., Guiver, M.D., Mikhailenko, S.D., Kaliaguine, S., Synthesis and characterization of poly(aryl ether ketone) copolymers containing (hexafluoroisopropylidene)-diphenol moiety as proton exchange membrane materials, *Polymer*, 2005, 46, 3257–3263.

64. Li, X.F., Zhang, G., Xu, D., Zhao, C.J., Na, H., Morphology study of sulfonated poly(ether ether ketone ketone)s (SPEEKK) membranes: The relationship between morphology and transport properties of SPEEKK membranes, *J. Power Sources*, 2007, 165, 701–707.

65. Sgreccia, E., Chailan, J.F., Khadhraoui, M., Di Vona, M.L., Knautha, P., Mechanical properties of proton-conducting sulfonated aromatic polymer membranes: Stress–strain tests and dynamical analysis, *J. Power Sources*, 2010, 195, 7770–7775.

66. Kundu, S., Simon, L.C., Fowler, M.W., Comparison of two accelerated Nafion™ degradation experiments, *Polym. Degrad. Stabil.*, 2008, 93, 214–224.

67. Benjamin, T.G., Membrane and MEA accelerated stress test protocols, presented at *High Temperature Membrane Working Group Meeting*, Washington, DC, May 14, 2007.

68. Fang, X., Shen, P.K., Song, S., Stergiopoulos, V., Tsiakaras, P., Degradation of perfluorinated sulfonic acid films: An in-situ infrared spectro-electrochemical study, *Polym. Degrad. Stabil.*, 2009, 94, 1707–1713.

69. Chen, C., Levitin, G., Hess, D.W., Fuller, T.F., XPS investigation of Nafion membrane degradation, *J. Power Sources*, 2007, 169, 288–295.

70. Li, H.T., Zhang, G., Ma, W.J., Zhao, C.J., Zhang, Y., Han, M.M., Zhu, J., Liu, Z.G., Wu, J., Na, H., Composite membranes based on a novel benzimidazole grafted PEEK and SPEEK for fuel cells, *Int. J. Hydrogen Energy*, 2010, 35, 11172–11179.

71. Kreuer, K.D., Paddison, S.J., Spohr, E., Schuster, M., Transport in proton conductors for fuel-cell applications: Simulations, elementary reactions, and phenomenology, *Chem. Rev.*, 2004, 104, 4637–4678.

72. Ochi, S., Kamishima, O., Mizusaki, J., Kawamura, J., Investigation of proton diffusion in Nafion 117 membrane by electrical conductivity and NMR, *Solid State Ionics*, 2009, 180, 580–584.

73. Liu, B.J., Hu, W., Robertson, G.P., Guiver, M.D., Sulphonated biphenylated poly(aryl ether ketone)s for fuel cell applications, *Fuel Cells*, 2010, 1, 45–53.

74. Lin, C.K., Tsai, J.C., The effect of the side-chain ratio on main-chain-type and side-chain-type sulfonated poly(ether ether ketone) for direct methanol fuel cell applications, *J. Mater. Chem.*, 2012, 22, 9244–9252.

75. Tsai, J.C., Lin, C.K., Acid–base blend membranes based on Nafion/aminated SPEEK for reducing methanol permeability, *J. Taiwan Inst. Chem. E*, 2011, 42, 281–285.

76. Bi, C., Zhang, H., Xiao, S., Zhang, Y., Mai, Z., Li, X., Grafted porous PTFE/partially fluorinated sulfonated poly(arylene ether ketone) composite membrane for PEMFC applications, *J. Membr. Sci.*, 2011, 376, 170–178.

77. Zhou, S., Hai, S.D., Kim, D., Cross-linked poly(arylene ether ketone) proton exchange membranes with high ion exchange capacity for fuel cells, *Fuel Cells*, 2012, 12, 589–598.

78. Kim, J., Kim, D., Pendant-sulfonated poly(arylene ether ketone) (PAEK) membranes cross-linked with a proton conducting reagent for fuel cells, *J. Membr. Sci.*, 2012, 405, 176–184.

79. Li, X.F., Liu, C.P., Xu, D., Zhao, C.J., Wang, Z., Zhang, G., Na, H., Xing, W., Preparation and properties of sulfonated poly(ether ether ketone)s (SPEEK)/polypyrrole composite membranes for direct methanol fuel cells, *J. Power Sources*, 2006, 162, 1–8.

80. Zhang, H., Zhang, T., Wang, J., Pei, F., He, Y., Liu, J., Enhanced proton conductivity of sulfonated poly(ether ether ketone) membrane embedded by dopamine-modified nanotubes for proton exchange membrane fuel cell, *Fuel Cells*, 2013, 13, 1155–1165.

81. Meenakshi, S., Bhat, S.D., Sahu, A.K., Sridhar, P., Pitchumani, S., Modified sulfonated poly(ether ether ketone) based mixed matrix membranes for direct methanol fuel cells, *Fuel Cells*, 2013, 13, 851–861.

82. Kumar, R., Mamlouk, M., Scott, K., Sulfonated polyether ether ketone-sulfonated graphene oxide composite membranes for polymer electrolyte fuel cells, *RSC Adv.*, 2014, 4, 617–623.

83. Luo, Q., Zhang, H., Chen, J., You, D., Sun, C., Zhang, Y., Preparation and characterization of Nafion/SPEEK layered composite membrane and its application in vanadium redox flow battery, *J. Membr. Sci.*, 2008, 325, 553–558.

84. Mai, Z., Zhang, H., Li, X., Bi, C., Dai, H., Sulfonated poly(tetramethyldiphenyl ether ether ketone) membranes for vanadium redox flow battery application, *J. Power Sources*, 2011, 196, 482–487.

85. Jia, C., Liu, J., Yan, C., A multilayered membrane for vanadium redox flow battery, *J. Power Sources*, 2012, 203, 190–194.

86. Jang, I.-Y., Kweon, O.-H., Kim, K.-E., Hwang, G.-J., Moon, S.-B., Kang, A.-S., Application of polysulfone (PSf)-and polyether ether ketone (PEEK)-tungstophosphoric acid (TPA) composite membranes for water electrolysis, *J. Membr. Sci.*, 2008, 322, 154–161.

87. Jang, I.-Y., Kweon, O.-H., Kim, K.-E., Hwang, G.-J., Moon, S.-B., Kang, A.-S., Covalently cross-linked sulfonated poly(ether ether ketone)/tungstophosphoric acid composite membranes for water electrolysis application, *J. Power Sources*, 2008, 181, 127–134.

6 Sulfonated Polyphenylenes and the Related Copolymer Membranes

Xuan Zhang and Shouwen Chen

CONTENTS

6.1 INTRODUCTION

Classical sulfonated poly(phenylene)s (SPPs) provide advantages as polymer electrolyte membranes (PEMs) due to the lack of hetero linkages in their polymer backbones. Owing to the high C–C bond dissociation energy of phenyl–phenyl group (478.6±6.3 KJ mol^{-1}, 298 K) [1,2], these chemically stable polymers have drawn many attentions as PEM candidates.

In the view of chemistry, there exist three types of linkages of the polymer main chains, that is, *ortho-*, *meta-*, and *para-*positions. Generally, poly(*o*-phenylene)s are rather rare due to the problem of reactivity and steric hindrance. In contrast, the *meta*-linkage makes polymer chains to be more flexible; the introduction of long alkoxyl side chain would effectively prevent the attainment of planar arrangement with longer conjugation lengths in the backbones of poly(*m*-phenylene)s [3], which provide some process abilities for molecular designings.

Extensive researches have been focused on poly(*para*-phenylene)s. Based on a coplanar molecular packing, these polymers generally exhibited favorable conjugation length in the solid-state results in a stiff, rodlike feature. Moreover, the wholly aromatic polymer main chains are exceptionally stable to hydrolysis and oxidation. However, the disadvantage is their low solubility in most of the organic solvents, which brings the difficulties in obtaining polymers with high molecular weight to form the robust membranes. Generally, two strategies have been used to overcome the aforementioned shortcomings. One is to directly introduce side chains with sulfonated acid groups. The other is to investigate some flexible oligomers by polycondensations.

This chapter gives an overview of SPPs and derivatives for cationic fuel cell (FC) and vanadium redox flow battery (VRFB) applications. The synthesis strategies, ex situ properties, morphology, and cell performance are briefly summarized and discussed.

6.2   MEMBRANE PREPARATION

Typical polyphenylenes can be obtained by means of the polymerization types. Early research on sulfonated alkoxyl PPs utilized dibromo- and diboronic acid-substituted precursors via Suzuki coupling reaction for the polymer synthesis, as shown in Figure 6.1 [4].

Another pathway was based on the coupling of substituted dichlorophenyl monomers using nickel catalysts, with the most popular monomer of 2,5-dichloro-benzophenone, followed by sulfonation (Figure 6.2) [5,6]. These polymers exhibited good solubility in polar aprotic solvents and could be further reacted by nucleophilic substitutions (Figure 6.3) [7]. Probably because of the difficulties in obtaining high molecular weight as well as some other functionalized modifications, these PPs could be compensated by incorporating them as sulfonated segments into block copolymers with poly(arylene ether sulfone)s (Figure 6.4) [8].

Atom transfer radical polymerization was investigated to construct triblock (TB) materials, as shown in Figure 6.5. Polyphenylene (PP) with benzyl chloride as the end-capped functional group was firstly prepared as the macroinitiator, and then styrene was added with CuCl catalysts [9]. TB polymers were obtained with very low polymer dispersity index and afforded to sulfonation afterward. Clear phase-separated morphology was an evidence of the as-prepared TB polymer and thus exhibited good potential for some distributed junction photovoltaic applications.

The Diels–Alder reaction is another straightforward way to synthesize poly (p-phenylene)s [10–13]. Figure 6.6 shows the synthesis and sulfonation process of DA-sulfonated poly(phenylene sulfone) (DA-SPP) ionomers. However, they might be mixed with a few flexible *meta*-linkages due to their sterically crowded systems.

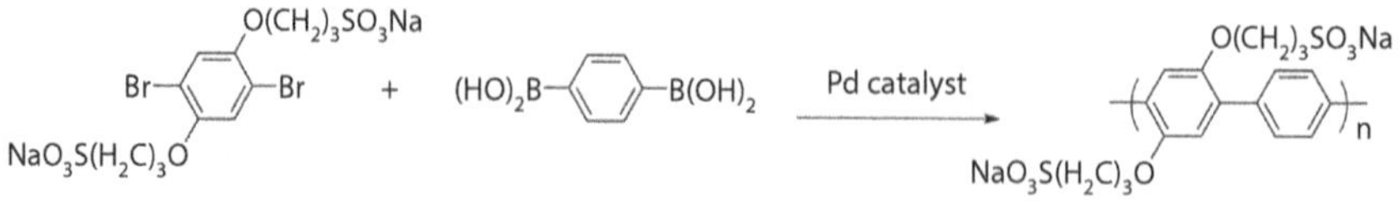

FIGURE 6.1   Synthetic route of sulfonated alkoxy-substituted poly(p-phenylene) by Suzuki coupling reaction. (From Child, A.D. and Reynolds, J.R., *Macromolecules*, 27(7), 1975, 1994.)

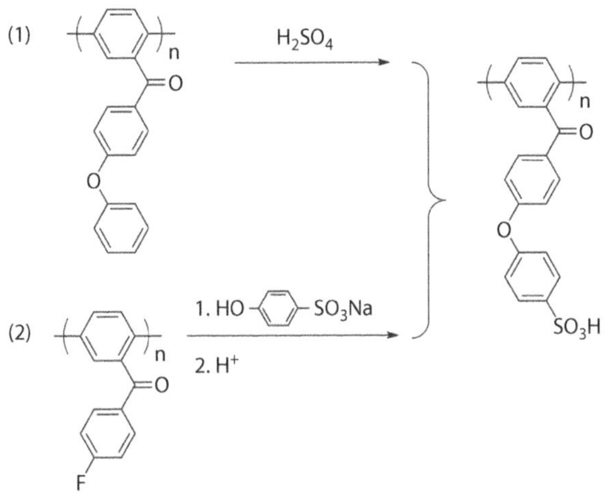

FIGURE 6.2 Synthetic route of typical sulfonated poly(4-phenoxybenzoyl-1,4-phenylene). (From Bae, J.M. et al., *Solid State Ionics*, 147(1–2), 189, 2002; Ghassemi, H. and McGrath, J.E., *Polymer*, 45, 5847, 2004.)

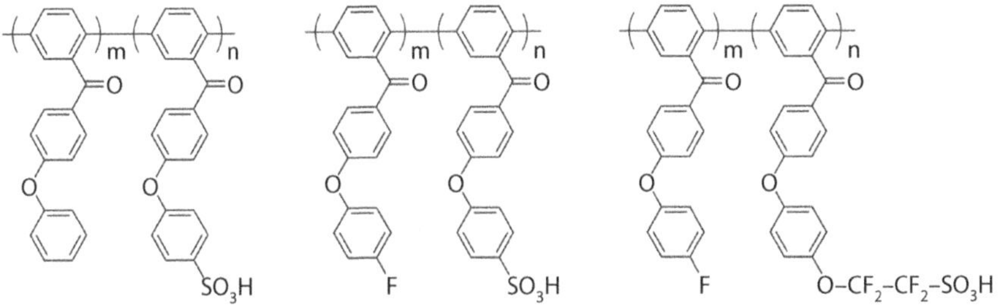

FIGURE 6.3 Chemical structures of the three types of SPP copolymers: sulfonated poly (*p*-benzoyl-phenylene) (sPPBP), grafted sulfonated poly(*p*-benzoyl-phenylene) (GS), and grafted perfluorinated sulfonated poly(*p*-benzoyl-phenylene) (GPS). (From Ninivin, C.L. et al., *J. Appl. Polym. Sci.*, 101, 944, 2006.)

Thus, these polymers may not be as strictly rodlike as the polymers prepared by nickel-catalytic aryl coupling. Their phenyl substituents provide a large number of potential sites for sulfonation, which can be also achieved by further sulfonation.

The fifth method to synthesize polyphenylenes is the Ullmann coupling of sulfonated dibromo aryl monomers. For this coupling reaction, it is needed to exchange the protons of the sulfonic acids for organic cations or metal ions, such as trimethyl benzylammonium or lithium ions before polymerizations [14,15]. Figure 6.7 shows the synthetic route of SPPs in the presence of copper catalyst. After polymerization, these groups were then deprotected by acid exchange process, and some of the resulting sulfonic acid groups could be used for cross-linking reaction by sulfone formation.

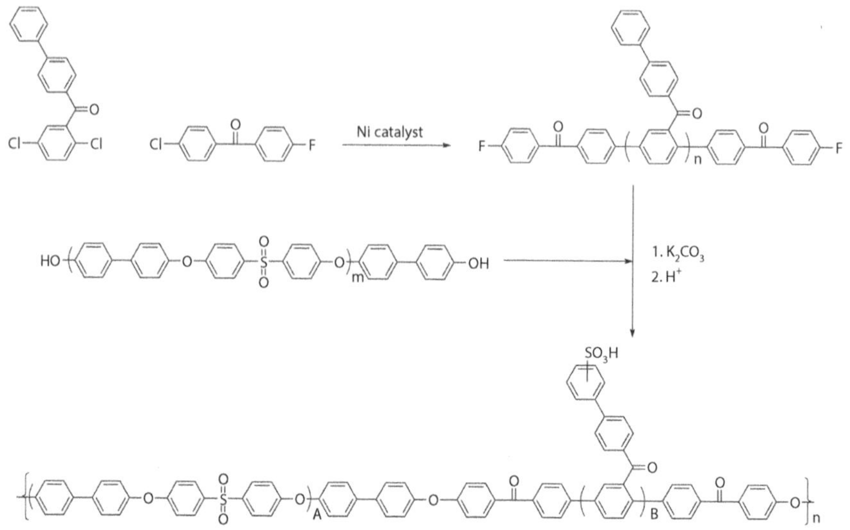

FIGURE 6.4 Synthesis of SPP-based copolymers. (From Ghassemi, H. et al., *Polymer*, 45, 5855, 2004.)

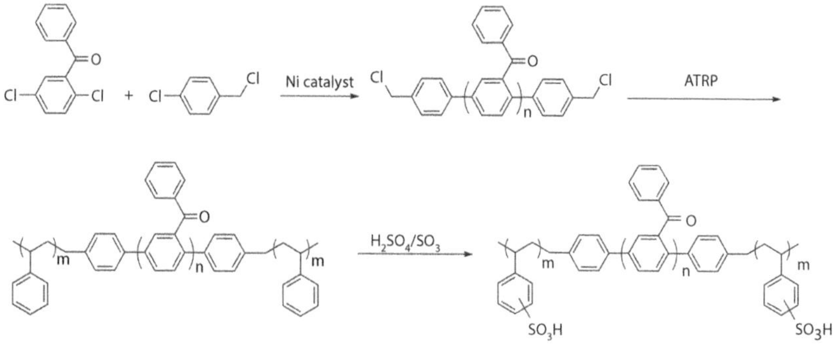

FIGURE 6.5 Synthetic route of PP-based sulfonated TB polymers. (From Hagberg, E.C. et al., *Macromolecules*, 37, 3642, 2004.)

In principle, all dichloro-aromatic monomers could be coupled by nickel-catalytic polymerization as described earlier. However, these catalysts are moisture sensitive, which bring some difficulties in the synthesis process and reagent storage. From this viewpoint, it is more reliable to get poly(*m*-phenylene)s via oxidation polymerization utilizing $FeCl_3$ or $VO(acac)_2$ as the common catalysts. Typically, alkoxyl groups or some ether linkages were preliminarily introduced onto the aromatic monomers so as to increase the electron density [16,17], as shown in Figure 6.8. After polymerization, excessive catalysts could be easily removed by rinsing with some common organic solvents. Note that it was a two-electron-transfer reaction; the molar ratio of the catalyst should be at least two equivalents to that of the total monomers.

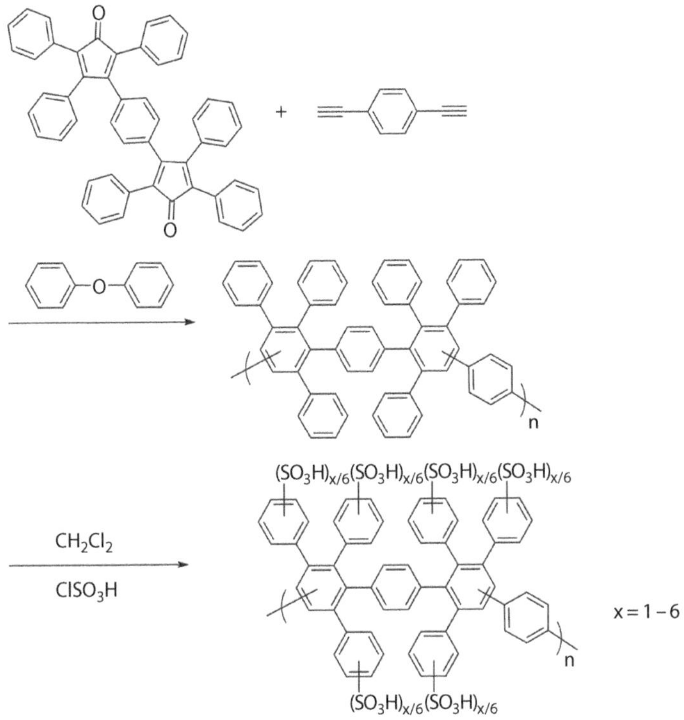

FIGURE 6.6 Synthetic route of DA-SPP from 1,4-*bis*(2,4,5-triphenyl cyclopentadienone) benzene and di(ethynyl)benzene. (From Fujimoto, C.H. et al., *Macromolecules*, 38, 5010, 2005.)

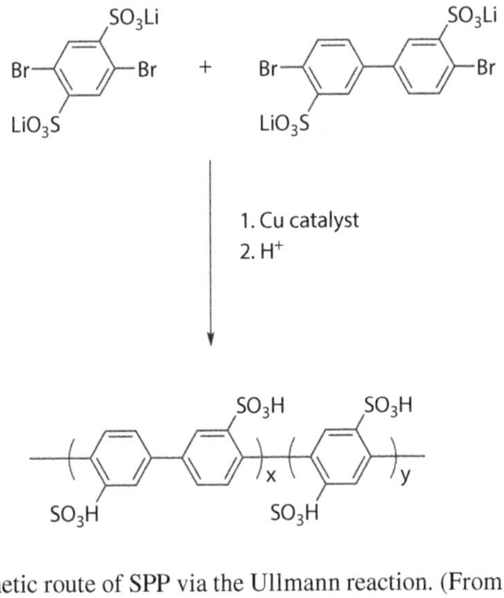

FIGURE 6.7 Synthetic route of SPP via the Ullmann reaction. (From Si, K. et al., *J. Mater. Chem.*, 22, 20907, 2012.)

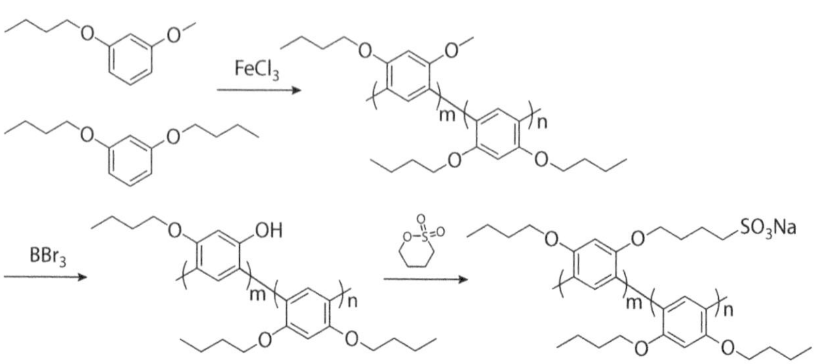

FIGURE 6.8 Synthetic routes for sulfonated poly(*m*-phenylene) copolymers. (From Zhang, X. et al., *Polym. Chem.*, 4, 1235, 2013.)

6.3 MEMBRANE PROPERTIES

Since SPPs were developed not long ago, systematic data on their electrochemical and physical properties have been rather limited. The first report was published by Rikukawa's group, in which two typical polymers of sulfonated poly(*p*-phenoxyben-zoyl-1,4-phenylene) (SPPBP) and sulfonated poly(ether ether ketone) were investigated for comparison [5,18]. At a similar degree of sulfonation, SPPBP showed better thermal stability and greater water uptake and proton conductivity under varied relative humidity (RH%) conditions than the former. Also, the structure anisotropy was studied by the same group using polarized microscopy measurement. SPPBP showed typical light-dark nematic textures in dimethyl sulfone (DMSO) solution (Figure 6.9), indicating a molecular organization of rigid-rod lamellar structure [19].

A comparison between sulfonated poly(*p*-benzoyl-1,4-phenylene) (SPBP) and SPPBP was carried out by Coutanceau's group [20]. Both SPBP and SPPBP are thermally stable up to at least 215°C. The permeability to methanol was even found to be 10 times lower than that of Nafion® 117 membrane, indicating a bright future for use in direct methanol fuel cell (DMFC) systems.

One-pot method was reported by Zhang et al., who directly introduced sulfonic acid group onto the 2,5-dichlorobenzophenone monomer to get sodium 3-(2,5-dichlorobenzoyl)benzenesulfonate (DCBS) and copolymerized them with the precursor or some functional dichloro-imide monomers via nickel-catalytic coupling [21–23]. Figure 6.10 lists the chemical structures of their dichloro-monomers. Compared with those ionomers obtained by postsulfonation, these SPP derivatives could be prepared with precisely controlled *ion exchange capacity (IEC)* values. Their polymer SPP-70 exhibited a high storage modulus over 2.0 GPa and loss modulus over 0.2 GPa ranging from 100°C to 450°C, respectively. The tensile strength and Young's modulus of the dry and hydrated SPP-70 membranes were more than 79 and 20 MPa and 1.7 and 0.5 GPa, respectively, which indicated that the SPP membranes were strong and tough enough for FC applications. The proton conductivity was also two to three times greater for SPP-70 than that of Nafion 117 in the hydrate state. The morphologies of these SPPs have also been characterized

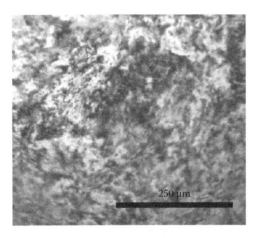

FIGURE 6.9 Optical polarization micrograph of a 200 mg mL^{-1} solution of SPPBP (2.81 mequiv g^{-1}) in DMSO. (Reprinted from *Polymer*, 52(26), Tonozuka, I., Yoshida, M., Kaneko, K., Takeoka, Y., and Rikukawa, M., Considerations of polymerization method and molecular weight for proton-conducting poly(*p*-phenylene) derivatives, 6020–6028, Copyright (2011), with permission from Elsevier.)

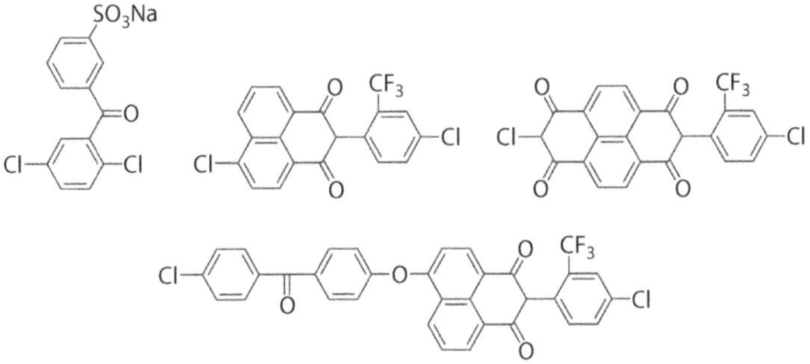

FIGURE 6.10 Chemical structures of DCBS and varied imide-based dichloro-monomers. (From Wu, S. et al., *Polymer*, 47, 6993, 2006; Qiu, Z. et al., *Macromolecules*, 39, 6425, 2006; Li, W. et al., *J. Membr. Sci.*, 315, 172, 2008.)

by high-resolution TEM and atom force microscopy (AFM). These techniques provide direct visualization of phase separation and the presence of ionic clusters in the membrane. For polybenzophenone-based SPPs, they displayed ionic domains with the size about 10–30 nm, and the width became larger with an increase in IEC values, as shown in Figure 6.11 [21].

Researcher of JSR Corporation, Japan, extensively investigated sulfonated 1,4-phenylene monomers for PP-based multiblock copolymers [24]. Esterification was initially introduced with varied aliphatic alcohols before the nickel-catalyzed copolymerization. Then, sulfonated polymers were obtained by a hydrolysis process. Their membrane has an elastic modulus of 1.8 GPa, a tensile stress at yield of

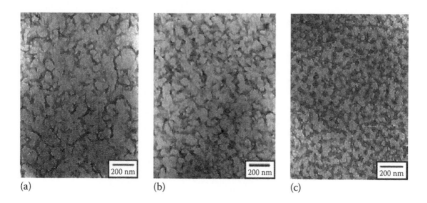

(a)   (b)   (c)

FIGURE 6.11 TEM images of (a) SPP-30, (b) SPP-50, and (c) SPP-70. (Reprinted from *Polymer*, 47, Wu, S., Qiu, Z., Zhang, S., Yang, X., Yang, F., and Li, Z., The direct synthesis of wholly aromatic poly(*p*-phenylene)s bearing sulfobenzoyl side groups as proton exchange membranes, 6993–7000, Copyright (2006), with permission from Elsevier.)

83 MPa (130 MPa at break), and an elongation at break of 100%, exceeding the per-fluorinated membrane 10 times in elastic modulus and 2.6 times in tensile strength. Meanwhile, since JSR electrolyte polymer has a rigid structure with high water resistance, it is expected to have 20%–50% higher conductivity than the fluorinated membrane of Nafion 112 from the results. Figure 6.12 shows the synthetic route of JSR membranes. Anyway, the exact chemical structure of their hydrophobic moiety was unpublished for the industrial concerns. Meanwhile, they comprehensively introduced the four strategies to control the film morphology [24], that is, (1) the mass ratio between sulfonated and nonsulfonated moiety, (2) the chemical structure of the hydrophilic block, (3) the lengths of each block, and (4) the film casting condition. As it is clearly seen in Figure 6.12, desired morphologies were obtained just like what it were designed according to those parameters.

Recently, Chen et al. prepared a series of SPP-*co*-PAEK random copolymers by coupling of two reactive functional monomers, 2,5-dichloro-3′-sulfobenzophenone and 2,2′-*bis*[4-(4-chlorobenzoyl)] phenoxyl perfluoropropane [25]. Later on, novel sulfonated ionomers were developed by the same group utilizing poly(phenylene ether ketone) (segment 1) and wholly sulfonated poly(arylene ether sulfone) (segment 2) as the polymer backbones and graft pendants, respectively [26]. The synthetic route is shown in Figure 6.13. Condensation occurred between functional fluorine groups in segment 1 and phenoxide-terminated groups in segment 2 and thus formed the graft-cross-linked (GC) architecture. The GC membranes exhibited reasonably high mechanical strength with Young's modulus, maximum stress, and elongation at break values that ranged from 0.9 to 1.3 GPa, 31 to 49 MPa, and 11% to 28% under 25°C/50% RH condition, respectively. It is worth noting that membranes showed good proton-conductive behavior with $\sigma_{\perp}/\sigma_{\parallel}$ (proton conductivity through-plane/in-plane) values in the range of 0.65–0.92. Taking the real proton conducting state in the FC system into account, the evaluation of $\sigma_{\perp}/\sigma_{\parallel}$ values was

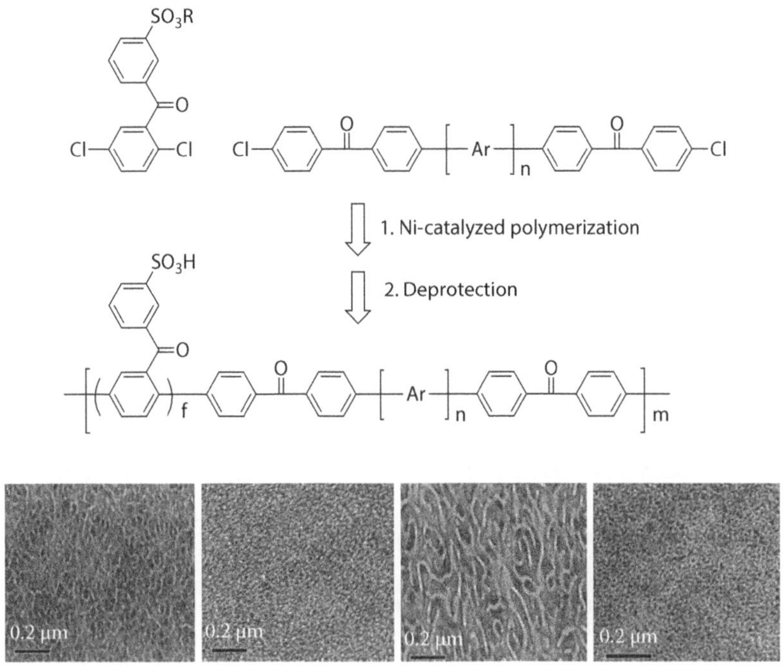

FIGURE 6.12 Chemical structures and TEM pictures of JSR membranes. (Reprinted by permission from Macmillan Publishers Ltd. *Polym. J.*, Goto, K., Rozhanskii, I., Yamakawa, Y., Otsuki, T., and Naito, Y., Development of aromatic polymer electrolyte membrane with high conductivity and durability for fuel cell, 41(2), 95–104, copyright 2009.)

quite essential and should be set as a rule in the future membrane characterizations. Morphological studies were carried out by the tapping mode AFM (TM-AFM) method. A well-defined nanoscale phase separation morphology was observed for all the obtained GC membranes, as shown in Figure 6.14. For instance, membrane GC(7/10) (*x/y* refers to molar ratio of the length of hydrophobic to hydrophilic moiety) with a low *IEC* (1.26 mequiv g^{-1}) exhibited a distinct phase separation morphology. The average width of hydrophilic domains is about 25–35 nm, which was some larger than that of Nafion® exposed to liquid water [27]. In addition, the discrepancies in phase separations were closely related to the block length of hydrophilic and hydrophobic units.

A series of fully rigid-rod aromatic polysulfonic acids based on biphenyl backbones were prepared by Si and Litt [28]. Cross-linking reaction was occurred in the presence of polyphosphoric acid to form a grafted structure by sulfone forming (Figure 6.15). The grafting degree was controlled by varying the reaction time and temperature. Due to the high local density of the architecture, the *IEC* value of their membrane reached extremely high to about 8.0 mequiv g^{-1}, which would be the highest value among the reported literatures on sulfonated PEMs. All the films showed elastic moduli between 0.8 and 2.2 GPa with the break at 6%–9% elongation under

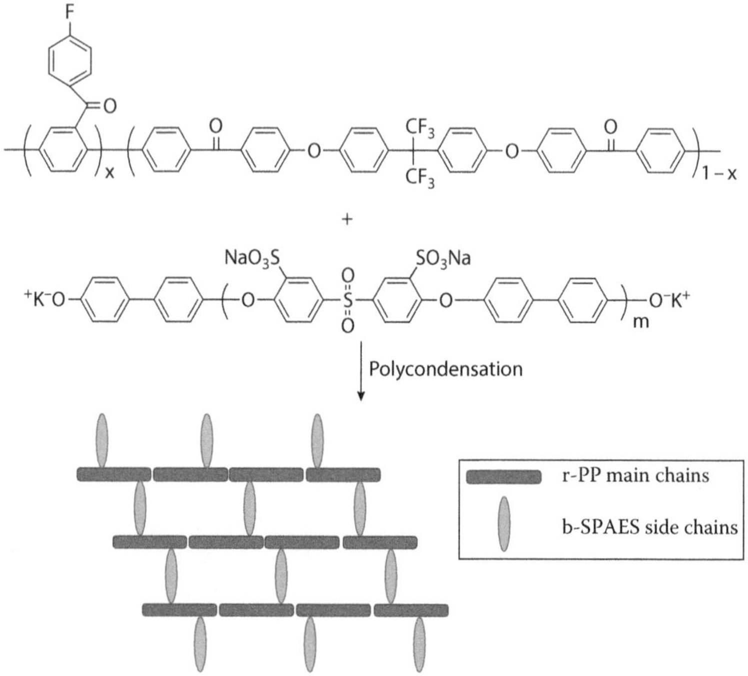

FIGURE 6.13 Schematic diagram of the GC membranes. (From Zhang, X. et al., *Macromol. Rapid Commun.*, 32(14), 1108, 2011.)

low RH condition. Based on the high IEC level, the conductivity was four to five times as high as that of Nafion over the entire RH range from 20% to 90%. Also, the through-plane conductivity was measured as a function of RH. The data were nearly 0.1 S cm^{-1} under 120°C/30% RH condition, which reached the target of the U.S. Department of Energy (U.S. DOE) 2015.

Table 6.1 summarizes a comprehensive collection of characterization data on SPPs and their derivatives, where Nafion is listed for comparison.

6.4 MEMBRANE APPLICATIONS

6.4.1 PEM H$_2$/O$_2$ Fuel Cells

Cell performance is the most important in situ test for evaluating the electrochemical properties of membranes in the FC system. Several efforts have been made on H$_2$/O$_2$ polymer electrolyte membrane fuel cell (PEMFC) performance of SPPs during the past decade. The first report on SPP-based H$_2$/O$_2$ FC was published by Bae's group [5]. Unfortunately, the conductivity of their membranes was too low to obtain an ideal cell performance; the maximum power output was only 300 mW cm^{-2}.

Recently, PEMFC performance of SPP-*co*-PAEKs membranes was reported by Chen et al., as shown in Figure 6.16 [25]. Under a mild operation condition (90°C,

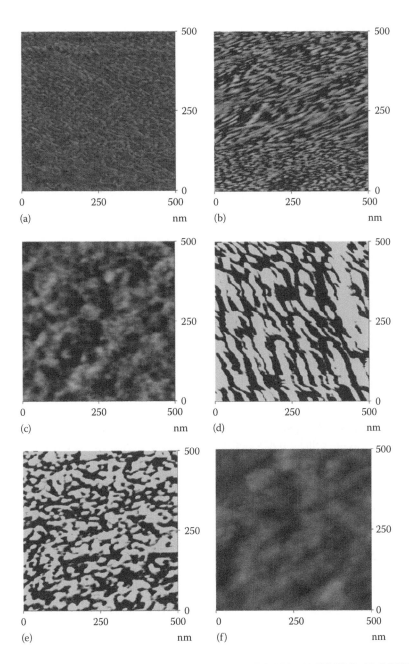

FIGURE 6.14 TM-AFM phase images: (a) GC(3/5), (b) GC(3/10), (c) GC(5/10), (d) GC(5/15), (e) GC(7/10), and (f) GC(7/15). (Zhang, X., Hu, Z., Luo, L., Chen, S., Liu, J., Chen, S., and Wang L.: Graft-crosslinked copolymers based on poly(arylene ether ketone)-*gc*-sulfonated poly(arylene ether sulfone) for PEMFC applications. *Macromol. Rapid Commun.* 2011. 32(14). 1108–1113. Copyright Wiley-VCH Verlag GmbH & Co. KGaA. Reproduced with permission.)

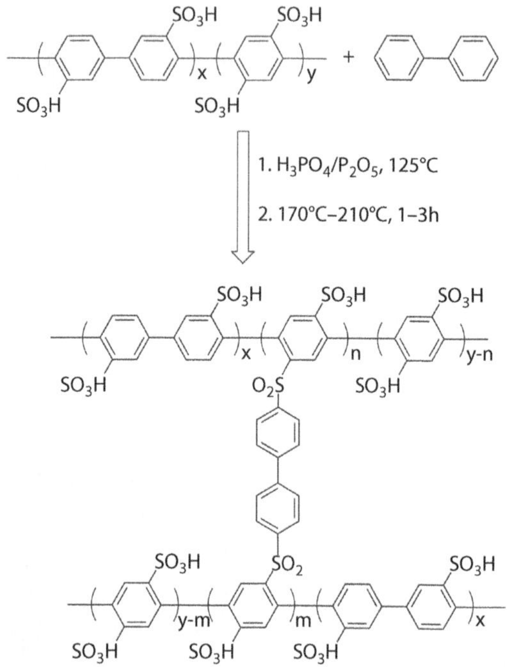

FIGURE 6.15 Synthetic route of high-*IEC* SPPs. (From Si, K. and Litt, M.H., *ECS Trans.*, 41(1), 1645, 2011.)

0.2 MPa, 82%/68% RH), SPP-*co*-PAEK(3/1) showed excellent PEM FC performance with open-circuit voltage (OCV) of 0.94 V, $V_{1.0}$ (cell voltage at current density of 1.0 A cm^{-2}) of 0.61 V, and P_{max} of 0.85 W cm^{-2}. The rigid poly(*p*-phenylene) bond made the hydrophilic moiety stacking packed, resulting in a continuous proton-conductive pathway and reasonably high cell performance.

Several extension researches were put through by Okamoto's group [32,35,36]. BT3 (a multiblocked copolymer [PSP-*b*-PAEK] membranes comprising of 2,5-dichloro-3′-sulfo-4′-((2,4-disulfo)phenoxy)benzophenone as the hydrophilic part) with a low *IEC* of 1.21 mequiv g^{-1} showed a rather high cell performance at 90°C, 0.2 MPa, and 82%/68% RH, that is, $V_{0.5}$ of 0.72 V and P_{max} of >0.95 W cm^{-2}. These might be the best ever reported performance on SPPs. A decrease in either gas pressure from 0.2 to 0.1 MPa (ambient pressure) or gas humidification from 82%/68% RH to 30% RH caused a small reduction in the PEMFC performance. Another comparable data were found for GC(3/10) (Figure 6.17) [36]; they showed $V_{0.5}$ of 0.75 V and P_{max} of 0.93 W cm^{-2} under the same operation condition as the aforementioned. It is encouraging to see that the $V_{0.3}$ reached to as high as 0.8 V, which has been already satisfied by the U.S. DOE target 2020 (0.8 V), whereas the P_{max} still need to be further improved (U.S. DOE, 1.0 W cm^{-2}) [37]. Figure 6.17d shows the durability test of GC(5/10) at 90°C, 0.2 MPa, and 50% RH under a constant load current density of 0.5 A cm^{-2} with monitoring the cell voltage and cell resistance. The cell voltage and OCV were held almost aptotic with only slight decreases (0.03–0.04 V) for 1000 h. Meanwhile,

TABLE 6.1
Physicochemical Properties of SPP Membranes

Polymer	IEC mmol g⁻¹	Water Uptake[a] %	Swelling State[b] Δl	Swelling State[b] Δt	Proton Conductivity[c] 50% RH mS cm⁻¹	Proton Conductivity[c] In Water mS cm⁻¹	Methanol Permeability 10⁻⁷ cm² s⁻¹	Young's Modulus GPa	References
Nafion	0.91	35–40	0.11–0.16	—	15–30	135–145	11–12	0.2–0.3	[5]
SPPBP 85%	2.38	—	—	—	—	10	—	—	[8]
6-f	1.20	50	—	—	—	36[a]	—	—	[20]
SPBP	0.7	43	—	—	—	—	1.4	—	[6]
SPPBP	1.3	65	—	—	—	—	1.0	—	[10]
SDAPP3	1.8	75	—	—	—	87[a]	—	1.2	[22]
SPP-50	2.19	42	0.09	0.21	—	60[a]	1.5	2.8	[24]
NPI(1)/SPP(70)	2.00	72	0.11	0.31	36	321	6.2	1.26	[29]
BC5	1.92	73	—	—	36	203[d]	—	—	[25]
JSR	—	—	—	—	—	110(80%)	—	1.8	

(Continued)

TABLE 6.1 (Continued)
Physicochemical Properties of SPP Membranes

Polymer	IEC mmol g^{-1}	Water Uptake^a %	Swelling State^b Δ$_l$	Δ$_t$	Proton Conductivity^c 50% RH mS cm^{-1}	In Water mS cm^{-1}	Methanol Permeability 10^{-7} cm^2 s^{-1}	Young's Modulus GPa	References
SPP-3	2.50	52(95%)^c	0.07	0.36	17	208(95%)	—	—	[30]
3	2.08	30(90%)^c	—	—	30	ca. 300(95%)	—	—	[31]
SPP-co-PAEK(3/1)	1.98	55	0.12	0.13	7	164^d	—	1.8	[26]
GC-5/10	1.55	64	0.07	0.15	12	186^d	—	1.33	[27]
BT3	1.21	25	0.05	0.09	14	146^d	—	1.5	[32]
B20P80-g-BP10%-210C-3h	5.9	81	0.12	0.33	ca. 200	ca. 600(95%)	—	1.56	[33]
3b	2.49	222^c	0.67	0.58	30	200(95%)	—	—	[34]
c-SPMP8-2	2.93	67	0.16	0.16	31	251	—	—	[16]

The data in the parenthesis refer to the relative humidity condition.

a   At 30°C in water.
b   R.T. in water.
c   At 80°C.
d   At 60°C.

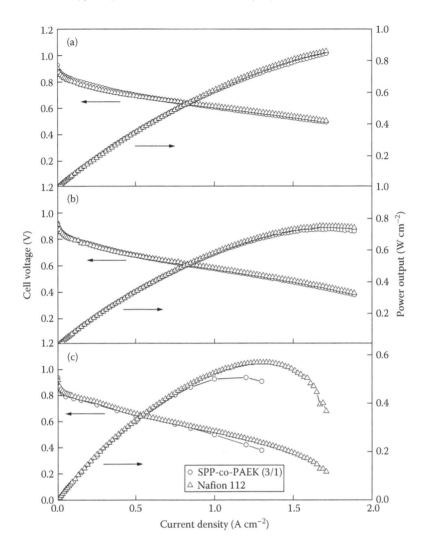

FIGURE 6.16 Performance of PEMFCs with SPP-*co*-PAEK (3/1) and Nafion 112 at 90°C/0.2 MPa with supply of H_2/air and gas humidification of (a) 82%/68% RH, (b) 48%/48%, and (c) 27%/27% RH. (Reprinted from *J. Power Sources*, 216, Zhang, X., Hu, Z., Pu, Y., Chen, S., Ling, J., Bi, H., Chen, S., Wang, L., and Okamoto, K., Preparation and properties of novel sulfonated poly(*p*-phenylene-*co*-aryl ether ketone)s for polymer electrolyte fuel cell applications, 261–268, Copyright (2012), with permission from Elsevier.)

the polarization and power output curves were charged with a slight reduction in the PEMFC performance occurred by 640 h, but no further reduction after 1000 h. The chemically stable backbones would effectively prevent the degradation during the test from the radical attack.

Miyatake and coworkers reported the cell performance of sulfonated poly(phenylene ether ketone) block copolymer membranes [31]. Figure 6.18 sorts out

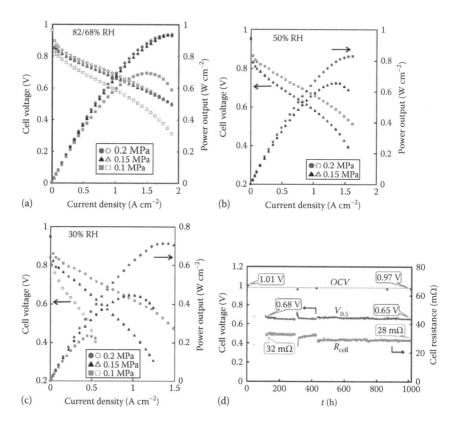

FIGURE 6.17   (a), (b), and (c) PEMFC performances for GC(3/10) at 90°C and (d) durability test for with GC(5/10) at 90°C, 0.15 MPa, 50% RH, and 0.5 A cm^{-2}. (Reprinted from *J. Power Sources*, 247, Hara, R., Endo, N., Higa, M., Okamoto, K., Zhang, X., Bi, H., Chen, S., Hu, Z., and Chen, S., Polymer electrolyte fuel cell performance of poly(arylene ether ketone)-*graft*-crosslinked-poly(sulfonated arylene ether sulfone), 932–938, Copyright (2014), with permission from Elsevier.)

their PEMFC and durability results. Good performance was obtained at high humidity (80% and 100% RH, 80°C, and ambient pressure); however, the performance became worse with decreasing humidity because of the rapid decrease in proton conductivity. For instance, $V_{0.5}$ decreased from 0.69 V (100% RH) to 0.39 V (40% RH) for membrane 3. The long-term test was also acceptable with OCV decreased from initially 1.07 to 0.94 V after 1100 h, indicating high durability.

It is quite interesting to evaluate the PEMFC performance of high-*IEC* membranes [33]. The membrane was somewhat brittle when operated in a membrane electrode assembly (MEA) fabrication. Yet, low OCV was obtained to 0.84 V by a non-hot-pressed membrane, indicating high gas permeation. Also, the voltage curve dropped significantly below 1.0 A cm^{-2}, which was inferior to Nafion. Obviously, high *IEC* values did not bring about the high electrochemical performance due to the extremely swelling state. These might be a huge error if the membrane sheets were

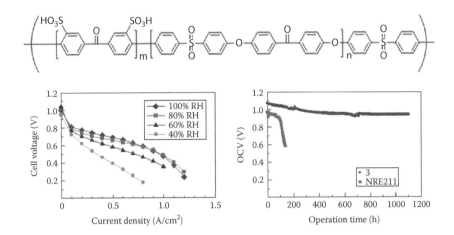

FIGURE 6.18 Chemical structures of sulfonated polybenzophenone/poly(arylene ether) block copolymer and their PEMFC performance. (Reprinted with permission from Miyahara, T., Hayano, T., Matsuno, S., Watanabe, M., and Miyatake, K., Sulfonated polybenzophenone/ poly(arylene ether) block copolymer membranes for fuel cell applications, *ACS Appl. Mater. Interfaces*, 4(6), 2881. Copyright 2012 American Chemical Society.)

too small, since the dimensional changes were somewhat imperceptible in the ex situ characterization. However, the issue became more serious when the membranes were conducted in the FC system. Too high *IEC*s caused too large swellings and seemed not proper to be adopted without dramatic improvement.

The cell performance of DA-SPPs was characterized by Sandia National Laboratories [38]. Under an idea condition (100% RH and pure O_2 as cathode gas), membrane sulfonated Diels-Alder poly(phenylene) (SDAPP4) with an *IEC* value of 2.3 mequiv g^{-1} achieved 1.02 A cm^{-2} at 0.65 V, which was 15% less than that of Nafion 212. Unfortunately, the membranes also displayed quite a large RH dependence with the same order for conductivity. Recently, the electrochemical properties were conducted by Kim and coworkers based on a series of DA-SPPSs [13]. Membrane DA-SPPS-3 ($IEC_{titr} = 1.88$ mequiv g^{-1}) exhibited slightly higher cell performance than that of Nafion in the entire range of current density, and P_{max} was obtained roughly to 0.69 W cm^{-2}, indicating a promising future. An overview comparison data on PEMFC performance is summarized in Table 6.2.

6.4.2 DIRECT METHANOL FUEL CELLS

DMFC performance on three typical polybenzophenone-typed SPPs was launched by Coutanceau's group [6]. Membrane GPS five times higher of the MeOH diffusion coefficient still exhibited better performance than the other two membranes (SPPBP and GS); the P_{max} reached to 22 mW cm^{-2} at 0.11 mA cm^{-2}, whereas 10 and 15 mW cm^{-2} for the latter two. Generally, these performances were not good enough in face of commercialization due to the too low conductivity of the as-prepared membranes.

TABLE 6.2

PEMFC Performance Data of Reported SPP Membranes

Conditions[a]	Membrane	OCV V^{-1}	$V_{0.5}$ V^{-1}	P_{max} W cm^{-2}	$\sigma_{\perp,FC}$[b] mS cm^{-1}	References
0.2/82[c]	GC(3/10)	1.00	0.75	0.93	95	[26]
	GC(5/10)	1.00	0.75	0.94	98	[26]
	BT3	0.95	0.72	>0.95	74	[32]
	BM1	0.95	0.72	>0.92	91	[35]
	SPP-*co*-PAEK(3/1)	0.95	0.69	0.85	54	[25]
	BC3	—	—	0.75[d]	—	[29]
	Nafion 112	0.95	0.71	>0.90	93	[26]
0.1/82[c]	GC(3/10)	0.96	0.69	0.69	73	[26]
	BT3	0.94	0.63	0.70	54	[32]
	BM1	0.95	0.65	0.68	68	[35]
	3[e]	1.00	0.69	—	—	[31]
	SDAPP4[e]	—	ca. 0.63	—	—	[38]
	Nafion 112	0.95	0.71	>0.90	93	[26]
0.1/50[c]	BT3	0.93	0.60	0.46	36	[32]
	BM1	0.95	0.59	0.42	39	[35]
	Nafion 112	0.95	0.60	0.39	36	[26]
0.1/30[c]	GC(3/10)	0.94	0.46	0.23	(15)	[26]
	3	—	0.39[f]	—	—	[31]

[a] PEMFC operation conditions: *x*/*y* refers to gas pressure (MPa) and RH.
[b] Recorded at 1.0 A cm^{-2}, the data in parentheses were measured at 0.5 A cm^{-2}.
[c] 82%/62% RH.
[d] At 0.3 MPa.
[e] At 80°C, 100% RH.
[f] At 80°C, 40% RH.

MEA based on DA-SPP ionomers was also tested in the DMFC system [38]. As shown in Figure 6.19, no obvious difference was found for the power output in the high-voltage and low-current-density region (>0.4 V, <0.1 A cm^{-2}, 1 M MeOH) for all the membranes; however, the conductivity effect became apparent in higher current densities, indicating the nonignorable current times resistance losses (iR losses). The DMFC performance decreased with increasing methanol concentration to 3 M, which might be attributed to the larger fuel poisoning effect at the anode. By calculating the membrane fuel crossover, a reduction of 75% MeOH permeability would be essentially required to mitigate the poisoning effect in the DMFC.

Although SPPs exhibited excellent MeOH blocking capacities (a magnitude of 10^{-7}) owing to their rigid polymer skeletons, the current DMFC performance was not qualified. A possible reason was due to the relatively low through-plane proton conductivities of those aforementioned polymers. Further studies on proton-conductive model in DMFC system are needed to reveal the mass transfer mechanism.

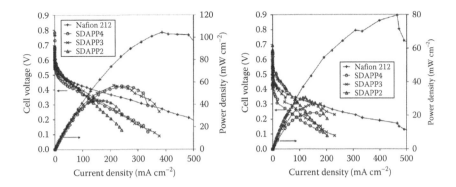

FIGURE 6.19 DMFC performance of DA-SPPs and Nafion (80°C, 200 sccm of air fed to the cathode at 15 psig backpressure; 1 mL min⁻¹ methanol fed to anode at 0 psig backpressure: 1 M left, 3 M right.). (Reprinted from *J. Power Sources*, 195, Stanis, R.J., Yaklin, M.A., Cornelius, C.J., Takatera, T., Umemoto, A., Ambrosini, A., and Fujimoto, C.H., Evaluation of hydrogen and methanol fuel cell performance of sulfonated Diels Alder poly(phenylene) membranes, 104–110, Copyright (2010), with permission from Elsevier.)

6.4.3 VANADIUM REDOX FLOW BATTERIES

Owing the similar roles as the separator and ion carrier, PEMs that severed in FCs were usually investigated for VRFB applications [39–44]. Sandia National Laboratory evaluated the VRFB performance of their DA-SPPs [45], and their VRFB properties were quite encouraging (Figure 6.20). All the samples showed good columbic efficiencies and energy efficiencies of greater than 92% and 88%, respectively. However, *IECs* had large influence on the operation cycles; a higher *IEC* value resulted in a lower cycle number as well as a faster degradation rate.

6.5 SUMMARY AND PERSPECTIVES

Currently, the critical bottleneck for PEMs is the durability problems. A lot of efforts have been made in developing suitable materials with superior overall property. As an important branch of PEMs, SPP and its derivatives are able to pave the road toward commercialization of PEMFCs due to their exceptionally stable chemical structures (phenyl–phenyl bond). The ex situ properties for these types of membranes are almost satisfied with the U.S. DOE target, 2015, as the recent achievements.

As far as the future research topics from lab to industrial scale are concerned, the cost needs to be taken into consideration. Although it is widely considered that the prices of hydrocarbon monomers are quite low, the investigation of too complicated polymer structures or noble metal catalysts would still inevitably increase the manufacturing costs. Meanwhile, postsulfonation needs to be possibly avoided due to use of large excessive amount of halogenated solvent, long reaction time, as well as the problem on acid treatment [46]. Therefore, sulfonated monomers are more preferred for the purpose of molecular designs; however, the purity is highly required indeed in order to obtain polymer with high molecular weight. Taking a look at the membrane

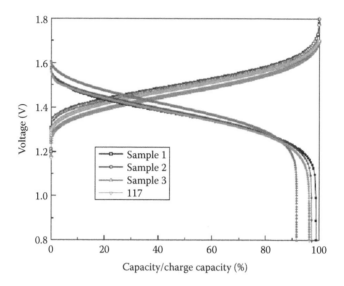

FIGURE 6.20 Charge–discharge curves of the SDAPP membranes. (Reprinted from *Electrochem. Commun.*, 20, Fujimoto, C., Kim, S., Stains, R., Wei, X., Li, L., and Yang, Z.G., Vanadium redox flow battery efficiency and durability studies of sulfonated Diels Alder poly(phenylene)s, 48–51, Copyright (2012), with permission from Elsevier.)

durability, the current progress (*ca.* 1000 h) is still needed to be improved (U.S. DOE milestone 2015, within a power degrades by 20%, durability over 5000 h).

In fact, it is rather difficult to design or expect a single type of SPP derivative to meet all the requirements. Molecular simulation seems to be an effective way as a guide for the materials designs: either to explore the structure–property relationship or to predict some molecular actions. Future works rely on omnifarious efforts, including membranes, catalysts, transport modeling study, and technical assessments.

ACKNOWLEDGMENTS

The authors appreciate the support of this research by the National Natural Science Foundation of China (21006052, 21276128), the Natural Science Foundation of Jiangsu Province (BK20140782, BK20141398), the Basic Research Program of Jiangsu Province of China (BK2011713), and the Fundamental Research Funds for the Central Universities (30920130121014).

REFERENCES

1. Luo Y. R., *Comprehensive Handbook of Chemical Bond Energies*, CRC Press, Boca Raton, FL, 2007.
2. David R. L., *CRC Handbook of Chemistry and Physics*, 90th edn., CRC Press, Boca Raton, FL, 2010.

3. Okada T., Fujiwara N., Ogata T., Haba O., Ueda M., Synthesis of regiocontrolled poly(4,6-di-*n*-butoxy-1,3-phenylene) by oxidative coupling polymerization, *Journal of Polymer Science: Part A: Polymer Chemistry*, 1997, 35(11), 2259–2266.

4. Child A. D., Reynolds J. R., Water-soluble rigid-rod polyelectrolytes: A new self-doped, electroactive sulfonatoalkoxy-substituted poly(*p*-phenylene), *Macromolecules*, 1994, 27(7), 1975–1977.

5. Bae J. M., Honma I., Murata M., Yamamoto T., Rikukawa M., Ogata N., Properties of selected sulfonated polymers as proton-conducting electrolytes for polymer electrolyte fuel cells, *Solid State Ionics*, 2002, 147(1–2), 189–194.

6. Ninivin C. L., Longeau A. B., Demattai D., Palmas P., Saillard J., Coutanceau C., Lamy C., Leger J. M., Determination of the physicochemical characteristics and electrical performance of postsulfonated and grafted sulfonated derivatives of poly(*para*-phenylene) as new proton-conducting membranes for direct methanol fuel cell, *Journal of Applied Polymer Science*, 2006, 101, 944–952.

7. Ghassemi H., McGrath J. E., Synthesis and properties of new sulfonated poly(*p*-phenylene) derivatives for proton exchange membranes. I, *Polymer*, 2004, 45, 5847–5854.

8. Ghassemi H., Ndip G., McGrath J. E., New multiblock copolymers of sulfonated poly(4′-phenyl-2,5-benzophenone) and poly(arylene ether sulfone) for proton exchange membranes. II, *Polymer*, 2004, 45, 5855–5862.

9. Hagberg E. C., Goodridge B., Ugurlu O., Chumbley S., Sheares V. V., Development of a versatile methodology for the synthesis of poly(2,5-benzophenone) containing coil-rod-coil triblock copolymers, *Macromolecules*, 2004, 37, 3642–3650.

10. Fujimoto C. H., Hickner M. A., Cornelius C. K., Loy D. A., Ionomeric poly(phenylene) prepared by Diels-Alder polymerization: Synthesis and physical properties of a novel polyelectrolyte, *Macromolecules*, 2005, 38, 5010–5016.

11. Stanis R. J., Yaklin M. A., Cornelius C. J., Takatera T., Umemoto A., Ambrosini A., Fujimoto C. H., Evaluation of hydrogen and methanol fuel cell performance of sulfonated Diels Alder poly(phenylene) membranes, *Journal of Power Sources*, 2010, 195(1), 104–110.

12. Fujimoto C. H., Hibbs M., Ambrosini A., Multi-block sulfonated poly(phenylene) copolymer proton exchange membranes, 2012, U.S. Patent 8110636 B1.

13. Lim Y., Lee H., Lee S., Jang H., Hossain M. A., Cho Y., Kim T., Hong Y., Kim W., Synthesis and properties of sulfonated poly(phenylene sulfone)s without ether linkage by Diels-Alder reaction for PEMFC application, *Electrochimica Acta*, 2014, 119, 16–23.

14. Litt M. H., Focil S. G., Liquid crystal poly(phenylene sulfonic acids), 2005, U.S. patent 20050239994 A1.

15. Si K., Dong D. X., Wycisk R., Litt M., Synthesis and characterization of poly(*para*-phenylene disulfonic acid), its copolymers and their *n*-alkylbenzene grafts as proton exchange membranes: High conductivity at low relative humidity, *Journal of Materials Chemistry*, 2012, 22, 20907–20917.

16. Zhang X., Sheng L., Higashihara T., Ueda M., Polymer electrolyte membranes based on poly(*m*-phenylene)s with sulfonic acid via long alkyl side chains, *Polymer Chemistry*, 2013, 4, 1235–1242.

17. Percec V., Okita S., Wang J. H., Synthesis of aromatic polyethers by Scholl reaction. VI. Aromatic polyethers by cation-radical polymerization of 4,4′-, 3,3′-, and 2,2′-*bis*(1-naphthoxy)biphenyls and of 1,3-*bis*(1-naphthoxy)benzene, *Macromolecules*, 1992, 25(1), 64–74.

18. Kobayashi T., Rikukawa M., Sanui K., Ogata N., Proton-conducting polymers derived from poly(ether-etherketone) and poly(4-phenoxybenzoyl-1,4-phenylene), *Solid State Ionics*, 1998, 106, 219–225.

19. Tonozuka I., Yoshida M., Kaneko K., Takeoka Y., Rikukawa M., Considerations of polymerization method and molecular weight for proton-conducting poly(*p*-phenylene) derivatives, *Polymer*, 2011, 52(26), 6020–6028.

20. Ninivin C. L., Longeau A. B., Demattel D., Coutanceau C., Lamy C., Leger J. M., Sulfonated derivatives of polyparaphenylene as proton conducting membranes for direct methanol fuel cell application, *Journal of Applied Electrochemistry*, 2004, 34, 1159–1170.

21. Wu S., Qiu Z., Zhang S., Yang X., Yang F., Li Z., The direct synthesis of wholly aromatic poly(*p*-phenylene)s bearing sulfobenzoyl side groups as proton exchange membranes, *Polymer*, 2006, 47, 6993–7000.

22. Qiu Z., Wu S., Li Z., Zhang S., Xing W., Liu C., Sulfonated poly(arylene-*co*-naphthalimide)s synthesized by copolymerization of primarily sulfonated monomer and fluorinated naphthalimide dichlorides as novel polymers for proton exchange membranes, *Macromolecules*, 2006, 39, 6425–6432.

23. Li W., Cui Z., Zhou X., Zhang S., Dai L., Xing W., Sulfonated poly(arylene-*co*-imide)s as water stable proton exchange membrane materials for fuel cells, *Journal of Membrane Science*, 2008, 315, 172–179.

24. Goto K., Rozhanskii I., Yamakawa Y., Otsuki T., Naito Y., Development of aromatic polymer electrolyte membrane with high conductivity and durability for fuel cell, *Polymer Journal*, 2009, 41(2), 95–104.

25. Zhang X., Hu Z., Pu Y., Chen S., Ling J., Bi H., Chen S., Wang L., Okamoto K., Preparation and properties of novel sulfonated poly(*p*-phenylene-*co*-aryl ether ketone)s for polymer electrolyte fuel cell applications, *Journal of Power Sources*, 2012, 216, 261–268.

26. Zhang X., Hu Z., Luo L., Chen S., Liu J., Chen S., Wang L., Graft-crosslinked copolymers based on poly(arylene ether ketone)-*gc*-sulfonated poly(arylene ether sulfone) for PEMFC applications, *Macromolecular Rapid Communications*, 2011, 32(14), 1108–1113.

27. McLean R. S., Doyle M., Sauer B. B., High-resolution imaging of ionic domains and crystal morphology in ionomers using AFM techniques, *Macromolecules*, 2000, 33(17), 6541–6550.

28. Si K., Litt M. H., Rigid rod poly(phenylene sulfonic acid) PEMs and MEAs with grafted and crosslinked biphenyl groups: Mechanical and electrical properties, *ECS Transactions*, 2011, 41(1), 1645–1656.

29. Bi H., Chen S., Chen X., Chen K., Higa M., Endo N., Okamoto K., Wang L., Poly(sulfonated phenylene)-*block*-polyimide copolymers for fuel cell applications, *Macromolecular Rapid Communications*, 2009, 30, 1852–1856.

30. Seesukphronrarak S., Ohira K., Kidena K., Takimoto N., Kuroda C. S., Ohira A., Synthesis and properties of sulfonated copoly(*p*-phenylene)s containing aliphatic alkyl pendant for fuel cell applications, *Polymer*, 2010, 51, 623–631.

31. Miyahara T., Hayano T., Matsuno S., Watanabe M., Miyatake K., Sulfonated polybenzophenone/poly(arylene ether) block copolymer membranes for fuel cell applications, *ACS Applied Materials & Interfaces*, 2012, 4(6), 2881–2884.

32. Chen S., Hara R., Chen K., Zhang X., Endo N., Higa M., Okamoto K., Wang L., Poly(phenylene) block copolymers bearing tri-sulfonated aromatic pendant groups for polymer electrolyte fuel cell applications, *Journal of Materials Chemistry A*, 2013, 1(28), 8178–8189.

33. Si K., Wycisk R., Dong D., Cooper K., Rodgers M., Brooker P., Slattery D., Litt M., Rigid-rod poly(phenylenesulfonic acid) proton exchange membranes with cross-linkable biphenyl groups for fuel cell applications, *Macromolecules*, 2013, 46(2), 422–433.

34. Nakabayashi K., Matsumoto K., Shibasaki Y., Ueda M., Synthesis and properties of sulfonated poly(2,5-diphenethoxy-*p*-phenylene), *Polymer*, 2007, 48(20), 5878–5883.

35. Chen S., Chen K., Zhang X., Hara R., Endo N., Higa M., Okamoto K., Wang L., Poly(sulfonated phenylene)-*block*-poly(arylene ether sulfone) copolymer for polymer electrolyte fuel cell application, *Polymer*, 2013, 54(1), 236–245.

36. Hara R., Endo N., Higa M., Okamoto K., Zhang X., Bi H., Chen S., Hu Z., Chen S., Polymer electrolyte fuel cell performance of poly(arylene ether ketone)-*graft*-crosslinked-poly(sulfonated arylene ether sulfone), *Journal of Power Sources*, 2014, 247, 932–938.

37. Department of Energy Hydrogen and Fuel Cells Program, Fuel Cell Technologies Office Multi-Year Research, Development and Demonstration Plan, 2012, http://www1.eere.energy.gov/hydrogenandfuelcells/mypp/pdfs/fuel_cells.pdf. Accessed June 08, 2012.

38. Stanis R. J., Yaklin M. A., Cornelius C. J., Takatera T., Umemoto A., Ambrosini A., Fujimoto C. H., Evaluation of hydrogen and methanol fuel cell performance of sulfonated Diels Alder poly(phenylene) membranes, *Journal of Power Sources*, 2010, 195, 104–110.

39. Kazacos M. S., Chakrabarti M. H., Hajimolana S. A., Mjalli F. S., Saleem M., Progress in flow battery research and development, *Journal of the Electrochemical Society*, 2011, 158(8), R55–R79.

40. Leung P., Li X., Leon C. P., Berlouis L., Lowa C. T. J., Walsh F. C., Progress in redox flow batteries, remaining challenges and their applications in energy storage, *RSC Advances*, 2012, 2, 10125–10156.

41. Schwenzer B., Zhang J., Kim S., Li L., Liu J., Yang Z., Membrane development for vanadium redox flow batteries, *ChemSusChem*, 2011, 4, 1388–1406.

42. Li X., Zhang H., Mai Z., Zhang H., Vankelecom I., Ion exchange membranes for vanadium redox flow battery (VRB) applications, *Energy & Environmental Science*, 2011, 4, 1147–1160.

43. Liu J., Zhang J. G., Yang Z., Lemmon J. P., Imhoff C., Graff G. L., Li L., Hu J., Wang C., Xiao J., Xia G. et al. Materials science and materials chemistry for large scale electrochemical energy storage: From transportation to electrical grid, *Advanced Functional Materials*, 2012, 23(8), 929–946.

44. Wang W., Luo Q., Li B., Wei X., Li L., Yang Z., Recent progress in redox flow battery research and development, *Advanced Functional Materials*, 2012, 23(8), 970–986.

45. Fujimoto C., Kim S., Stains R., Wei X., Li L., Yang Z. G., Vanadium redox flow battery efficiency and durability studies of sulfonated Diels Alder poly(phenylene)s, *Electrochemistry Communications*, 2012, 20, 48–51.

46. Bae B., Miyatake K., Watanabe M., Alternative hydrocarbon membranes by step growth, In: Matyjaszewski K., Möller M., (eds.), *Polymer Science: A Comprehensive Reference*, Elsevier, Amsterdam, The Netherlands, 2012, Chapter 10.34, p. 647.

7 Polyphosphazene Membranes

Jun Lin and Shaojian He

CONTENTS

7.1 INTRODUCTION

Polymer electrolyte membranes are one of the key components in many electrochemical energy conversion and storage devices, with the main functions of conducting ions and acting as separators [1–3]. They are inherently safer than those classical salt-liquid solvent electrolytes in lithium batteries, as there will be no problem of internal shorting or electrolyte leakage using these polymer electrolytes. Unlike the flammable organic solvents that always raise the combustion concerns, these polymer electrolytes are nonflammable. In fuel cells, the application of solid polymer electrolytes could overcome the leakage problem associated with liquid electrolytes.

Among the solid polymer electrolytes for lithium batteries, poly(ethylene oxide) (PEO) has been the first developed and most extensively investigated [4–7]. The oxygen atoms on the backbone of PEO could help with the salt solvation and ion-pair separation via the coordination with lithium ions. PEO has relatively low glass-transition temperature (T_g) so that the polymer chain is flexible enough to facilitate ion conduction. However, as the ions can only migrate through the amorphous phase in a polymer, the presence of crystalline phase in PEO significantly impedes the ion transport, and the ionic conductivity only ranges from 10^{-8} to 10^{-4} S cm^{-1} between room temperature and 100°C [8].

In hydrogen/air fuel cells that operate at a temperature lower than 80°C, a per-fluorosulfonated polymer developed by DuPont, Nafion®, is currently the polymer electrolyte membrane of choice [9–12]. Despite the fact that Nafion has relatively high proton conductivity under high humidity condition with good chemical and mechanical stability, Nafion suffers a few critical problems such as low proton con-ductivity at relatively low humidity and high temperature (>100°C) and high metha-nol permeability (unsuitable for direct methanol fuel cells [DMFCs]). In addition, Nafion is expensive and poses potential threats to the environment with hydrofluoric acid release upon the decomposition under typical fuel cell operating conditions. There have been a variety of membrane materials investigated as the alternatives for Nafion, including sulfonated poly(aryl ether ketones), sulfonated polyimides, sulfo-nated poly(aryl ether sulfones), and acid-doped polybenzimidazole (PBI) [10,13–16]. However, most of these membranes suffer from poor oxidative stability, and long-term operation in fuel cell remains questionable.

Polyphosphazenes have emerged as a class of promising solid polymer electrolyte materials for energy applications due to their inherently high stability and a wide range of synthetic variability. Herein, a summary is presented on the synthesis of polyphosphazenes, membrane fabrication and characterization, and applications for lithium batteries, fuel cells, and dye-sensitized solar cells (DSSCs).

7.2 POLYPHOSPHAZENES

Polyphosphazenes are inorganic polymers that consist of a backbone with alternat-ing phosphorus–nitrogen atoms [17,18]. Each phosphorus atom is attached with two side groups, which could be organic, inorganic, or organometallic (Figure 7.1). Polyphosphazenes with a broad range of properties can be obtained by changing

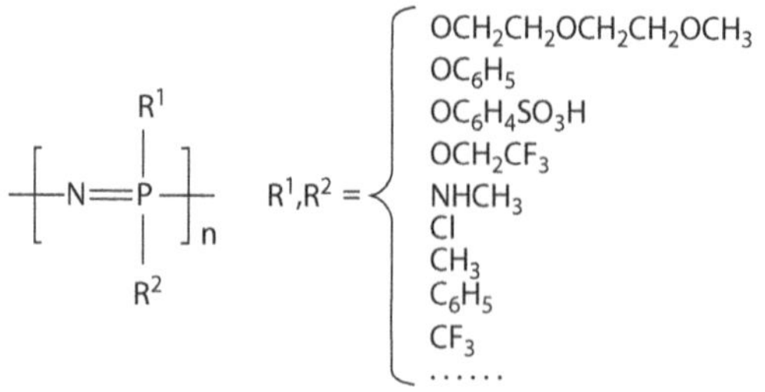

$$R^1, R^2 = \begin{cases} OCH_2CH_2OCH_2CH_2OCH_3 \\ OC_6H_5 \\ OC_6H_4SO_3H \\ OCH_2CF_3 \\ NHCH_3 \\ Cl \\ CH_3 \\ C_6H_5 \\ CF_3 \\ \cdots\cdots \end{cases}$$

FIGURE 7.1   General structures of polyphosphazenes.

the side groups with various functionalities. In particular, polyphosphazenes with small or flexible side groups can be fully amorphous, with a T_g as low as $-80°C$, which allows them to be potential candidate for lithium batteries [19]. As compared to classical hydrocarbon materials, polyphosphazenes are more resistant to thermo-oxidative decomposition due to the fact that both phosphorus and nitrogen atoms in the backbone are in their highest oxidation states. This feature makes polyphospha-zenes especially attractive for fuel cell applications where the oxidative stability of most hydrocarbon materials has been unsatisfactory.

7.2.1 SYNTHESIS OF POLYPHOSPHAZENES

In general, the synthesis of polyphosphazenes can follow three main routes as follows:

1. Preparation of a macromolecular precursor, poly(dichlorophosphazene) (PDCP), which is then reacted with nucleophilic reagents to replace all the chlorine atoms
2. *Living* cationic condensation polymerization of substituted phosphoranimines
3. Ring opening polymerization of substituted cyclophosphazenes

Among these methods, the first one has been the most widely used since it was first developed in 1960s by Allcock et al. (Figure 7.2) [20–22]. In most of cases, an important step involved in this process is the thermal ring opening polymerization of commercially available hexachlorocyclophosphazene (HCCP) to produce PDCP under strictly controlled experimental conditions (250°C, vacuum of 10^{-2} torr, and reaction time of 24–72 h). Usually, the molecular weight of PDCP is quite high with relatively wide molecular weight distribution. Immediate nucleophilic substitution or appropriate stabilization of PDCP is always necessary because of the extreme

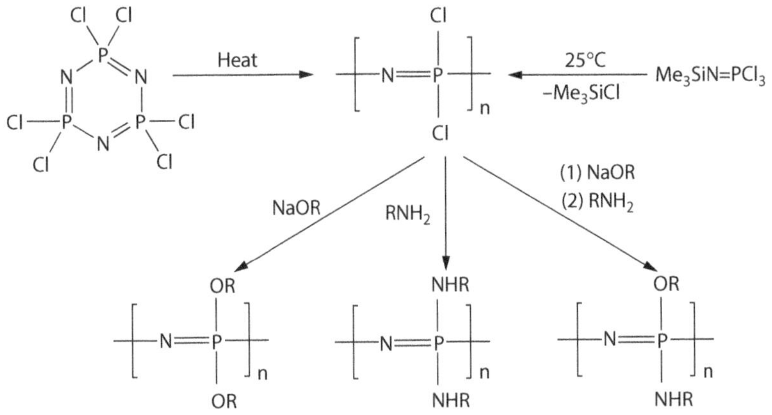

FIGURE 7.2 Scheme of polyphosphazene synthesis via macromolecular substitution.

reactivity of the chlorine atoms toward water and nucleophilic groups. It was found that even 0.0009% water resulted in the cross-linking of PDCP within days [23]. Therefore, to avoid cross-linking and branching, extreme caution must be taken during the final stages of ring opening polymerization step, and complete replacement of chlorine atoms during the PDCP substitution step should be achieved. There are also other approaches to synthesize PDCP [17], among which a *living* cationic polymerization of trichloro(trimethylsilyl)phosphoranimine, $Cl_3P=NSiMe_3$, with small amounts of PCl_5 as the initiator (Figure 7.2) is the most interesting [24,25]. PDCP with controllable molecular weight and narrow polydispersities can be prepared via such synthesis route. Another advantage of this approach is that the synthesis can take place at ambient temperature with much better reproducibility, as compared to high-temperature requirement in ring opening polymerization or other condensation processes with low reproducibility.

Later on, such ambient temperature synthesis approach was extended to a variety of organo-substituted phosphoranimines, to directly synthesize polyphosphazenes with controlled molecular weight and low polydispersities, so that PDCP preparation and following chlorine substitution steps were eliminated [26]. Such *living* cationic condensation polymerization method also allows the preparation of polyphosphazenes with complex structures such as comb, star, and dendritic architectures, as well as block and graft copolymers with organic macromolecules [18].

7.2.2 Polyphosphazenes Containing Ethyleneoxy Side Groups

7.2.2.1 Poly[*bis*(2-(2-methoxyethoxy)ethoxy)phosphazene]

The first polyphosphazene electrolyte was discovered in 1984, when Blonsky et al. synthesized poly[*bis*(2-(2-methoxyethoxy)ethoxy)phosphazene] (MEEP) with the reaction between the sodium salt of 2-(2-methoxyethoxy)ethanol and the polymer precursor, PDCP, in the presence of tetra-*n*-butylammonium bromide (Figure 7.3) [19]. Solvent-free complexes of $(LiSO_3CF_3)_{0.25}$·MEEP (the molar ratio of Li^+ to MEEP repeating unit) had a room temperature ionic conductivity in the range of 10^{-5} S cm^{-1}, one to three orders of magnitude higher than that of complexes of PEO and $LiSO_3CF_3$. Such high ionic conductivity is mainly attributed to the flexible phosphorus–nitrogen backbone and the oligomeric etheric side chains of MEEP, which is an amorphous polymer with a T_g as low as −84°C [27]. Further study found that the treatment of $(LiSO_3CF_3)_{0.25}$·MEEP films with acetonitrile vapor resulted in the film plasticization and the improvement of ionic conductivity by ca. 300 times [28].

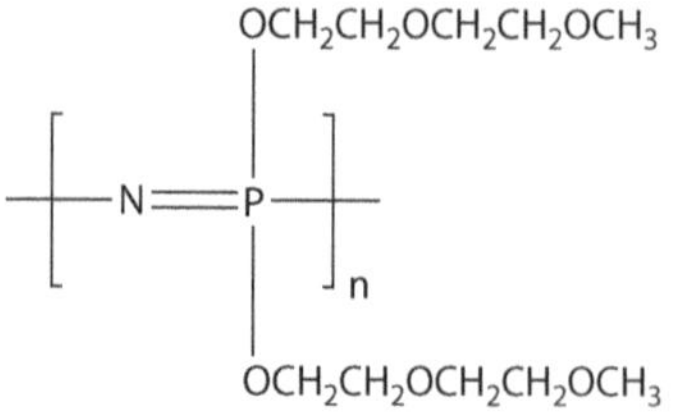

$$OCH_2CH_2OCH_2CH_2OCH_3$$

$$OCH_2CH_2OCH_2CH_2OCH_3$$

FIGURE 7.3   Structure of MEEP.

In the attempt to understand the ionic conduction mechanism in MEEP–lithium triflate complexes, Allcock et al. used ^{1}H, ^{13}C, ^{15}N, and ^{31}P nuclear magnetic resonance (NMR) spectroscopy as well as molecular dynamics (MD) simulations to probe the interaction between the polymer and Li$^+$ ions [29]. It was found that Li$^+$ ions were loosely bound to the oxygen atoms of the polymer, and this facilitated the ion transport through the polymer matrix. Based on the ^{15}N natural abundance NMR results, the authors also concluded that the nitrogen atoms on the polymer backbone were not involved in coordination to Li$^+$ ions. MD simulations, however, showed that Li$^+$ ions coordinate with both the oxygen and nitrogen atoms in the polymer with no preferred binding sites existing between the ions and the polymers.

In another study, Luther et al. took a different approach to investigate the ion transport in MEEP–lithium triflate complexes by using ^{15}N-labeled MEEP as they claimed natural abundance deficiency of ^{15}N nuclei would inhibit effective study of normal MEEP by NMR spectroscopy [30]. During the preparation process of ^{15}N-labeled MEEP, the polymer precursor, ^{15}N-labeled PDCP, was synthesized directly from the reaction between ^{15}N-labeled ammonium sulfate and PCL$_5$ instead of ring opening polymerization of HCCP. The differences in the P–N stretching and P–O bending observed in the IR and Raman spectra indicated the association of Li$^+$ ions with both nitrogen and oxygen atoms in the polymer. Nitrogen atoms were found to play a significant role in Li$^+$ ion complexation and transport, as indicated by the chemical shift changes in NMR (^{15}N and ^{31}P) and significant decrease in NMR temperature-dependent spin–lattice relaxation minimum values ($T_{1\,min}$) upon the addition of lithium triflate to the polymer. The experimental results concurred with MD simulation prediction, suggesting a *pocket* structure where a Li$^+$ ion coordinates with one backbone nitrogen atom and the neighboring ether oxygen atoms (Figure 7.4).

Wang and Balbuena employed ab initio quantum mechanics (QM) calculations and classical MD simulations to study the microscopic structure of MEEP–lithium triflate complexes [31]. They found that the backbone nitrogen atoms have stronger

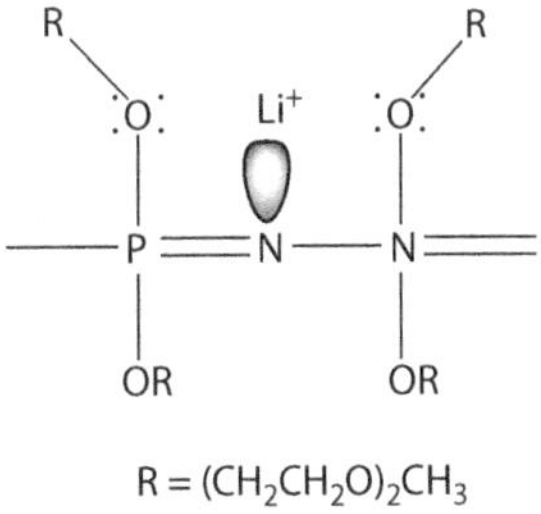

R = (CH$_2$CH$_2$O)$_2$CH$_3$

FIGURE 7.4 A *pocket* structure formed by the lithium association with the backbone nitrogen atom and the neighboring ether oxygen atoms. (Reprinted with permission from Luther, T.A., Stewart, F.F., Budzien, J.L., LaViolette, R.A., Bauer, W.F., Harrup, M.K., Allen, C.W., and Elayan, A., On the mechanism of ion transport through polyphosphazene solid polymer electrolytes: NMR, IR, and Raman spectroscopic studies and computational analysis of ^{15}N-labeled polyphosphazenes, *J. Phys. Chem. B*, 107, 3168. Copyright 2003 American Chemical Society.)

affinity for Li$^+$ ions as compared to the oxygen atoms in the pendent groups and that a more stable *pocket* structure would be the one where Li$^+$ ions coordinate with multiple nitrogen atoms and one methoxy oxygen.

By using Raman spectroscopy, Frech et al. found that the maximum conductivity of MEEP–lithium triflate complexes occurred at a composition consisting exclusively of contact ion pairs and triple anions without free triflate ions [32]. The authors inferred that dissociation of these immobile, ionically associated species would generate transient triflate ions, which move via a series of dissociation–reassociation steps through the liquid-like domains of highly disordered ethylene oxide side chains. Further study by Lee and Allcock examined the effect of the cations (monovalent and divalent) and anions on ionic conduction in MEEP [33]. The anions selected in this study were *bis*(trifluoromethanesulfonyl)imidate (TFSI) and triflate anions, between which TFSI has more extensive charge delocalization and a higher degree of dissociation. It was found that MEEP electrolytes with the TFSI anions had higher ionic conductivity than that with the triflate anions, even though MEEP–TFSI complexes also had higher T_g. Additionally, the maximum ionic conductivity was reached for MEEP with magnesium salts at a salt molar fraction half of that with lithium salts. These results suggest that anions are the dominant charge carriers for the ion conduction in MEEP, while cations interact with the polymer strongly via coordination with nitrogen or oxygen atoms, resulting in the reduction of the backbone flexibility.

7.2.2.2 MEEP Derivatives

Despite the fact that MEEP has high ionic conductivity when complexed with lithium salts, the potential of MEEP as a solid membrane electrolyte is significantly limited by an inherent lack of dimensional stability as it flows slowly even under light pressure. The last 30 years have seen a large number of polyphosphazenes developed and investigated as alternative candidates for solid polymer electrolytes, most of which have been based on MEEP, containing ethyleneoxy side groups.

7.2.2.2.1 Backbone and Side-Chain Modification

Allcock et al. synthesized a series of mixed-substituent polyphosphazenes with both alkoxy and alkoxy ether side groups (Figure 7.5) [34]. All the synthesized mixed-substituent polyphosphazenes had lower T_g than MEEP, indicating higher polymer flexibility. However, these polymers all had lower conductivity than MEEP, because there were fewer ion coordination sites available with the introduction of noncoordinating alkoxy side groups that overweighed the increase of free volume due to the enhanced polymer flexibility. The maximum ionic conductivity was found to decline for the polyphosphazene–lithium triflate complexes with the increase in the length of the alkyl component of the alkoxy groups.

Allcock's group further studied the effect of polymer side-chain structure on the ionic conductivity and dimensional stability, by comparing polyphosphazenes bearing linear and branched oligoethyleneoxy side chains with MEEP [35]. For the polymers with linear oligoethyleneoxy side chains, the maximum ionic conductivity of polymer–lithium triflate complexes was very close to that of MEEP

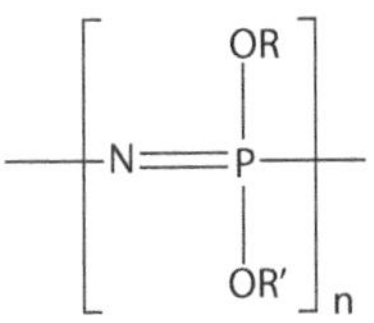

$R' = CH_2CH_2OCH_2CH_2OCH_3$

(1) $R = (CH_2)_2CH_3$
(2) $R = (CH_2)_3CH_3$
(3) $R = (CH_2)_4CH_3$
(4) $R = (CH_2)_5CH_3$
(5) $R = (CH_2)_6CH_3$
(6) $R = (CH_2)_7CH_3$
(7) $R = (CH_2)_8CH_3$
(8) $R = (CH_2)_9CH_3$

FIGURE 7.5 Mixed-substituent polyphosphazenes with both alkoxy and alkoxy ether side groups. (Reprinted with permission from Allcock, H.R., Napierala, M.E., Cameron, C.G., and O'Connor, S.J.M., Synthesis and characterization of ionically conducting alkoxy ether/alkoxy mixed-substituent poly(organophosphazenes) and their use as solid solvents for ionic conduction, *Macromolecules*, 29, 1951. Copyright 1996 American Chemical Society.)

and increased with the increase of side-chain length. Similar to MEEP, these linear side-chain polymers also have poor dimensional stability. In comparison, the branched side-chain polymers have much better dimensional stability, probably due to the improved side-chain entanglement. Meanwhile, these branched side-chain polymers still have similar conductivity as MEEP, with similar T_g in the range of $-82°C$ to $-79°C$.

A highly branched polymer containing 5 branches from a cyclotriphosphazene pendent side group (with 26 ethyleneoxy units per repeat unit) was synthesized to further investigate the effect of complex branching on physical properties and ionic conductivity of polymers (Figure 7.6) [36]. Slightly higher conductivity than MEEP was achieved for this highly branched polymer, with improved dimensional stability due to the much enhanced chain entanglements.

Aryloxy side groups were introduced in another attempt to prepare solid polyphosphazene electrolytes with good mechanical properties [37]. Two types of polyphosphazenes were synthesized, mixed-substituent phenoxy/oligoethyleneoxy polyphosphazenes and single-substituent aryloxy polyphosphazenes with oligoethyleneoxy units attached to the aromatic rings in the para position, as shown in Figure 7.7. All the polymers have higher T_g than MEEP, with the improved mechanical properties. The optimum ionic conductivity of polymer–lithium TFSI (LiTFSI) complexes was in the range of $\sim 10^{-5}$ S cm^{-1}, close to that of MEEP system. With the same 50:50 molar ratio of aryloxy/oligoethyleneoxy groups, the single-substituent polymers had higher T_g than cosubstituent polymers, with lower ionic conductivity.

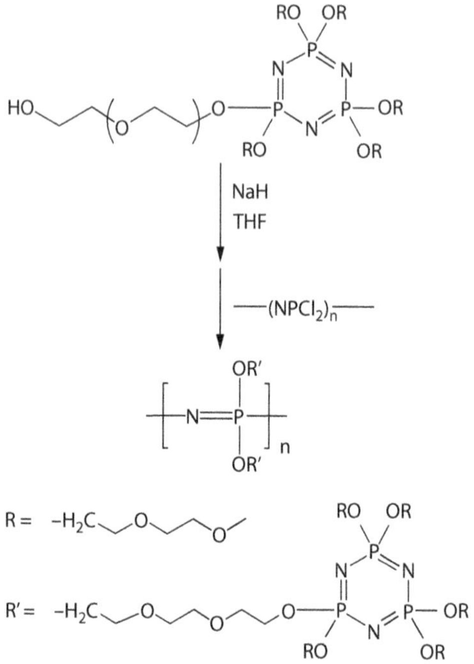

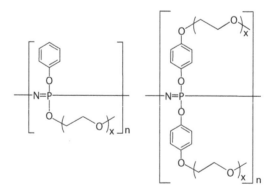

FIGURE 7.6 Synthesis scheme of a highly branched polyphosphazene. (Reprinted with permission from Allcock, H.R., Sunderland, N.J., Ravikiran, R., and Nelson, J.M., Polyphosphazenes with novel architectures: Influence on physical properties and behavior as solid polymer electrolytes, *Macromolecules*, 31, 8026. Copyright 1998 American Chemical Society.)

FIGURE 7.7 Structure of mixed-substituent phenoxy/oligoethyleneoxy polyphosphazenes and single-substituent aryloxy polyphosphazenes with oligoethyleneoxy units attached to the aromatic rings in the para position. (Reprinted from *Solid State Ionics*, 156, Allcock, H.R. and Kellam, E.C., The synthesis and applications of novel aryloxy/oligoethyleneoxy substituted polyphosphazenes as solid polymer electrolytes, 401–414, Copyright 2003, with permission from Elsevier.)

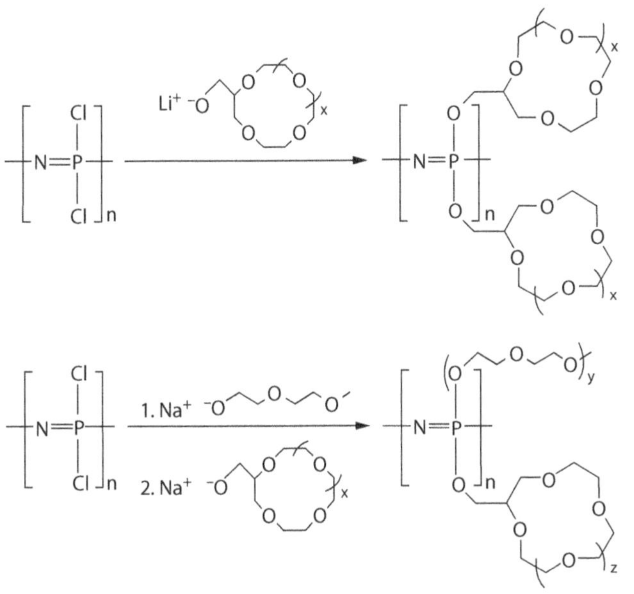

FIGURE 7.8 Synthesis of single-substituent and mixed-substituent polyphosphazenes with crown ether and MEE groups. (Reprinted with permission from Allcock, H.R., Olmeijer, D.L., and O'Connor, S.J.M., Cation complexation and conductivity in crown ether bearing polyphosphazenes, *Macromolecules*, 31, 753. Copyright 1998 American Chemical Society.)

An interesting series of MEEP derivatives containing crown ether side groups was reported by Allcock et al. (Figure 7.8) [38]. Single-substituent and mixed-substituent polyphosphazenes with crown ether and 2-(2-methoxyethoxy)ethoxy (MEE) groups (1:3 molar ratio) were synthesized, in which three types of crown ether groups with different cavity sizes were used. All the polymers containing crown ether side groups have higher T_g than MEEP, with the single-substituent polymers having higher T_gs (~40°C) than mixed-substituent ones (~64°C). No measurable ionic conductivity was obtained for single-substituent polymer–LiCLO$_4$ complexes. For the mixed-substituent polymers, the highest ionic conductivity was in the range of 10^{-5} S cm^{-1}, and great reduction in ionic conductivity was observed with the formation of a 1:1 or 2:1 crown ether–cation complex. All these results suggest significant contribution to the ion conduction by the cations. The authors believed the strong inhibition of cation migration due to stable crown ether–cation complexation was much more dominant than the facilitation of solvation and ion-pair separation that also resulted from such complexation.

Conner et al. incorporated 2-(2-phenoxyethoxy)ethoxy (PEE) side groups to the polymer backbone, with the intention of improving dimensional stability without reducing the number of coordination sites available [39]. Increase in T_g was found to correlate with increasing content of PEE side groups. Gel polymer electrolytes (GPEs) formed by polymer, LiTFSI, and propylene carbonate (PC) in a molar ratio of 8:1:6.3 showed the ionic conductivity close to MEEP GPEs, ~10^{-4} S cm^{-1}.

The incorporation of strongly polar functional groups such as sulfones or sulfoxides raised the polymer T_g to almost ambient temperature, and the polymer–lithium triflate complexes had conductivity lower than 10^{-8} S cm^{-1} [40]. These polar groups also interfered strongly with the ion transport, even in the presence of PC.

Single-ion conductive polyphosphazenes were also synthesized and studied as solid polymer electrolyte [41,42]. By reacting lithium bisulfite with allyl side groups, polyphosphazenes containing sulfopropyl oligo(oxyethylene) side chains were obtained [41]. However, room temperature conductivity of only 10^{-8} to 10^{-7} S cm^{-1} was achieved with these polymers, as the salt dissociation energy was still too high to obtain high conductivity. Arylsulfonimide groups were attached to the polyphosphazene backbone in order to increase free volume and facilitate cation transport [42]. With the increase of lithium sulfonimide substituent component in the polymer, the ionic conductivity decreased in spite of the increasing amount of charge carrying species. Maximum conductivity at ambient temperature was ~10^{-6} S cm^{-1}, which was lower than that of the model system with MEE and phenoxy side groups complexed with free lithium sulfonimide. Such behavior was related with the macromolecular motion that was faster in the model system.

Poly(aminophosphazenes) were studied by several groups for the potential applications for solid polymer electrolytes [43–48]. Dimensionally stable polyphosphazene electrolyte membranes, poly[*bis*(pentylamino)phosphazene] (PPAP) and poly[*bis*(hexylamino)phosphazene] (PHAP), were first synthesized and complexed with LiCLO$_4$, respectively [43]. However, the room temperature ionic conductivity of these materials was unsatisfactory, in the range of only 10^{-9} and 10^{-7} S cm^{-1}. Therefore, in the later study, the MEE side groups were incorporated to the polymer backbone to prepare a series of mixed-substituent polyphosphazenes [44]. Combination of MEE and alkylamine groups produced freestanding films with enhanced conductivity. The higher the MEE substituent content, the higher the ionic conductivity. The maximum room temperature conductivity reached 2.2×10^{-5} S cm^{-1}, comparable with that of MEEP.

Wiemhöfer and coworkers synthesized a series of poly(aminophosphazene) electrolytes via the replacement of chlorine atoms on the polymer precursor, PDCP, which was prepared by *living* cationic polymerization [45–48]. Polyphosphazene [NP(NHR)$_2$]n [R = oligo(propylene oxide) units] was cross-linked by UV irradiation to give a mechanically stable polymer electrolyte membrane, with the highest ionic conductivity reaching 6.5×10^{-6} S cm^{-1} at ambient temperature [45]. They also synthesized a series of mixed-substituent polyphosphazenes with both (2-methoxyethyl)amino and *n*-propylamino side groups, which were mechanically stable after being complexed with lithium triflate [46,47]. Room temperature conductivity of only ~3×10^{-7} S cm^{-1} was obtained for these polymers complexed with lithium triflate. The addition of Al$_2$O$_3$ nanoparticles led to an almost 100 times increase of ionic conductivity, up to 10^{-5} S cm^{-1}. Based on the results from solid-state NMR spectroscopy, the authors found that 10%–15% of the Li$^+$ ions were quite mobile and 25%–30% of the Li$^+$ ions were located within the pockets formed by the side chains, with the rest 60% of the Li$^+$ ions rather immobile and unavailable for ion transport [48]. With the addition of Al$_2$O$_3$ particles, the coordination of Li$^+$ ions to the alumina surface was observed, which may contribute to the ion transport as indicated by the significant increase of ionic conductivity upon the formation of composites.

7.2.2.2.2 Block Copolymers

Block copolymers were expected to combine the best properties of MEEP with those of traditional organic polymers. *Living* cationic polymerization was employed to synthesize polyphosphazenes-*co*-PEO block copolymers with controlled molecular weight and narrow polydispersities [49]. By using PEO-based phosphoranimines either as a terminator or to be converted to macroinitiators via reaction with PCl_5, triblock or diblock copolymers of polyphosphazene with PEO could be prepared. The room temperature ionic conductivities of these materials, after being complexed with varying ratios of lithium triflate, were in the range of 10^{-6} and 10^{-5} S cm^{-1}, with the maximum conductivity ~100 times higher than that of PEO homopolymer-based electrolytes and very close to that of MEEP.

7.2.2.2.3 Cross-Linking

Cross-linking, including chemical cross-linking, ^{60}Co gamma radiation, and UV radiation, has been employed in MEEP to overcome its problem of low dimensional stability [50–53]. The ionic conductivity was practically not changed after cross-linking, along with much improved dimensional stability. Among these cross-linking methods, chemical cross-linking involves the introduction of a difunctional reagent into the system, thus resulting in impurities, while gamma irradiation has its limited accessibility and relatively high cost. Therefore, UV cross-linking seems to be the most promising, as it was found to be possible even in the absence of a photoinitiator.

An interesting type of polymer electrolytes based on cross-linked cyclotriphosphazenes was reported by Kaskhedikar et al. [54]. They used terminal vinyl units in the oligoether side chains of the cyclotriphosphazenes to prepare cross-linked lithium salt-containing polymer membranes. The polymers containing 50% of vinyl units did not form a complete interpenetrating network, leading to dimensional instability, even though the maximum room temperature conductivity for these polymers reached 3.2×10^{-5} S cm^{-1}, the same as MEEP. The highly cross-linked polymers containing 100% vinyl units have excellent mechanical properties; however, the conductivity was only in the range of 10^{-9} and 10^{-8} S cm^{-1}. The addition of MEEP to this highly cross-linked polymer network (1:1 mass ratio) resulted in a polymer electrolyte with a high room temperature conductivity of 1.3×10^{-5} S cm^{-1} and good mechanical properties.

7.2.2.2.4 Blends

Blending is also a common practice to improve certain properties in polymers. There have been many reports in the literature on MEEP blends, to enhance the dimensional stability without compromising the ionic conduction [55–60].

Abraham et al. conducted a series of pioneering work on MEEP blends to improve the dimensional stability of MEEP, using high-molecular-weight PEO, poly(propylene oxide) (PPO), in situ photopolymerized poly(ethylene glycol diacrylate) (PEGDA), or polyvinylpyrrolidone (PVP) [55–58]. The blends showed significantly improved mechanical properties as compared to MEEP lithium salt systems, even though most of the blends complexed with lithium salts were amorphous according to x-ray diffraction (XRD) results (MEEP/PEO blends complexed with lithium salts except LiTFSI were semicrystalline). Such improvement was attributed to the reinforcement of mechanical strength by the dispersion of mechanically strong second polymer component into the MEEP matrix. MEEP/PEO (55:45 in weight) blends complexed with TFSI attained the room temperature conductivity of 6.7×10^{-5} S cm^{-1}, which was about the same as MEEP–TFSI complex and higher than all other MEEP–lithium salt complexes. More detailed study on MEEP/PEO–lithium salt (LiClO$_4$ or LiBF$_4$) complexes was accomplished by differential scanning calorimetry (DSC) and ^{7}Li NMR spectroscopy [59]. DSC results were consistent with early results on XRD, showing the existence of both amorphous MEEP and crystalline PEO phases. Based on ^{7}Li NMR results such as line width and spin–lattice relaxation time (T_1), Li$^+$ ions present in the amorphous and crystalline phases of the MEEP/PEO blends could be characterized separately. It was found that significant Li$^+$ ion mobility in the MEEP-like phase occurs only above T_g.

Landry et al. investigated the blending miscibility of MEEP with a series of polymer components [60]. Polymers containing acidic or proton-donating groups were found to have good miscibility with MEEP due to hydrogen-bonding-type interaction between the blend components, while polymers such as polystyrene (PS), poly(methyl methacrylate) (PMMA), and polyvinyl acetate (PVAc) were not miscible with MEEP. The dimensional stability of MEEP was improved in the miscible blends, and T_g remained significantly lower below additivity indicating ion transport may remain fast. However, no ionic conductivity results were reported.

7.2.2.2.5 Composites

Another approach employed to improve the dimensional stability of MEEP was to form composites with rigid inorganic components, which could be particles or a network.

Kim et al. used sol-gel method to prepare mechanically strong MEEP–silica composite electrolyte from a mixture of MEEP, tetraethyl orthosilicate (TEOS), and lithium triflate [61]. There was practically no change in T_g with the change in the amount of TEOS, indicating the segmental motion of MEEP was not affected by the incorporation of silica network, and there were no covalent connections between MEEP and silica. Meanwhile, the amount of TEOS content did affect the ionic conductivity of the composite materials. The increasing TEOS component always resulted in the decrease of ionic conductivity, presumably due to the longer conduction pathway resulting from the phase separation between MEEP and TEOS. The highest ionic conductivity reached 4.6×10^{-5} S cm^{-1} for the composites with 10 wt% TEOS and 15 wt% lithium sulfate.

To improve the interaction between MEEP and inorganic components so that dimensional stability could be further enhanced, Brusatin et al. functionalized MEEP with free hydroxyls to the end of side groups of MEEP so that a 3D network could be formed due to covalent bonding between MEEP and silica [62]. Homogeneous and mechanically stable hybrid materials were obtained. No phase separation was observed. The highest conductivity of 3×10^{-5} S cm^{-1} was achieved for the samples doped with lithium triflate at 60°C. One of the major drawbacks for this type of hybrid materials, however, was the lack of control over the hydroxyl density and the possible presence of residual hydroxyl groups that will prevent their use in combination with Li electrodes.

The further study on sol-gel preparation of MEEP–silicate composite materials involved the synthesis of a polyphosphazene precursor via the covalent linkage of MEEP with an organometallic alkoxide (triethoxysilane group) [63]. The following hydrolysis and condensation produced a covalently interconnected hybrid material with controlled morphologies and physical properties. A maximum ionic conductivity of 7.69×10^{-5} S cm^{-1} was achieved for the composite materials complexed with LiTFSI salt.

Besides silicate, aluminum oxide (α-Al$_2$O$_3$) was also studied as the inorganic components to be dispersed in MEEP–LiClO$_4$ complexes to form the composites [64]. The ambient temperature conductivity was increased with the addition of α-Al$_2$O$_3$, with a maximum conductivity of 9.7×10^{-5} S cm^{-1} obtained for the composite electrolytes with 2.5 wt% α-Al$_2$O$_3$. The significant increase in cation transference number suggested the major contribution of Li$^+$ ions to the ionic conduction and the enhancement of Li$^+$ ion transport due to the α-Al$_2$O$_3$ addition. Similar to the early study on MEEP–silicate composites [61], the Arrhenius temperature dependence of composite conductivity was observed with the addition of α-Al$_2$O$_3$, indicating the ion transport in the composites follows a thermally activated process by a diffusion mechanism. In comparison, the ion transport in MEEP–lithium salt complexes was believed to migrate through the free volumes that provided the segmental motion of the polymer chains, indicated by a Vogel–Tamman–Fulcher (VTF) temperature dependence of conductivity.

7.2.2.3 Polymers with Pendent Cyclotriphosphazenes

Inoue et al. first reported a series of polycascade polymers with pendent oligo(oxyethylene)cyclotriphosphazene side groups for the application for solid polymer electrolytes [65–68]. In spite of relatively rigid backbone of PS or its derivatives (Figure 7.9), conductivity as high as 10^{-4} S cm^{-1} was reached for these polymers, suggesting the backbone flexibility is not an important factor in ion transport. Instead, the segmental mobility of side chains and the ion–dipole interactions were believed to play an important role in determining the complex conductivity, and the ions migrate through a continuous conducting phase formed by ethyleneoxy side groups without being affected by the backbone mobility. It was found that the conductivity was even higher for the polymers with longer spacer between cyclotriphosphazene side groups and aromatic rings, indicating the formation of continuous conducting phase became more favorable with the introduction of longer spacer [68].

Another series of norbornene-based polymer electrolytes bearing pendent cyclotriphosphazene side groups were synthesized and characterized by Allcock et al. [69–73]. In comparison with styrene-based polymers prepared by Inoue et al. [65–68], most of the norbornene-based polymers exhibit no crystallization

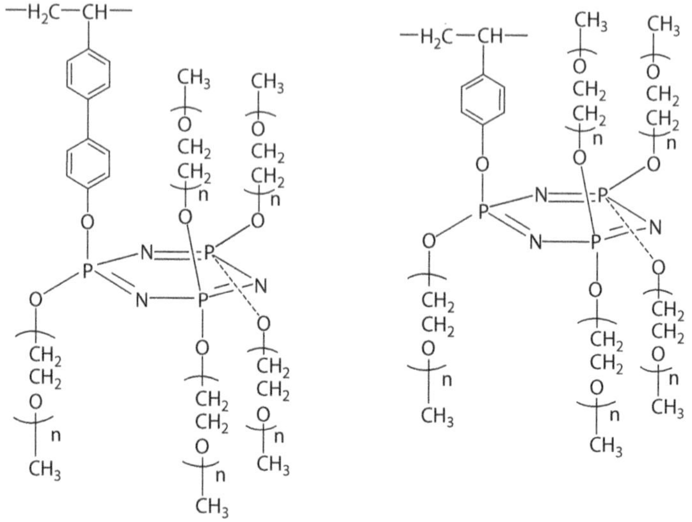

FIGURE 7.9   Structure of polycascade polymers. (Reprinted with permission from Inoue, K., Nishikawa, Y., and Tanigaki, T., High-conductivity electrolytes composed of polystyrene carrying pendant oligo(oxyethylene)cyclotriphosphazenes and lithium perchlorate, *J. Am. Chem. Soc.*, 113, 7609. Copyright 1991 American Chemical Society.)

or melting transitions above the ambient temperature. In general, the complexes of lithium salts with these norbornene-based polymers have ionic conductivity in the range of 10^{-5} S cm^{-1}, similar as that of MEEP–salt complexes. With the increase of the lengths of the oligoethyleneoxy side units attached to the cyclotriphosphazene rings, the conductivities of the polymer–salt complexes increased generally due to the increasing amount of available cation coordination sites [70]. Oxygen atoms were also introduced to the backbone of polynorbornenes in one study, and it was found that ionic conductivity was increased despite the higher glass-transition temperature [71]. This was related with the additional coordination sites provided by the backbone oxygen atoms. Another two studies on these norbornene-based materials deal with improving water resistance for operation in a humid environment [72,73]. Polynorbornene copolymers with two different types of pendent cyclotriphosphazene side groups, one of which bore PEE and the other with 4-(lithium carboxalato)phenoxy (LiOOCPh) units, were prepared [72]. Room temperature conductivity of 5.9×10^{-6} S cm^{-1} was achieved for the polymer with 90% PEE and 10% LiOOCPh, with surface contact angle at 76°. To further improve the polymer hydrophobicity, 2,2,2-trifluoroethoxy (TFE) side groups were introduced to the cyclotriphosphazene rings in addition to MEE side groups [73]. Of the three systems investigated, including composite blends of all TFE and all MEE groups and copolymers from all TFE and all MEE monomers, the homopolymer system with 40% MEE and 60% TFE units on every cyclotriphosphazene exhibited room temperature conductivity of 1.2×10^{-5} S cm^{-1} with a modest hydrophobic surface (water contact angle of 77.7°).

7.2.3 SULFONIC ACID FUNCTIONALIZED POLYPHOSPHAZENES

The most well-studied polyphosphazene-based membrane materials for fuel cells are those bearing sulfonic acid groups. They have been synthesized by either a post-sulfonation approach or direct synthesis from the macromolecular precursor, PDCP.

7.2.3.1 Postsulfonation of Polyphosphazenes

Until now, most of sulfonated polyphosphazenes have been prepared by electrophilic sulfonation of poly[(aryloxy)phosphazenes]. Montoneri et al. achieved the sulfonated poly[*bis*(phenoxy)phosphazene] (SPBPP) by applying sulfur trioxide (SO_3) in chlorinated solvents [74–76]. They showed that only when the molar ratio of SO_3 to poly[*bis*(phenoxy)phosphazene] was higher than one did sulfonation of aromatic rings occur. Otherwise, the insufficient amount of SO_3 could only attack the backbone nitrogen atoms to form $\equiv N \rightarrow SO_3$ complexes. Kurachi and Kajiwara used chlorosulfonic acid to sulfonate poly(anilinophosphazenes) in tetrachloroethane [77]. Poly[(anilinosulfamic)phosphazenes] of two different sulfonation degrees were obtained. Allcock and coworkers conducted sulfonation of a variety of aryloxy- or arylamino-containing small-molecule phosphazene cyclic trimmers and polyphosphazenes using chlorosulfonic acid, concentrated or fuming sulfuric acid [78]. The sulfonation was found to mainly take place at the para position, while meta-substitution was observed when there are alkyl groups in the para position. During the sulfonation process, chain cleavage and polymer degradation occurred, with the higher stability for polymers bearing the bulkier aryloxy side groups. Allcock's group later attempted to use 1,3-propanesultone as the sulfonating reagent to prepare sulfonated poly(aminophosphazenes), but the reaction yields were quite low even for the polymers with amine groups attached to the backbone with alkyl spacer groups (Figure 7.10) [79]. Most of sulfonated polyphosphazenes from the aforementioned early work were water soluble when they have moderately high sulfonation degrees, and therefore, no results were reported on their fuel cell–related properties, such as proton conductivity and water uptake.

The first study on sulfonated polyphosphazenes for proton exchange membrane (PEM) applications was performed by Wycisk and Pintauro [80]. They chose a total of four poly[(aryloxy)phosphazenes], poly[(3-methylphenoxy)(phenoxy)phosphazene], poly[(4-methylphenoxy)(phenoxy)phosphazene], poly[(3-ethylphenoxy)(phenoxy)phosphazene], and poly[(4-ethylphenoxy)(phenoxy)phosphazene], to be sulfonated with SO_3 in dichloroethane (Figure 7.11). It was found that ethylphenoxy derivatives underwent significant degradation during the sulfonation process, but the methylphenoxy derivatives were resistant to degradation, and the sulfonation degree was easy to be controlled by varying the molar ratio of SO_3 to the polymers. Water-insoluble sulfonated polymers were obtained for poly[(3-methylphenoxy)(phenoxy)phosphazene] with an IEC up to 2.1 mmol g^{-1} and poly[(4-methylphenoxy)(phenoxy)phosphazene] with an IEC up to 1.3 mmol g^{-1}. *N,N*-Dimethylacetamide or 1-methyl-2-pyrrolidinone was used as the solvent to prepare the polymer solution from which the membranes of sulfonated poly[(methylphenoxy)(phenoxy)phosphazene] were cast. For membranes with an IEC in the range of 1–1.5 mmol g^{-1}, water uptake less than 40 wt% and specific resistivity lower than 10 ohm m in 0.1 N NaCl were observed.

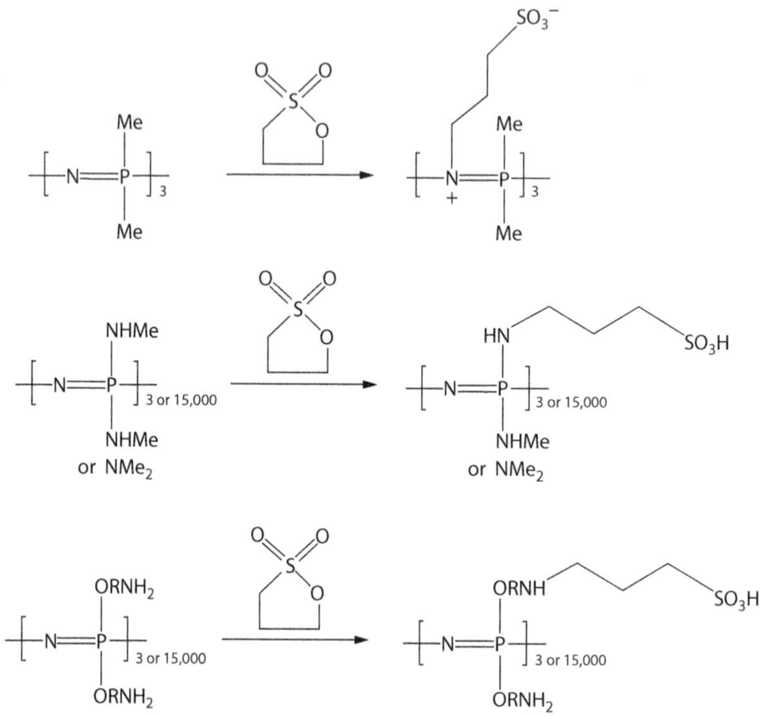

FIGURE 7.10 Sulfonation of poly(aminophosphazenes) with 1,3-propanesultone. (Reprinted with permission from Allcock, H.R., Klingenberg, E.H., and Welker, M.F., Alkanesulfonation of cyclic and high polymeric phosphazenes, *Macromolecules*, 26, 5512. Copyright 1993 American Chemical Society.)

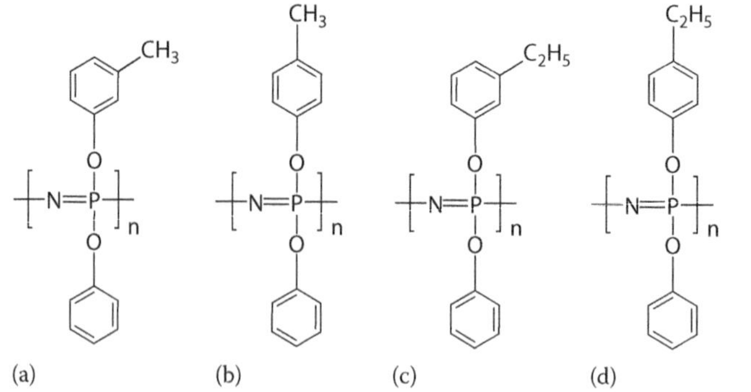

FIGURE 7.11 The structures of the four poly[(aryloxy)phosphazenes]: (a) poly[(3-methylphenoxy)(phenoxy)phosphazene], (b) poly[(4-methylphenoxy)(phenoxy)phosphazene], (c) poly[(3-ethylphenoxy)(phenoxy)phosphazene], and (d) poly[(4-ethylphenoxy)(phenoxy) phosphazene]. (Reprinted from *J. Membr. Sci.*, 119, Wycisk, R. and Pintauro, P.N., Sulfonated polyphosphazene ion-exchange membranes, 155–160, Copyright 1996, with permission from Elsevier.)

Further work on sulfonated poly[bis(3-methylphenoxy)phosphazene] (SPBMPP) confirmed that methyl groups promoted aromatic ring sulfonation, which started when the ratio of SO_3 to the polymer was higher than 0.64 [81]. For the sulfonated polymers with an IEC lower than 1.0 mmol g^{-1}, the sulfonation mainly occurred at the para position, while all available carbon atoms at the aromatic ring could be attacked by SO_3 when the polymer IEC was higher than 1.0 mmol g^{-1}. It was also found that during the sulfonation process, there was always the formation of $\equiv N \rightarrow SO_3$ complexes and $\equiv N^+H\cdots\cdots HSO_4^-$ (Figure 7.12), which would increase the polymer hydrophilicity and water uptake. Wide-angle XRD results indicated 3D crystal structure was lost after sulfonation, but a 2D ordered phase was retained in sulfonated polymers (Figure 7.13) [82]. Guo et al. furthered their study to prepare cross-linked SPBMPP of an IEC of 1.4 mmol g^{-1} [83]. Polymer membranes containing the photoinitiator (benzophenone) were cast before being exposed to UV light to be cross-linked. Cross-links were believed to be created by the recombination of macroradicals, which were formed by benzophenone-derived diradicals abstracting hydrogen atoms from the methyl carbons of the methylphenoxy side groups (Figure 7.14) [84,85]. Both water and methanol sorptions were reduced after cross-linking,

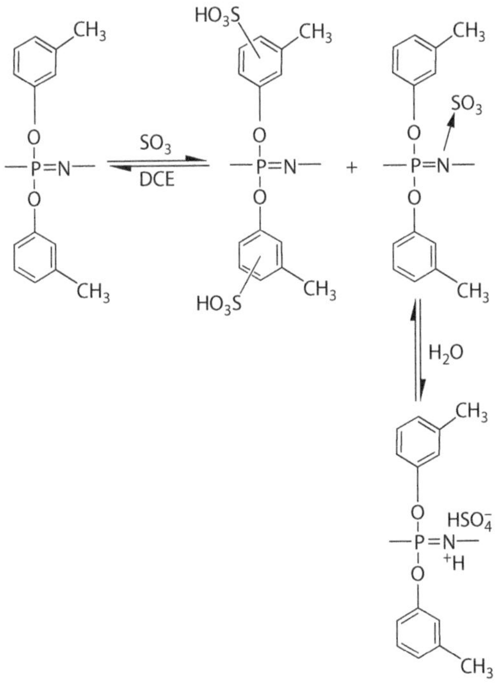

FIGURE 7.12 Sulfonation of poly[bis(3-methylphenoxy)phosphazene]. (Tang, H., Pintauro, P.N., Guo, Q., and O'Connor, S.: Polyphosphazene membranes. III. Solid-state characterization and properties of sulfonated poly[bis(3-methylphenoxy)phosphazene]. J. Appl. Polym. Sci. 1999. 71. 387–399. Copyright Wiley-VCH Verlag GmbH & Co. KGaA. Reprinted with permission.)

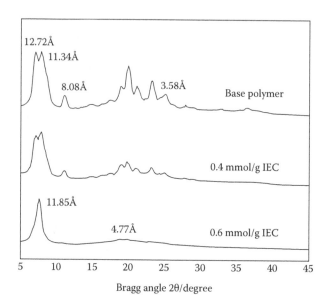

FIGURE 7.13 Wide-angle XRD of dry poly[*bis*(3-methylphenoxy)phosphazenes] at 25°C. (Tang, H. and Pintauro, P.N.: Polyphosphazene membranes. IV. Polymer morphology and proton conductivity in sulfonated poly[*bis*(3-methylphenoxy)phosphazene] films. *J. Appl. Polym. Sci.* 2000. 79. 49–59. Copyright Wiley-VCH Verlag GmbH & Co. KGaA. Reprinted with permission.)

whereas proton conductivity was not affected by cross-linking. Methanol diffusion coefficient of 1.62×10^{-8} cm^2 s^{-1} and water diffusion coefficient less than 1.2×10^{-8} cm^2 s^{-1} at 30°C were both significantly lower than that in a Nafion 117 membrane, and proton conductivity of 0.040 S cm^{-1} at 25°C was only 30% lower than that in Nafion 117. In addition, the cross-linked membrane demonstrated excellent oxidative stability, with less than 5% weight loss after being soaked in Fenton's reagent for 24 h at 68°C. In comparison, commercial styrenic cation exchange membranes suffered almost complete degradation under the same testing condition.

Besides cross-linking, blending technique was also adopted by Pintauro's group in order to control the water/methanol uptake properties of sulfonated polyphosphazene membranes [86–89]. An inert polymer, polyacrylonitrile (PAN) or polyvinylidene fluoride (PVDF), was first used to blend with SPBMPP [86–88]. Blended membranes were prepared by casting the polymer solution and then UV cross-linking the dried membranes with benzophenone as the photoinitiator. The SEM pictures of blended membranes revealed a phase-separated morphology with micrometer-sized domains, indicating the incompatibility between the inert polymer and SPBMPP (Figure 7.15a and b). For SPBMPP/PAN blended membranes, room temperature proton conductivity varied from 0.008 to 0.05 S cm^{-1}, as compared to 0.06 S cm^{-1} for Nafion 117. Methanol permeability of SPBMPP/PAN blended membranes was much lower than that of Nafion 117, and the highest membrane selectivity (the ratio of proton conductivity to methanol permeability) was 2.8 times higher than that of Nafion 117. Promising

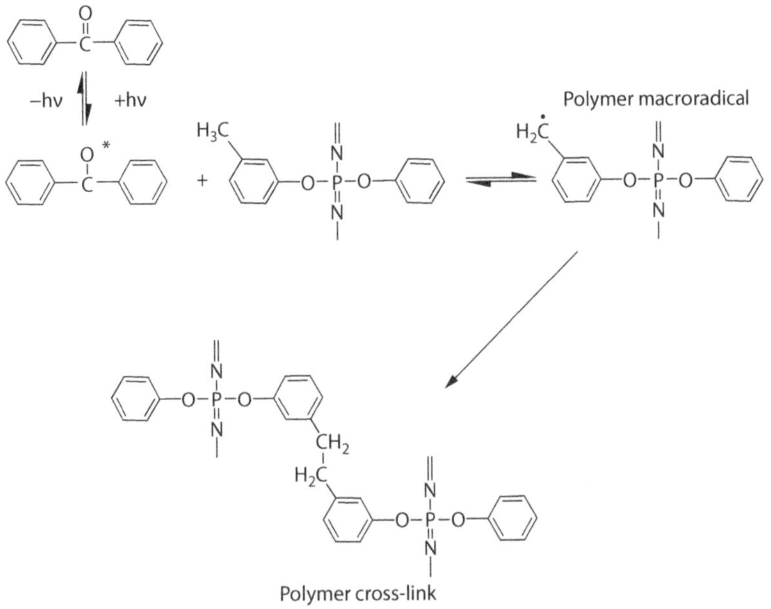

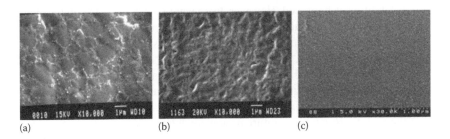

FIGURE 7.14 UV cross-linking of poly[*bis*(3-methylphenoxy)phosphazenes]. (Graves, R. and Pintauro, P.N.: Polyphosphazene membranes. II. Solid-state photocrosslinking of poly[(alkylphenoxy)(phenoxy)phosphazene] films. *J. Appl. Polym. Sci.* 1998. 68. 827–836. Copyright Wiley-VCH Verlag GmbH & Co. KGaA. Reprinted with permission.)

FIGURE 7.15 SEM images of sulfonated polyphosphazene blends: (a) SPBMPP-PVDF, (b) SPBMPP-PAN, and (c) SPBPP-PBI. (Reprinted with kind permission from Springer Science+Business Media: *Fuel Cells II*, Polyphosphazene membranes for fuel cells, 2008, 157, Wycisk, R. and Pintauro, P.N./Scherer, G., ed.)

DMFC performance was achieved with SPBMPP/PAN blended membrane–based membrane electrode assembly (MEA) as will be discussed in Section 7.3.2.

Miscible blends were prepared when PBI was used to blend with SPBPP of an IEC of 1.4 mmol g^{-1} [89]. In this work, concentrated sulfuric acid was used as both solvent and sulfonating agent to prepare SPBPP. In SPBPP/PBI blends, PBI also serves as the cross-linking agent due to its basicity, resulting in the interactions between the acidic protons of the sulfonic groups and the basic nitrogen atoms of the imidazole units of PBI. Therefore, good compatibility was achieved in the membrane, as indicated by the nano-sized domain morphology in the SEM analysis (Figure 7.15c) [88]. The T_g of blended membranes decreased at first with PBI content not greater than 10 wt% and then increased significantly when PBI content was more than 20 wt%. Such interesting behavior was ascribed to the superposition of two opposite effects [88]. On the one hand, the lower membrane IEC due to blending with PBI results in weaker hydrogen-bonding network and higher flexibility in phosphazene backbone movement, which will lower the T_g. On the other hand, the energetically favorable interaction with PBI of high T_g (399°C) leads to higher T_g of the blend. The room temperature proton conductivities of membranes ranged from 0.005 to 0.08 S cm^{-1}, with water uptake of between 18 and 75 wt% and methanol permeability of between 1.0×10^{-7} and 2.0×10^{-6} cm^2 s^{-1} (60°C in 1 M methanol), depending on the membrane IEC and PBI content (Figure 7.16). The decrease of proton conductivity with PBI content was attributed to the combined effects from three aspects: (1) a dilution effect due to the addition of less conductive component (PBI), (2) lower water uptake, and (3) lower effective IEC resulting from the acid–base interaction between SPBPP and PBI. The calculated membrane selectivity was more than 50% higher than Nafion, suggesting methanol crossover in a blended SPBPP/PBI membrane–based MEA would be much lower than that in Nafion MEA of the same thickness (DMFC test results will be discussed in Section 7.3.2).

A self-cross-linkable hydroxymethylphenoxy group was introduced by Song et al. to the polyphosphazene backbone before the polymer sulfonation with fuming sulfuric acid [90]. Cross-linked structures were formed after the sulfonated polymer membrane was heated at 80°C under vacuum, with methyol groups reacting with the neighboring aromatic rings or between themselves. After cross-linking, membrane thermal stability was improved. However, there were no data reported on the change of proton conductivity and methanol permeability of membranes upon cross-linking.

Burjanadze et al. took another approach in order to stabilize the water content in the SPBPP membranes [91]. SPBPP was trapped in an interpenetrating hydrophilic network of hexa(vinyloxyethoxyethoxy)cyclotriphosphazene (CVEEP), which was cross-linked during the membrane casting process (Figure 7.17). Membranes exhibited good thermal and mechanical properties with relatively high IEC of between 1.62 and 1.79 mmol g^{-1}. Room temperature proton conductivity of up to 0.013 S cm^{-1} was obtained for these membranes. The same group also reported the synthesis of sulfonated poly[*bis*(2-phenoxyethoxy)phosphazene] (PhEP–SO$_3$H) with concentrated sulfuric acid [92]. The sulfonation was confirmed by Fourier transform infrared (FTIR) spectroscopy, as shown in Figure 7.18. Before the sulfonation, the –P=N– stretching of the backbone and the –P–O–C– stretching of the side chains were at ~1250 and 1200 cm^{-1}, respectively. However, both peaks were overlapped by the broad peaks centered at 1130 cm^{-1} for –SO$_3$Na in PhEP–SO$_3$H, which also exhibited two peaks between 600 and 640 cm^{-1} due to the existence of the free sulfonic acid groups. In this work, the polymer precursor, PDCP, was prepared via *living* cationic polymerization of Cl$_3$P=NSi(CH$_3$)$_3$ monomer. The sulfonated polymer in sodium salt form was pressed to pellets and tested in a water-saturated nitrogen atmosphere for the ionic conductivity. Depending on the test temperature from 20°C to 60°C, conductivity ranged from 0.018 to 0.071 S cm^{-1}, with the highest conductivity obtained at 40°C.

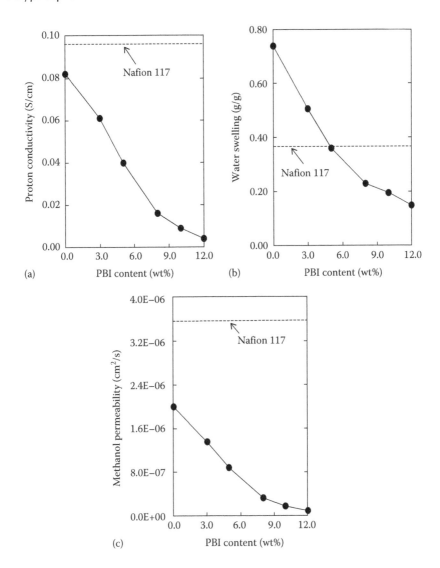

FIGURE 7.16 Effect of PBI content on the properties of SPBPP-PBI blended membranes: (a) room temperature proton conductivity, (b) room temperature water uptake, and (c) methanol permeability at 60°C with 1.0 M methanol. (Reprinted with kind permission from Springer Science+Business Media: *Fuel Cells II*, Polyphosphazene membranes for fuel cells, 2008, 157, Wycisk, R. and Pintauro, P.N./Scherer, G., ed.)

However, membranes were not prepared from solution casting due to the crystalline form of the sulfonated polymer. In another report by Fei et al., a hybrid polymer with a polynorbornene backbone and pendent sulfonated cyclic phosphazene side groups was synthesized [93]. ^{60}Co gamma radiation cross-linking was applied after the polymer film was cast, in order to restrict water swelling. The prepared membranes had relatively low IEC ranging from 0.267 to 0.49 mmol g^{-1}. It was also found that there

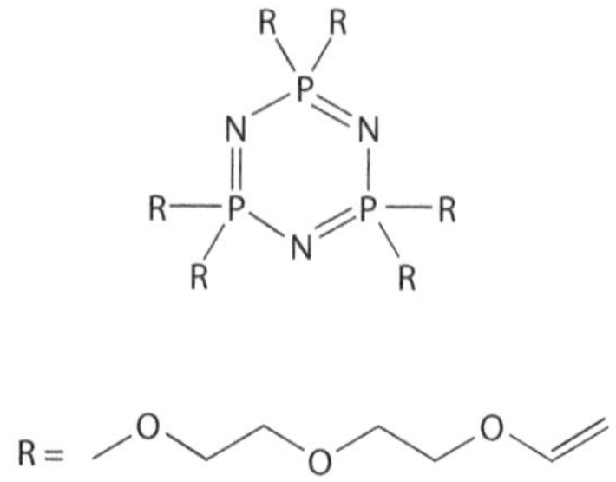

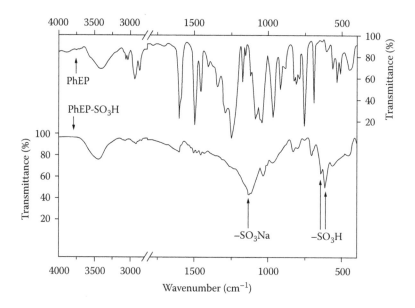

FIGURE 7.17 Structure of CVEEP. (Reprinted from *Solid State Ionics*, 177, Burjanadze, M., Paulsdorf, J., Kaskhedikar, N., Karatas, Y., and Wiemhöfer, H.D., Proton conducting membranes from sulfonated poly[*bis*(phenoxy)phosphazenes] with an interpenetrating hydrophilic network, 2425–2430, Copyright 2006, with permission from Elsevier.)

FIGURE 7.18 Characteristic FTIR spectra of PhEP and PhEP–SO$_3$H. (Reprinted from *Solid State Ionics*, 169, Paulsdorf, J., Burjanadze, M., Hagelschur, K., and Wiemhöfer, H.D., Ionic conductivity in polyphosphazene polymer electrolytes prepared by the living cationic polymerization, 25–33, Copyright 2004, with permission from Elsevier.)

was unusually high water and methanol uptake in the membranes, even though both proton conductivity and methanol permeability are quite low (more than 100 times smaller than that for Nafion). However, all the calculated membrane selectivity was lower than that of Nafion.

Recently, an interesting modified postsulfonation procedure was developed by He et al. to prepare a series of perfluoroalkyl sulfonic acid–functionalized polyphosphazenes (PSA-P) [94]. A polymer precursor, poly[(4-bromophenoxy) (phenoxy)phosphazene], was synthesized first before reacting with potassium

1,1,2,2-tetrafluoro-2-(1,1,2,2-tetrafluoro-2-iodoethoxy) ethanesulfonate by copper-catalyzed coupling reaction. The prepared PSA-P membranes demonstrated quite low water uptake (ranging from 5 to 12 wt% at 25°C), mainly due to relatively low IEC (ranging from 0.3 to 0.6 mmol g^{-1}). Compared with SPBPP with similar IEC, the room temperature conductivity of PSA-P was 50% higher (0.009 S cm^{-1} for PSA-P vs. 0.006 S cm^{-1} for SPBPP), which could be attributed to the stronger acidity of perfluoroalkyl sulfonic acid groups in PSA-P. In addition, PSA-P membranes also exhibited better thermal and oxidative stability compared to SPBPP, probably due to the presence of C–F bonds in PSA-P.

7.2.3.2 Direct Synthesis of Sulfonated Polyphosphazenes

Difficulty has long been posed for single-step replacement of chlorine atoms of the PDCP precursor with a sulfonic acid–containing nucleophile to achieve direct sulfonation, because the reaction between sulfonic acid groups and PDCP led to unstable derivatives that later underwent degradation.

One of the earliest works on realizing direct sulfonation of polyphosphazenes was reported by Ganapathiappan et al. (Figure 7.19) [95]. The disodium salt of 2-hydroxy-ethanesulfonic acid was reacted with an excess of linear PDCP in the presence of 15-crown-5, a phase-transfer catalyst, to obtain a partially substituted intermediate, which might also be cross-linked. Then the sodium salt of 2-(2-methoxyethoxy) ethanol or low-molecular-weight poly(ethylene glycol methyl ether) was added to replace the sulfonate groups and remaining chlorine atoms, during which any

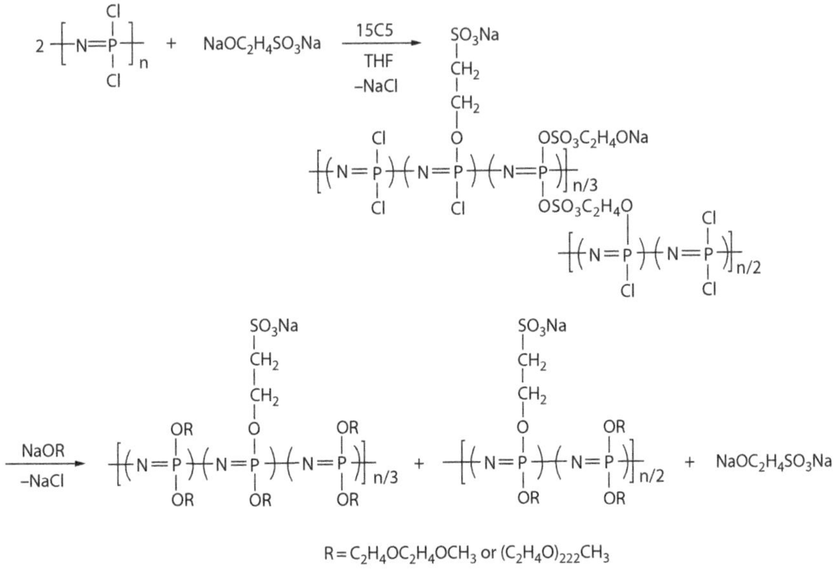

FIGURE 7.19 Direct synthesis of sulfonated polyphosphazene with disodium salt of 2-hydroxyethanesulfonic acid. (Reprinted with permission from Ganapathiappan, S., Chen, K., and Shriver, D.F., A new class of cation conductors: Polyphosphazene sulfonates, *Macromolecules*, 21, 2299. Copyright 1988 American Chemical Society.)

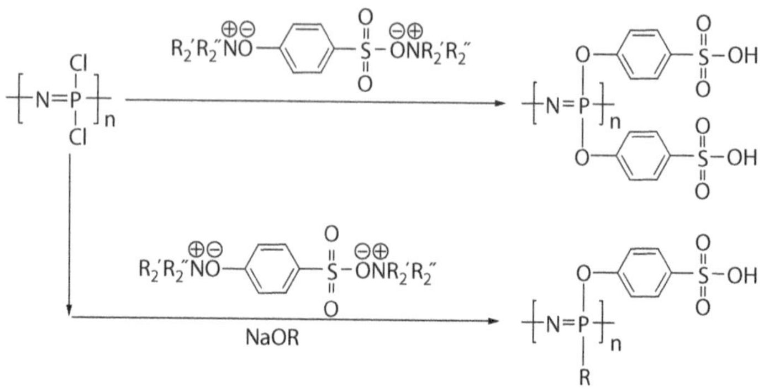

FIGURE 7.20 Direct synthesis of sulfonated polyphosphazenes using the *noncovalent* protection method. (Reprinted with permission from Andrianov, A.K., Marin, A., Chen, J., Sargent, J., and Corbett, N., Novel route to sulfonated polyphosphazenes: Single-step synthesis using "noncovalent protection" of sulfonic acid functionality, *Macromolecules*, 37, 4075. Copyright 2004 American Chemical Society.)

cross-linking by the alkoxysulfonate was also removed. Through such synthetic procedure, the length of the ether side groups and the ratio of the sulfonate groups to ether groups can be easily varied. In general, about 50% of sulfonate groups in the reactants were incorporated in the final sulfonated polymer.

A *noncovalent protection* method was developed by Andrianov et al. to prepare directly sulfonated polyphosphazenes (Figure 7.20) [96]. The formation of unstable intermediates was prevented by protecting hydroxybenzenesulfonic acid with hydrophobic dimethyldipalmitylammonium ions, which were not reactive against PDCP. The protective dimethyldipalmitylammonium groups can be easily removed by treatment with potassium hydroxide and hydrochloric acid sequentially. It was found that the conversion of sodium phenoxide into a hydrophobic ammonium salt did not lower its reactivity against PDCP. High-molecular-weight sulfonated polyphosphazene homopolymers and copolymers were synthesized. However, no membrane was prepared from these polymers, and no characterization results were reported on ionic conductivity, water uptake, or methanol permeability.

7.2.4 Phosphonic Acid–Functionalized Polyphosphazenes

As an alternative to the sulfonic acid–functionalized polyphosphazenes, polyphosphazenes with phosphonic acid groups have been investigated by Allcock's group [97–102]. Their first attempt was to attach phosphonate side groups to the polymer backbone by treatment of halogenated phosphate esters with hydroxyl-functionalized phosphazenes [97]. However, such polymers with phosphorus–oxygen–carbon linkages were prone to hydrolysis and thermal cleavage. Further studies on the introduction of phosphonate units to form phosphorus–carbon linkages have involved both reactions between bromomethylene-phenoxy side groups and a sodium dialkyl phosphite and between a dialkyl chlorophosphate and lithiophenoxy side groups that were formed by treating bromophenoxy groups with *n*-butyllithium [98].

Diphenyl chlorophosphate was used in a later study to react with lithio-functionalized aryloxyphosphazenes that were formed by treatment of bromophenoxy aryloxyphosphazenes with more efficient *tert*-butyllithium [99]. About half of bromophenoxy side groups were converted to aryloxy diphenyl phosphonate ester units. Following hydrolysis with sodium hydroxide and subsequent acidification with hydrochloric acid, phenylphosphonic acid–functionalized poly(aryloxyphosphazenes) were obtained, with about half of phosphonate ester groups were converted to hydroxyl groups (Figure 7.21). Membranes of phosphonated polyphosphazenes with either 3-methylphenoxy or 4-methylphenoxy units were casted from N,N-dimethylformamide. For 4-methylphenoxy substituted polymers with an IEC between 1.17 and 1.43 mmol g^{-1}, membrane proton conductivity varied from 0.038 to 0.054 S cm^{-1}, and water uptake ranged from 19 to 32 wt%. In comparison, membranes of 3-methylphenoxy derivatives had slightly lower T_g (38.5°C vs. 42.4°C), higher proton conductivity (0.056 vs. 0.045 S cm^{-1}), and lower

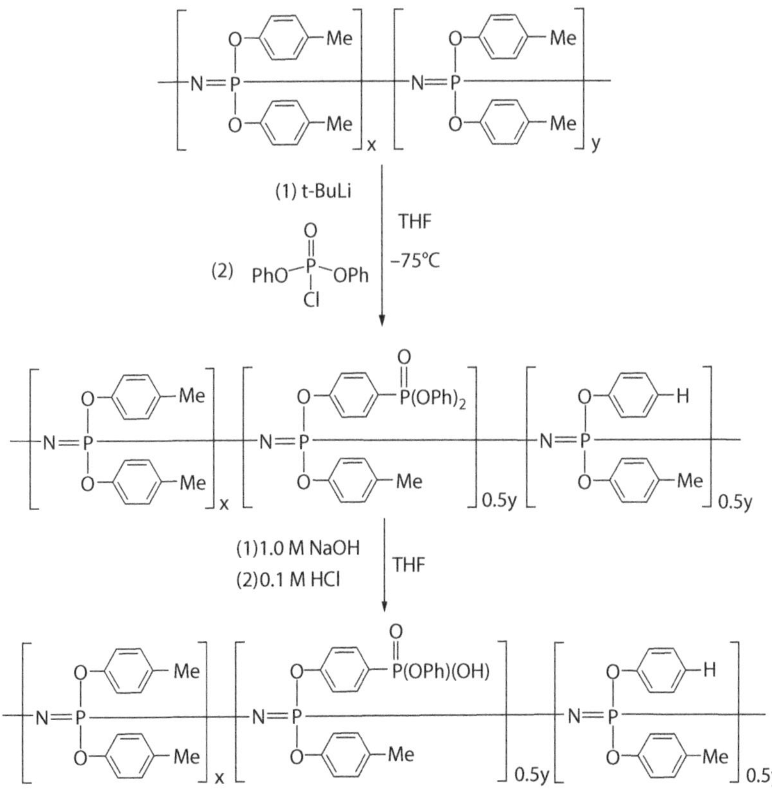

FIGURE 7.21 Synthesis of phenylphosphonic acid–substituted polyphosphazenes. (Reprinted from *J. Membr. Sci.*, 201, Allcock, H.R., Hofmann, M.A., Ambler, C.M., Lvov, S.N., Zhou, X.Y., Chalkova, E., and Weston, J., Phenyl phosphonic acid functionalized poly[aryloxyphosphazenes] as proton-conducting membranes for direct methanol fuel cells, 47–54, Copyright 2002, with permission from Elsevier.)

water uptake (12 vs. 25 wt%) with a similar IEC (1.35 vs. 1.34 mmol g^{-1}). Upon exposure to ^{60}Co gamma radiation, these membranes of 3-methylphenoxy derivatives were cross-linked, although it was found that various radiation doses had little effect on membrane proton conductivity or water uptake. The authors also evaluated the methanol diffusion in these cross-linked phosphonated polyphosphazenes with 3-methylphenoxy units [100,101]. At 80°C and 2.8 bar, methanol diffusion coefficients were found to be between 0.14×10^{-6} and 0.27×10^{-6} cm^2 s^{-1} with 3 M aqueous methanol solution, which was more than 10 times lower than that for Nafion 117. With 12 M aqueous methanol solution, there was three to five times increase in methanol diffusion coefficients for cross-linked phosphonated polyphosphazene membranes than that with 3 M methanol, while there was practically no change in methanol diffusion coefficient for Nafion 117, indicating phosphonated polymers were much more sensitive to the methanol concentration. At 120°C and 6.2 bar, methanol diffusion coefficients in phosphonated polyphosphazene membranes with 50% (v/v) methanol solution were approximately nine times lower than that in Nafion 117. With the measured proton conductivity, the calculated membrane selectivity for phosphonated polyphosphazenes was more than two times higher than that for Nafion 117 over the temperature range of between 22°C and 125°C.

7.2.5 SULFONIMIDE-FUNCTIONALIZED POLYPHOSPHAZENES

The incorporation of highly acidic sulfonimide side groups to polyphosphazenes was investigated by Hofmann et al. via the reaction scheme shown in Figure 7.22 [103]. They first synthesized the sulfonimide-containing phenoxide nucleophiles, of which sulfonimide side group is not nucleophilic and allows for the displacement of chlorine atoms in PDCP without the formation of unstable intermediates or cross-linked products. During the polymer synthesis process, sodium 4-methylphenoxide was first used to replace 50% of the chlorine atoms in PDCP to ensure polymer solubility in tetrahydrofuran for the next step, during which the sulfonimide-containing phenoxide nucleophiles reacted with the partially substituted PDCP to attach sulfonimide groups onto the polymer backbone. Sodium 4-methylphenoxide was used again during the final step to replace the remaining chlorine atoms. The acid form of the polymer was obtained by multiple precipitations into concentrated hydrochloric acid. Membranes were cast from dioxane and cross-linked by ^{60}Co gamma radiation. Water uptake was 119 wt% for the uncross-linked membrane with an IEC of 0.99 mmol g^{-1}, much higher than that of Nafion 117 (30 wt%). Cross-linking had a significant effect on reducing membrane swelling, with 73 and 42 wt% water uptake for membranes exposed to radiation doses of 20 and 40 Mrad, respectively. Room temperature proton conductivity was measured to be 0.049 S cm^{-1} for the uncross-linked membrane and 0.071 S cm^{-1} for the cross-linked membrane with 20 Mrad radiation dosages. Such improvement in proton conductivity by cross-linking was probably due to the facilitation of proton transport with smaller distance between the acid groups after cross-linking. Further cross-linking reduced the conductivity to 0.058 S cm^{-1}, even though the water content in the membrane was not low, ~23 H$_2$O/SO$_3$H.

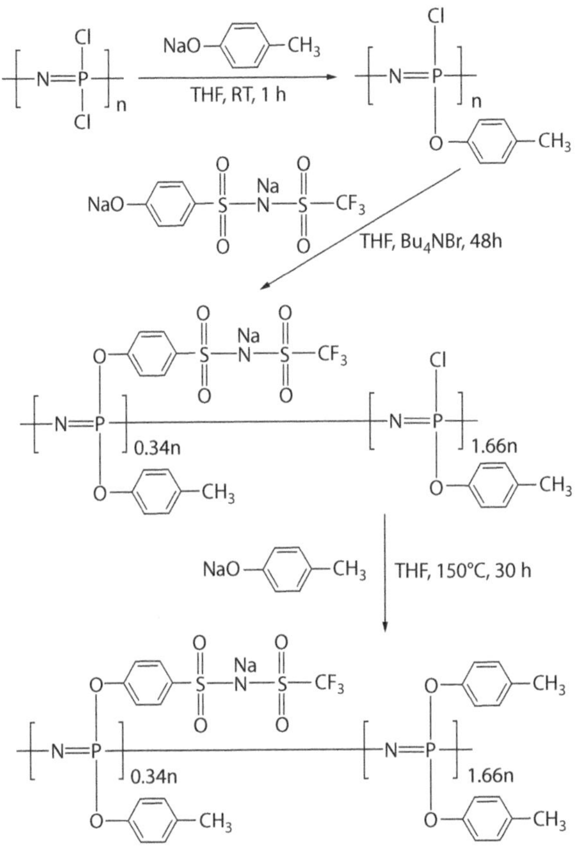

FIGURE 7.22 Synthesis of sulfonimide-functionalized polyphosphazenes. (Reprinted with permission from Hofmann, M.A., Ambler, C.M., Maher, A.E., Chalkova, E., Zhou, X.Y., Lvov, S.N., and Allcock, H.R., Synthesis of polyphosphazenes with sulfonimide side groups, *Macromolecules*, 35, 6490. Copyright 2002 American Chemical Society.)

Blended membranes of the sulfonimide polyphosphazene and PVDF in a 75%/25% w/w ratio were cast from *N,N*-dimethylacetamide. No phase separation was observed, with the membrane water uptake reduced to 41 wt% from 119 wt%. It was surprising that the proton conductivity of blended membrane increased more than 20% to 0.060 S cm^{-1}, even though PVDF is an inert polymer.

7.2.6 Azole-Functionalized Polyphosphazenes

Nitrogen-containing aromatic heterocycle-based polymer systems have been investigated to meet the challenges for operation under anhydrous fuel cell conditions, as the heterocycles perform the functions of water in proton transport with the basic nitrogen sites on the heterocycles acting as strong proton acceptors [104]. Hacivelioglu et al. developed two types of azole-functionalized polyphosphazenes as the anhydrous PEMs, which were synthesized by replacing the chlorine atoms

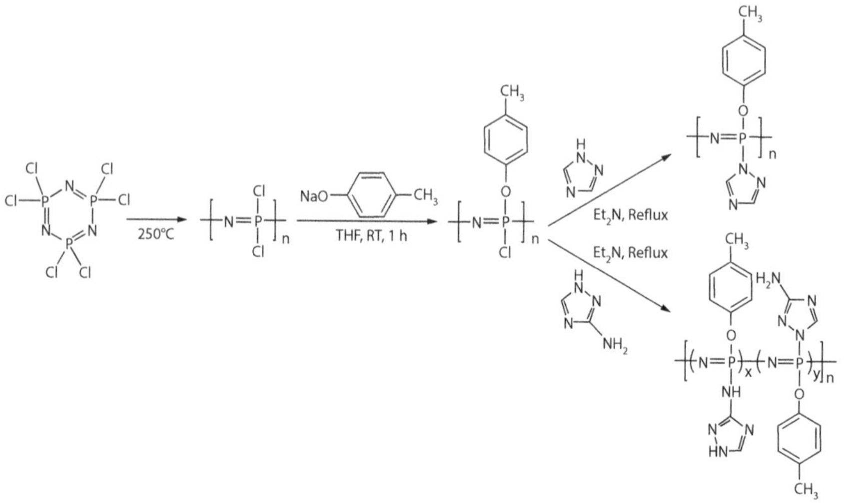

FIGURE 7.23 Synthesis scheme of azole-functionalized polyphosphazenes. (Hacivelioglu, F., Oezden, S., Celik, S.U., Yesilot, S., Kilic, A., and Bozkurt, A., Azole substituted polyphosphazenes as non-humidified proton conducting membranes, *J. Mater. Chem.*, 21, 1020–1027, Copyright 2011. Reprinted by permission of The Royal Society of Chemistry.)

of poly[(4-methylphenoxy)chlorophosphazene] with 1*H*-1,2,4-triazole and 3-amino-1*H*-1,2,4-aminotriazole in the presence of triethylamine, respectively (Figure 7.23) [105]. Flexible membranes were cast from the solution of trifluoromethanesulfonic acid doped in azole-functionalized polyphosphazenes. It was found that proton conductivity increased with dopant concentration and temperature. When the molar ratio of dopant to the azole unit in the polymer was 2, the membrane proton conductivity under 0% relative humidity reached the maximum at 0.003 S cm^{-1} at 50°C and 0.0412 S cm^{-1} at 130°C for triazole and aminotriazole-substituted polymers, respectively. However, further doping of trifluoromethanesulfonic acid reduced membrane proton conductivity, probably due to the unavailability of unprotonated triazole units in the presence of excess acid.

In another study (Figure 7.24) [106], a PS backbone hybrid polymer with pendent 1-oxy-[4-(1,2,4-triazol-1-yl)]phenyl–substituted cyclophosphazene (PVTP) was synthesized by reacting an excess amount of sodium salt of [4-(1,2,4-triazol-1-yl)] phenol with a polymer precursor, (4-vinylphenoxy)pentachlorocyclotriphosphazene (poly-VPCP), which was prepared according to the literature [66]. The complete replacement of the chlorine atoms in poly-VPCP by 1-oxy-[4-(1,2,4-triazol-1-yl)]phenyl groups was confirmed by ^{31}P NMR, indicated by the disappearance of the peaks at $\delta = 22.2$ ppm and $\delta = 11.8$ ppm, which correspond to the resonances of the phosphorous atoms in –PCl$_2$ and –P(ClOR) units of poly-VPCP, respectively (Figure 7.25). Instead in PVTP, only a single peak appeared at $\delta = 8.6$ ppm, since all the phosphorous atoms have the similar chemical environment. Trifluoromethanesulfonic acid was doped in polymer to form proton-conducting membranes. Maximum proton conductivity of 0.043 S cm^{-1} at 150°C was realized in anhydrous state when the molar ratio of dopant

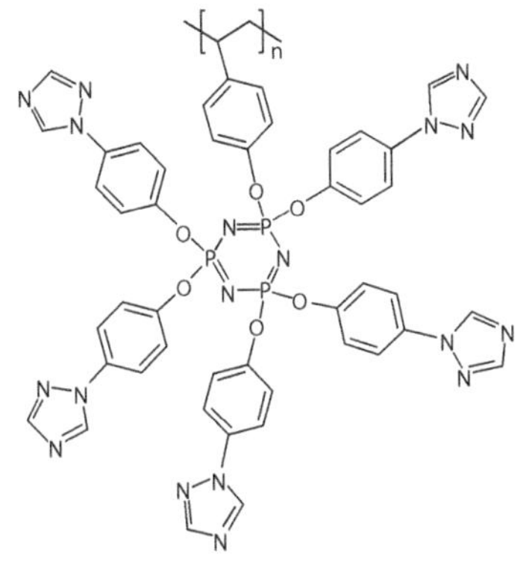

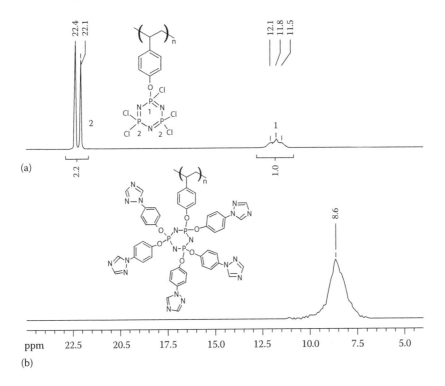

FIGURE 7.24 Structure of the hybrid polymer with PS backbone and pendent azole-functionalized cyclophosphazene (PVTP). (Reprinted from *Polymer*, 54, Alidagi, H.A., Girgic, O.M., Zorlu, Y., Hacivelioglu, F., Celik, S.U., Bozkurt, A., Kilic, A., and Yesilot, S., Synthesis and proton conductivity of azole-substituted cyclic and polymeric phosphazenes, 2250–2256, Copyright 2013, with permission from Elsevier.)

FIGURE 7.25 ^{31}P NMR spectrum of poly-VPCP (a) in CDCl$_3$ and PVTP (b) in DMSO-d$_6$. (Reprinted from *Polymer*, 54, Alidagi, H.A., Girgic, O.M., Zorlu, Y., Hacivelioglu, F., Celik, S.U., Bozkurt, A., Kilic, A., and Yesilot, S., Synthesis and proton conductivity of azole-substituted cyclic and polymeric phosphazenes, 2250–2256, Copyright 2013, with permission from Elsevier.)

to the azole unit in the polymer was 2. Cross-linking would be necessary for PVTP to be applied in the practical DMFC operation due to its solubility in methanol.

7.2.7 POLYPHOSPHAZENES DOPED WITH PHOSPHORIC ACID

Polymers doped with phosphoric acid (PA) are another type of PEMs that have potential working in fuel cells under low relative humidity conditions [107]. PA could act as both proton donor and proton acceptor, and dynamic hydrogen bond networks could be formed so that protons could move via hydrogen bond breaking and forming processes [108]. Dotelli et al. have examined a series of polyphosphazenes doped with PA for proton conduction at anhydrous state [109–111]. Due to the fragility of acid-doped poly(dipropyl)phosphazene (PDPrP) and poly(diethyl, dipropyl)phosphazene copolymer (PDEt, DPrP), composite membrane materials were fabricated with polyphenylene sulfide (PPS), the sulfonated poly[(hydroxy)propyl, pehenyl]ether (SPHPE) and sulfonated polyimide (SPI), respectively. Among these materials, the SPHPE–PDPrP–PA composite exhibited the highest proton conductivity of 0.0071 S cm^{-1} in the dry state at 127°C, which was attributed by the authors to the higher acid doping level (1.7 H$_3$PO$_4$/N) as compared to the other two composites (1.0 H$_3$PO$_4$/N). It was surprising that, at the same acid doping level, the PPS–PDPrP–PA composite had higher proton conductivity than the SPI–PDEt, DPrP–PA composite, which should have larger amount of proton donor due to the presence of sulfonic acid groups in SPI.

7.3 MEMBRANE APPLICATIONS IN ELECTROCHEMICAL DEVICES FOR ENERGY STORAGE AND CONVERSION

7.3.1 APPLICATIONS FOR LITHIUM BATTERIES

As compared to the numerous literature reports on preparation and characterization of polyphosphazene electrolytes for lithium ion conduction, there have been only a few studies on lithium battery using polyphosphazene electrolyte membranes.

Nazri et al. fabricated a Li/TiS$_2$ cell using MEEP lithium triflate as the electrolyte [112]. The open-circuit voltage was found to be 2.8–2.9 V after long-term stand (>10 h). The midpoint of the charge and discharge occurred at ~3.2 and 1.8 V, respectively, when the charging and discharging current was 30 and 50 μA cm^{-2}, respectively. Therefore, the cell had a power density of ~0.09 mW cm^{-2} on discharge. Electrode polarization was clearly observed during deintercalation of lithium from TiS$_2$ electrode. The resistance of lithium/MEEP interface increased with time due to the accumulation of anions at the interface.

Abraham et al. studied a few types of polyphosphazenes as the electrolyte to prepare Li/TiS$_2$ cells, including MEEP, MEEP/PEO, and MEEP/PVP blends [55,57]. For a cell with 50:50 MEEP/PEO complexed with LiClO$_4$ as the electrolyte, the cathode utilization of 0.3 mA h cm^{-2} was achieved with a discharging rate of 0.25 mA cm^{-2} at 50°C, and the utilization was 0.9 mA h cm^{-2} at 2.0 mA cm^{-2} at 100°C [55]. When MEEP–LiClO$_4$ immobilized within a fiberglass filter paper was used as the electrolyte, a Li/TiS$_2$ cell exhibited an open-circuit voltage of 2.8–2.9 V [57]. When the

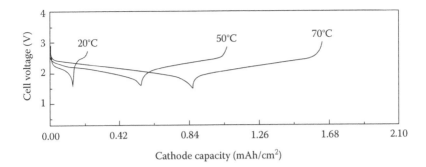

FIGURE 7.26 Constant current discharges of a Li/(MEEP/PVP–LiClO$_4$)/TiS$_2$ cell at C/20 and different temperatures: i_d=0.08 mA cm^{-2}, i_c=0.04 mA cm^{-2}, voltage limits = 1.6–3.0 V. (Reprinted with permission from Abraham, K.M. and Alamgir, M., Dimensionally stable MEEP-based polymer electrolytes and solid-state lithium batteries, *Chemistry of Materials*, 3, 339. Copyright 1991 American Chemical Society.)

discharging rates were 0.05, 0.1, and 0.2 mA cm^{-2} at 50°C, the cell delivered the capacities of 1.10, 0.45, and 0.28 mA h cm^{-2}, respectively. The midpoint of charge and discharge was 2.5 and 2.1 V, respectively. The coulombic efficiency during the cycling was between 90% and 100%. When MEEP/PVP blend complexed with LiClO$_4$ was used as the electrolyte, the cell had a capacity of 0.14 mA h cm^{-2} at room temperature with a discharging rate of 0.05 mA cm^{-2}, ~17% of its theoretical capacity (Figure 7.26). The 100% cathode utilization was accomplished when the temperature was raised to 100°C. For cells using MEEP/PEO–LiClO$_4$ electrolytes, the rate capabilities were quite low at room temperature, as the cell could not be discharged at a practical rate. Raising the temperature could significantly improve the cell performance, with a cathode capacity of 1.2 mA h cm^{-2} achieved at 65°C.

7.3.2 Applications for PEM Fuel Cells

Among the proton-conducting polyphosphazene membranes prepared for conductivity studies, only a few were fabricated into the MEA to be tested in fuel cells.

Chalkova et al. investigated the performance of sulfonimide polyphosphazene membrane–based MEAs in an H$_2$/O$_2$ fuel cell [113]. The membrane of 0.01 cm thick had proton conductivity of 0.058 S cm^{-1} with 42 wt% water uptake. The electrodes were prepared by applying the catalyst ink-containing carbon-supported platinum (20% Pt on Vulcan XC-72R) and Nafion binder onto the carbon cloth. The catalyst loading was 0.33 mg cm^{-2} for both the anode and cathode. The MEA was hot pressed at 65°C and 400 psi for 30 s. In comparison, Nafion-based MEA (with catalyst loading of 0.26 mg cm^{-2} for the anode and 0.48 mg cm^{-2} for the cathode) was pressed at a higher temperature and pressure for longer period of time, at 125°C and 1400 psi for 2 min. A maximum power density of 0.36 W cm^{-2} at 0.87 A cm^{-2} and a limiting current of 1.12 A cm^{-2} was achieved for the sulfonimide polyphosphazene MEA at 22°C with unhumidified hydrogen and oxygen (Figure 7.27a). When the test temperature was raised to 80°C

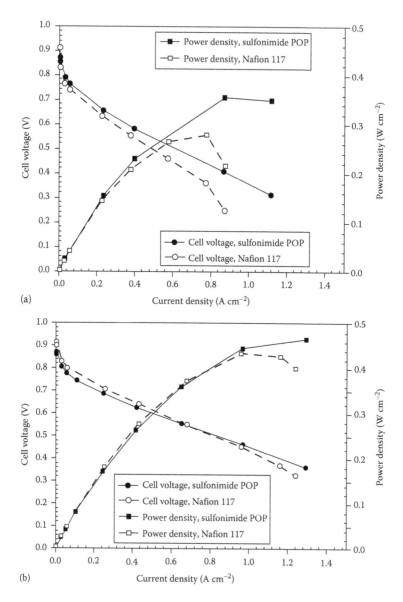

FIGURE 7.27 H$_2$/O$_2$ fuel cell performance with sulfonimide-functionalized polyphosphazene membranes: (a) 22°C, 2 bar H$_2$ and 3 bar O$_2$, 0% RH and (b) 80°C, 3 bar H$_2$ and 3.3 bar O$_2$, 100% RH. (Reprinted with permission from Chalkova, E. et al., *Electrochem. Solid-State Lett.*, 5, A221. Copyright 2002 The Electrochemical Society.)

and full humidification was realized for hydrogen and oxygen, the fuel cell performance was improved, with a maximum power density of 0.47 W cm^{-2} at the limiting current of 1.29 A cm^{-2} (Figure 7.27b). Such performance was superior to that of Nafion 117 MEA, as can be seen in Figure 7.27a and b. Since the thickness of the sulfonimide polyphosphazene membrane was half that of Nafion 117 (0.02 cm) and the conductivity was approximately 60% that of Nafion 117 (0.1 S cm^{-1}), the areal resistance (the ratio of membrane thickness to proton conductivity) of the sulfonimide polyphosphazene membrane would be ~17% lower than that of Nafion 117. The smaller slopes for the I–V curves also suggest lower areal resistance of sulfonimide polyphosphazene MEA as compared to that of Nafion 117 MEA. There was no long-term test of these polyphosphazene membranes, as it may raise concern on the MEA durability at 80°C with MEA hot pressing temperature of only 65°C during MEA preparation. The authors also noted that due to the solubility of uncross-linked sulfonimide polyphosphazene membranes in methanol, the application in DMFCs would require a high level of cross-linking and fuel feed in low methanol concentration [114].

Pintauro's group has studied a series of blended polyphosphazene membranes for their application in DMFCs [86–89]. MEAs were prepared by pressing UV-cross-linked SPBMPP/PAN blended membranes at 120°C and 125 psi for 5 min, between two electrodes, each of which has a catalyst loading of 4.0 mg cm^{-2} (1:1 platinum–ruthenium alloy and platinum black as the anode and cathode catalysts, respectively) [86]. A total of four membranes of similar thickness (0.0137–0.0158 cm) with different blend compositions were selected for the MEA preparation, with SPBMPP (IEC = 2.1 mmol g^{-1}) content ranging from 45 to 55 wt% and corresponding IEC ranging from 0.95 to 1.15 mmol g^{-1}. A Nafion 117–based MEA was also prepared under the same conditions for comparison. Figure 7.28 shows the steady-state current density–voltage curves at 60°C with 1.0 M methanol at a flow rate of 20 mL min^{-1} and humidified air at 150 sccm and 30 psi back pressure. The fuel cell performance increased with membrane IEC, mainly due to the lower areal resistance of higher IEC membranes. The MEA based on the SPBMPP/PAN membrane with an IEC of 1.15 mmol g^{-1} achieved a power output very close to that with Nafion 117 MEA at current density less than 0.15 A cm^{-2}, with methanol crossover three times lower than that in Nafion 117 MEA. For a three-layer composite MEA (a membrane with an IEC of 0.95 mmol g^{-1} sandwiched between two membranes with an IEC of 1.15 mmol g^{-1}), very high open-circuit voltage (0.822 V) was observed because of the low methanol crossover (~10 times lower than that with Nafion 117). The power output was slightly lower than that with single-layer MEA based on the membrane with an IEC of 1.15 mmol g^{-1}, probably due to the higher areal resistance of membranes and contact resistance between the membrane layers.

Blended membranes of PVDF and either sulfonated poly[(3-methylphenoxy) (4-ethylphenoxy)phosphazene] (SP3MP4EPP) or sulfonated poly[(4-ethylphenoxy) (phenoxy)phosphazene] (SP4EPPP) were also investigated for their performance in DMFCs [87,88]. Three-layer composite MEAs were fabricated, in which two electrode-pressed Nafion 112 membranes were attached to both sides of blended

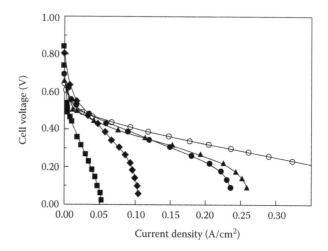

FIGURE 7.28 DMFC performance with SPBMPP/PAN blended membranes: 60°C, 1.0 M methanol at 20 mL min^{-1}, and 50°C humidified air at 150 sccm with 30 psi back pressure. Nafion 117 (○); SPBMPP/PAN blended membranes with different IECs: 1.15 mmol g^{-1} (▲); 1.10 mmol g^{-1} (●); 1.0 mmol g^{-1} (◆); 0.95 mmol g^{-1} (■). (Reprinted with permission from Carter, R. et al., *Electrochem. Solid-State Lett.*, 5, A195. Copyright 2002 The Electrochemical Society.)

membranes, in order to avoid the attachment problems when the electrode was directly hot pressed to the blended membranes. For the SP3MP4EPP/PVDF membrane–based MEA, the electrochemical performance was almost the same as that of Nafion 117 at current density lower than 0.15 A cm^{-2}, with slightly lower power output than that of Nafion 117 at higher current density. Methanol crossover was four times lower than that of Nafion 117. While for SP4EPPP/PVDF membrane–based MEA, the methanol crossover was further reduced to 10 times lower than that of Nafion 117, but the power output was much smaller with much higher ohmic loss, presumably due to the higher areal resistance of SP4EPPP/PVDF blended membranes.

Further work by Pintauro and coworkers involved the DMFC application of a series of SPBPP/PBI blended membranes, which were prepared by adding 5–12 wt% PBI to SPBPP with two different IECs (1.2 and 1.4 mmol g^{-1}) [89]. Similar MEA structures as that in SP3MP4EPP/PVDF study, three-layer composite MEAs, were fabricated, in which a blended membrane was sandwiched between two Nafion 112 half MEAs (only one side of a Nafion 112 membrane hot pressed with an electrode). Such MEA design is quite useful in identifying promising membrane materials, as finding proper catalyst/binder composition and hot pressing conditions poses a significant challenge for the MEA preparation. As shown in Figure 7.29a, the fuel cell performance improved with the increasing content of proton-conducting SPBPP, with SPBPP1.4-PBI03 (blended SPBPP of 1.4 mmol g^{-1} IEC with 3 wt% PBI) reaching nearly the same power output as a sandwiched Nafion 112 MEA with more than 30% lower methanol crossover.

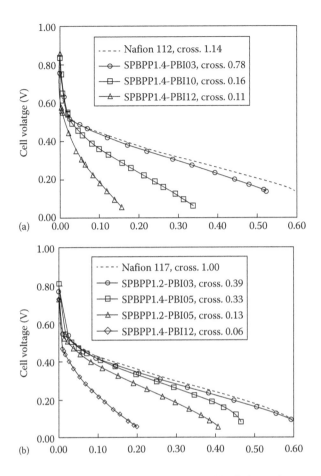

FIGURE 7.29 DMFC performance with SPBPP/PBI blended membranes: 60°C, 1.0 M methanol, and ambient pressure air at 500 sccm. (a) Sandwiched MEA and (b) direct hot pressing of electrodes. Cross denotes the methanol crossover flux (mol cm^{-2} min^{-1}) at open circuit, relative to that in Nafion 117. (Reprinted with kind permission from Springer Science+Business Media: *Fuel Cells II*, Polyphosphazene membranes for fuel cells, 2008, 157, Wycisk, R. and Pintauro, P.N./Scherer, G., ed.)

Another set of SPBPP/PBI-based MEAs, which were prepared by hot pressing the electrodes directly to the membrane at 80°C and 125 psi for 3 min, were also examined for their performance in DMFCs operating at 60°C with 1.0 M aqueous methanol feed and 500 sccm ambient air [89]. As compared to a Nafion 117 MEA (also prepared by direct electrode attachment to the membrane), practically the same fuel cell performance was achieved by SPBPP1.2-PBI03 (blended SPBPP of 1.2 mmol g^{-1} IEC with 3 wt% PBI) with a maximum power density of 89 mW cm^{-2} (vs. 96 mW cm^{-2} for Nafion 117) (Figure 7.29b). The methanol crossover of SPBPP1.2-PBI03 MEA was 2.5 times lower than that of Nafion 117, as indicated by its higher open-circuit voltage. It can be seen from the *I–V* curves on both Figure 7.29a and b that relatively inferior performance was recorded for the MEAs based on membranes with lower proton conductivity and lower methanol permeability (as suggested by the lower methanol crossover for the MEAs tested in the fuel cells). This suggests that membrane resistance instead of membrane methanol permeability would be the major factor in determining the MEA performance in a DMFC under the aforementioned testing conditions. However, with the higher methanol feed concentration, methanol barrier properties of membranes become more dominant on influencing the fuel cell performance. With the methanol feed concentration increasing from 1.0 to 10.0 M, the maximum power density of more methanol-permeable Nafion 117 MEA decreased substantially from 96 to only 9 mW cm^{-2}, while SPBPP1.2-PBI03 MEA still retained more than 60% power (89 mW cm^{-2} at 1.0 M and 60 mW cm^{-2} at 10.0 M). It should also be noted that MEA performance in a DMFC might be affected by the methanol flow rate significantly, especially at high methanol feed concentration [115], but the methanol flow rate was kept constant at 2 mL min^{-1} for all methanol feed concentrations in this study.

Long-term fuel cell tests were also performed for the SPBPP/PBI MEAs in a load cycling mode (cycling of 59 min at 0.1 A cm^{-2} and 1 min at open circuit). For an SPBPP1.4-PBI08 MEA (blended SPBPP of 1.4 mmol g^{-1} IEC with 8 wt% PBI), it was found that the cell voltages decrease at a rate of about 0.8 mV h^{-1}, from 0.39 to 0.31 V at 0.1 A cm^{-2} after 100 h of load cycling. There was no change in methanol crossover throughout the 100 h cycling, but there was constant increase in cell resistance, as measured by a current interrupt method. Such voltage loss during the cycling experiment was attributed to the increasing interfacial resistance between the membrane and the Nafion binder in the electrodes because of growing difference in swelling between Nafion and SPBPP/PBI blend.

7.3.3 Applications for Dye-Sensitized Solar Cells

Allcock and coworkers have examined the application of MEEP as the polymer electrolyte in DSSCs [116,117].

A small molecule plasticizer, PC, was mixed with MEEP at a weight ratio of 50:50, to which 1-methyl-3-propylimidazolium (PMII) and I$_2$ with a molar ratio of 10:1 were added to form MEEP/PC electrolytes. In comparison with PEO/PC electrolyte with the similar composition, MEEP/PC electrolytes gave better performance in a DSSC mainly due to their lower internal resistance (Figure 7.30) [116]. MEEP/PC was also more stable than PEO/PC, as the homogenous MEEP/PC mixture could be stable for at least 1 month, while PEO/PC did not appear to be homogeneous within only a few days.

For the DSSCs based on MEEP electrolytes without the addition of plasticizer, heat treatment played an important role in filling electrolyte into the nanoporous electrodes and determining the cell performance [117]. Longer heating time led to higher degree of infiltration. Solvent assistance also had significant effect on the extent of the electrolyte infiltration. Unfortunately, the power conversion efficiency of these MEEP-based cells was still relatively low, less than 1%, probably due to the insufficient electrolyte infiltration and high electrolyte resistance.

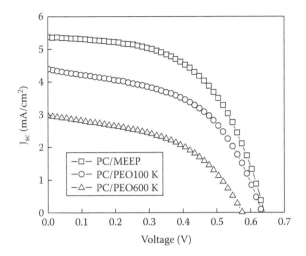

FIGURE 7.30 *J–V* characteristics of DSSCs with MEEP/PC and PEO/PC electrolytes under AM 1.5 G illumination. All electrolytes contain 1.0 M PMII, 0.1 M I_2, and 0.4 M 4-*tert*-butylpyridine. (Reprinted with permission from Lee, S.-H.A., Jackson, A.-M.S., Hess, A., Fei, S.-T., Pursel, S.M., Basham, J., Grimes, C.A., Horn, M.W., Allcock, H.R., and Mallouk, T.E., Influence of different iodide salts on the performance of dye-sensitized solar cells containing phosphazene-based nonvolatile electrolytes, *J. Phys. Chem. C*, 114, 15234. Copyright 2010 American Chemical Society.)

7.4 CONCLUSIONS AND PERSPECTIVES

In this chapter, the development of polyphosphazene electrolyte membranes has been discussed on topics including the polymer synthesis, characterization, and applications for the electrochemical energy conversion and storage devices. Over the past decades, much effort has been devoted to the enhancement of ion transport together with the improvement of dimensional stability in polyphosphazenes. With the changes of side group and/or backbone structure, polyphosphazenes with high ionic conductivity and reasonable mechanical properties can be obtained. Forming composites with inorganic nanoparticles, such as aluminum oxide or silica, can lead to further improvement in lithium ion conductivity. As for fuel cell applications, the most promising materials have been aryloxy-substituted polyphosphazenes functionalized with sulfonic acid groups, as they exhibit good proton conductivity with low water uptake and methanol permeability. Fuel cell performance with these sulfonated polyphosphazene-based MEAs was comparable to that with Nafion.

Despite much progress, challenges remain in improving reproducibility and reducing the cost of polyphosphazene electrolytes. *Living* cationic condensation polymerization method seems to be a more promising way to obtain polymers with complex structures, as well as controllable molecular weight and small molecular weight distributions.

Theoretical modeling of polyphosphazene electrolytes is another research area yet to be explored in more detail. As there are almost unlimited possibilities for the side group functionalization of polyphosphazenes, how to identify the structure characteristics for the optimization of both ion conductivity and dimensional stability would be a daunting task for the synthetic and physical chemists. Modern computational tools and molecular modeling could be of great help in accomplishing such task.

REFERENCES

1. MacCallum, J. R., Vincent, C. A., *Polymer Electrolyte Reviews*, Vol. 1, 1987, Springer, London.
2. Agrawal, R. C., Pandey, G. P., Solid polymer electrolytes: Materials designing and all-solid-state battery applications: An overview, *Journal of Physics D: Applied Physics*, 2008, 41, 223001.
3. Hallinan, D. T., Balsara, N. P., Polymer electrolytes, *Annual Review of Materials Research*, 2013, 43, 503–525.
4. Fenton, D. E., Parker, J. M., Wright, P. V., Complexes of alkali metal ions with poly(ethylene oxide), *Polymer*, 1973, 14, 589.
5. Berthier, C., Gorecki, W., Minier, M., Armand, M. B., Chabagno, J. M., Rigaud, P., Microscopic investigation of ionic conductivity in alkali metal salts-poly(ethylene oxide) adducts, *Solid State Ionics*, 1983, 11, 91–95.
6. Ratner, M. A., Shriver, D. F., Ion transport in solvent-free polymers, *Chemical Reviews*, 1988, 88, 109–124.
7. Perrier, M., Besner, S., Paquette, C., Vallee, A., Lascaud, S., Prud'homme, J., Mixed-alkali effect and short-range interactions in amorphous poly(ethylene oxide) electrolytes, *Electrochimica Acta*, 1995, 40, 2123–2129.
8. Dias, F. B., Plomp, L., Veldhuis, J. B. J., Trends in polymer electrolytes for secondary lithium batteries, *Journal of Power Sources*, 2000, 88, 169–191.
9. Kreuer, K. D., On the development of proton conducting polymer membranes for hydrogen and methanol fuel cells, *Journal of Membrane Science*, 2001, 185, 29–39.
10. Hickner, M. A., Ghassemi, H., Kim, Y. S., Einsla, B. R., McGrath, J. E., Alternative polymer systems for proton exchange membranes (PEMs), *Chemical Reviews*, 2004, 104, 4587–4612.
11. Mauritz, K. A., Moore, R. B., State of understanding of Nafion, *Chemical Reviews*, 2004, 104, 4535–4586.
12. Peighambardoust, S. J., Rowshanzamir, S., Amjadi, M., Review of the proton exchange membranes for fuel cell applications, *International Journal of Hydrogen Energy*, 2010, 35, 9349–9384.
13. Neburchilov, V., Martin, J., Wang, H., Zhang, J., A review of polymer electrolyte membranes for direct methanol fuel cells, *Journal of Power Sources*, 2007, 169, 221–238.
14. Peckham, T. J., Holdcroft, S., Structure–morphology–property relationships of non-perfluorinated proton-conducting membranes, *Advanced Materials*, 2010, 22, 4667–4690.
15. Laberty-Robert, C., Valle, K., Pereira, F., Sanchez, C., Design and properties of functional hybrid organic–inorganic membranes for fuel cells, *Chemical Society Reviews*, 2011, 40, 961–1005.
16. Zhang, H., Shen, P. K., Recent development of polymer electrolyte membranes for fuel cells, *Chemical Reviews*, 2012, 112, 2780–2832.
17. Gleria, M., De Jaeger, R., Polyphosphazenes: A review, in *New Aspects in Phosphorus Chemistry V*, Majoral, J.-P., ed., 2005, Springer, Berlin, Germany, pp. 165–251.
18. Allcock, H. R., Recent developments in polyphosphazene materials science, *Current Opinion in Solid State and Materials Science*, 2006, 10, 231–240.
19. Blonsky, P. M., Shriver, D. F., Austin, P., Allcock, H. R., Polyphosphazene solid electrolytes, *Journal of the American Chemical Society*, 1984, 106, 6854–6855.

20. Allcock, H. R., Kugel, R. L., Synthesis of high polymeric alkoxy- and aryloxyphosphonitriles, *Journal of the American Chemical Society*, 1965, 87, 4216–4217.

21. Allcock, H. R., Kugel, R. L., Phosphonitrilic compounds. VII. High molecular weight poly(diaminophosphazenes), *Inorganic Chemistry*, 1966, 5, 1716–1718.

22. Allcock, H. R., Kugel, R. L., Valan, K. J., Phosphonitrilic compounds. VI. High molecular weight poly(alkoxy- and aryloxyphosphazenes), *Inorganic Chemistry*, 1966, 5, 1709–1715.

23. Andrianov, A. K., Chen, J., LeGolvan, M. P., Poly(dichlorophosphazene) as a precursor for biologically active polyphosphazenes: Synthesis, characterization, and stabilization, *Macromolecules*, 2003, 37, 414–420.

24. Honeyman, C. H., Manners, I., Morrissey, C. T., Allcock, H. R., Ambient temperature synthesis of poly(dichlorophosphazene) with molecular weight control, *Journal of the American Chemical Society*, 1995, 117, 7035–7036.

25. Allcock, H. R., Crane, C. A., Morrissey, C. T., Nelson, J. M., Reeves, S. D., Honeyman, C. H., Manners, I., "Living" cationic polymerization of phosphoranimines as an ambient temperature route to polyphosphazenes with controlled molecular weights, *Macromolecules*, 1996, 29, 7740–7747.

26. Allcock, H. R., Nelson, J. M., Reeves, S. D., Honeyman, C. H., Manners, I., Ambient-temperature direct synthesis of poly(organophosphazenes) via the "living" cationic polymerization of organo-substituted phosphoranimines, *Macromolecules*, 1997, 30, 50–56.

27. Allcock, H. R., Austin, P. E., Neenan, T. X., Sisko, J. T., Blonsky, P. M., Shriver, D. F., Polyphosphazenes with etheric side groups: Prospective biomedical and solid electrolyte polymers, *Macromolecules*, 1986, 19, 1508–1512.

28. Reed, R. A., Wooster, T. T., Murray, R. W., Yaniv, D. R., Tonge, J. S., Shriver, D. F., Solid state voltammetry in ionically conducting phosphazene-$LiSO_3CF_3$ films, *Journal of the Electrochemical Society*, 1989, 136, 2565–2570.

29. Allcock, H. R., Napierala, M. E., Olmeijer, D. L., Best, S. A., Merz, K. M., Ionic conduction in polyphosphazene solids and gels: ^{13}C, ^{31}P, and ^{15}N NMR spectroscopy and molecular dynamics simulations, *Macromolecules*, 1999, 32, 732–741.

30. Luther, T. A., Stewart, F. F., Budzien, J. L., LaViolette, R. A., Bauer, W. F., Harrup, M. K., Allen, C. W., Elayan, A., On the mechanism of ion transport through polyphosphazene solid polymer electrolytes: NMR, IR, and Raman spectroscopic studies and computational analysis of ^{15}N-labeled polyphosphazenes, *The Journal of Physical Chemistry B*, 2003, 107, 3168–3176.

31. Wang, Y., Balbuena, P. B., Combined ab initio quantum mechanics and classical molecular dynamics studies of polyphosphazene polymer electrolytes: Competitive solvation of Li^+ and $LiCF_3SO_3$, *The Journal of Physical Chemistry B*, 2004, 108, 15694–15702.

32. Frech, R., York, S., Allcock, H., Kellam, C., Ionic transport in polymer electrolytes: The essential role of associated ionic species, *Macromolecules*, 2004, 37, 8699–8702.

33. Lee, D. K., Allcock, H. R., The effects of cations and anions on the ionic conductivity of poly *bis*(2-(2-methoxyethoxy)ethoxy)phosphazene doped with lithium and magnesium salts of trifluoromethanesulfonate and *bis* (trifluoromethanesulfonyl)imidate, *Solid State Ionics*, 2010, 181, 1721–1726.

34. Allcock, H. R., Napierala, M. E., Cameron, C. G., O'Connor, S. J. M., Synthesis and characterization of ionically conducting alkoxy ether/alkoxy mixed-substituent poly(organophosphazenes) and their use as solid solvents for ionic conduction, *Macromolecules*, 1996, 29, 1951–1956.

35. Allcock, H. R., O'Connor, S. J. M., Olmeijer, D. L., Napierala, M. E., Cameron, C. G., Polyphosphazenes bearing branched and linear oligoethyleneoxy side groups as solid solvents for ionic conduction, *Macromolecules*, 1996, 29, 7544–7552.

36. Allcock, H. R., Sunderland, N. J., Ravikiran, R., Nelson, J. M., Polyphosphazenes with novel architectures: Influence on physical properties and behavior as solid polymer electrolytes, *Macromolecules*, 1998, 31, 8026–8035.

37. Allcock, H. R., Kellam, E. C., The synthesis and applications of novel aryloxy/oligoethyleneoxy substituted polyphosphazenes as solid polymer electrolytes, *Solid State Ionics*, 2003, 156, 401–414.

38. Allcock, H. R., Olmeijer, D. L., O'Connor, S. J. M., Cation complexation and conductivity in crown ether bearing polyphosphazenes, *Macromolecules*, 1998, 31, 753–759.

39. Conner, D. A., Welna, D. T., Chang, Y., Allcock, H. R., Influence of terminal phenyl groups on the side chains of phosphazene polymers: Structure–property relationships and polymer electrolyte behavior, *Macromolecules*, 2007, 40, 322–328.

40. Allcock, H. R., Olmeijer, D. L., Polyphosphazenes functionalized with sulfone or sulfoxide groups: Synthesis, characterization, and possible polymer electrolyte applications, *Macromolecules*, 1998, 31, 8036–8046.

41. Tada, Y., Sato, M., Takeno, N., Nakacho, Y., Shigehara, K., Attempts at lithium single-ionic conduction by anchoring sulfonate anions as terminating groups of oligo(oxyethylene) side chains in comb-type polyphosphazenes, *Chemistry of Materials*, 1994, 6, 27–30.

42. Allcock, H. R., Welna, D. T., Maher, A. E., Single ion conductors—Polyphosphazenes with sulfonimide functional groups, *Solid State Ionics*, 2006, 177, 741–747.

43. Chen-Yang, Y. W., Hwang, J. J., Chang, F. H., Polyphosphazene electrolytes. 1. Preparation and conductivities of new polymer electrolytes based on poly[*bis*(amino) phosphazene] and lithium perchlorate, *Macromolecules*, 1997, 30, 3825–3831.

44. Chen-Yang, Y. W., Hwang, J. J., Huang, A. Y., Polyphosphazene electrolytes. 2. Synthesis and properties of new polymer electrolytes based on poly((amino) (2-methoxyethoxy) ethoxy)phosphazenes, *Macromolecules*, 2000, 33, 1237–1244.

45. Kaskhedikar, N., Paulsdorf, J., Burjanadze, M., Karatas, Y., Wilmer, D., Roling, B., Wiemhöfer, H. D., Ionic conductivity of polymer electrolyte membranes based on polyphosphazene with oligo(propylene oxide) side chains, *Solid State Ionics*, 2006, 177, 703–707.

46. Paulsdorf, J., Kaskhedikar, N., Burjanadze, M., Obeidi, S., Stolwijk, N. A., Wilmer, D., Wiemhöfer, H. D., Synthesis and ionic conductivity of polymer electrolytes based on a polyphosphazene with short side groups, *Chemistry of Materials*, 2006, 18, 1281–1288.

47. Kaskhedikar, N., Paulsdorf, J., Burjanadze, A., Karatas, Y., Roling, B., Wiemhöfer, H. D., Polyphosphazene based composite polymer electrolytes, *Solid State Ionics*, 2006, 177, 2699–2704.

48. van Wullen, L., Koster, T. K. J., Wiemhöfer, H.-D., Kaskhedikar, N., Local cation coordination motifs in polyphosphazene based composite electrolytes, *Chemistry of Materials*, 2008, 20, 7399–7407.

49. Allcock, H. R., Prange, R., Hartle, T. J., Poly(phosphazene-ethylene oxide) di- and triblock copolymers as solid polymer electrolytes, *Macromolecules*, 2001, 34, 5463–5470.

50. Tonge, J. S., Shriver, D. F., Increased dimensional stability in ionically conducting polyphosphazenes systems, *Journal of the Electrochemical Society*, 1987, 134, 269–270.

51. Bennett, J. L., Dembek, A. A., Allcock, H. R., Heyen, B. J., Shriver, D. F., Radiation crosslinking of poly[*bis*(2-(2-methoxyethoxy)ethoxy)phosphazene]: Effect on solid-state ionic conductivity, *Chemistry of Materials*, 1989, 1, 14–16.

52. Nazri, G. A., Meibuhr, S. G., Effect of γ-radiation on the structure and ionic conductivity of 2-(2-methoxy-ethoxy-ethoxy) polyphosphazene+LiCF$_3$SO$_3$, *Journal of the Electrochemical Society*, 1989, 136, 2450–2454.

53. Nelson, C. J., Coggio, W. D., Allcock, H. R., Ultraviolet radiation-induced crosslinking of poly[*bis*(2-(2-methoxyethoxy)ethoxy)phosphazene], *Chemistry of Materials*, 1991, 3, 786–787.

54. Kaskhedikar, N., Burjanadze, M., Karatas, Y., Wiemhoefer, H. D., Polymer electrolytes based on cross-linked cyclotriphosphazenes, *Solid State Ionics*, 2006, 177, 3129–3134.

55. Abraham, K. M., Alamgir, M., Perrotti, S. J., Rechargeable solid-state Li batteries utilizing polyphosphazene-poly(ethylene oxide) mixed polymer electrolytes, *Journal of the Electrochemical Society*, 1988, 135, 535–536.

56. Abraham, K. M., Alamgir, M., Reynolds, R. K., Polyphosphazene-poly(olefin oxide) mixed polymer electrolytes: I. Conductivity and thermal studies of MEEP/PEO(LiX)$_n$, *Journal of the Electrochemical Society*, 1989, 136, 3576–3582.

57. Abraham, K. M., Alamgir, M., Dimensionally stable MEEP-based polymer electrolytes and solid-state lithium batteries, *Chemistry of Materials*, 1991, 3, 339–348.

58. Abraham, K. M., Alamgir, M., Moulton, R. D., Polyphosphazene-poly(olefin oxide) mixed polymer electrolytes: II. Characterization of MEEP/PEO(LiX)$_n$, *Journal of the Electrochemical Society*, 1991, 138, 921–927.

59. Adamic, K. J., Greenbaum, S. G., Abraham, K. M., Alamgir, M., Wintersgill, M. C., Fontanella, J. J., Lithium-7 nmr study of polymer electrolytes based on composites of poly[*bis*((methoxyethoxy)ethoxy)phosphazene] and poly(ethylene oxide), *Chemistry of Materials*, 1991, 3, 534–538.

60. Landry, C. J. T., Ferrar, W. T., Teegarden, D. M., Coltrain, B. K., Novel miscible blends of etheric polyphosphazenes with acidic polymers, *Macromolecules*, 1993, 26, 35–46.

61. Kim, C., Kim, J. S., Lee, M.-H., Ionic conduction of sol-gel derived polyphosphazene/silicate hybrid network, *Synthetic Metals*, 1998, 98, 153–156.

62. Brusatin, G., Guglielmi, M., De Jaeger, R., Facchin, G., Gleria, M., Musiani, M., Sol-gel hybrid materials based on hydroxylatedpoly[*bis*(methoxy-ethoxy-ethoxy)phosphazene] and silica: A ceramic ionic conductor, *Journal of Materials Science*, 1997, 32, 4415–4420.

63. Allcock, H. R., Chang, Y., Welna, D. T., Ionic conductivity of covalently interconnected polyphosphazene—Silicate hybrid networks, *Solid State Ionics*, 2006, 177, 569–572.

64. Chen-Yang, Y. W., Chen, H. C., Lin, F. J., Liao, C. W., Chen, T. L., Preparation and conductivity of the composite polymer electrolytes based on poly[*bis*(methoxyethoxyethoxy) phosphazene], LiClO$_4$ and α-Al$_2$O$_3$, *Solid State Ionics*, 2003, 156, 383–392.

65. Inoue, K., Nishikawa, Y., Tanigaki, T., Ionic conductivity of polymer complexes formed by polystyrene derivatives with a pendant oligo(oxyethylene)cyclotriphosphazene and LiClO$_4$, *Macromolecules*, 1991, 24, 3464–3465.

66. Inoue, K., Nishikawa, Y., Tanigaki, T., High-conductivity electrolytes composed of polystyrene carrying pendant oligo(oxyethylene)cyclotriphosphazenes and lithium perchlorate, *Journal of the American Chemical Society*, 1991, 113, 7609–7613.

67. Inoue, K., Nishikawa, Y., Tanigaki, T., Ionic conductivities in polystyrene carrying a pendant oligo (oxyethylene)cyclotriphosphazene-alkali thiocyanate complexes, *Solid State Ionics*, 1992, 58, 217–220.

68. Inoue, K., Tanigaki, T., Yuan, Z., Synthesis of novel polystyrene derivatives with pendant oligo(oxyethylene)cyclo-triphosphazenes and ionic conductivities of their LiClO$_4$ complexes, *Polymers for Advanced Technologies*, 1993, 4, 74–79.

69. Allcock, H. R., Laredo, W. R., Morford, R. V., Polymer electrolytes derived from polynorbornenes with pendent cyclophosphazenes: Poly(ethylene glycol) methyl ether (PEGME) derivatives, *Solid State Ionics*, 2001, 139, 27–36.

70. Allcock, H. R., Laredo, W. R., Kellam, E. C., Morford, R. V., Polynorbornenes bearing pendent cyclotriphosphazenes with oligoethyleneoxy side groups: Behavior as solid polymer electrolytes, *Macromolecules*, 2001, 34, 787–794.

71. Allcock, H. R., Bender, J. D., Morford, R. V., Berda, E. B., Synthesis and characterization of novel solid polymer electrolytes based on poly(7-oxanorbornenes) with pendent oligoethyleneoxy-functionalized cyclotriphosphazenes, *Macromolecules*, 2003, 36, 3563–3569.

72. Welna, D. T., Stone, D. A., Allcock, H. R., Lithium-ion conductive polymers as prospective membranes for lithium-seawater batteries, *Chemistry of Materials*, 2006, 18, 4486–4492.

73. Stone, D. A., Welna, D. T., Allcock, H. R., Synthesis and characterization of lithium-ion conductive membranes with low water permeation, *Chemistry of Materials*, 2007, 19, 2473–2482.

74. Montoneri, E., Gleria, M., Ricca, G., Pappalardo, G. C., On the reaction of catenapoly(diphenoxy-λ^5-phosphazene) with sulfur trioxide, *Die Makromolekulare Chemie*, 1989, 190, 191–202.

75. Montoneri, E., Gleria, M., Ricca, G., Pappalardo, G. C., New acid-polyfunctional water-soluble phosphazenes: Synthesis and spectroscopic characterization, *Journal of Macromolecular Science: Part A—Chemistry*, 1989, 26, 645–661.

76. Montoneri, E., Ricca, G., Gleria, M., Gallazzi, M. C., Complexes of hexaphenoxycyclotriphosphazene and sulfur trioxide, *Inorganic Chemistry*, 1991, 30, 150–152.

77. Kurachi, Y., Kajiwara, M., Synthesis and properties of poly (organophosphazenes) ionomers $[NP(HNC_6H_5)_{2-(x+y)}(HNC_6H_4SO_3H)_x(HNC_6H_4SO_3Li)_y]_n$, *Journal of Materials Science*, 1991, 26, 1799–1802.

78. Allcock, H. R., Fitzpatrick, R. J., Salvati, L., Sulfonation of (aryloxy)- and (arylamino) phosphazenes: Small-molecule compounds, polymers, and surfaces, *Chemistry of Materials*, 1991, 3, 1120–1132.

79. Allcock, H. R., Klingenberg, E. H., Welker, M. F., Alkanesulfonation of cyclic and high polymeric phosphazenes, *Macromolecules*, 1993, 26, 5512–5519.

80. Wycisk, R., Pintauro, P. N., Sulfonated polyphosphazene ion-exchange membranes, *Journal of Membrane Science*, 1996, 119, 155–160.

81. Tang, H., Pintauro, P. N., Guo, Q., O'Connor, S., Polyphosphazene membranes. III. Solid-state characterization and properties of sulfonated poly[*bis*(3-methylphenoxy) phosphazene], *Journal of Applied Polymer Science*, 1999, 71, 387–399.

82. Tang, H., Pintauro, P. N., Polyphosphazene membranes. IV. Polymer morphology and proton conductivity in sulfonated poly[*bis*(3-methylphenoxy)phosphazene] films, *Journal of Applied Polymer Science*, 2000, 79, 49–59.

83. Guo, Q., Pintauro, P. N., Tang, H., O'Connor, S., Sulfonated and crosslinked polyphosphazene-based proton-exchange membranes, *Journal of Membrane Science*, 1999, 154, 175–181.

84. Wycisk, R., Pintauro, P. N., Wang, W., O'Connor, S., Polyphosphazene membranes. I. Solid-state photocrosslinking of poly[(4-ethylphenoxy)(phenoxy)phosphazene], *Journal of Applied Polymer Science*, 1996, 59, 1607–1617.

85. Graves, R., Pintauro, P. N., Polyphosphazene membranes. II. Solid-state photocrosslinking of poly[(alkylphenoxy)(phenoxy)phosphazene] films, *Journal of Applied Polymer Science*, 1998, 68, 827–836.

86. Carter, R., Wycisk, R., Yoo, H., Pintauro, P. N., Blended polyphosphazene/polyacrylonitrile membranes for direct methanol fuel cells, *Electrochemical and Solid-State Letters*, 2002, 5, A195–A197.

87. Pintauro, P. N., Wycisk, R., Sulfonated polyphosphazene membranes for direct methanol fuel cells, in *Phosphazenes: A Worldwide Insight*, Gleria, M., De Jaeger, R., eds., 2004, Nova Science Publishers, New York, pp. 591–620.

88. Wycisk, R., Pintauro, P. N., Polyphosphazene membranes for fuel cells, in *Fuel Cells II*, Scherer, G., ed., 2008, Springer, Berlin, Germany, pp. 157–183.

89. Wycisk, R., Lee, J. K., Pintauro, P. N., Sulfonated polyphosphazene-polybenzimidazole membranes for DMFCs, *Journal of the Electrochemical Society*, 2005, 152, A892–A898.

90. Song, H., Lee, S. C., Heo, H. Y., Kim, D. I., Lee, D.-H., Lee, J. H., Chang, J. Y., Improvement of thermal stability of sulfonated polyphosphazenes by introducing a self-crosslinkable group, *Journal of Polymer Science Part A: Polymer Chemistry*, 2008, 46, 5850–5858.

91. Burjanadze, M., Paulsdorf, J., Kaskhedikar, N., Karatas, Y., Wiemhöfer, H. D., Proton conducting membranes from sulfonated poly[*bis*(phenoxy)phosphazenes] with an interpenetrating hydrophilic network, *Solid State Ionics*, 2006, 177, 2425–2430.

92. Paulsdorf, J., Burjanadze, M., Hagelschur, K., Wiemhöfer, H. D., Ionic conductivity in polyphosphazene polymer electrolytes prepared by the living cationic polymerization, *Solid State Ionics*, 2004, 169, 25–33.

93. Fei, S.-T., Wood, R. M., Lee, D. K., Stone, D. A., Chang, H.-L., Allcock, H. R., Inorganic–organic hybrid polymers with pendent sulfonated cyclic phosphazene side groups as potential proton conductive materials for direct methanol fuel cells, *Journal of Membrane Science*, 2008, 320, 206–214.

94. He, M.-L., Xu, H.-L., Dong, Y., Xiao, J.-H., Liu, P., Fu, F.-Y., Hussain, S., Zhang, S.-Z., Jing, C.-J., Hao, X., Zhu, C.-J., Synthesis and characterization of perfluoroalkyl sulfonic acid functionalized polyphosphazene for proton-conducting membranes, *Journal of Macromolecular Science, Part A*, 2013, 51, 55–62.

95. Ganapathiappan, S., Chen, K., Shriver, D. F., A new class of cation conductors: Polyphosphazene sulfonates, *Macromolecules*, 1988, 21, 2299–2301.

96. Andrianov, A. K., Marin, A., Chen, J., Sargent, J., Corbett, N., Novel route to sulfonated polyphosphazenes: Single-step synthesis using "noncovalent protection" of sulfonic acid functionality, *Macromolecules*, 2004, 37, 4075–4080.

97. Allcock, H. R., Taylor, J. P., Phosphorylation of phosphazenes and its effects on thermal properties and fire retardant behavior, *Polymer Engineering & Science*, 2000, 40, 1177–1189.

98. Allcock, H. R., Hofmann, M. A., Wood, R. M., Phosphonation of aryloxyphosphazenes, *Macromolecules*, 2001, 34, 6915–6921.

99. Allcock, H. R., Hofmann, M. A., Ambler, C. M., Morford, R. V., Phenylphosphonic acid functionalized poly[aryloxyphosphazenes], *Macromolecules*, 2002, 35, 3484–3489.

100. Allcock, H. R., Hofmann, M. A., Ambler, C. M., Lvov, S. N., Zhou, X. Y., Chalkova, E., Weston, J., Phenyl phosphonic acid functionalized poly[aryloxyphosphazenes] as proton-conducting membranes for direct methanol fuel cells, *Journal of Membrane Science*, 2002, 201, 47–54.

101. Fedkin, M. V., Zhou, X., Hofmann, M. A., Chalkova, E., Weston, J. A., Allcock, H. R., Lvov, S. N., Evaluation of methanol crossover in proton-conducting polyphosphazene membranes, *Materials Letters*, 2002, 52, 192–196.

102. Zhou, X., Weston, J., Chalkova, E., Hofmann, M. A., Ambler, C. M., Allcock, H. R., Lvov, S. N., High temperature transport properties of polyphosphazene membranes for direct methanol fuel cells, *Electrochimica Acta*, 2003, 48, 2173–2180.

103. Hofmann, M. A., Ambler, C. M., Maher, A. E., Chalkova, E., Zhou, X. Y., Lvov, S. N., Allcock, H. R., Synthesis of polyphosphazenes with sulfonimide side groups, *Macromolecules*, 2002, 35, 6490–6493.

104. Celik, S. U., Bozkurt, A., Hosseini, S. S., Alternatives toward proton conductive anhydrous membranes for fuel cells: Heterocyclic protogenic solvents comprising polymer electrolytes, *Progress in Polymer Science*, 2012, 37, 1265–1291.

105. Hacivelioglu, F., Oezden, S., Celik, S. U., Yesilot, S., Kilic, A., Bozkurt, A., Azole substituted polyphosphazenes as non-humidified proton conducting membranes, *Journal of Materials Chemistry*, 2011, 21, 1020–1027.

106. Alidagi, H. A., Girgic, O. M., Zorlu, Y., Hacivelioglu, F., Celik, S. U., Bozkurt, A., Kilic, A., Yesilot, S., Synthesis and proton conductivity of azole-substituted cyclic and polymeric phosphazenes, *Polymer*, 2013, 54, 2250–2256.

107. Li, Q., Jensen, J. O., Savinell, R. F., Bjerrum, N. J., High temperature proton exchange membranes based on polybenzimidazoles for fuel cells, *Progress in Polymer Science*, 2009, 34, 449–477.

108. Steininger, H., Schuster, M., Kreuer, K. D., Kaltbeitzel, A., Bingol, B., Meyer, W. H., Schauff, S., Brunklaus, G., Maier, J., Spiess, H. W., Intermediate temperature proton conductors for pem fuel cells based on phosphonic acid as protogenic group: A progress report, *Physical Chemistry Chemical Physics*, 2007, 9, 1764–1773.

109. Dotelli, G., Gallazzi, M. C., Mari, C. M., Greppi, F., Montoneri, E., Manuelli, A., Polyalkylphosphazenes as solid proton conducting electrolytes, *Journal of Materials Science*, 2004, 39, 6937–6943.

110. Dotelli, G., Gallazzi, M. C., Perfetti, G., Montoneri, E., Proton conductivity of poly(dipropyl)phosphazene-sulfonated poly[(hydroxy)propyl, phenyl]ether-H_3PO_4 composite in dry environment, *Solid State Ionics*, 2005, 176, 2819–2827.

111. Dotelli, G., Gallazzi, M. C., Bagatti, M., Montoneri, E., Boffa, V., Proton conductivity of poly(dialkyl)phosphazenes-phosphoric acid composites at low humidity, *Solid State Ionics*, 2007, 178, 1442–1450.

112. Nazri, G., MacArthur, D. M., Ogara, J. F., Polyphosphazene electrolytes for lithium batteries, *Chemistry of Materials*, 1989, 1, 370–374.

113. Chalkova, E., Zhou, X., Ambler, C., Hofmann, M. A., Weston, J. A., Allcock, H. R., Lvov, S. N., Sulfonimide polyphosphazene-based H_2/O_2 fuel cells, *Electrochemical and Solid-State Letters*, 2002, 5, A221–A222.

114. Allcock, H. R., Wood, R. M., Design and synthesis of ion-conductive polyphosphazenes for fuel cell applications: Review, *Journal of Polymer Science Part B: Polymer Physics*, 2006, 44, 2358–2368.

115. Lin, J., Wu, P. H., Wycisk, R., Trivisonno, A., Pintauro, P. N., Direct methanol fuel cell operation with pre-stretched recast Nafion®, *Journal of Power Sources*, 2008, 183, 491–497.

116. Lee, S.-H. A., Jackson, A.-M. S., Hess, A., Fei, S.-T., Pursel, S. M., Basham, J., Grimes, C. A., Horn, M. W., Allcock, H. R., Mallouk, T. E., Influence of different iodide salts on the performance of dye-sensitized solar cells containing phosphazene-based nonvolatile electrolytes, *The Journal of Physical Chemistry C*, 2010, 114, 15234–15242.

117. Fei, S.-T., Lee, S.-H. A., Pursel, S. M., Basham, J., Hess, A., Grimes, C. A., Horn, M. W., Mallouk, T. E., Allcock, H. R., Electrolyte infiltration in phosphazene-based dye-sensitized solar cells, *Journal of Power Sources*, 2011, 196, 5223–5230.

8 Phosphoric Acid–Doped Polybenzimidazole Membranes for High-Temperature Proton Exchange Membrane Fuel Cells

Jianhua Fang and Xiaoxia Guo

CONTENTS

8.1 INTRODUCTION

High-temperature (typically 120°C–200°C) polymer electrolyte membrane (PEM) fuel cells (HT-PEMFCs) have attracted considerable attention in the past decades because they possess many advantages such as high CO tolerance of electrodes, easy water management, and fast electrochemical kinetics [1,2]. For use in HT-PEMFCs, an ideal PEM must meet the following requirements: (1) high proton conductivity at elevated temperatures without humidification (low humidities), (2) excellent chemical and electrochemical stability at elevated temperatures, (3) high mechanical strength and toughness, and (4) high thermal stability. Sulfonated polymers can hardly be used as the PEMs in HT-PEMFCs because of their very low proton conductivities ($<10^{-2}$ S cm^{-1}) at low relative humidities (e.g., <30%) resulting from the insufficient water molecules as the carrier for proton transport. In contrast, phosphoric acid (PA)–doped polybenzimidazole (PBI) membranes have been reported to have high proton conductivities (>0.01 S cm^{-1}) at low relative humidities or even under complete anhydrous condition at elevated temperatures, which makes them very suitable for use in HT-PEMFCs. The PA-doped PBI membranes as PEMs were first reported by Litt, Savinell and coworkers [1], and since then, a huge number of research articles have been published.

PBIs refer to a class of heterocyclic polymers consisting of benzimidazole linkages in polymer backbone. Poly(2,2′-m-phenylene-5,5′-bibenzimidazole) is almost the only commercial PBI (Celazole®). Its chemical structure is shown in Figure 8.1. This polymer is often simply named as PBI or m-PBI (m refers to the meta-substitution positions of the phenylene moiety).

Besides the commercial PBI, poly(2,5-benzimidazole) is another kind of frequently studied PBI in academic research. The chemical structure of this polymer is shown in Figure 8.2. Because it is synthesized from an AB-type monomer, 3,4-diaminobenzoic acid, it is also called AB-PBI.

By changing the chemical linkages (e.g., m-phenylene in m-PBI) between benzimidazole groups, various PBIs with different fine chemical structures may be

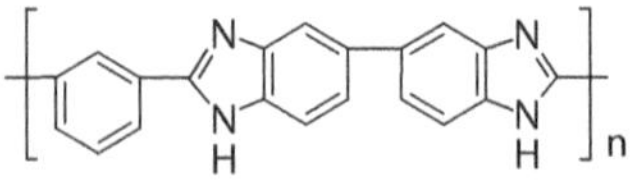

FIGURE 8.1 Chemical structure of poly(2,2′-m-phenylene-5,5′-bibenzimidazole).

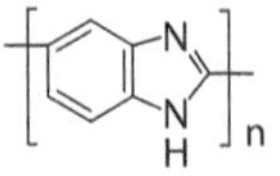

FIGURE 8.2 Chemical structure of poly(2,5-benzimiodazole).

obtained. In this chapter, the synthesis and characterization, acid doping, proton conductivity, and fuel cell performances of various PBIs will be briefly introduced.

8.2 SYNTHESIS AND CHARACTERIZATION OF POLYBENZIMIDAZOLES

8.2.1 SYNTHESIS OF NON-SULFONATED POLYBENZIMIDAZOLES

8.2.1.1 Synthesis from Diphenyl Esters of Dicarboxylic Acids and Tetraamines

In 1961, Vogel and Marvel first reported the synthesis of various high-molecular-weight PBIs by melt polymerization of diphenyl esters of dicarboxylic acids and tetraamines (Figure 8.3) [3]. This is the earliest successful synthetic method for PBIs. The polymerization is performed by primarily preheating the mixture of a dicarboxylic acid diphenyl ester and a tetraamine at 220°C–290°C for a short period of time (10–30 min) under inert (nitrogen) atmosphere, then evacuating to 0.1 mmHg pressure at this temperature to remove phenol and water to yield a glassy foam, and finally gradually raising the reaction temperature to 400°C and keeping at this temperature for a long period of time (1.5–9 h) under high vacuum (0.1 mmHg). They found that the use of free acids and the corresponding dimethyl esters gave inferior results. This is likely related to the insufficiently thermal stability of the free acids and the dimethyl esters, which decomposed prior to the commencement of polymerization. Under similar conditions, the polymerization of an AB type of monomer, phenyl 3,4-diaminobenzoate, also yielded high-molecular-weight PBI (AB-PBI, Figure 8.4).

It should be noted that the polymerization must be performed under inert atmosphere in combination with high-vacuum procedure because the tetraamines are very sensitive to oxidation especially at elevated temperatures. In addition, though various high-molecular-weight PBIs have been successfully synthesized via this method,

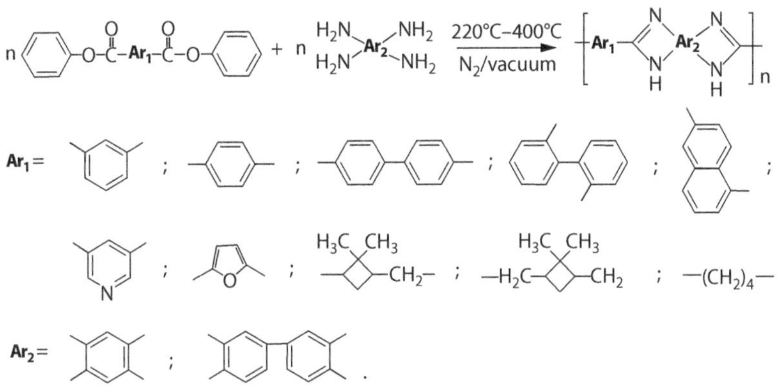

FIGURE 8.3 Synthesis of various PBIs via condensation polymerization of dicarboxylic acid diphenyl esters and tetraamines. (From Vogel, H. and Marvel, C.S., *J. Polym. Sci.*, L, 511, 1961.)

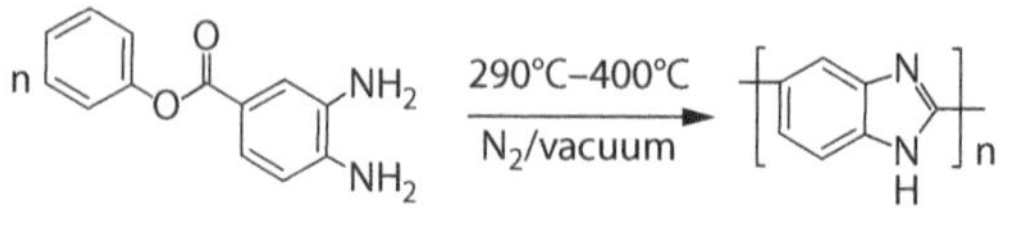

FIGURE 8.4 Synthesis of the AB-PBI via condensation polymerization of phenyl 3,4-diaminobenzoate. (From Vogel, H. and Marvel, C.S., *J. Polym. Sci.*, L, 511, 1961.)

the resulting polymers might have broad-molecular-weight distribution because of the rapid solidification of the polymerization system, which restricts the free motion of the mass.

8.2.1.2 Synthesis from Dicarbixylic Acids and Tetraamines in Polyphophoric Acid

Imai and coworkers developed a modified method for PBI synthesis, which can be conducted under relatively moderate conditions [4,5]. A dicarboxylic acid is directly used to polymerize with the hydrochloric acid salt of a tetraamine in polyphosphoric acid (PPA) at ~200°C (Figure 8.5). This method is almost applicable to the synthesis of various PBIs. The merits of this method are obvious: (1) dicarboxylic acids rather than the esters of dicarboxylic acids are directly used as monomers, and therefore, the monomer resources are much more abundant and (2) the polymerization temperature is much lower than that of the foregoing melt polymerization resulting in less consumption of energy.

It should be noted that the hydrochloric acid salt of a tetraamine can be replaced with the free tetraamine despite that the latter is much less stable to oxidation than the former. Based on the authors' experience, it is helpful to perform the polymerization according to the following procedures. First, a tetraamine is mixed with PPA at room temperature under nitrogen flow. Then the mixture is slightly heated till the tetraamine is completely dissolved. Finally, stoichiometric amount dicarboxylic acid is added and the reaction mixture is successively heated at ~150°C for 2–4 h and 200°C for 5–20 h. As the tetraamine is dissolved in PPA, it is immediately protonated leading to greatly improved oxidative stability of this monomer.

8.2.1.3 Synthesis from Dibasic Acids and Tetraamines in Eaton's Reagent

Ueda et al. reported that in some cases, PBIs could be synthesized by polymerization of dicarboxylic acids and tetraamine hydrochloride salts in Eaton's reagent (a solution mixture of phosphorus pentoxide [P_2O_5] in methanesulfonic acid [MSA] at the weight ratio of P_2O_5/MSA = 1/10, it is also denoted as PPMA) [6]. They also

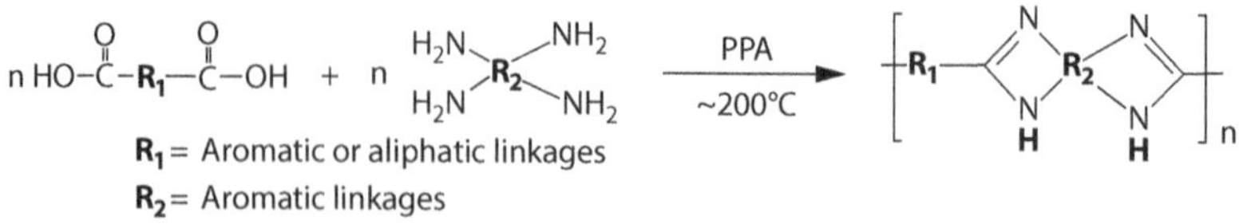

R$_1$ = Aromatic or aliphatic linkages
R$_2$ = Aromatic linkages

FIGURE 8.5 Synthesis of various PBIs via condensation polymerization in PPA. (From Iwakura, Y. et al., *J. Polym. Sci. Part A: General Papers*, 2, 2605, 1964.)

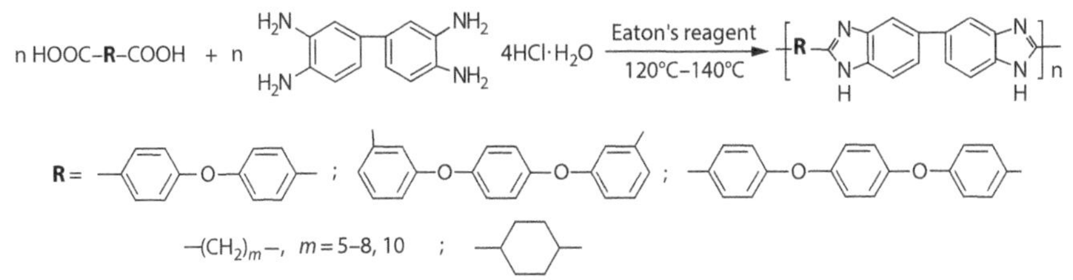

FIGURE 8.6 Synthesis of various PBIs via condensation polymerization in Eaton's reagent. (Reprinted with modification from Ueda, M. et al., *Macromolecules*, 18, 2723, 1985. With Permission.)

found that 5 mL of Eaton's reagent was appropriate for the polymerization on a 1.0 mmol scale judging from the highest molecular weight (inherent viscosity) of the resulting PBIs [6]. This approach has the merits of fast polymerization rate (10 min to a couple of hours) and lower reaction temperature (120°C–140°C) in comparison with the foregoing diphenyl esters of dicarboxylic acids method and the PPA method. This approach is applicable to both the aliphatic dicarboxylic acids and the aromatic dicarboxylic acids with low electron affinity (low positivity of the carbonyl carbon). Figure 8.6 shows some successful examples synthesized by this method.

Just like the situation of the polymerization in PPA, here, the tetraamine hydrochloride salts can also be replaced with the free tetraamines and it is preferentially to completely dissolve the free tetraamines in Eaton's reagent prior to the addition of the dicarboxylic acid monomers. However, it should be noted that as the dicarboxylic acids with high electron affinity such as isophthalic acid (*i*PTA), terephthalic, and 4,4′-dicarboxyldiphenyl sulfone are used as the acidic monomers, the resulting products are only low-molecular-weight polymers [6].

In some special cases, Ueda's approach is more useful for PBI synthesis. For example, the authors found that the copolymerization of disodium 4,6-*bis*(4-carboxyphenoxy) benzene-1,3-disulfonate (BCPOBDS-Na), 4,4′-dicarboxydiphenyl ether (DCDPE), and 3,3′-diaminobenzidine (DABz) in Eaton's reagent at 140°C yielded organo soluble linear sulfonated polybenzimidazoles (SOPBIs) [7], whereas insoluble gel was obtained when the polymerization was conducted in PPA probably because of the occurrence of cross-linking reactions between sulfonate groups and activated benzene rings.

8.2.1.4 Synthesis from Aromatic Dialdehydes and Tetraamines

In 1983, Neuse and Loonat first reported a two-stage synthesis approach for the preparation of various PBIs that involves the low-temperature solution polymerization of aromatic *bis*(*o*-diamines) with aromatic dialdehydes and subsequent conversion of the resulting azomethine-type prepolymers to all-aromatic PBIs [8]. The reactions are illustrated in Figure 8.7. The first-stage polycondensation is conducted in dipolar aprotic media at −18°C to +25°C under anaerobic conditions to yield linear and soluble prepolymers. The prepolymers are converted to the desired PBIs by the second-stage oxidative cyclodehydrogenation reactions in the presence

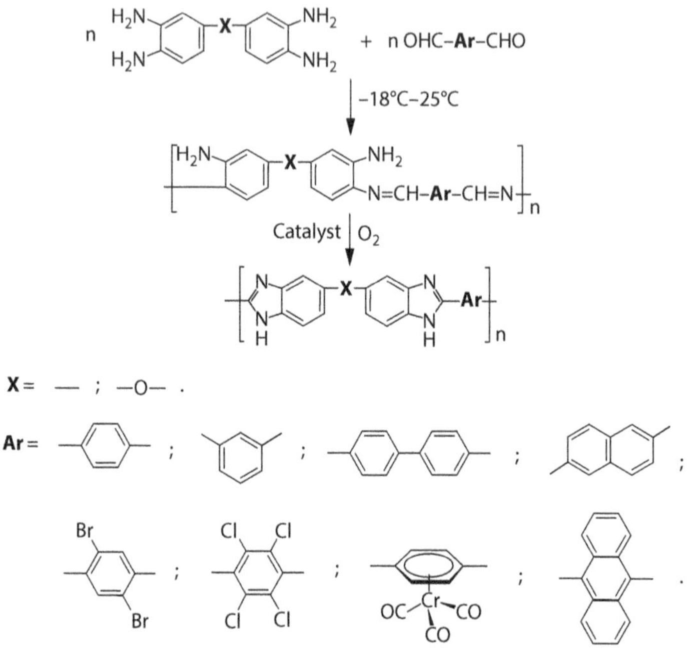

FIGURE 8.7 Polycondensation of aromatic tetraamines with aromatic dialdehydes. (Reprinted with modification from Neuse, E.W. and Loonat, M.S., *Macromolecules*, 16, 128, 1983. With permission.)

of catalytic amounts of certain transition-metal compounds such as anhydrous iron (III) chloride at temperatures below 100°C. The merits of this approach are obvious: (1) the prepolymers are generally organo soluble which is favorable for processing and (2) the reaction temperatures are much lower than that of the foregoing mentioned approaches, which is energy saving. However, in the first-stage polycondensation, there is a risk of cross-linking because the *o*-diamino groups both are readily to react with the aldehyde groups. This can be avoided by slow addition (dropwise) of a highly diluted (0.1 M) dialdehyde solution to a diluted (0.5 M) tetraamine solution when doing the polycondensation [8].

8.2.1.5 Synthesis from Benzimidazole-Based Monomers

In 1996, Hedrick and coworkers reported various poly(aryl ether benzimidazole)s, which were synthesized via condensation polymerization of a benzimidazole-based difluoro monomer with various bisphenols in aprotic dipolar solvents in the presence of K_2CO_3 at elevated temperatures up to 180°C (Figure 8.8) [9]. The benzimidazole-based difluoro monomer was synthesized via condensation reaction of 4-fluorobenzoic acid and DABz at 175°C in PPA (Figure 8.9). They also synthesized two interesting benzimidazole-based AB-type monomers (A and B refer to fluoro and hydroxyl groups) via multi step reactions (Figure 8.10). The formation of imidazole rings is achieved via the ring-closure reaction of the *o*-amino amide intermediates

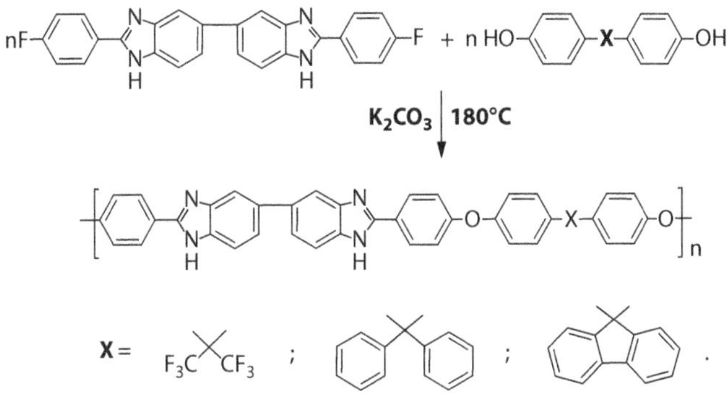

FIGURE 8.8 Synthesis of poly(ether benzimidazole)s from a difluoro monomer and diphenols. (Reprinted with modification from Twieg, R. et al., *Macromolecules*, 29, 7335, 1996. With permission.)

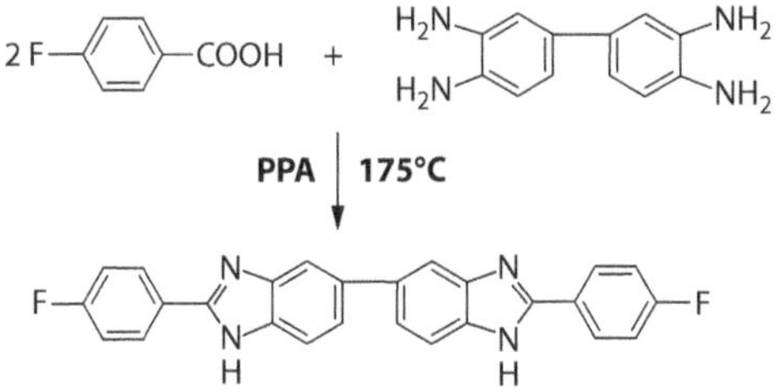

FIGURE 8.9 Synthesis of a benzimidazole-based difluoro monomer. (Reprinted from Twieg, R. et al., *Macromolecules*, 29, 7335, 1996. With permission.)

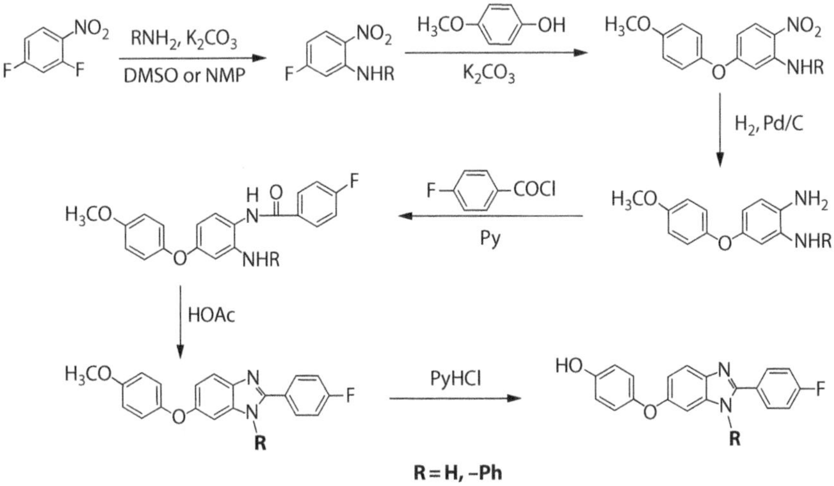

FIGURE 8.10 Synthesis of benzimidazole-based AB-type monomers. (Reprinted with modification from Twieg, R. et al., *Macromolecules*, 29, 7335, 1996. With permission.)

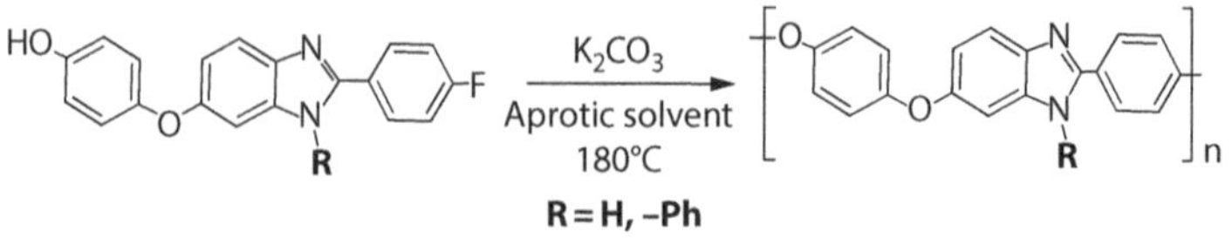

$$R = H, -Ph$$

FIGURE 8.11 Synthesis of polybenzimidazoles from the AB-type monomers. (Reprinted with modification from Twieg, R. et al., *Macromolecules*, 29, 7335, 1996. With permission.)

in acetic acid. Self-condensation polymerization of these two monomers gives the corresponding poly(aryl ether benzimidazole)s (Figure 8.11).

8.2.2 Synthesis of Sulfonated Polybenzimidazoles

8.2.2.1 Direct Polymerization of Sulfonated Monomers

In principle, the foregoing mentioned synthetic approaches for non-SOPBIs are also applicable to SOPBIs. However, one must be aware that the sulfonic acid groups may decompose at high temperatures (e.g., >250°C) due to their thermal instability and the sulfonic acid salts are usually infusible. This is why no sulfonated PBI (SOPBI) has been reported to be synthesized via melt polymerization method. In addition, in some cases, sulfonic acid groups may undergo side reactions when the polymerization of sulfonated monomers is conducted in PPA or in Eaton's reagent. And therefore, it is essential to carefully select the polymerization approach and to control the polymerization conditions when performing the polymerization in PPA or in Eaton's reagent. For example, the copolymerization of 4,4′-dicarboxyldiphenyl ether-2,2′-disulfonic acid disodium salt (DCDPEDS), DCDPE, and DABz in PPA at 190°C for 20 h yielded insoluble gel (Figure 8.12) [10]. This is probably because an unclear side reaction might have occurred on the sulfonic acid groups. Replacing PPA with Eaton's reagent, the polymerization conducting at 140°C for a short period of time (<4 h) yielded only low-molecular polymer because of the low reactivity of DCDPEDS resulting from the strong electron-withdrawing effect of the sulfonic acid group, while longer reaction time (20 h) also resulted in insoluble gel [10].

Gomez-Romero and coworkers reported that the condensation polymerization of 5-sulfoisophthalic acid mono sodium salt and 1,2,4,5-tetraaminobenzenetetrahydrochloride in PPA at 200°C for 5 h under nitrogen successfully yielded the desired product, poly[*m*-(5-sulfo)-phenylenebenzobisimidazole] [11]. Here, the dicarboxylic acid monomer, 5-sulfoisophthalic acid mono sodium salt, has the structural feature that the carbonyl groups possess high electron affinity due to the presence of electron-withdrawing substituents. In this case, PPA is a proper medium for the polymerization, whereas Eaton's reagent is unsuitable. In contrast, the following example (Figure 8.13) demonstrates that Eaton's reagent is an appropriate medium for synthesis of sulfonated poly(ether benzimidazoles) [7]. In PPA, however, insoluble gels readily formed probably because of the occurrence of cross-linking reactions between sulfonate groups and activated benzene rings.

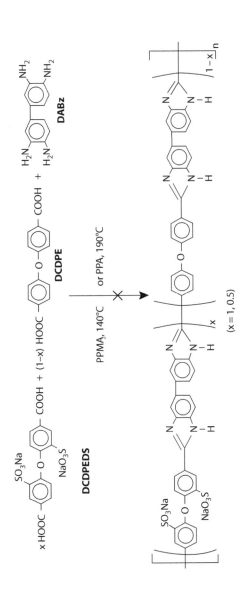

FIGURE 8.12 Synthesis of SOPBIs from a sulfonated dicarboxylic acid monomer. (Reprinted from Xu, H. et al., *Polymer*, 48, 5556, 2007. With permission.)

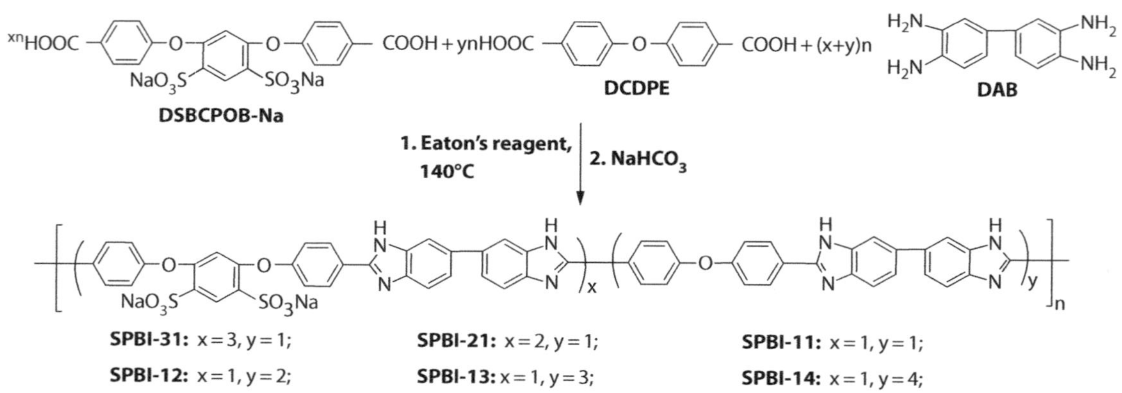

FIGURE 8.13 Synthesis of various SOPBIs from BCPOBDS-Na. (Reprinted from Sheng, L. et al., *J. Power Sources*, 196, 3039, 2011. With permission.)

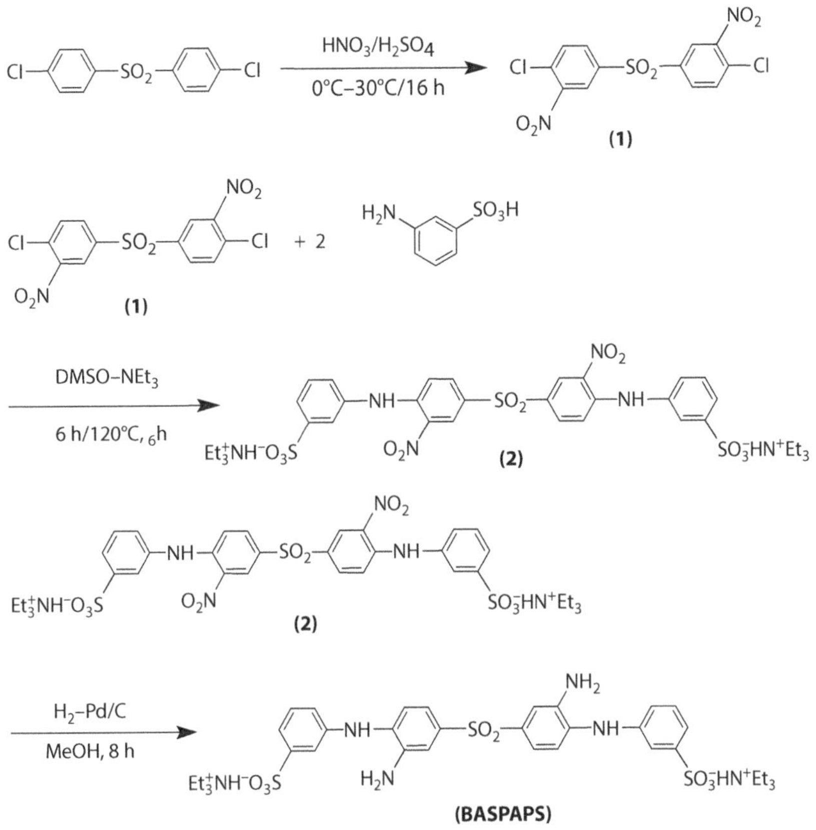

FIGURE 8.14 Synthesis of a disulfonated tetraamine monomer. (Reprinted from Jouanneau, J. et al., *Macromolecules*, 40, 983, 2007. With permission.)

SOPBIs can also be synthesized from sulfonated tetraamines and dicarboxylic acids. Mercier and coworkers reported the synthesis of various N-substituted PBIs by using Ueda's approach [12]. As illustrated in Figures 8.14 and 8.15, a sulfonated N-substituted tetraamine is synthesized by three-step reactions and subsequently copolymerized with DCDPE and a nonsulfonated tetraamine to yield the SOPBIs with varied degree of sulfonation (DS).

8.2.2.2 Postsulfonation

Postsulfonation of a parent PBI is, in principle, a facile approach for the synthesis of SOPBIs comparing with the approaches of direct polymerization of sulfonated monomers. It may avoid complex synthetic procedures for monomers. Concentrated sulfuric acid and fuming sulfuric acid are frequently used as the sulfonating reagent. For the commercial PBI (*m*-PBI) and the AB-PBI, the postsulfonation reaction must be performed at extremely high temperatures (~450°C) because the benzene rings in the polymer backbones are extremely unreactive resulting from the

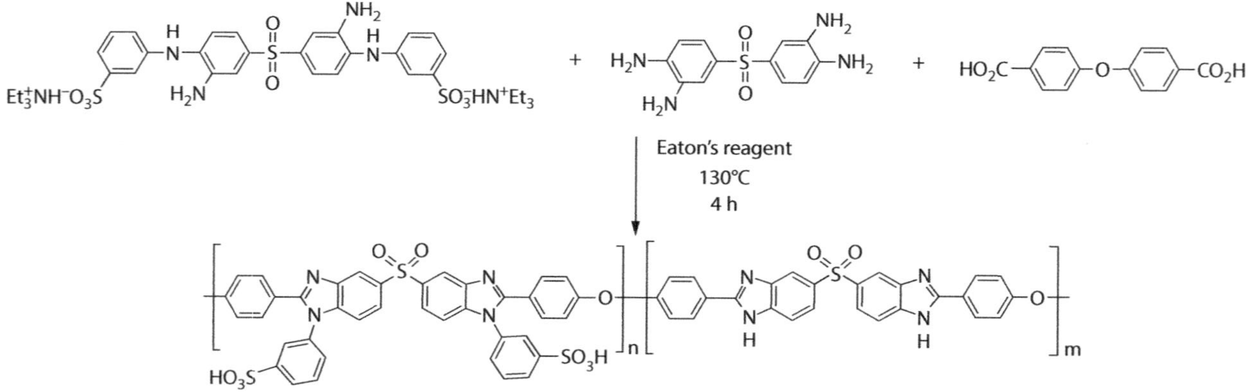

FIGURE 8.15 Synthesis of SOPBIs from a disulfonated tetraamine monomer. (Reprinted from Jouanneau, J. et al., *Macromolecules*, 40, 983, 2007. With permission.)

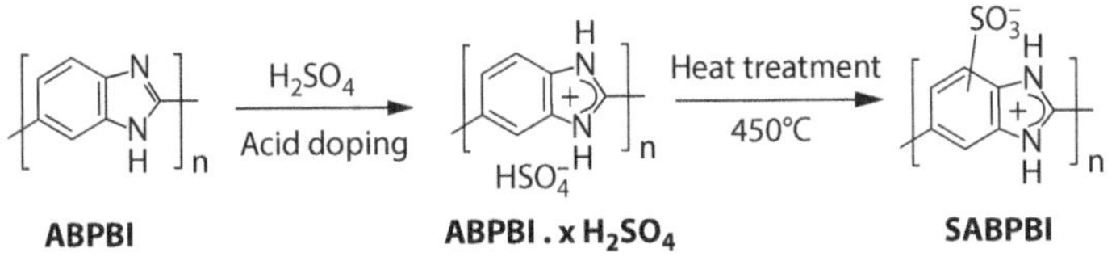

FIGURE 8.16 Synthesis of sulfonated AB-PBI by postsulfonation. (Reprinted from Asensio, J.A. et al., *Electrochim. Acta*, 49, 4461, 2004. With permission.)

strong electron-withdrawing effect of the protonated imidazole rings [13–15]. The substitution position is not fixed (Figure 8.16) and the DS cannot be precisely controlled, which are two major drawbacks associated with the postsulfonation method. Moreover, for the *m*-PBI, the sulfonation may occur at any of the three benzene rings of a repeat unit.

Unlike the *m*-PBI and AB-PBI, the poly(ether benzimidazole) (OPBI) synthesized from DCDPE and DABz can be readily sulfonated with concentrated sulfuric acid under mild conditions (80°C) because of the high reactivity of the ether bond-linked benzene rings of the OPBI (Figure 8.17). The sulfonation sites occurred in the acid moiety (DCDPE) not in the tetraamine moiety (DABz) because the nitrogen atoms were protonated in sulfuric acid leading to great deactivation effect on the benzene rings of DABz moiety, whereas the electron-donating effect of ether bond activated the benzene rings of the DCDPE moiety.

The DS increases with prolonging reaction time. Replacing concentrated sulfuric acid with fuming sulfuric acid is also very effective for enhancing the DS. A high value of DS (154%) was achieved as the sulfonation was conducted in fuming sulfuric acid (20% SO_3) at 80°C for 5 h [10].

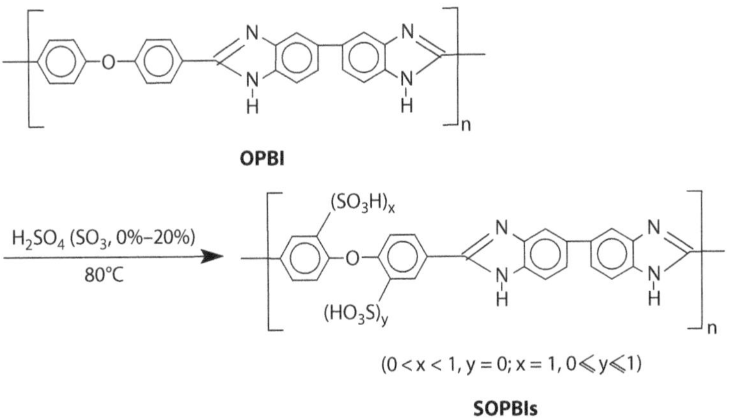

FIGURE 8.17 Synthesis of SOPBIs by postsulfonation under mild reaction conditions. (Reprinted from Xu, H. et al., *Polymer*, 48, 5556, 2007. With permission.)

The postsulfonation can also be achieved via grafting reactions. The N–H bond of the imidazole ring is a good site for grafting reaction. Roziere and coworkers [16] reported a two-step reaction procedure to graft sulfonated aryl groups onto *m*-PBI. In the first step reaction, the *m*-PBI is reacted with LiH in DMAc to yield the lithium benzimidazol-1-ide. Then, the intermediate product is reacted with sodium 4-bromomethylbenzenesulfonate to give the SOPBI (Figure 8.18). Replacing sodium 4-bromomethylbenzenesulfonate with propanesultone or butanesultone yields the sulfopropyl- or sulfobutyl-grated PBI (Figure 8.19) [17].

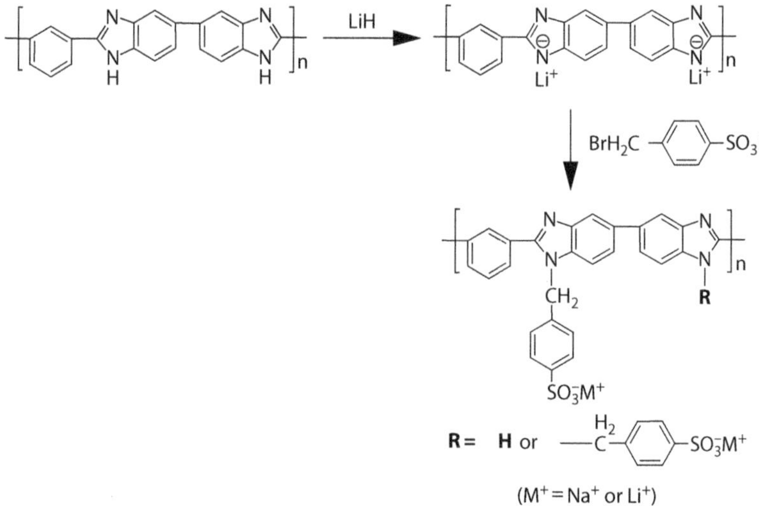

FIGURE 8.18 Synthetic route for the sulfonated PBI reported in Ref. [16]. (Reprinted with modification from Glipa, X. et al., *Solid State Ionics*, 97, 323, 1997. With permission.)

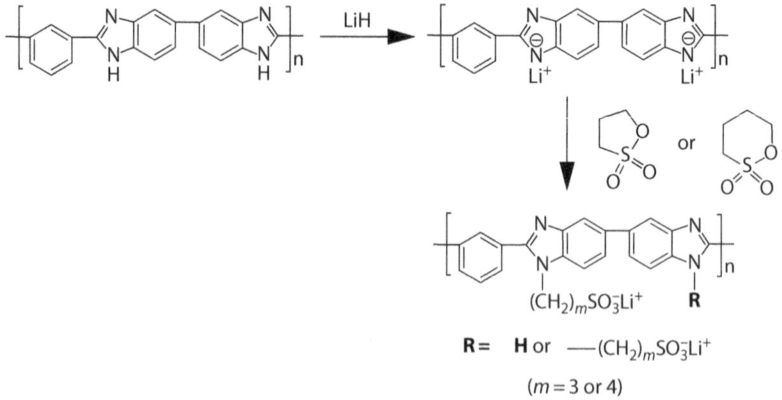

FIGURE 8.19 Synthetic route for the sulfonated PBI reported in Ref. [17]. (Reprinted with modification from Bae, J.-M. et al., *Solid State Ionics*, 147, 189, 2002. With permission.)

8.2.3 SPECTRAL CHARACTERIZATION

The chemical structures of PBIs are usually characterized by Fourier transform infrared (FT-IR) spectroscopy. The key structural feature of PBIs is the benzimidazole groups. A benzimidazole group is built up of C=N, C–N, N–H, C=C (benzene ring backbone) and =C–H (benzene ring) bonds that show characteristic absorption bands in FT-IR spectra. Although a slight shift may occur depending on the fine chemical structures of the polymers, the characteristic absorption bands owing to N–H stretch usually appear at 3450–3400 cm^{-1} (sharp, isolated N–H) and 3300–2500 cm^{-1} (broad, self-associated N–H) [11,18]. The characteristic absorption band due to =C–H stretch usually appears at 3020–3100 cm^{-1} (weak). The characteristic absorption bands due to C=N and C=C stretches usually appear around 1640–1500 cm^{-1}, while the strong absorption band assigned to in-plane deformation of benzimidazole rings usually appears at 1495–1395 cm^{-1} [11].

It should be noted that the characteristic absorption bands of benzimidazole groups may be largely changed due to the presence of acid in the polymers. Bouchet and Siebert investigated the FT-IR spectra of the pure m-PBI and the m-PBI doped with different acids [18]. Figure 8.20 is their reported FT-IR spectra of the pure m-PBI and the PA-doped m-PBI with different acid content reported. It can be seen that the absorption band at 3412 cm^{-1} assigned to isolated N–H stretch shifted to lower wave number due to increase in concentration of hydrogen bonded N–H. Another significant feature is the increasing intensities of peaks at 1565 and 1630 cm^{-1} with the acid content in the polymer due to protonation of benzimidazole groups [18].

Besides FT-IR spectra, nuclear magnetic resonance (NMR) spectra, in particular, proton NMR (^{1}H NMR) spectra are also frequently used to characterize PBI structure if the polymer is soluble in an appropriate deuterated solvent. Figure 8.21

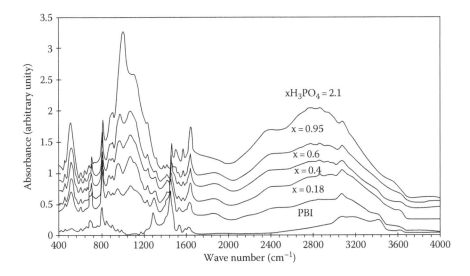

FIGURE 8.20 FT-IR spectra of the m-PBI with different content of PA. (Reprinted from Bouchet, R. and Siebert, E., *Solid State Ionics*, 118, 287, 1999. With permission.)

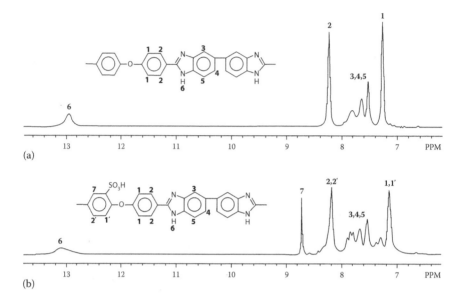

FIGURE 8.21 ¹H NMR spectra of OPBI (a) and SOPBI (b) prepared by postsulfonation in concentrated sulfuric acid at 80°C for 10 h. (Reprinted from Xu, H. et al., *Polymer*, 48, 5556, 2007. With permission.)

is the ¹H NMR spectra of the OPBI, a kind of PBI synthesized from DCDPE and 3,3′-diaminobenzimdine (DABz), and the corresponding SOPBI [10]. For the OPBI, the peak at ~13 ppm is assigned to the proton (H6) of the benzimidazole ring N–H. The peak at 8.25 ppm is assigned to the protons (H2) of benzene rings of the DCDPE moiety adjacent to imidazole rings. The three peaks (H3–5) at approximately 7.8–7.5 ppm are assigned to the benzene rings of DABz moiety. The peak at ~7.3 ppm is assigned to the protons (H1) of benzene rings of the DCDPE moiety adjacent to ether bond. For the SOPBI, the peak assigned to N–H of imidazole group is shifted to ~13.1 ppm. A new peak at 8.74 ppm is assigned to the protons adjacent to sulfonic acid group.

8.3 SOLUBILITY

Because of the stiff backbone and strong interchain hydrogen bond interaction that may cause aggregation of polymers, PBIs generally show poor solubility in common organic solvents. For example, the AB-PBI is only soluble in concentrated sulfuric acid, MSA, and formic acid (FA) but is insoluble in common organic solvents such as *N,N*-dimethylformamide (DMF), *N,N*-dimethylacetamide (DMAc), 1-methyl-2-pyrrolidinone (NMP) and dimethylsulfoxide (DMSO) [3]. It is greatly desirable to develop organosoluble PBIs, while the key performances such as high thermal stability, high mechanical strength and modulus, excellent chemical stability, and good flame retardance are maintained. Since most PBIs are synthesized from dicarboxylic acids and tetraamines or their hydrogen chloride salts, the solubility of PBIs is strongly related to the chemical structures of both the dicarboxylic acid moieties and the tetraamine moieties. The structural parameters include the configuration of moieties, polar groups, bulky groups, flexible linkages, density of imidazole rings, copolymerization, etc. Table 8.1 cites the solubility behaviors of the PBIs derived from DABz and various dicarboxylic acids with different structural features in some organic solvents [3,19–22]. *i*PTA and terephthalic acid (*t*PTA) are isomers to each other; however, the *m*-PBI shows significantly better solubility than the *p*-PBI. This indicates that with the same tetraamine moiety (DABz), the dicarboxylic acid moiety with nonlinear configuration is favorable for improving the solubility of the PBI. Replacing *i*PTA with twist dicarboxylic acid (2,2′-biphenyldicarboxylic acid) or the ones with bulky groups (4-(4,5-diphenyl-imidazole-1,2-yl)-dibenzoic acid, 5-(4-aminophenoxy)isophthalic acid) or even 5-aminoisophthalic acid (APTA) causes great improvements in solubility of the polymers (2,2′-BPPBI, IDBA-PBI, PBI-A, DABz-APTA). These polymers show excellent solubility in all the solvents listed in the table. This is likely attributed to the weakened interchain interaction resulting from the loose packing of polymer chains and/or the low density of benzimidazole groups. Replacing *i*PTA with 3,5-pyridinedicarboxylic acid or 2,5-furandicarboxylic acid results in similar solubility behavior to that of the *m*-PBI. The PBIs (OPBI and BDA-PBI) derived from 4,4′-oxybisbenzoic acid and 4,4′-biphenyldicarboxlic acid are another interesting pair of example. The former are well soluble in DMF, DMAc, NMP and DMSO, whereas the latter is insoluble in DMF and DMSO. The difference in solubility of these two polymers should be ascribed to the different configuration of the dicarboxylic acid moieties as well as the different chain flexibility (Figure 8.22).

Changing the structure of the tetraamine moiety also changes the solubility (Table 8.2). With the same dicarboxylic acid moiety, the 1,2,4,5-tetraamine (TAB)-based PBIs show poor solubility than the DABz-based ones. For example, the *i*PTA-TAB is insoluble in DMAc, DMF, or DMSO [3], whereas the *m*-PBI is soluble or partially soluble in these solvents. In contrast, the PBIs derived from 2,6-*bis*(3,4-diaminophenyl)-4-phenyl-pyridine are well soluble in FA, DMAc, and NMP, which are superior than the ones derived from DABz [22].

SOPBIs in their proton form are generally insoluble in organic solvents such as DMSO, NMP, DMAc, and DMF resulting from the ionic cross-linking (sulfonic acid–benzimidazole interaction) [7,10]. To the authors' experience even trace amount of residual acid may greatly decrease the solubility of the non-SOPBIs. The solubility of the SOPBIs can be significantly improved by transforming the proton form into the triethylammonium form or the metallic ion form.

Besides the structural feature of the PBIs, molecular weight also affects the solubility. It is well known that the polymers with higher molecular weight tend to have lower solubility because of the stronger interchain interaction. Such an effect is fairly pronounced for PBIs. For example, Cabasso and coworkers [23] reported that low-molecular-weight ($M_w = 35,800$) *m*-PBI is soluble in refluxing DMF, whereas high-molecular-weight ($M_w = 70,800$) *m*-PBI is only soluble in DMSO at 180°C but is insoluble in refluxing DMF. The solubility can be further enhanced by adding lithium chloride to the polymer solutions. Akpalu and coworkers reported that the high-molecular-weight *m*-PBI of which $M_w = 199,000$ is soluble in DMAc containing 4% lithium chloride [24]. The incorporation of lithium chloride into a PBI solution

TABLE 8.1

Solubility Behaviors of Various Polybenzimidazoles Derived from 3,3'-Diaminobenzidine and Various Dicarboxylic Acid Monomers

Acronym	Repeat Unit	Solubility					Refs.
		FA	DMF	DMAc	NMP	DMSO	
m-PBI		5–6[a]	+–	+	N/A	+	[3]
p-PBI		2–3[a]	–	N/A	N/A	0.5–1	[3]
Py-PBI		10–15[a]	+–	N/A	N/A	+	[3]
FPBI		3–4[a]	+–	N/A	N/A	+	[3]
2,8-NPBI		8–10[a]	–	N/A	N/A	+	[3]
2,2'-BPPBI		+	+	N/A	N/A	+	[3]
IDBA-PBI		N/A	+	+	+	+	[19]
PBI-A		+	+	+	+	+	[20]
DABz-APTA		+	+	+	+	+	[21]

(Continued)

TABLE 8.1 (*Continued*)

Solubility Behaviors of Various Polybenzimidazoles Derived from 3,3'-Diaminobenzidine and Various Dicarboxylic Acid Monomers

Acronym	Repeat Unit	Solubility					Refs.
		FA	DMF	DMAc	NMP	DMSO	
OPBI	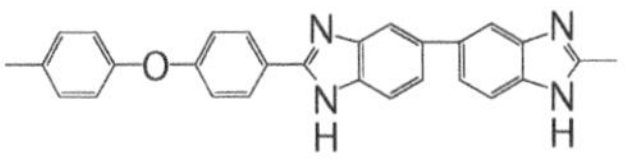	+	+	+	+	+	[10]
BDA-PBI	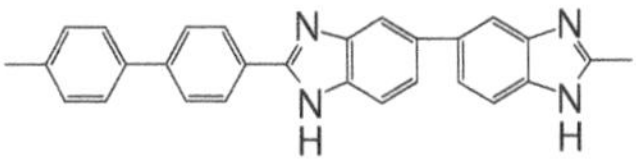	+–	–	+	+	–	[22]
BuDA-DABz	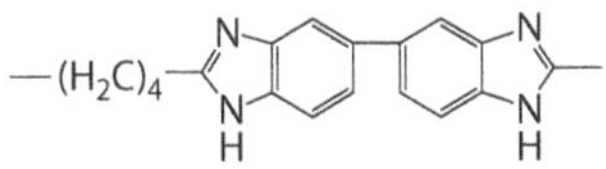	+	+–	N/A	N/A	+	[3]

Keys: "+", soluble; "+–", partially soluble; "–", insoluble.

N/A: not available from literature.

[a] Unit: g/100 mL solvent.

Rigid, linear
(a)

Flexible, angled
(b)

FIGURE 8.22 Difference in configuration of the dicarboxylic acid moieties. (a) Rigid, linear and (b) Flexible, angled.

results in association of lithium chloride with the imidazole –NH and –N=C– groups and the dissociation of interchain hydrogen bonds, and therefore, the PBI solubility is improved [25].

Since the imidazole –NH is a reactive group, PBIs can be chemically modified via N-substitution (grafting). The modified PBIs generally show much improved solubility comparing with the parent polymers because the hydrogen bond interactions are weakened due to the reduction of imidazole –NH content. The modification can be achieved by reacting the parent polymers with LiH or NaH to yield the lithium or sodium benzimidazol-1-ide intermediate products in the first step followed by reacting with proper electrophilic reagents to produce the desired N-substituted PBIs. Kumbharkar and Kharul [26] reported the synthesis (Figure 8.23) and solubility of various N-substituted PBIs. Most N-substituted PBIs are well soluble in aprotic solvents such as DMF, DMAc, and NMP at room temperature and some of them are even soluble in low-boiling-point solvents such as chloroform and THF [26]. Jana's group [27] synthesized a series of *n*-alkyl-grafted PBIs with varied alkyl chain length (C_2–C_{14}) with the degree of N-substitution of 42%–65% (Figure 8.24). It appears that the alkyl chain length hardly affects the solubility of the N-substituted polymers.

TABLE 8.2

Solubility Behaviors of Various Polybenzimidazoles Derived from Two Tetraamine Monomers, 1,2,4,5-Benzenetetraamine (TAB) and 2,6-*bis*(3′,4′-Diaminophenyl)-4-Phenylpyridine (Py-TAB), and Various Dicarboxylic Acid Monomers

Acronym	Repeat Unit	Solubility					Refs.
		FA	DMF	DMAc	NMP	DMSO	
*i*PTA-TAB	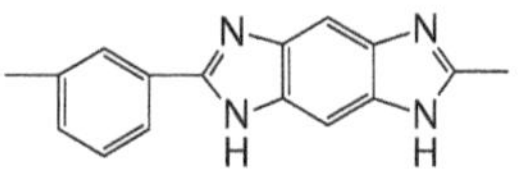	5–6[a]	–	N/A	N/A	–	[3]
*t*PTA-TAB	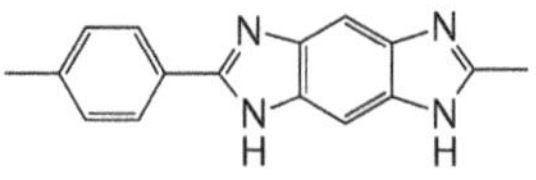	2–3[a]	–	N/A	N/A	–	[3]
BuDA-TAB	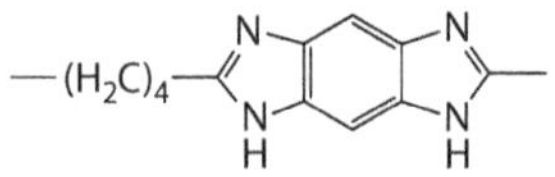	+	–	N/A	N/A	+–	[3]
*i*PTA-PyPBI	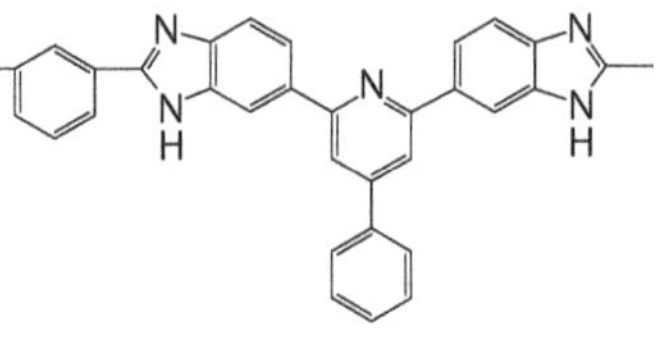	+	N/A	+	+	N/A	[22]
TPA-PyPBI	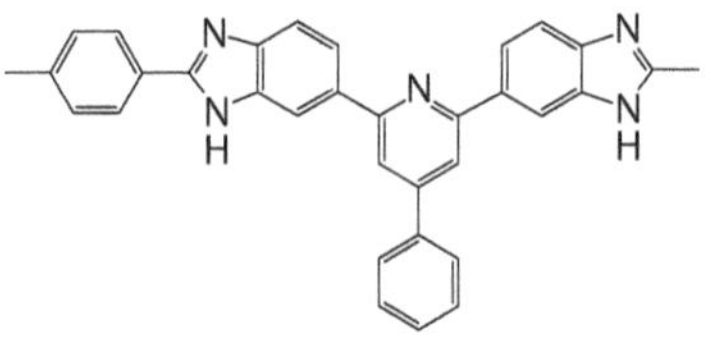	+	N/A	+	+	N/A	[22]
BDA-PyPBI	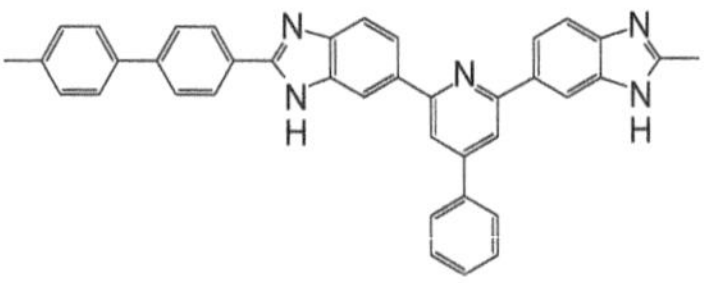	+	N/A	+	+	N/A	[22]
OBA-PyPBI	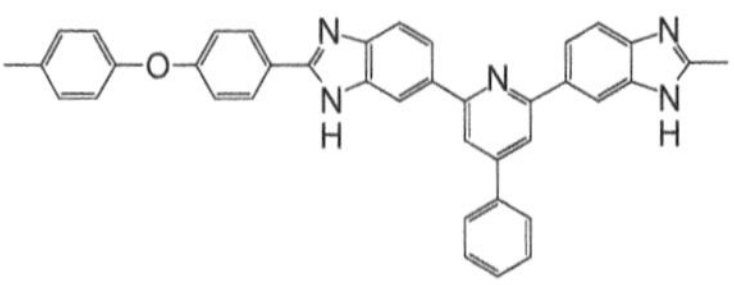	+	N/A	+	+	N/A	[22]

(*Continued*)

TABLE 8.2 (*Continued*)

Solubility Behaviors of Various Polybenzimidazoles Derived from Two Tetraamine Monomers, 1,2,4,5-Benzenetetraamine (TAB) and 2,6-*bis*(3',4'-Diaminophenyl)-4-Phenylpyridine (Py-TAB), and Various Dicarboxylic Acid Monomers

Acronym	Repeat Unit	Solubility					Refs.
		FA	DMF	DMAc	NMP	DMSO	
BPDA-PyPBI	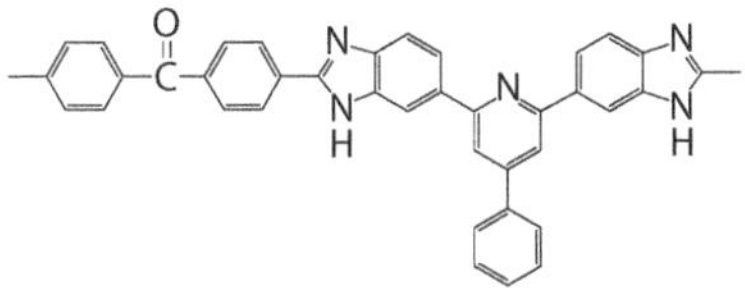	+	N/A	+	+	N/A	[22]
HFIPA-PyPBI	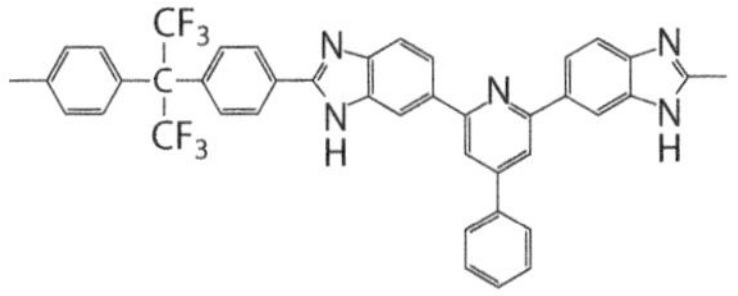	+	N/A	+	+	N/A	[22]

Keys: "+", soluble; "+−", partially soluble; "−", insoluble.

N/A: not available from literature.

[a] Unit: g/100 mL solvent.

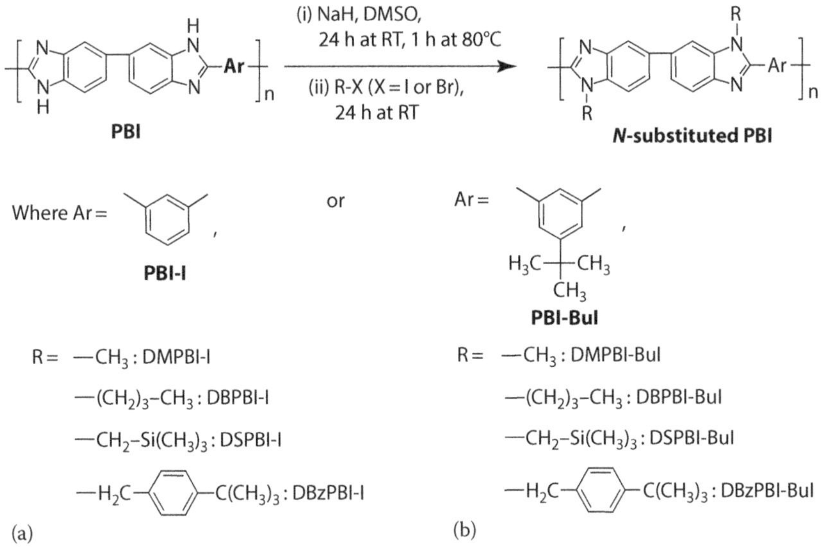

FIGURE 8.23 N-substitution of (a) PBI-I and (b) PBI-BuI. (Reprinted from Kumbharkar, S.C. and Kharul, U.K., *Eur. Polym. J.*, 45, 3363, 2009. With permission.)

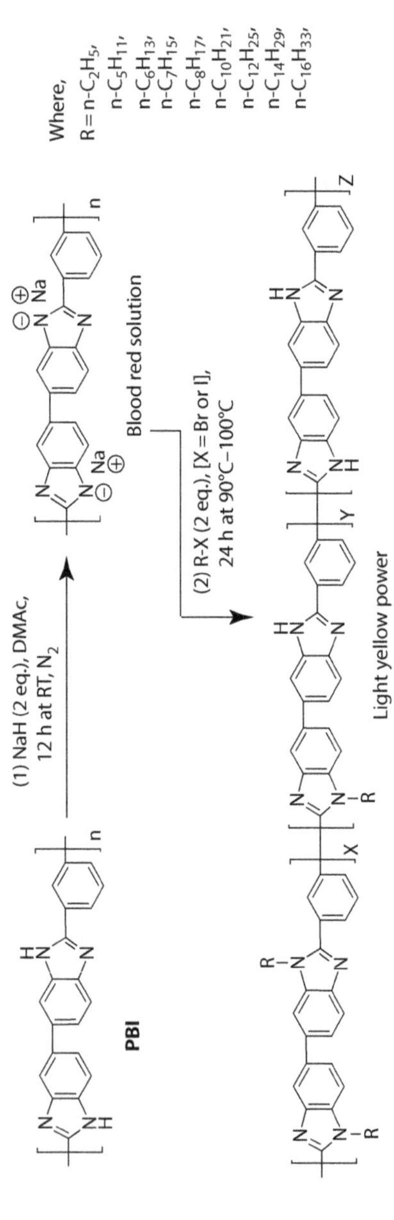

FIGURE 8.24 Synthesis of N-substituted PBI. (Reprinted from Maity, S. et al., *Eur. Polym. J.*, 49, 2280, 2013. With permission.)

8.4 ACID DOPING, PROTON CONDUCTIVITY, AND MECHANICAL PROPERTIES

8.4.1 ACID DOPING

The nitrogen atom of an imidazole ring possesses a pair of nonbonded electrons, and therefore, imidazole groups are a kind of Lewis base. Since both the carbon and nitrogen of the –C=N– bond are sp^2 hybridized, the nitrogen atom of the –C=N– bond is more basic than the one of the –C–NH– bond. As a result, protonation preferentially occurred on the nitrogen atom of the –C=N–, and PBIs are generally soluble in or have strong interactions with strong acids such as concentrated sulfuric acid, MSA and PA. This is why the PBI membranes can be readily doped with various acids. The PA-doped PBI membranes are especially of more interests because of their important applications in HT-PEMFCs as will be introduced in the following sections. The PA-doping is usually performed by soaking the PBI membranes in aqueous PA solutions at room temperature or at elevated temperatures for a given period of time. The sorption equilibrium can be reached after 50 h at room temperature for the commercial PBIs (m-PBI) [2]. It can be shorter or longer depending on the fine chemical structure and morphology of the PBI membranes, the concentration of PA and temperature.

The degree of PA doping is usually characterized by the parameter, doping level (DL) or uptake (S_{PA}). The DL is defined as a number equal to the number of PA molecules per repeat unit of PBI, while the S_{PA} is termed as the weight of the absorbed PA in grams per 100 g of PBI. The S_{PA} is usually measured according to the following procedures [2,7,21]:

A sheet of preweighed dry PBI membrane is immersed in aqueous PA solution of desired concentration at a given temperature till sorption equilibrium is achieved. Then the membrane is taken out, blotted with tissue paper, and dried at 110°C–120°C for a period of time (typically 10 h) in vacuo to remove the absorbed water in the membranes. The S_{PA} is calculated from the following equation:

$$S_{PA} = \frac{W_1 - W_0}{W_0}$$

where W_0 and W_1 refer to the weight of the undoped dry membrane and the doped dry membrane, respectively.

This method is quite simple and easy to operate in the lab. However, it should be noted that the experimental conditions are very crucial to ensure accurate measurement. Usually the measurement must be repeated at least three times for a PBI and the averaged value is considered as the S_{PA} to minimize the experimental error as possible. Moreover, vacuum is essential to quickly remove the absorbed water in the membranes, while 110°C–120°C seems to be the suitable temperature for membrane drying [2]. Too high temperatures (>120°C) and/or too long time may cause self-condensation of PA, whereas too low temperatures (<100°C) or too short time (less than a few hours) may cause insufficient removal of water.

The PA content in PA-doped membranes can also be analyzed by titration with standard sodium hydroxide solution [28]. In principle, titration method is more accurate than the weighing method provided that the absorbed PA in the samples has been completely neutralized.

For a given polybenezimidazole, both DL and S_{PA} are useful for characterization of the degree of acid doping. However, in some cases, S_{PA} is more useful for comparison of the sorption capacities between different PBIs. For example, the commercial PBI possesses two imidazole rings per repeat unit, whereas the AB-PBI possesses only one imidazole ring per repeat unit. As a result, when the DL is the same for both polymers, the S_{PA} of the commercial PBI is only less than half of that of the AB-PBI.

The DL and the S_{PA} are affected by many factors such as the chemical structure of polymers, PA concentration, and temperature. Li and coworkers studied in detail the variation of DL of m-PBI as a function of PA concentration at room temperature. As shown in Figure 8.25, in the acid concentration ranging from 2 to 10 mol L^{-1} (~18–66 wt%), the DL increases only slightly, whereas the DL increases rapidly above 13 mol L^{-1} (~79 wt%) [29].

The PBIs with high density of benzimidazole groups or other basic groups and/or loose packing of polymer chains generally show high S_{PA}. Leykin et al. [30] reported that the PBI with benzimidazol-2-yl substituents (chemical structure is shown in Figure 8.26) showed a high DL of 17.2 (read from Figure 8.27) in 60% PA solution at room temperature which is much higher than that (~3.5 [29]) of the m-PBI under the same conditions. This should be attributed to the high benzimidazole group density and loose chain packing of the polymer. Benicewicz and coworkers reported that the incorporation of pyridine groups as an additional nitrogen containing heterocycle enhanced the interaction of PA and the polymer [31]. The authors found that the PBI prepared from 3-aminoisophthalic acid and DABz (chemical structure is shown in Figure 8.28) was soluble in 85% PA at room temperature because of the high density

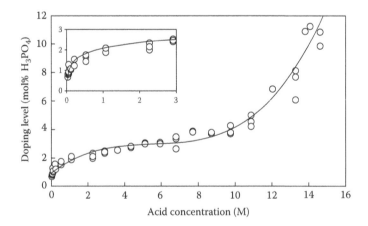

FIGURE 8.25 Variation of acid DL of the m-PBI membrane at room temperature as a function of acid concentration. (Reprinted from Li, Q. et al., *Solid State Ionics*, 168, 177, 2004. With permission.)

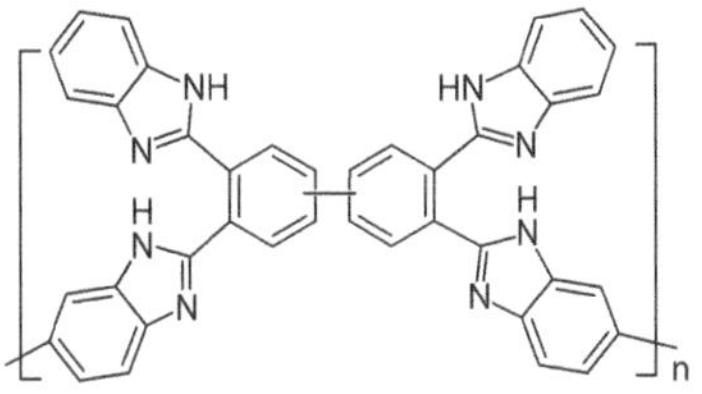

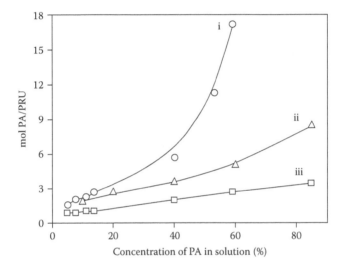

FIGURE 8.26 Chemical structure of the PBI with benzimidazol-2-yl substituents. (Reprinted from Leykin, A.Y. et al., *J. Membr. Sci.*, 347, 69, 2010. With permission.)

FIGURE 8.27 Variation of acid DL of various PBI membrane at room temperature as a function of acid concentration. (Reprinted from Leykin, A.Y. et al., *J. Membr. Sci.*, 347, 69, 2010. With permission.)

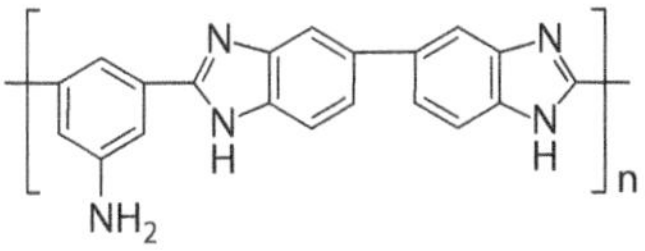

FIGURE 8.28 Chemical structure of the PBI derived from APTA and DABz. (Reprinted from Xu, N. et al., *J. Polym. Sci. Part A: Polym. Chem.*, 47, 6992, 2009. With permission.)

of basic groups (primary amino groups and benzimidazole groups) as well as the loose chain packing resulting from the bulky amino groups [21].

Benicewicz's group [28,31] developed a sol–gel method by which highly doped PBI membranes (DL = 20–40) were prepared. The as-synthesized PBI solutions in PPA are directly cast into films while hot without isolation and subsequently the films are exposed to a humidified atmosphere (40% ± 5%) at 25°C for 24 h.

Moisture-sorption-induced hydrolysis of the solvent from PPA (a good solvent for PBI) to PA (a poor solvent for PBI) yields the highly PA-doped membranes. The resulting membranes show exceptionally high proton conductivity at elevated temperatures as will be introduced in the following section.

8.4.2 PROTON CONDUCTIVITY

The blank PBIs are protonic insulator of which proton conductivity is in the order of $\sim10^{-9}$–10^{-10} S cm^{-1} [32], and therefore, acid doping is essential to ensure sufficiently high proton conductivity. As foregoing mentioned, for the PA-doped PBI membranes the protonation preferentially occurs on the –N=C– nitrogen of a benzimidazole ring. Once the –N=C– nitrogen is protonated, the positive charge will be delocalized in the whole imidazole ring and the other nitrogen atom (–C–NH–) of this imidazole ring will no longer be protonated. As a result, the maximum protonation is one PA molecule per benzimidazole groups. As the doping exceeds the maximum protonation, besides the protonated (bonded) PA molecules, *free* (nonbonded) PA molecules will also exist in the membranes. It should be noted that deprotonation (the reverse process of protonation) may also occur though the protonation process is dominant. It is believed that such a protonation–deprotonation process is one of the mechanisms for proton transport [33]. For the *m*-PBI/PA membrane, at a DL=2 (one PA molecule per benzimidazole ring) the proton conductivity is about 2.5×10^{-2} S cm^{-1} at 200°C [2]. In this case, the protonation–deprotonation process and the dihydrogen phosphate anion (H$_2$PO$_4^-$) should be mainly responsible for the ionic conductance. At a higher DL of 5.6 the conductivity increases to 6.8×10^{-2} S cm^{-1} at 200°C indicating a significant contribution from the *free acid* [2].

PA is a good proton conductor even in completely anhydrous state because of its reversible self-dissociation property (2H$_3$PO$_4$ $\rightleftharpoons$ H$_4$PO$_4^+$+H$_2$PO$_4^-$). Such a unique future makes it very suitable for use in high-temperature PEMFCs. Since conductivity is closely related to the mobility of ionic species and the mobility of the *free acid* is much higher than the bonded acid, raising the *free acid* content in membranes should be an effective way for enhancement of proton conductivity, that is, high DL is necessary to achieve high proton conductivity (e.g., > 0.1 S cm^{-1}). Benicewicz's group reported that the highly PA-doped (DL: ~32) PBI membranes prepared by sol–gel method exhibited exceptionally high proton conductivity (0.26 S cm^{-1}) under anhydrous condition at 200°C [28]. However, too high DL often causes excess swelling and unacceptable mechanical strength. This issue will be discussed in detail in the following section.

Although PA can undergo dissociation under anhydrous condition, the proton transport is still facilitated with the aid of water. Thus, the proton conductivity increases with increasing relative humidity. It is reported that, at 200°C, an increase in relative humidity from 0.15% to 5% resulted in an increase in conductivity from 0.038 to 0.068 S cm^{-1} [2]. This clearly indicates the significant effect of water content in the membranes on conductivity. Therefore, when doing comparison of proton conductivity reported in different literature, one must pay attention to the drying conditions for conductivity measurement. To eliminate the effect of water on proton conductivity of the PA-doped membranes, the authors proposed the following procedures:

A sheet of membrane and two pairs of blackened platinum plate electrodes were set in a Teflon cell. The cell was placed in a thermo-controlled chamber with an inlet and an outlet for continuous nitrogen flow throughout the measurement. Before starting the conductivity measurement, the chamber was heated at 170°C for 1 day to remove any water vapor so that the relative humidity could approach to 0%. After that, the temperatures of the chamber was set at the desired values and kept constant for 30 min at each point [7,21].

The dependence of proton conductivity on temperature can be interpreted by the Arrhenius equation: $\sigma = \sigma_0 \exp(-E_a/RT)$, where σ, E_a, R, and T refer to conductivity, activation energy, gas constant and absolute temperature, while σ_0 is the preexponential factor. The plot of natural logarithm of conductivity ($\ln \sigma$) versus the reciprocal of temperature (1/T) gives a straight line ($\ln \sigma = -E_a/RT + \ln \sigma_0$). The slope of the line equals to the term $-E_a/R$. Thus, the activation energy (E_a) equals the negative product of the slope and gas constant (R). Figure 8.29 shows the plot of $\ln \sigma$ of three PA-doped PBI membranes versus 1000/T (unpublished data of our group). From the slope of each line, the activation energy values are calculated to be 23.4, 20.8, and 22.2 kJ mol^{-1} from top to bottom in the figure. Li's group reported that the activation energy values of the PA-doped m-PBI with DL of 2.0–5.6 are in the range of 18–25 kJ mol^{-1} [2].

The proton-conducting mechanism of PA-doped PBI membranes has been well studied by many research groups [33–37], and it has been commonly accepted that the proton conduction is mainly determined by the Grotthuss mechanism (proton hopping) especially at temperatures below 100°C [36]. The proton hopping may occur between acid–acid or acid–water (in case of hydration) or acid–benzimidazole ring. The presence of a lot of free PA molecules in PBI membranes is very crucial to achieve high proton conductivity because the free acid (PA) is the major contributing factor for proton transport [2].

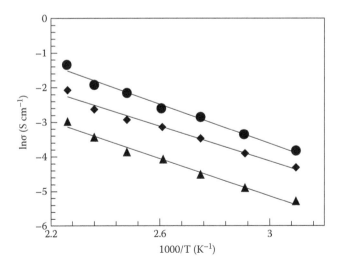

FIGURE 8.29 Temperature dependence of proton conductivity of three PA-doped PBI membranes. (From unpublished data of our group.)

8.4.3 MECHANICAL PROPERTIES

PA is a liquid at room temperature. The PA absorbed in PBI membranes functions as not only a proton-conducting source but also a plasticizer. The PA-induced plasticization effect causes reduction in glass transition temperature (T_g) and mechanical strength and modulus of the membranes. For a given membrane, the higher the DL (or S_{PA}), the larger the magnitude of reduction in mechanical strength. As a membrane is highly doped, it may excess swell and turn to be a rubbery material leading to a drastic reduction or completely loss of mechanical strength. Therefore, usually, the PA-doping must be controlled at a reasonable level to well balance proton conductivity and mechanical strength. An optimal DL of 5–6 has been suggested for the m-PBI membrane [2]. From the viewpoint of practical applications, it is strongly desirable to develop membranes with high proton conductivity while high mechanical strength is maintained. It is found that besides acid uptake, many other factors such as polymer molecular weight, chemical structure of polymers, and cross-linking also have large impact on the mechanical properties of PBI membranes. For the PBI membranes with exactly the same chemical structure and the same DL, those with higher molecular weight tend to show higher mechanical strength. For example, Lobato et al. [38] reported that at a DL of 6.7 the high-molecular-weight ($M_w = 105,100$) m-PBI membrane exhibited a stress at break of 15.5 MPa, which is about 2.5 as large as that (6.2 MPa) of the low-molecular-weight ($M_w = 38,400$) m-PBI membrane indicating that increasing polymer molecular weight is favorable for enhancement of membrane mechanical strength. Without acid doping, the high-molecular-weight membrane also shows much higher stress at break (~120 MPa) than the low-molecular-weight one (~45 MPa). This is because higher molecular weight corresponds to longer polymer chains and thus higher degree of chain entanglement which enhances interchain interaction. Benicewicz's group [28] reported that the PA-doped p-PBI membranes prepared by sol–gel method exhibit tensile stress in the range ~1.0–3.5 MPa and elongation at break in the range of 150%–390%. Though the mechanical strength of these membranes is quite low due to their extremely high acid DLs (20–40), the difference in tensile stress among the membranes with different polymer molecular weight (inherent viscosity) is still significant. The membrane with an inherent viscosity of 3.0 dL g^{-1} exhibits a tensile stress of about 3.5 MPa, which is nearly twice of that of the membrane with the polymer inherent viscosity of 1.7 dL g^{-1}.

Theoretically, the molecular weight of condensation polymers is affected by many factors such as the purity of monomers, the molar ratio between the reactive functional groups (e.g., carboxylic group and o-diamino group), monomer concentration, monomer reactivity, reaction time, and any possible side reactions. To achieve high molecular weight, the purity of monomers must be as high as possible (e.g., >99%), and the molar ratio between the reactive functional groups must be strictly controlled at 1:1 (occasionally, one may use the excess of one of the monomers to produce functional group [from the excess monomer] terminated oligomers, which can be used for the next step reactions). In addition, increasing monomer concentration is usually favorable for achieving high-molecular-weight polymers. Dilute polymerization system tends to form macrocyclic compounds, which can hardly form tough films.

Polymerization reaction time is also a critical factor to condensation polymerization. In the initial stage, only low-molecular-weight oligomers (dimer, trimer, etc.) are formed and the chain growth continues with time going. To the authors' experience, for the synthesis of PBIs from diacids and tetraamine monomers in PPA, the reaction time of about 20 h is usually long enough for achieving high-molecular-weight polymers. Of course, in many cases, raising reaction temperature can accelerate the reaction rate and thus the reaction time may be shortened.

The chemical structure of PBIs also affects their mechanical properties. Leykin et al. reported that the benzimidazol-2-yl-substituted PBI membrane (BPBI, Figure 8.26) cross-linked with sulfuric acid exhibits high storage modulus at elevated temperatures which is comparable to that of the m-PBI membrane despite the higher DL of the former [30]. Benicewicz's group [28] and Lee's group [39] found that under similar DL the p-PBI exhibits higher tensile strength than the m-PBI.

Covalent cross-linking is an effective and widely used approach for improving the mechanical strength of acid-doped PBI membranes. Li's group reported that the m-PBI membranes could be covalently cross-linked by thermal treatment (160°C–300°C) using α,α'-dibromo-p-xylene as the cross-linker and the cross-linking density could be controlled by regulating the weight fraction of the loaded cross-linker [40]. The relevant cross-linking reaction is illustrated in Figure 8.30. The cross-linked membrane (theoretical cross-linking density: 13.5%) with a DL of 8.5 shows almost the same mechanical strength as the non-cross-linked one with a significantly lower DL (6.8) [40]. Na's group [41] reported that the m-PBI membranes could also be cross-linked by using 1,3-bis(2,3-epoxypropoxy)-2,2-dimethylpropane (NGDE) as cross-linker and the cross-linking was based on the chemical reaction between the imidazole –NH– groups and the epoxy groups at 150°C under vacuum. The cross-linking density could be controlled by regulating the loading level of the cross-linker (0, 5, 10, 15, 20 wt%). For the plain (without acid doping) membranes, the tensile strength increased with increasing the loading level of the cross-linker and reached maximum at 15 wt%. Further increase in cross-linker loading caused drastic decrease in tensile strength and elongation at break because the membrane became brittle resulting from the too high cross-linking density. The cross-linking also improved the mechanical strength of the PA-doped membranes although the acid DL slightly decreased for the cross-linked membranes under the same doping

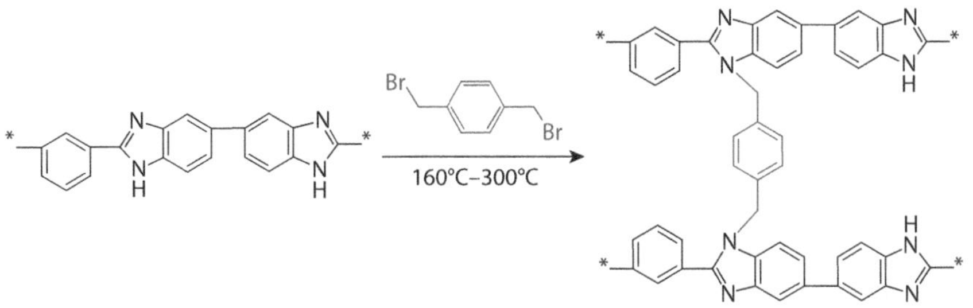

FIGURE 8.30 Cross-linking reaction between the m-PBI and α,α'-dibromo-p-xylene (cross-linker).

conditions. A facile cross-linking approach has been recently developed by the authors' group. This approach involves the synthesis of a series of random and sequential PBI copolymers with pendant amino groups via condensation polymerization of APTA, *i*PTA, and DABz in PPA (Figure 8.31) [21]. The pendant amino groups are utilized for covalent cross-linking using ethylene glycol diglycidyl ether (EGDE) and 1,3-dibromopropane (DBP) as the cross-linkers. The cross-linking was achieved during the process of membrane formation by casting and drying polymer solutions containing stoichiometric amount of a cross-linker. Because of the high reactivity of the pendant amino groups as well as the cross-linkers, the cross-linking reaction was conducted under mild conditions (at 80°C for a few hours). The cross-linking reactions are illustrated in Figure 8.32. As the DBP was used as cross-linker, excess triethylamine was added as a base to neutralize the produced hydrogen bromide. It was found that cross-linked membranes generally showed high mechanical properties even at high PA DLs, whereas the uncross-linked membranes highly swelled or even dissolved in PA indicating that the cross-linking is very effective for enhancing membrane mechanical strength. The sequenced copolymer membrane with the composition of APTA/*i*PTA = 1/2 cross-linked with DBP exhibits a tensile stress of 17 MPa despite its high PA uptake (330 wt%) [21]. The membrane displayed very high proton conductivity at elevated temperature at completely anhydrous conditions (0.14 S cm^{-1} at 170°C) indicating excellent performance of this membrane.

Hyperbranched polymers are a class of dendritic polymers, which possess unique 3D branched architecture with many open and accessible cavities inside the polymer structure [42,43]. The cavities may be utilized to absorb/accommodate PA molecules if the polymers have high affinity to PA. In addition, hyperbranched polymers have a large number of terminal functional groups, which may be utilized for cross-linking. Because the cross-linking mainly occurs at the outside (terminal groups) of the quasi-global-like macromolecules, the cavities inside the polymer structure are hardly affected by cross-linking. As a result, high PA uptake and good mechanical strength may be achieved with hyperbranched PBI (HBPBI) membranes if they are properly cross-linked. Based on these considerations, the authors' group have developed various cross-linked HBPBI membranes for fuel cells [44,45]. The HBPBIs were synthesized via the A$_2$ + B$_3$ approach. Here, A$_2$ and B$_3$ refer to a difunctional monomer and a trifunctional monomer, respectively. Condensation polymerization of 1,3,5-benzenetricarboxylic acid (BTA, B$_3$) and DABz (A$_2$, note that an *ortho*-diamino groups are regarded as one functional group) resulted in carboxyl or amine-terminated HBPBI depending on the monomer molar ratio and the monomer addition manner (Figure 8.33). Simultaneous addition of BTA and DABz with the molar ratio of 1:1 (manner 1) gave carboxyl-terminated HBPBI (HBPBI-1), whereas the addition of BTA portion-wise to DABz solution in PPA at the molar ratio of DABz/BTA = 2:1 (manner 2) yielded amine-terminated HBPBI (HBPBI-2). The free carboxyl and amino groups of HBPBI-1 and HBPBI-2 could further react with *o*-diaminobenzene and benzoic acid, respectively, to form the chemically modified polymers (Figure 8.34). Although hyperbranched polymers generally have poor film-forming ability because of lack of chain entanglement, the HBPBI free-standing membranes with good mechanical properties (tensile strength, ~70 MPa; elongation at break, 1.5%–1.8%) were obtained by cross-linking treatment of the amine-terminated and

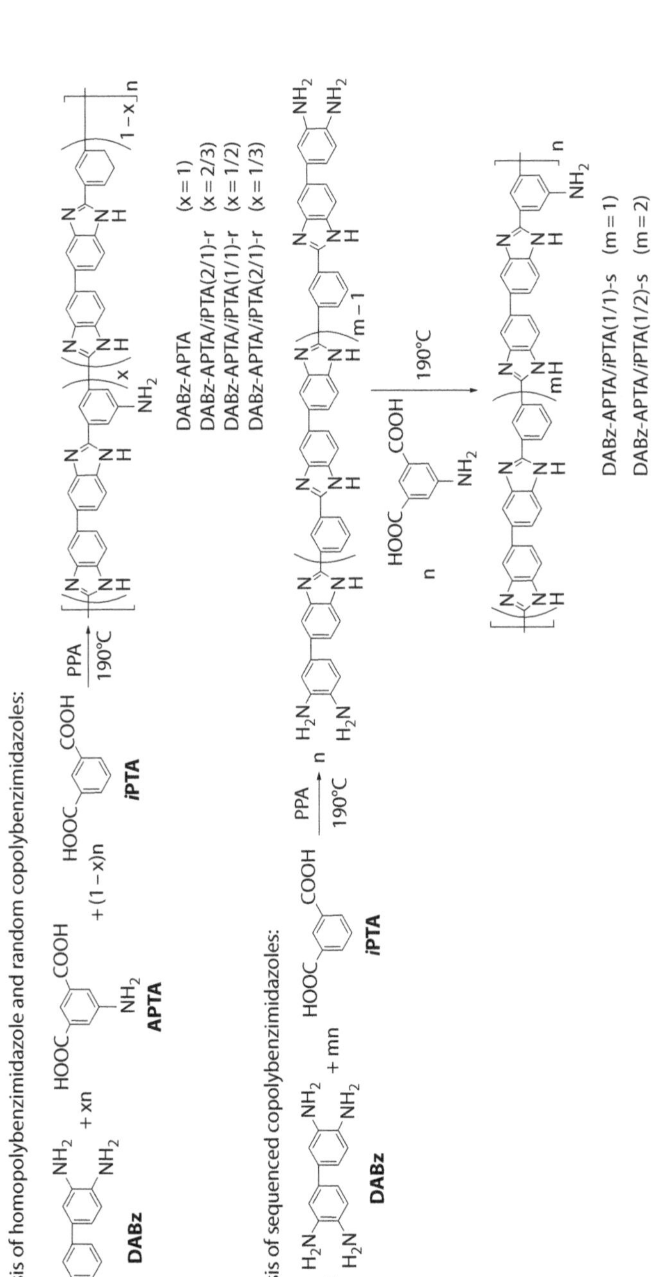

FIGURE 8.31 Synthesis of various random and sequenced copolybenzimidazoles with pendant amino groups. (Reprinted from Xu, N. et al., *J. Polym. Sci. Part A: Polym. Chem.*, 47, 6992, 2009. With permission.)

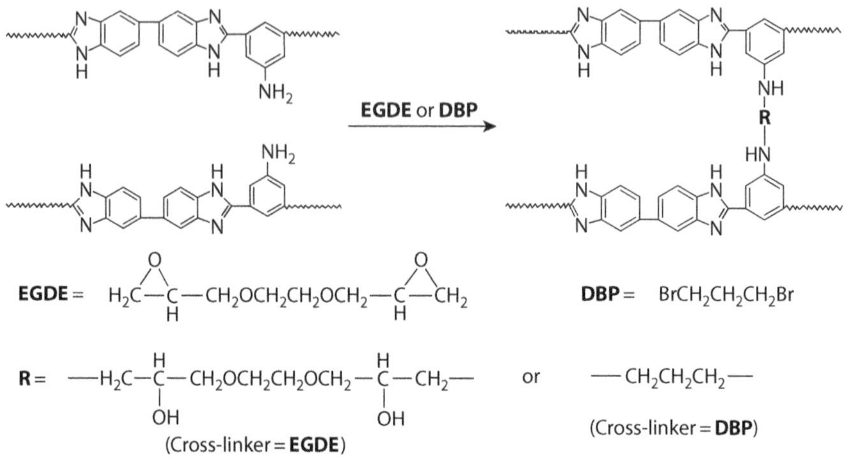

FIGURE 8.32 Cross-linking reaction between the PBI copolymers with pendant amino groups and the cross-linkers EGDE and DBP. (Reprinted from Xu, N. et al., *J. Polym. Sci. Part A: Polym. Chem.*, 47, 6992, 2009. With permission.)

partially chemically modified HBPBIs using terephthaldehyde (TPA) as cross-linker (Figure 8.35). The HBPBI membranes exhibited high PA uptakes (266–376 wt%) in 85% PA solution and the PA-doped HBPBI-6 (40% *o*-diamino groups have been reacted with benzoic acid) membrane showed higher tensile strength (8.1 MPa) than the PA-doped *m*-PBI (5.0 MPa) despite the higher DL of the former (354 vs. 330 wt%) [44]. This is likely attributed to the covalent cross-linking as well as the hyper-branched architecture of the HBPBI-6.

Another route for the synthesis of various amine-terminated HBPBIs is shown in Figure 8.36 [45]. It involves the condensation polymerization of aromatic dicarboxylic acids (*t*PTA, *i*PTA, 2,2-*bis*(4-carboxyphenyl)hexafluoropropane [6FA], 4,4-benzophenonedicarboxylic acid [BPDA], and DCDPE) with an in situ synthesized aromatic *tris*(*o*-diamino) intermediate product (hexamine) from BTA and DABz (BTA/DABz = 1/3, by mole) in PPA at 190°C for 20 h at the molar ratio of acid/*tris*(*o*-diamino) of 1:1. The HBPBI free-standing membranes were fabricated by solution cast method in the presence of a cross-linker (EGDE or TPA) on the basis of the cross-linking reaction between the terminal amino groups of the HBPBIs and the epoxy or formaldehyde groups of the cross-linkers. The resulting HBPBI membranes displayed good mechanical properties (tensile strength, 52–98 MPa; elongation at break, 3.2%–6.2%) and high thermal stability. In 85% PA solution, the HBPBI membranes cross-linked with PTA exhibited much higher PA uptakes (404–572 wt%) than the ones cross-linked with EGDE (150–205 wt%). High proton conductivity up to 0.064 S cm^{-1} was obtained with the PA-doped and TPA-cross-linked HBPBI (synthesized from *i*PTA and the *tris*(*o*-diamino) compound) membranes at 170°C and 0% relative humidity.

Besides the aforementioned approaches, incorporation of inert (little/none affinity to PA) segments into PBI backbone may be another effective way to enhance the

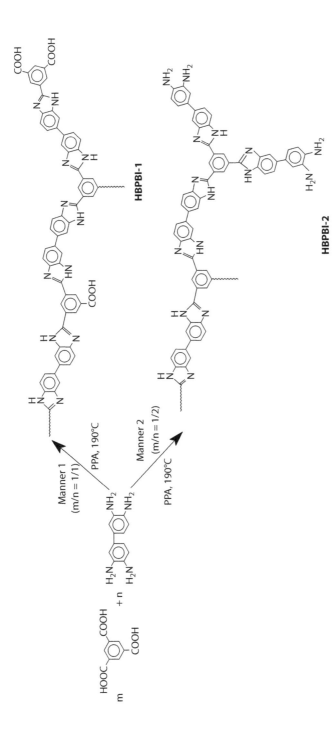

FIGURE 8.33 Synthesis of HPBPIs using BTA as B_3 type of monomer and DABz as A_2 type of monomer. (Reprinted from Xu, H. et al., *J. Polym. Sci. Part A: Polym. Chem.*, 45, 1150, 2007. With permission.)

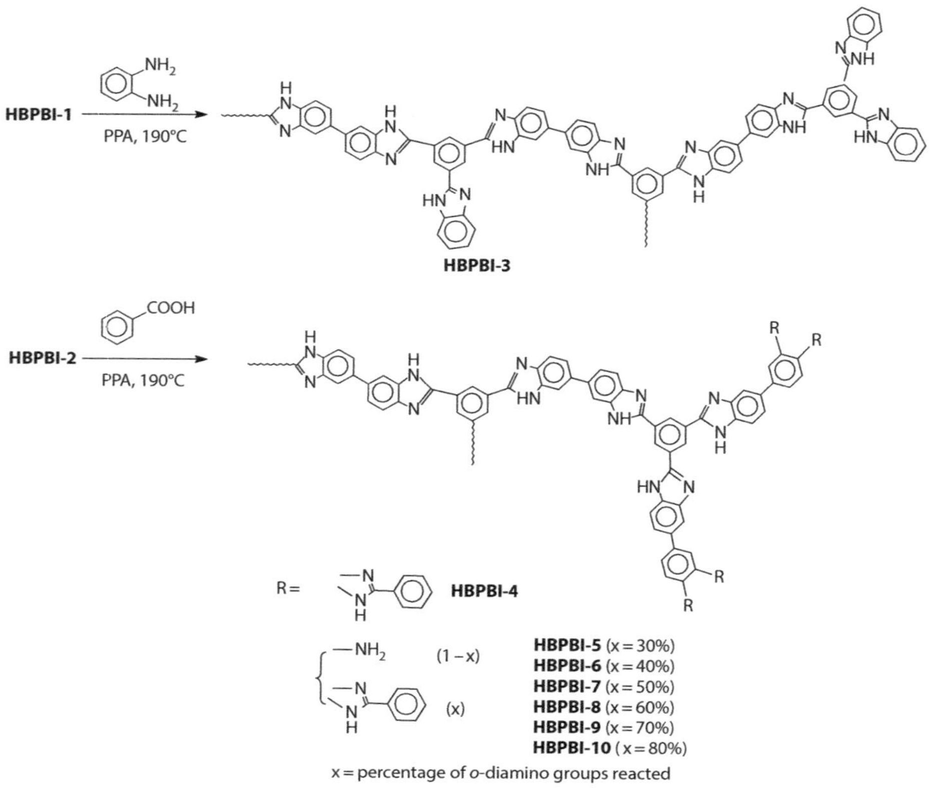

FIGURE 8.34 Preparation of chemically modified HPBPIs. (Reprinted from Xu, H. et al., *J. Polym. Sci. Part A: Polym. Chem.*, 45, 1150, 2007. With permission.)

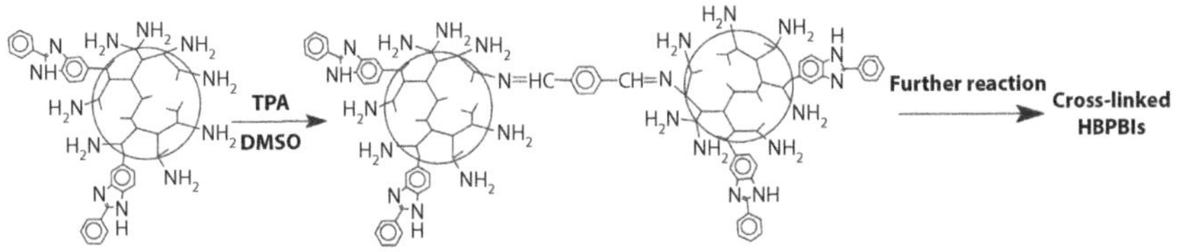

FIGURE 8.35 Cross-linking reactions of partially modified HPBPIs. (Reprinted from Xu, H. et al., *J. Polym. Sci. Part A: Polym. Chem.*, 45, 1150, 2007. With permission.)

mechanical strength of the PA-doped membranes. McGrath's group [46] reported a series of multiblock copolymers, poly(arylene ether sulfone)-*b*-PBI (BPS-PBI), as potential PEMs for high-temperature fuel cells. The poly(arylene ether sulfone) block has little affinity to PA molecules, and thus it may have the functions to suppress membrane swelling during the process of acid doping and to maintain membrane mechanical strength. Although no detailed comparison of mechanical properties between the BPS-PBI copolymer membranes and the *m*-PBI is available from this literature, it is still worth introducing this work. The copolymer synthesis involves three-step reactions. In the first step, *o*-diamine-terminated PBI oligomers with controlled molecular weights ($M_n = 5,400–16,100$, measured by ^1H NMR) were synthesized by condensation polymerization of *i*PTA with excess DABz in PPA at 200°C for 24 h (Figure 8.37a). Secondly, benzoic acid–terminated poly(arylene ether sulfone) oligomers (BPS) with controlled molecular weights ($M_n = 5,400–14,700$, measured by ^1H NMR) were synthesized by condensation polymerization of dichlorodiphenyl sulfone, 4,4′-biphenol, and 3-hydroxybenzoic acid in DMAc in the presence of potassium carbonate at 180°C for 48 h (Figure 8.37b). Finally, the PBI oligomers were further polymerized with the poly(arylene ether sulfone) oligomers in NMP at 200°C for 48 h based on the reaction between the terminal *o*-diamine and the terminal carboxyl groups (Figure 8.37c). The resulting multiblock copolymers, BPS 5-PBI 5, BPS10-PBI 10, and BPS15-PBI 15, have almost the same composition but different block lengths (note: the numbers of the copolymer names refer to the target molecular weights of the oligomers divided by 1000). These block copolymer membranes exhibited quite similar PA absorption behavior and a maximum DL of ~12 was achieved when 14.6 M PA solution was used, which is quite close to that (~11.5) of the *m*-PBI under the same doping conditions. However, because of the presence of the BPS block the PA uptakes (145–174 w/w%, calculated from the acid content in the sample membranes) of these block copolymer membranes are much lower than that (~350 w/w% [29]) of the *m*-PBI. The BPS15-PBI 15 (DL, 11.5) showed a considerably high tensile strength (36 MPa), which is likely attributed to the inert poly(ether sulfone) block as well as the moderate PA uptake.

It is well known that aromatic polyimides (PIs) are a class of high-performance polymers with extremely high thermal stability, high mechanical strength and modulus, good chemical resistance, and excellent film-forming ability. These outstanding properties are mainly attributed to the strong interchain interactions resulting from the intramolecular and intermolecular charge–transfer complex as well as the stiff backbone of the PIs. The incorporation of imide segments into PBI backbone is

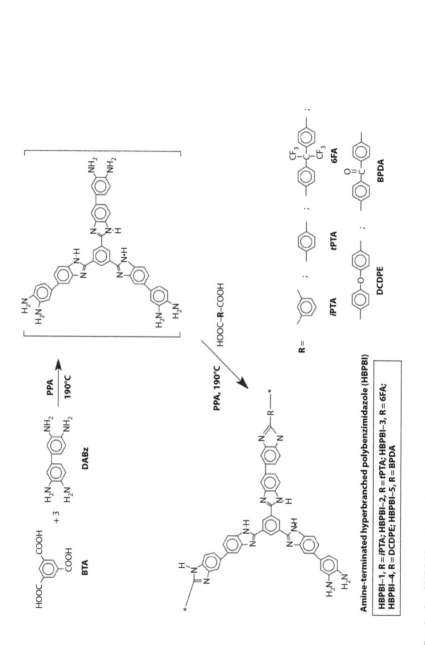

FIGURE 8.36 Synthesis of HPBPIs using an in situ synthesized aromatic tris(*o*-diamino) intermediate as B$_3$ type of monomer and various dicarboxylic acid as A$_2$ type of monomers. (Reprinted from Xu, H. et al., *J. Membr. Sci.*, 288, 255, 2007. With permission.)

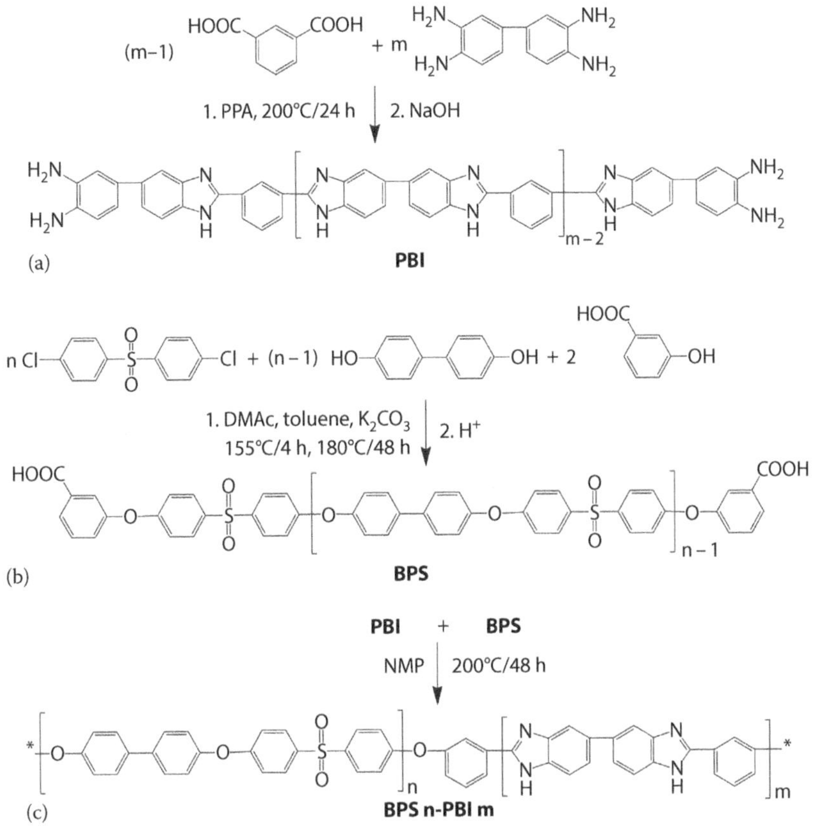

FIGURE 8.37 Synthesis of poly(arylene ether sulfone)-*b*-PBI (BPS-PBI) multiblock copolymers: (a) synthesis of the PBI oligomer, (b) synthesis of the carboxyl-terminated poly(ethyl sulfone) oligomer (BPS), and (c) block copolymerization. (Reprinted with modification from Lee, H.-S. et al., *Polymer*, 49, 5387, 2008. With permission.)

expected to be efficient to improve the membrane mechanical properties. Based on this consideration, we have recently synthesized a series of poly(imide benzimidazole) s (PIBIs) as a new type of PEMs for high-temperature fuel cells [47]. The synthesis of the PIBIs was performed by random copolymerization of biphenyl-4,4′-diyldi(oxo)-4,4′-bis(1,8-naphthalenedicarboxylic anhydride) (BPNDA), (2-(4-aminophenyl)-5-aminobenzimidazole) (APABI), and 4,4′diaminodiphenyl ether (ODA) in *m*-cresol in the presence of benzoic acid and isoquinoline at 180°C for 20 h (Figure 8.38). At room temperature, the resulting PIBIs exhibited quite low capacity to absorb PA because of their rather low densities of benzimidazole groups. However, very high acid uptakes could be achieved by raising temperature to 180°C in PPA or in 85 wt% PA under pressure. Among the PIBIs, the PIBI-1/0 that was synthesized from BPNDA and APABI (in the absence of ODA) exhibited the highest acid uptake (780 w/w%, in PPA) and the highest anhydrous proton conductivity (0.26 S cm^{-1}) at 170°C owing to its highest content of benzimidazole groups. Moreover, despite their very

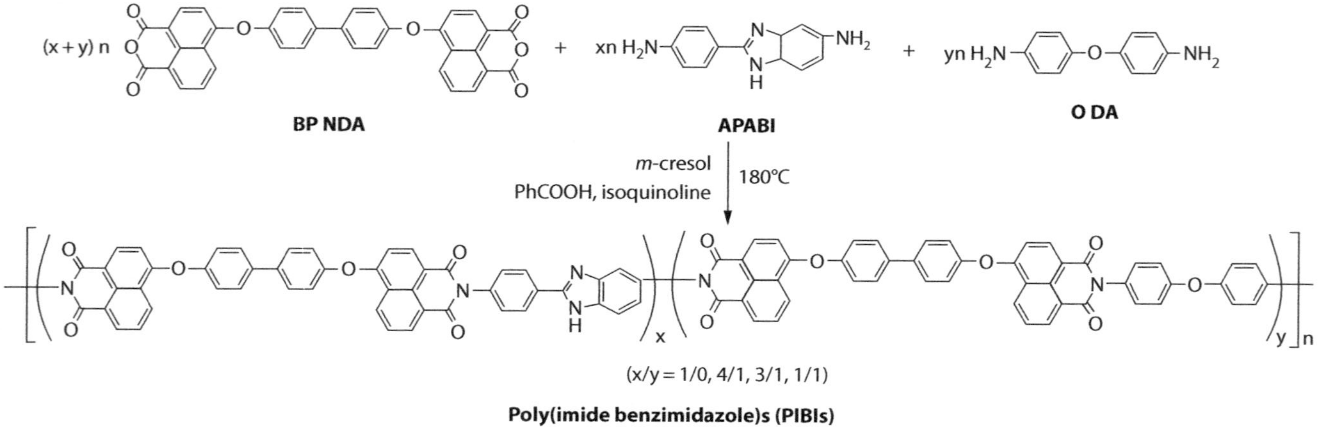

FIGURE 8.38 Synthesis of the PIBIs. (Reprinted from Yuan, S. et al., *J. Membr. Sci.*, 454, 351, 2014. With permission.)

high acid uptakes, the PIBI membranes exhibited significantly higher mechanical strength than many PBI membranes with similar acid uptakes. The PIBI-1/1 membrane with a 1:1 molar ratio between APABI and ODA, for example, had similar acid uptake (450 w/w%, in PPA) but much higher tensile stress (28 MPa) than a membrane based on high-molecular-weight PBI (6.2 MPa) [38]. This indicates that the incorporation of imide segments into PBI backbone is really very effective for enhancing the mechanical strength of the PA-doped membranes.

8.5 CHEMICAL STABILITY

Fuel cell durability is closely dependent on the chemical stability of PEMs. Radical-oxidation-induced degradation of polymer main chains has been identified to be one of the major reasons that cause deterioration of proton exchange membranes. Peroxide radicals (hydroxyl and hydroperoxy radicals, etc.) are formed in fuel cells due to the oxygen permeation through the membrane from the cathode side and reduction at the anode side [48]. Therefore, it is very important to develop proton exchange membranes with high radical oxidative stability. Usually, the radical oxidative stability is evaluated by Fenton's test, a kind of accelerated aging test. Fenton's reagent, known as an extremely strong oxidant, is a mixture of Fe^{2+} and hydrogen peroxide. As trace amount of Fe^{2+} in combination of a lot of hydrogen peroxide is used, Fe^{2+} is rapidly oxidized by hydrogen peroxide yielding Fe^{3+}, a hydroxyl radical, and a hydroxide anion. Fe^{3+} is then reduced back to Fe^{2+} by another molecule of hydrogen peroxide, forming a hydroperoxy radical and a proton:

$$Fe^{2+} + H_2O_2 \rightarrow Fe^{3+} + HO^{\cdot} + HO^{-}$$

$$Fe^{3+} + H_2O_2 \rightarrow Fe^{2+} + HO^{\cdot}_2 + H^{+}$$

For sulfonated polymer membranes, Fenton's test is usually performed by immersing the samples in dilute hydrogen peroxide containing trace amount of Fe^{2+} (typically, 3 wt% H_2O_2 + 3 ppm Fe^{2+}, Fe^{2+} is added in its form of $FeSO_4$) at 80°C for a period of time. The radical oxidative stability is evaluated either by the elapsed time when the samples started to break into pieces (τ_1) and completely dissolved (τ_2) or by the weight loss after the samples were soaked in Fenton's reagent for a given period of time. The method of measuring τ_1 and τ_2 is very simple and easy to operate but is less accurate than the weighing method. In general, sulfonated hydrocarbon polymer membranes exhibit a τ_1 and a τ_2 from tens of minutes to a few hours depending on their fine chemical structures. PBI membranes, however, are generally much more stable than common sulfonated hydrocarbon polymer membranes. The m-PBI and the OPBI (prepared from 4,4′-dicarboxylic acid and DABz), for example, showed only small weight loss values of 9.9% and 6.0%, respectively, after Fenton's test in 3 wt% H_2O_2 containing 3 ppm $FeSO_4$ at 80°C for 12 h indicating very good radical oxidative stability [10]. This is likely attributed to the high stability of benzimidazole groups as well as the low water uptake of these two membranes.

It should be noted that there are no strict standard operation conditions for Fenton's test and different testing conditions may be employed by different researchers.

As a result, one must be aware of the detailed operation conditions when doing comparison of Fenton's test data from different publications. Besides the foregoing described conditions, Li et al. [40] proposed a modified method for Fenton's test by using a solution of 3 wt% H_2O_2 containing 4 ppm Fe^{2+} (Fe^{2+} was added in its form of $(NH_4)_2Fe(SO_4)_2 \cdot 6H_2O$ for accelerating the effect to produce hydroxide radicals [40]) as Fenton's reagent and decreasing the test temperature to 68°C. The authors examined Fenton's test in acidic environment by directly immersing the PA-doped PBI membranes in 3 wt% H_2O_2 containing 3 ppm $FeSO_4$ at 80°C [21]. The PA concentration was controlled at 0.01 mol L^{-1} assuming that the PA completely diffused into Fenton's reagent from the membrane samples. Under the weak acidic condition, the m-PBI exhibited a durability τ (elapsed time when the membranes became brittle [broken when they were bent]) of 12 h, which is nearly one-third of that (32 h) of the acid-free membrane [21]. This indicates that the oxidation is accelerated by the slight acidification of Fenton's reagent. Pu's group investigated the effects of acidification conditions on Fenton's test results in details. They found that at low PA concentration, around 0.001 mol L^{-1}, the m-PBI exhibited a shorter τ (defined as the floccules time) and a larger weight loss than in PA-free solution. However, at high PA concentration of 10 mol L^{-1}, a reverse effect was observed, that is, much longer τ and lower weight loss were obtained indicating that the radical oxidative stability was improved by adding a lot of PA. This is probably because the protonated imidazole rings are more stable to radical attack [49].

It has been identified that the radical oxidative stability of PBI membranes is affected by many factors such as the polymer molecular weight, the fine chemical structure, and the presence/absence of covalent cross-linking. Li, He and coworkers [50] evaluated the radical oxidative stability of the m-PBI membranes with different molecular weights by Fenton's test in 3 wt% H_2O_2 containing 4 ppm Fe^{2+} (in its form of $(NH_4)_2Fe(SO_4)_2 \cdot 6H_2O$) at 68°C. They found that the molecular weight of the m-PBI significantly affected the radical oxidative stability, and the high molecular weight, the higher the stability. This is because higher-molecular-weight polymers tend to have higher density of chain entanglement and thus stronger interchain interactions, which is favorable for enhancing the antioxidation property of the polymers. Moreover, longer polymer chains (higher molecular weight) needs longer time to degrade into low-molecular-weight materials than the short polymer chains.

Pu's group compared the radical oxidative stability of four kinds of PBI membranes (Figure 8.39) via Fenton's test in 3 w/v% H_2O_2 containing 2 ppm Fe^{2+} (in its form of $FeSO_4 \cdot 7H_2O$) at 80°C. They reported that the stabilities of these polymers are in the order: m-PBI > ABPBI > PBI-hex > PBI-imi [49]. They ascribed the stability difference to the difference in π-conjugation of benzene rings and imidazole rings. On the basis of FT-IR and ^1H NMR analysis, they proposed a possible degradation mechanism as shown in Figure 8.40.

Li and coworkers [51] also proposed a mechanism, which is somewhat different from that proposed by Pu's group. They found that the weak point of the polymer chain is not the N–H group of an imidazole ring but the carbon atom linking benzenoid ring and imidazole ring. The degradation of acid-doped PBI membranes under Fenton's test conditions starts by the attack of hydroxyl radicals at the carbon atom linking imidazole ring and benzenoid ring, which would eventually lead to the

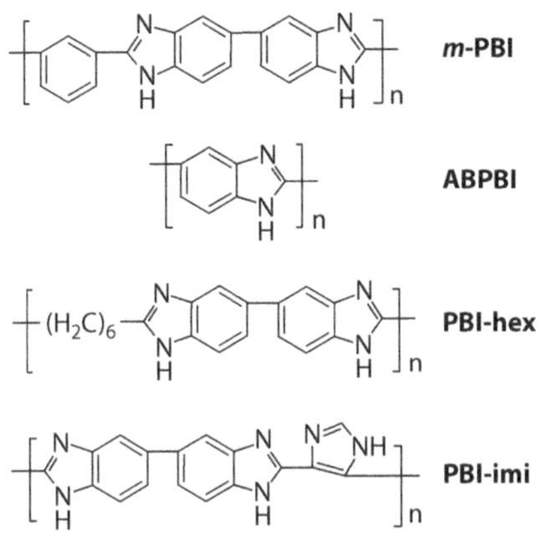

FIGURE 8.39 Chemical structures of four kinds of polybenzimidazoles. (Reprinted from Chang, Z. et al., *Polym. Degrad. Stabil.*, 95, 2648, 2010. With permission.)

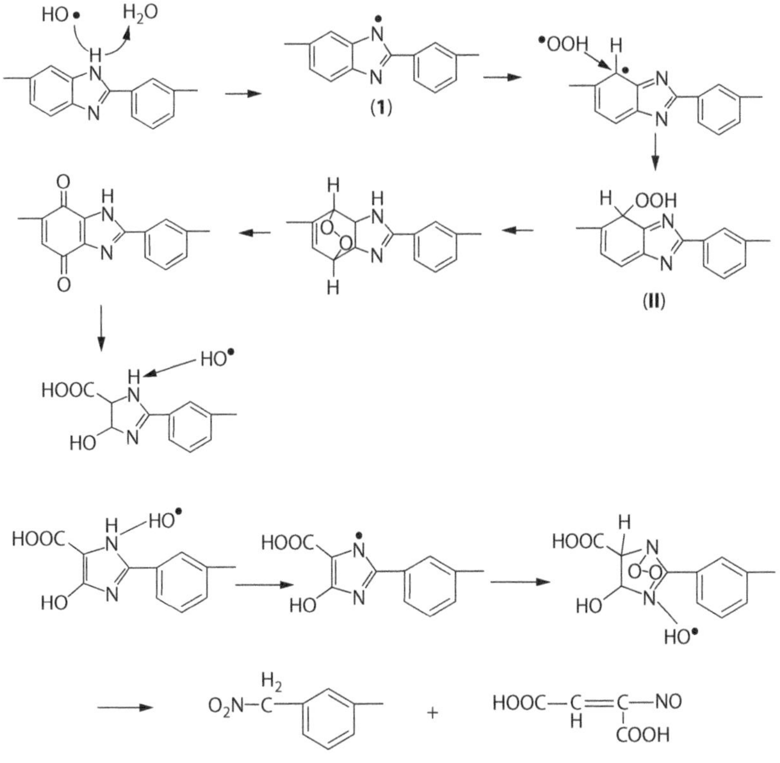

FIGURE 8.40 Degradation mechanism of polybenzimidazoles proposed by Pu et al. (Reprinted from Chang, Z. et al., *Polym. Degrad. Stabil.*, 95, 2648, 2010. With permission.)

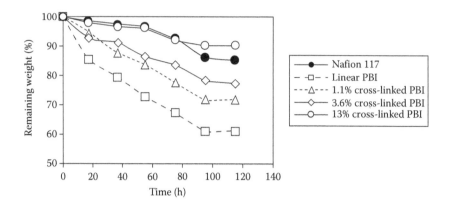

FIGURE 8.41 Membrane degradation in 3% H_2O_2 containing 4 ppm Fe^{2+} at 68°C. Solid lines indicate that the samples remained in a whole membrane form, whereas dashed lines indicate that samples were broken into small pieces. (Reprinted from Li, Q. et al., *Chem. Mater.*, 19, 350, 2007. With permission.)

imidazole ring opening. The chain scission generates small molecules and terminal groups that are further oxidized by an endpoint oxidation.

Covalent cross-linking has been identified to be an effect approach for improving the radical oxidative stability of PBI membranes. Li and coworkers reported that the *m*-PBI showed a weight loss of ~40% after 100 h testing in 3 wt% H_2O_2 containing 4 ppm Fe^{2+} (in its form of $(NH_4)_2Fe(SO_4)_2 \cdot 6H_2O$) at 68°C, whereas the *p*-xylene dibromide cross-linked *m*-PBI (cross-linking density: 13%) exhibited only a weight loss of ~10% under the same conditions, which is even slightly lower than that of Nafion (Figure 8.41) [40]. Na's group reported that the cross-linked membranes prepared from the *m*-PBI and an epoxy resin cross-linker, NDGE, exhibited significantly better stability than the pristine membrane (uncross-linked *m*-PBI) and the stability increased with increasing the cross-linking density [41]. In Fenton's reagent (3% H_2O_2 solution containing 4 ppm Fe^{2+}) at 70°C, the cross-linked PBI-NDGE-20% membrane (cross-linker weight content: 20%) did not break into pieces and kept its shape for more than 480 h and its weight loss was approximately 35%, whereas the pristine membrane broke into pieces after 19 h and the weight loss was about 49% after 480 h test. The authors' group examined the effect of different cross-linkers on the radical oxidative stability of various cross-linked PBI membranes and found that the membranes cross-linked with DBP exhibited significantly better stability than the ones cross-linked with EGDE [21].

8.6 SINGLE-CELL PERFORMANCE

The performance of a fuel cell is affected by many factors such as PEM conductivity, catalyst (Pt) loading level, membrane electrode assembly (MEA) fabrication techniques, fuel (hydrogen, methanol, etc.) and its feeding conditions (pressure, flow rate, humidification, etc.), oxygen/air feeding conditions (partial pressure, flow rate, humidification, etc.), and operation temperature. The fabrication techniques of MEA are very crucial to fuel cell performance. The key issues associated with the fabrication of MEA are as follows: (1) keeping ohmic contact between the catalyst layer and the membrane and (2) keeping high porosity of the catalyst layer to allow fast mass (hydrogen, oxygen, etc.) transfer. In general, the MEA fabrication techniques can be classified into two types: catalyst-coated membrane (CCM) and catalyst-coated substrate (CCS). For the CCM method, a catalyst ink, which is a mixture of the catalyst (platinum supported on carbon) suspended in an ionomer solution (typically, the m-PBI solution in DMAc), is sprayed onto both sides of a PBI membrane and is subsequently dried at elevated temperatures to remove the residual solvent. The ionomer of ink functions as the binder to fix the Pt/C catalyst nanoparticles as well as the proton conductor. The membrane can be either first acid doped or undoped. When the acid-free PBI membrane is used as the PEM, the MEA needs to be acid doped by soaking in PA solution. When the acid-doped membrane is used as the PEM, the soaking procedure may be omitted because the acid molecules may migrate from the membrane into the catalyst layer and thus the binder is doped with acid. Of course, it is also reported that further doping treatment may achieve better fuel cell performance [52]. For the CCS method, the foregoing mentioned ink is sprayed onto one side of a gas diffusion layer (GDL) and is subsequently dried at elevated temperatures to remove the residual solvent. The resulting catalyst-coated GDL is called gas diffusion electrode (GDE). The GDE is usually acid doped by soaking in a PA solution and subsequently assembled with PEM (two sheets of the GDE and a sheet of the acid-doped PBI membrane are sandwiched and hot pressed to form an MEA).

The characterization of fuel cell performance is usually performed by recording the variation of cell voltage (V) as a function of current density (I) and the I–V curve is called the polarization curve. As shown in Figure 8.42, for a typical polarization curve, the cell drop is composed of three parts: kinetic, ohmic, and mass transfer [53]. The initial rapid voltage drop of the fuel cell is attributed to oxygen reduction and hydrogen oxidation kinetics. Since oxygen reduction is much more difficult than hydrogen oxidation, the former should be the main reason for the initial voltage drop. In the ohmic part, the cell voltage drop is mainly due to the internal resistance of the fuel cell, including electrolyte membrane resistance, catalyst layer resistance, and contact resistance. In the mass transfer part, the voltage drop is due to the transfer speed of hydrogen and oxygen to the electrode surface [53]. The cell voltage drop due to mass transfer occurs in the highest current density region. High current density means large amount of hydrogen and oxygen are reacted per unit time, which may cause local starvation of the reactants due to the slow mass transfer rate. For a given fuel cell, generally increasing operation temperature results in better fuel cell performance because of the faster electrochemical kinetics and the higher proton conductivity at higher temperatures [7,31,54].

Figure 8.43 shows the polarization curves of a H_2/O_2 and H_2/air single cell using the PBI membrane (DL, ~32, prepared by solgel method) as the PEM at 160°C, which is reported by Benicewicz's group [28]. The Pt-loading level is 1.0 mg cm^{-2} for both anode and cathode, and the fuel cell is operated at ambient pressure without external humidification. The fuel cell performance of H_2/O_2 system is significantly better than that of H_2/air system due to the higher oxygen partial pressure of the former. Because of the exceptionally high PA DL and thus very high proton conductivity of

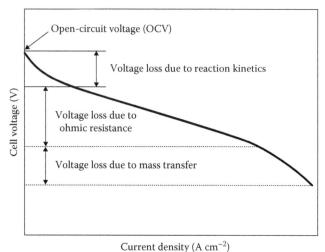

FIGURE 8.42   Diagram of polarization curve of a fuel cell.

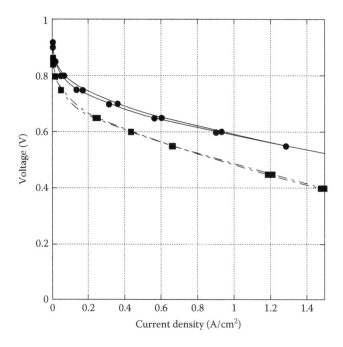

FIGURE 8.43   Fuel cell performance for the PBI membranes from the solgel process. Performance curves of fuel cells under H_2/air (squares) and H_2/O_2 (circles), without any feed gas humidification. The membrane PA DL was approximately 32 mol of PA per mole PBI repeat unit. The catalyst loading in both electrodes was 1.0 mg cm^{-2} Pt, and the cell was operated at 160°C at constant stoichiometry of 1.2 stoic and 2.5 stoic at the anode and the cathode, respectively. (Reprinted from Xiao, L. et al., *Chem. Mater.*, 17, 5328, 2005. With permission.)

the membrane, the fuel cell exhibits very high performance with the highest power output of approximately 0.9 W cm^{-2} at 2.5 A cm^{-2} [28].

The authors' group prepared a kind of SOPBIs (Figure 8.14), which exhibited PA uptakes in the range 180–240 w/w% and tensile strength in the range 13–20 MPa [7]. The SPBI-11 membrane (PA uptake, 240 w/w%; tensile strength, 20 MPa) was used as the PEM, and its H$_2$/O$_2$ and H$_2$/air fuel cell performance was evaluated at ambient pressure and 170°C without external humidification. The Pt-loading level is 0.5 mg cm^{-2}. As shown in Figure 8.44, the fuel cell exhibited fairly good performance despite the relatively lower PA uptake. This might be related to the presence of two kinds of proton-conducting sources, sulfonic acid groups and PA. Further studies are needed to clarify the exact mechanism.

Li's group investigated the effect of feed gas pressure on fuel cell performance. They found that as expected, the fuel cell exhibited better performance with increasing feed gas pressure [2].

Besides performance, durability is another important concern of fuel cells. To evaluate fuel cell durability, two modes, steady-state (continuous running) and shut-down–restart cycling, are often used. In the steady-state mode, a fuel cell is operated continuously at a fixed current density to measure the voltage decay. Quite different durability data from a few hundred hours to nearly twenty thousand hours have been reported in the literature [2,28,47,55–58]. In principle, the durability is affected by many factors such as the chemical stability and conductivity stability of PEMs and the stability of catalyst layer. Oono and coworkers [56] studied the long-term cell degradation mechanism with a PA-doped PBI membrane (DL is unavailable from the literature). The H$_2$/air fuel cell was continuously tested at 150°C at 0.2 A cm^{-2} without external humidification. They found that the fuel cell exhibited extremely

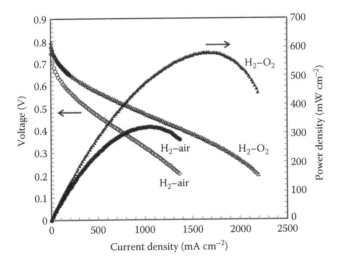

FIGURE 8.44 Comparison of performance curves of the fuel cell with the PA-doped sulfonated PBI membrane PEM to H$_2$–O$_2$ and H$_2$–air fuel cell at 170°C. (Reprinted from Sheng, L. et al., *J. Power Sources*, 196, 3039, 2011. With permission.)

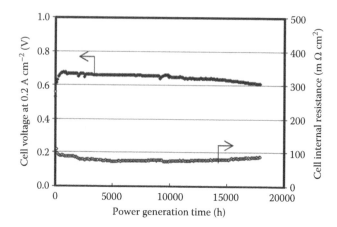

FIGURE 8.45 Long-term power generation test results for Cell E operated until its cell voltage dropped by 10% from its peak value (17,860 h). (Reprinted from Oono, Y. et al., *J. Power Sources*, 210, 366, 2012. With permission.)

good long-term durability (17,860 h, voltage decay: ≤10%, Figure 8.45). Moreover, the cell voltage declined very gradually in the initial 14,000 h test, and thereafter a more rapid decrease in cell voltage occurred. The voltage decay within the initial 14,000 h was probably caused by the active area decrease due to catalyst agglomeration and so on, while the voltage decay after 14,000 h was found to correlate well with decreases in the membrane thickness, which is likely attributed to the chemical degradation of the PBI [56].

Benicewicz's group evaluated the durability of the fuel cell with the PBI membranes prepared by solgel method [55]. They also reported a very good steady-state long-term durability of greater than 14,000 h at 120°C at 0.2 A cm^{-2} without external humidification [55].

Considering that the PA-doped PBI membrane fuel cells always have a risk of acid leakage owing to the possible formation of water droplets in the cathode side when the fuel cells are shut down, the evaluation of fuel cell durability by shutdown–restart cycling mode is of more significance. Li's group evaluated the durability of the fuel cell with the PA-doped *m*-PBI membrane by measuring the current density decay at a fixed cell voltage at 150°C. They reported that over a period of 1400 h, a slight decrease in current density was observed after 47 cycles [2]. Benicewicz's group systematically evaluated the fuel cell durability of two kinds of PBI membranes, *p*-PBI (synthesized from *t*PTA and DABz) and 2OH-PBI (synthesized from 2,5-dihydroxy*t*PTA and DABz) by both the shutdown–restart cycling mode (100 cycles) and the thermal cycling mode (180°C–120°C, 100 cycles; 120°C–80°C, 100 cycles) at a fixed current density (0.2 A cm^{-2}). They found that the cell voltage degradation rate was closely dependent on the test conditions, and in most cases, the maximal voltage degradation rate was less than 300 μV cycle^{-1} (50 μV h^{-1}), which was comparable to those measured under steady state [55]. Moreover, under many typical operating conditions, PA loss from the membranes did not appear to be a major factor of fuel cell failure due to abundance (acid DL, 26.2) of PA in membranes prepared by solgel process and the low PA loss rates under these operating conditions [55]. However, they also reported that a sharp increase in PA loss was occasionally observed during fuel cell operation probably due to liquid water formation at the cathode side [55].

8.7 SUMMARY

In this chapter, the synthetic approaches, structural characterization, solubility properties, acid doping, proton conductivity, mechanical properties, chemical stability, and fuel cell performance of PBIs have been introduced. Solution polymerization from dicarboxylic acids and aromatic tetraamines (bis(*o*-diamino) monomers) in PPA at 190°C–200°C or Eaton's reagent at 120°C–140°C is the most widely used method for synthesis of various PBIs. FT-IR spectroscopy is a useful tool to characterize PBI structure. The commercial PBI (*m*-PBI) and the AB-PBI exhibit poor solubility in organic solvents due to the strong interchain interactions resulting from hydrogen bond. Structural modifications such as incorporation of bulky groups into polymer backbone, introduction of moieties with nonlinear configuration, N-substitution, and polymer molecular weight control can effectively improve the polymer solubility. PA-doped PBI membranes may exhibit high proton conductivity even at completely anhydrous conditions. Proton conductivity is strongly dependent on PA uptake or acid DL and the higher PA uptake tends to give higher proton conductivity. Moreover, the free acid in PBI membranes is the main contributing factor to the high proton conductivity. However, too high PA uptake causes poor mechanical strength of the membranes due to the plasticization effect of PA. Covalent cross-linking, structural modification, and increasing polymer molecular weight are effective approaches to improve the mechanical strength of the highly doped membranes. Long-term durability up to nearly 18,000 h of the fuel cells based on PA-doped PBI membranes under steady-state operation has been reported. Good durability of the fuel cells operated under shutdown–start-up mode with the highly doped PBI membranes (prepared by solgel method) has also been reported. However, some challenges still remain in this area. The biggest challenge is to find an effective way to diminish/eliminate PA leakage, which is essential to assure long-term durability of fuel cells for many applications. Mechanical strength and modulus are still one of the major concerns associated with the highly doped PBI membranes. Though recently great progresses have been made in solving this problem, further enhancement in mechanical strength and modulus is desirable. Finally, some other issues such as ionomers and the related technologies for MEA fabrication and carbon erosion problem are also needed to be further studied.

REFERENCES

1. Wainright, J. S., Wang, J.-T., Weng, D., Savinell, R. F., Litt, M., Acid-doped polybenzimidazoles: A new polymer electrolyte, *J. Electrochem. Soc.* 1995, 142, L121–L123.
2. Li, Q., He, R., Jensen, J. O., Bjerrum, N. J., PBI-based polymer membranes for high temperature fuel cells—Preparation, characterization and fuel cell demonstration, *Fuel Cells* 2004, 4, 147–159.

3. Vogel, H., Marvel, C. S., Polybenzimidazoles, new thermally stable polymers, *J. Polym. Sci.* 1961, L, 511–539.
4. Iwakura, Y., Uno, K., Imai, Y., Polyphenylenebenzimidazoles, *J. Polym. Sci. Part A: General Papers* 1964, 2, 2605–2615.
5. Iwakura, Y., Uno, K., Imai, Y., Polybenzimidazoles. II. Polyalkylenebenzimidazoles, *Die Makromolekulare Chemie* 1964, 77, 33–40.
6. Ueda, M., Sato, M., Mochizuki, A., Poly(benzimidazole) synthesis by direct reaction of diacids and tetraamine, *Macromolecules* 1985, 18, 2723–2726.
7. Sheng, L., Xu, H., Guo, X., Fang, J., Fang, L., Yin, J., Synthesis and properties of novel sulfonated polybenzimidazoles from disodium 4,6-bis(4-carboxyphenoxy)benzene-1, 3-disulfonate, *J. Power Sources* 2011, 196, 3039–3047.
8. Neuse, E. W., Loonat, M. S., Two-stage polybenzimidazole synthesis via poly(azomethine) intermediates, *Macromolecules* 1983, 16, 128–136.
9. Twieg, R., Matray, T., Hedrick, J. L., Poly(aryl ether benzimidazoles), *Macromolecules* 1996, 29, 7335–7341.
10. Xu, H., Chen, K., Guo, X., Fang, J., Yin, J., Synthesis of novel sulfonated polybenzimidazole and preparation of cross-linked membranes for fuel cell application, *Polymer* 2007, 48, 5556–5564.
11. Asensio, J. A., Borros, S., Gomez-Romero, P., Proton-conducting polymers based on benzimidazoles and sulfonated benzimidazoles, *J. Polym. Sci. Part A: Polym. Chem.* 2002, 40, 3703–3710.
12. Jouanneau, J., Mercier, R., Gonon, L., Gebel, G., Synthesis of sulfonated polybenzimidazoles from functionalized monomers: Preparation of ionic conducting membranes, *Macromolecules* 2007, 40, 983–990.
13. Ariza, M. J., Jones, D. J., Rozikre, J., Role of post-sulfonation thermal treatment in conducting and thermal properties of sulfuric acid sulfonated poly(benzimidazole) membranes, *Desalination* 2002, 147, 183–189.
14. Asensio, J. A., Borros, S., Gomez-Romero, P., Sulfonated poly(2,5-benzimidazole) (SABPBI) impregnated with phosphoric acid as proton conducting membranes for polymer electrolyte fuel cells, *Electrochim. Acta* 2004, 49, 4461–4466.
15. Asensio, J. A., Borros, S., Gomez-Romero, P., Enhanced conductivity in polyanion-containing polybenzimidazoles. Improved materials for proton-exchange membranes and PEM fuel cells, *Electrochem. Commun.* 2003, 5, 967–972.
16. Glipa, X., Haddad, M. E., Jones, D. J., Roziere, J., Synthesis and characterization of sulfonated polybenzimidazole: A highly conducting proton exchange polymer, *Solid State Ionics* 1997, 97, 323–331.
17. Bae, J.-M., Honma, I., Murata, M., Yamamoto, T., Rikukawa, M., Ogata, N., Properties of selected sulfonated polymers as proton-conducting electrolytes for polymer electrolyte fuel cells, *Solid State Ionics* 2002, 147, 189–194.
18. Bouchet, R., Siebert, E., Proton conduction in acid doped polybenzimidazole, *Solid State Ionics* 1999, 118, 287–299.
19. Liang, Z., Jiang, X., Xu, H., Chen, D., Yin, J., Polybenzimidazoles (PBIs) derived from non-coplanar dibenzoic acid containing imidazole (IDBA): Synthesis, characterization and properties, *Macromol. Chem. Phys.* 2009, 210, 1632–1639.
20. Kulkarni, M., Potrekar, R., Kulkarni, R. A., Vernekar, S. P., Synthesis and characterization of novel polybenzimidazoles bearing pendant phenoxyamine groups, *J. Polym. Sci. Part A: Polym. Chem.* 2008, 46, 5776–5793.
21. Xu, N., Guo, X., Fang, J., Xu, H., Yin, J., Synthesis of novel polybenzimidazoles with pendant amino groups and the formation of their crosslinked membranes for medium temperature fuel cell applications, *J. Polym. Sci. Part A: Polym. Chem.* 2009, 47, 6992–7002.

22. Maity, S., Jana, T., Soluble polybenzimidazoles for PEM: Synthesized from efficient, inexpensive, readily accessible alternative tetraamine monomer, *Macromolecules* 2013, 46, 6814–6823.

23. Yuan, Y., Johnson, F., Cabasso, I., Polybenzimidazole (PBI) molecular weight and Mark-Houwink equation, *J. Appl. Polym. Sci.* 2009, 112, 3436–3441.

24. Shogbon, C. B., Brousseau, J.-L., Zhang, H., Benicewicz, B. C., Akpalu, Y. A., Determination of the molecular parameters and studies of the chain conformation of polybenzimidazole in DMAc/LiCl, *Macromolecules* 2006, 39, 9409–9418.

25. Su, P.-H., Lin, H.-L., Lin, Y.-P., Yu, T. L., Influence of catalyst layer polybenzimidazole molecular weight on the polybenzimidazole-based proton exchange membrane fuel cell performance, *Int. J. Hydrogen Energ.* 2013, 38, 13742–13753.

26. Kumbharkar, S. C., Kharul, U. K., N-substitution of polybenzimidazoles: Synthesis and evaluation of physical properties, *Eur. Polym. J.* 2009, 45, 3363–3371.

27. Maity, S., Sannigrahi, A., Ghosh, S., Jana, T., N-alkyl polybenzimidazole: Effect of alkyl chain length, *Eur. Polym. J.* 2013, 49, 2280–2292.

28. Xiao, L., Zhang, H., Scanlon, E., Ramanathan, L. S., Choe, E.-W., Rogers, D., Apple, T., Benicewicz, B. C., High-temperature polybenzimidazole fuel cell membranes via a sol-gel process, *Chem. Mater.* 2005, 17, 5328–5333.

29. Li, Q., He, R., Berg, R. W., Hjuler, H. A., Bjerrum, N. J., Water uptake and acid doping of polybenzimidazoles as electrolyte membranes for fuel cells, *Solid State Ionics* 2004, 168, 177–185.

30. Leykin, A. Y., Askadskii, A. A., Vasilev, V. G., Rusanov, A. L., Dependence of some properties of phosphoric acid doped PBIs on their chemical structure, *J. Membr. Sci.* 2010, 347, 69–74.

31. Xiao, L., Zhang, H., Jana, T., Scanlon, E., Chen, R., Choe, E.-W., Ramanathan, L. S., Yu, S., Benicewicz, B. C., Synthesis and characterization of pyridine-based polybenzimidazoles for high temperature polymer electrolyte membrane fuel cell applications, *Fuel Cells* 2005, 5, 287–295.

32. Xing, B., Savadogo, O., The effect of acid doping on the conductivity of polybenzimidazole (PBI), *J. New Mater. Electrochem. Syst.* 1999, 2, 95–101.

33. Hughes, C. E., Haufe, S., Angerstein, B., Kalim, R., Mahr, U., Reiche, A., Baldus, M., Probing structure and dynamics in poly[2,2'-(m-phenylene)-5,5'-bibenzimidazole] fuel cells with magic-angle spinning NMR, *J. Phys. Chem. B* 2004, 108, 13626–13631.

34. Ma, Y.-L., Wainright, J. S., Litt, M. H., Savinell, R. F., Conductivity of PBI membranes for high-temperature polymer electrolyte fuel cells, *J. Electrochem. Soc.* 2004, 151, A8–A16.

35. Schechter, A., Savinell, R. F., Wainright, J. S., Ray, D., ^{1}H and ^{31}P NMR study of phosphoric acid–doped polybenzimidazole under controlled water activity, *J. Electrochem. Soc.* 2009, 156, B283–B290.

36. Asensio, J. A., Sanchez, E. M., Gomez-Romero, P., Proton-conducting membranes based on benzimidazole polymers for high-temperature PEM fuel cells. A chemical quest, *Chem. Soc. Rev.* 2010, 39, 3210–3239.

37. He, R., Li, Q., Xiao, G., Bjerrum, N. J., Proton conductivity of phosphoric acid doped polybenzimidazole and its composites with inorganic proton conductors, *J. Membr. Sci.* 2003, 226, 169–184.

38. Lobato, J., Canizares, P., Rodrigo, M. A., Linares, J. J., Aguilar, J. A., Improved polybenzimidazole films for H_3PO_4-doped PBI-based high temperature PEMFC, *J. Membr. Sci.* 2007, 306, 47–55.

39. Kim, T.-H., Kim, S.-K., Lim, T.-W., Lee, J.-C., Synthesis and properties of poly(aryl ether benzimidazole) copolymers for high-temperature fuel cell membranes, *J. Membr. Sci.* 2008, 323, 362–370.

40. Li, Q., Pan, C., Jensen, J. O., Noye, P., Bjerrum, N. J., Cross-linked polybenzimidazole membranes for fuel cells, *Chem. Mater.* 2007, 19, 350–352.

41. Wang, S., Zhang, G., Han, M., Li, H., Zhang, Y., Ni, J., Ma, W. et al., Novel epoxy-based cross-linked polybenzimidazole for high temperature proton exchange membrane fuel cells, *Int. J. Hydrogen Energ.* 2011, 36, 8412–8421.

42. Fang, J., Kita, H., Okamoto, K.-I., Hyperbranched polyimides for gas separation applications. 1. Synthesis and characterization, *Macromolecules* 2000, 33, 4639–4646.

43. Fang, J., Kita, H., Okamoto, K.-I., Gas permeation properties of hyperbranched polyimide membranes, *J. Membr. Sci.* 2001, 182, 245–256.

44. Xu, H., Chen, K., Guo, X., Fang, J., Yin, J., Synthesis and properties of hyperbranched polybenzimidazoles via $A_2 + B_3$ approach, *J. Polym. Sci. A: Polym. Chem.* 2007, 45, 1150–1158.

45. Xu, H., Chen, K., Guo, X., Fang, J., Yin, J., Synthesis of hyperbranched polybenzimidazoles and their membrane formation, *J. Membr. Sci.* 2007, 288, 255–260.

46. Lee, H.-S., Roy, A., Lane, O., McGrath, J. E., Synthesis and characterization of poly(arylene ether sulfone)-*b*-polybenzimidazole copolymers for high temperature low humidity proton exchange membrane fuel cells, *Polymer* 2008, 49, 5387–5396.

47. Yuan, S., Guo, X., Aili, D., Pan, C., Li, Q., Fang, J., Poly(imide benzimidazole)s for high temperature polymer electrolyte membrane fuel cells, *J. Membr. Sci.* 2014, 454, 351–358.

48. Kinumoto, T., Inaba, M., Nakayama, Y., Ogata, K., Umebayashi, R., Tasaka, A., Iriyama, Y., Abe, T., Ogumi, Z., Durability of perfluorinated ionomer membrane against hydrogen peroxide, *J. Power Sources* 2006, 258, 1222–1228.

49. Chang, Z., Pu, H., Wan, D., Jin, M., Pan, H., Effects of adjacent groups of benzimidazole on antioxidation of polybenzimidazoles, *Polym. Degrad. Stabil.* 2010, 95, 2648–2653.

50. Yang, J. S., Cleemann, L. N., Steenberg, T., Terkelsen, C., Li, Q. F., Jensen, J. O., Hjuler, H. A., Bjerrum, N. J., He, R. H., High molecular weight polybenzimidazole membranes for high temperature PEMFC, *Fuel Cells* 2014, 14, 7–15.

51. Liao, J. H., Li, Q. F., Rudbeck, H. C., Jensen, J. O., Chromik, A., Bjerrum, N. J., Kerres, J., Xing, W., Oxidative degradation of polybenzimidazole membranes as electrolytes for high temperature proton exchange membrane fuel cells, *Fuel Cells* 2011, 11, 745–755.

52. Cho, Y.-H., Kim, S.-K., Kim, T.-H., Cho, Y.-H., Lim, J. W., Jung, N., Yoon, W.-S., Lee, J.-C., Sung, Y.-E., Preparation of MEA with the polybenzimidazole membrane for high temperature PEM fuel cell, *Electrochem. Solid-State Lett.* 2011, 14, B38–B40.

53. Yuan, X.-Z., Song, C., Wang, H., Zhang, J. 2009. *Electrochemical Impedance Spectroscopy in PEM Fuel Cells.* Springer, London, U.K.

54. Li, Q., Hjuler, H. A., Bjerrum, N. J., Phosphoric acid doped polybenzimidazole membranes: Physiochemical characterization and fuel cell applications, *J. Appl. Electrochem.* 2001, 31, 773–779.

55. Yu, S., Xiao, L., Benicewicz, B. C., Durability studies of PBI-based high temperature PEMFCs, *Fuel Cells* 2008, 8, 165–174.

56. Oono, Y., Sounai, A., Hori, M., Long-term cell degradation mechanism in high-temperature proton exchange membrane fuel cells, *J. Power Sources* 2012, 210, 366–373.

57. Jung, G.-B., Tseng, C.-C., Yeh, C.-C., Lin, C.-Y., Membrane electrode assemblies doped with H_3PO_4 for high temperature proton exchange membrane fuel cells, *Int. J. Hydrogen Energ.* 2012, 37, 13645–13651.

58. Galbiati, S., Baricci, A., Casalegno, A., Carcassola, G., Marchesi, R., On the activation of polybenzimidazole-based membrane electrode assemblies doped with phosphoric acid, *Int. J. Hydrogen Energ.* 2012, 37, 14475–14481.

9 Hybrid/Composite Membranes

Yan Yin

CONTENTS

9.1   INTRODUCTION

As a representative of the electrochemical polymer electrolyte membranes, the proton exchange membrane (PEM) has been developed extensively as both a separator and an electrolyte in the operating fuel cell. In order to realize the potential of fuel cells, there is a need for smart and rational design of novel electrolyte membrane materials based on a fundamental understanding of membrane morphology, proton and mass transfer, and chemical and mechanical properties. However, additional advancements will still be necessary to meet aggressive operating conditions of higher temperatures and/or lower humidities, as well as longer operating longevity demanded in both automotive and stationary applications. The current PEM fuel cell utilizes perfluorosulfonic acid (PFSA) polymer membranes, for example, Nafion®, as electrolyte that is limited to low-temperature and high methanol crossover. The state-of-the-art membrane research is focused on developing electrolyte materials that provide good conductivity in the absence of water. Hybrid/composite membranes exhibit the capability of both reinforcement of a thin membrane to provide mechanical durability and improvement of water retention to promote proton transport.

9.2   MEMBRANE SYNTHESIS

9.2.1   Materials Requirement and Selection

The heart of a PEM fuel cell is a polymer electrolyte membrane that separates reactant gases and provides path for proton transport. The desired membrane properties for PEM fuel cells include high proton conductivity and poor electron conductivity, low permeability for reactant gases, minimal crossover of fuel and water especially for direct methanol fuel cells (DMFCs), good mechanical strength, chemical and thermal stabilities, compatibility with electrode materials, dimensional stability (minimal shrinking and swelling) during fuel cell operation, durability under prolonged operation (~5000 h for transportation applications[1]) at elevated temperatures and during freeze–thaw cycles, and low cost. Up to date, none of the available membrane materials meets all of these requirements.[2]

The operating conditions of future PEM fuel cells will be more stringent than they are today. The need for system simplification to save cost and minimize parasitic energy losses drives the conditions to less desirable environments for fuel cell membranes. The markets and applications today and for the near future can be divided into three areas: stationary, automotive, and portable.[3] Each of these has different operating requirements, which in turn lead to different membrane selections. Although the exact conditions and membrane selections are still evolving today, the trends are becoming clear.

The stationary market will require long durability, high efficiency, stable power output, and long lifetime under relatively constant operating conditions. Fewer if any on–off cycles are anticipated in many stationary applications. Higher operating temperatures are desirable to utilize the waste heat of the process more efficiently and to be less sensitive to fuel impurities, which in turn could lead to simplified reforming technologies. Drier operating conditions are not an absolute requirement, and full humidification of the gas streams may be acceptable. Customers in the stationary market today are demanding 40,000 h life and are beginning to discuss 60,000 h lifetimes or longer.[4] The longevity of membrane materials is driven by the need to reduce the cost of power generated, which is a strong function of lifetime.

In automotive market, on the other hand, the primary requirements will be high temperature and drier operating conditions.[5] This is driven by the available radiator technology used to transport excess heat away from the fuel cell stack, easier water management, and increased rates of reaction and diffusion. To maintain the current size of automotive radiators in PEM fuel cell vehicles, temperature above 100°C is necessary for the fuel cell stack to increase the heat transfer efficiency of the radiator. An additional advantage of higher temperatures is lower sensitivity to fuel impurities, particularly carbon monoxide.[6] System size and weight constrains associated with water management are driving lower-humidity conditions as well. Operating under automotive driving cycles, with frequent on–off cycles including freeze tolerance, high temperature tolerance and reasonable high conducting performance must also be part of the membrane material considerations.

The portable market is perhaps the least well defined and hardest to predict. Methanol fuel is currently being favored for this application because of its high energy density and portability.[7,8] Usually, these methanol-fueled systems also run dry, typically with ambient air on the cathode. The DMFCs have a challenging set of membrane materials issues in relation to methanol crossover.[9–11] Methanol crossover is closely related to several factors including membrane structure and morphology, membrane thickness, and fuel cell operating conditions such as temperature, pressure, and methanol concentration.[9] Hydrocarbon and composite fluorinated membranes currently show the most potential for low-cost membranes with low methanol permeability and high durability.[10] To date, there has been continuous extensive research on developing membranes that can fulfill all of the essential characteristics to yield the desired performance in DMFCs.[11]

Nafion-based hybrid membranes are traditionally used electrolytes for both PEM fuel cell and DMFCs.[12–17] Besides Nafion, many composite engineering thermoplastic polymers based on poly(etheretherketone) (PEEK),[18] polyvinyl alcohol (PVA),[19] polysulfone,[20,21] polybenzimidazole (PBI),[22] polyimide,[23] and other organic–inorganic composite membranes[24,25] have been employed as alternative membranes for both PEM fuel cells and DMFCs due to their lower cost, comparable conductivity, high mechanical and thermal stabilities, and easy modification as well.

9.2.2 POLYMER SYNTHESIS METHODS AND PROCESSES

Hybrid/composite polymer electrolyte membranes are generally composed of proton-conducting polymer and inorganic/organic filler. The proton-conducting polymers (usually as sulfonated polymers) are basically used as conduction matrix, in most cases, and sometimes are used as pore-filling materials. The syntheses of polymer matrix have been introduced in other chapters and will not be described in this chapter. For hybrid/composite polymer electrolytes, the synthetic method can be mainly classified into (1) physical mixing of inorganic filler,[26–34] (2) covalent cross-linking,[35,36] (3) graft polymerization,[37,38] and (4) acid–base or polymer blending.[39–43] Most of the hybrid/composite polymers are synthesized via sol–gel process. Attempts to prevent the loss of water through modification of the chemical composition of Nafion membranes have been performed widely.[26–29,37,44] These studies are generally carried out by dispersion of nanoparticles or by direct growth of inorganic phase using SO_3^- groups of Nafion to catalyze sol–gel polymerization of metal oxide precursor.

In order to realize the optimum fuel cell performance, perfluorinated polymer-based composite membranes modified with ceramic/inorganic fillers, namely, SiO_2, TiO_2, ZrO_2, clay, and activated carbon, are extensively used to promote proton conduction in the membranes at elevated temperatures or under low relative humidity (RH).[26,27,45–51] Most of the composite materials are synthesized via physical mixing and doping of nanometer metal ions.

Silica-filled Nafion-based composite membrane is one of the most studied hybrid membranes for PEM fuel cells. These composite membranes are usually fabricated by embedding silica particles as inorganic fillers in PFSA ionomer by water hydrolysis process. The use of Nafion as acid helps in forming silica/siloxane polymer within the membrane. The inorganic filler materials have high affinity to water and assist proton transport across the electrolyte membrane. Sahu et al.[26] reported the synthesis of silica-containing Nafion hybrid membranes via sol–gel method. They used tetraethoxysilane (TEO), isopropyl alcohol, and deionized water as the starting materials to prepare silica sol. The mixture is sonicated for 15 min to obtain a visibly homogeneous and transparent colloidal suspension. The required amount of sol is then impregnated with 5 wt% commercial Nafion solution, and the resultant admixture is sonicated for another 15 min before casting into membrane. Generally, the hydrolysis procedure has a significant influence on the size and distribution of SiO_2 nanoparticles, which are responsible for the quality and performance of the composite membranes. Sometimes, silica with surface modified with hydrophobic perfluorododecyl groups is used as the additive for PFSA membrane. The hybrid materials are prepared by membrane casting from alcohol polymer solutions containing precursors for the particle formation followed by the hydrolysis of the precursors. It was found that incorporation of silica with hydrophobic surface results in the decrease of water uptake (WU) in comparison with membrane doped with pure SiO_2, while the conductivity is improved.[30] Mesoporous silica-containing sulfonic acid groups is also employed as a filler material to improve the proton conductivity and water retention at high temperature and low RH. Tsai et al.[33] synthesized mesoporous phenyl silica (PS) using TEO and phenyltriethoxysilane as the Si precursors.

The PS is sulfonated with concentrated sulfuric acid to prepare the mesoporous sulfonated PS. These hybrid membranes are prepared via in situ cocondensation of TEO and chlorosulfonylphenethylsilane via self-assembly route using organic surfactants as templates for the tuning of the architecture of the silica or hybrid organo-silica components.[44]

Another type of metallic physical filler widely used for hybrid nanocomposite polymer electrolytes is TiO_2.[29,30,49,50] The hygroscopic metal oxide particles are incorporated into the hydrophobic domains of the PEM to enhance the water retention of Nafion. Titanium dioxide is a good candidate as hydrophilic filler for Nafion because it provides suitable hydration of the membrane under fuel cell operation conditions, especially at elevated temperatures.[29] Nanometer TiO_2 powder is firstly synthesized via sol–gel method from $Ti(OiPr)_4$ and then calcined at around 400°C before mixing with Nafion solution.[30] In order to avoid modifications in the intrinsic properties of the polymeric matrix, the synthetic methodology must occur at low temperature (<100°C) and produce crystalline particles with average size compatible with ionic clusters.[49] Usually, the titanium dioxide produced by conventional sol–gel process is amorphous, and the crystallization of stable nanoparticles requires heat treatments at high temperature (e.g., 400°C), as mentioned earlier, which increases the average particle size.

Recently, zirconium modified Nafion composites have attracted attention due to their retention of water under high temperature and low RH conditions.[27,31,45,50] In general, during preparation of the hybrid materials, zirconium oxide particles with a size of more than 10 nm are simply mixed with the polyelectrolyte solution. Pan and Zhang et al. synthesized crystallized zirconia nanoparticles with diameters of 6 nm in Nafion solution through sol–gel process.[27] Nafion molecules are self-assembled onto zirconia nanoparticles through electrostatic interactions and prevent the further growth of initial formed particles. Nakajima et al. developed surface-modified zirconium oxide with a nanodispersion into polymer electrolyte by utilizing the *capping phenomenon*, which is the multipoint adsorption of a polyelectrolyte to the zirconium oxide.[31] In order to improve the proton-conducting capability, a sulfated-zirconia filler is developed by Giffin et al.,[45] and the presence of acidic functionalities on the surface of the filler endued a reasonably high conductivity (3×10^{-3} S cm^{-1} at 120°C) even in completely dry conditions. Di Noto et al. reported two families of hybrid inorganic–organic proton-conducting membranes based on Nafion and a different *core–shell* nanofiller.[50] The nanofillers are based on either a ZrO_2 *core* covered with a HfO_2 *shell* or a HfO_2 *core* solvated by a *shell* of SiO_2 nanoparticles.

Besides the typical metal oxides as mentioned earlier, other inorganic fillers such as clay and activated carbon are also studied.[46–48,51] Giannelis groups[46] synthesized a series of Nafion–clay nanocomposite membranes using proton exchanged clay nanoparticles as the mixing filler. Well-dispersed, mechanically robust, free-standing nanocomposite membranes were prepared by casting from a water suspension at 180°C under pressure. Chang and coworkers reported a highly porous activated carbon incorporated into Nafion to obtain unprecedented level of WU, resulting in dramatic enhancements in proton conductivity at low RH.[47,48] The cost-effective and high-throughput method for producing composite membranes with high WU is developed by combining high-porosity and superior surface area activated carbon with Nafion. Fatyeyeva et al. synthesized Nafion composites with modified layered inorganic clay (Laponite) into Nafion membranes, which contribute to an increase of the proton conductivity at 85°C.[51] The modification of Laponite particles was performed by the plasma activation process and the chemical grafting of sulfonic groups as well as by the direct plasma sulfonation.

Fabrication of hybrid/composite polyelectrolytes can also be carried out by cross-linking. Na group prepared hybrid PEMs with silane cross-linking and thiol-ene click chemistry.[35,36] During silane cross-linking process, two silane monomers, namely, 3-glycidoxypropyl-trimethoxysilane (GPTMS) and 3-mercaptopropyl-trimethoxysilane (MPTMS), are first blended with a sulfonated poly(arylene ether ketone) (SPAEK). Then the blended membrane is heated to induce the grafting of GPTMS onto SPAEK. Finally, a hydrolysis–condensation is performed on the grafted membrane to induce cross-linking. The –SH groups of MPTMS are oxidized to sulfonic acid groups, which are attributed to enhance the proton conductivity.[35] On the other hand, a combination of silane cross-linking and thiol-ene click chemistry based on SPAEK with propenyl groups has been reported to further increase membrane stability.[36] Figure 9.1 shows an illustration of the preparation for cross-linked hybrid membranes using silane cross-linking and thiol-ene click chemistry. Propenyl, containing a double bond, is an excellent cross-linking agent, because it can initiate a radical reaction by heating or ultraviolet irradiation. It could be involved in thermally and photochemically initiated thiol-ene click reactions using thiol and allyl functionalized polymers. As shown in Figure 9.1, the hydrophilic SPAEK could be blocked in the Si–O–Si network, making a contribution to decrease swelling and methanol permeation.

Grafting method is sometimes coupled with cross-linking or inorganic filling. Joseph et al. reported organically modified mesoporous silica/Nafion membranes prepared by a surfactant template sol–gel process involving Nafion solution and silica precursors.[37] Phosphonic acid functions were used as an alternative for the sulfonic acid groups in fuel cell membranes and synthesized Nafion/phosphonic acid-grafted mesostructured silica hybrid membranes. The silica network is embedded in the polymer and the $-PO_3H_2$ moieties, resulting in effective water retention at low RH. The main attractions of using $-PO_3H_2$ groups over $-SO_3H$ groups are due to the stability of hydrogen bonding. It is anticipated that blending of commonly employed sulfonated materials with low contents of phosphonic acid–functionalized phases can ensure the required high conductivities in the entire temperature range of PEM fuel cell. Pereira et al. synthesized Nafion–mesoporous silica hybrid membranes by the in situ generation of inorganic silica using a surfactant-assisted sol–gel process.[44] The sulfonic acid moieties were grafted onto the mesoporous silica by postoxidation reactions with a view to improve conductivity. During the preparation of hybrid composites, two individual homogeneous inorganic and organic solutions are created, followed by mixing and solvent evaporation under carefully controlled conditions.

Chi et al. reported the preparation of poly(vinylidene fluoride) (PVdF) nano-composite membranes based on graft polymerization and sol–gel process.[38] In their study, poly(vinylidene fluoride-co-chlorotrifluoroethylene) (P(VDF-co-CTFE)) was grafted with poly(styrene sulfonic acid) (PSSA) via atom transfer radical

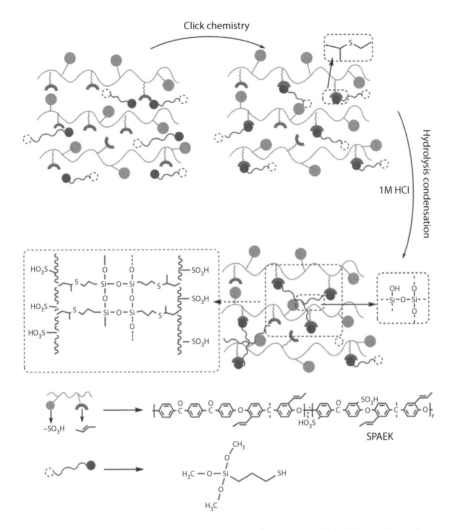

FIGURE 9.1 Scheme for illustrating the preparation of cross-linked hybrid membranes using silane cross-linking and thiol-ene click chemistry. (From Gao, S. et al., *J. Power Sources*, 214, 285, 2012.)

polymerization to prepare a proton-conducting P(VDF-co-CTFE)-g-PSSA graft polymer. The corresponding graft copolymer was combined with a silica precursor, MPTMS, through a sol–gel reaction and a subsequent oxidation process.

Acid–base and blend membranes are alternative candidates for hybrid/composite polyelectrolytes.[39–43] Fully aromatic polymers are currently widely used polymer blenders except for Nafion itself. The purpose for acid–base blending is to find an optimal compromise between enhancement of proton conductivity by sulfonation and improvement of thermal and morphological stability by composition. In blend composite membranes, the majority partner should ensure good proton-conducting behavior, and the minority partner should improve the mechanical properties and stabilize the morphology of the polymer and should decrease methanol crossover as well especially when being employed in DMFCs.

Molla et al. developed Nafion-based composite membranes using PVA as a nanofiber substrate to suppress methanol permeation.[39–41] By using the ultrathin PVA nanofiber substrate, the membrane thickness is greatly decreased while maintaining good mechanical properties. During the synthesis of the Nafion/PVA composite, porous PVA mats were produced first by a standard electrospinning setup through the feeding of a water-based solution of PVA. After chemical functionalization of PVA nanofibers with ion exchange and cross-linking with glutaraldehyde, the resultant PVA nanofibers were impregnated with the 5 wt% Nafion solution in isopropanol and water with a certain weight ratio. The impregnation process is usually repeated several times to ensure complete infiltration of Nafion. Very recently, Lin and Wang[42] reported Nafion/PVA blend composite membrane that is somewhat different from their composite fibers[43] with similar fabrication method as reported in literatures.[39] Lin et al. synthesized their blend composites by mixing Nafion/PVA solutions with different weight ratios between Nafion and PVA, followed by casting to get composite membranes. They observed agglomeration and nonhomogeneous distribution of PVA in the blend membranes.

In order to decrease reactant crossover, such as methanol, fully aromatic polymers are widely studied, among which sulfonated poly(ether ether ketone) (SPEEK) is a typical representative.[52–54] Xue et al. reported an acid–base hybrid polymer electrolyte membranes based on SPEEK and 3-aminopropyltriethoxy silane (APTES).[52] The addition of APTES to SPEEK may ensure high dimensional stability and membrane swelling at high temperature due to the electrostatic interaction between the basic amine groups of APTES and the acidic sulfonic groups of SPEEK. Yang developed blend membranes consisting of SPEEK and PVA with flexible PVA content to realize low methanol permeability, since PVA possesses excellent methanol resistance.[53] The SPEEK/PVA blend membranes were prepared from solvent evaporation of the mixed solution. Di Vona et al. developed hybrid blend polymer composites from SPEEK and substituted polyphenylsulfone (PPSU).[54] Before the blending process, PPSU was generally modified either by silylation via metalation reaction with butyllithium followed by electrophilic substitution with phenyl-trichlorosilane or by sulfonation from hydrolysis and reaction with concentrated sulfuric acid.

Yilnaztuerk et al. reported self-assembled polyelectrolyte complexes from layer-by-layer (LbL) method with highly charged polyvinyl sulfate potassium salt and poly(allylamine hydrochloride) on Nafion membrane.[55] The multilayered composite membranes possess both high proton conductivity and methanol blocking properties. The preparation of ultrathin polyelectrolyte films starts with the immersion of a positively charged substrate in an aqueous solution of an anionic polyelectrolyte so that a thin layer of this compound is adsorbed and the surface charge of the substrate reverted. Subsequent dipping of this substrate into a solution of a cationic electrolyte again leads to adsorption of a thin layer and the surface charge is rendered positive again. Multiple repetition of the adsorption steps results in a multilayer membrane with alternation positive and negative excess charges.

9.2.3 Membrane Formation

For inorganic hybrid membranes, the membrane formation can be divided into two ways: (1) the pristine polymer membrane is impregnated into the inorganic ion solution to combine metal ions; (2) the modified inorganic filler is added into the polymer solution and then cast into membrane via solvent evaporation. Obviously, the former method is very simple, while the inorganic particle content and distribution will be uneasy to control. Hence, the latter method is widely preferred and a sol composed of inorganic ion and corresponding solvent is prepared first via hydrolysis, followed by impregnation with Nafion solution.[26,30,32]

A typical Nafion hybrid membrane containing heteropolyacid modified inorganic filler is prepared as follows.[56] Phosphotungstic acid (PWA) supported on silica gel is prepared from commercial PWA and silica gel with an appropriate amount of PWA and SiO_2. The mixture is ultrasonicated and then the suspension is dried at room temperature (RT) to obtain the resulting solid material that is crushed into a fine powder. Then an appropriate amount of 5 wt% Nafion solution is mixed with the PWA-modified silica (or unmodified SiO_2, TiO_2, WO_3) in an ultrasonic bath for 30 min, followed by solution casting in a glass tray and heated at 70°C for 30 min. Then, the recast composite membranes are detached from the glass tray by adding some deionized water. Finally, the membranes are purified by heating at 80°C in 3% H_2O_2 solution and deionized water and 0.5 M H_2SO_4 and deionized water for 2 h, respectively.

Usually, the membrane formation and fabrication are correlated with the hybrid/composite polymer synthesis procedure. Figure 9.2 shows an illustration of the preparation procedure of hybrid membranes containing inorganic fillers such as silica.[44] A sol is prepared first using absolute ethanol, Pluronic P123, and TEOS. In an attempt to increase the conductivity of the silica phase and the film flexibility, part of the TEOS (10% molar) is substituted by organoalkoxysilane (2-(4-chlorosulfonylphenyl) ethylsilane) in some cases. The organosilane solution is added drop by drop continuously in a stirred Nafion alcoholic solution containing 20 wt% water. The resulting homogeneous sol is then stirred at RT for 20 h. The hybrid membranes are obtained by pouring the mother solution onto a glass support using a hand coater (UNICOATER 409 Erichsen equipped with an Autonics temperature controller TZ4SP Series) followed by drying at 30°C for 10 h through laminar flow. During slow solvent evaporation, a polymeric mesoporous silicon oxide phase is grown by the sol–gel process. The condensation reactions and cooperative self-assembly organizations take place during solvent evaporation. The hybrid-cast membranes are dried at 120°C for 10 h to remove trapped volatiles and promote further inorganic condensation inside the Nafion network. The hybrid membranes are then separated from the glass plate by dipping them in deionized water. Finally, the membranes are purified by boiling in deionized water and 3% H_2O_2 solution to remove surfactant and residual organic impurities, and boiling in 0.5 M sulfuric acid solution to remove any possible inorganic contaminants, and finally rinsing with deionized water. The activated membranes in their acid form are stored in deionized water prior to use.

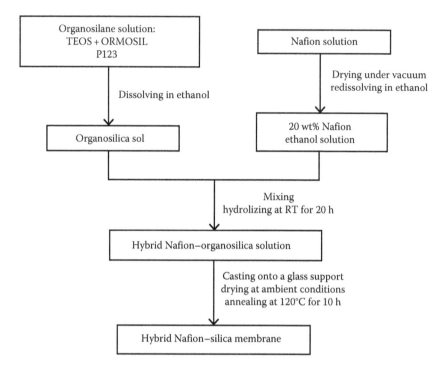

FIGURE 9.2 Preparation protocol of Nafion/mesoporous silica hybrid membranes. (From Pereira, F. et al., *Chem. Mater.*, 20, 1710, 2008.)

Fu et al. reported a covalent organic/inorganic hybrid membrane with a semi-interpenetrating polymer network structure.[57] The membrane preparation procedures include four steps. The first step is thermal polymerization. The polyvinyl chloride (PVC) film was cleaned in *n*-hexane for 4 h before being soaked in a monomer solution of styrene (St), *p*-vinylbenzyl chloride (*p*-VBC), and divinylbenzene (DVB) and benzoyl peroxide for 3 h at RT. The monomer swollen PVC film was placed between two glass plates and sealed with aluminum foil to prevent evaporation of the monomers. Thermal polymerization was carried out in an oven at 80°C for 8 h. During the second step, the film was soaked in 3-(methylamino)propyl-trimethoxysilane (MAPTMS)/benzene solution for 24 h, allowing diffusion of MAPTMS into the benzene-swollen film and the amine groups reacted with the chloromethyl group of VBC. After being washed with benzene to remove excess MAPTMS, the third step was carried out under acidic conditions to promote a sol–gel reaction. The film was soaked in 0.05 M HCl for 3 h to partially hydrolyze the methoxy group, followed by full hydrolyzation–polycondensation at 80°C for 12 h and 100°C for 1 h, respectively. Finally, the preswollen film in 1,2-chloroethane was sulfonated in a mixed solution of chlorosulfonic acid and 1,2-dichloroethane (1:9, v/v) at RT for 12 h, reacidified in 1 M HCl for 12 h, and washed completely in purified water prior to use.

Thanganathan and Bobba prepared a hybrid membrane via the sol–gel technique using PWA mixed with PVA and phosphosilicate (P_2O_5/SiO_2).[58] PVA (average

Mw = 100,000) was dissolved in deionized water at 90°C for 5 h. The solution was then cooled to RT. At the same time, a PWA solution was obtained by dissolving the PWA in deionized water. The two PVA and PWA solutions were mixed one by one with the phosphosilicate solution and stirred at RT for 6 h. TEOS (orthoethylsilicate) was reacted with H_2O and C_2H_5OH (as 0.1 M HCl aqueous solution) under the hydrolysis and condensation reaction of three ethoxy groups to produce an Si–O–Si network structure. The hybrid solution was kept in a flask equipped with a stirrer, and slow stirring was maintained at RT for 12 h. The reaction mixture was cast onto clean glass plates or Petri dishes and kept at RT until the membrane was optically dry. The resultant membrane was then peeled off and heated in an oven at 120°C for 24 h.

The fabrication of blend membranes is relatively simple compared with that of inorganic filler mixed or cross-linked composite membranes. Bi et al.[59] prepared a series of cross-linked sulfonated poly(arylene ether sulfone)/sulfonated polyimide (cSPAES/SPI) blend membranes by mixing a certain amount of cSPAES and SPI in their triethylamine (TEA) salt forms using m-cresol as a solvent, followed by filtration and casting onto a Petri dish and dried at 80°C for 2 h, 100°C for 2 h, and 120°C for 15 h, respectively, followed by proton exchange with 2 M HCl to obtain the blend hydrocarbon polyelectrolytes.

9.3   MEMBRANE CHARACTERIZATION

9.3.1   SPECTRAL STUDIES

Spectral studies such as Fourier transform infrared/attenuated total reflectance (FTIR/ATR) analysis, solid-state or pulse field gradient (PFG) nuclear magnetic resonance (NMR), x-ray diffraction (XRD) or small angle x-ray scattering (SAXS) analysis, and Raman study are basically the characterization methods for hybrid/composite membranes. Sometimes, the multiple spectral analyses are necessary to better understand the structure–property relationship for a kind of composite polyelectrolytes.

ATR–FTIR spectra for Nafion and silica–Nafion composite membranes have been investigated.[26] For silica–Nafion composite membranes, most of the characteristic absorption bands due to Nafion are generally suppressed. The absorption band at 950 cm^{-1} is attributed to Si–OH stretching vibration, and the absorption peak at 800 cm^{-1} is assigned to the asymmetric stretching vibration due to the Si–O–Si group. These absorption bands confirm the existence of silica in the composite membrane. Kurniawan et al. recently reported durability properties of Nafion–hydrophilic silica hybrid membrane against trace radical species in PEM fuel cells.[34] Their IR spectra indicated that for neat Nafion, the peaks of C–F antisymmetric stretching at 1209 cm^{-1} and symmetric stretching shift to higher wave numbers with decreases in intensity after durability test, while for silica-containing nanocomposite membranes with an addition of 1 wt% hydrophilic silica to the Nafion matrix, the peak bands located at 1207, 1151, 1057, and 968 cm^{-1} arising from the ionomer decrease in intensity with the immersion aging time simultaneously without the significant spectra shape variation, implying the stability of the nanocomposite membranes against durability test.

The characterization of pure Nafion and modified membranes with different percentages of titanium tetraisopropoxide (TIP) shows the presence of IR bands of Ti–O and TiO_2.[29] The field of low-wave-number band between 400 and 700 cm^{-1} and that between 990 and 1200 cm^{-1} characterize the vibration of the groups O–R isopropoxide that directly related to titanium. The crystallization of TiO_2 nanoparticles is characterized by the presence of a spectral band between 400 and 500 cm^{-1}, indicating the formation of the bridge Ti–O–. The peaks at 437.6, 434.9, and 437.2 cm^{-1} are also investigated for the presence of O–Ti band. The strong and wide peak at 3494.7 cm^{-1} and peak at 1631.5 cm^{-1} are due to the hydroxyl groups of Ti–OH with which physisorbed water molecules are bound by weak hydrogen bonds. In general, lattice water absorbs at 2884–3420 cm^{-1} (antisymmetric and symmetric OH stretching) and at 1630–1600 cm^{-1} (HOH bending). Whereas for pure Nafion membrane, a broad band at 2899 cm^{-1} attributed to vibrations of the O–H groups is observed.

Giffin et al. reported FTIR–ATR spectra of the sulfonated-zirconia/Nafion composite membranes compared with pristine Nafion.[45] A significant difference in the FTIR spectra of the composite membrane and pristine Nafion is found in the Nafion acid mode region found above the strong C–F and C–C stretching modes. In dry Nafion, a weak band can be found at approximately 1475 cm^{-1} that has been attributed to a sulfonate group with a covalently bound (undissociated) proton.[60] This band disappears upon hydration of the pristine Nafion as the proton dissociates forming the sulfonated anion and a proton associated with the water present in the system. Unlike the spectrum of pristine Nafion, the absence of the band in the spectrum of composite membrane indicates that the protons are dissociated from Nafion's pendant side chains even in completely dry conditions. This result is different from that found in some other composite systems. For example, in Nafion composite membranes doped with *core–shell* nanofiller, the acid band of Nafion remains in the presence of the nanofiller indicating that the proton is still covalently bonded to the pendant side chain.[60] The fact that there are no undissociated protons on the Nafion side chains in the composite membrane has important implications for its structure, suggesting that the filler in the composite increases the number of charge carriers present in the dehydrated system as compared to pristine Nafion.

Di Noto et al. reported the spectra of both sides of the $[Nafion/(TiO_2F)_x]$ membranes as a function of the mass fraction of the TiO_2F nanofiller.[61] With respect to the top side of the membrane (Side A), Side B became more enriched in the TiO_2F nanofiller, giving rise to an opaque texture. At the same time, the top side of the membrane (Side A) remained smooth and glossy. It is observed that the spectral profiles are very similar in comparison to those reported elsewhere for a similar class of hybrid inorganic–organic Nafion-based materials.[60] This makes the correlative assignment of the FTIR–ATR spectra of the hybrid $[Nafion/(TiO_2F)_x]$ membranes very easy. It is noted that all the spectra of the Side A of the membranes are similar to the spectrum of pristine Nafion. On the other hand, all the spectra of the Side B of the $[Nafion/(TiO_2F)_x]$ membranes are quite different from the spectrum of the pristine recast Nafion membrane. This evidence is interpreted admitting that during the solvent-casting process, the TiO_2F nanofiller developed a concentration profile.[61]

XRD analysis is an important tool for determining the phase changes (crystalline and amorphous) of the proton-conducting hybrid membranes as well as their microstructure. XRD usually includes wide-angle XRD and small-angle x-ray scattering, among which SAXS is widely utilized to investigate the structural changes in hydrated Nafion membranes upon introduction of the inorganic phase.

The SAXS results for recast Nafion, SiO_2 composite, and TiO_2 composite membranes in relation to the Bragg spacing, which is correlated with the peak scattering angle and in turn represents the average interionic cluster distance, as a function of water content, were reported by Adjemian et al.[16] The Bragg spacing is reported to depend on the water content and equivalent weight of the membranes. The Bragg spacing associated with metal oxide–containing membranes shows closer ionic clusters than the unmodified Nafion membrane. It has been suggested that decreases in cluster-to-cluster distance correlate with cluster size, smaller separations being associated with larger cluster sizes. Therefore, hydrated silica- and titania-containing membranes support larger water clusters, consistent with the conclusion that these materials are mechanically stiffer at the molecular level. It is observed that at hydration levels below ~7 waters/sulfonate group (relatively dehydrated), the Bragg spacing is invariant with metal oxide composition, further supporting the proposed relationship between water cluster size and presence of the metal oxide phase. That is, below ~7 waters/sulfonate, the available water is a limiting reagent and improved structural membrane properties do not enhance the membrane aqueous phase. Once sufficient water is available, the ability of the Nafion phase to host larger water clusters with a closer contact becomes an enabling issue. The ability to support larger water clusters is likely associated with the improved apparent cell conductivity under elevated temperature operational conditions in which the polymer is placed under significant external stress by the kinematics of the cell test frame.[16]

The clustering of ionic groups in Nafion is usually indicated by the existence of a scattering vector maximum at $q \sim 0.11$ Å^{-1} or $q \sim 0.13$ Å^{-1}, often called as *ionomer peak* in SAXS.[37] The crystallites in the fluorocarbon matrix give rise to another prominent feature in the SAXS patterns of Nafion 117, which is commonly termed as *matrix knee*. This broad shoulder peak appears at $q < 0.05$ Å^{-1} and it corresponds to an intercrystalline repeat length of 16–18 nm. In Joseph's study,[37] the SAXS profiles obtained for predried, unmodified Nafion clearly show the ionomer peak centered around $q \sim 0.14$ Å^{-1} and a broad peak due to the crystallites at around $q \sim 0.05$ Å^{-1}. Although the ion cluster size in the Nafion polymer network depends on the hydration levels of the polymer, the ionic clusters exist even in dry Nafion membranes. For the dehydrated samples, the cluster size may be small with a small number of ionic SO_3H groups in each cluster and a small characteristic separation. Interestingly, the composite membranes show a shift in the ionomer peak position to larger q values, at slightly less intensity than the Nafion peak. The higher q values obtained for the composite membranes indicate that either the number density of ionic clusters increases or their cluster size decreases in these systems, resulting in the reduction of the center-to-center cluster distance. In addition to the shift in ionomer peak, the SAXS patterns of surfactant-extracted composite membranes reveal the presence of mesoporous silica embedded in the polymer. Broad and intense Bragg peaks can be observed for all composite membranes with a characteristic d-spacing of about 11 nm. A single diffraction peak in the SAXS pattern with no higher-order Bragg peaks reveals that the pore organization does not exhibit mesoscopic order.[37]

Recently, Hammami et al. reported the XRD studies of Nafion–TiO$_2$ hybrid membranes.[29] Their XRD patterns show peaks corresponding to the TiO$_2$ anatase and rutile phases and the characteristic peak of Nafion. The diffraction peaks of the inorganic phase were detected in the XRD samples prepared with 7% and 9% of TIP at $2\theta = 54.4°$ and 56.7°, which could be indexed to the (211) and (220) crystal planes of TiO$_2$ rutile phase. In addition, three peaks appear at $2\theta = 26.1°$, 38.6°, 48.13°, respectively, and 54.27°, which could be indexed to the (101), (004), (200), and (105) planes of TiO$_2$ anatase phase. More importantly, the crystalline TiO$_2$ nanoparticles were inserted in the hydrophilic sites, and Nafion acts as a template for the inorganic phase. The combined results of the microstructural characterization indicated that the samples are composed of TiO$_2$ anatase phase. Such findings are relevant for the proton transport and are reflected in the protonic properties of the hybrid electrolytes.

The XRD diffraction patterns of nanocomposite membranes containing 0–20 wt% MPTMS indicate that pristine P(VDF-co-CTFE)-g-PSSA graft copolymer exhibited broad amorphous halos centered at 18.2° and 20.2°. In general, the intensities of the peaks were reduced as the concentration of MPTMS increased, implying that silica perturbed the microstructures of the graft copolymer membranes. In addition to microstructural changes, the nanostructural changes in the membranes as a result of the addition of silica were also investigated using SAXS analysis, as shown in Figure 9.3.[51] The pristine graft copolymer exhibited a maximum peak at $q = 0.207$ nm^{-1}, suggesting a well-developed, nanophase-separated structure existed. Upon the addition of MPTMS, the nanocomposite membranes remained in a microphase-separated state, but their q value maxima shifted to higher values. At the same time, the peaks became broader with increasing MPTMS concentration. The domain spacing of the graft polymer was estimated from the peak maximum using the Bragg relationship. Figure 9.3b shows that the domain spacing of P(VDF-co-CTFE)-g-PSSA decreased from 30 to 17 nm with increasing MPTMS concentration up to 5 wt%, above which it did not change greatly. Instead, additional domain spacing was observed near 40 nm at concentrations greater than 5 wt% MPTMS, which increased linearly with increasing MPTMS concentration. This suggests that the microphase-separated structure of the graft copolymer was maintained up to 5 wt% MPTMS, above which the organic graft copolymer and the inorganic silica were macroscopically phase separated due to a miscibility limit between them. These morphological characteristics affect the properties, including both the WU and proton conductivity, of P(VDF-co-CTFE)-g-PSSA/SiO$_2$–SO$_3$H membranes.

Shao et al. reported the XRD patterns of recast Nafion membrane and recast composite Nafion membrane containing the additives SiO$_2$, TiO$_2$ WO$_3$, and SiO$_2$/PWA.[56] The patterns of composite membrane with TiO$_2$ and WO$_3$ show all the characteristic peaks of the respective inorganic additives, indicating the formation of enough crystallites on the outer surface of the membrane. The diffractogram of the SiO$_2$ and SiO$_2$-/PWA-impregnated membrane exhibits no peak characteristics of SiO$_2$ and PWA, indicating that the presence of SiO$_2$ and PWA in the membrane is an amorphous nature. Furthermore, it is noted that the XRD peaks at about 18° of 2θ become sharper by incorporation of the inorganic additives than that of recast Nafion membrane, indicating the enhancement of crystallinity of the incorporated membranes.

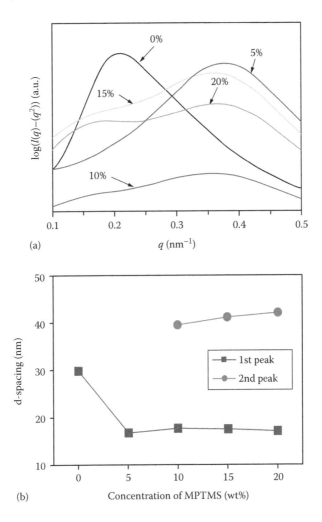

FIGURE 9.3 (a) SAXS patterns and (b) d-spacing of P(VDF-*co*-CTFE)-*g*-PSSA/SiO$_2$-SO$_3$H nanocomposite membranes with different MPTMS concentrations. (From Fatyeyeva, K. et al., *J. Membr. Sci.*, 369, 155, 2011.)

Thanganathan and Bobba investigated the XRD patterns for pure PVA, pure PMA, and the PVA/PMA/P$_2$O$_5$/SiO$_2$ hybrid composite membranes, as shown in Figure 9.4.[58] The peaks in the pure materials (Figure 9.4I and II) disappeared in the composite membrane (Figure 9.4III), confirming that the hybrid membrane had a semicrystalline phase and that hydrogen bonding occurred between the polymer chains. Moreover, the hybrid composite membranes showed a broad peak at about 20°–22° corresponding to 2θ.

Pulsed field gradient NMR is a smart technique to analyze water in PEMs and evaluate diffusion coefficients of proton nuclei, regardless of the environmental state of water in the PEM and under a wide range of sample conditions, without damaging

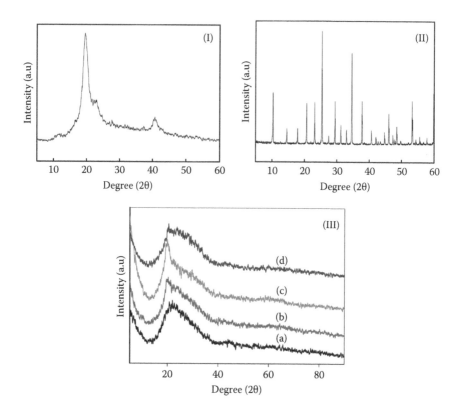

FIGURE 9.4 XRD patterns of (I) pure PVA, (II) pure PWA, and (III) the PVA/PWA/P₂O₅/SiO₂ hybrid composite membranes: (a) 50 wt.%/20/10/20 (mol%), (b) 60 wt.%/15/10/15 (mol%), (c) 79 wt.%/10/10/10 (mol%), and (d) 80 wt.%/5/10/5 (mol%). (From Thanganathan, U. and Bobba, R., *J. Alloy. Compd.*, 540, 184, 2012.)

the sample. During NMR measurements, an isolated signal is typically seen for protons and water in the PEM. This signal is completely distinct from that of liquid water, making NMR a good candidate method for evaluating water diffusion behavior in PEMs conditioned at determined values of temperature and humidity.[2,20,62] Li and Madsen et al. have reported anisotropic diffusion and morphology studies in perfluorosulfonated ionomers investigated by NMR.[63] It was found that for extruded membranes with stronger alignment, a faster in-plane diffusion than through-plane diffusion is investigated, while diffusion anisotropy is minimal for weakly aligned membranes.

Ye et al. studied proton mobilities in Nafion and Nafion/SiO₂ composites using high-resolution solid-state magic angle spinning (MAS) NMR.[64] From their ¹H NMR spectra, it is found that low concentrations of TEOS or short permeation times (PTs) are necessary to allow complete hydrolysis of TEOS in Nafion. Incomplete hydrolysis of TEOS leaves residual ethyl groups on the surface of silica, which not only reduces the amount of water adsorbed by silica but also blocks the pathway of proton transport in the Nafion/SiO₂ composites. The diffusion coefficients established using

PFG NMR show that the best Nafion/SiO_2 composite can be obtained from synthesis with a low concentration of TEOS in a methanol solution.

The residual sulfuric acid protons are generally shown as a high-frequency proton resonance with a broad peak. Because of the complete hydrolysis of TEOS under these conditions, no residual ethyl group signals are observed. The absence of Si–OH proton signal could be attributed to the insignificant amount of surface Si–OH groups compared with the large number of sulfuric acid protons. After silica is doped into Nafion, water adsorbed on the silica surface forms hydrogen bonds with Si–OH groups. The presence of water will shift the sulfonic acid proton resonance to a lower chemical shift. When a higher concentration of TEOS is used to dope silica into Nafion, new peaks will be observed accordingly. The peaks at 1.0 and 3.5 ppm are generally assigned to CH_3 and CH_2 groups of the residual ethyl groups of incompletely hydrolyzed TEOS, respectively. A broad resonance spanning from 5 to 8 ppm is basically assigned to hydrogen-bonded –SO_3H and Si–OH protons. Wide range of local environments in the membrane causes significant spectral overlap. A broad resonance at 4 ppm is assigned to water physisorbed on silica.[64]

The ^{29}Si NMR spectra as shown in Figure 9.5 displayed that the Nafion/SiO_2 composite has a high ratio of Q_3/Q_4 sites, consistent with a small particle size and many surface hydroxyl groups.[64] The NMR research demonstrates the role of high-surface-area SiO_2 particles in trapping water and building a pathway for structural (Grotthuss mechanism) proton diffusion. As shown in Figure 9.5a, there is no silicon resonance observed at −80 ppm in all ^{29}Si spectra, confirming that there is no unreacted TEOS in these samples. The majority of silicon nuclei in the silica made from TEOS and sulfuric acid are in the form of Q_4. This indicates that a full

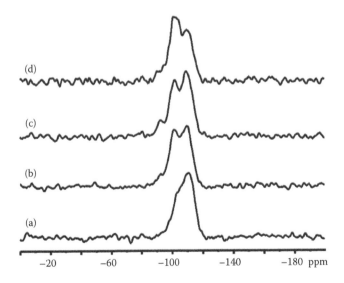

FIGURE 9.5 ^{29}Si MAS NMR of SiO_2 and Nafion/SiO_2 composites dried at 105°C for 24 h, MAS = 5 kHz: (a) SiO_2 from TEOS and H_2SO_4 (aq); (b) Nafion/SiO_2 (0:1:1, PT = 120 min); (c) Nafion/SiO_2 (0:1:1, PT = 24 h); (d) Nafion/SiO_2 (2.3:1:1, PT = 24 h). (From Ye, G. et al., *Macromolecules*, 40, 1529, 2007.)

condensation reaction has occurred among the TEOS molecules and that the average silica particle size is large. The Nafion/SiO_2 composite prepared using a short PT (120 min) gives an increased ratio of Q_3 to Q_4 as shown in Figure 9.5b, which suggests smaller silica particles in the composite compared with neat SiO_2. This is likely due to the restricted motion of silica in water channels of Nafion. Compared with silica made from a mixture of TEOS and sulfuric acid, such a narrow channel will certainly restrict the motion of the precursor units within the membrane. Therefore, the condensation reaction among these silica units will be limited. This maintains a large number of surface hydroxyl groups of silica. Thus, more silicon atoms have the Q_3 coordination environment in the composite. Nevertheless, when a longer PT is allowed (24 h), the ratio of Q_3 to Q_4 decreases slightly, as shown in Figure 9.5c. This may be attributed to the condensation reaction dominating over the hydrolysis reaction of TEOS. Interestingly, this composite also shows a peak at −90 ppm, which is attributed to silicon in the form of $(HO)_2–Si–(O–Si)_2$, denoted as Q_2. The Q_2 sites of silicon could be also due to the restricted condensation reaction between newly formed small silica particles, since the initially formed silica particles would be large and further prevent the fast motion of the subsequently formed small particles. As expected, because of the reduced concentration of TEOS, the composite Nafion/SiO_2 (2.3:1:1) shows an even higher ratio of Q_3 to Q_4, as shown in Figure 9.5d, suggesting a further decrease in the condensation among silica particles. This is desirable because silica with more surface hydroxyl groups will provide more sites to retain water.[64]

Nicotera et al. studied water transport properties and the effects of the addition of SiO_2, TiO_2, or $Zr(HPO_4)_2$ to Nafion matrix via NMR characterization.[65] They found that each of these additions resulted in improved WU and improved cell performance compared to unmodified Nafion, as a result of the membrane pore structure, which has larger cavities for water clustering. Two peaks were observed in their NMR spectra, about which the primary peak at 1–3 kHz is attributed to the bulk water and the second peak (<0 kHz) is assigned to the tightly bound water, which is more pronounced at high temperatures. This indicates that most of the water is associated with cluster filling the pore volumes, and some of the water solvates the sulfonated groups, the latter portion being strongly confined by the relatively large electrostatic and hydrogen-bonding forces in the vicinity of the inner-pore surfaces. It is clear that the effect of the particles is to improve the WU of the membranes, about which the enhanced absorption is not merely a consequence of the fact that SiO_2, TiO_2, or $Zr(HPO_4)_2$ particles are hygroscopic but is also attributable to their effect on the pore structure, creating larger cavities where more water molecules can be hosted in cluster form.[65]

Kannan et al. investigated WU properties of phosphosilicate Nafion hybrid membranes using solid-state NMR spectra.[66] The ^{29}Si CP/MAS NMR spectrum of the hybrid sample shows two characteristic peaks at −58.9 and −68.4 ppm. The ^{31}P CP/MAS NMR spectrum of Nafion hybrids shows two peaks at 3.7 and −27.6 ppm, respectively. The major peak at 3.7 ppm can be assigned to $O=P–(OR)(OH)_2$, which confirms that the major part of phosphate-containing groups has infiltrated to the membrane and remained unreacted. The other peak at −27.6 ppm indicates the presence of phosphorus atom with two bridging oxygens such as Si–O–P and P–O–P bonds. These results demonstrate the formation of phosphosilicate structures and the presence of −POH, which can form strong hydrogen bonding with water molecule.

Vibrational spectroscopic methods have higher sensitivity toward water molecules especially at a low vapor pressure. Confocal micro-Raman spectroscopy is a noninvasive, nondestructive technique with a high spatial resolution of micrometers, higher than that of the NMR techniques. With Raman spectroscopy, the chemical composition of membranes can be obtained, which is a major advantage over scattering techniques. Micro-Raman spectroscopy hence can be a powerful tool for the direct, simultaneous measurement of the chemical compositions in PEMs and the water content.[67,68]

Watanabe group[68] investigated the interior properties of water distribution inside Nafion membranes at various temperatures and humidities using in situ confocal micro-Raman spectroscopy. A basic Raman spectrum of the pure Nafion membrane at RT displayed peaks at around 1058, 970, and 804 cm^{-1} originates from the stretching vibrations of S–O, C–O, and S–C in the Nafion side chains. The peaks of the asymmetric and symmetric stretching vibrations of CF_2, which occupies a large portion of the membrane, were observed around 1211 and 731 cm^{-1}, respectively. The C–C single bond peaks were observed around 1297 and 1376 cm^{-1}.

Di Noto et al. studied the effect of SiO_2 on relaxation phenomena and mechanism of ion conductivity of $Nafion/(SiO_2)_x$ composite membrane spectral analyses including FTIR and FT-Raman.[69] Their FT-Raman studies indicated that the fluorocarbon chains of Nafion hydrophobic domains assume the typical helical conformation structure with a $D(14\pi/15)$ symmetry. At the 400–1400 cm^{-1} region of the FT-Raman profiles, vibrational modes diagnostic for the PTFE backbone chain were confirmed. Table 9.1 illustrates the frequencies of Nafion PTFE domain vibrational modes together with their symmetry class, which coincides with that of a fluorocarbon chain in a $D(14\pi/15)$ factor group. It should be pointed out that as φ (the molar ratio of $SiO_2/–SO_3H$) increases, the narrow bands typical of crystalline PTFE spectra exhibit peak broadening owing to the crystallinity decrease in Nafion and in $[Nafion/(SiO_2)_x]$ composites. Furthermore, the location of fluorocarbon vibrations is unaffected by the SiO_2 concentration. The bands peaking at 986 and 970 cm^{-1} are attributed to the $\nu_{as}(C–O–C)$ and $\nu_s(C–O–C)$ modes, respectively, of ether side chains. It is noted that the profiles of these peaks are independent of the $[Nafion/(SiO_2)_x]$ composition. The band measured at ca. 1059 cm^{-1} in the FT-Raman spectra is ascribed to the symmetric stretchings of $\nu_s(SO_3^-)$ acid groups. The Raman spectra in the 1640–1590 cm^{-1} region present the bending vibrations typical of low associated water species. In accordance with the investigation carried out for water in vapor phases, the doublet of bands at 1594 and 1625 cm^{-1} is assigned correlatively to water species in dimer form, $(H_2O)_2$. As expected, this latter species of water aggregates has the lowest possible dipolar moment. These characteristics lead us to hypothesize that these water dimer species represent the water aggregates located in the bulk environments of the material where the dielectric constant is low, that is, near the hydrophobic Nafion host domains. Therefore, it is expected that water dimer species are dispersed in the hydrophobic channels with a diameter of 1–1.5 nm, which interconnect the polar hydrophilic cages of $[Nafion/(SiO_2)_x]$ composites.[69]

Di Noto group[70] recently also reported on the PEMs based on Nafion, SiO_2, and a protic ionic liquid (PIL), triethylammonium trifluoromethanesulfonate (TEATF), using vibrational studies of FTIR and FT-Raman. Their vibrational studies showed that (1) the neutralization of membranes by TEA and their impregnation with TEATF influence the conformational composition of fluorocarbon backbone chains

TABLE 9.1

FT-IR and FT-Raman Band Assignments of $[Nafion/(SiO_2)_x]$ Composite Membranes

Observed Band Spectra Detected Freq. $(cm^{-1})^a$

Nafion		Composite Membranes		Band Assignments[b]	Symmetry Class[c]
FT-IR	FT-Raman	FT-IR	FT-Raman		
—	—	465	—	$\gamma(Si-O)$	—
526 (m)	—	526 (m)	—	$\delta(CF_2)$	E_2
555 (m)	—	545 (m)	568 (w)	$t(CF_2)$	A_2
633 (m)	622 (vs,b)	638 (m)	622 (w,b)	$\omega,\delta(CF_2)$; (H_2O)	A_2
659 (sh,vw)	—	—	—	—	—
723 (w,b)	732 (vs)	717 (w,b)	732 (vs)	$\nu_s(CF_2)$	A_1
779 (vw,b)	—	779 (vw,b)	—	$\nu_s(CF_2)$	E_2
804 (w)	805 (m)	801 (w)	805 (m)	$\nu(C-S)$	—
970 (sh)	969 (w)	970 (sh)	969 (w)	$\nu_s(C-O-C)$	—
987 (s)	—	986 (s)	—	$\nu_{as}(C-O-C)$	—
1060 (vs)	1059 (m)	1062 (s)	1059 (m)	$\nu_s(SO_3^-)$	—
—	—	1088 (sh)	1089 (vw)	$\nu(Si-O)$	—
1160 (s)	1172 (vw)	1150 (s)	1172 (vw)	$\nu_s(CF_2)$	E_1
1207 (s)	—	1207 (s)	—	$\nu_{as}(CF_2)$, $\nu_{as}(SO_3^-)$	A_2
1216 (sh)	1217 (w)	1221 (sh)	1217 (w)	$\nu_{as}(CF_2)$	E_1
1300 (sh)	1297 (w)	1302 (sh)	1297 (w)	$\nu(C-C)$	E_2
1319 (sh)	—	1321 (sh)	—	$\nu(C-C)$	—
1384 (m)	1377 (m)	1382 (m)	1377 (m)	$\nu_s(C-C)$	A_1
1405 (sh)	—	1405 (m)	—	$\nu(S=O)$	—
—	1594 (m)	—	1594 (m)	$\delta(H_2O)_2$	—
—	1636 (sh)	—	1625 (sh)	$\delta(H_2O)_2$	—
1633 (w,b)	1649 (m)	1633 (w,b)	1649 (m)	$\delta(H_2O)_n$	—
1722 (sh)	—	1722 (sh)	—	$\delta[H_3O^+(H_2O)_n]$	—
1760 (sh)	—	1760 (sh)	—	$\delta[H_3O^+-SO_3^-](H_2O)_n$	—
2860 (w,b)	—	2863 (vw)	—	$\nu_3(H_3O^+)$	E
3400 (s,b)	3300 (vs,b)	3400 (s,b)	3311 (vs,b)	$\nu_{hy}(H_2O)$	—

Source: Di Noto, V. et al., *J. Phys. Chem. B*, 110, 24972, 2006.

[a] Relative intensities of observed bands are reported in parentheses: vs, very strong; s, strong; m, medium; w, weak; vw, very weak; b, broad; sh, shoulder.

[b] ν, stretching; δ, bending; ω, wagging; t, twisting; r, rocking; as, antisymmetric mode; s, symmetric mode.

[c] $D(14\pi/15)$ symmetry class of the helical perfluorocarbon backbone.

in hydrophobic domains of Nafion and (2) in the hybrid materials, triflate anions are arranged in micelle-like nanoparticles with a *core* consisting of interacting $-CF_3$ groups and a *shell* of hydrophilic sulfonic anion groups.

9.3.2 MORPHOLOGICAL STUDIES

Morphological study for hybrid/composite membranes is a very important method to investigate the microstructure and thus correlate with the properties of the composite polyelectrolytes. To better understand the *structure–property* relationship for PEMs, except for spectral analysis as mentioned earlier, microscopic studies such as field emission (FE) scanning electron microscopy (SEM), transmission electron microscopy (TEM), and atomic force microscopy (AFM) technologies are basically extensively utilized.

SEM or FE SEM are probably the most widely used microscopic method to analyze the surface or cross-sectional microstructure for composite polymer membranes.[26,46,55,56,64] Kannan et al. investigated the microstructure of phosphosilicate-modified Nafion membranes, and the typical SEM images are shown in Figure 9.6a and b, respectively.[66] The morphology of the blank Nafion as shown in Figure 9.6a is uniform and smooth. In contrast, the surface of the Nafion 1:1 hybrid membrane (Figure 9.6b) shows a phase-separated morphology with a near-uniform distribution of oxide particles in one phase and almost negligible amount in the other phase. The phase-separated morphology arises from the use of polar solvent such as ethanol, which is selectively compatible with only the hydrophilic (ionic domain) part of the copolymer, whereas the fluorinated polymer backbone in Nafion is noncompatible with this solvent. Also, a qualitative compositional analysis of the pure Nafion and Nafion hybrids was carried out to determine the elements present in the membrane. The energy-dispersive x-ray analysis (EDAX) spectrum of Nafion shows the presence of fluorine, carbon, oxygen, and sulfur. In hybridized Nafion membrane, the intensity of oxygen peak increased significantly showing the presence of organic methacrylate and inorganic oxides. The EDAX spectrum confirms the presence of P, Si, C, and O peaks and a lower intensity F peak than Nafion, indicating phosphosilicate inorganic particles reside in the membrane. The cross-sectional image of the hybrid membrane in comparison to pure Nafion indicates the presence of cavities due to the removal of particulate structures from the membrane during cross sectioning.[66]

Yao et al. synthesized a highly proton-conductive inorganic fiber/polymer hybrid PEM taking advantage of sulfated-zirconia ($S-ZrO_2$) fibers as long-range ionic channel inducer.[71] $S-ZrO_2$, recognized as the strongest solid super acid with a Hammett acid strength H_0 of -16.03 (whereas both Nafion and 100% H_2SO_4 show only -12), was selected as the inorganic component of hybrid PEMs. A complicated microscopic and spectral analysis is integrated to give an insight into the structure–property relationship. The calcined $S-ZrO_2$ mat has a compact structure with a fiber diameter of about 155 nm. All strong reflection peaks resulting from phase composition can be readily indexed to tetragonal ZrO_2, indicating the successful formation of $S-ZrO_2$. The formation of $S-ZrO_2$ was also confirmed by FTIR spectroscopy. $S-ZrO_2$ fiber mats were then used to construct $S-ZrO_2$ fiber/CPAMPS hybrid PEMs by immersing the mats in AMPS monomer, followed by in situ polymerization with AIBN as

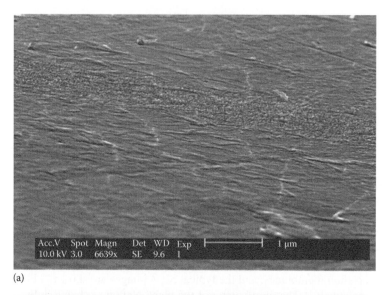

(a)

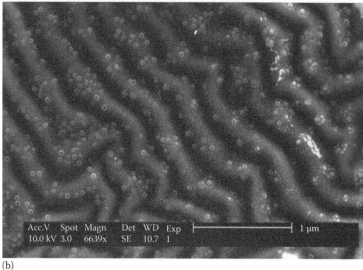

(b)

FIGURE 9.6 SEM images of surface morphology for (a) pure Nafion and (b) Nafion 1:1 hybrid. (From Kannan, A.G. et al., *J. Membr. Sci.*, 333, 50, 2009.)

the initiator and EGD as the cross-linker. FTIR spectrum and SEM images were obtained to verify the formation of S-ZrO$_2$/C-PAMPS hybrid PEMs.[71]

A group of FE SEM images for silica-containing Nafion hybrid composites are shown in Figure 9.7 to elucidate the silica distribution in Nafion,[72] where *US* means unmodified silica and *ILS* means ionic liquid–functionalized silica. As shown in Figure 9.7a through i, the rough surface in the virgin Nafion (VN) is attributed to the detachment of Nafion chains during cryofracture. Similar types of morphology can

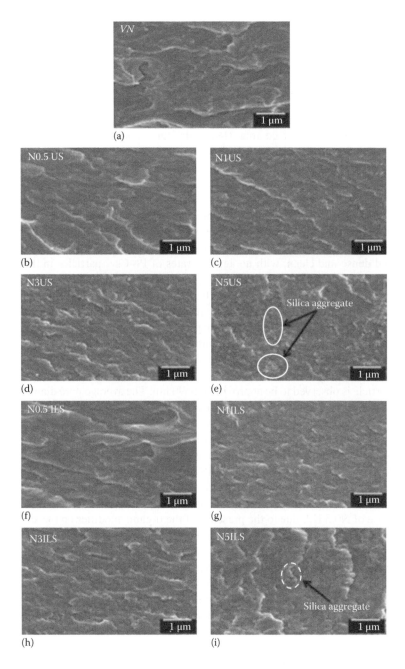

FIGURE 9.7 FE SEM photomicrographs of the functionalized silica-containing Nafion hybrid membranes. (a) Virgin Nafion, (b) Nafion membrane with 0.5% US, (c) Nafion membrane with 1% US, (d) Nafion membrane with 3% US, (e) Nafion membrane with 5% US, (f) Nafion membrane with 0.5% ILS, (g) Nafion membrane with 1% ILS, (h) Nafion membrane with 3% ILS, and (i) Nafion membrane with 5% ILS. (From Mishra, A.K. et al., *J. Mater. Chem.*, 22, 24366, 2012.)

be seen in all the nanocomposites. In the case of Nafion with US, the silica nanoparticles are uniformly distributed at low silica content (up to 1 wt% US content), with
an increasing aggregation tendency for N3US and N5US. However, functionalization
of silica with ionic liquid leads to an increased compatibility between the Nafion
and ILS. Hence, uniform distribution of ILS can be seen for N3ILS with a loosely
aggregated morphology for N5ILS compared to the tightly aggregated morphology
of N5US. It is worth mentioning here that the aggregation tendency of nanofillers
reduces their exposed surface area. Hence, the surface area of interaction between
silica and Nafion in N5US and N5ILS is expected to be quite low compared to the
respective composites with a well-dispersed morphology.[72]

Zhang et al. developed an inorganic/organic self-humidifying membrane based
on SPEEK hybrid with $Cs_{2.5}H_{0.5}PW_{12}O_{40}$ catalyst ($Pt–Cs_{2.5}$).[73] To determine the
$Pt–Cs_{2.5}$ catalyst distribution along the membrane cross section, the Cs/S elemental profiles across the sample thicknesses were also carried out simultaneously by
energy-dispersive x-ray detector. The cross-sectional $SPEEK/Pt–Cs_{2.5}$ membrane
appeared dense and clean, with no agglomerates of $Pt–Cs_{2.5}$ particles in the whole
membrane cross section, implying that $Pt–Cs_{2.5}$ particles were not recrystallized into
large particles after incorporating with SPEEK, but were highly dispersed throughout the polymer matrix. The Cs element also showed a uniform distribution in the
whole membrane from cross-sectional image.[73]

The microstructure of pure US (as synthesized) was also investigated by Mishra
and Lee et al. via FE TEM characterization.[72] The formation of highly ordered hexagonal silica with cylindrical nanochannels is evident, and the size of the individual
silica particle is observed to be approximately 90 nm. The average pore diameter was
calculated to be approximately 3.4 nm. The FETEM images of N0.5 US and N0.5
ILS, which were chosen in order to observe the nanoscale distribution of silica in the
Nafion matrix at very low silica content, represent that even at a low silica content,
most of the US nanoparticles can be seen aggregated together. The size of the silica
particles in N0.5 US varies from 90 to 187.4 nm. However, the silica particles are
well separated from each other in the case of N0.5 ILS with a particle size range of
90–140 nm. This is possibly due to the increased interaction between ILS and Nafion
in the case of N0.5 ILS due to the presence of the organic modifier on the silica surface. However, due to the dominance of intraparticle interaction in Nafion/US, most
of the silica particles form aggregates.[72]

Recently, Wang et al. synthesized a series of $Nafion/CeO_2$ hybrid composites with
different CeO_2 loading via a self-assemble route, which is different from traditional
sol–gel method.[74] Figure 9.8 exhibits the high-resolution TEM images of the resulting
composite membranes in comparison with that by conventional sol–gel method. The
average particle size was about 3–6 nm for the $Nafion/CeO_2$ (1 wt%), $Nafion/CeO_2$
(3 wt%), and $Nafion/CeO_2$ (5 wt%) nanoparticles. The $Nafion/CeO_2$ nanoparticles
prepared by the self-assembly route show a uniform distribution and are comparable
to sol–gel-derived CeO_2 nanoparticles. The formation of uniformly distributed and
nanosized $Nafion/CeO_2$ particles is most likely due to the strong stabilization and
steric hindrance effects of the self-assembled Nafion ionomers on the CeO_2 nanoparticles (Figure 9.8f). In the presence of Nafion ionomers, self-assembly would occur
between the positively charged CeO_2 and the negatively charged SO_3^- end groups of

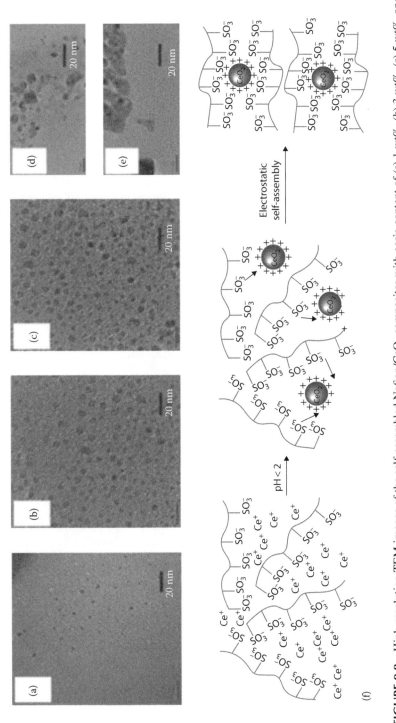

FIGURE 9.8 High-resolution TEM images of the self-assembled Nafion/CeO$_2$ composites with ceria content of (a) 1 wt%, (b) 3 wt%, (c) 5 wt%, (d) 10 wt%; image of the Nafion/CeO$_2$ composites prepared from conventional sol–gel method with ceria content of (e) 5 wt%; schematic diagram of the formation of (f) self-assembled Nafion/CeO$_2$ composites. (From Wang, Z. et al., *J. Membr. Sci.*, 421–422, 201, 2012.)

Nafion ionomers by the electrostatic force. In the case of sol–gel method, because water acts both as reaction reagent and solvent in the hydrolysis reaction, water is essential for the hydrolysis reaction and for the cerium polymerization. However, excess water in the solution would dilute the Nafion content, reducing the self-assembly between the ceria particles and Nafion ionomers. This leads to the significant reduction in the steric hindrance effect of Nafion ionomers, resulting in the significant grain growth of the ceria particles.[74]

Jiang group[75] developed silica/poly(divinylbenzene)-based polymeric microcapsules (PMCs) modified with three kinds of functional groups as carboxylic acid (PMC-C), sulfonic acid (PMC-S), and pyridyl groups (PMC-N), about which the PMCs displayed enhanced water retention capability even under low RH of 20%. The PMCs were well designed to have core–shell structure with controllable shell thickness, and then composite polyelectrolytes were fabricated from PMCs and chitosan (CS). Figure 9.9 displays the TEM images of the functionalized PMCs.

As shown in Figure 9.9, all the PMCs have well-defined capsular structure, dense shell, and large lumen. The PMCs are quite uniform with a lumen diameter of around 400 nm. The shell thicknesses of PMC-C, PMC-S, and PMC-N are around 127,

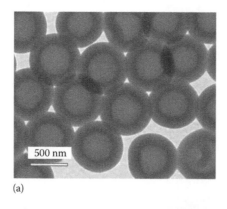

(a)

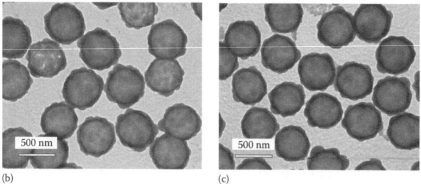

(b)   (c)

FIGURE 9.9 TEM images of the PMCs: (a) PMC-C, (b) PMC-S, and (c) PMC-N. (From Wang, J. et al., *Adv. Funct. Mater.*, 21, 971, 2011.)

80, and 90 nm, respectively, all of which are robust enough to preserve the hollow structure. The PMCs were then embedded into the CS matrix to fabricate composite membranes. It is considered that the PMCs maintained their pristine structure and homogeneous dispersion within the membrane matrix owing to their hydrogen bonding and electrostatic interactions and hence their favorable compatibility with CS.[75]

AFM is considered an appropriate technique to analyze the phase distribution of polymer/inorganic composite materials. The roughness of the Nafion and the Nafion–TiO_2 hybrid membranes from AFM images can be characterized via root-mean-square (RMS) value. Nafion hybrid membranes with TiO_2 usually exhibit a surface topography composed of groups of varying size TiO_2 with irregular shapes. The irregular shape of nanoparticles is caused by low temperatures, and the kinetic energy is not sufficient to induce the coalescence of grains. Surface morphological changes were observed in the hybrid Nafion membranes with 7% of TIP nanoparticles, which are well separated and become more visible.[29] The high contrast of the color distribution in narrow diameters suggests that the *hard* TiO_2 is well dispersed in the polymer electrolyte, while the relatively featureless boundary implies that many of the TiO_2 nanoparticles are covered by the *soft* polymer ionomers. The harmonization between the inorganic phase and polymer phase decreases the contrast of the inorganic domains. The roughness value of pure Nafion membrane is around 0.559 mm, which increases to 0.630 mm for hybrid Nafion with 5% TIP and corresponding 1.087 mm for 7% TIP. The surface of the composite membrane is clearly very smooth. In the presence of Nafion ionomers, self-assembly would occur between the positively charged TiO_2 and the negatively charged SO_3^- end groups of Nafion ionomers by the electrostatic force. The self-assembly of the Nafion and TiO_2 nanoparticles also gives the composite structure a good compatibility on the interface.[29]

Xing and Na group[76] reported on surface morphologies of SPEEK and aminopropyltriethoxysilane (KH550) hybrid membranes doped with PWA. The AFM spectra of their hybrid membranes are shown in Figure 9.10. The roughness of the pristine SPEEK membrane surface indicated that the height contrast between hydrophilic and hydrophobic domains is prominent. Comparing the AFM images of different

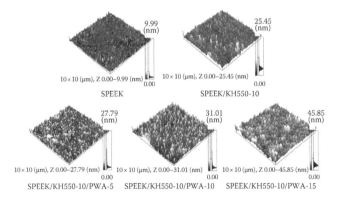

FIGURE 9.10 AFM images of SPEEK and their hybrid membranes doped with PWA. (From Fu, T. et al., *Solid State Ionics*, 179, 2265, 2008.)

membranes, the surface morphology of the membranes changed significantly after incorporation with KH550 and PWA. With the increment of the weight ratio of KH550 and PWA, the surface became gradually rougher.

Xu et al. studied the effect of acids and water addition on morphology and proton conduction in sol–gel-derived acid–base polysiloxane composites based on poly(N-(2-aminoethyl)-3-aminopropyl-trimethoxysilane) (PAAS).[77] Figure 9.11 shows the AFM topographic images, and Table 9.2 lists the RMS roughness of the PAAS-R membranes prepared by varied water addition. RMS roughness represents the inhomogeneity of the membrane. With water addition increasing, the morphology of PAAS–HSO$_4$ was greatly changed as follows: (a) large particles (about 1 μm) → (b) small particles (about 200 nm) → (c and d) (network with branching points) and

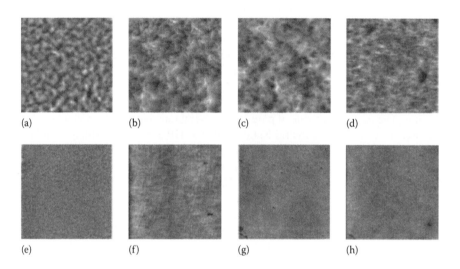

(a)   (b)   (c)   (d)

(e)   (f)   (g)   (h)

FIGURE 9.11 AFM topographic images of (a–d) PAAS–HSO$_4$ and (e–h) PAAS–H$_2$PO$_4$ membranes formed by different water additions (a, e, 25 mL; b, f, 50 mL; c, h, 250 mL; d, h, 500 mL). (From Zeng, S. et al., *Solid State Ionics*, 181, 1408, 2010.)

TABLE 9.2

RMS Roughness of the PAAS-R Membrane Prepared by Different Water Additions

	RMS Roughness (Å)	
Water Addition (mL)	PAAS–HSO$_4$	PAAS–H$_2$PO$_4$
25	417	28
50	288	6
250	166	3
500	142	3

Source: Zeng, S. et al., *Solid State Ionics*, 181, 1408, 2010.

thus led to RMS roughness decreasing visibly. In contrast, that of PAAS–H$_2$PO$_4$ was always flat and clean, as the pH of the PAAS–H$_2$PO$_4$ sol was insensitive to the water addition.

9.3.3 Physical Properties

The solubility and viscosity of the hybrid/composite membranes are generally dependent on the polymer matrix and the degree of compatibility and dispersibility between the polymer and inorganic components. Usually, the polymer matrix is soluble in isopropanol (for Nafion families) or dimethyl sulfoxide, dimethylformamide, dimethylacetamide, N-methyl-2-pyrrolidone, or m-cresol, for most of the hydrocarbon polymers such as SPEEK, sulfonated polyether sulfone (SPS), sulfonated poly(arylene ether sulfone) (SPAES), SPAEK, and SPI.

Thermal and mechanical properties for hybrid/composite membranes are very important in relation to membrane stability, which will affect the lifetime under practical operating conditions. Thermogravimetry analysis (TGA) is usually employed to investigate thermal property, and dynamic mechanical analysis (DMA) is generally used to characterize mechanical property of membranes. Nafion families are well known to be unstable at temperatures higher than 100°C due to their low glass transition temperature (T_g); however, modifications with inorganic filler showed increased mechanical stability due to the increased T_gs resulting from the interactions between sulfonic acid group and inorganic fillers.[16] Table 9.3 lists the T_g values for various composite membranes as determined using DMA. Other candidates based on hydrocarbon polymer matrix composited with inorganic fillers such as Al$_2$O$_3$, SiO$_2$, ZrO$_2$, and TiO$_2$ generally display reasonable high thermal and mechanical properties. Table 9.4 illustrates an overview of thermal and mechanical properties for various hybrid/composite membranes based on the results reported in literatures.

A typical thermogravimetric profile of polymer membranes is usually represented by the relationship between polymer weight loss/residue and temperature.[61] It is generally observed that Nafion has three obvious weight losses during heating process, namely I, II, and III. At the lowest temperatures ($T < 100$°C), a slight weight loss occurred gradually, which is ascribed to desorption of the residual water from the

---

TABLE 9.3
Glass Transition Temperatures (T_g) for Various Composite Membranes Determined by DMS

Membrane Material	T_g (°C)
Nafion (recast)	92
Nafion/Al$_2$O$_3$	95
Nafion/SiO$_2$	100
Nafion/ZrO$_2$	112
Nafion/TiO$_2$	120

Source: Adjemian, K.T. et al., *Chem. Mater.*, 18, 2238, 2006.

TABLE 9.4
Thermal and Mechanical Properties of Hybrid/Composite Membranes

Hybrid/Composite Membrane	Polymer Matrix	Inorganic Filler and Content (wt%)	Decomposition Temperature (°C)[a]	Elastic/Storage Modulus (MPa)[b]	References
Nafion	Nafion	—	320	300	[56,61]
Nafion/SiO$_2$	Nafion	SiO$_2$ (10%)	350	—	[56]
Nafion/SiO$_2$/PWA	Nafion	SiO$_2$/PWA (10%)	340	—	[56]
Nafion/WO$_3$	Nafion	WO$_3$ (10%)	280	—	[56]
Nafion/TiO$_2$	Nafion	TiO$_2$ (10%)	315	—	[56]
Nafion/TiO$_2$	Nafion	TiO$_2$ (15%)	170	500	[61]
Nafion/(ZrHf)$_x$	Nafion	ZrHf (10%)	200	—	[49]
Nafion/(ZrHf)$_x$	Nafion	ZrHf (15%)	200	—	[49]
Nafion/(SiHf)$_x$	Nafion	SiHf (10%)	200	—	[49]
Nafion/(SiHf)$_x$	Nafion	SiHf (15%)	200	—	[49]
Nafion/SiO$_2$	Nafion	SiO$_2$ (15%)	170	200	[70]
Nafion/SiO$_2$(TEA)/(TEATF)[c]	Nafion	SiO$_2$ (15%)	330	200	[70]
SPEEK/KH550[d]	SPEEK	KH550 (10%)	299	—	[76]
SPEEK/KH550/PWA	SPEEK	KH550/PWA(10%/5%)	298	—	[76]
SPEEK/THP-TiO$_2$[e]	SPEEK	THP-TiO$_2$ (5%)	215	1400±500	[78]
SPEEK/Soil-TiO$_2$[e]	SPEEK	Soil-TiO$_2$ (5%)	210	880±20	[78]
SPEEK/IMCs-HPW[f]	SPEEK	IMCs-HPW (15%)	270	—	[79]
SPS/Si	SPS	SiO$_2$ (20%)	250	—	[20]
SPS/Si (P)	SPS	SiO$_2$ (P) (20%)	250	—	[20]
SPAES/SiO$_2$	SPAES	SiO$_2$ (6%)	320	835	[80]
SPAES/SiO$_2$	SPAES	SiO$_2$ (10%)	280	—	[81]

(Continued)

TABLE 9.4 (Continued)
Thermal and Mechanical Properties of Hybrid/Composite Membranes

Hybrid/Composite Membrane	Polymer Matrix	Inorganic Filler and Content (wt%)	Decomposition Temperature (°C)[a]	Elastic/Storage Modulus (MPa)[b]	References
SPI/SMCM[g]	SPI	SMCM (20%)	—	2500	[82]
		SMCM (40%)	—	1600	[82]
SPI/SMSN[h]	SPI	SMSN (7%)	290	667	[83]
SPI/mSiO$_2$[i]	SPI	mSiO$_2$ (10%)	250	—	[84]
PVA/PWA/(P$_2$O$_5$/SiO$_2$)	PWA/(P$_2$O$_5$/SiO$_2$)	PVA (20%)	250	—	[58]
PVA/Si	PVA	SiO$_2$ (50%)	250	—	[85]
PGA/xH$_3$PO$_4$[j]	PGA	H$_3$PO$_4$ (x=1,2,3[j])	200	—	[86]
PMMA/SiO$_2$-P$_2$O$_5$[k]	PMMA	SiO$_2$-P$_2$O$_5$ (40%)	130–310	—	[87]
Poly(VDF-*ter*-PFSVE-*ter*-VTEOS)/SiO$_2$[l]	terpolymer	SiO$_2$ (35%)	185–220	—	[88]
PBI/Nafion/SiO$_2$	PBI/Nafion	SiO$_2$ (5%)	300	8.38[m]	[89]

[a] From TGA reported in literatures, usually refers to the first decomposition of sulfonic acid group.

[b] From dynamic mechanical analysis reported in literatures at RT.

[c] TEA means TEA and TEATF refers to PILs of triethylammonium trifluoromethanesulfonate.

[d] KH550 means aminopropyltriethoxysilane.

[e] THP means tri(hydroxymethyl)-propane and Soil refers to silicone oil.

[f] IMCs-HPW means imidazole microcapsules loaded with PWA.

[g] SMCM means sulfonated Si-MCM-41.

[h] SMSN means sulfonated mesoporous silica nanoparticle.

[i] mSiO$_2$ means mesoporous silica.

[j] PGA/xH$_3$PO$_4$ means H$_3$PO$_4$-doped polysioxane-amide-1,2,4-triazole (x=H$_3$PO$_4$/3-amino-1,2,4-triazole, mole ratio).

[k] PMMA means poly(methyl methacrylate).

[l] Poly(VDF-*ter*-PFSVE-*ter*-VTEOS) means poly(vinylidene fluoride-*ter*-perfluoro(4-methyl-3,6-dioxaoct-7-ene sulfonyl fluoride)-*ter*-vinyltriethoxysilane).

[m] The data mean the tensile strength.

membranes. The initial water desorption is more pronounced in the pristine recast Nafion, in accordance with the larger WU of this membrane in comparison with the hybrid [Nafion/(TiO$_2$F)$_x$] materials. I, II, and III are attributed to the degradation of sulfonic acid groups ($100°C < T < 250°C$), the degradation of the perfluoroetheral side chains of Nafion ($300°C < T < 380°C$), and the decomposition of the perfluorinated backbone chains of Nafion ($T > 400°C$), respectively. With respect to pristine recast Nafion, I and II of hybrid [Nafion/(TiO$_2$F)$_x$] displayed a relatively earlier weight loss or lower temperature tolerance (I, 170°C of hybrid materials versus 210°C of pristine Nafion; II, 320°C of hybrid materials versus 355°C of pristine Nafion). This evidence is interpreted as follows: the TiO$_2$F nanofiller acts as a catalyst, promoting the I and II thermal degradation events. Thus, it can be assumed that the TiO$_2$F nanofiller is interacting with the −SO$_3$H-tipped perfluoroethereal side chains of the Nafion host polymer.[61] Similar results in relation to TiO$_2$ containing hybrid/composite membranes are also found in other literatures as listed in Table 9.4.[56,78]

The TGA thermograms of the pure Nafion and inorganic additive–incorporated membranes are shown in Figure 9.12.[56] It is seen that the decomposition temperature of the composite membranes shifts with the nature of inorganic content. The pure Nafion, Nafion/SiO$_2$, and Nafion/SiO$_2$/PWA started to decompose at temperatures of 320°C, 350°C, and 340°C (see Table 9.4), respectively, whereas the TiO$_2$ and WO$_3$ composite membranes decompose relatively at lower temperatures, that is, at 280°C and 315°C (see Table 9.4), respectively. This indicates that the TiO$_2$ and WO$_3$ fillers have accelerated the decomposition of the membrane at an early temperature than that of the pure Nafion membrane, which is inconsistent with that in literatures.[61] On the other hand, the desulfonation temperature of the hybrid membranes shifted to higher temperatures as the SiO$_2$ loading increased due to the Si−O−Si cross-linking structure, thus increasing the thermal stability of the membrane, which is in good agreement with those reported in literatures.[20,70,80–84]

Di Noto et al. give an example of DMA analysis for Nafion/TiO$_2$ hybrid membranes, which illustrates the relationship between storage modulus (E') and temperature.[61] The typical Nafion behavior is characterized by (1) a decrease in both E' and E'' as T is raised and (2) a collapse of the mechanical properties of both pristine recast Nafion and [Nafion/(TiO$_2$F)$_{0.05}$] at $T > 100°C$; on the other hand, [Nafion/(TiO$_2$F)$_{0.10}$] and [Nafion/(TiO$_2$F)$_{0.15}$] maintain appreciable mechanical properties at temperatures as high as 210°C. With respect to the pristine recast Nafion, the hybrid [Nafion/(TiO$_2$F)$_x$] membranes are characterized by higher E' values at both $T = 25°C$ and $T = 100°C$. For all the hybrid [Nafion/(TiO$_2$F)$_x$] membranes, E' values are around 500 MPa (see Table 9.4), while the corresponding E' value for pristine Nafion is much lower and equal to ca. 300 MPa.[61]

The hybrid membranes based on hydrocarbon polymer matrix with a cross-linking network structure are also reported to show improved tensile strength and Young's modulus compared with the pristine membrane.[80,81] For example, the tensile strength of the SPASE/SiO$_2$ (10 wt%) hybrid membrane increased by 31% compared to the pristine SPAES. With the increase in SiO$_2$ content, both the tensile strength and modulus of the corresponding hybrid membranes also increased.[80] However, the content of the inorganic filler should also be controlled carefully because a large

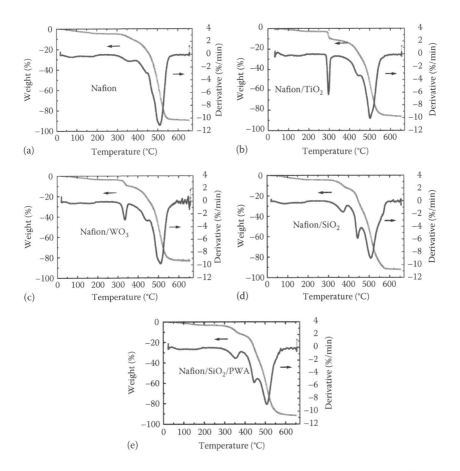

FIGURE 9.12 TGA thermograms and associated derivative curves, taken at a heating rate of 10°C/min, under N_2, for (a) Nafion, (b) Nafion/TiO_2, (c) Nafion/WO_3, (d) Nafion/SiO_2, and (e) Nafion/SiO_2/PWA. (From Shao, Z.G. et al., *Solid State Ionics*, 177, 779, 2006.)

excess of addition of inorganic filler will also bring mechanical loss or brittleness for the hybrid membrane in some cases.[20,82]

9.3.4 Ion-Exchange Capacity, Water Uptake, and Swelling Ratio

Addition of inorganic fillers to the pristine membranes will generally result in decrease in ion-exchange capacity (IEC) except that the inorganic fillers are also sulfonated before mixing with PEMs. However, the most attractive property of inorganic fillers should be their high water retention capability, which has a significant influence on the proton conductivity. Table 9.5 lists the IEC, WU, and proton conductivity (σ) for various hybrid/composite membranes.

The WU is generally reported using water weight percent of dry membranes according to Equation 9.1:

TABLE 9.5

IEC, WU, and Proton Conductivity (σ) for Hybrid/Composite Membranes

Hybrid/ Composite Membrane	Inorganic Filler Content (wt%)	IEC[a] (mequiv. g^{-1})	WU[b] (g 100 g^{-1})	Swelling Ratio (%)[c]	σ[d] (mS cm^{-1})	References
Nafion 115	—	0.91	26	—	8.1[e]	[56]
Nafion/SiO$_2$	SiO$_2$ (10%)	—	34	—	10.7[e]	[56]
Nafion/SiO$_2$/ PWA	SiO$_2$/PWA (10%)	—	38	—	26.7[e]	[56]
Nafion/WO$_3$	WO$_3$ (10%)	—	37	—	10.1[e]	[56]
Nafion/TiO$_2$	TiO$_2$ (10%)	—	34	—	8.9[e]	[56]
Nafion/TiO$_2$	TiO$_2$ (15%)	0.80	19	—	—	[61]
Nafion/(ZrHf)$_x$	ZrHf (10%)	0.80	35	—	—	[49]
Nafion/(ZrHf)$_x$	ZrHf (15%)	0.78	36	—	—	[49]
Nafion/(SiHf)$_x$	SiHf (10%)	0.82	40	—	—	[49]
Nafion/(SiHf)$_x$	SiHf (15%)	0.80	43	—	—	[49]
Nafion/SiO$_2$	SiO$_2$ (15%)	—	—	—	40[h]	[70]
Nafion/ SiO$_2$(TEA)/ (TEATF)[f]	SiO$_2$ (15%)	—	33[g]	—	4.7[h]	[70]
SPEEK/KH550[i]	KH550 (10%)	1.90	46	—	155[j]	[76]
SPEEK/KH550/ PWA	KH550/ PWA(10%/5%)	1.85	46	—	164[j]	[76]
SPEEK/ THP-TiO$_2$	THP-TiO$_2$ (5%)	2.20	77	—	—	[78]
SPEEK/ Soil-TiO$_2$	Soil-TiO$_2$ (5%)	2.20	46	—	200[k]	[78]
SPEEK/ IMCs-HPW	IMCs-HPW (15%)	1.55	78	—	31.6	[79]
SPS/Si	SiO$_2$ (20%)	0.85	23	—	51.0	[20]
SPS/Si (P)	SiO$_2$ (P) (20%)	0.96	29	—	63.6	[20]
SPAEK/SiO$_2$	SiO$_2$ (5%)	1.93	28 (43[l])	7.4 (1.05[l])	72.0 (125[l])	[36]
	SiO$_2$ (10%)	1.52	27 (35[l])	6.7 (10.2[l])	67.0 (118[l])	[36]
	SiO$_2$ (20%)	1.31	22 (31[l])	3.2 (4.8[l])	48.5 (1[l])	[36]
SPAES/SiO$_2$	SiO$_2$ (6%)	—	28	9	47.0	[80]
		—	43[m]	13.2[m]	96.0[l]	[80]
	SiO$_2$ (10%)	—	25	8	44.0	[80]
		—	38[m]	12.5[m]	90.0[l]	[80]
SPAES/SiO$_2$	SiO$_2$ (10%)	1.63	44	—	56.0	[81]
SPI/SMCM	SMCM (20%)	1.90	90	4.4	152	[82]
	SMCM (40%)	1.93	96	4.7	124	[82]
SPI/SMSN	SMSN (7%)	2.62	51	—	26.5 (42.5[l])	[83]

(Continued)

TABLE 9.5 (*Continued*)

IEC, WU, and Proton Conductivity (σ) for Hybrid/Composite Membranes

Hybrid/ Composite Membrane	Inorganic Filler Content (wt%)	IEC[a] (mequiv. g⁻¹)	WU[b] (g 100 g⁻¹)	Swelling Ratio (%)[c]	σ[d] (mS cm⁻¹)	References
SPI/mSiO$_2$	mSiO$_2$ (10%)	1.64	39	—	153 (187[l])	[84]
	mSiO$_2$ (20%)	1.55	42	—	171 (204[l])	[84]
PVA/PWA/ (P$_2$O$_5$/SiO$_2$)	PVA (15%)	2.56	56	71	—	[58]
	PVA (20%)	1.38	71	89	—	[58]
PVA/Si	SiO$_2$ (50%)	1.86	91	—	52.2	[85]
	SiO$_2$ (70%)	1.95	123	—	55.0	[85]
PGA/xH$_3$PO$_4$	H$_3$PO$_4$ ($x=1$)	—	—	—	1.48[n]	[86]
	H$_3$PO$_4$ ($x=2$)	—	—	—	10.7[n]	[86]
	H$_3$PO$_4$ ($x=3$)	—	—	—	14.3[n]	[86]
PMMA/ SiO$_2$-P$_2$O$_5$	SiO$_2$-P$_2$O$_5$ (35%+5%)	—	3.2	—	186[o]	[87]
	SiO$_2$-P$_2$O$_5$ (30%+10%)	—	6.0	—	380[o]	[87]
Poly(VDF-*ter*-PFSVE-*ter*-VTEOS)/SiO$_2$	SiO$_2$ (35%)	0.50	85	—	21 (11[p])	[88]
PBI/Nafion/SiO$_2$	SiO$_2$ (5%)	—	42[q]	7	—	[89]

[a] Determined by titration.

[b] Measured for fully hydrated membranes at RT.

[c] Measured at RT until others mentioned.

[d] Measured at RT at 100% RH.

[e] Measured at 110°C and 70% RH.

[f] TEA means TEA and TEATF refers to PILs of triethylammonium trifluoromethanesulfonate.

[g] Measured for fully hydrated membranes absorbed with TEATF ionic liquid.

[h] Measured at 105°C and 100% RH.

[i] KH550 means aminopropyltriethoxysilane.

[j] Measured at 80°C and 100% RH.

[k] Measured at 100°C and 90% RH.

[l] Measured at 80°C.

[m] Measured at 100°C.

[n] Measured at 200°C and anhydrous condition.

[o] Measured at 50°C and 90% RH.

[p] Measured at 120°C and 90% RH.

[q] Measured at 60°C.

$$\text{Water uptake (wt\%)} = \frac{W_{\text{wet}} - W_{\text{dry}}}{W_{\text{dry}}} \times 100\% \qquad (9.1)$$

where W_{wet} and W_{dry} refer to the weight of the swelled and dry membrane, respectively.

The swelling ratio is usually calculated from the change of either membrane diameter or membrane length according to Equation 9.2:

$$\text{Swelling ratio (\%)} = \frac{L_{\text{wet}} - L_{\text{dry}}}{L_{\text{dry}}} \times 100\% \qquad (9.2)$$

where L_{wet} and L_{dry} refer to the length/diameter of the swelled and dry membrane, respectively.

The incorporation of inorganic fillers containing silica, titanium, and zirconium generally leads to a slight decrease in IEC due to the dilution effect and the decrease in proportion to the filler content.[36,49,84] However, the IEC variation versus inorganic filler content changes oppositely for PVA-based hybrid membranes.[58,85] For example, the phosphonic acid–functionalized PVA–silica composite membranes displayed a slightly IEC increase from 1.90 to 1.95 mequiv. g^{-1} when the silica content ranged from 40 to 70 wt%.[85] On the other hand, most of the hybrid membranes exhibit an increase in WU with increasing inorganic filler content.[82,84,85] The increase of WU can be attributed to (1) hygroscopic interior for the inorganic filler themselves and (2) the plasticization effect of the nanoparticles, which will result in loose packing of the polymer chains. It is known that besides the sulfonic acid groups in matrix, the hygroscopic Si–OH groups could also absorb and hold water molecules, which increased the content of bound water. In addition, the nanosized SiO_2 particles had large specific surface areas, which induced high WU of hybrid membranes compared with that of the pristine membrane. However, there are also studies claiming the existence of an optimized content of inorganic fillers that the hybrid membranes thus display a peak WU. As shown in Table 9.5, the SPAES/SiO_2 hybrid membrane containing 6 wt% SiO_2 displayed a slightly larger WU (28%) than that (25%) for hybrid membrane containing 10 wt% SiO_2.[80] And the difference becomes larger under high temperatures. It is considered that when the SiO_2 content was up to 10wt%, the cross-linking density was increased also, which increased the interaction of the polymers and restricted the mobility/flexibility of the polymer chains.[80]

Membrane swelling is an important parameter with respect to membrane stability and fuel cell durability under practical operations. It is no doubt that larger WU is favorable for proton transport but generally leads to more serious membrane swelling (see Table 9.5). It is very interesting that a general trend of the reduction in swelling ratio with the increasing SiO_2 loading from 0 to 10 wt% was found, but an opposite trend was observed for the WU, for SPAES/SiO_2 hybrid membranes as reported by Ren et al.[80] The reason is that the cross-linking structure increased the interaction of the polymer molecules and restricted the movement of polymer chains in the hybrid membranes. When the membrane was soaked in the water, it swelled due to the hygroscopic effect of the sulfonic acid groups. During the membrane preparation process, the chains of the polymer were fixed up when the solvents were removed at high temperature, but a certain gap existed between the polymer and inorganic particles due to the further condensation between the residual Si–OH of the SiO_2 particles in the hybrid membranes, which became the water storage space. Therefore, the hybrid membranes exhibited higher WU but lower swelling ratio.[80]

9.3.5 PROTON CONDUCTIVITY

The proton conductivities of Nafion and inorganic additive–containing Nafion composite membranes under various temperature and RH are also partly summarized in Table 9.5.

A typical relationship between proton conductivity and temperature for fully hydrated Nafion-based membranes is shown in Figure 9.13a, and the related RH

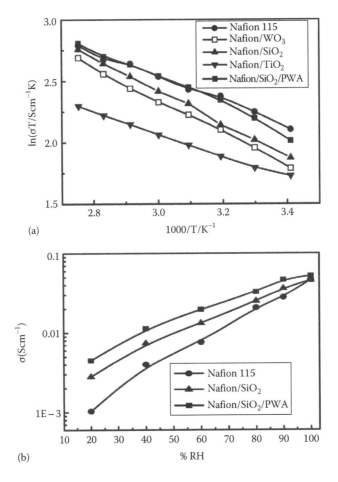

(a)

(b)

FIGURE 9.13 Proton conductivity of Nafion, Nafion/TiO_2, Nafion/WO_3, Nafion/SiO_2, and Nafion/SiO_2/PWA composite membranes as functions of (a) temperature under 100% RH and (b) RH at 100°C. (From Shao, Z.G. et al., *Solid State Ionics*, 177, 779, 2006.)

dependence of proton conductivity is shown in Figure 9.13b.[56] As shown in Figure 9.13a, the proton conductivity of the inorganic impregnated membranes is lower than that of the Nafion 115 membrane in the lower temperature range. In contrast, the proton conductivity of the SiO_2, WO_3, and SiO_2-/PWA-containing membranes is found to be close to that of the Nafion 115 at the high temperature range. On the other hand, the proton conductivity of TiO_2-impregnated membrane is found to be low even at this high temperature range. The activation energies of Nafion 115, Nafion/SiO_2, Nafion/WO_3, Nafion/TiO_2, and Nafion/SiO_2/PWA are calculated to be as 8.3, 9.6, 11.0, 8.4, and 11.2 kJ mol^{-1}, respectively. The higher activation energy values observed for the composite membranes might be caused by the presence of inorganic additives in the membrane. As shown in Figure 9.13b, the proton conductivity of the Nafion/SiO_2 composite membrane is higher than that of Nafion 115 under low RH. This can be explained by the fact that the silica particles can retain water even at high temperature and at low RH; this property can help to prevent the dehydration of the membrane during the operation of the fuel cell. It can also be noted that the proton conductivity of the Nafion/SiO_2/PWA composite membrane is higher than that of the Nafion/SiO_2 composite membrane under low RH. It can be explained by the higher uptake of water by the Nafion/SiO_2/PWA composite membrane (see Table 9.5).

Shahi also reported on the proton conductivities for SPS/SiO_2 hybrid membranes compared with those of pristine membrane as well as Nafion 117.[20] The Arrhenius plots for various membranes are shown in Figure 9.14. The activation energy of these

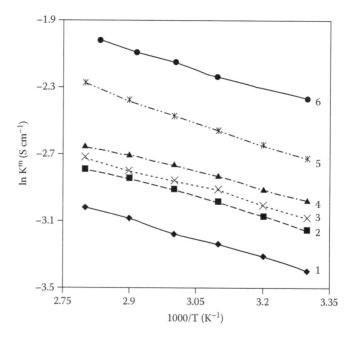

FIGURE 9.14 Arrhenius plots for different membranes. (1) SPS, (2) SPS-Si/10, (3) SPS-Si(P)/10, (4) SPS-Si/20, (5) SPS-Si(P)/20, and (6) Nafion 117. (From Shahi, V.K., *Solid State Ionics*, 177, 3395, 2007.)

hybrid membranes ranged from 5.43 to 7.47 kJ mol^{-1}, which is lower than those for Nafion-based hybrid membranes.[56]

Okamoto et al.[82] studied the proton conductivities for SPI/SiO$_2$ hybrid membranes at different humidification conditions at 60°C. They found that under 100% RH, the hybrid membrane showed a decrease in proton conductivity compared with that of pristine SPI (from 178 to 124 mS cm^{-1}); however, the conductivity at low RHs of 30%–50% hardly changed. It is considered that in the highly swollen hybrid membranes, the doped silica particles act as an insulating filler to reduce the conductivity by the tortuosity effect. On the other hand, under the low humidification of 30%–50% RH, the silica particles may act as water adsorbate to compensate the conductivity.

Recently, Ren et al. developed a cross-linked SPAES/SiO$_2$ hybrid membrane and the proton conductivity as a function of SiO$_2$ content.[80] At the same temperature, the proton conductivities of the hybrid membranes increased with the increasing SiO$_2$ content in the range of 0–6 wt%, which is consistent with the WU as mentioned earlier (see Table 9.5). In general, there are two primary conductive mechanisms describing proton diffusion through the membranes, namely, vehicle mechanism and Grotthus mechanism (hopping). It is probable that the bound water participates via the Grotthus mechanism, and the free water takes part mostly via the vehicle mechanism.[91,92] When SiO$_2$ is introduced into the hybrid membranes, the IEC is reduced due to the dilution effect of SiO$_2$ on the sulfonic acid groups. However, the hygroscopic Si–OH group and cross-linking networks also held some bound water that could compensate for the loss of free water to a certain degree. The bound water facilitates proton transport through the Grotthus mechanism, which contributes to enhancing the proton conductivity. In addition, the positively charged –Si–OH groups in the acidic medium perhaps also endow the SiO$_2$ particles with proton conduction capability. As a result, the proton conductivities of the hybrid membranes are improved. However, when SiO$_2$ content is increased to 10 wt%, the proton conductivities of the hybrid membranes begin to decrease. When excess SiO$_2$ particles are introduced into the membranes, the –SO$_3$H groups decrease obviously due to the dilution effect, which results in a lower IEC and a reduction of free water so as to lower the membrane's proton conductivity. The bound water held by the hygroscopic Si–OH could not compensate for the loss of free water enough. Therefore, the proton conductivity is dominated by the number of sulfonic groups. In addition, the larger SiO$_2$ content blocks the hydrophilic pathways and hence suppresses the transmission channel of proton, which also leads to the lower proton conductivity.[80]

9.3.6 Chemical Stability

Oxidative stability is one of the important factors for evaluating the lifetime of PEMs under harsh fuel cell conditions of oxidation–reduction reactions. Free radicals, such as oxygen, hydroxide, and peroxide, usually attack the hydrophilic domain of the polymer resulting in the degradation of polymer chains. In general, a higher sulfonation degree value (in proportion to IEC) leads to a lower oxidation stability. Table 9.6 lists the oxidative stability of a series of SPAES/SiO$_2$ hybrid membranes at 80°C in Fenton's reagent.[80] Oxidative stability is represented by the retained weight of membrane after the treatment. It was found that the chemical stability increased with

TABLE 9.6

Oxidative Stability of SPAES/SiO$_2$-x Hybrid Membranes at 80°C in Fenton's Reagent[a]

	Oxidative Stability	
Sample (%)	RW (%)[b]	T (h)[c]
SPAES/SiO$_2$-0	90	2.5
SPAES/SiO$_2$-3	95	>3
SPAES/SiO$_2$-6	96	>4
SPAES/SiO$_2$-10	98	>4

Source: Ren, J. et al., *J. Membr. Sci.*, 434, 161, 2013.

[a] x means the silica content (wt%).

[b] Retained weights of membranes after treating in Fenton's reagent for 1 h.

[c] The time the membranes started to disappear.

increasing SiO$_2$ content, and all the hybrid membranes displayed improved oxidative stability than the pristine membrane.

Gao et al. reported on the oxidative stability of SPAEK cross-linked by MPTMS-containing −SH groups (Trademark: KH590) via the thiol-ene click reaction.[36] Their cross-linked hybrid membranes also displayed a significant improvement in the oxidative stability than the pristine membrane and the stability is in proportion to the silica content.

Chen et al. reported chemically stable hybrid PEMs, namely, StSi-g-ETFE, from poly(ethylene-co-tetrafluoroethylene) (ETFE) grafted with p-styryltrimethoxysilane (StSi) by γ−ray preirradiation and the successive hydrolysis and sulfonation.[93] For comparison, the St/DVB (95/5 by weight) monomers instead of StSi were also grafted to the ETFE films under the same conditions. Figure 9.15 shows the weight changes as a function of immersing time for the resulting hybrid PEMs with and without cross-linking from HCl and the St/DVB-grafted PEM in a 3% H$_2$O$_2$ aqueous solution at 60°C. It was found that the degradation (determined as a weight loss) began after about 70 h for the StSi-grafted PEMs not given HCl treatment and the St/DVB-grafted PEM, but after about 250 h for the StSi-grafted PEMs given HCl treatment. In this case, the time elapsed before onset of degradation is defined as the durability time, which is a parameter of the chemical stability of the PEM. Thus, the StSi-grafted PEM without HCl treatment displayed chemical stability similar to the St/DVB-grafted PEMs. With HCl treatment, −Si−O−Si− cross-links were introduced into the PEMs, leading to the significantly improved chemical stability.[93]

Recently, Wang et al. synthesized Nafion/CeO$_2$ hybrid composites through a self-assemble route, and the hybrid membranes displayed a significant improvement of oxidative stability.[74] Figure 9.16 illustrates the fluoride generation from Nafion series membranes exposed to Fenton's reagents. For the self-assembled Nafion/CeO$_2$ hybrid membranes, the fluoride emission rate (FER) was reduced with the increase in the ceria content. The simulated FER value at the initial 12 h was approximately

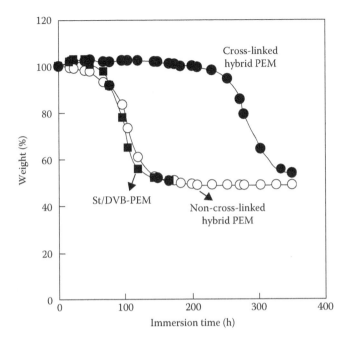

FIGURE 9.15 Weight change of the cross-linked and non-cross-linked StSi-grafted hybrid PEMs and the St/DVB-grafted PEM in 3% H_2O_2 aqueous solution at 60°C. (From Chen, J. et al., *J. Polym. Sci. A: Polym. Chem.*, 46, 5559, 2008.)

0.68–1.96 mg h^{-1}, for the self-assembled Nafion/CeO_2 hybrid membranes with ceria content of 1, 3, 5, 10 wt%, and 3.05 mg h^{-1}, for the pristine Nafion membrane, respectively. The FER value of the sol–gel-derived Nafion/CeO_2 hybrid membrane (5 wt%) was 1.56 mg h^{-1}, much higher than that of the self-assembled Nafion/CeO_2 hybrid membranes with the same ceria content (1.01 mg h^{-1}), implying that the large particles, low surface area, and low particle dispersion of the ceria oxide have a negative influence on the radical scavenge. After the initial 12 h, the FER improved with increased fluoride emission/corrosion time slope for all samples. As a result, the total FER values for the 24 h corrosion were approximately 43.05, 8.67, 6.01, and 4.47 mg h^{-1}, respectively, for the self-assembled Nafion/CeO_2 hybrid membranes with ceria content of 1, 3, 5, 10 wt%, in comparison to 55.78 mg h^{-1} for the pristine Nafion membrane and 11.64 mg h^{-1} for the sol–gel-derived Nafion/CeO_2 hybrid membrane (5 wt%). PFSA degradation processes via an unzipping mechanism through the carboxylic acid groups have been proposed in recent years. There is current debate concerning where these carboxylic acid end groups originated from. One widely accepted mechanism points to the weak polymer end groups as the initial source of carboxylic acid end groups. Another mechanism suggests that the PFSA side chain cleavage is the alternative source for the defect end groups.[94] Thus, the side chain cleavage creates additional carboxylic acid end groups on the polymer and accelerates the degradation rate of the radical corrosion test. After 24 h radical corrosion, the fluoride ion emission was approximately 72.1 mg for 1 g self-assembled

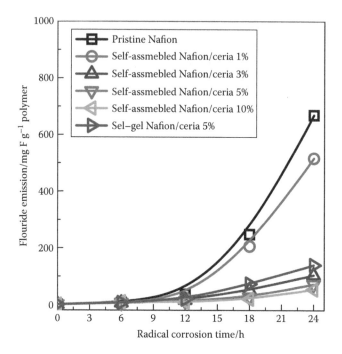

FIGURE 9.16 Fluoride evolution from the PEMs as a function of immersion time under Fenton's degradation test. (From Wang, Z. et al., *J. Membr. Sci.*, 421–422, 201, 2012.)

Nafion/CeO$_2$ hybrid membrane with ceria content of 5 wt%, in comparison to 669.34 mg for the pristine Nafion membrane. The low fluoride emission acceleration of the Nafion/CeO$_2$ hybrid membranes suggests an improved stability because of the anti-oxidation of ceria nanoparticles.[74]

9.4 MEMBRANE CONDUCTION MECHANISM AND THEORETICAL MODELING

9.4.1 MEMBRANE CONDUCTION MECHANISM MODELING

It is well known that PEMs exhibit unique ion-selective transport that has enabled a breakthrough in high-performance PEM fuel cells. The morphology and proton transport mechanisms of the commercially available Nafion and potential alternative hydrocarbon polymers have been extensively studied with respect to proton-conductive channels.[95–97]

Hybrid/composite membranes based on Nafion and hydrocarbon polyelectrolytes have also been investigated for the proton conduction mechanism.[90,98–100] Di Noto group has made extensive research on inorganic–organic proton-conducting membranes based on Nafion and various metal ionic fillers.[14,45,61,69,70,98] They concluded that M$_x$O$_y$ (M = Ti, Zr, Hf, Ta, and W) influences (1) the relaxations of both the hydrophobic and the hydrophilic domains of Nafion polymer matrix and (2) the thermal stability range of conductivity and the conductivity value of membranes. The nano-fillers affect the macromolecular dynamics of Nafion-based polymer host owing to the formation of dynamic cross-links, $R-SO_3H \cdots M_xO_y \cdots HSO_3-R$, in hydrophilic polar cages. The membranes doped with HfO_2 and WO_3 oxoclusters present a stability range of conductivity of $5°C \leq T \leq 135°C$ and give rise to σ values of 2.8×10^{-2} and 2.5×10^{-2} S cm^{-1}, respectively, at 135°C and 100% RH.[98]

It has been reported that inorganic filler promotes water sorption, and the superior water retention capability is favorable for proton transport in hybrid membranes; hence, higher proton conductivity will be obtained (see Table 9.5). WUs of the Nafion–mesoporous silica membranes increased with silica content (see Table 9.5), and the inclusion of sulfonated silica to Nafion has further enhanced WU.[82,83] Distribution of small nanoparticles of silica in the ionic cluster of Nafion leads to an increased water retention ability of the composite membrane compared to that of silica with larger particle sizes. This might be explained by the increased surface area of interaction between water and silica due to increased hydrogen bonding. The high water retention capacity of the hybrid membrane is more significant at temperatures higher than 80°C.[90]

Incorporation of phosphosilicate to Nafion is found to improve WU of the nano-composite membrane due to the high hydrophilicity of the phosphosilicate.[66] WU of the Nafion–silica–PWA membrane was 38%, compared to 32% and 34% of the recast Nafion and Nafion–silica nanocomposite, respectively.[56] However, higher filler content reduced the WU of the Nafion–silica–PWA nanocomposite due to the increased particle size, which in turn prevents the particles from remaining within the ionic cluster.

Figure 9.17 shows a schematic diagram representing the ion conduction channel in (a) Nafion and blockage of the ion conduction channel in the hybrid membranes due to the presence of aggregated (b) silica and (c) nanaoclays.[90] Inferior proton conductivity of the Nafion–clay nanocomposite membrane in comparison to the VN was reported by several researchers even after the addition of sulfonated clay.[46,51] The lower proton conductivities of the composite membranes were ascribed to the blockage of the proton conduction channel (Figure 9.17b) by the silica aggregates at low temperature and to the assistance in proton conduction due to the hydrophilic nature of silicates at low humidity and high temperature.

Proton conductivity and WU, studied as function of RH, can be combined to calculate the effective proton mobility in ionomers. The plot of proton mobility versus proton concentration shows common features for various ionomers, which can be related to percolation and tortuosity of hydrated nanometric channels. Proton conductivity can even be enhanced in cross-linked ionomers, possibly due to a reduction of channel tortuosity.[99]

Ogawa et al. performed theoretical studies on proton transfer by investigation on surface of zirconium phosphate with adsorbed water molecules.[100] They claimed that the interactions between the phosphate groups at the ZrP surface and the adsorbed water molecules are relatively large, and a strong hydrogen-bonding network is generated locally. Because of the strong interactions, water molecules can be attached to the ZrP surface, and the O–O distance becomes shorter than that in bulk water systems. Because of the short O–O distances and the delocalized charge of each atom,

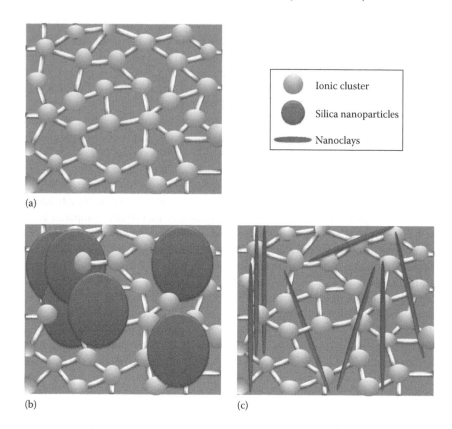

FIGURE 9.17 Schematic diagram representing the ion conduction channel in (a) Nafion and blockage of the ion conduction channel in the hybrid membranes due to the presence of (b) aggregated silica and (c) nanoclays. (From Mishra, A.K. et al., *Prog. Polym. Sci.*, 37, 842, 2012.)

the activation energy of proton transfer at the ZrP surface decreases and causes high proton conductivity even under conditions of high temperature and low humidity.

Figure 9.18 shows the simplest path, in which only a single proton transfers from H_3O^+ to H_2O with the activation energy, $E_a = 3.51$ kJ mol^{-1}, through a Zundel structure $(H_5O_2^+)$.[100] The initial geometry (state), the transition state, and the final geometry (state) are shown in Figure 9.18a through c, respectively. This type of proton-transfer path has been proposed by theoretical and experimental studies on the mechanism of proton transfer in bulk water and ice. It should be noted that, in contrast to proton transfer on Pt (111) and polar surfaces such as Al_2O_3 and SiC (100), in which both separated protons and hydroxyl groups can be easily stabilized, hydroxyl groups are unstable at the ZrP surface and no hydroxyl groups can be found through Ogawa's calculations.[100]

The interaction energies of the water molecules that make a hydrogen bond with POH are relatively high. It is found that one of the water molecules on the ZrP surface makes three hydrogen bonds with three individual POHs, as shown in Figure 9.19a.

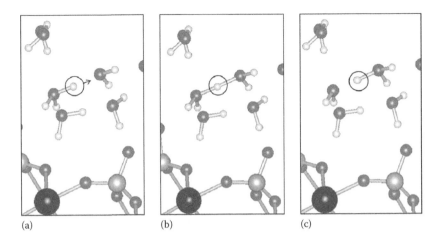

(a)   (b)   (c)

FIGURE 9.18 Example of proton-transfer paths modeling water-adsorbed ZrP surfaces with an activation energy of 3.51 kJ mol^{-1}: (a) initial state, (b) transition state, and (c) final state. Along this path, the following proton transfer takes place: $H_3O^+ + H_2O \rightarrow$ transition state $\rightarrow H_2O + H_3O^+$. Atoms shown in this figure in ascending order in size refer to hydrogen, oxygen, phosphorus, and zirconium, respectively. (From Ogawa, T. et al., *J. Phys. Chem. C*, 115, 5599, 2011.)

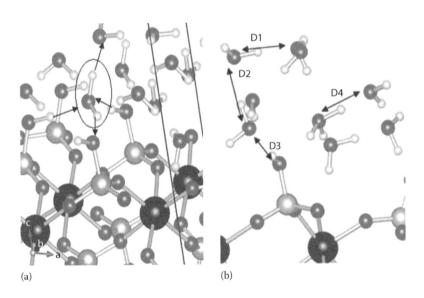

(a)   (b)

FIGURE 9.19 (a) Example of the hydrogen-bonding network in water-adsorbed ZrP surfaces. One of the water molecules (circled) forms three hydrogen bonds with three individual POHs indicated by the arrows and (b) atoms and molecules labeled according to the classifications of the hydrogen bonds. Atoms shown in this figure in ascending order in size refer to hydrogen, oxygen, phosphorus, and zirconium, respectively. (From Ogawa, T. et al., *J. Phys. Chem. C*, 115, 5599, 2011.)

In this case, the ZrP surface fixes water molecules with strong interaction energies exceeding 170 kJ mol^{-1}. The averaged interaction energies between water molecules are calculated to be 108 kJ mol^{-1}.[100] The interaction energy through the hydrogen-bonding network between water and POH in that structure is more than 1.6 times stronger than that of bulk water systems. Although the example in Figure 9.19a may be a special case, this structure, where one water molecule makes several hydrogen bonds with hydrophilic groups, shows the importance of forming several hydrogen bonds (hydrogen-bonding network) with POH. Although the *acidity* of POH is less than that of PFSA, the hydrophilicity of ZrP is higher than that of PFSA.

To investigate precisely the strong interaction between water and ZrP, Ogawa et al. also examined the effect of the interaction on the O–O distances between oxygen atoms that donate or accept protons in the reactant states.[100] In this case, the O–O distances between water–water or water–POH in the initial states are classified into four groups (Figure 9.19b), and the O–O distances in each group are examined as follows (defined as *bound water* by being bound by a hydrogen bond from POH):

- Group D1: Both oxygen atoms compose the water molecules, which are not affected by POH in the ZrP system.
- Group D2: One of the oxygen atoms composes bound water and the other composes the water that is not bound by a hydrogen bond from POH in the ZrP system.
- Group D3: Both oxygen atoms compose the POH or bound water in the ZrP system.
- Group D4: Either of two oxygen atoms composes H3Ot in the ZrP system.

9.4.2 MEMBRANE STRUCTURE–RELATED MODELING

The incorporation of inorganic fillers into sulfonated aromatic polymers can be carried out through two kinds of connection mode, namely, Van der Waals bonds or covalent bonds (cross-links). Van der Waals bonds are considered in composite materials, where a second phase is added and dispersed inside the polymer matrix. Di Vona and Knauth have studied second phase hybrid organic–inorganic materials (such as Si-PPSU) and inexpensive nanocrystalline metal oxides (such as TiO$_2$).[54,78,99] Covalent cross-linking bonds by SO$_2$ bridges can be formed by thermal treatment (solvothermal cross-linking)[101] or by clever chemical synthesis (giving hybrid ionomers called SOSiPEEK)[102]. The bond energy for hybrid/composite membranes is in the order of cross-linked ionomers (SOSiPEEK) > composite ionomers (SPEEK-SiPPSU) > SPEEK-TiO$_2$.

Modeling of polymer membrane structure through dynamics simulations is generally coupled with the water/proton-transfer properties. Yana et al. conducted molecular dynamics (MD) simulations of Krytox-Silica-Nafion composite membranes in comparison to the pure Nafion.[103] A 5 wt% of carboxylic acid terminated perfluoropolyether hybrid with silica (Krytox-Silica) in Nafion composite polymer was used in the modification of a PEM in order to improve its efficiency at high operating temperatures. MD simulations were carried out in order to understand the microscopic properties of two systems, Krytox-Silica in Nafion and pure

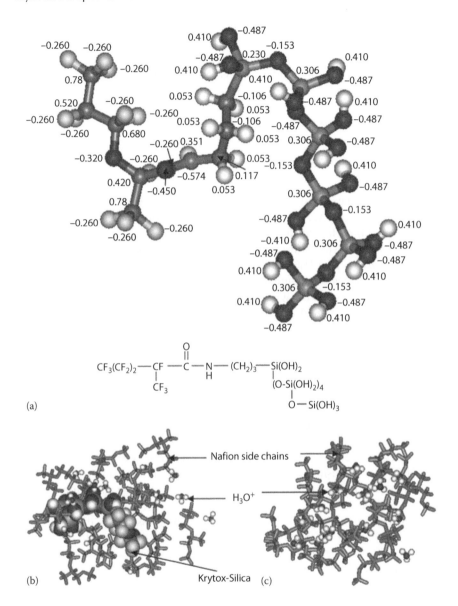

FIGURE 9.20 Initial structures of (a) Krytox-Silica molecule and charges, (b) 5% wt of Krytox-Silica in Nafion system, and (c) pure Nafion system. (From Yana, J. et al., *Polymer*, 51, 4632, 2010.)

Nafion. Models of the initial structure of Krytox-Silica as shown in Figure 9.20a and of a composition structure incorporating 5 wt% Krytox-Silica in a Nafion composite polymer consisting of 15 Nafion side chains, 15 hydronium ions, and 1 of Krytox-Silica, as shown in Figure 9.20b, were established. In another system, pure Nafion was modeled without Krytox-Silica (model structure shown in Figure 9.20c).

Models with various amounts of water molecules and temperatures were simulated to study the water content and temperature effects. All structures of each model were randomly posited for simulation of small amorphous hydrophilic pore.

The interaction at the molecular level is investigated via the radial distribution function (RDF), which is the function of the spherically averaged distribution of interatomic distance (r) between two species A and B, totaling N_A and N_B, in a volume of v, as calculated by[104]

$$g_{AB}(r) = \frac{N_{AB}(r)v}{4\Pi r^2 dr N_A N_B} \tag{9.3}$$

Figure 9.21 shows the diffusion coefficients of H_3O^+ from MD simulations: (a) variation of temperature, (b) variation of water content, (c) variation of temperature and water content, (d) proton conductivity of experimental data,[105] (e) MSD average of H_3O^+ in 5% wt of Krytox-Silica in Nafion system, and (f) MSD average of H_3O^+ in pure Nafion system. The H_3O^+ diffusion coefficients as a function of temperature at 100% water molecules (Figure 9.21a) showed that the hydronium ions' mobility increases at high temperature, which is consistent with the tangential direction conductivities of Nafion, with 91 and 102 mS cm^{-1} at 25°C and 30°C, respectively.[103,106]

It has been observed via experimental study that the conductivity of Nafion strongly depends on the water content.[95–97,106] Based on this model, the effect of water content at the operating temperature of 353 K is investigated with the water content varied from 30%, 50%, and 100% RH against the fully solvated system. The higher the percent of water molecules, the higher the diffusion coefficient of H_3O^+ (shown in Figure 9.21b). The effect of the amount of water molecules on the diffusion coefficient of 5 wt% of Krytox-Silica in Nafion is less than that for the pure Nafion system, especially at low water content. Under the assumption that at high temperatures the water can evaporate, the percentage of water should be smaller at higher temperatures. The plot of diffusion coefficient and temperature shows a similar trend to that found from experimental measurements, reported by Chirachanchai and coworkers (Figure 9.21c and d).[105]

Distribution of water around SO_3^- of each system is essentially the same. This is due to an excess of water in both systems. Concerning the effect of temperature, the H_3O^+ diffusion coefficients of Krytox-Silica in Nafion increased until 353 K and then kept constant, whereas the pure Nafion system had a positive slope until 333 K (Figure 9.21a). Concerning water content (Figure 9.21b), the H_3O^+ diffusion coefficient of 5 wt% of Krytox-Silica in Nafion system was constant beyond 50% water molecules, while in the case of pure Nafion system, the same coefficient was decreasing with a higher slope below that point. This indicates that the selected diffusion coefficients for the 5 wt% of Krytox-Silica in Nafion system are increased from 333 to 353 K and then decreased from 353 to 373 K as a result of the temperature and water content effects, respectively (Figure 9.21c and d). However, the diffusion coefficient of H_3O^+ in pure Nafion is mainly influenced by the water content.

The distribution of water around SO_3^- in the models is shown in terms of RDF of $O(SO_3^-)$-$O(H_2O)$ and the distribution of water around SO_3^- in Krytox-Silica in Nafion system higher than pure Nafion system.[103] Based on Yana's work, the sharp

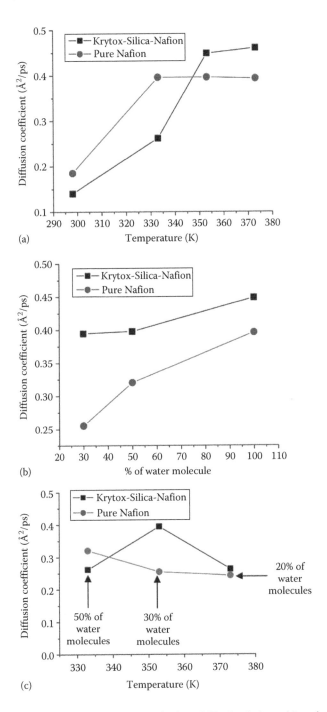

FIGURE 9.21 Diffusion coefficients of H_3O^+ from MD simulations: (a) variation of temperature, (b) variation of water content, (c) variation of temperature and water content.

(*Continued*)

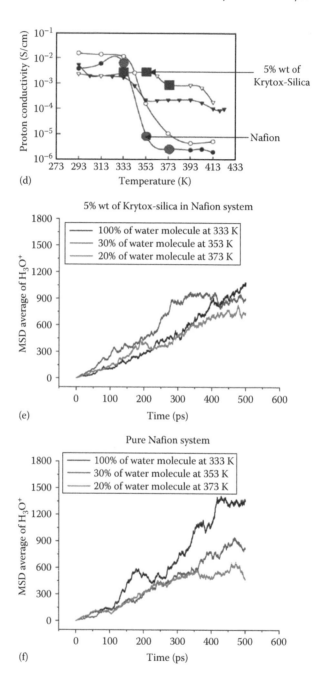

FIGURE 9.21 (Continued) Diffusion coefficients of H_3O^+ from MD simulations: (d) proton conductivity of experimental data,[105] (e) MSD average of H_3O^+ in 5 wt% of Krytox-Silica in Nafion system, and (f) MSD average of H_3O^+ in pure Nafion system. (From Yana, J. et al., *Polymer*, 51, 4632, 2010.)

peak at 2.5 Å in the RDF plot of $O(H_3O^+)$-$O(H_2O)$ for the constant 100% water model for the Krytox-Silica in Nafion system and the pure Nafion system indicated the potentially Zundel ion formation, and this complex was found related to H_3O^+ movement or proton conduction. The peak integration of the RDF plot shows that the number of water molecules around H_3O^+ in the pure Nafion system is always lower than the Krytox-Silica in Nafion system. The interaction of SO_3^- and H_3O^+ can be investigated from the RDF of $O(SO_3^-)$-$O(H_3O^+)$. As a result, the distribution around SO_3^- from the peak integration at 2.8 Å of the RDF shows that the number of water molecules around SO_3^- in the pure Nafion system is less than those for the 5 wt% Krytox-Silica in Nafion system at 353 K (30% of water molecule) and 373 K (20% of water molecule). Again, the number of water molecules distributed around the SO_3^- group in all systems is not significantly different, because the amounts of water per sulfonate group are sufficient for each of the systems. The distribution of H_3O^+ around H_3O^+ from the peak integration of RDF plot shows that the number of H_3O^+ around H_3O^+ in both systems was not significantly different, indicating that H_3O^+ ions move between side chains only and are stable in this interaction. Almost no water molecules from the peak integration indicate that silica did not form a strong interaction with water, and thus proton conductivities are minimally affected by the 5 wt% Krytox-Silica in the Nafion. Krytox-Silica material should act as a water absorbent in the hybrid polymer membrane system, while this will not interrupt the proton diffusion process.[103]

The results were in good agreement with the experiments and could be used to describe the application of Krytox-Silica-Nafion composite at high temperatures. The effect of the amount of water molecules on the diffusion coefficient or proton conductivity showed more deviations between 5% wt of Krytox-Silica-Nafion composite and pure Nafion system at lower water content (or higher temperature) than at high water content (or low temperature). According to the diffusion coefficient results, the percentage of water molecules at each temperature corresponds to the known experimental trend. Silica, as the water absorbent in the hybrid polymer membrane, does not have a strong interaction with water molecules or H_3O^+ ions; thus, the proton conductivities will not be highly affected by adding Krytox-Silica to the Nafion.[103]

9.4.3 OTHER THEORETICAL STUDIES

Although theoretical studies about water/proton transport in pristine Nafion and alternative hydrocarbon PEMs have been widely carried out, the modeling and simulation for hybrid/composite membranes are relatively seldom reported. The proton conduction mechanism for hybrid/composite membranes is considered similar as that for regular PEMs based on the so-called hopping (Grotthus), where protons hop from one hydrolyzed ionic site ($SO_3^-H_3O^+$) to another across the membrane, and electroosmotic drag (*vehicle*), where protons bound to water ($H^+(H_2O)_x$) drag one or more water molecules across the membrane.[107] The more important role for inorganic fillers for hybrid/composite membranes should still focus on their water retention capability, as mentioned earlier.

9.5 MEMBRANE APPLICATIONS IN ELECTROCHEMICAL DEVICES FOR ENERGY STORAGE AND CONVERSION

9.5.1 APPLICATIONS FOR PEM FOR FUEL CELLS

The final application for most of the hybrid/composite membranes is focused on PEM fuel cells, especially for DMFCs. Cell performance study, about which power density can be calculated at a particular voltage, RH, and temperature, is the main evaluation method for membrane final potentials. Table 9.7 gives an overview of the current densities measured for various membranes under different experimental conditions.

Most of the silica-based Nafion nanocomposite membranes have exhibited superior cell performance compared to that of the pristine Nafion.[90] The addition of TBS to Nafion–silica nanocomposite significantly increased the electrochemical properties of the membrane. Nafion–silica–TBS nanocomposite has a current density of 1600 mA cm^{-2} at 0.6 V and 1100 mW cm^{-2} (maximum power density), compared to 999 mA cm^{-2} of Nafion at 0.6 V and 838 mW cm^{-2}. Nafion–silica–ZrO$_2$ composite membranes have superior cell performance at 120°C and low RH compared to those of the VN and Nafion–ZrO$_2$ membrane.[90] Addition of HPMC to SPEEK–silica nanocomposites improves the current density of the nanocomposites irrespective of the change in RH. The current densities of SPEEK–silica–HPMC, SPEEK–silica nanocomposites, and Nafion were 785, 598, and 176 mA cm^{-2}, respectively, under 70% RH and 0.4 V cell potential, with power densities of 126, 161, and 27 mW cm^{-2}, respectively, under 10% RH and 80°C.[90] Addition of Laponite to SPPEK was noted to achieve the current density of 370 mA cm^{-2} compared to 80 mA cm^{-2} for SPEEK at 0.6 V (see Table 9.7).

Another important advantage for incorporating inorganic fillers or composites into polymer matrix might be the excellent barrier capacity toward methanol, which will improve cell performance in DMFCs.[108–111] Sahu et al. reported on Nafion-based hybrid membranes composited with silica, mesoporous zirconium phosphate (MZP), and mesoporous titanium phosphate (MTP), respectively.[108] They found that methanol crossover generally decreases with increasing current density as the concentration of methanol at the anode/membrane interface decreases. The methanol crossover decreases significantly for the composite membranes compared to pristine Nafion 117. Nafion–MTP exhibits a lower methanol crossover in relation to other types of composite membranes, which could be due to the presence of relatively bigger particles and lower surface area of the filler. Nafion–MZP shows an intermediate value compared to Nafion–silica and Nafion–MTP composite membranes. In general, the hydrophilic nature of silica, MZP, and MTP particles in the composite membranes prevents the formation of nonselective voids for methanol transport without affecting the migration of protons as evidenced from the proton conductivity.[108]

The DMFC performances for a series of Nafion-based hybrid membranes compared with that of pristine Nafion are shown in Figure 9.22a through c, respectively. Most of the hybrid membranes displayed better cell performance compared with that of Nafion. For Nafion–MZP and Nafion–MTP composite membranes,

TABLE 9.7
WU and PEM Fuel Cell Performance of Various Polyelectrolyte Membranes

Membranes	WU (%)	Operating Temperature (°C)	Cell Voltage (V)	Current Density (mA cm^{-2})
Nafion 117	23	80	0.2	600
Nafion–silica (3 nm)	28			1130
Nafion–silica (90 nm)	25			1030
Nafion–silica (1 μm)	21			500
Nafion	—	140	0.4	133
Nafion–silica	—			225
Nafion–mesoporous silica	48	120	0.6	540
Recast Nafion	14	60	0.4	540
Nafion–sulfonated silica	25			1040
Nafion–silica-TBS[a]	—	80	0.6	1600
Nafion–115	32	110	0.4	95
Nafion–silica	34			320
Nafion–silica-PWA	38			540
Recast Nafion	50	80	0.6	600
Nafion–Laponite	87			—
Nafion–sulfonated Laponite	70			720
Recast Nafion	14	60	60	450
Nafion-H$^+$ MMT[b]	13			—
Nafion–sulfonated MMT	20			800
SPEEK	30	120	0.1	150
SPEEK–silica	39			250
SPEEK–silica–HPMC[c]	49			700
SPEEK (70°C)	100	60	0.6	80
SPEEK-Laponite clay	30			370
SDF	100–190	93	0.3	1200
SDF–silica	187–385			1700

Source: Mishra, A.K. et al., *Prog. Polym. Sci.*, 37, 842, 2012.

[a] TBS refers to a commercial fluorosurfactant containing a sulfonic acid group (Zonyl TBS).

[b] MMT refers to montmorillonite.

[c] HPMC refers to hydroxypropyl methyl cellulose.

the combination of robust mesostructure with strong acid functionality of MZP and MTP helps in furthering the DMFC performance.[108]

It should be noted that the PVA-based composite/blend membranes generally showed high potential for DMFC applications due to the better methanol barrier capability for PVA itself resulting from the dense molecular packing structure caused by inter- and intramolecular hydrogen bonding.[109–111] Kim group developed PVA-based nanocomposites modified with sulfonated mesoporous benzene-silicas and obtained

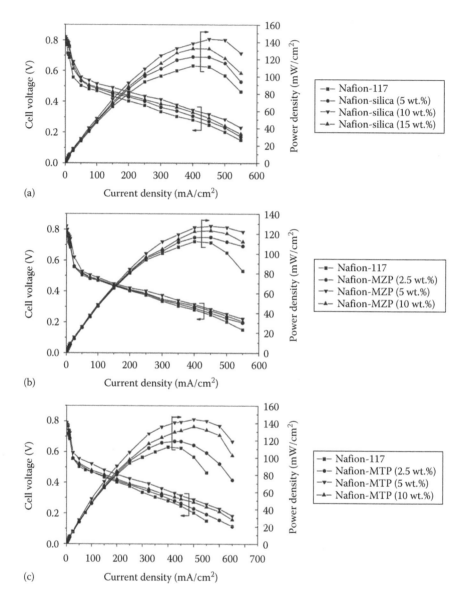

FIGURE 9.22 Performance curves for DMFCs with Nafion 117 and (a) Nafion–silica, (b) Nafion-MZP, and (c) Nafion-MTP composited membranes at 70°C using aqueous methanol and oxygen reactants at atmospheric pressure. Anode, 2 M aqueous methanol with a flow rate of 2 sccm; cathode, oxygen with a flow rate of 200 sccm. (From Sahu, A.K. et al., *J. Membr. Sci.*, 345, 305, 2009.)

TABLE 9.8

Ionic Conductivity, Methanol Permeability, and Selectivity of Various Membranes

Membranes	Conductivity ($\times 10^{-3}$ S cm^{-1})	Methanol Permeability ($\times 10^{-7}$ cm^2 s^{-1})	Selectivity[c] ($\times 10^3$ S cm^{-3} s)
PVA	28	24.35	11.5
PCS91-G1h[a]	20	16.24	12.5
PCS91-G2h	13	9.42	13.5
PCS91-G3h	10	7.42	13.3
Nafion 115	14	18	7.8
QAPVA/SiO$_2$[b]	6.8	8.45	8.05
QAPVA	7.3	10	1.8

Source: Yang, J.M. and Chiu, H.C., *J. Membr. Sci.*, 419–420, 65, 2012.

[a] PCS91-G1h means the PCS was cross-linked with glutaraldehyde for 1 h.

[b] QAPVA means QCS with quaternized polyvinyl alcohol blends.

[c] Selectivity equals to conductivity/methanol permeability.

good performance in lowering the methanol crossover (about 68% reduction in comparison with the Nafion 117), and mesoporous benzene–silica with smaller particle morphology and pores (2–3 nm) was observed to be a more effective additive.[109]

Yun et al. synthesized nanocomposite membranes (SPVA) based on cross-linked sulfonated PVA and sulfonated multiwalled carbon nanotubes (s-MWNTs) for DMFC applications.[110] The composite membranes exhibited excellent proton conductivity ranging from 0.032 to 0.075 S cm^{-1} as well as low methanol permeability ranging from 1.12×10^{-8} to 3.32×10^{-9} cm^2 s^{-1} at 60°C. Recently, Yang and Chiu prepared a series of novel cross-linked PVA/chitosan (PCS) blend membranes and investigated the ionic conductivity, methanol permeability, and selectivity of various membranes (see Table 9.8).[111]

9.5.2 APPLICATIONS FOR BATTERIES

Lithium batteries are acquiring a leading role in various important markets with particular impact on that of the consumer electronics and, in prospect, on that of electric or hybrid cars. However, considering the evolution of these markets and their associated energy and power demands, new and improved battery configurations are requested. Progress in lithium battery technology may be achieved by passing from a conventional liquid electrolyte structure to a solid-state polymer configuration.[112] Inorganic fillers cannot only be utilized as proton-conductive additives for PEMs but also be selected as nanocomposites in polyelectrolytes for lithium battery.

Croce et al. developed a series of composited membranes based on poly(ethylene oxide) (PEO) and lithium salts with the dispersion of functionalized ZrO$_2$ ceramic filler.[112] The composite polyelectrolyte containing 5 wt% zirconium displayed improved ionic conductivity than the unmodified membranes by one order of

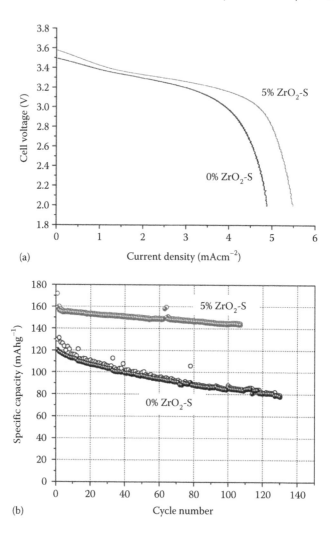

FIGURE 9.23 Performances of the Li/P(EO)$_{20}$LiClO$_4$+5% S-ZrO$_2$/LiFePO$_4$ battery and the Li/P(EO)$_{20}$LiClO$_4$/LiFePO$_4$ battery at 90°C. (a) Typical current–voltage polarization curves and (b) capacity versus charge–discharge cycles. (From Croce, F. et al., *J. Power Sources*, 162, 685, 2006.)

magnitude. Figure 9.23a shows a typical comparison of current–voltage polarization curves for Li/P(EO)$_{20}$LiClO$_4$/LiFePO$_4$ battery with and without ZrO$_2$ nanofillers. The resistance of the composite electrolyte battery, as determined by the linear part of the curve in the middle current range, is lower than that of the battery using a standard, ceramic-free electrolyte. More significant is the difference in the value of the limiting current, which is consistently higher for the battery based on the composite electrolyte. The cycling capabilities of the Li/P(EO)$_{20}$LiClO$_4$+5% ZrO$_2$/LiFePO$_4$ battery are shown in Figure 9.23b, which reports the capacity versus cycle number. The battery operates over 100 cycles with a very good capacity retention, thus confirming its long life expectance. Also important is the observed high charge–discharge efficiency (ratio between charge and discharge capacity), a value that confirms the good interfacial stability between the lithium metal electrode and the lithium transference number of the associated composite polymer electrolytes.[112]

Chiappone et al. prepared methacrylic-based thermo-set gel-PEMs reinforced with microfibrillated cellulose (MFC) particles and obtained ductile composited membranes with excellent mechanical properties, as well as high ionic conductivity (10^{-3} S cm^{-1} at RT).[113] In general, for a lithium battery, the anodic reaction occurs in the vicinity of 0 V versus Li/Li$^+$, while the cathode potentials can approach values as high as 4.5 V versus Li/Li$^+$, implying that the electrochemical stability window (ESW) is a fundamental parameter regarding cycling reversibility. The ESW of the MFC-3 composite polymer electrolyte (representative for all the samples prepared) was measured at ambient temperature.[113] In fact, independently of the amount of MFC added, all samples exhibited a wide ESW ranging from the lithium plating to around 5.0 V versus Li/Li$^+$. The resulting current–voltage traces showed in the first sweeping cycle a current flow at about 2.7 V versus Li/Li$^+$, followed by a large current flow starting at around 1.5 V. The first event is most probably due to some decomposition of the cellulosic material, while the latter is associated with a multistep decomposition process, very likely due to the reduction of the carbonate solution component, with the consequent formation of a passivating film on the testing electrode. The increase of the current during the first anodic scan, which is due to the decomposition of the electrolyte, was taken in correspondence to the onset of a low current peak at approximately 4.3 V versus Li/Li$^+$. However, even if the current rises at a lower voltage, after few cycles, its trend consistently deviates from that observed in the first cycle, showing a sort of *passivation* phenomenon, which extends the anodic stability up to a higher voltage, that is, above 5 V versus Li/Li$^+$.[113]

The Tang group[114] and the He group[115] developed organic/inorganic hybrid membranes based on PVdF using titania-poly(methyl methacrylate) (PMMA-TiO$_2$) as nanocomposites. He et al. incorporated hexafluoropropylene (HFP) to the PMMA matrix and investigated the charge–discharge cycling performance of LiCoO$_2$/Li button cells with different electrolytes. The cell with PVdF-HFP/TiO$_2$-PMMA blend composite polymer electrolyte displayed better cyclic performance than that with liquid electrolyte and pristine PVdF–HFP gel polymer electrolyte.[115]

Prasanth et al. studied the effect of nanoclay on ionic conductivity and electrochemical properties of PVdF-based nanocomposite porous polymer membranes.[116] They found that the incorporation of 2 wt% nanoclay in the hybrid polymer gel electrolyte (PGE) enhances its ionic conductivity from 0.78 to 3.08 mS cm^{-1} at RT. Li-ion half cell comprising Li/PVdF-clay(2 wt%)-PGE/LiMn$_2$O$_4$ delivers high charge–discharge capacity (about 120 mA h g^{-1}) and shows a stable cycle performance. Angulakhsmi et al. incorporated nanochitin in PEO-LiPF$_6$ matrix and found that the incorporation of chitin whiskers significantly improves the ionic conductivity, thermal stability, and mechanical integrity along with the interfacial properties.[117] The galvanostatic cycling behavior in an LiFePO$_4$/CeLi cell at 70°C displayed an improved specific capacity and outstanding cycling stability.

9.5.3 APPLICATIONS FOR SOLAR CELLS

Proton-conducting composite membranes functionalized by SiO_2 particles with ordered mesoporosity can be used in photocatalysis and dye-sensitized solar cells (DSSCs) and showed more stable photovoltaic performance than that of the DSSC assembled with liquid electrolyte.[118,119] Choi and Kim reported on hybrid/composite membranes based on polypropylene nonwoven matrix used in DSSCs.[119] During their preparation of hybrid membrane, PEO and SiO_2 particles were coated onto the porous polypropylene nonwoven matrix to get gel polymer electrolyte, which exhibit ionic conductivity higher than 1.1×10^{-3} S cm^{-1} at RT. DSSC employing the hybrid composite membrane with PEO and 10 wt% SiO_2 exhibited an open circuit voltage (VOC) of 0.77 V and a short circuit current of 10.78 mA cm^{-2} at an incident light intensity of 100 mW cm^{-2}, yielding a conversion efficiency of 5.2%.

Choi and Kim reported the photocurrent density–voltage curves of the DSSCs assembled with the hybrid composite membranes coated by PEO and different content of SiO_2.[119] For comparison, the photocurrent density–voltage curve of the DSSC prepared with liquid electrolyte was also measured. The DSSCs assembled with the hybrid composite membranes exhibit a lower short circuit current density (J_{sc}) and higher VOC than the liquid electrolyte-based DSSC. Lower value of J_{sc} in the DSSC with hybrid composite membrane originates from its lower ionic conductivity. Higher resistance for ion migration reduces the supply of I_3^- to the Pt counterelectrode, which causes the depletion of I_3^- and also retards the kinetics of the dye regeneration, therefore decreasing the J_{sc}. An increase of VOC in the DSSC is related to the reduction of charge recombination on the TiO_2 and electrolyte interface. The injected electron in TiO_2 is known to recombine with I_3^- in the electrolyte. If I_3^- diffuses in a slower media-like gel polymer matrix, the recombination reaction will be depressed, resulting in an increase of VOC. With the addition of SiO_2 into the electrolyte, the pH value increased from 5.94 (liquid electrolyte without SiO_2) to values ranging from 7.25 to 7.64. Both the J_{sc} and the conversion efficiency η increased with SiO_2 content up to 10 wt% and decreased with the further increase. An increase in J_{sc} with SiO_2 content is related to the improved ionic conductivity of the electrolyte.[119]

9.5.4 APPLICATIONS FOR WATER ELECTROLYSIS

Since hydrogen is one of the most promising energy carriers because of its unharmful emissions in almost every application where fossil fuels are being used today, hydrogen production and storage become more and more important. Synergy between hydrogen and electricity and renewable energy sources is particularly promising from a practical viewpoint. PEMs such as Nafion and alternative hydrocarbon polyelectrolyte can also be employed in PEM electrolyzers, and the half-cell reactions during electrolysis are reversible of those during fuel cell operation.[120,121]

Nafion-based composite membranes might be one of the early studied PEMs as solid polymer electrolyte (SPE) water electrolyzer. Baglio et al. reported on the performances for Nafion/TiO_2 hybrid membranes in an SPE electrolyzer.[122] They found that the hybrid membranes displayed higher performance with respect to pristine

Nafion 115 due to the water retention capability of the TiO_2 filler. A promising increase in electrical efficiency was recorded at low current densities for the composite membrane-based SPE electrolyzer at high temperature compared to conventional membrane-based devices. Hansen and Aili group developed phosphoric acid-doped membranes based on Nafion, PBI, and their blends for steam electrolysis at temperatures above 100°C.[123,124] A membrane electrode assembly (MEA) based on phosphoric acid–doped Nafion was operated at 130°C at ambient pressure with a current density of 300 mA cm^{-2} at 1.75 V, with no membrane degradation observed during a test of 90 h. The PBI-based MEAs showed better polarization curves (500 mA cm^{-2} at 1.75 V) but poor durability. The durability problems were apparently attributed to severe membrane failure since a decrease in the cell voltage under a constant current, accompanied by a reduction of the gas evolution (both hydrogen and oxygen), was observed.[123] When replacing Nafion with another type of polymer in the PFSA family, namely, Aquivion™, with a shorter side chain, the resulting phosphoric acid–doped hybrid membrane exhibited a much better performance in the steam electrolysis test at 130°C, with current densities up to 775 mA cm^{-2} at 1.8 V.[124]

Except for Nafion-based PEMs, other hydrocarbon-based PEMs are also studied in water electrolysis applications.[125–127] Wang group developed blend composite membranes from SPEEK and poly(ether sulfone) (PES) as SPE membranes for hydrogen production via water electrolysis.[125] Their SPEEK/PES hybrid blends displayed an electrolytic current of 1655 mA cm^{-2} at 2.0 V and 80°C, indicating a great potential for this kind of SPEs as candidates for water electrolysis application. Lee et al. synthesized a series of cross-linked SPEEK/HPA composited membranes with different HPA loadings including tungstophosphoric acid (TPA), molybdophosphoric acid (MoPA), and tungstosilicic acid (TSiA) and investigated the effect of cross-linking agent and HPA content on the performance of the resulting composited membranes.[126] They obtained an optimum condition of cross-linked SPEEK/HPAs with 0.01 mL of cross-linking agent content and achieved the ion conductivity in the order of magnitude: SPEEK/TPA30 (30 wt%) < SPEEK/MoPA40 < SPEEK/TSiA30. It is expected that the cross-linked SPEEK/TPA30 was suitable as an alternative membrane in large-scale polymer electrolyte membrane electrolysis system taking both electrocatalytic activity and mechanical stability into account. The dual effect of higher proton conductivity and electrocatalytic activity with the addition of HPAs causes a synergy effect.

Recently, Song et al. developed covalently cross-linked SPEEK/Cs-TPA/CeO$_2$ composite membrane for PEM water electrolysis.[127] In this case, TPA with a cesium was added to the SPEEK to increase proton conductivity. CeO$_2$ was used to scavenge free radicals that attack the membrane in the water electrolysis and to improve the durability of the membrane. The composite membrane featured the electrochemical characteristics, such as 0.13 S cm^{-1} of proton conductivity at 80°C. Pt(NH$_3$)$_4$Cl$_2$, Pd(NH$_3$)$_4$Cl$_2$, RhCl$_3$, and Co(NH$_6$)$_4$Cl$_3$ were used to prepare a variety of the MEAs as electrocatalytic precursors. Consequently, cross-linked SPEEK/Cs-TPA/CeO$_2$ (1%) composite membrane is expected to be a promising one that may replace Nafion 117 membrane based on integration of the mechanical characteristics, oxidation durability, and electrochemical properties. The Pd/Nafion 117/Pt-Pd MEA showed better performance than that of Pt/Nafion 117/Pt MEA. In addition, Pd/PEM/Pt-Pd

MEA with cross-linked SPEEK/Cs-TPA/CeO$_2$ (1%) showed a possibility of practical application for PEM water electrolysis cell due to the optimal comprehensive performance, with an electrochemical activity surface area of 26.2 m^2 g^{-1}, cell voltage of 1.82 V, and voltage efficiency of 81.3%.

9.5.5 Applications for Vanadium Redox Flow Batteries

Redox flow batteries operate on the principle of using the redox chemistry of two electrochemically active materials dissolved in solution. Many different redox couples such as vanadium–bromide, iron–chromium, zinc–bromine, and all-vanadium have been investigated for use in redox flow batteries. Vanadium redox flow battery (VRFB) offers excellent qualities for large energy storage systems such as a long lifespan, quick response time, deep-discharge capability, and low maintenance cost. Nafion-based PEMs have been used as proton-selective membranes because of their high chemical and mechanical stabilities as well as good proton conductivity.

Teng et al. prepared Nafion/PTFE blend membranes through solution casting and investigated the membrane properties with respect to VRFB application.[128] It was found that the addition of hydrophobic PTFE reduces the WU, IEC, and conductivity of blend membranes despite the high miscibility of PTFE with Nafion. The addition of PTFE can otherwise increase the crystallinity and thermal stability of Nafion/PTFE membranes and reduce the vanadium permeability. The energy efficiency of this VRFB with N$_{0.7}$P$_{0.3}$ blend membrane (with 30 wt% of PTFE) was 85.1% at a current density of 50 mA cm^{-2}, which was superior to that of recast Nafion membrane (80.5%). Self-discharge test shows that the decay of open circuit potential of N$_{0.7}$P$_{0.3}$ membrane is much lower than that of pristine Nafion membrane. More than 50 cycles of charge–discharge test proved that the N$_{0.7}$P$_{0.3}$ membrane possesses high stability in long time running, suggesting that the addition of PTFE is a simple and effective way to improve the performance of Nafion for VRFB application.

Many attempts have been made to block the cross-diffusion of vanadium cations through polyelectrolyte membranes, in order to modify the water-filled channels by introducing inorganic materials such as SiO$_2$ and TiO$_2$.[129–131] Vijayakumar et al. investigated the properties of Nafion/SiO$_2$ hybrid membranes in VRFB application via spectral analyses including NMR, x-ray photoelectron spectroscopy (XPS), and FTIR.[129] The XPS study reveals the chemical identity and environment of vanadium cations accumulated at the surface. On the other hand, the ^{19}F and ^{29}Si NMR measurement explores the nature of the interaction between the silica particles, Nafion side chains, and diffused vanadium cations. The ^{29}Si NMR shows that the silica particles interact via hydrogen bonds with the sulfonic groups of Nafion and the diffused vanadium cations. Chen et al. synthesized a series of sulfonated poly(fluorenyl ether ketone)/(sulfonated) SiO$_2$ (SFPEK/SiO$_2$) hybrid membranes for VRFB applications.[130] Table 9.9 illustrates the proton conductivity, VO^{2+} permeability, and Φ of the hybrid membranes, pristine SPFEK, and Nafion 117.[130]

It can be seen in Table 9.9 that both the proton conductivity and VO^{2+} permeability increase with the temperature. The SPFEK/3%SiO$_2$ hybrid membrane has higher proton conductivity and VO^{2+} permeability than pristine SPFEK membrane. This is

TABLE 9.9

Proton Conductivity, VO²⁺ Permeability, and Φ of the Hybrid Membranes, Pristine SPFEK, and Nafion 117

Membranes	Conductivity ($\times 10^{-2}$ S cm⁻¹)		VO²⁺ Permeability ($\times 10^{-12}$ m² s⁻¹)		Φ^a ($\times 10^6$ S cm⁻³ s)	
	20°C	40°C	20°C	40°C	20°C	40°C
Nafion 117	3.22	4.73	4.90	12.32	0.66	0.38
SPEEK	2.19	3.54	1.05	2.91	2.09	1.22
SPEEK/3%SiO₂	3.18	4.47	1.40	3.60	2.27	1.24
SPEEK/9%SiO₂	2.09	3.29	0.87	2.35	2.40	1.40
SPEEK/15%SiO₂	1.38	1.98	0.47	1.33	2.94	1.49
SPEEK/3%SiO₂–SO₃H	2.24	3.61	0.98	2.65	2.29	1.36
SPEEK/9%SiO₂–SO₃H	2.37	3.76	1.03	2.74	2.30	1.37
SPEEK/15%SiO₂–SO₃H	2.48	3.82	1.05	2.79	2.36	1.37

Source: Chen, D. et al., *J. Power Sources*, 195, 7701, 2010.

ᵃ Φ means selectivity of various membranes (proton conductivity/VO²⁺ permeability).

because the incompatible SiO_2 particles expanded the network of SPFEK resulting in larger cavities for the transportation of hydrated ions. The larger content of SiO_2 will fill the cavities itself and block the hydrophilic pathways, which is validated by the declines of proton conductivity and VO²⁺ permeability of SPFEK/9% SiO_2 and SPFEK/15% SiO_2 hybrid membranes. For the SPFEK/SiO_2–SO₃H hybrid membranes, the proton conductivities increased, while the VO²⁺ permeabilities decreased slightly. This is due to the good adhesion between SiO_2–SO₃H particles and SPFEK matrix, while the in situ–formed silica-rich layer with higher sulfonic acid group concentration favors the transportation of protons but suppresses the permeation of VO²⁺. In conclusion, all the hybrid membranes have comparable proton conductivity and dramatically lower VO²⁺ permeability than Nafion.[130]

The curves of the discharge voltages versus discharge times of the VRFBs assembled with the hybrid membranes and the SPFEK membrane at the current densities of 20 and 60 mA cm⁻² were measured by Chen et al.[130] The voltage efficiency, which is represented by the discharge voltage, is mainly influenced by the resistance of the membrane. However, the coulombic efficiency of different cells, which is showed by the discharge time when the discharge current is the same, is affected by the resistance as well as the vanadium ion permeability of the membrane. It is observed that for the VRFB-SPFEK/SiO_2 and VRFB-SPFEK, the changing tendency of discharge voltage was as follows: VRFB-SPFEK/3% SiO_2 > VRFB-SPFEK/9% SiO_2 > VRFB-SPFEK > VRFB-SPFEK/15% SiO_2. There was no obvious difference in the discharge characteristics of the VRFBs assembled with different SPFEK/SiO_2–SO₃H hybrid membranes. The voltage efficiency and coulombic efficiency of VRFB-SPFEK/15% SiO_2–SO₃H were little higher than that of VRFB-SPFEK at the current density of 20 mA cm⁻², and the discrepancy in voltage efficiency was more significant at the current density of 60 mA cm⁻².[130]

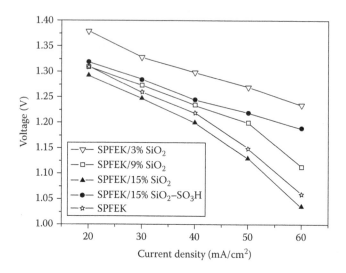

FIGURE 9.24 Effect of current density on average discharge voltage of the VRFB assembled with SPFEK and hybrid membranes. (From Chen, D. et al., *J. Power Sources*, 195, 7701, 2010.)

The effect of average discharge voltage on current density of the VRFB assembled with SPFEK and hybrid membranes is shown in Figure 9.24. It is apparent that the average discharge voltage decreases with the increase in membrane resistance and discharge current density. The VRFB-SPFEK/3% SiO_2, VRFB-SPFEK/9% SiO_2, and VRFB-SPFEK/SiO_2–SO_3H possess higher average discharge voltage than the VRFB-SPFEK at all the tested current densities.[130]

9.5.6 APPLICATIONS FOR OTHER ENERGY POWERS

Recently, hybrid/composite membranes have also been used as anion-exchange membranes (AEMs) for fuel cell applications.[131–137] AEMs are also involved in a wide range of applications such as cleanup, liquid separation, and decontamination processes.[131,138–141] An anionic membrane should fulfill stringent mechanical, chemical, and thermal requirements, and the required properties are summarized as follows: (1) an efficient transfer of the hydroxyl ions ensuring high ionic conductivity, (2) high membrane barrier capability to fuel crossover ensuring high fuel cell performance and safety, and (3) superior chemical stability in a strong acidic or basic medium and good membrane durability.[131]

Wu et al. synthesized a series of anion-exchange organic/inorganic hybrid membranes based on quaternizing of the copolymer of vinylbenzyl chloride (VBC) and γ-methacryloxypropyl trimethoxy silane (γ-MPS), followed by a sol–gel reaction to the copolymer and monophenyltriethoxysilane (EPh).[132] The oxidative resistance can be enhanced with a proper dosage of EPh added (in the range of 25%–50%), and the durability time in Fenton's reagent can reach 12 h. TGA analysis indicates that the membranes have thermal degradation temperatures higher than 257°C.

The membranes showed a tensile strength 65 MPa with an elongation at break of more than 30%. The conductivity of the fully hydrated membranes is in the range of 22.7–43.3 mS cm^{-1} at RT. Wu et al. also prepared a series of silica/poly(2,6-dimethyl-1,4-phenylene oxide) (PPO) anion-exchange hybrid membranes with heat treatment at 120°C–140°C.[133] They found that the physicochemical properties of the membranes, including ion-exchange property, hydrophilicity, OH$^-$ conductivity, and tensile property, can be easily controlled by adjusting the heating temperature and time.

Up to date, AEMs have been reported to show high potential for direct alcohol fuel cells (DAFCs), including both DMFC and direct ethanol fuel cells (DEFCs). Shahi group developed SPEEK-zeolite-zirconium hydrogen phosphate (ZrP) nanocomposite PEM by in situ infiltration and precipitation, focusing on reduction in the methanol permeability at moderate temperature of 70°C without sacrificing proton conductivity.[134] Very recently, Wang and Wang synthesized a series of organic/inorganic hybrid membranes from cross-linked quaternized chitosan (QCS, the quaternization degree [DQ] of the QCS was 80.8% ± 3.5%) with different contents of TEOS.[135] It was found that the anionic conductivities of the hybrid membranes showed a level of 10^{-2} S cm^{-1} at 80°C. The methanol permeability of the membranes was in a range from 7.5 × 10^{-6} to 1.5 × 10^{-6} cm^2 s^{-1}. However, the hydroxyl conductivity of the hybrid membranes treated with 1 mol L^{-1} KOH generally decreased by 12.8%–45.5% compared with the untreated one.

Proton-conducting hybrid membranes composed of PVA, HPW, and (diethylenetriamine)pentaacetic acid (DTPA) have been reported attractive for DEFC applications.[136] Ethanol permeabilities obtained for PVA/HPW/DTPA was about two orders of magnitude smaller than Nafion 117, and the proton conductivity in ethanol was in the order of 10^{-3} S cm^{-1} with added 4 wt% of DTPA and generally increases with the addition of HPW.[136] Zhao group synthesized cross-linked PVA hybrid membranes composited with layered double hydroxide (LDH).[137] They found that incorporating 20 wt% LDH into the PVA resulted in not only a higher ionic conductivity but also a lower ethanol permeability. The performance test of the DEFC using the PVA/LDH hybrid membrane shows that the fuel cell can yield a power density of 82 mW cm^{-2} at 80°C, which is much higher than that of the AEM DEFC employing the quaternary ammonium group functionalized membrane. The ionic conductivity is significantly increased after the addition of LDHs, from 6.2 mS cm^{-1} of neat PVA to 17.0, 26.6, and 27.8 mS cm^{-1}, corresponding to PVA/10 LDH, PVA/20 LDH, and PVA/30 LDH at 30°C, respectively. The reason why the addition of LDHs into PVA improves ionic conductivity can be explained as follows.[137] First, the Mg-AlCO$_3$$^{2-}$ LDH was testified to be a hydroxide conductor by a concentration cell as mentioned in Section 9.1. Meanwhile, alkaline-doped PVA was also demonstrated to conduct the OH$^-$. Therefore, there exists synergic effect between PVA and LDHs due to the strong interfacial adhesion. Secondly, the crystallinity of PVA polymer was inhibited by the LDHs, and the amorphous phase is increased as discussed earlier. The improvement of amorphous domain created more free volumes, which are favorable for the increase in the ionic conductivity.

Except for applications as AEMs utilized in DAFC, hybrid/composite membranes can also be employed for pervaporation of liquid separation such as dehydration of ethanol/water azeotrope,[138] for acid recovery,[139] and for diffusion dialysis process.[140]

Based on Pandey and Shahi research, CS/silica hybrid membranes cross-linked with formaldehyde displayed high selectivity of 5282 and total flux of 0.59 L m^{-2} at 30°C in ethanol/water mixture (90%), which gives an insight into new nonhazardous and environmentally benign materials with low cost.[138] Wu et al. prepared anion-exchange hybrid membranes for diffusion dialysis process, from PVA and multi-alkoxy silicon copolymer with high thermal stability, mechanical strength, and swelling resistance.[139] They also developed hybrid membranes containing both $-N^+(CH_3)_3 Br^-$ and $-OH$ groups from quaternized poly(2,6-dimethyl-1,4-phenyleneoxide)(QPPO) and PVA. The hybrid membranes displayed a good acid recovery capability such as HCl; the sorbed HCl concentration increased from 0.36 to 2.9 mol L^{-1} with an increase in $-OH$ group content. The dialysis coefficient of HCl (U_H) increased from 0.021 to 0.049 m h^{-1}, but the separation factor decreased from 39 to 26.[140]

Utilizations of hybrid/composite membranes can also be found for ultrafiltration membranes toward protein separation[141] and functionalized polyelectrolyte capsules using filamentous *Escherichia coli* cells,[142] which will further extend the application fields for functionalized hybrid/composite membranes.

9.6 TYPICAL EXAMPLE ANALYSIS FROM MATERIAL SELECTION, SYNTHESIS, CHARACTERIZATION, AND APPLICATIONS

9.6.1 EXAMPLE OF HYBRID/COMPOSITE MEMBRANES FOR PEMFCs[143]

9.6.1.1 Materials Selection

It has been demonstrated that proton exchange membrane fuel cells (PEMFC) operation at high temperature (120°C–150°C) could enhance the kinetic reaction rates at both electrodes and improve the carbon monoxide tolerance and simplified heat and water managements. Incorporation of hydrophilic inorganic fillers to enhance the water retention has been reported to enable the PEMs with high conductivity at high temperature. Inorganic oxide fillers such as silica, titania, and zeolite have been applied to modify humidified Nafion membranes to improve the water retention ability at elevated temperatures. Since silica nanoparticles have a strong tendency to aggregate particularly at high particle concentrations, the effective incorporation and homogeneous dispersion of silica nanoparticles within the polymer matrix are the key challenges to be addressed. Hence, silica is selected as a typical inorganic filler to modify commercially available Nafion. 1-Methyl-3-[(triethoxysilyl)propyl] imidazolium chloride-surface functionalized silica (Im-SiO$_2$) nanoparticles were synthesized and used as the additives. A PIL, N-ethylimidazolium trifluoromethane sulfonate ([EIm][TfO]), was chosen as a nonaqueous proton carriers. Im-SiO$_2$-doped cross-linked poly(styrene-co-acrylonitrile) hybrid proton-conducting membranes were prepared by in situ polymerization of a mixture containing monomer oils, PIL, and well-dispersed Im-SiO$_2$ nanoparticles.

9.6.1.2 Synthesis

First, monodispersed SiO$_2$ spherical nanoparticles with the mean diameter of 130 nm were synthesized as follows: a mixture of TEOS (5 mL) and ethanol

(5 mL) was added into the flask at a rate of about 1 mL min^{-1} via a dropping funnel. The concentrations of NH_3, H_2O, and TEOS in the final mixture were ca. 0.3, 1.7, and 0.2 M, respectively. After stirring for 4 h at RT, the nanoparticles were obtained by centrifugation. The nanoparticles were washed with ethanol several times and dried under vacuum for 24 h.

Second, 1-methyl-3-[(triethoxysilyl)propyl]imidazolium chloride (TMICl) was synthesized as follows in a typical synthetic procedure: a mixture of 1-methylimidazole and 3-triethoxy-silylpropyl chloride (equalmolaramount) was heated at 80°C for 24 h under nitrogen atmosphere to synthesize TMICl. After cooling down, the pale-yellow viscous liquid was washed with ether for three times and dried under vacuum at 80°C for 24 h.

Third, TMICl functionalized silica nanoparticles were prepared as follows in a typical synthetic procedure: 0.2 g TMICl was added into a suspension of 2.0 g SiO_2 nanoparticles in 50 mL dry chloroform. The mixture was refluxed under nitrogen atmosphere at 80°C for 24 h. The product was centrifuged and washed with chloroform to remove the unreacted TMICl. The obtained TMICl functionalized silica nanoparticle is referred as Im-SiO_2 nanoparticle as shown in Figure 9.25.

Fourth, N-ethylimidazolium trifluoromethanesulfonate ([EIm][TfO]) was synthesized as follows in a typical procedure: N-ethylimidazole was dissolved in CH_2Cl_2 and cooled with an ice bath, and then trifluoromethane sulfonic acid was added dropwisely. The molar ratio of N-ethylimidazole/trifluoromethane sulfonic acid was kept at 1:1. The reaction mixture was allowed to warm to RT slowly and stirred continuously for 2 days. CH_2Cl_2 was evaporated and the resultant viscous yellow oil was washed with ethyl acetate and ether twice and then dried in dynamic vacuum at 80°C for 24 h before use.

And finally, the resulting hybrid membranes were prepared as follows in a typical procedure: a mixture containing styrene/acrylonitrile (1:3 weight ratio, 60 wt%), [EIm][TfO] (40 wt%), DVB (2 wt% to the formulation based on the weight of monomer), and 1 wt% of benzoin isobutyl ether (photoinitiator), with or without Im-SiO_2 nanoparticles was stirred and ultrasonicated to obtain a homogeneous solution,

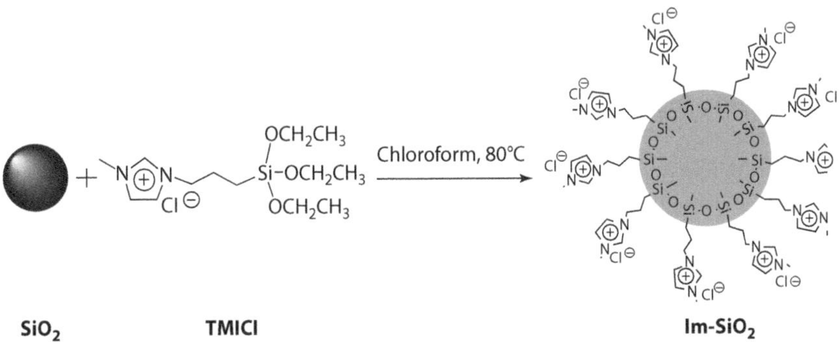

SiO$_2$ TMICl   Im-SiO$_2$

FIGURE 9.25 General synthesis procedures for the preparation of Im-SiO_2 nanoparticles. (From Lin, B. et al., *Fuel Cells*, 13, 72, 2013.)

which was then cast into a glass mold and photo cross-linked by irradiation with UV light of 250 nm wavelength for 30 min at RT. Poly(styrene-co-acrylonitrile) (PSAN) has been widely used as membrane materials due to their good chemical stability, mechanical properties, and ease of processing. Therefore, PASN was chosen as a polymer matrix for the preparation of hybrid PEMs, and [EIm][TfO] was used as the anhydrous proton conductor.

9.6.1.3   Characterization

Im-SiO$_2$ nanoparticles were applied as the additive for the preparation of PIL-based hybrid PEMs. Compared with the unmodified SiO$_2$ nanoparticles, floccules on the surface of Im-SiO$_2$ nanoparticles might be due to the tethered of TMICl were observed from TEM images.

The FTIR spectra display a broad stretching band centered at 3470 cm^{-1}, which is ascribed to the hydroxyl bond (OH) groups of SiO$_2$ nanoparticles. An absorption peak at 1634 cm^{-1} can be assigned as hydrogen-bonding bending vibrations. The absorption peaks at 1243 and 803 cm^{-1} can be assigned as Si–O–Si stretching vibrations. The broad peak at 471 cm^{-1} is the characteristic absorption peak of Si–O rocking vibrations. The spectrum of TMICl confirms the presence of absorption at about 780 cm^{-1} related to the vibration of Si–O. A band about 1290 cm^{-1} is related to the vibration of C–N bond of imidazole. The absorption peaks at 1443 and 1575 cm^{-1} are related to C=N and C=C vibrations. The peaks at about 2977 and 3100 cm^{-1} are assigned to the stretching of methylene and methyl (–CH$_2$– and –CH$_3$). The results of FTIR confirm the chemical structure of TMICl. The new bands corresponding to TMICl are observed at 3102, 2944, 1569, and 1458 cm^{-1}, indicating a successful surface modification of SiO$_2$. The nanoparticle surface modification of SiO$_2$ is further confirmed by the mass loss in the TGA curves. The main mass loss of SiO$_2$ nanoparticles at 650°C can be assigned to the condensation of –OH on the surface of nanoparticles. There was no mass loss observed from 650°C to 750°C, and the mass loss is 7.8% for Im-SiO$_2$ nanoparticles due to the degradation of TMICl attached on the surface of Im-SiO$_2$ nanoparticles. All the membranes lose less than 2% in weight up to 200°C, and the temperature at 5% weight loss was higher than 300°C. These results confirm that this type of PIL-based membranes indeed confer a high thermal stability, far beyond the range of interest for application in PEMFCs. However, addition of Im-SiO$_2$ nanoparticles could not dramatically improve the decomposition temperature of the composite membranes.

The SEM images of the cross section of the PIL-based hybrid membranes containing 15 wt% Im-SiO$_2$ nanoparticles are investigated. The defects observed in the hybrid membranes were caused by the cutting damage. It can be found that the Im-SiO$_2$ nanoparticles are homogenously dispersed in the hybrid membranes, and no agglomeration of Im-SiO$_2$ nanoparticles was observed in the sample. Although the PIL component could not be distinguished from the SEM images in the present work, we believe that the homogenous distribution of the Im-SiO$_2$ nanoparticles favors the formation of PIL conductive networks in the membrane, which therefore could enhance the conductivity of the hybrid membrane.

9.6.1.4   Performance/Application

Figure 9.26a exhibits the conducting performance for PIL-based hybrid membranes containing 40 wt% of [EIm][TfO], incorporated with or without Im-SiO_2 nanoparticles, respectively, at a wide range of temperature under dry condition. The pristine membrane shows the conductivity of 5.01×10^{-5} S cm^{-1} at 40°C. With the

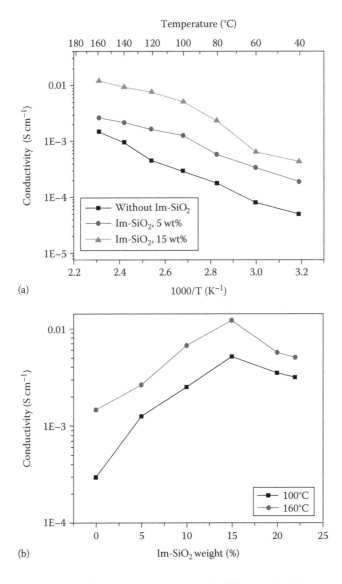

(a)

(b)

FIGURE 9.26   (a) Conductivity Arrhenius plots of the PIL-based hybrid membranes as a function of temperature. (b) Conductivity of the PIL-based hybrid membranes as a function of the weight fraction of Im-SiO_2 nanoparticles at 100°C and 160°C. (From Lin, B. et al., *Fuel Cells*, 13, 72, 2013.)

incorporation of 5 wt% Im-SiO$_2$ nanoparticles, the conductivity was considerably improved, reaching the value of 10^{-4} S cm^{-1} under the same experimental conditions. All the samples show no decay in conductivity even at the temperature as high as 160°C, reaching the values in the order of 10^{-3} S cm^{-1}. The enhancement of the conductivity is probably due to the PIL continuous networks or interconnected channels formed in the membrane, which caused by the distribution of Im-SiO$_2$ nanoparticles

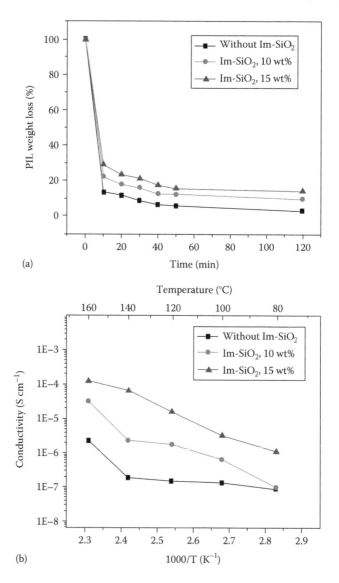

FIGURE 9.27 (a) Results of ionic liquid retention ability test carried out in distilled water; (b) conductivity Arrhenius plots of membranes after water-extraction treatment for 2 h. (From Lin, B. et al., *Fuel Cells*, 13, 72, 2013.)

and the capillary interaction between the Im-SiO$_2$ nanoparticles. Figure 9.26b shows the effect of Im-SiO$_2$ variation on the conductivity of the hybrid membranes. The conductivity increased with increasing the amount of Im-SiO$_2$ nanoparticles, however, dramatically decreased with the addition of excess Im-SiO$_2$. This is probably because that further addition of excess Im-SiO$_2$ nanoparticles could block the formed ion transport channels. In addition, Im-SiO$_2$ themselves are less conductive than PIL; therefore, it is not surprising that excess addition of Im-SiO$_2$ nanoparticles declined the conductivity of the corresponding hybrid membranes.

The progressive release of PILs during the operation in the fuel cells is a great challenge. The retention ability of the PIL in the composite membrane was confirmed by determining the weight loss of membrane samples after being immersed in distilled water, as shown in Figure 9.27a. It can be seen that the pristine membranes without the Im-SiO$_2$ nanoparticles lost most of the ionic liquid component (about 90 wt%) in 10 min after being immerged in water, while the incorporation of Im-SiO$_2$ nanoparticles could slow the release of ionic liquid component. It is worth noting that there is still about 14.2 wt% of PILs kept in the hybrid membrane doped with 15 wt% Im-SiO$_2$ nanoparticles, even after being immersed in distilled water for 120 min, which is more effective than the silica nanoparticles without surface functionalized. It is considered that the enhancement of the PIL-holding strength can be ascribed to the ionic forces between Im-SiO$_2$ nanoparticles and PIL. The enhanced retention ability of the PILs in the composite membrane was further confirmed by conductivity measurements of the water-extracted membranes. Figure 9.27b shows the conductivity Arrhenius plots of membranes after 120 min water-extraction treatment. The conductivity of each samples dramatically decreased after the extraction of PIL by water. However, the membrane containing Im-SiO$_2$ nanoparticles still shows much higher conductivity than pristine membrane. This result further confirmed that Im-SiO$_2$ nanoparticles are effective in holding ionic liquids.

9.6.2   EXAMPLE OF HYBRID/COMPOSITE MEMBRANES FOR DMFCs

9.6.2.1   Materials Selection

Since the late 1960s, Nafion has been the most utilized proton-conductive membrane in DMFCs. However, due to temperature limit of Nafion membrane and methanol crossover, DMFC can only be operated below 100°C. As observed through previous studies, DMFC operated up to 100°C can achieve higher power output compared to that operated at lower temperature. In order to achieve higher power output, membrane cannot only withstand high temperature but also provide higher proton conductivity and lower methanol crossover when operated at high temperatures.[144] Up to date, many kinds of composite materials and inorganic fillers have found applications to modify Nafion for fabrication of organic/inorganic or acid/base membranes, as elaborated in the former sections.

Zirconium phosphate (ZrP) has been widely used as inorganic fillers to modify Nafion to get high performances in both PEMFC and DMFC at high temperatures.[144–146] It was observed that a higher loading of ZrP made the composite membrane less sensitive to the dehydration effect and reduced methanol crossover.

In this section, zirconium phosphate is selected as the filling materials to form ZrP/Nafion hybrid membranes.

9.6.2.2 Synthesis

The composite membrane was prepared using Nafion 115 as a substrate. The treated Nafion 115 with a given dry weight was swollen in ethanol–water solution (2:3) at 80°C for an hour, and then it was immersed in 0.06 M $ZrOCl_2$ solution for at least an hour at 80°C to exchange H^+ in Nafion 115 with Zr^{4+}. After being rinsed with deionized water, the membrane was dipped in 1 M H_3PO_4 overnight at 80°C. Finally, the resulting membrane was repeatedly rinsed with deionized water to remove excessive H_3PO_4. The resulting composite membrane was dried in the vacuum oven at 80°C for 24 h and then weighed. The content of ZrP in the composite membrane was about 23 wt%.

9.6.2.3 Characterization

Liquid uptake (A) was measured by immersing the membranes into methanol solution with various concentrations and calculated as follows:

$$A = \frac{W_{wet} - W_{dry}}{W_{dry}} \times 100\% \qquad (9.4)$$

where
W_{wet} is the weight of wet membrane
W_{dry} is the weight of dry membrane

It was found that the liquid uptakes of both membranes increased linearly with increasing methanol concentration. For example, the liquid uptake of Nafion 115 increased from 34.3 wt% in 0 M methanol solution to 58.6 wt% in 10 M methanol solution, while that of the composite membrane increased from 28.3 to 37.5 wt% in the corresponding methanol solution. This suggests that Nafion 115 absorbed more liquids and swelled more seriously than ZrP/Nafion 115 in high concentration methanol solution. It is considered that the mechanical stability for ZrP/Nafion 115 was improved due to the incorporation of ZrP. The reason may be as follows: when Nafion 115 was soaked into the methanol solution, methanol molecules could diffuse easily into the ion clusters within Nafion 115 like water molecules. However, for ZrP/Nafion 115, ZrP in the pores possibly occupied an amorphous region that was originally filled with water and methanol molecules in Nafion 115, resulting in the decrease in free volume, cluster size, and the percent sorption capacity of methanol.

The TGA curve of composite membranes indicated that there was a gradual weight loss up to about 280°C, due to the evaporation of a little water adsorbed into the composite membrane during the course of preservation. Upon further heating, the composite membrane exhibited a rapid weight loss from 280°C to 310°C, corresponding to the decomposition of sulfonic acid groups. From 310°C to 350°C, the weight loss is attributed to the decomposition of polar perfluorosulfonic vinyl ether segments $(P = CF_2 - CF(O - CF_2 \cdots))$ and after which, occurs the decomposition of nonpolar terafluoroethylene segments $(N = CF_2 - CF_2)$ of Nafion 115.

The last remainder should be inorganic phase-ZrP. Both the pristine and composite membranes exhibited comparable thermal stability. It can be concluded that ZrP did not deteriorate the thermal stability of the composite membrane. In addition, the TGA results confirmed the existence of ZrP in the composite membrane, and the estimated ZrP's content agreed with the nominal amount.

The XRD pattern of ZrP/Nafion 115 composite membrane was investigated. The interlayer distance d is calculated from Bragg diffraction equation:

$$2d \sin \theta = n\lambda \tag{9.5}$$

where

d is the distance between atomic layers in a crystal
λ is the wavelength of the x-ray beam
n is an integer
θ is half of the diffraction angle

It is considered that two diffraction peaks at d-spacings of 0.52 nm (100) and 0.23 nm (101) are attributed to the crystalline perfluorinated backbone of Nafion 115. Others (0.44, 0.365, 0.264, and 0.170 nm) are assigned to ZrP crystal in the composite membrane, because these peaks matched up with the typical peaks of crystalline ZrP in literatures.[147] At the same time, these diffraction peaks also agreed well with the XRD pattern of ZrP powder. According to the Debye–Sherrer formula, the mean particle size of ZrP can be calculated to be about 5.0 nm from two peaks at 0.264 and 0.17 nm. The EDX pattern effectively verified the presence of ZrP. It was also found that the atomic ratio of P and Zr was 1.74, which is close to 2, the value in the molecular formula $Zr(H_2PO_4)_2$.

The IECs of Nafion 115 and ZrP/Nafion 115 were determined by back titration method. The IEC of Nafion 115 was 0.909 mequiv. g^{-1}, which is identical to the reported values. When about 23 wt% ZrP was incorporated into Nafion 115, the IEC of the resulting membrane increased significantly from 0.909 to 1.94 mequiv. g^{-1}. This favorable result is reasonable because it was reported that ZrP is a well-known ion exchanger[148] and its total IEC was 6.64 mequiv. g^{-1}.[149] Thus, the IEC of composite membrane can be calculated to be 2.2 mequiv.g^{-1}, which is close to the experimental value of 1.94 mequiv. g^{-1}. This suggests that as expected, the number of acid sites in the composite membranes was increased due to the filling of ZrP, thus avoiding large loss of the proton conductivity.

9.6.2.4   Performance/Application

The proton conductivity of the membrane has been attributed to a *liquid-like* proton conductivity mechanism. According to the mechanism, the protons are transported through hydrogen-bonding network in the water-rich ionic pores and channels of the membrane. It was found that the proton conductivity of Nafion 115 was 0.10 S cm^{-1}, which agreed well with the results published.[145] While for the composite membrane, the proton conductivity was 0.084 S cm^{-1}, which was slightly lower than that of pristine Nafion. To some degree, as a well-known ion exchanger and solid acid, ZrP entrapped in the ion cluster could make up for the loss of the proton conductivity, which resulted in the decrease of the free volume in the ion cluster. The reason is that the structure of α-ZrP is a layered one, and each layer consists of zirconium atoms lying in a plane and bridging through phosphate groups, which locate alternately above and below this plane. This layered structure is advantageous to the proton transport because it has many pendants (P–O–H groups) and forms a hydrogen-bonding network with water. However, it was reported that zirconium phosphate glasses had a proton conductivity of 10^{-2} S cm^{-1} at ambient temperature under the condition of full humidification[150]; thus, this inherent lower proton conductivity still resulted in mild loss of the composite membrane's proton conductivity.

The methanol crossover of ZrP/Nafion 115 increased with temperature but decreased by about 50% compared with that of Nafion 115 within the entire temperature range from 30°C to 90°C. Arrhenius plots of limiting current density (j_{lim}) of the methanol crossover as a function of temperature for Nafion 115 and the composite membrane from 30°C to 90°C can be drawn. The activation energy (E_a) of the methanol crossover through the membrane is calculated by fitting to the Arrhenius equation:

$$j_{lim} = A \cdot \exp\left(\frac{-E_a}{RT}\right) \tag{9.6}$$

where j_{lim}, A, E_a, R, and T denote the limiting current density of the methanol crossover, the frequency factor, the activation energy for the methanol crossover, the gas constant, and the absolute temperature, respectively. It is noted that the values of E_a of the composite membrane is 18.7 kJ mol^{-1}, which is higher than that (16.3 kJ mol^{-1}) of Nafion 115. Higher activation energy suggests that it is difficult to diffuse through the composite membrane for methanol molecules. It is obvious that the composite membrane is endowed with a tortuous diffusion pathway for methanol across the membrane due to the filling of ZrP. As a result, methanol crossover through the composite membrane is suppressed.

For DMFC applications, a higher methanol concentration is strongly desirable due to higher specific energy if the problem of methanol crossover could be resolved. The behaviors of Nafion 115 and composite membranes are evaluated when high-concentration methanol solutions (5 or 10 M) were employed in DMFC at 75°C. The performances of single cells are tested. When the cell was pumped with 5 M methanol solution, the composite membrane showed better performance than Nafion 115 mainly due to improved mass transportation polarization. Their peak power densities were 96.3 and 91.6 mW cm^{-2}, respectively. When the methanol concentration was further increased to 10 M, the peak power density of DMFC with composite membrane was 76.19 mW cm^{-2}, which was much higher than that (42.4 mW cm^{-2}) of DMFC with Nafion 115. The satisfying results should be owed to lower methanol crossover of the composite membrane.

Considering the importance of the PEM's durability for the real application and commercialization of fuel cells, the life test of single cell with composite membrane was performed with feed methanol concentration of 1 and 10 M, respectively. It can be found that DMFC with ZrP/Nafion membrane can run stably 100 h above 0.4 V, when 1 M methanol solution was used as fuel. Although the durability of 100 h is small, the preliminary result is inspiring. The reason is due to that the durability test was performed after high-methanol solution (5 and 10 M) was fed into the anode again and again in order to determine the polarization curves of DMFC at different temperatures. The lifetime test of high-concentration DMFC with composite membrane was also performed at 75°C. It is suggested that DMFC with composite membrane can run stably at least 7 h even at very high concentrations.

9.7 CHALLENGES AND PERSPECTIVES

Hybrid/composite membranes have shown great potential for not only PEMFC and DAFC but also Li$^+$ battery and VRFB applications. Addition of functional inorganic fillers to the polymeric membranes has not only improved the performance of the polymer at higher operating temperature but also led to increased thermal, mechanical, and WU properties of the nanocomposite membrane. Despite the fact that inorganic-based nanocomposite membranes for fuel cells are under investigation for many years, no worthwhile/breakthrough membrane material is found yet and interest in the field is waning in recent years. The improvement in properties of the inorganic nanoparticle-based polymeric membranes highly depends on the degree of dispersion of the nanofillers in the polymer matrix. Surface modifications of silicates along with the addition of proton-conducting and hydrophilic fillers result in increased conductivity of the nanocomposite membranes. Leaching of PWA, sulfonated POSS, and highly sulfonated silica during continuous usage may restrict the end applications in MEA. The optimum degree of sulfonation of silica and addition of PWA and sulfonated POSS along with mesoporous silica particles may be beneficial to avoid leaching.[90]

Mesoporous silica in combination with ionic liquid provides an interconnecting PIL channel, which can operate without external RH, and thus produced greater improvement in proton conductivity of the hybrid/composite membrane. Hence, the present research could be directed to the utilization of mesoporous silica as a potential polymer matrix filler for high-temperature PEMFC. Fine tuning of the surface area and pore size of the mesoporous silica in combination with the ionic liquid can lead to synergistic improvement in the technical properties, the proton conductivity, and cell performance of the nanocomposite membrane.[90]

Sulfonated aromatic polymers (SAPs) are promising inexpensive polymer electrolyte membranes for fuel cells, but there still exist some challenges about this kind of materials including excessive swelling in water, poor mechanical strength, and low dimensional stability, especially for highly sulfonated SAP. Cross-linking is widely proposed as an efficient strategy to deal with these challenges due to the reticulation effect of entangled polymer chains. Composite SAPs could be cross-linked (1) via covalent bonds, including esterification, addition, Friedel–Crafts reactions, and formation of $-SO_2-$ bridges, (2) ionic bonds, and (3) combined covalent and ionic cross-linking.[151] In recent in-depth investigations, some strategies against decrease of proton conductivity of cross-linked SAP were explored: (1) a high initial degree of sulfonation, (2) the introduction of additional carboxylic or sulfonated acid groups, (3) an optimized cross-linking degree or chain length of cross-linkers that can tune the WU and free volume, (4) an optimized microstructure that can contribute to a high proton conductivity by forming fine nanoscale phase separation, and (5) the introduction of additional inorganic solid acids into the cross-linked SAP matrix. However, further careful design and optimization are necessary in order to improve the overall performances.[151]

DMFCs based on methanol aqueous solution do not need humidifying system and special thermal management ancillary devices. Hence, the power and energy densities are superior, even when compared with indirect fuel cell and newly developed lithium-ion batteries.[152] One of the challenges in current DMFC research is the development of novel proton-conductive membranes for improving the performance of DMFCs by improving proton conductivity and reducing methanol permeability. The polymer membrane should simultaneously maintain a large proton conductivity (typically 0.1 S cm^{-1} at 90°C) and have sufficient chemical, thermal, and mechanical stabilities, low permeability to reactants, low cost, and ready availability.[9,144,152] Since the aim of membrane properties for DMFC is to increase the proton conductivity and decrease the methanol permeability, the selectivity factor can be regarded as a guide to develop the better characteristic of membrane for DMFC usage. The higher selectivity factor means the membrane can provide the better performance in DMFC. Upon researches up to date, the selectivity factor sometimes not depend on temperature but more depending on material used in the membrane preparation itself. However, for practical applications, the selectivity factor is not enough. The main parameter that provides an indication on how far this membrane technology from the practical application is the DMFC power density. The maximum power density is not only influenced by the material used as the composite membrane but also usually affected by the temperature of DMFC operation. So, in order to create the membrane with better performance, extensive researches must be conducted in the future using various types of materials that not only have high mechanical and thermal durability but also can produce high power density. In order to get the higher power density, several approaches can be tried in the future such as stacking the DMFC, using bigger active area, or using heater to maintain the high temperature, whereas some other obstacles such as water management, heat management, and expenditure need to be taken into consideration.[153]

To date, different kinds of energy storage techniques have been developed for different applications. Each technology has its own advantages and limitations and is suitable for a limited range of applications, among which vanadium redox flow battery has enormous impact on the stabilization and smooth output of renewable energy. Considering the performance requirement and cost, electrochemical systems or rechargeable batteries are becoming an ideal option for electricity energy storage due to their advantages such as not being limited by geographical requirement, the ability to be installed anywhere, the quite large control window for power rate and capacity, and the fast response time and low environmental footprint.[154] The most challenges for VRFB development might be the energy density and the cycle life and durability of electrolyte membranes, as well as the carbon electrode materials. The concept of the porous membrane overcomes the stability restriction caused by ion-exchange groups from traditional ion-exchange membranes, leading to great progress for metal oxide composited hybrid membranes.[154,155] Chemical and physical modifications of Nafion membranes to improve ion selectivity therefore receive a continuous research interest. And the development of nonfluorinated hybrid membranes has focused on sulfonated aromatic polymers in a hope that the rigid chain and less connected ionic cluster will prevent vanadium ion crossover.

9.8   CHAPTER SUMMARY

Hybrid/composite membranes, which typically contain high-performance inorganic materials (fillers) dispersed in a polymeric matrix, have gained ground as a means of combining the best characteristics of polymeric and inorganic materials while potentially overcoming their individual limitations. Hence, there comes the review of this chapter in order to give a glimpse on the development of the hybrid/composite membranes during this decade. It is important to incorporate the microstructure with the mechanism as well as the characteristics of the related materials to better understand the nature of the electrolyte membranes. This chapter gives an overview on membrane synthesis (Section 9.2), characterization (Section 9.3), and stability analysis, membrane conduction mechanism, and theoretical modeling (Section 9.4), together with performance analyses (Section 9.5), which are applicable for electrochemical devices for energy storage and conversion. Specifically, typical example analyses (Section 9.6) with respect to PEMFC and DMFC applications are introduced based on material selection, synthesis, characterization, and applications. Finally, challenges and perspectives for hybrid/composite membranes are highlighted briefly.

REFERENCES

1. Marcinkoski, J., Kopasz, J.P., Benjamin, T.G., Progress in the US DOE fuel cell subprogram efforts in polymer electrolyte fuel cells, *Int. J. Hydrogen Energy*, 2008, 33, 3894–3902.
2. Devanathan, R., Recent developments in proton exchange membranes for fuel cells, *Energy Environ. Sci.*, 2008, 1, 101–119.
3. Beuscher, U., Cleghorn, S.J.C., Johnson, W.B., Challenges for PEM fuel cell membranes, *Int. J. Energy Res.*, 2005, 29, 1103–1112.
4. Barbir, F., System design for stationary power generation, *Handbook of Fuel Cells—Fundamentals, Technology and Applications*, Vielstich, W., Lamm, A., Gasteiger, H.A. (eds.), Wiley, 2003, Chapter 51, p. 683.
5. Li, Q., He, R., Jensen, J.O., Bjerrum, N.J., Approaches and recent development of polymer electrolyte membranes for fuel cells operating above 100°C, *Chem. Mater.*, 2003, 15, 4896–4915.
6. Lakshminarayana, G., Vijayaraghavan, R., Nogami, M., Kityk, I.V., Anhydrous proton conducting hybrid membrane electrolytes for high temperature (>100°C) proton exchange membrane fuel cells, *J. Electrochem. Soc.*, 2011, 158, B376–B383.
7. Li, X., *Principles of Fuel Cells*, Taylor & Francis, New York, 2006, Chapter 10, p. 507.
8. Kamaruddin, M.Z.F., Kamarudin, S.K., Daud, W.R.W., Masdar, M.S., An overview of fuel management in direct methanol fuel cells, *Renew. Sust. Energ. Rev.*, 2013, 24, 557–565.
9. Ahmed, M., Dincer, I., A review on methanol crossover in direct methanol fuel cells: Challenges and achievements, *Int. J. Energy Res.*, 2011, 35, 1213–1228.
10. Neburchilov, V., Martin, J., Wang, H., Zhang, J., A review of polymer electrolyte membranes for direct methanol fuel cells, *J. Power Sources*, 2007, 169, 221–238.

11. Perghambardoust, S.J., Rowshanzamir, S., Amjadi, M., Review of the proton exchange membranes for fuel cell applications, *Int. J. Hydrogen Energy*, 2010, 35, 9349–9384.

12. Korin, E., Siton, O., Bettelheim, A., Fuel cells and ionically conductive membranes: An overview, *Rev. Chem. Eng.*, 2007, 23, 35–63.

13. Lufrano, F., Baglio, V., Staiti, P., Antonucci, V., Arico, A.S., Performance analysis of polymer electrolyte membranes for direct methanol fuel cells, *J. Power Sources*, 2013, 243, 519–534.

14. Di Noto, V., Boaretto, N., Negro, E., Pace, G., New inorganic-organic proton conducting membranes based on Nafion and hydrophobic fluoroalkylated silica nanoparticles, *J. Power Sources*, 2010, 195, 7734–7742.

15. Kwak, S.H., Yang, T.H., Kim, C.S., Yoon, K.H., Nafion/mordenite hybrid membrane for high-temperature operation of polymer electrolyte membrane fuel cell, *Solid State Ionics*, 2003, 160, 309–315.

16. Adjemian, K.T., Domieney, R., Krishnan, L., Ota, H., Majsztrik, P., Zhang, T., Mann, J. et al., Function and characterization of metal oxide-Nafion composite membranes for elevated-temperature H_2/O_2 PEM fuel cells, *Chem. Mater.*, 2006, 18, 2238–2248.

17. Sahu, A.K., Meenakshi, S., Bhat, S.D., Shahid, A., Sridhar, P., Pitchumani, S., Shukla, A.K., Meso-structured silica-Nafion hybrid membranes for direct methanol fuel cells, *J. Electrochem. Soc.*, 2012, 159, F702–F710.

18. Iulianelli, A., Basile, A., Sulfonated PEEK-based polymers in PEMFC and DMFC applications: A review, *Int. J. Hydrogen Energy*, 2012, 37, 15241–15255.

19. Maiti, J., Kakati, N., Lee, S.H., Jee, S.H., Viswanathan, B., Yoon, Y.S., Where do poly(vinyl alcohol) based membranes stand in relation to Nafion for direct methanol fuel cell applications, *J. Power Sources*, 2012, 216, 48–66.

20. Shahi, V.K., Highly charged proton-exchange membrane: Sulfonated poly(ether sulfone)-silica polyelectrolyte composite membranes for fuel cells, *Solid State Ionics*, 2007, 177, 3395–3404.

21. Seol, J.H., Won, J.H., Yoon, K.S., Hong, Y.T., Lee, S.Y., SiO_2 ceramic nanoporous substrate-reinforced sulfonated poly(arylene ether sulfone) composite membranes for proton exchange membrane fuel cells, *Int. J. Hydrogen Energy*, 2012, 37, 6189–6198.

22. Asensio, J.A., Sanchez, E.M., Gomez-Romero, P., Proton-conducting membranes based on benzimidazole polymers for high-temperature PEM fuel cells. A chemical quest, *Chem. Soc. Rev.*, 2010, 39, 3210–3239.

23. Yin, Y., Yamada, O., Tanaka, K., Okamoto, K., On the development of naphthalene-based sulfonated polyimide membranes for fuel cell applications, *Polym. J.*, 2006, 38, 197–219.

24. Li, S., Liu, M., Synthesis and conductivity of proton-electrolyte membranes based on hybrid inorganic–organic copolymers, *Electrochim. Acta*, 2003, 48, 4271–4276.

25. Honma, I., Nakajima, H., Nishikawa, O., Sugimoto, T., Nomura, S., Organic/inorganic nano-composites for high temperature proton conducting polymer electrolytes, *Solid State Ionics*, 2003, 162–163, 237–245.

26. Sahu, A.K., Selvarani, G., Pitchumani, S., Sridhar, P., Shukla, A.K., A sol-gel modified alternative Nafion-silica composite membrane for polymer electrolyte fuel cells, *J. Electrochem. Soc.*, 2007, 154, B123–B132.

27. Pan, J., Zhang, H., Chen, W., Pan, M., Nafion–zirconia nanocomposite membranes formed via in situ sol–gel process, *Int. J. Hydrogen Energy*, 2010, 35, 2796–2801.

28. Liu, D., Geng, L., Fu, Y., Dai, X., Qi, B., Lv, C., In situ sol–gel route to novel sulfonated polyimide–SiO_2 hybrid proton-exchange membranes for direct methanol fuel cells, *Polym. Int.*, 2010, 59, 1578–1585.

29. Hammami, R., Ahamed, Z., Charradi, K., Beji, Z., Ben Assaker, I., Ben Naceur, J., Auvity, B., Squadrito, G., Chtourou, R., Elaboration and characterization of hybrid polymer electrolytes Nafion-TiO_2 for PEMFCs, *Int. J. Hydrogen Energy*, 2013, 38, 11583.

30. Safronova, E.Y., Yaroslavtsev, A.B., Nafion-type membranes doped with silica nanoparticles with modified surface, *Solid State Ionics*, 2012, 221, 6–10.
31. Nakajima, T., Tamaki, T., Ohashi, H., Yamaguchi, T., Introduction of size-controlled Nafion/ZrO$_2$ nanocomposite electrolyte into primary pores for high Pt utilization in PEFCs, *J. Electrochem. Soc.*, 2013, 160, F129–F134.
32. Tang, H., Pan, M., Synthesis and characterization of a self-assembled Nafion/silica nanocomposite membrane for polymer electrolyte membrane fuel cells, *J. Phys. Chem. C*, 2008, 112, 11556–11568.
33. Tsai, C.H., Lin, H.J., Tsai, H.M., Hwang, J.T., Chang, S.M., Chen-Yang, Y.W., Characterization and PEMFC application of a mesoporous sulfonated silica prepared from two precursors, tetraethoxysilane and phenyltriethoxysilane, *Int. J. Hydrogen Energy*, 2011, 36, 9831–9841.
34. Kurniawan, D., Morita, S., Kitagawa, K., Durability of Nafion-hydrophilic silica hybrid membrane against trace radial species in polymer electrolyte fuel cells, *Microchem. J.*, 2013, 108, 60–63.
35. Lin, H., Zhao, C., Jiang, Y., Ma, W., Na, H., Novel hybrid polymer electrolyte membranes with high proton conductivity prepared by a silane-crosslinking technique for direct methanol fuel cells, *J. Power Sources*, 2011, 196, 1744–1749.
36. Gao, S., Zhao, C., Na, H., Chemically stable hybrid polymer electrolyte membranes prepared by silane-crosslinking and thiol-ene click chemistry, *J. Power Sources*, 2012, 214, 285–291.
37. Joseph, J., Tseng, C.Y., Hwang, B.J., Phosphonic acid-grafted mesostructured silica/Nafion hybrid membranes for fuel cell applications, *J. Power Sources*, 2011, 196, 7363–7371.
38. Chi, W.S., Patel, R., Hwang, H., Shul, Y.G., Kim, J.H., Preparation of poly(vinylidene fluoride) nanocomposite membranes based on graft polymerization and sol–gel process for polymer electrolyte membrane fuel cells, *J. Solid State Electrochem.*, 2012, 16, 1405–1414.
39. Molla, S., Compan, V., Gimenez, E., Blazquez, A., Urdanpilleta, I., Novel ultrathin composite membranes of Nafion/PVA for PEMFCs, *Int. J. Hydrogen Energy*, 2011, 36, 9886–9895.
40. Molla, S., Compan, V., Performance of composite Nafion/PVA membranes for direct methanol fuel cells, *J. Power Sources*, 2011, 196, 2699.
41. Molla, S., Compan, V., Lafuente, S.L., Prats, J., On the methanol permeability through pristine Nafion® and Nafion/PVA membranes measured by different techniques. A comparison of methodologies, *Fuel Cells*, 2011, 11, 897–906.
42. Lin, H.L., Wang, S.H., Nafion/poly(vinyl alcohol) nano-fiber composite and Nafion/poly(vinyl alcohol) blend membranes for direct methanol fuel cells, *J. Membr. Sci.*, 2014, 452, 253–262.
43. Lin, H.L., Wang, S.H., Chiu, C.K., Yu, T.L., Chen, L.C., Huang, C.C., Cheng, T.H., Lin, J.M., Preparation of Nafion/poly(vinylalcohol) electro-spun fiber composite membranes for direct methanol fuel cells, *J. Membr. Sci.*, 2010, 365, 114–122.
44. Pereira, F., Valle, K., Belleville, P., Morin, A., Lambert, S., Sanchez, C., Advanced mesostructured hybrid silica-Nafion membranes for high-performance PEM fuel cell, *Chem. Mater.*, 2008, 20, 1710.
45. Giffin, G.A., Piga, M., Lavina, S., Navarra, M.A., D'Epifanio, A., Scrosati, B., Di Noto, V., Characterization of sulfated-zirconia/Nafion composite membranes for proton exchange membrane fuel cells, *J. Power Sources*, 2012, 198, 66–75.
46. Alonso, R.H., Estevez, L., Lian, H., Kelarakis, A., Giannelis, E.P., Nafion–clay nanocomposite membranes: Morphology and properties, *Polymer*, 2009, 50, 2402–2410.
47. Chien, H.C., Tsai, L.D., Lai, C.M., Lin, J.N., Zhu, C.Y., Chang, F.C., Highly hydrated Nafion/activated carbon hybrids, *Polymer*, 2012, 53, 4927–4930.

48. Chien, H.C., Tsai, L.D., Lai, C.M., Lin, J.N., Zhu, C.Y., Chang, F.C., Characteristics of high-water-uptake activated carbon/Nafion hybrid membranes for proton exchange membrane fuel cells, *J. Power Sources*, 2013, 226, 87–93.

49. Santiago, E.I., Isidoro, R.A., Dresch, M.A., Matos, B.R., Linardi, M., Fonseca, F.C., Nafion-TiO$_2$ hybrid electrolytes for stable operation of PEM fuel cells at high temperature, *Electrochim. Acta*, 2009, 54, 4111–4117.

50. Di Noto, V., Boaretto, N., Negro, E., Giffin, G.A., Lavina, S., Polizzi, S., Inorganic-organic membranes based on Nafion, [(ZrO$_2$)·(HfO$_2$)$_{0.25}$] and [(SiO$_2$)·(HfO$_2$)$_{0.28}$]. Part I: Synthesis, thermal stability and performance in a single PEMFC, *Int. J. Hydrogen Energy*, 2012, 37, 6199–6214.

51. Fatyeyeva, K., Chappey, C., Poncin-Epaillard, F., Langevin, D., Valleton, J.M., Marais, S., Composite membranes based on Nafion and plasma treated clay charges: Elaboration and water sorption investigations, *J. Membr. Sci.*, 2011, 369, 155–166.

52. Xue, Y., Fu, R., Wu, C., Lee, J.Y., Xu, T., Acid-base hybrid polymer electrolyte membranes based on SPEEK, *J. Membr. Sci.*, 2010, 350, 148–153.

53. Yang, T., Preliminary study of SPEEK/PVA blend membranes for DMFC applications, *Int. J. Hydrogen Energy*, 2008, 33, 6772–6779.

54. Di Vona, M.L., Sgreccia, E., Licoccia, S., Khadhraoui, M., Denoyel, R., Knauth, P., Composite proton-conducting hybrid polymers: Water sorption isotherms and mechanical properties of blends of sulfonated PEEK and substituted PPSU, *Chem. Mater.*, 2008, 20, 4327–4334.

55. Yilnaztuerk, S., Deligoez, H., Yilmazoglu, M., Damyan, H., Oeksuezoemer, F., Koc, S.N., Durmus, A., Guerkaynak, M.A., Self-assembly of highly charged polyelectrolyte complexes with superior proton conductivity and methanol barrier properties for fuel cells, *J. Power Sources*, 2010, 195, 703–709.

56. Shao, Z.G., Xu, H.F., Li, M., Hsing, I.M., Hybrid Nafion-inorganic oxides membrane doped with heteropolyacids for high temperature operation of proton exchange membrane fuel cell, *Solid State Ionics*, 2006, 177, 779–785.

57. Fu, R.Q., Woo, J.J., Seo, S.J., Lee, J.S., Moon, S.H., Covalent organic/inorganic hybrid proton-conductive membrane with semi-interpenetrating polymer network: Preparation and characterizations, *J. Power Sources*, 2008, 179, 458–466.

58. Thanganathan, U., Bobba, R., Enhanced conductivity and electrochemical properties for class of hybrid systems via sol-gel techniques, *J. Alloy. Compd.*, 2012, 540, 184–191.

59. Bi, H., Wang, J., Chen, S., Hu, Z., Gao, Z., Wang, L., Okamoto, K., Preparation and properties of cross-linked sulfonated poly(arylene ether sulfone)/sulfonated polyimide blend membranes for fuel cell application, *J. Membr. Sci.*, 2010, 350, 109–116.

60. Di Noto, V., Piga, M., Lavina, S., Negro, E., Yoshida, K., Ito, R., Furukawa, T., Structure, properties and proton conductivity of Nafion/[(TiO$_2$)·(WO$_3$)0.148]$_{\psi TiO2}$ nanocomposite membranes, *Electrochim. Acta*, 2010, 55, 1431–1444.

61. Di Noto, V., Bettiol, M., Bassetto, F., Boaretto, N., Negro, E., Lavina, S., Bertasi, F., Hybrid inorganic–organic nanocomposite polymer electrolytes based on Nafion and fluorinated TiO$_2$ for PEMFCs, *Int. J. Hydrogen Energy*, 2012, 37, 6169–6181.

62. Zhang, J., Giotto, M.V., Wen, W.Y., Jones, A.A., An NMR study of the state of ions and diffusion in perfluorosulfonate ionomer, *J. Membr. Sci.*, 2006, 269, 118–125.

63. Li, J., Wilmsmeyer, K.G., Madsen, L.A., An isotropic diffusion and morphology in perfluorosulfonate ionomers investigated by NMR, *Macromolecules*, 2009, 42, 255–262.

64. Ye, G., Janzen, N., Goward, G.R., Proton dynamics of Nafion and Nafion/SiO$_2$ composites by solid state NMR and pulse field gradient NMR, *Macromolecules*, 2007, 40, 1529–1537.

65. Nicotera, I., Zhang, T., Bocarsly, A., Greenbaum, S., NMR characterization of composite polymer membranes for low-humidity PEM fuel cells, *J. Electrochem. Soc.*, 2007, 154, B466–B473.

66. Kannan, A.G., Choudhury, N.R., Dutta, N.K., In situ modification of Nafion membranes with phospho-silicate for improved water retention and proton conduction, *J. Membr. Sci.*, 2009, 333, 50–58.

67. Huguet, P., Morin, A., Gebel, G., Deabate, S., Sutor, A.K., Peng, Z., In situ analysis of water management in operating fuel cells by confocal Raman spectroscopy, *Electrochem. Commun.*, 2011, 13, 418–422.

68. Hara, M., Inukai, J., Miyatake, K., Uchida, H., Watanabe, M., Temperature dependence of the water distribution inside a Nafion membrane in an operating polymer electrolyte fuel cell. A micro-Raman study, *Electrochim. Acta*, 2011, 58, 449–455.

69. Di Noto, V., Gliubizzi, R., Negro, E., Pace, G., Effect of SiO_2 on relaxation phenomena and mechanism of ion conductivity of [Nafion/(SiO_2)x] composite membranes, *J. Phys. Chem. B*, 2006, 110, 24972–24986.

70. Thayumanasundaram, S., Piga, M., Lavina, S., Negro, E., Jeyapandian, M., Ghassemzadeh, L., Müller, K., Di Noto, V., Hybrid inorganic–organic proton conducting membranes based on Nafion, SiO_2 and triethylammonium trifluoromethanesulfonate ionic liquid, *Electrochim. Acta*, 2010, 55, 1355–1365.

71. Yao, Y., Guo, B., Ji, L., Jung, K.H., Lin, Z., Alcoutlabi, M., Hamouda, H., Zhang, X., Highly proton conductive electrolyte membranes: Fiber-induced long-range ionic channels, *Electrochem. Commun.*, 2011, 13, 1005–1008.

72. Mishra, A.K., Kuila, T., Kim, D.Y., Kim, N.H., Lee, J.H., Protic ionic liquid-functionalized mesoporous silica-based hybrid membranes for proton exchange membrane fuel cells, *J. Mater. Chem.*, 2012, 22, 24366–24372.

73. Zhang, Y., Zhang, H., Bi, C., Zhu, X., An inorganic/organic self-humidifying composite membranes for proton exchange membrane fuel cell application, *Electrochim. Acta*, 2008, 53, 4096–4103.

74. Wang, Z., Tang, H., Zhang, H., Lei, M., Chen, R., Xiao, P., Pan, M., Synthesis of Nafion/CeO_2 hybrid for chemically durable proton exchange membrane of fuel cell, *J. Membr. Sci.*, 2012, 421–422, 201–210.

75. Wang, J., Zhang, H., Yang, X., Jiang, S., Lv, W., Jiang, Z., Qiao, S., Enhanced water retention by using polymeric microcapsules to confer high proton conductivity on membranes at low humidity, *Adv. Funct. Mater.*, 2011, 21, 971–978.

76. Fu, T., Wang, J., Ni, J., Cui, Z., Zhong, S., Zhao, C., Na, H., Xing, W., Sulfonated poly(ether ether ketone)/aminopropyltriethoxysilane/phosphotungstic acid hybrid membranes with non-covalent bond: Characterization, thermal stability, and proton conductivity, *Solid State Ionics*, 2008, 179, 2265–2273.

77. Zeng, S., Hu, S., Pan, S., Wu, G., Xu, W., Effects of acids and water addition on morphology and proton conduction in sol-gel derived acid-base polysiloxane, *Solid State Ionics*, 2010, 181, 1408–1414.

78. Di Vona, M.L., Sgreccia, E., Donnadio, A., Casciola, M., Chailan, J.F., Auer, G., Knauth, P., Composite polymer electrolytes of sulfonated poly-ether-ether-ketone (SPEEK) with organically functionalized TiO_2, *J. Membr. Sci.*, 2011, 369, 536–544.

79. Wu, H., Shen, X., Gao, Y., Li, Z., Jiang, Z., Composite proton conductive membranes composed of sulfonated poly(ether ether ketone) and phosphotungstic acid-loaded imidazole microcapsules as acid reservoirs, *J. Membr. Sci.*, 2014, 451, 74–84.

80. Ren, J., Zhang, S., Liu, Y., Wang, Y., Pang, J., Wang, Q., Wang, G., A novel crosslinking organic–inorganic hybrid proton exchange membrane based on sulfonated poly(arylene ether sulfone) with 4-amino-phenyl pendant group for fuel cell application, *J. Membr. Sci.*, 2013, 434, 161–170.

81. Chun, J.H., Kim, S.G., Lee, J.Y., Hyeon, D.H., Chun, B.H., Kim, S.H., Park, K.T., Crosslinked sulfonated poly(arylene ether sulfone)/silica hybrid membranes for high temperature proton exchange membrane fuel cells, *Renew. Energ.*, 2013, 51, 22–28.

82. Okamoto, K., Yaguchi, K., Yamamoto, H., Chen, K., Endo, N., Higa, M., Kita, H., Sulfonated polyimide hybrid membranes for polymer electrolyte fuel cell applications, *J. Power Sources*, 2010, 195, 5856.

83. Liu, D., Geng, L., Fu, Y., Dai, X., Lv, C., Novel nanocomposite membranes based on sulfonated mesoporous silica nanoparticles modified sulfonated polyimides for direct methanol fuel cells, *J. Membr. Sci.*, 2011, 366, 251–257.

84. Geng, L., He, Y., Liu, D., Dai, X., Lv, C., Facile in situ template synthesis of sulfonated polyimide/mesoporous silica hybrid proton exchange membrane for direct methanol fuel cells, *Micropor. Mesopor. Mat.*, 2012, 148, 8–14.

85. Binsu, V.V., Nagarale, R.K., Shahi, V.K., Phosphonic acid functionalized aminopropyl triethoxysilane–PVA composite material: Organic–inorganic hybrid proton-exchange membranes in aqueous media, *J. Mater. Chem.*, 2005, 15, 4823–4831.

86. Yan, S., Zeng, S., Su, X., Yin, H., Xiong, Y., Xu, W., H_3PO_4-doped 1,2,4-triazole-polysiloxane proton conducting membrane prepared by sol–gel method, *Solid State Ionics*, 2011, 198, 1–5.

87. Chen, B., Li, G., Wang, L., Chen, R., Yin, F., Proton conductivity and fuel cell performance of organic–inorganic hybrid membrane based on poly(methyl methacrylate)/silica, *Int. J. Hydrogen Energy*, 2013, 38, 7913–7923.

88. Sel, O., Soules, A., Ameduri, B., Boutevin, B., Laberty-Robert, C., Gebel, G., Sanchez, C., Original fuel-cell membranes from crosslinked terpolymers via a "Sol–Gel" strategy, *Adv. Funct. Mater.*, 2010, 20, 1090–1098.

89. Wang, L., Advani, S.G., Prasad, A.K., PBI/Nafion/SiO_2 hybrid membrane for high-temperature low-humidity fuel cell applications, *Electrochim. Acta*, 2013, 105, 530–534.

90. Mishra, A.K., Bose, S., Kuila, T., Kim, N.H., Lee, J.H., Silicate-based polymer-nanocomposite membranes for polymer electrolyte membrane fuel cells, *Prog. Polym. Sci.*, 2012, 37, 842–869.

91. Pivovar, B.S., Wang, Y.X., Cussler, E.L., Pervaporation membranes in direct methanol fuel cells, *J. Membr. Sci.*, 1999, 154, 155–162.

92. Kreuer, K.D., Proton conductivity: Materials and applications, *Chem. Mater.*, 1996, 8, 610–641.

93. Chen, J., Asano, M., Maekawa, Y., Yoshida, M., Chemically stable hybrid polymer electrolyte membranes prepared by radiation grafting, sulfonation, and silane-crosslinking techniques, *J. Polym. Sci. A: Polym. Chem.*, 2008, 46, 5559–5567.

94. Danilczuk, M., Perkowski, A.J., Schlick, S., Ranking the stability of perfluorinated membranes used in fuel cells to attack by hydroxyl radicals and the effect of Ce(III): A competitive kinetics approach based on spin trapping ESR, *Macromolecules*, 2010, 43, 3352–3358.

95. Wu, L., Zhang, Z., Ran, J., Zhou, D., Li, C., Xu, T., Advances in proton-exchange membranes for fuel cells: An overview on proton conductive channels (PCCs), *Phys. Chem. Chem. Phys.*, 2013, 15, 4870–4887.

96. Pozuelo, J., Riande, E., Saiz, E., Compan, V., Molecular dynamics simulations of proton conduction in sulfonated poly(phenylsulfone)s, *Macromolecules*, 2006, 39, 8862–8866.

97. Schmidt-Rohr, K., Chen, Q., Parallel cylindrical water nanochannels in Nafion fuel-cell membranes, *Nat. Mater.*, 2008, 7, 75–83.

98. Di Noto, V., Lavina, S., Negro, E., Vittadello, M., Conti, F., Piga, M., Pace, G., Hybrid inorganic–organic proton conducting membranes based on Nafion and 5 wt% of M_xO_y (M = Ti, Zr, Hf, Ta and W). Part II: Relaxation phenomena and conductivity mechanism, *J. Power Sources*, 2009, 187, 57–66.

99. Knauth, P., Di Vona, M.L., Sulfonated aromatic ionomers: Analysis of proton conductivity and proton mobility, *Solid State Ionics*, 2012, 225, 255–259.

100. Ogawa, T., Ushiyama, H., Lee, J.M., Yamaguchi, T., Yamashita, K., Theoretical studies on proton transfer among a high density of acid groups: Surface of zirconium phosphate with adsorbed water molecules, *J. Phys. Chem. C*, 2011, 115, 5599–5606.

101. Di Vona, M.L., Sgreccia, E., Licoccia, S., Alberti, G., Tortet, L., Knauth, P., Analysis of temperature-promoted and solvent-assisted cross-linking in sulfonated poly(ether ether ketone) (SPEEK) proton-conducting membranes, *J. Phys. Chem. B*, 2009, 113, 7505–7512.

102. Di Vona, M.L., Marani, D., D'Ottavi, C., Trombetta, M., Traversa, E., Beurroies, I., Knauth, P., Licoccia, S., A simple new route to covalent organic/inorganic hybrid proton exchange polymeric membranes, *Chem. Mater.*, 2006, 18, 69–75.

103. Yana, J., Nimmanpipug, P., Chirachanchai, S., Gosalawit, R., Dokmaisrijan, S., Vannarat, S., Vilaithong, T., Lee, V.S., Molecular dynamics simulations of Krytox-Silica-Nafion composite for high temperature fuel cell electrolyte membranes, *Polymer*, 2010, 51, 4632–4638.

104. Elliott, J.A., Hanna, S., Elliott, A.M.S., Cooley, G.E., Atomistic simulation and molecular dynamics of model systems for perfluorinated ionomer membranes, *Phys. Chem. Chem. Phys.*, 1999, 1, 4855–4863.

105. Gosalawit, R., Chirachanchai, S., Manuspiya, H., Traversa, E., Krytox-Silica-Nafion® composite membrane: A hybrid system for maintaining proton conductivity in a wide range of operating temperatures, *Catal. Today*, 2006, 118, 259–265.

106. Silva, R.F., De Francesco, M., Pozio, A., Tangential and normal conductivities of Nafion® membranes used in polymer electrolyte fuel cells, *J. Power Sources*, 2004, 134, 18–26.

107. Deluca, N.W., Elabd, Y.A., Polymer electrolyte membranes for the direct methanol fuel cell: A review, *J. Polym. Sci., Polym. Phys. Ed.*, 2006, 44, 2201–2225.

108. Sahu, A.K., Bhat, S.D., Pitchumani, S., Sridhar, P., Vimalan, V., George, C., Chandrakumar, N., Shukla, A.K., Novel organic–inorganic composite polymer-electrolyte membranes for DMFCs, *J. Membr. Sci.*, 2009, 345, 305–314.

109. Cho, E., Kim, H., Kim, D., Effect of morphology and pore size of sulfonated mesoporous benzene-silicas in the preparation of poly(vinyl alcohol)-based hybrid nanocomposite membranes for direct methanol fuel cell application, *J. Phys. Chem. B*, 2009, 113, 9770–9778.

110. Yun, S., Im, H., Heo, Y., Kim, J., Crosslinked sulfonated poly(vinyl alcohol)/sulfonated multi-walled carbon nanotubes nanocomposite membranes for direct methanol fuel cells, *J. Membr. Sci.*, 2011, 380, 208–215.

111. Yang, J.M., Chiu, H.C., Preparation and characterization of polyvinyl alcohol/chitosan blended membrane for alkaline direct methanol fuel cells, *J. Membr. Sci.*, 2012, 419–420, 65–71.

112. Croce, F., Sacchetti, S., Scrosati, B., Advanced lithium batteries based on high-performance composite polymer electrolytes, *J. Power Sources*, 2006, 162, 685–689.

113. Chiappone, A., Nair, J.R., Gerbaldi, C., Jabbour, L., Bongiovanni, R., Zeno, E., Beneventi, D., Penazzi, N., Microfibrillated cellulose as reinforcement for Li-ion battery polymer electrolytes with excellent mechanical stability, *J. Power Sources*, 2011, 196, 10280–10288.

114. Cui, W., Tang, D., Gong, Z., Electrospun poly(vinylidene fluoride)/poly(methyl methacrylate) grafted TiO_2 composite nanofibrous membrane as polymer electrolyte for lithium-ion batteries, *J. Power Sources*, 2013, 223, 206–213.

115. Gao, J., Wang, L., Fang, M., He, X., Li, J., Gao, J., Deng, L., Wang, J., Chen, H., Structure and electrochemical properties of composite polymer electrolyte based on poly vinylidene fluoride-hexafluoropropylene/titania-poly(methyl methacrylate) for lithium-ion batteries, *J. Power Sources*, 2014, 246, 499–504.

116. Prasanth, R., Shubha, N., Hng, H.H., Srinivasan, M., Effect of nano-clay on ionic conductivity and electrochemical properties of poly(vinylidene fluoride) based nanocomposite porous polymer membranes and their application as polymer electrolyte in lithium ion batteries, *Eur. Polym. J.*, 2013, 49, 307–318.

117. Angulakhsmi, N., Thomas, S., Nair, J.R., Bongiovanni, R., Gerbaldi, C., Manuel Stephan, A., Cycling profile of innovative nanochitin-incorporated poly (ethylene oxide) based electrolytes for lithium batteries, *J. Power Sources*, 2013, 228, 294–299.

118. Marschall, R., Sharifi, M., Wark, M., Proton-conducting composite membranes for future perspective applications in fuel cells, desalination facilities and photocatalysis, *Chem. Ing. Tech.*, 2011, 83, 2177–2187.

119. Choi, Y.J., Kim, D.W., Photovoltaic performance of dye-sensitized solar cells assembled with hybrid composite membrane based on polypropylene non-woven matrix, *Bull. Korean Chem. Soc.*, 2011, 32, 605–608.

120. Ito, H., Maeda, T., Nakano, A., Takenaka, H., Properties of Nafion membranes under PEM water electrolysis conditions, *Int. J. Hydrogen Energy*, 2011, 36, 10527–10540.

121. Siracusano, S., Baglio, V., Lufrano, F., Staiti, P., Aricò, A.S., Electrochemical characterization of a PEM water electrolyzer based on a sulfonated polysulfone membrane, *J. Membr. Sci.*, 2013, 448, 209–214.

122. Baglio, V., Ornelas, R., Matteucci, F., Martina, F., Ciccarella, G., Zama, I., Arriaga, L.G., Antonucci, V., Aricò, A.S., Solid polymer electrolyte water electrolyser based on Nafion-TiO$_2$ composite membrane for high temperature operation, *Fuel Cells*, 2009, 9, 247–252.

123. Aili, D., Hansen, M.K., Pan, C., Li, Q., Christensen, E., Jensen, J.O., Bjerrum, N.J., Phosphoric acid doped membranes based on Nafion, PBI and their blends—Membrane preparation, characterization and steam electrolysis testing, *Int. J. Hydrogen Energy*, 2011, 36, 6985–6993.

124. Hansen, M.K., Aili, D., Christensen, E., Pan, C., Eriksen, S., Jensen, J.O., Von Barner, J.H., Li, Q., Bjerrum, N.J., PEM steam electrolysis at 130°C using a phosphoric acid doped short side chain PFSA membrane, *Int. J. Hydrogen Energy*, 2012, 37, 10992–11000.

125. Wei, G., Xu, L., Huang, C., Wang, Y., SPE water electrolysis with SPEEK/PES blend membrane, *Int. J. Hydrogen Energy*, 2010, 35, 7778–7783.

126. Lee, K.M., Woo, J.Y., Jee, B.C., Hwang, Y.K., Yun, C., Moon, S.B., Chung, J.H., Kang, A.S., Effect of cross-linking agent and heteropolyacid (HPA) contents on physico-chemical characteristics of covalently cross-linked sulfonated poly(ether ether ketone)/HPAs composite membranes for water electrolysis, *J. Ind. Eng. Chem.*, 2011, 17, 657–666.

127. Song, M.A., Ha, S.I., Park, D.Y., Ryu, C.H., Kang, A.S., Moon, S.B., Chung, J.H., Development and characterization of covalently cross-linked SPEEK/Cs-TPA/CeO$_2$ composite membrane and membrane electrode assembly for water electrolysis, *Int. J. Hydrogen Energy*, 2013, 38, 10502–10510.

128. Teng, X., Sun, C., Dai, J., Liu, H., Su, J., Li, F., Solution casting Nafion/polytetrafluoroethylene membrane for vanadium redox flow battery application, *Electrochim. Acta*, 2013, 88, 725–734.

129. Vijayakumar, M., Schwenzer, B., Kim, S., Yang, Z., Thevuthasan, S., Liu, J., Graff, G.L., Hu, J., Investigation of local environments in Nafion–SiO$_2$ composite membranes used in vanadium redox flow batteries, *Solid State Nucl. Mag.*, 2012, 42, 71–80.

130. Chen, D., Wang, S., Xiao, M., Han, D., Meng, Y., Sulfonated poly (fluorenyl ether ketone) membrane with embedded silica rich layer and enhanced proton selectivity for vanadium redox flow battery, *J. Power Sources*, 2010, 195, 7701–7708.

131. Couture, G., Alaaeddine, A., Boschet, F., Ameduri, B., Polymeric materials as anion-exchange membranes for alkaline fuel cells, *Prog. Polym. Sci.*, 2011, 36, 1521–1557.

132. Wu, Y., Wu, C., Xu, T., Yu, F., Fu, Y., Novel anion-exchange organic-inorganic hybrid membranes: Preparation and characterizations for potential use in fuel cells, *J. Membr. Sci.*, 2008, 321, 299–308.

133. Wu, Y., Wu, C., Xu, T., Lin, X., Fu, Y., Novel silica/poly(2,6-dimethyl-1,4-phenylene oxide) hybrid anion-exchange membranes for alkaline fuel cells: Effect of heat treatment, *J. Membr. Sci.*, 2009, 338, 51–60.

134. Tripathi, B.P., Kumar, M., Shahi, V.K., Highly stable proton conducting nanocomposite polymer electrolyte membrane (PEM) prepared by pore modifications: An extremely low methanol permeable PEM, *J. Membr. Sci.*, 2009, 327, 145–154.

135. Wang, J., Wang, L., Preparation and properties of organic–inorganic alkaline hybrid membranes for direct methanol fuel cell application, *Solid State Ionics*, 2014, 255, 96–103.

136. Gomes, A.S., Dutra Filho, J.C., Hybrid membranes of PVA for direct ethanol fuel cells (DEFCs) applications, *Int. J. Hydrogen Energy*, 2012, 37, 6246–6252.

137. Zeng, L., Zhao, T.S., Li, Y.S., Synthesis and characterization of crosslinked poly (vinyl alcohol)/layered double hydroxide composite polymer membranes for alkaline direct ethanol fuel cells, *Int. J. Hydrogen Energy*, 2012, 37, 18425–18432.

138. Pandey, R.P., Shahi, V.K., Functionalized silica–chitosan hybrid membrane for dehydration of ethanol/water azeotrope: Effect of cross-linking on structure and performance, *J. Membr. Sci.*, 2013, 444, 116–126.

139. Wu, C., Wu, Y., Luo, J., Xu, T., Fu, Y., Anion exchange hybrid membranes from PVA and multi-alkoxy silicon copolymer tailored for diffusion dialysis process, *J. Membr. Sci.*, 2010, 356, 96–104.

140. Wu, Y., Luo, J., Zhao, L., Zhang, G., Wu, C., Xu, T., QPPO/PVA anion exchange hybrid membranes from double crosslinking agents for acid recovery, *J. Membr. Sci.*, 2013, 428, 95–103.

141. Kumar, M., Ulbricht, M., Novel antifouling positively charged hybrid ultrafiltration membranes for protein separation based on blends of carboxylated carbon nanotubes and aminated poly(arylene ether sulfone), *J. Membr. Sci.*, 2013, 448, 62–73.

142. Lederer, F., Gunther, T.J., Weinert, U., Raff, J., Pollmann, K., Development of functionalised polyelectrolyte capsules using filamentous *Escherichia coli* cells, *Microb. Cell Fact.*, 2012, 11, 163.

143. Lin, B., Qiu, B., Qiu, L., Si, Z., Chu, F., Chen, X., Yan, F., Imidazolium-functionalized SiO_2 nanoparticle doped proton conducting membranes for anhydrous proton exchange membrane applications, *Fuel Cells*, 2013, 13, 72–78.

144. Ahmad, H., Kamarudin, S.K., Hasran, U.A., Daud, W.R.W., Overview of hybrid membranes for direct-methanol fuel-cell applications, *Int. J. Hydrogen Energy*, 2010, 35, 2160–2175.

145. Hou, H., Sun, G., Wu, Z., Jin, W., Xin, Q., Zirconium phosphate/Nafion 115 composite membrane for high-concentration DMFC, *Int. J. Hydrogen Energy*, 2008, 33, 3402–3409.

146. Costamagna, P., Yang, C., Bocarsly, A.B., Srinivasan, S., Nafion 115/zirconium phosphate composite membranes for operation of PEMFCs above 100°C, *Electrochim. Acta*, 2002, 47, 1023–1033.

147. Bauer, F., Willert-Porada, M., Microstructural characterization of Zr-phosphate-Nafion® membranes for direct methanol fuel cell (DMFC) applications, *J. Membr. Sci.*, 2004, 233, 141–149.

148. Jones, D.J., Roziere, J., Recent advances in the functionalisation of polybenzimidazole and polyetherketone for fuel cell applications, *J. Membr. Sci.*, 2001, 185, 41–58.

149. Alberti, G., Casciola, M., Costatino, U., Levi, G., Inorganic ion exchange membranes consisting of microcrystals of zirconium phosphate supported by Kynar®, *J. Membr. Sci.*, 1978, 3, 179–190.

150. Abe, Y., Li, G., Nogami, M., Kasuga, T., Hench, L.L., Superprotonic conductors of glassy zirconium phosphates, *J. Electrochem. Soc.*, 1996, 143, 144–147.

151. Hou, H., Di Vona, M.L., Knauth, P., Building bridges: Crosslinking of sulfonated aromatic polymers—A review, *J. Membr. Sci.*, 2012, 423–424, 113–127.

152. Zhao, T.S., Kreuer, K.D., Nguyen, T.V., *Advances in Fuel Cells*, Vol. 1. Elsevier Ltd., New York, 2007.

153. Kamarudin, S.K., Achmad, F., Daud, W.R.W., Overview on the application of direct methanol fuel cell (DMFC) for portable electronic devices, *Int. J. Hydrogen Energy*, 2009, 34, 6902–6916.

154. Ding, C., Zhang, H., Li, X., Liu, T., Xing, F., Vanadium flow battery for energy storage: Prospects and challenges, *J. Phys. Chem. Lett.*, 2013, 4, 1281–1294.

155. Wang, W., Luo, Q., Li, B., Wei, X., Li, L., Yang, Z., Recent progress in redox flow battery research and development, *Adv. Funct. Mater.*, 2013, 23, 970–986.

10 Aliphatic Polymer Electrolyte Membranes

Jinli Qiao, Junkun Tang, and Yuyu Liu

CONTENTS

10.1 INTRODUCTION

The demand for high-performance energy conversion materials/devices has risen
substantially due to the increased concerns in natural resources deficiency and environmental protection. In this way, electrochemical energy storage and conversion
technologies such as fuel cells, batteries, supercapacitors, water electrolysis, and
CO_2 electrochemical reduction have been recognized as the most feasible, highly
efficient, and clean electrochemical means to help in reducing greenhouse effect and
the contamination in our environment [1]. Polymer electrolyte membrane (PEM)
materials, as the core component of these technologies, have aroused much interest
in fundamental research as well as in the field of practical applications because of
their possible use as thin-film solid electrolytes in various electrochemical devices.
The key factors for a successful PEM should include sufficient ionic conductivity,
long-life cell performance, and acceptable production cost. PEMs can be divided
into two categories: one is the alkaline PEMs, which conduct hydroxide ions (OH^-),
and the other is the acidic PEMs, which conduct protons (H^+) [1]. Compared with
the Li-ion solid polymer electrolytes, the alkaline PEMs have been introduced to the
consumers in the last decade due to the rapid market growth in portable electronic
devices such as notebook computers, mobile phones, personal digital assistants,

digital cameras, and video camcorders. These include alkaline rechargeable bat-teries, especially Ni–MH (metal hydride), Zn–Ni, and other hybrid systems such as air metal [2,3]. These materials are also of interest for application in the alkaline fuel cells (AFCs) [1] and supercapacitors [4]. This is because compared with Li-ion solid polymer electrolytes, the alkaline PEMs have a number of distinct character-istics, such as easy preparation, low cost, abundance of the basic components, and high ionic conductivity at room temperature (10^{-3} S cm^{-1}). Besides, there are sev-eral advantages when the operational environment is alkaline. Here are examples: (1) Faster kinetics can be reached under alkaline conditions. This fast kinetics could reduce or remove the need for precious metal catalysts such as Pt-based catalysts as for acidic conditions. (2) Corrosion problems can be minimized under alkaline conditions, and (3) the usage of small organic fuels is favored due to their fast kinet-ics of electrooxidation in alkaline medium. As a result of these advantages, alkaline PEM fuel cells, for example, have in recent years been considered the next genera-tion of fuel cell technology, and great effort has been put into their research and development [1].

With respect to the demand for low cost, easy preparation, and simple struc-ture of PEM membranes, the aliphatic membranes are apparently one of the most important kinds of candidates. In this chapter, we will provide a particular com-prehensive description of the fundamentals of aliphatic PEMs in the literature, including material selection, design and synthesis, characterizations, their conduc-tion behavior, chemical stability, and mechanisms. Most attention was focused on the polymer membranes with aliphatic main chains, not on perfluorostructured, polyaromatic, or polyheterocyclic membranes, as the latter have been well dis-cussed in other chapters. But the aliphatic PEMs containing a fluorostructured component such as poly(vinylidene fluoride) (PVDF) and an aromatic component such as poly(styrene sulfonic acid) (PSSA) may be involved for a good funda-mental understanding. In addition, the most recent significant progress in both research and applications of these aliphatic PEMs is also summarized, including their possible applications in fuel cells, Zn–air batteries, and rechargeable Ni–MH batteries as well as carbon dioxide electroreductions. In fact, so far, many types of aliphatic polymer materials were utilized as PEMs, such as poly(vinyl alcohol) (PVA), poly(acrylic acid) (PAA), poly(methyl methacrylate) PMMA, poly(ethylene oxide) (PEO), PEO–propylene oxide, poly(acrylonitrile) (PAN), poly(vinyl chlo-ride) (PVC), and even cellulose (bacterial cellulose [BC]) and chitosan (CS) as well as their composites. In this chapter, for a good fundamental understanding, the applications of these aliphatic PEMs in both acidic and alkaline media are provided together.

10.2 MEMBRANE SYNTHESIS

10.2.1 Materials Requirement and Selection

To be an applicable PEM, the membrane should have appropriate ionic conductiv-ity in the first place, which requires enough ionic groups in the membrane system as ionic group donators. Unlike many traditional PEMs like Nafion® and sulfonate aromatic polymers (which contain sulfonic acid groups directly attached to the main chain or carrying short pendant side chains with terminal sulfonic acid units), most aliphatic polymers like PVA, PMMA, PEO, PAN, and PVC do not have ionic groups in the polymer structure. Therefore, they are frequently used as base materials in PEMs, and the ionic groups can be incorporated through introduction of ion-containing polymers or solution. As base materials, aliphatic polymers have the advantage of low fuel permeability, good mechanical property, and enough chemical stability. Inorganic materials doping, polymer blending, and cross-linking procedure are usually used to improve the stability and fuel resistance of PEMs. In this way, the polymer acid/alkali solutions with proper ionic groups would be composited with base materials for ionic and hydrophilic group donating. Therefore, the ionic group content in the polymers or ionic concentration in the solutions is important for the PEM system to ensure enough ionic conductivity. Additionally, the ratio of the components in the composite can also affect the overall performance of the corresponding PEMs.

10.2.1.1   Poly(Vinyl Alcohol)

PVA is a water-soluble synthetic polymer with polyhydroxy functional groups. In comparison with the membranes with aromatic skeletons, PVA has a number of distinguished characteristics such as low cost and easy preparation as well as biodegradability. Therefore, PVA is commonly used in papermaking, textiles, and a variety of coatings. In addition, it has good film-forming capacity, adequate hydrophilic properties, and a high density of reactive chemical functions that are favorable for cross-linking by irradiation, chemical, or thermal treatments. Alkaline PEMs based on PVA have been studied for applications on Zn–Ni [5–8], Ni–MH [9], and Zn–air batteries [10].

In recent years, PVA has been used as PEM for fuel cell applications [11–16]. Furthermore, it shows unique characterizations of good resistance to organic solvent and superior methanol barriers to Nafion, where the alcohol cross leak is a key issue in practical uses for direct methanol fuel cells (DMFCs). Since no ionic group is contained in the side chains, acid or basic groups should be introduced through ionic polymers or ionic solution.

Ionic polymers were most commonly used for ionic groups introduced. Qiao et al. have introduced poly(acrylamide-co-diallyldimethylammonium chloride) (PAADDA) as anion charge carriers into PVA for anion fuel cell applications [15,16]. PSSA was selected as cationic polymer by Yang et al. and Kin et al. for DMFC systems [11,14,17]. On the other hand, ionic group can also been introduced by acid or basic solution such as KOH or H_3PO_4 [18–20]. Besides, some inorganic materials have also acted as acid group carriers in PVA systems such as silicotungstic acid (STA) [21], sulfonated polyhedral oligosilsesquioxane (sPOSS) [22], and hydroxyapatite (HAP) [23].

Introduced with proper ionic group as well as some cross-linking if needed [19,24,25], PVA have been used for a variety of fuel cells such as DMFC [24,26,27], AFCs [15,16], and direct ethanol fuel cell (DEFC) [23,28]. Research shows that PVA membrane is a promising base material for ion-exchange membrane for fuel cells.

10.2.1.2 Poly(Vinylidene Fluoride)

PVDF is a highly nonreactive and pure thermoplastic fluoropolymer produced by the polymerization of vinylidene difluoride. Because of its high mechanical strength, good film-forming ability, as well as its high resistance to solvent, acid, base, and heat, PVDF has aroused the researchers' interest as PEM for fuel cell applications in recent years. Similar to PVA, there is no ionic group in PVDF, and the researchers should introduce some either by blending or by grafting in which PVDF acts as the base material.

As to the fabrication of PEMs, PVDF was usually used as a copolymer with hexafluoropropylene, indicated as PVDF–HFP. Kumar et al. have done much work on PVDF–HFP-based PEMs [29–33]. Since then, the copolymer was blended with other polymers or inorganic materials for ion introduction like Nafion, montmorillonite (MMT), and $AlO[OH]_n$. Besides, PVDF has been grafted with PSSA or vinylimidazole (VIm) in some research works [34,35], and Al_2O_3, poly(ethylene glycol) (PEG), or poly(2-acrylamido-2-methyl-1-propanesulfonic acid) (PAMPS) have been used to blend with PVDF system [35,36]. In these researches, PVDF-based PEMs are designed mainly for DMFC application as well as DEFC.

10.2.1.3 Poly(Ethylene Oxide)

PEO is a polyether compound with many applications, which refers to a polymer of ethylene oxide. With high tensile strength and flexibility and adequate film-forming ability, PEO is another common base material for PEMs, except for PVA and PVDF. Similarly, since PEO does not have ionic groups like PVA and PVDF, it was usually blended with ionic polymers, for example, sulfonated polyimide (SPI) [37], or impregnated with KOH solutions [38–40]. The former can be used for PEM fuel cells, while the latter for secondary batteries frequently.

10.2.1.4 Polyolefin

Polyolefins like polyethylene (PE) and PE terephthalate (PET) are the most widely used polymers in many commercial applications. Owing to its excellent chemical and electrochemical stability, low cost, and easy preparation, in the recent decade, some of the researchers have shifted attention to these polymers. Sulfonated poly(arylene ether sulfone), poly(vinylbenzyl chloride) (PVBC), and polystyrene (PS) have been grafted onto PE [41–43], which were designed for DMFC, proton-exchange membrane fuel cell (PEMFC), or AFC. In addition, PET has also been utilized for DMFC through Nafion ionomer impregnating by Lee and coworkers [44].

10.2.1.5 Poly(Methyl Methacrylate)

PMMA is another base material used as PEMs for electrochemical applications, due to its excellent thermal and mechanical stabilities, as well as its ability to significantly increase ion conductivity of PEMs. Mostly PMMA was synthesized as the main component of block copolymers, and another component can be PS and PSSA [45–47]. As one of the most promising aliphatic polymers, PMMA has also been blended with other components like PVDF and SiO_2 [48–50]. These PEMs prepared from PMMA have the potential to be used in a variety of applications such as fuel cells, lithium batteries, and even electrical double-layer capacitors.

10.2.1.6 Bacterial Cellulose

BC, which is synthesized by *Acetobacter xylinum*, is a natural and low-cost biopolymer. It is distinguished by its special 3D nanostructure morphology, high water holding capacity, biodegradability, high tensile strength, low gas permeation, and good biocompatibility. Hence, BC has been studied in various fields, such as food additives, high-quality audio membranes, functional papers, pervaporation membranes, and biomedical materials. BC membranes can retain its chemical and thermal stability up to 275°C with high mechanical strength [51]. Furthermore, the hydroxyl groups on its backbone can provide BC with a high hydrophilicity, which is crucial for the operation of PEMs, such as in fuel cells. Also because of the lack of ionic groups, BC was impregnated with phosphoric acid (H_3PO_4/BC) or phytic acid (PA/BC) for H_2/O_2 fuel cell applications [52]. By in situ chemical-reduction method, platinum nanoparticles were also deposited on the BC membrane surface for membrane electrode assembly (MEA) fabrications [53].

10.2.1.7 Chitosan

Unlike the aforementioned base materials, CS is a cationic polyelectrolyte. The hydrophilic nature of CS ensures a high selectivity for water in the pervaporative separation of alcohol–water mixtures. Therefore, CS was usually chosen as the cationic polyelectrolyte for the fabrication of PEM membranes for its dual function as an alcohol barrier and a proton conductor. In addition to its high hydrophilicity, the good film-forming property, high mechanical strength, and chemical resistance make CS very promising material. In addition, CS is a cycloaliphatic polymer; it contains active $-NH_2$ groups and $-OH$ groups that can be utilized in constructing an ionic cross-linked interpenetrating network (IPN) structure. In this way, PEMs with an IPN structure formed by CS and an anionic polyelectrolyte, such as poly(acrylic acid) (PAA) [54] and PAMPS [55], offered a feasible solution to obtain both low methanol crossover and high proton conductivity, which is of growing interest in DMFC membrane development [54,56]. Besides, CS, a deacetylation product of chitin obtained from crab and shrimp shells, is considered to be an extremely cheap, nonhazardous, and environmentally benign polymer. Figure 10.1 illustrated these typical aliphatic polymers selected for PEM fabrication.

10.2.2 Polymer Synthesis Methods and Processes

As described earlier, aliphatic polymers that are used as base material for PEMs are the commonly used membranes such as PVA, PVDF, and PMMA. In general, there are two means to increase the ionic conductivity of the PEMs: (1) suppression of crystallization of polymer chains to improve polymer chain mobility and (2) increase in the carrier concentration. To realize these processes, they can be composited with other polymers or inorganic materials. Besides, copolymerization and cross-linking techniques are also two important procedures for membrane synthesis/preparation.

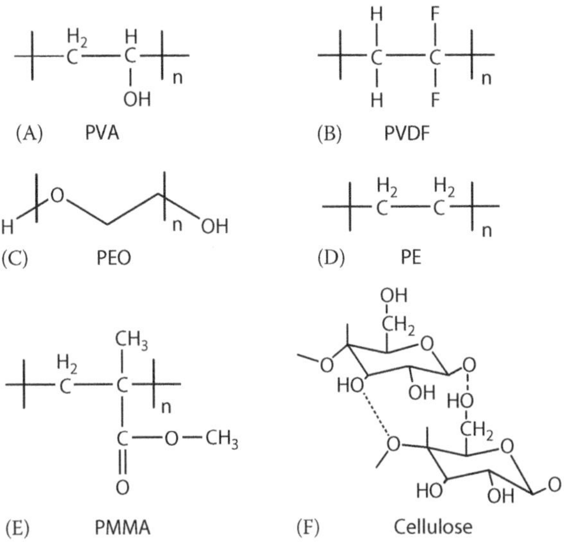

FIGURE 10.1 Typical aliphatic polymers selected for PEM fabrication. (A) PVA, (B) PVDF, (C) PEO, (D) PE, (E) PMMA, and (F) Cellulose.

1. *Organic–inorganic hybrid composites*: The organic–inorganic hybrid composites are also called the inorganic filler doped with polymer composites, which have been extensively studied since Wright's discovery of PEO complexes with alkali metal salts [57] and the recognition by Armand et al. [58] of their potential use in batteries. The composites can usually be prepared by a simple physical mixing approach (PM), that is, the inorganic filler like KOH, TiO_2, Al_2O_3, and SiO_2 powders or their solutions are directly mixed with polymer matrix by mechanical stir. This direct mixing can achieve the improved conductivity. The disadvantage is that most of the inorganic filler powders cannot uniformly disperse into the polymer matrix due to the aggregation together and the formation of conglomerations. Another method is the in situ synthesis (ISS) approach, that is, the inorganic filler powders are prepared in the doping process of composites, for example, hydroxyapatite (HA, $Ca_{10}(PO_4)_6(OH)_2$); thus, more porous structure can be obtained by ISS process than by PM process due to the relative small size and the uniform distribution of the tiny HA particle in base material, such as in the PVA matrix synthesized by ISS process [6].

2. *Copolymerization*: The copolymerization of the preceding aliphatic polymers with other polymers is another method to fabricate PEMs, in which block or graft copolymerization has been used. This method is easily used for designing different main chain structures. For the composition method, the base polymers were usually obtained from commercial products directly, whereas for the latter method, the copolymers should be synthesized in laboratory by the researchers. Via sequential atom transfer radical

polymerization, for example, the proton-exchange membranes based on partially sulfonated poly(styrene-block-methyl methacrylate) (SPS-b-PMMA) can be obtained by postsulfonation of poly(styrene-b-methyl methacrylate) PS-b-PMMA. By changing the content of PMMA, various block copolymers can be synthesized. Then the block copolymers can be converted to their ionomers by sulfonation using acetyl sulfate as sulfonating agent. Generally, the sulfonation reaction is performed at 30°C [47]. Sulfonic PS is a commonly used material for sulfonic group introduction in PEMs [43,46,59]. In addition, quaternary amine for anion conduction in anion-exchange membrane (AEM) can also be obtained by introducing PVBC [42].

3. *Cross-linking*: For membrane synthesis using cross-linking technique, the base polymer material should usually have –OH and –NH$_2$ groups that can react with cross-linker, for example, glutaraldehyde (GA) or diethylenetriaminepentaacetic acid to form a cross-linking network, which reinforces the strength of the membranes obtained [27,28]. According to the cross-linking processed, the cross-linking method can be divided into thermal cross-linking, chemical cross-linking, and UV radiation. This method in fact belongs to chemical cross-linking method since the cross-linker is added from the outside as chemical reaction reagent. Usually, for obtaining high density of cross-linkages of base polymer (e.g., PVA) matrix, the thermal cross-linking technique was often applied along with the chemical cross-linking technique. By this, the second polymer (usually the polymer containing conduction ions) can be well *trapped* into the base polymer network. The chemical and thermal cross-linking reactions for blended CS and PVA membranes can be seen in Figure 10.2 [60].

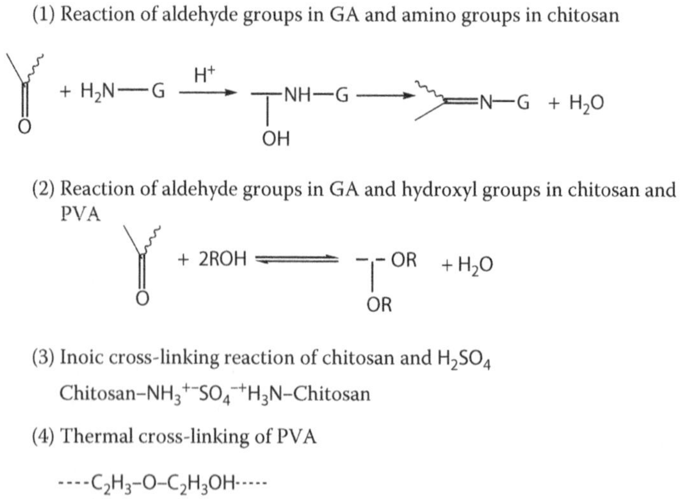

(1) Reaction of aldehyde groups in GA and amino groups in chitosan

(2) Reaction of aldehyde groups in GA and hydroxyl groups in chitosan and PVA

(3) Inoic cross-linking reaction of chitosan and H$_2$SO$_4$

Chitosan–NH$_3^{+-}$SO$_4^{-+}$H$_3$N–Chitosan

(4) Thermal cross-linking of PVA

----C$_2$H$_3$–O–C$_2$H$_3$OH-----

FIGURE 10.2 Chemical and thermal cross-linking reactions for blended CS and PVA membranes. (From Svang-Ariyaskul, A. et al., *J. Membr. Sci.*, 280, 815, 2006.)

According to the sequence of cross-linker that was added by Qiao et al., direct and indirect chemical cross-linking methods are further divided [16]. For direct chemical cross-linking method, usually appropriate amount of cross-linker was directly added into a polymer mixed solution to react, and the pH of the solution was adjusted by adding one drop of HCl (37.5%) or 5–10 wt% sulfuric acid, which was usually used as catalyst. Then, the resulting solution was poured into plastic Petri dishes/or coated onto the surface of a glass plate and dried, and water was evaporated under ambient conditions or at certain temperature. For indirect chemical cross-linking method, on the other hand, the mixed solution was at first poured into plastic dishes/ or coated onto the surface of a glass plate after removing the air under vacuum. The membrane was peeled off from the plastic substrate and then were soaked (usually square pieces of membranes of ca. 1.5×2 cm) in a reaction solution containing cross-linking reagent and small amount of HCl or H_2SO_4.

In fact, the cross-linking can often be finished by UV radiation, which is also called photo-induced cross-linking technique. During the process, the monomer with small amount (usually 1 wt% benzoin ethyl ether) of a photoinitiator dissolved in distilled water was mixed with a base polymer solution under stirring. Then, the mixture solutions were irradiated by UV light with or without degassing at room temperature to initiate the polymerization. After this, the mixture solutions were poured onto a clean Teflon surface and maintained in a vacuum oven to keep the polymerization going and then allowed the solvent to evaporate until membranes were formed. Finally, the membranes were washed with deionized (DI) water to remove the unreacted monomer and the untrapped polymer in base polymer, then followed by drying in a vacuum oven before use.

10.2.3 Membrane Formation

Generally, there are mainly two methods for PEM formation: solution cast and melt exclusion. For aliphatic PEM preparation, the former method is more suitable as the precursor system for membrane preparation is usually the composite of two or more components.

For solution casting [32,61], the base polymer material and the ionic group containing polymer are separately dissolved into appropriate solvent at first. Efforts should be made to ensure that the dissolution is completed, for example, by heating, agitation, or refluxing. After that, the two polymer solutions are then mixed together until a homogeneous solution is formed. If needed, inorganic component should be added into the solution. Afterward, the corresponding mixed solution would be either casted onto a flat plate using a doctor blade or poured into a flat dish. The plate or dish is usually made of glass, occasionally polytetrafluoroethylene (PTFE) or metal, and adjusted even using the level gauge. Then the solution would undergo a heating procedure for solvent evaporation, which varies depending on the researchers and solution systems. Finally, the membrane would form on the plate/ dish. Some solvent, usually water, would be utilized to facilitate membrane peeling from the plate/dish, based on the swelling of membrane by the solvent, particularly PVDF- or PMMA-based composite membranes. For PVA-based membranes, plastic Petri dishes are always used for solution casting, and no any auxiliary procedure is needed. The membranes could be directly peeled from the Petri dishes very easily [15,16,27]. Finally, the obtained membranes would be stored either in water or in dry state before use.

10.3 MEMBRANE CHARACTERIZATION

The structure and properties of aliphatic PEMs are of vital importance for electrochemical device performances, which are, of course, the main concern of the researchers. Consequently, there are plenty of characterizations to evaluate aliphatic PEM structure and properties, such as spectral studies, morphological studies, physical properties, and ion conductivity.

10.3.1 SPECTRAL STUDIES (IR, RAMAN SPECTROSCOPY, NMR, XRD, ETC.)

10.3.1.1 Infrared Radiation

Molecular structure is of vital importance for aliphatic PEMs. As one of the most important spectral studies, infrared radiation (IR) is often used by the researchers for functional group characterization, composition analysis, as well as structure identification [33,46,48]. As PEMs are almost entirely film form, the IR tests could be conducted directly, without any additional sample preparation, through either transmitting or reflecting model. From IR tests, some slight change of functional group in aliphatic polymer molecular structure could be detected. As reported by Yun et al., for the sulfonation of polymers, the introduced sulfonate group, for example, in SPVA, could be well identified through the two new peaks at 1266 and 1718 cm^{-1}, which were assigned to the asymmetric stretching vibrations of O=S=O in the sulfonic acid group and the C=O stretching peak of carboxyl acid groups as shown in Figure 10.3 [62]. Bai and Ho have used IR to determine the structure change of SPI membranes with the amounts of PEO soft segments though the variation of the two peaks assigned to the C–O–C and C–N groups in the PEO-imide soft segments [37]. In addition, the structure change through cross-linking could also be detected by IR tests from an increase or decrease of some peaks of the related groups in the cross-linking process [15,27,63], for example, in preparing PVA-based membrane materials by *trapping* PAMPS as proton-conduction membrane composites or PAADDA as AEM. In other words, IR test is a very useful tool to determine the structure or structure change of aliphatic PEMs.

10.3.1.2 Micro-Raman Spectroscopy

Except for IR test, micro-Raman measurement was also used to determine the structure or structure change of aliphatic PEMs. A typical candidate is the use of micro-Raman spectroscopy to analyze the structure change of PVA after introducing HAP or MMT, and the increase in amorphous domains of the composite membranes was demonstrated [23,26]. Besides, micro-Raman spectroscopy has also been used for the confirmation of the existence of ionic groups ($-SO_3^-$) and in particular for the degradation of PEMs after fuel cell tests [14,64].

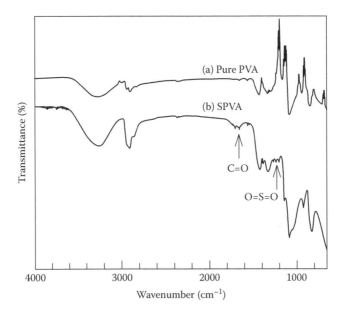

FIGURE 10.3 FT-IR spectra of (a) pure PVA and (b) SPVA membrane. (From Yun, S. et al., *J. Membr. Sci.*, 380, 208, 2011.)

10.3.1.3 Nuclear Magnetic Resonance

Since many aliphatic PEMs were prepared from some new synthesized polymers in researches, nuclear magnetic resonance (NMR) is a useful tool to characterize the molecular structure of these polymers. In most cases, it was used to verify the coherence of the designed and actual structure of the new polymer [41,47]. Sometimes, it has also been used to identify a series of new polymers, when the researchers need to study on the effect of polymer structure on the properties of PEMs [65]. Confirmation of quaternization was also done by 1H NMR spectroscopy (Figure 10.4), such as in cross-linked quaternized PVA (QPVA) [66] and cross-linked quaternized-CS [67] AEMs by observing the chemical shifts (ppm) of related functional groups (OH, CH, CH_3) or increase in these peak intensities.

10.3.1.4 X-Ray Diffraction

Due to not having ionic groups in the polymer structure, most aliphatic polymers like PVA, PMMA, PEO, PAN, and PVC are frequently used as base materials in PEMs, and the ionic groups should be incorporated through introduction of ion-containing polymers or solution. At this stage, the x-ray diffraction (XRD) measurement was usually performed to examine the crystallinity of these composite membranes [66,68]. As observed by Wang et al., the PVA polymer exhibits a semicrystalline structure with a large peak at a 2θ angle of $19°–20°$ and a small peak of $39°–40°$ [68]. However, the peak intensity of the poly(vinyl alcohol)/3-(trimethylammonium) propyl-functionalized silica (PVA–TMAPS) membranes with different weight ratios of PVA to TMAPS was reduced as compared with that of pure PVA polymer, which

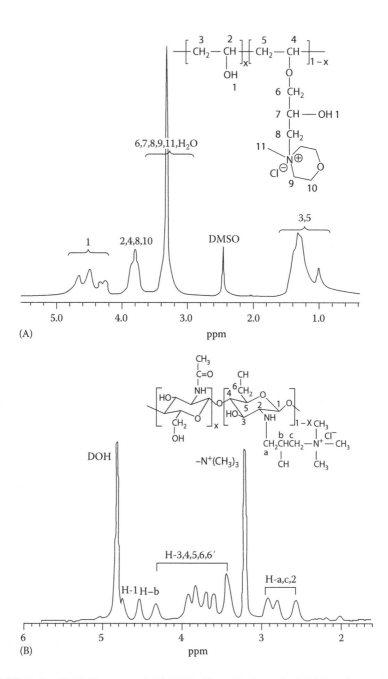

FIGURE 10.4 1H NMR spectra of (A) QPVA (From Ye, L. et al., *Solid State Ionics*, 240, 1, 2013) and (B) HTCC (DQ, 78.3%). (From Wan, Y. et al., *J. Power Sources*, 195, 3785, 2010.)

means that this membrane has the lowest crystallinity. The low crystallinity reveals that a more amorphous phase exists in this membrane. This means that the structure of the membrane is more disordered, that is, the PVA and the TMAPS are mixed more uniformly. Better mixing between the base polymer and introduced ion-containing polymers or solution is helpful for the improvement of ionic conductivity of the membrane.

10.3.2 Morphological Studies (SEM, TEM, EM, XRD, Etc.)

10.3.2.1 Scanning Electron Microscope

As functional membranes, the morphology of PEMs is very important for their performances in electrochemical applications, and scanning electron microscope (SEM) is a powerful tool for morphological studies. Mostly, SEM was utilized to observe the surface morphology of PEMs, which may confirm the uniformity of membrane surface or study the morphology variation of membranes with structure or composition change [19,31,36]. Cross-sectional SEM is also significant for PEMs, which can observe the morphology in the membrane through the thickness direction. It has been used to examine membranes' morphology distribution from one side to the other, especially for the composite membranes with both organic and inorganic phases [42,47,70].

Kumar et al. have observed both the surface and cross-sectional images of PVDF–HFP membranes before and after PSSA grafting and sulfonation (Figure 10.5A and found the original highly porous PVDF–HFP membrane becoming less porous after grafting and almost smooth after sulfonation [32].

The cross view of the SEM pictures can also be used to examine the phase separation phenomena in the membranes; therefore, the compatibility between the two composites can be verified [16]. Zhou et al. have observed the cross-sectional images of poly(vinyl alcohol) modified quaternized hydroxyethylcellulose ethoxylate (PVA/QHECE) composite membranes both for OH$^-$ form and for Cl$^-$ form (Figure 10.5B). Small shallow folded cavities were observed to be irregularly distributed in PVA/QHECE membrane and became obvious with increasing QHECE content in polymers. Therefore, the compatibility between PVA and QHECE is poor when in so high content of QHECE [71].

10.3.2.2 Transmission Electron Microscopy

For PEMs, transmission electron microscopy (TEM) was usually used to study the microstructure of PEMs according to the contrast between domains. For most PEMs, there are at least two different domains: hydrophilic and hydrophobic domains. To improve the contrast between the two domains, heavy metal salt like ruthenium tetroxide and lead acetate was often used to stain the membrane [43,46]. In TEM images, the bright portions are attributed to the hydrophobic domain, while the darker portions represent the hydrophilic domain. Choi and Jo have conducted TEM tests for sulfonated PS-b-PMMA membranes with different degrees of sulfonation and blend ratios [46]. They have found that the hydrophilic domains become larger as the degree of sulfonation is increased and the morphology of the membranes

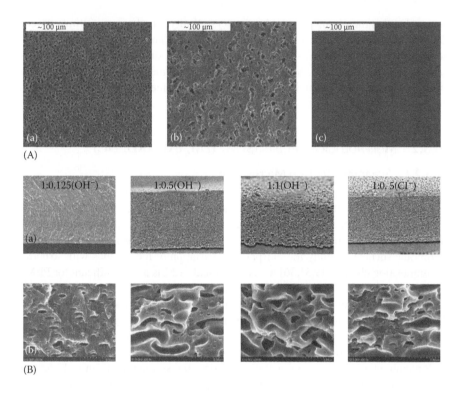

FIGURE 10.5 (A) The surface morphological images of PVDF–HFP membrane: (a) before grafting, (b) after grafting, and (c) after sulfonation. (From Gnana kumar, G. et al., *J. Membr. Sci.*, 350, 92, 2010.) (B) SEM pictures of the cut view of PVA/QHECE alkaline membranes: (a) (1,000×) and (b) (20,000×). (From Zhou, T.C. et al., *J. Power Sources*, 227, 291, 2013.)

changes from a well-ordered lamellar structure to a disordered form as the PVDF content in the blend is increased.

10.3.2.3 Elemental Mapping

In addition to morphology image, elemental mapping (EM) is also commonly used to determine the element distribution, so as to study the distribution of the corresponding components [33,72], which is a supplementary mode of morphology image test. In the energy dispersive X-ray (EDX) mapping image, the highlighted bright dots revealed high element concentration, which can let us know whether the distributions of elements were homogeneous or not [73]. Fu et al. prepared the alkaline solid PEMs by incorporating KOH in PVA and found that there was no obvious change in potassium element before and after conditioned PVA membrane in 4 M KOH, but a much higher concentration in oxygen element particular to conditioned PVA membrane at 80°C (Figure 10.6). This may be due to the fact that OH⁻ in KOH may react with acetal group in PVA matrix to generate –OH and water; thus, the combined KOH molecules by long-distance interaction may exist in the PVA matrix, which was helpful for the ionic conductivity.

FIGURE 10.6 EDX mapping of (A, D) potassium element and (B, C) oxygen element within PVA/KOH, respectively. Doping KOH concentration in solution: 4 M, (A and B) conditioned at RT and (C and D) conditioned at 80°C. Condition time 24 h, followed by complete removal of free KOH prior to testing. (From Fu, J. et al., *Synthetic Met.*, 160, 193, 2010.)

10.3.3 PHYSICAL PROPERTIES (SOLUBILITY, VISCOSITY, THERMAL STABILITY, MECHANICAL PROPERTIES, ETC.)

10.3.3.1 Thermal Stability

Thermal stability is an important property for PEMs, especially in high-temperature systems, and the corresponding characterization is thermogravimetric analysis (TGA). As most of the prepared aliphatic PEMs reported in the literature are block copolymers, graft copolymers, and composites of polymers, with frequently some inorganic materials included, TGA was used to study the differences between PEMs with different compositions [24,37,42,47].

For example, Guiver et al. have reported the TGA thermogram of ultrahigh-molecular-weight PE (UHMWPE), styrene-grafted UHMWPE (UHMWPE-g-PS), and styrene-grafted and sulfonated UHMWPE (UHMWPE-g-PSSA) with different degrees of grafting, as shown in Figure 10.7 [43]. Results showed that UHMWPE-g-PS behaved identically to nongrafted UHMWPE only with the onset point due to

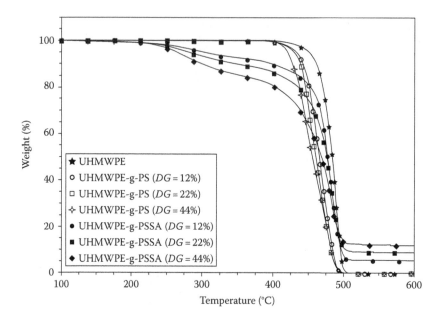

FIGURE 10.7   TGA curves of control UHMWPE, styrene-grafted UHMWPE (UHMWPE-g-PS), and UHMWPE-g-PSSA. (From Sherazi, T.A. et al., *J. Membr. Sci.*, 333, 59, 2009.)

weight loss of UHMWPE-g-PS slightly lower than the onset point of UHMWPE. While for UHMWPE-g-PSSA membrane, there was a gradual weight loss in the temperature range of 150°C–260°C, which is due to loss of water associated with sulfonic acid groups.

According to the literature about the thermostability of aliphatic PEMs, it is concluded that thermostability of PEMs is a comprehensive property of all components in the composite membrane. Usually, the incorporation of sulfonate group would give two additional stages in a TGA curve from 40°C to 300°C, which may, respectively, stem from the elimination of water and sulfonic acid groups [32,46]. Besides, the introduction of cross-linking structure and inorganic materials (ceramic effect) would improve the thermostability apparently, which were commonly applied in aliphatic PEM studies [14,19,32,33,74].

10.3.3.2   Mechanical Properties

Mechanical properties of aliphatic PEMs were conducted by universal test instrument and dynamic mechanical analysis [14,75]. From the former method, stress–strain curves were obtained as well as parameters like tensile strength, Young's modulus, and elongation at break [14,26,41,75,76]. For the latter method, except for the stress–strain curves, we can get the curves of storage modulus (E′) and tan (δ) value versus temperature. From the literature, we know that introduction of high-T_g chains into aliphatic membrane may increase its tensile strength and modulus and decrease its elongation. And high molecular weight is beneficial to tensile strength and Young's modulus, while sulfonation of aliphatic membranes may somewhat reduce the overall mechanical strength. However, sulfonation of high-molecular-weight aliphatic membrane would almost have no effect on the mechanical strength [41]. Research showed that incorporation of inorganic filter, cross-linking, and annealing treatment would enhance the mechanical properties of aliphatic PEMs.

10.3.4 ION-EXCHANGE CAPACITY, SOLVENT UPTAKE, AND SWELLING RATIO

10.3.4.1 Ion-Exchange Capacity

As PEM materials, aliphatic polymers should have ionic groups that are essential for electrochemical applications. Thus, the concentration of ionic groups in membrane system is one of the parameters to which the researchers are concerned, which is represented as ion-exchange capacity (IEC). In most literatures, IEC was measured by the classical back titration method.

For proton-exchange membrane, the membrane should be immersed in dilute sulfuric acid solution for enough time to convert it into the H$^+$ form completely. The membrane was washed with DI water to remove excess H$^+$ and then equilibrated with NaCl salt solution. The IEC value was determined by titration with NaOH solution using phenol red as the endpoint indicator and defined as mequiv of cationic groups per gram of dried sample [33,47,65,77]. While for AEM, the membrane was soaked in a dilute KOH solution to convert it into the OH^{-1} form. The membrane was washed with DI water to remove excess KOH and then equilibrated with HCl solution. The IEC value was determined from the reduction in acid measured using back titration and defined as mequiv of quaternary ammonium groups per gram of dried sample [15,42].

10.3.4.2 Water Uptake and Swelling Ratio

Since PEMs are usually immersed in water solution environment in electrochemical applications, water uptake (WU) and swelling ratio (SR) in water need to be paid close attention. WU of the membranes was evaluated by the mass differences between the wet and dry membranes. It should be noted that the membrane weighting order, weighting dry membrane before or after wet membrane, varied depending on the researchers' preference. The weight percentage uptake was determined by the following equation:

$$WU\,(\%) = \frac{W_w - W_d}{W_d} \qquad (10.1)$$

where W_d and W_w are the weights of the dry and fully wet membranes, respectively.

SR can be measured similar to WU, in which a membrane volume, not weight, is identified by its length, width, and thickness. The SR was determined by the following equation:

$$SR\,(\%) = \frac{V_w - V_d}{V_d} \qquad (10.2)$$

where V_d and V_w are the volumes of the dry and fully wet membranes, respectively.

10.3.5　Ionic Conductivity

As PEMs are mainly used in electrochemical applications, ionic conductivity is one of the most important properties. That is to say, a good PEM should have appropriate ionic conductivity first. Therefore, the membrane materials are usually polymer electrolytes with negatively or positively charged groups on the polymer backbone or side chains or with charged group carried by some inorganic materials. For a new PEM, ionic conductivity is an indispensable characterization that needs to be studied.

Without absorbing water, these PEMs tend to be rather rigid and are poor ionic conductors. The ionic conductivity would dramatically increase with water content [78]. Therefore, ionic conductivity tests are mostly conducted in water or in water vapor with adequate related humidity. Two types of ionic conductivity for PEMs were used: in-plane and through-plane conductivities. The former represents the conductivity along the membrane surface direction, and the latter refers to the conductivity across the membrane thickness direction. In addition, there are two methods for conductivity measurement: two-point probe electrode and four-point probe electrode. The latter method is more accurate but the former uses a simpler device. Therefore, comparison of ionic conductivities between membranes must be of the same type and measured through the same method. The aforementioned conductivity measurements are suitable for both proton conductivity and anion conductivity. Proton conductivity (σ) is calculated by the following equation [79–82]:

$$\sigma = \frac{L}{RA} \tag{10.3}$$

For in-plane test, L is the distance between the two electrodes and A is the cross-sectional area of the membrane; for through-plane test, L is the thickness of the membrane and A is the overlap area of the two electrodes. R is the impedance of membrane from the Nyquist plot.

10.3.6　Permeability

During the operation of fuel cells, fuels like hydrogen or methanol would permeate through the PEMs, simultaneously with the ion conduction. The permeation would reduce the cell voltage, with fuel cell efficiency decreased accordingly. In addition, for hydrogen fuel cells, the permeation of hydrogen would result in radical generation, for example, peroxide and hydroperoxide radicals, which may lead to chemical degradation of the membrane and ionomer contained in the catalyst layer in fuel cell operation [83]. Therefore, fuel permeation should be as low as possible for PEMs.

Two main methods were used for fuel permeability determination: ex situ and in situ methods. The former method, also named diffusion cell method, was mostly used for methanol permeability measurements [22,84,85]. In this measurement, a diffusion cell was used, consisting of two reservoirs of distinct composition, with a sample membrane between them. The membrane acts as a parting plane to prevent liquid passing through directly and only permit permeation of the molecules in solution. The two reservoirs were injected with aqueous methanol solution with a certain concentration and pure distilled water, respectively. Diffusion would start right after the injection of solution in the two reservoirs, with the concentration of methanol in the water reservoir, measured by gas chromatography, increased with time, which can be described by the following equation [22,86]:

$$C_B(t) = \frac{A}{V_B} \frac{DK}{L} C_A t \qquad (10.4)$$

where

C_B and C_A are the real-time methanol concentrations of the water reservoir and initial methanol concentration of the methanol solution reservoir

A, L, and V_B are the effective area of the membrane, the membrane thickness, and the volume of the permeated reservoir, respectively

D and K are the methanol diffusivity and partition coefficient between the membrane and the product DK that represents the membrane permeability (P in cm² s⁻¹), which can be calculated from the slope of the straight line obtained from plots of methanol concentration versus time, as the typical charts shown in Figure 10.8 [87].

However, this diffusion cell method was just a simulation of the situation of fuel cell operation. In situ measurement is a more accurate method that determines fuel concentration during fuel cell operation. Hydrogen or oxygen permeation was usually detected using this method by chronoamperometry [88–91]. Before the measurement,

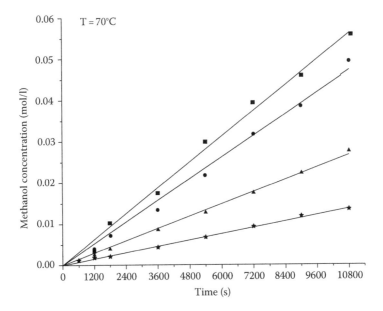

FIGURE 10.8 Methanol concentration versus time at the permeability experiments for composite Nafion/PVA membranes with different thicknesses. (From Mollá, S. and Compañ, V., *J. Power Sources*, 196, 2699, 2011.)

MEA should be fabricated from the sample PEMs and then assembled into the single-cell system, which is the same as fuel cell. Potential step experiments were performed to evaluate fuel permeability using an electrochemical interface with a certain flow rate of humidified H_2 at the anode and humidified N_2 at the cathode. The anode served as both the counter and reference electrodes. The cathode potential was increased gradually, and the steady-state current density corresponds to the H_2 cross-over current density. Although more complicated than ex situ method, the in situ method is much more accurate and closer to the real situation of fuel cell operation.

10.3.7 CHEMICAL STABILITY

Chemical stability of aliphatic PEM is also an important factor that affects membrane durability. During electrochemical device's operation, for example, PEMFC, radicals like HO·· and HOO·· would be formed in the catalytic process, which may induce membrane degradation [47,77]. In this case, the Fenton test (H_2O_2 aqueous solution containing Fe^{2+}) is applied as an accelerated test, which is mostly used for the evaluation of oxidative stability of membranes.

As to AFCs, AEM has poor tolerance to strong alkaline media, which may affect the transport of hydroxide anions and AFC performances [92]. Thus, the tolerance of membrane toward alkaline media should be conducted for AEMs, in addition to the common Fenton tests.

10.4 MEMBRANE CONDUCTION MECHANISM AND THEORETICAL MODELING

During fuel cell operation, ions would conduct through PEMs, which leads to current flow in the cell system. Thus, the conductivity of PEMs plays an *important role* in fuel cell performances, and membrane conduction mechanism has drawn much attention in fuel cell researches. There are mainly two kinds of conduction mediums in PEMs according to the fuel cell type: proton (H^+) and hydroxyl (OH^-). As aliphatic membranes were usually used for anion-conduction membrane and proton-conduction mechanism has been discussed already in the preceding chapters, we will focus our attention mainly to hydroxyl conduction mechanism in this chapter, although they have a lot in common.

In AFCs, hydroxyl anions transport from the cathode to the anode, suffering a certain obstruction from the PEM, catalyst layer, and diffusion layer. The anion conductivity of PEM is based on many factors such as material component of a membrane, molecular structure of a polymer, membrane morphology, anion concentration and location in a membrane, and circumstance of a membrane. Grew and Chiu have put forward transport mechanisms that are possible for hydroxyl anion in the AEMs through a dusty fluid model [93]. Similar to the classic proton-conduction mechanism, which included combinations of the Grotthuss mechanism, en masse diffusion, migration, surface site hopping, and convective processes [94–102], they predicted that the hydroxyl anion-conduction mechanism can be the product of any combination of these mechanisms including the Grotthuss mechanism, surface site hopping, diffusion, migration, and/or convection (Figure 10.9). The Grotthuss

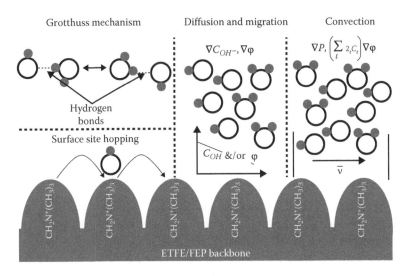

FIGURE 10.9 Transport mechanisms that are possible for hydroxyl anion in the AEM. (From Grew, K.N. and Chiu, W.K.S., *J. Electrochem. Soc.*, 157, B327, 2010.)

mechanism, an important contribution mechanism for proton conduction in aqueous solution, was also suggested to be suitable for hydroxyl anion transport. It represents that the hydroxyl anion would move via structural diffusion in water channels of the AEMs. As shown in Figure 10.9, OH$^-$ can grab a proton from the adjacent water molecular through hydrogen bonds and form H$_2$O, with the original H$_2$O transforming to OH$^-$, which results in OH$^-$ diffusion. Similar to the transport through hopping on the sulfonic side chains for proton conduction, hydroxyl anion can also hope on cationic side chains, for example, quaternary ammonium on the internal surface of water channels. Owing to the electrode reactions, concentrations of hydroxyl anion display obvious difference in the two electrodes, resulting in the diffusion/migration of the hydroxyl anions. In addition, convection processes stem from a pressure gradient and/or potential gradient, which may lead to net motion of hydroxyl anions. The importance of these mechanisms in hydroxyl anion conduction, especially the proportion of these processes, varies greatly with many factors like membrane types, humidity, and temperature. However, there are very few researches about anion-conduction mechanisms [103–108], and many aspects of them are still unclear.

10.5 MEMBRANE APPLICATIONS IN ELECTROCHEMICAL DEVICES FOR ENERGY STORAGE AND CONVERSION

10.5.1 APPLICATIONS FOR PROTON-EXCHANGE MEMBRANE FUEL CELLS

PEMFC is one of the most important fuel cells that were studied intensively by the researchers. Due to their high electrochemical properties as well as excellent chemical resistance, Nafion is the reference material for preparing proton-exchange membranes. But Nafion has the drawback of high cost, difficulty in synthesis, and environmental problems, while aliphatic membranes were applied as alternative PEMs for PEMFC. Among these membranes, PMMA and PVDF were the most commonly used as base polymers [47,50,109]. Besides, a few literatures also reported the application of PEO and PE as base polymer of PEMs for PEMFC [43,76]. The base aliphatic polymer would be utilized to synthesize copolymers with other polymers, for example, PS or composite with inorganic materials like silica, followed by sulfonation with proper sulfonating agent. However, there is a problem for aliphatic membranes and the lack of systematical study on PEMFC performances, which needs further research (Table 10.1).

10.5.2 Applications for Direct Methanol Fuel Cells

DMFC is another type of fuel cells, which has a reasonably stable energy density and is targeted especially for portable applications. Most of the aliphatic PEMs reported in the literature are designed for DMFC, since the main advantage of which is low fuel permeability. Among these aliphatic membranes, most effort has been put in PVA- and PVDF-based PEMs by the researchers [17,22,29,111]. Moreover, other aliphatic polymers like PE, PEG, and PET were studied in the literature occasionally [43,44,74]. Many polymers and inorganic materials have been introduced to prepare the composite membranes, through copolymerization or blending. The introduced polymers included Nafion, PS, HFP, and MA, while the inorganic materials consist of SiO_2, POSS, and AlO(OH) (Table 10.2).

10.5.3 Applications for Anion-Exchange Membrane Fuel Cells

AFC is another type of fuel cell with OH^- as current conduction medium. As the traditional liquid electrolyte AFC has many drawbacks, for example, CO_2 sensitivity, AEM fuel cell (AEMFC) was more attractive to the researchers in recent years [92]. Nafion, the reference material for proton-exchange membranes, was no longer suitable for AEMFC. Therefore, many studies about aliphatic membranes were conducted for AEMFC. Most of them were focused on PVA-based membrane. In these studies, PVA polymers were composited with other polymers like poly(12-acryloylaminododecanoic acid) (PAADA), PVP, and PS [15,16,63,118]. For the membranes without anion like PVA/PAP and PVA/PS, they need to be immersed in alkali solution, for example, KOH solution, for anion conduction in fuel cell. While for the membrane with anion already, for example, PVA/PAADA, alkali solution is no longer required. In addition, to fix alkali in the membrane, PVA pure or composite membranes were usually cross-linked before use [19,25] (Table 10.3).

10.5.4 Applications for Direct Ethanol Fuel Cells

Apart from the common fuel cells discussed earlier, aliphatic membranes can also be applied to DEFCs, for example, PVA- and PVDF-based composite membranes [23,28,33]. Gnana kumar has used irradiated PVdF–HFP–MMT composite membranes for DEFC application [33]. It was reported that the fabricated composite exhibited higher thermal stability and ensured the higher-temperature applications

TABLE 10.1
Properties of Aliphatic PEMs for PEMFC Utilization Described in the Literature

	IEC (mequiv g^{-1})	T (°C)	WU (wt%)	Conductivity (mS cm^{-2})	Perm.$_{MeOH}$ (cm^2 s^{-1})	Power Density$_{Max}$ (mW cm^{-2})	Stability
PVdF/PES [65]	0.16	RT	20	1.2			
PVdF/DPES [65]	0.27	RT	30	3.3			
PVdF/SPES [65]	0.38	RT	23	4.5			
PEO/GO [76]		60		90		53	
UHMWPE-g-PSSA-58 [43]	2.62	30	172.39	211 (35°C)	2.8×10^{-6}		
SPS-b-PMMA[47]	1.63	80	43.6	151.0			
P(S-co-SSA)-b-PMMA/PVDF [46]	1.87	RT	51	32			
PVDF-CSPS [110]	1.7	22	45			155	

TABLE 10.2
Properties of Aliphatic PEMs for DMFC Use Described in the Literature

	IEC (mequiv g^{-1})	T (°C)	WU (wt%)	Conductivity (mS cm^{-2})	Perm.$_{MeOH}$ (cm^2 s^{-1})	Power Density$_{Max}$* (mW cm^{-2})	Stability
Nafion-PE (DE2020) [112]		25		35	2.2×10^{-6}	250 (80°C 6 M MeOH)	
PVDF-g-PSSA [34]	2.5	25	85	45		~116 (70°C 1 M MeOH)	
PVDF-coated Nafion [113]		rt		5.5	9.5×10^{-8}	58.3 (30°C 1 M MeOH)	
PSSA-PVDF [111]		25		87.4		~52 (55°C 0.5 M MeOH)	
P(VdF-co-HFP)/Nafion [114]		30	26.4	1.52	1.24×10^{-7}		
AMPS, asymmetric based acrylic [115]		90	~60	42	1×10^{-8}		
Radiation-grafted LDPE/PTFE [116]	1.97	20		16.7	1.34×10^{-6}	70 (90°C 2 M MeOH)	
PVDF-SiO$_2$ [117]		25		70 mS cm^{-1}	22 mA cm^{-2a} (60°C)		
PVDF-SiO$_2$ gel [117]		25		200 mS cm^{-1}	53 mA cm^{-2a} (60°C)	85 (80°C 1 M MeOH)	

^{a} The methanol permeability was tested by in situ chronoamperometry.

TABLE 10.3

Properties of Aliphatic PEMs for AEMFC Use Described in the Literature

	IEC (mequiv g⁻¹)	T (°C)	WU (wt%)	Conductivity (mS cm⁻²)	Perm.$_{MeOH}$ (cm² s⁻¹)	Power Density$_{Max}$ (mW cm⁻²)	Stability
PE-g-PVBC-TOH [42]	0.58	30	23.58	23.1 (90°C)	5.92×10^{-8}		60°C/1 M NaOH/120 h
PVA/PAADDA-d (1:1) [16]	1.61	25	121.1	3.03	4.16×10^{-7}		60°C/water/350 h
PVA/PAADDA (1:1) [15]	1.61	25	80.0	0.114			
PVA/PDDA-OH⁻ [119]	0.85	25	96	25		32.7	80°C/8 M KOH/360 h
PVA$_{15}$ [19]		20	~45	~190		72 (80°C)	
γ-Grafted ETFE [120]	2.11	40	64.4	45		48	60°C/10 M KOH/120 h
AQPVBH [121]	0.232	rt	10.42	11	1.35×10^{-8}		60°C/3 M KOH/120 h
Quaternized/cross-linked PCMS [122,123]	1.14	50		9.2		55	50°C/methanol/air fuel cell/233 h
ETFE-g-PVB trimethyl ammonium [124,125]	1.42	50	40	34	5.4×10^{-7}	130	
Poly(epichlorohydrin)/DABCO/ quinuclidine [126]	1.3	60		13		100 (25°C)	
Poly(epichlorohydrin)/DABCO/ TEA [127]	0.54	25	31–45	10		42 (60°C)	
Poly(MMA-co-BA-co-VBC) [128]	1.25	rt	239	5.3		35 (60°C)	

by ceramic efforts of the included filler. For DEFC performances, the maximum power density has reached to 93 mW cm^{-2} for PVdF–PSSA at 60°C using 4 M concentration of ethanol.

10.5.5 APPLICATION FOR BATTERIES

Aliphatic polymers have also been applied to batteries, especially nickel/metal secondary batteries. PEO-based polymers were used as solid polymer electrolytes in batteries [38–40]. However, these studies have been gradually out of the researchers' sight because of some environment problems.

10.6 TYPICAL EXAMPLE ANALYSIS FROM MATERIAL SELECTION, SYNTHESIS, CHARACTERIZATION, AND APPLICATIONS

10.6.1 CROSS-LINKED PVA/PDDA AS ANION-EXCHANGE MEMBRANE FOR FUEL CELL APPLICATIONS

PVA has been often used as ion-exchange membrane for fuel cell applications. Poly(diallyldimethylammonium chloride) (PDDA) is a water-soluble quaternized copolymer with conductive anions (OH$^-$) as charge carriers, which also have a certain tolerance to alkaline environment. In this part, we will introduce a typical example of cross-linked PVA/PDDA blend membrane as AEM for fuel cell applications conducted by Qiao and coworkers.

10.6.1.1 Membrane Preparation

The PVA/PDDA membranes were formed by a solution cast method as described in Figure 10.10. Typically, a stock PVA (Mw = 86,000–89,000) aqueous solution was first prepared by dissolving PVA (50 g) in distilled water (500 mL) at 90°C

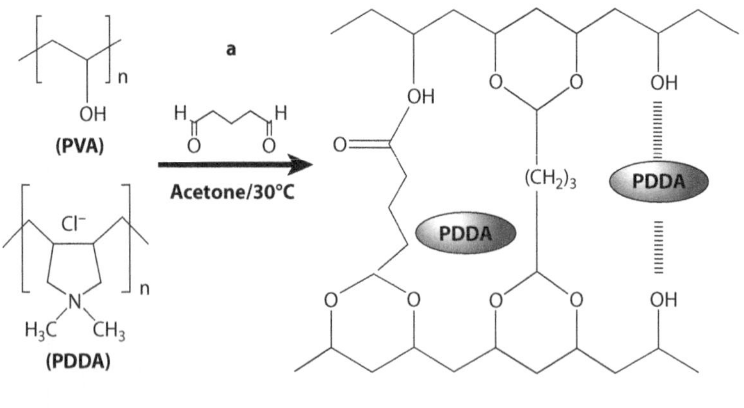

FIGURE 10.10 Chemical structure of cross-linked PVA/PDDA composite. (From Qiao, J. and Zhang, J., *J. Power Sources*, 237, 1, 2013; Zhang, J. et al., *J. Power Sources*, 240, 359, 2013.)

with stirring. Appropriate amounts of PDDA (20% water solution, Mw = 400,000–500,000) were then mixed with the aforementioned PVA under mechanical agitation, in a controlled PVA/PDDA mass ratio. The resulting solutions were poured into plastic Petri dishes, and water was evaporated under ambient conditions. When visually dry, the membrane was peeled from the plastic substrate with a thickness of about 60–80 μm [119,129].

10.6.1.2   Cross-Linking of PVA/PDDA Membranes

The membranes were treated by both thermal and chemical cross-linking methods. The thermal cross-linking refers to the membranes that were annealed at different temperatures (from 110°C to 150°C) for an hour to induce physical cross-linking between PVA chains. Then samples of polymer membranes (1 × 1.5 cm) were soaked in reaction solution consisting of 10 wt% GA and 0.2 wt% hydrochloric acid in acetone for further chemical cross-linking. Due to the reaction catalyzed by H^+, the cross-linking and acetal took place between the PVA hydroxyl and the GA aldehyde groups in the membrane.

10.6.1.3   Characterizations and Applications

For conductivity characterization, the membranes were rendered conducting by immersion of PVA/PDDA membranes in 2 M KOH and equilibrated for 24 h to convert it from Cl^- form into OH^- form. Then the membranes were taken out and rinsed repeatedly with DI water to remove the absorbed KOH on membrane surfaces for final conductivity measurements. The OH^- conductivity of the formed membrane was measured by an AC impedance technique using an electrochemical impedance analyzer. The maximum OH^- conductivity value measured at 25°C with AC impedance spectroscopy reaches as high as 0.025 S cm^{-1} at a PVA/PDDA polymer composition of 1:0.5 by mass.

Alkaline stability was tracked by immersing PVA/PDDA membrane (1:0.5 by mass) into 8 M KOH solution at 80°C. It was then washed with DI water for several times to remove free KOH on the membrane surface, and then the OH^- conductivity of the membrane was measured at room temperature. Result shows that no appreciable decay of the OH^- conductivity could be observed even after 360 h of immersion, and so was the WU. Oxidative durability of PVA/PDDA membranes was also carried out through time-dependent measurements of the weight changes in 3% H_2O_2 solution at an elevated temperature of 60°C. The weight of the membrane samples tends to maintain a constant value (about 80 wt%) with no further weight losses again up to 240 h at 60°C, even higher than Tokuyama A901 although the PVA/PDDA membrane is just made of aliphatic skeletons.

H_2/O_2 fuel cell performance was measured using PVA/PDDA membrane-based MEA (1:0.5 by mass). The AFC exhibits an open circuit voltage (OCV) of 0.81 V and a peak power density of 32.7 mW cm^{-2} at a current density of 72.9 mA cm^{-2} (Figure 10.11). Although it is much lower than some alkaline exchange membranes, TMA functionalized (LDPE-co-VBC) [130] and poly(ETFE-g-VBC) [131] membrane by introducing nonfluorinated or partially fluorinated groups; such excellent

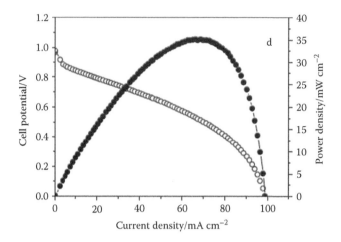

FIGURE 10.11 H_2/O_2 fuel cell performance of PVA/PDDA membrane at 25°C. (From Qiao, J. and Zhang, J., *J. Power Sources*, 237, 1, 2013.)

electrochemical performance in the AEMFC has rarely been observed for PVA-based cation-free hydrocarbon membranes, as PVA/PDDA membranes are all hydrocarbon chains.

10.6.2 HYDROXIDE-ION-CONDUCTION MEMBRANE USED FOR ELECTROCHEMICAL CONVERSION OF CARBON DIOXIDE IN ALKALINE POLYMER ELECTROLYTE MEMBRANE CELLS

The electrochemical reduction of carbon dioxide to produce fuel has often been termed by some as *artificial* photosynthesis. Compared to the traditional researches using alkaline solution electrolytes, porous separators, and solid metallic electrode structures, there are numerous benefits to using a cell design based on a solid polymeric ion-conduction MEA with porous catalytic electrodes. In this part, we will introduce a typical example of hydroxide-ion-conduction membrane used for electrochemical conversion of carbon dioxide in alkaline PEM cells, conducted by Valdez and coworkers [132].

10.6.2.1 Membrane Preparation

The hydroxide-ion-conduction membrane (AMI-7001S) was obtained from American Membranes International Inc. (Ringwood, NJ). The AMI (Applied Membranes, Inc.) membrane consisted of PS with quaternary amine side groups, and the polymer had been cross-linked with divinylbenzene similar to the anion-exchange resins and electrodialysis membranes. The AMI membrane had a thickness of 450 μm. The *as-received* membranes from AMI contained chloride ions. Therefore, these membranes were soaked overnight in a solution of 1 M sodium bicarbonate to exchange the chloride ions for hydroxide ions (Figure 10.12).

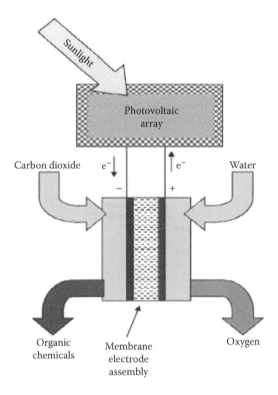

FIGURE 10.12   Polymer membrane cell configuration for the electrochemical reduction of carbon dioxide. (From Narayanan, S.R. et al., *J. Electrochem. Soc.*, 158, A167, 2011.)

10.6.2.2   Characterization and Applications

The performances of the Nafion and the hydroxide-ion-conduction membrane cells were compared for two cathode solutions: (1) DI water (18 MΩ cm) saturated with carbon dioxide by continuous bubbling of the pure gas at a pressure of 1 atm and (2) 1 M sodium bicarbonate solution. As shown in Figure 10.13, the cell voltages with the hydroxide-ion membrane were found to decrease substantially upon changing the electrolyte from CO_2-saturated DI water to 1 M sodium bicarbonate. However, the performance of the Nafion-based cell did not alter significantly for the same change in the cathode solution. This phenomenon can be ascribed to differences in ionic contact at the electrode/membrane interface, as hydroxide-ion membrane is cross-linked and could not be hot pressed to improve the ionic contact. Thus, the use of a liquid electrolyte could substantially improve the ionic contact between the catalyst layer and the hydroxide-ion-conduction membrane.

The cumulative faradaic efficiency can describe the conversion efficiency of carbon dioxide, which is defined as the total charge used for producing formate divided by the total charge passed through the cell during the experiment. As shown in Figure 10.14, the initial efficiencies for formate production from 1 M bicarbonate solutions are found to be as high as 80%, but it begins to decline over a period of 1 h because

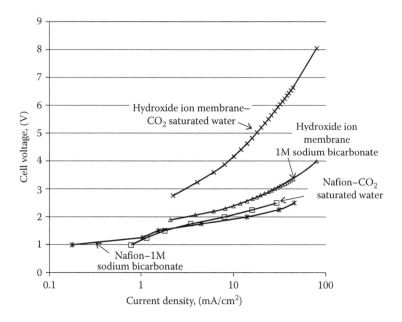

FIGURE 10.13 Performance of cells with the alkaline Nafion membrane and hydroxide-ion-conduction membrane using various cathode solutions indicated. (From Narayanan, S.R. et al., *J. Electrochem. Soc.*, 158, A167, 2011.)

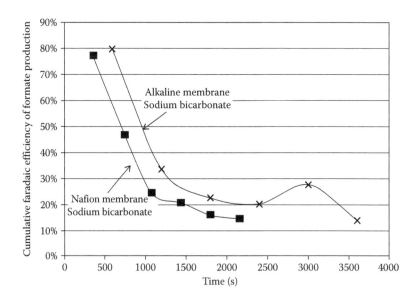

FIGURE 10.14 Comparison of the cumulative efficiency of formate production in the Nafion membrane cell and hydroxide-ion-conduction membrane cell with the sodium bicarbonate solution in the cathode. (From Narayanan, S.R. et al., *J. Electrochem. Soc.*, 158, A167, 2011.)

of the buildup of diffusion barriers from bicarbonate and carbon dioxide reduction at the surface of the electrode. This situation could be improved by introducing pulsed mode electrolysis instead of steady-state electrolysis.

10.6.3 HYBRID DUAL-NETWORK CHITOSAN/PVA MEMBRANES FOR DIRECT METHANOL FUEL CELLS

CS and PVA membranes have suffered from high degree of swelling and less stability in fuel cell operation. In this part, we will introduce a hybrid dual-network solid PEM (Figure 10.15) with interconnected networks of natural polymer CS and a synthetic polymer PVA cross-linked with GA/SSA and incorporated with stabilized silicotungstic acid (SWA) used for DMFCs [133].

10.6.3.1 Membrane Preparation

CS–PVA–CS–SWA hybrid membranes were prepared by solution casting technique. In brief, 0.75 g of CS was dissolved in 75 mL of 2 wt% acetic acid with stirring to form a homogeneous viscous solution. A total of 2.5 g of PVA was dissolved in 25 mL of DI water at 60°C. PVA and CS are cross-linked with 0.6 mL of SSA and 0.5 mL of GA solution, respectively. Both CS and PVA solutions are mixed together and allowed to stir for 24 h. The required amount of stabilized SWA was ultrasonicated for 2 h and then added to CS–PVA solution to form a hybrid solution. The solutions were cast in a plexiglass plate and the solvent was evaporated at 30°C to form a membrane. The membrane was peeled off from the glass plate and washed with DI water repeatedly to remove the residual impurities and be used for further studies.

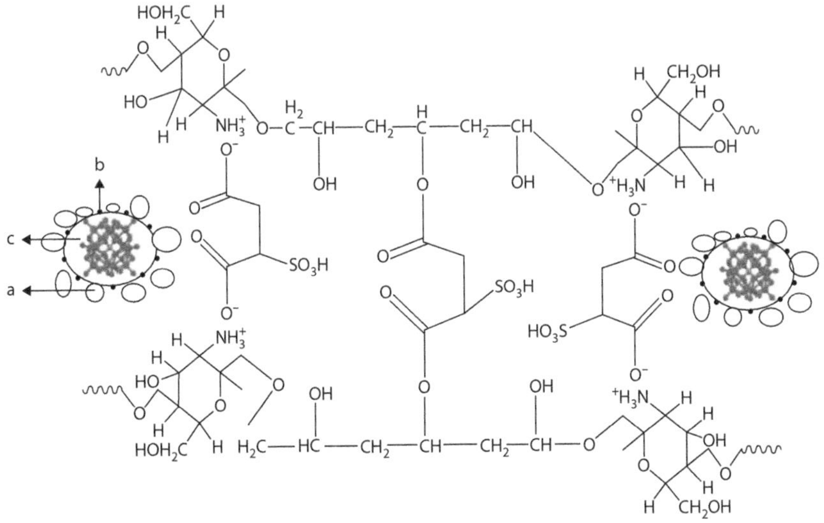

FIGURE 10.15 Dual network of CS–PVA–SSA–SWA. Note: (a) Methanol molecule, (b) Hydrated phase of SWA, and (c) Keggin cage of SWA. (From Meenakshi, S. et al., *J. Solid State Electrochem.*, 16, 1709, 2012.)

10.6.3.2 Characterization and Applications

Proton conductivity measurements were performed by AC impedance technique. The proton conductivities for all the membranes increased with increase in temperature from 30°C to 100°C as seen in Figure 10.16. It is noteworthy that the proton conductivity increases with increase in SWA content from 5 to 10 wt% and decreases at 15 wt% SWA. Higher content of SWA (>10 wt%) in cross-linked CS–PVA blend disrupts the proton conduction path by blocking the voids of polymer matrix and decreases its proton conductivity, which may be also attributed to the water sorption data. The methanol crossover for the membranes is measured in situ in fuel cells under OCV condition at 70°C. The methanol crossover flux is lower for CS–PVA–SSA–SWA (10 wt%) (about 3.3×10^{-7} mol s^{-1} cm^{-2}) hybrid membranes in comparison with Nafion 117 and other membranes. In addition, the electrochemical selectivity of CS–PVA–SSA–SWA (10 wt%) has reached to 2.69×10^{4} S cm^{-3} s.

MEAs of membranes were made and DMFC tests were conducted. The voltage and power density versus current density plots of membranes were shown in

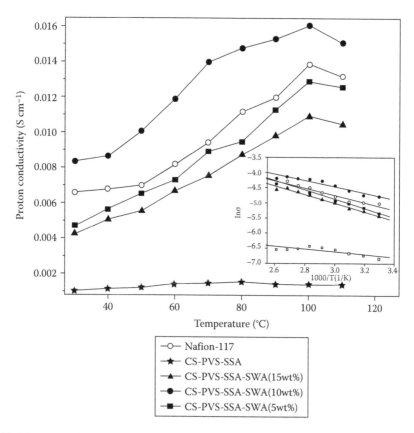

FIGURE 10.16 Proton conductivity versus temperature and ln σ versus 1000/T plot of Nafion 117, CS–PVA–SSA blend, and CS–PVA–SSA–SWA (5–15 wt%) hybrid membranes. (From Meenakshi, S. et al., *J. Solid State Electrochem.*, 16, 1709, 2012.)

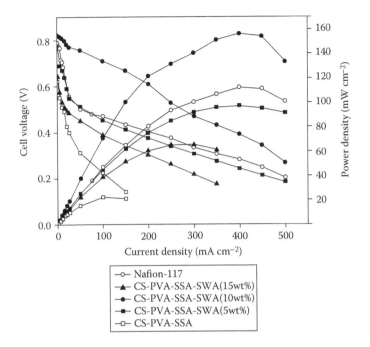

FIGURE 10.17   Voltage and power density versus current density plot for several membranes. (From Meenakshi, S. et al., *J. Solid State Electrochem.*, 16, 1709, 2012.)

Figure 10.17. Peak power densities of 156 mW cm^{-2} at a load current density of 400 mA cm^{-2} were obtained for the MEA based on CS–PVA–SSA–SWA (10 wt%) dual-network hybrid membranes. The higher DMFC performance observed for the hybrid membranes is due to the lower valence and electronegativity and increased stability of the Keggin type of anion in SWA structure that enhances its acid strength, thereby increasing proton conductivity.

10.7   CHALLENGES AND PERSPECTIVES

In comparison with the membranes with common aromatic skeletons such as poly-sulfane (PS), poly(ether ether ketone), poly(phthalazinon ether sulfone ketone), poly(etherimide), poly(benzimidazole), poly(phenylene oxide), polysiloxane, poly(oxyethylene) methacrylate, poly(arylene ether sulfone), and polyethersulfone Cardo, which are generally at high price and of complicated synthesis processes, the most important advantages of the aliphatic polymer materials as PEM membranes are their low cost, easy preparation, and simple structure. These aliphatic PEMs are particularly environmentally friendly (e.g., if the quaternization process is proceeded when these aromatic membranes are used for alkaline PEM fuel cells, the synthesis route uses chloromethyl ether for chloromethylation, which is very toxic and carcinogenic). However, the stability of the aliphatic PEMs is not very good. This is probably the biggest challenge when they are used in electrochemical devices.

This is because for the PEMs with aromatic skeletons, the bond energy of C–H bond is 435 kJ mol^{-1}, which is very close to the bond energy of C–F bond (485 kJ mol^{-1}) for the perfluorostructured membrane such as Nafion. However, for PEMs with aliphatic skeletons, the bond energy of C–H is only 350 kJ mol^{-1}, which is much lower than the membranes with both aromatic skeleton and perfluoroskeleton. The stability of the aliphatic PEMs on the electrode processes is too low to be practical.

In addition, early PEM research mainly focused on using polymer membranes as a solid matrix to hold alkaline solution. Hydrophilic polymers, such as PEO [134,135], CS [136], PVA, and polyacrylamide (PAM), or their copolymers [137–139] have been used to absorb the alkaline solution, and these polymers were generally cross-linked to obtain the required mechanical properties. However, these polymer membranes are not cation- or anion-free OH$^-$ or H$^+$ conductive, since they would not work without alkaline or acid solution. Some strong acids (including sulfuric acid, phosphoric acid, and heteropolyacid) and alkaline (e.g., KOH) are often doped into the polymer matrix to increase the H$^+$ conductivity and/or OH$^-$ conductivity. However, the accompanying decreased conductivity and the acid/KOH leaching-induced corrosion constitute the two inherent drawbacks. But this could be controlled effectively by adjusting the modification procedures, and the membranes can be thus designed with expectations of moderate swelling, good mechanical strength and flexibility, and high H$^+$/OH$^-$ conductivity and stability. All the aforementioned properties could be greatly improved even in a highly corrosive environment. Some typical techniques for membrane modifications are summarized as follows:

1. *Inorganic–Organic Hybrid Modifications*: Early PEM research mainly focused on using polymer membranes as a solid matrix to hold alkaline solution, particular to hydrophilic polymers including PEO [4,38–40], PVA [2,6,7], and PAA, or their copolymers [3,9,13]. The composites such as PVA/KOH/H$_2$O, PAA/Me$_4$NOH, PEO/PVA/KOH/H$_2$O, PEO/KOH/H$_2$O, PPA/KOH/H$_2$O, and poly(epichlorohydrin) PECH/PEO/KOH/H$_2$O (here PECH: polyerepichlorhydride) also have been investigated in the last several years, where KOH and H$_2$O were also used as plasticizers. Therefore, the electrical conductivity (usually 10^{-3}–10^{-2} S cm^{-1}) of these composite alkaline polymer electrolytes is greatly influenced by the contents of KOH and H$_2$O and the amorphous structure of the polymer matrix. However, in the preparation process, when KOH content is over the limitation of its solubility in a specific polymer matrix, the gel composite electrolyte phase, polymer + KOH + H$_2$O, is separated from the solution phase, KOH + H$_2$O. The phase separation makes the preparation of homogeneous and stoichiometric composite electrolyte impossible and also makes a poor mechanical property of the composite. By incorporating the inorganic filler, all the preceding properties could be improved greatly. Its most significant advantages are their lower glass transition temperature (T_g) and more amorphous domain of matrix due to the organic filler's incorporation; therefore, the conductivity and the mechanical property of these organic–inorganic hybrid composites were improved. The most typical candidate can be seen from the ZrO$_2$-doped PVA/ZrO$_2$/KOH/H$_2$O system [140], α-Al$_2$O$_3$-doped PVA/KOH/α-Al$_2$O$_3$/H$_2$O composite

system with a conductivity increased by three orders at room temperature [141], and TiO_2-, Al_2O_3-, and SiO_2-doped PEO [142]. The inorganic powder doped alkaline solid polymer electrolytes (ASPEs) with greatly improved conductivity by mixing nanosized TiO_2 and Al_2O_3.

2. *Blending Technique*: As one of the chemical modification methods, the blending technique is an extremely promising approach due to its potential for combining the attractive features of each blend component while at the same time reducing their deficient characteristics [143,144]. In other words, the blend properties of polymer possess the intrinsic chemical, physical, mechanic, and morphological properties of each polymer. The advantages from each polymer can be combined by blending two polymers. However, the blending of polyelectrolyte materials causes a precipitate or coacervate in an aqueous medium resulting from electrostatic interaction between two charges [18]. Therefore, the formation of heterogeneous solution prohibits the possibility of uniform membrane synthesis; in particular, when the mass ratio of the two polymers is high, the phase separation becomes more obvious. In addition, only by blending, the membrane may not be formed if the blending polymer has high hydrophilic property. At this stage, the membrane structure is usually finished by blending and then chemical cross-linking technique, such as PVA-based PEMs.

3. *Copolymerization Modification*: The choice of copolymerization is to cover the shortage of base polymers like high water solubility, low mechanical property, and low ionic group concentration, through introducing the other polymer chain segment. For instance, PMMA is a polymer without ionic group; Erdogan et al. has introduced sulfonic PS into PEM synthesis by block copolymerization [47]. The fabricated membrane has proton conductivity close to Nafion 117.

4. *Cross-Linking Technique Modifications*: In PEM fabrication, cross-linking has been used frequently to improve the physical or chemical stability or decrease fuel permeability. After cross-linking, membrane swelling could also be decreased significantly. The mechanical properties could also be improved, only with a small decrease in conductivity. Many PVA-based PEMs have been cross-linked, as this polymer is water soluble. For PEO-based PEMs, for example, cross-linking is found to improve the mechanical properties but does not degrade the electrochemical properties, which compare favorably with those of similar solid polymer electrolytes prepared with pure base polymer. In addition, introduction of a few oxypropylene oxide units in the PEO chains can prevent their recrystallization while keeping the good solvating properties of PEO toward lithium ions [145].

5. *Interpenetrating Polymer Network*: IPN usually constructed within polyelectrolyte complex membranes leads to an enhanced compatibility between the two polymers and thus a synergistic effect to suppress methanol crossover [11,12,18,19]. Membranes with an IPN structure formed by a cationic polyelectrolyte and an anionic polyelectrolyte may offer a feasible solution to obtain both low methanol crossover and high proton conductivity. For example, the incorporation of PAM with cross-linking agent provides a

semi-IPN with PVA. A decrease in polymer crystallinity can improve chain flexibility for more efficient ionic transport in the amorphous region.

6. *Membrane Fabrication by Inner Structure and Molecular Weight*: It has been demonstrated that the structure and the molecular weight of quaternized polymer introduced have a great influence on the alkaline stability of the membranes prepared. Introducing different functional polymer chains can be an effective way to create a microscopic phase-separated structure in ion-conduction polymer membranes, and thus the ionic conductivity could be improved. Qiao et al. has demonstrated that by applying PVA as a polymer matrix, the quaternized polymers with cyclic structure or side chain and high molecule weight show much higher alkaline stability than those with only main chain or low molecule weight [42,146,147]. Therefore, the method simply distinguishes it from the existing techniques, allowing the PEMs to be cost effective.

10.8   CHAPTER SUMMARY

To facilitate the research and development of aliphatic PEMs, this chapter gives a comprehensive overview of several decades of development and recent trends in the material selection, design and synthesis, characterizations, their conduction property and water absorption behavior, conduction mechanisms and chemical stability, as well as their possible applications in fuel cells, Zn–air batteries, and rechargeable Ni–MH batteries and even in carbon dioxide electroreduction. Various aliphatic PEMs explored and reported in the literature and their modification techniques are summarized. Challenges and perspectives are presented in depth with respect to membrane conductivity and stability. Some typical aliphatic PEMs and their associated data from material selection, synthesis, characterization, and applications are also summarized and presented to help readers quickly locate the information they are looking for.

It should be mentioned that all the PEM systems in this chapter are based on conduction mechanisms, that is, the proton conduction is based on the migration of hydronium ions (H_3O^+) or hydroxyl ions (OH^-) through the hydrophilic clusters of sulfonate aggregates or hopping between the *gaps* of the polymer matrix or the hydrophilic functional groups such as C–O, C=O, and C–N bonds [148], which require water supply for retaining their proton or hydroxyl ion conductivity. However, the presence of water causes some problems in the devices containing moisture-sensitive materials (e.g., electrochromic windows). Also, degradation of the polymer membrane will occur under strong acidic/alkaline conditions with considerable amounts of water [149], although the electrode flooding/drying issues would be improved when in alkaline conditions [150,151]. It has been demonstrated that dissolution of H_3PO_4 in PMMA plasticized by dimethylformamide (PMMA/DMF) or poly(glycidyl methacrylate) with DMF or propylene carbonate (PGMA/DMF and PGMA/PC) and PEO-modified poly(methacrylate) containing poly(ethylene glycol dimethyl ether) as an organic plasticizer [20,152–154] produces the so-called polymeric gel electrolytes with high proton conductivities around 10^{-4} S cm^{-1} at ambient temperatures. Because these kinds of polymeric gels are synthesized using anhydrous solvents under a dry atmosphere, the drawbacks mentioned earlier are avoided. In addition, the absence of water simplifies, in principle, the study of the conduction behavior of the system and is likely to provide better electrochemical and thermal stability of the material.

Another point is insufficient fundamental understanding. The literature contains very short of attempts to fundamentally understand the (H_3O^+)/hydroxyl ions (OH^-) conduction process on aliphatic PEMs through both experimental and theoretical modeling approaches to membrane material down selection with respect to new membrane design and membrane operation optimization. Computational modeling is an important aspect in membrane performance optimization. A fundamental understanding of the electrochemical and transport processes in aliphatic PEMs should be necessary.

REFERENCES

1. Y.-J. Wang, J. Qiao, R. Baker, J. Zhang, Alkaline polymer electrolyte membranes for fuel cell applications, *Chem. Soc. Rev.*, 42 (2013) 5768–5787.
2. A.A. Mohamad, N.S. Mohamed, M.Z.A. Yahya, R. Othman, S. Ramesh, Y. Alias, A.K. Arof, Ionic conductivity studies of poly(vinyl alcohol) alkaline solid polymer electrolyte and its use in nickel–zinc cells, *Solid State Ionics*, 156 (2003) 171–177.
3. N. Vassal, E. Salmon, J.F. Fauvarque, Electrochemical properties of an alkaline solid polymer electrolyte based on P(ECH-co-EO), *Electrochim. Acta*, 45 (2000) 1527–1532.
4. A. Lewandowski, M. Zajder, E. Frąckowiak, F. Béguin, Supercapacitor based on activated carbon and polyethylene oxide–KOH–H_2O polymer electrolyte, *Electrochim. Acta*, 46 (2001) 2777–2780.
5. Q.M. Wu, J.F. Zhang, S.B. Sang, Preparation of alkaline solid polymer electrolyte based on PVA–TiO_2–KOH–H_2O and its performance in Zn-Ni battery, *J. Phys. Chem. Solids*, 69 (2008) 2691–2695.
6. S.B. Sang, Q.M. Wu, Z. Gan, Influences of doping approach on conductivity of composite alkaline solid polymer electrolyte PVA–HA–KOH–H_2O, *Electrochim. Acta*, 53 (2008) 5065–5070.
7. A. Lewandowski, K. Skorupska, J. Malinska, Novel poly(vinyl alcohol)–KOH–H_2O alkaline polymer electrolyte, *Solid State Ionics*, 133 (2000) 265–271.
8. H.B. Yang, H.C. Zhang, X.D. Wang, J.H. Wang, X.L. Meng, Z.X. Zhou, Calcium zincate synthesized by ballmilling as a negative material for secondary alkaline batteries, *J. Electrochem. Soc.*, 151 (2004) A2126–A2131.
9. C.C. Yang, Polymer Ni-MH battery based on PEO-PVA-KOH polymer electrolyte, *J. Power Sources*, 109 (2002) 22–31.
10. C.C. Yang, S.J. Lin, Alkaline composite PEO-PVA-glass-fibre-mat polymer electrolyte for Zn-air battery, *J. Power Sources*, 112 (2002) 497–503.
11. D.S. Kim, T.I. Yun, M.Y. Seo, H.I. Cho, Y.M. Lee, S.Y. Nam, J.W. Rhim, Preparation of ion-exchange membranes for fuel cell based on crosslinked PVA/PSSA-MA/silica hybrid, *Desalination*, 200 (2006) 634–635.
12. D.S. Kim, M.D. Guiver, M.Y. Seo, H.I. Cho, D.H. Kim, A.W. Rhim, G.Y. Moon, S.Y. Nam, Influence of silica content in crosslinked PVA/PSSA-MA/silica hybrid membrane for direct methanol fuel cell (DMFC), *Macromol. Res.*, 15 (2007) 412–417.
13. C.C. Yang, C.T. Lin, S.J. Chiu, Preparation of the PVA/HAP composite polymer membrane for alkaline DMFC application, *Desalination*, 233 (2008) 137–146.

14. C.C. Yang, W.C. Chien, Y.J.J. Li, Direct methanol fuel cell based on poly(vinyl alcohol)/titanium oxide nanotubes/poly(styrene sulfonic acid) (PVA-nt-TiO$_2$/PSSA) composite polymer membrane, *J. Power Sources*, 195 (2010) 3407–3415.

15. J. Qiao, J. Fu, R. Lin, J. Liu, J. Ma, Poly (vinyl alcohol)/poly (acrylamide-co-diallyldimethylammonium chloride)(PVA/PAADDA) composite as novel anion-exchange membranes for potential use in fuel cells, *ECS Trans.*, 33 (2010) 1915–1922.

16. J. Qiao, J. Fu, L. Liu, Y. Liu, J. Sheng, Highly stable hydroxyl anion conducting membranes poly(vinyl alcohol)/poly(acrylamide-co-diallyldimethylammonium chloride) (PVA/PAADDA) for alkaline fuel cells: Effect of cross-linking, *Int. J. Hydrogen Energy*, 37 (2012) 4580–4589.

17. D. Kim, M. Guiver, M. Seo, H. Cho, D. Kim, J. Rhim, G. Moon, S. Nam, Influence of silica content in crosslinked PVA/PSSA-MA/silica hybrid membrane for direct methanol fuel cell (DMFC), *Macromol. Res.*, 15 (2007) 412–417.

18. F. Ahmad, E. Sheha, Preparation and physical properties of (PVA)0.7(NaBr)0.3(H$_3$PO$_4$) xM solid acid membrane for phosphoric acid—Fuel cells, *J. Adv. Res.*, 4 (2013) 155–161.

19. G. Merle, S.S. Hosseiny, M. Wessling, K. Nijmeijer, New cross-linked PVA based polymer electrolyte membranes for alkaline fuel cells, *J. Membr. Sci.*, 409–410 (2012) 191–199.

20. J. Qiao, N. Yoshimoto, M. Ishikawa, M. Morita, A proton conductor based on a polymeric complex of poly(ethylene oxide)-modified poly(methacrylate) with anhydrous H$_3$PO$_4$, *Chem. Mater.*, 15 (2003) 2005–2010.

21. A. Anis, A.K. Banthia, S. Bandyopadhyay, Synthesis & characterization of PVA/STA composite polymer electrolyte membranes for fuel cell application, *J. Mater. Eng. Perform.*, 17 (2008) 772–779.

22. Y.-W. Chang, E. Wang, G. Shin, J.-E. Han, P.T. Mather, Poly(vinyl alcohol) (PVA)/sulfonated polyhedral oligosilsesquioxane (sPOSS) hybrid membranes for direct methanol fuel cell applications, *Polym. Adv. Technol.*, 18 (2007) 535–543.

23. C.-C. Yang, Y.-J. Lee, S.-J. Chiu, K.-T. Lee, W.-C. Chien, C.-T. Lin, C.-A. Huang, Preparation of a PVA/HAP composite polymer membrane for a direct ethanol fuel cell (DEFC), *J. Appl. Electrochem.*, 38 (2008) 1329–1337.

24. D.S. Kim, H.B. Park, J.W. Rhim, Y. Moo Lee, Preparation and characterization of crosslinked PVA/SiO$_2$ hybrid membranes containing sulfonic acid groups for direct methanol fuel cell applications, *J. Membr. Sci.*, 240 (2004) 37–48.

25. V.M. Nikolic, A. Krkljes, Z.K. Popovic, Z.V. Lausevic, S.S. Miljanic, On the use of gamma irradiation crosslinked PVA membranes in hydrogen fuel cells, *Electrochem. Commun.*, 9 (2007) 2661–2665.

26. C.-C. Yang, Y.-J. Lee, Preparation of the acidic PVA/MMT nanocomposite polymer membrane for the direct methanol fuel cell (DMFC), *Thin Solid Films*, 517 (2009) 4735–4740.

27. J. Qiao, T. Hamaya, T. Okada, New highly proton-conducting membrane poly(vinylpyrrolidone)(PVP) modified poly(vinyl alcohol)/2-acrylamido-2-methyl-1-propanesulfonic acid (PVA–PAMPS) for low temperature direct methanol fuel cells (DMFCs), *Polymer*, 46 (2005) 10809–10816.

28. A.d.S. Gomes, J.C. Dutra Filho, Hybrid membranes of PVA for direct ethanol fuel cells (DEFCs) applications, *Int. J. Hydrogen Energy*, 37 (2012) 6246–6252.

29. G. Gnana Kumar, D.N. Lee, P. Kim, K.S. Nahm, R. Nimma Elizabeth, Characterization of PVdF-HFP/Nafion/AlO[OH]n composite membranes for direct methanol fuel cell (DMFC), *Eur. Polym. J.*, 44 (2008) 2225–2230.

30. G.G. Kumar, P. Uthirakumar, K.S. Nahm, R.N. Elizabeth, Fabrication and electro chemical properties of poly vinyl alcohol/para toluene sulfonic acid membranes for the applications of DMFC, *Solid State Ionics*, 180 (2009) 282–287.

31. G.G. Kumar, P. Kim, K.S. Nahm, R.N. Elizabeth, Structural characterization of PVdF-HFP/PEG/Al$_2$O$_3$ proton conducting membranes for fuel cells, *J. Membr. Sci.*, 303 (2007) 126–131.

32. G. Gnana kumar, J. Shin, Y.-C. Nho, I.S. Hwang, G. Fei, A.R. Kim, K.S. Nahm, Irradiated PVdF-HFP–tin oxide composite membranes for the applications of direct methanol fuel cells, *J. Membr. Sci.*, 350 (2010) 92–100.

33. G. Gnana kumar, Irradiated PVdF-HFP-montmorillonite composite membranes for the application of direct ethanol fuel cells, *J. Mater. Chem.*, 21 (2011) 17382–17391.

34. V. Saarinen, O. Himanen, T. Kallio, G. Sundholm, K. Kontturi, Current distribution measurements with a free-breathing direct methanol fuel cell using PVDF-g-PSSA and Nafion® 117 membranes, *J. Power Sources*, 163 (2007) 768–776.

35. K.A. Stewart, M. Singh, H.P.S. Missan, Novel PVdF-graft-VIm based nano-composite polymer electrolyte membranes for fuel cell applications, *ECS Trans.*, 25 (2009) 1459–1468.

36. J. Shen, J. Xi, W. Zhu, L. Chen, X. Qiu, A nanocomposite proton exchange membrane based on PVDF, poly(2-acrylamido-2-methyl propylene sulfonic acid), and nano-Al_2O_3 for direct methanol fuel cells, *J. Power Sources*, 159 (2006) 894–899.

37. H. Bai, W.S.W. Ho, New poly(ethylene oxide) soft segment-containing sulfonated polyimide copolymers for high temperature proton-exchange membrane fuel cells, *J. Membr. Sci.*, 313 (2008) 75–85.

38. S. Guinot, E. Salmon, J.F. Penneau, J.F. Fauvarque, A new class of PEO-based SPEs: Structure, conductivity and application to alkaline secondary batteries, *Electrochim. Acta*, 43 (1998) 1163–1170.

39. N. Vassal, E. Salmon, J.F. Fauvarque, Nickel/metal hydride secondary batteries using an alkaline solid polymer electrolyte, *J. Electrochem. Soc.*, 146 (1999) 20–26.

40. J.F. Fauvarque, S. Guinot, N. Bouzir, E. Salmon, J.F. Penneau, Alkaline poly(ethylene oxide) solid polymer electrolytes. Application to nickel secondary batteries, *Electrochim. Acta*, 40 (1995) 2449–2453.

41. H.K. Kim, M. Zhang, X. Yuan, S.N. Lvov, T.C.M. Chung, Synthesis of polyethylene-based proton exchange membranes containing PE backbone and sulfonated poly(arylene ether sulfone) side chains for fuel cell applications, *Macromolecules*, 45 (2012) 2460–2470.

42. T.A. Sherazi, J.Y. Sohn, Y.M. Lee, M.D. Guiver, Polyethylene-based radiation grafted anion-exchange membranes for alkaline fuel cells, *J. Membr. Sci.*, 441 (2013) 148–157.

43. T.A. Sherazi, S. Ahmad, M.A. Kashmiri, D.S. Kim, M.D. Guiver, Radiation-induced grafting of styrene onto ultrahigh-molecular-weight polyethylene powder for polymer electrolyte fuel cell application: II. Sulfonation and characterization, *J. Membr. Sci.*, 333 (2009) 59–67.

44. J.H. Shim, I.G. Koo, W.M. Lee, Nafion-impregnated polyethylene-terephthalate film used as the electrolyte for direct methanol fuel cells, *Electrochim. Acta*, 50 (2005) 2385–2391.

45. P. Piboonsatsanasakul, J. Wootthikanokkhan, S. Thanawan, Preparation and characterizations of direct methanol fuel cell membrane from sulfonated polystyrene/poly(vinylidene fluoride) blend compatibilized with poly(styrene)-b-poly(methyl methacrytlate) block copolymer, *J. Appl. Polym. Sci.*, 107 (2008) 1325–1336.

46. W.H. Choi, W.H. Jo, Preparation of new proton exchange membrane based on self-assembly of poly(styrene-co-styrene sulfonic acid)-b-poly(methyl methacrylate)/poly(vinylidene fluoride) blend, *J. Power Sources*, 188 (2009) 127–131.

47. T. Erdogan, E.E. Unveren, T.Y. Inan, B. Birkan, Well-defined block copolymer ionomers and their blend membranes for proton exchange membrane fuel cell, *J. Membr. Sci.*, 344 (2009) 172–181.

48. S. Rajendran, O. Mahendran, R. Kannan, Characterisation of [(1−x)PMMA–xPVdF] polymer blend electrolyte with Li+ ion, *Fuel*, 81 (2002) 1077–1081.

49. S.S. Sekhon, H.P. Singh, Proton conduction in polymer gel electrolytes containing chloroacetic acids, *Solid State Ionics*, 175 (2004) 545–548.

50. T. Uma, F. Yin, High proton conductivity of poly (methyl methacrylate)-based hybrid membrane for PEMFCs, *Chem. Phys. Lett.*, 512 (2011) 104–107.

51. J. George, K.V. Ramana, S.N. Sabapathy, J.H. Jagannath, A.S. Bawa, Characterization of chemically treated bacterial (*Acetobacter xylinum*) biopolymer: Some thermo-mechanical properties, *Int. J. Biol. Macromol.*, 37 (2005) 189–194.

52. G.P. Jiang, J.L. Qiao, F. Hong, Application of phosphoric acid and phytic acid-doped bacterial cellulose as novel proton-conducting membranes to PEMFC, *Int. J. Hydrogen Energy*, 37 (2012) 9182–9192.

53. J.Z. Yang, D.P. Sun, J. Li, X.J. Yang, J.W. Yu, Q.L. Hao, W.M. Liu, J.G. Liu, Z.G. Zou, J. Gu, In situ deposition of platinum nanoparticles on bacterial cellulose membranes and evaluation of PEM fuel cell performance, *Electrochim. Acta*, 54 (2009) 6300–6305.

54. M. Matsuguchi, H. Takahashi, Methanol permeability and proton conductivity of a semi-interpenetrating polymer networks (IPNs) membrane composed of Nafion((R)) and cross-linked DVB, *J. Membr. Sci.*, 281 (2006) 707–715.

55. Z.Y. Jiang, X.H. Zheng, H. Wu, J.T. Wang, Y.B. Wang, Proton conducting CS/P(AA-AMPS) membrane with reduced methanol permeability for DMFCs, *J. Power Sources*, 180 (2008) 143–153.

56. P. Mukoma, B.R. Jooste, H.C.M. Vosloo, Synthesis and characterization of cross-linked chitosan membranes for application as alternative proton exchange membrane materials in fuel cells, *J. Power Sources*, 136 (2004) 16–23.

57. D.E. Fenton, J.M. Parker, P.V. Wright, Complexes of alkali-metal ions with poly(ethylene oxide), *Polymer*, 14 (1973) 589.

58. M.B Armand, J.M Chabagno, M.J Duclot. In: P. Vashishta, J.N Mundy, G.K Shenoy, eds. *Fast Ion Transport in Solids Electrodes and Electrolytes*. Amsterdam, the Netherlands: North-Holland Elsevier, 1979, pp. 131–136.

59. G. Guo, R. Jie, S. An, Performance of the self-breathing air direct methanol fuel cell with modified poly (vinylidene fluoride) grafted onto a blended polystyrene sulfonated acid (m-PVDF-g-PSSA) membranes, *Adv. Mater. Res.*, 152–153 (2011) 149–153.

60. A. Svang-Ariyaskul, R.Y.M. Huang, P.L. Douglas, R. Pal, X. Feng, P. Chen, L. Liu, Blended chitosan and polyvinyl alcohol membranes for the pervaporation dehydration of isopropanol, *J. Membr. Sci.*, 280 (2006) 815–823.

61. D.S. Kim, I.C. Park, H.I. Cho, D.H. Kim, G.Y. Moon, H.K. Lee, J.W. Rhim, Effect of organo clay content on proton conductivity and methanol transport through crosslinked PVA hybrid membrane for direct methanol fuel cell, *J. Ind. Eng. Chem.*, 15 (2009) 265–269.

62. S. Yun, H. Im, Y. Heo, J. Kim, Crosslinked sulfonated poly(vinyl alcohol)/sulfonated multi-walled carbon nanotubes nanocomposite membranes for direct methanol fuel cells, *J. Membr. Sci.*, 380 (2011) 208–215.

63. J. Fu, J. Qiao, H. Lv, J. Ma, X.-Z. Yuan, H. Wang, Alkali doped poly(vinyl alcohol) (PVA) for anion-exchange membrane fuel cells: Ionic conductivity, chemical stability and FT-IR characterizations, *ECS Trans.*, 25 (2010) 15–23.

64. P. Gode, J. Ihonen, A. Strandroth, H. Ericson, G. Lindbergh, G. Sundholm, F. Sundholm, N. Walsby, Membrane durability in a PEM fuel cell studied using PVDF based radiation grafted membranes, *Fuel Cells*, 3 (2003) 21–27.

65. F. Pereira, A. Chan, K. Vallé, P. Palmas, J. Bigarré, P. Belleville, C. Sanchez, Design of interpenetrated networks of mesostructured hybrid silica and nonconductive poly(vinylidene fluoride)–cohexafluoropropylene (PVdF–HFP) polymer for proton exchange membrane fuel cell applications, *Chem. Asian J.*, 6 (2011) 1217–1224.

66. L. Ye, L.F. Zhai, J.H. Fang, J.H. Liu, C.H. Li, R. Guan, Synthesis and characterization of novel cross-linked quaternized poly(vinyl alcohol) membranes based on morpholine for anion exchange membranes, *Solid State Ionics*, 240 (2013) 1–9.

67. Y. Wan, B. Peppley, K.A.M. Creber, V.T. Bui, Anion-exchange membranes composed of quaternized-chitosan derivatives for alkaline fuel cells, *J. Power Sources*, 195 (2010) 3785–3793.

68. E.D. Wang, T.S. Zhao, W.W. Yang, Poly (vinyl alcohol)/3-(trimethylammonium) propyl-functionalized silica hybrid membranes for alkaline direct ethanol fuel cells, *Int. J. Hydrogen Energy*, 35 (2010) 2183–2189.

69. G.G. Kumar, D.N. Lee, P. Kim, K.S. Nahm, R.N. Elizabeth, Characterization of PVdF-HFP/Nafion/AlO[OH](n) composite membranes for direct methanol fuel cell (DMFC), *Eur. Polym. J.*, 44 (2008) 2225–2230.

70. T.A. Sherazi, M.D. Guiver, D. Kingston, S. Ahmad, M.A. Kashmiri, X. Xue, Radiation-grafted membranes based on polyethylene for direct methanol fuel cells, *J. Power Sources*, 195 (2010) 21–29.

71. T.C. Zhou, Z. Jing, J.L. Qiao, L.L. Liu, G.P. Jiang, J. Zhang, Y.Y. Liu, High durable poly(vinyl alcohol)/quaterized hydroxyethylcellulose ethoxylate anion exchange membranes for direct methanol alkaline fuel cells, *J. Power Sources*, 227 (2013) 291–299.

72. J. Byun, J. Sauk, H. Kim, Preparation of PVdF/PSSA composite membranes using supercritical carbon dioxide for direct methanol fuel cells, *Int. J. Hydrogen Energy*, 34 (2009) 6437–6442.

73. J. Fu, J.L. Qiao, X.Z. Wang, J.X. Ma, T. Okada, Alkali doped poly(vinyl alcohol) for potential fuel cell applications, *Synthetic Met.*, 160 (2010) 193–199.

74. A. Saxena, B.P. Tripathi, V.K. Shahi, Sulfonated poly(styrene-co-maleic anhydride)–poly(ethylene glycol)–silica nanocomposite polyelectrolyte membranes for fuel cell applications, *J. Phys. Chem. B*, 111 (2007) 12454–12461.

75. S. Alwin, S.D. Bhat, A.K. Sahu, A. Jalajakshi, P. Sridhar, S. Pitchumani, A.K. Shukla, Modified-pore-filled-PVDF-membrane electrolytes for direct methanol fuel cells, *J. Electrochem. Soc.*, 158 (2011) B91–B98.

76. Y.-C. Cao, C. Xu, X. Wu, X. Wang, L. Xing, K. Scott, A poly (ethylene oxide)/graphene oxide electrolyte membrane for low temperature polymer fuel cells, *J. Power Sources*, 196 (2011) 8377–8382.

77. T. Inan, H. Doğan, A. Güngör, PVdF-HFP membranes for fuel cell applications: Effects of doping agents and coating on the membrane's properties, *Ionics*, 19 (2013) 629–641.

78. S.J. Zaidi, T. Matsuura, *Polymer Membranes for Fuel Cells*, (2009) Springer-Verlag Inc., New York.

79. S. Ma, Z. Siroma, H. Tanaka, Anisotropic conductivity over in-plane and thickness directions in Nafion-117, *J. Electrochem. Soc.*, 153 (2006) A2274–A2281.

80. R.F. Silva, M. De Francesco, A. Pozio, Tangential and normal conductivities of Nafion® membranes used in polymer electrolyte fuel cells, *J. Power Sources*, 134 (2004) 18–26.

81. Z. Siroma, R. Kakitsubo, N. Fujiwara, T. Ioroi, S.-i. Yamazaki, K. Yasuda, Depression of proton conductivity in recast Nafion film measured on flat substrate, *J. Power Sources*, 189 (2009) 994–998.

82. J. Tang, W. Yuan, J. Wang, J. Tang, H. Li, Y. Zhang, Perfluorosulfonate ionomer membranes with improved through-plane proton conductivity fabricated under magnetic field, *J. Membr. Sci.*, 423–424 (2012) 267–274.

83. J. Peron, A. Mani, X. Zhao, D. Edwards, M. Adachi, T. Soboleva, Z. Shi, Z. Xie, T. Navessin, S. Holdcroft, Properties of Nafion® NR-211 membranes for PEMFCs, *J. Membr. Sci.*, 356 (2010) 44–51.

84. L. Su, Preparation of polysiloxane modified perfluorosulfonic acid composite membranes assisted by supercritical carbon dioxide for direct methanol fuel cell, *J. Power Sources*, 194 (2009) 220–225.

85. L. Su, Perfluorosulfonic acid membranes treated by supercritical carbon dioxide method for direct methanol fuel cell application, *J. Membr. Sci.*, 335 (2009) 118–125.

86. L. Su, Preparation of polysiloxane/perfluorosulfonic acid nanocomposite membranes in supercritical carbon dioxide system for direct methanol fuel cell, *Int. J. Hydrogen Energy*, 34 (2009) 6892–6901.

87. S. Mollá, V. Compañ, Performance of composite Nafion/PVA membranes for direct methanol fuel cells, *J. Power Sources*, 196 (2011) 2699–2708.

88. S. Zhang, X.-Z. Yuan, J.N.C. Hin, H. Wang, J. Wu, K.A. Friedrich, M. Schulze, Effects of open-circuit operation on membrane and catalyst layer degradation in proton exchange membrane fuel cells, *J. Power Sources*, 195 (2010) 1142–1148.

89. X.-Z. Yuan, S. Zhang, H. Wang, J. Wu, J.C. Sun, R. Hiesgen, K.A. Friedrich, M. Schulze, A. Haug, Degradation of a polymer exchange membrane fuel cell stack with Nafion® membranes of different thicknesses: Part I. In situ diagnosis, *J. Power Sources*, 195 (2010) 7594–7599.

90. J. Catalano, M.G. Baschetti, M.G. De Angelis, G.C. Sarti, A. Sanguineti, P. Fossati, Gas and water vapor permeation in a short-side-chain PFSI membrane, *Desalination*, 240 (2009) 341–346.

91. X. Cheng, J. Zhang, Y. Tang, C. Song, J. Shen, D. Song, J. Zhang, Hydrogen crossover in high-temperature PEM fuel cells, *J. Power Sources*, 167 (2007) 25–31.

92. G. Merle, M. Wessling, K. Nijmeijer, Anion exchange membranes for alkaline fuel cells: A review, *J. Membr. Sci.*, 377 (2011) 1–35.

93. K.N. Grew, W.K.S. Chiu, A dusty fluid model for predicting hydroxyl anion conductivity in alkaline anion exchange membranes, *J. Electrochem. Soc.*, 157 (2010) B327–B337.

94. T.J. Peckham, S. Holdcroft, Structure–morphology–property relationships of non-perfluorinated proton-conducting membranes, *Adv. Mater.*, 22 (2010) 4667–4690.

95. D. Marx, Proton transfer 200 years after von Grotthuss: Insights from ab initio simulations, *ChemPhysChem*, 7 (2006) 1848–1870.

96. P. Choi, N.H. Jalani, R. Datta, Thermodynamics and proton transport in Nafion: II. Proton diffusion mechanisms and conductivity, *J. Electrochem. Soc.*, 152 (2005) E123–E130.

97. P. Choi, N.H. Jalani, R. Datta, Thermodynamics and proton transport in Nafion: I. Membrane swelling, sorption, and ion-exchange equilibrium, *J. Electrochem. Soc.*, 152 (2005) E84–E89.

98. K.-D. Kreuer, S.J. Paddison, E. Spohr, M. Schuster, Transport in proton conductors for fuel-cell applications: Simulations, elementary reactions, and phenomenology, *Chem. Rev.*, 104 (2004) 4637–4678.

99. S.J. Paddison, R. Paul, The nature of proton transport in fully hydrated Nafion®, *PCCP*, 4 (2002) 1158–1163.

100. P. Commer, A.G. Cherstvy, E. Spohr, A.A. Kornyshev, The effect of water content on proton transport in polymer electrolyte membranes, *Fuel Cells*, 2 (2002) 127–136.

101. K.D. Kreuer, On the development of proton conducting polymer membranes for hydrogen and methanol fuel cells, *J. Membr. Sci.*, 185 (2001) 29–39.

102. M. Eikerling, A.A. Kornyshev, A.M. Kuznetsov, J. Ulstrup, S. Walbran, Mechanisms of proton conductance in polymer electrolyte membranes, *J. Phys. Chem. B*, 105 (2001) 3646–3662.

103. T.-H. Tsai, A.M. Maes, M.A. Vandiver, C. Versek, S. Seifert, M. Tuominen, M.W. Liberatore, A.M. Herring, E.B. Coughlin, Synthesis and structure–conductivity relationship of polystyrene-block-poly(vinyl benzyl trimethylammonium) for alkaline anion exchange membrane fuel cells, *J. Polym. Sci. B: Polym. Phys.*, 51 (2013) 1751–1760.

104. G. Sudre, S. Inceoglu, P. Cotanda, N.P. Balsara, Influence of bound ion on the morphology and conductivity of anion-conducting block copolymers, *Macromolecules*, 46 (2013) 1519–1527.

105. C.G. Arges, J. Parrondo, G. Johnson, A. Nadhan, V. Ramani, Assessing the influence of different cation chemistries on ionic conductivity and alkaline stability of anion exchange membranes, *J. Mater. Chem.*, 22 (2012) 3733–3744.

106. M. Zhang, H.K. Kim, E. Chalkova, F. Mark, S.N. Lvov, T.C.M. Chung, New polyethylene based anion exchange membranes (PE–AEMs) with high ionic conductivity, *Macromolecules*, 44 (2011) 5937–5946.

107. K.N. Grew, X. Ren, D. Chu, Effects of temperature and carbon dioxide on anion exchange membrane conductivity, *Electrochem. Solid-State Lett.*, 14 (2011) B127–B131.

108. S. Castañeda Ramírez, C.I. Sánchez Sáenz, Modeling and analysis of ion transport through anion exchange membranes used in alkaline fuel cells, *ECS Trans.*, 50 (2013) 2091–2107.

109. T. Lehtinen, G. Sundholm, S. Holmberg, F. Sundholm, P. Björnbom, M. Bursell, Electrochemical characterization of PVDF-based proton conducting membranes for fuel cells, *Electrochim. Acta*, 43 (1998) 1881–1890.

110. E.F. Abdrashitov, V.C. Bokun, D.A. Kritskaya, E.A. Sanginov, A.N. Ponomarev, Y.A. Dobrovolsky, Synthesis and properties of the PVDF-based proton exchange membranes with incorporated cross-linked sulphonated polystyrene for fuel cells, *Solid State Ionics*, 251 (2013) 9–12.

111. G.K.S. Prakash, M.C. Smart, Q.-J. Wang, A. Atti, V. Pleynet, B. Yang, K. McGrath, G.A. Olah, S.R. Narayanan, W. Chun, T. Valdez, S. Surampudi, High efficiency direct methanol fuel cell based on poly(styrenesulfonic) acid (PSSA)–poly(vinylidene fluoride) (PVDF) composite membranes, *J. Fluorine Chem.*, 125 (2004) 1217–1230.

112. M.H. Yildirim, D. Stamatialis, M. Wessling, Dimensionally stable Nafion–polyethylene composite membranes for direct methanol fuel cell applications, *J. Membr. Sci.*, 321 (2008) 364–372.

113. K.-Y. Cho, H.-Y. Jung, K.A. Sung, W.-K. Kim, S.-J. Sung, J.-K. Park, J.-H. Choi, Y.-E. Sung, Preparation and characteristics of Nafion membrane coated with a PVdF copolymer/recast Nafion blend for direct methanol fuel cell, *J. Power Sources*, 159 (2006) 524–528.

114. K.-Y. Cho, J.-Y. Eom, H.-Y. Jung, N.-S. Choi, Y.M. Lee, J.-K. Park, J.-H. Choi, K.-W. Park, Y.-E. Sung, Characteristics of PVdF copolymer/Nafion blend membrane for direct methanol fuel cell (DMFC), *Electrochim. Acta*, 50 (2004) 583–588.

115. H. Pei, L. Hong, J.Y. Lee, Embedded polymerization driven asymmetric PEM for direct methanol fuel cells, *J. Membr. Sci.*, 270 (2006) 169–178.

116. K. Scott, W.M. Taama, P. Argyropoulos, Performance of the direct methanol fuel cell with radiation-grafted polymer membranes, *J. Membr. Sci.*, 171 (2000) 119–130.

117. E. Peled, T. Duvdevani, A. Aharon, A. Melman, A direct methanol fuel cell based on a novel low-cost nanoporous proton-conducting membrane, *Electrochem. Solid-State Lett.*, 3 (2000) 525–528.

118. C.-C. Yang, S.-J. Chiu, W.-C. Chien, Development of alkaline direct methanol fuel cells based on crosslinked PVA polymer membranes, *J. Power Sources*, 162 (2006) 21–29.

119. J. Qiao, J. Zhang, Anion conducting poly(vinyl alcohol)/poly(diallyldimethylammonium chloride) membranes with high durable alkaline stability for polymer electrolyte membrane fuel cells, *J. Power Sources*, 237 (2013) 1–4.

120. J. Fang, Y. Yang, X. Lu, M. Ye, W. Li, Y. Zhang, Cross-linked, ETFE-derived and radiation grafted membranes for anion exchange membrane fuel cell applications, *Int. J. Hydrogen Energy*, 37 (2012) 594–602.

121. Y. Zhang, J. Fang, Y. Wu, H. Xu, X. Chi, W. Li, Y. Yang, G. Yan, Y. Zhuang, Novel fluoropolymer anion exchange membranes for alkaline direct methanol fuel cells, *J. Colloid Interface Sci.*, 381 (2012) 59–66.

122. J.R. Varcoe, R.C.T. Slade, E. Lam How Yee, An alkaline polymer electrochemical interface: A breakthrough in application of alkaline anion-exchange membranes in fuel cells, *Chem. Commun.*, (2006) 1428–1429.

123. T.N. Danks, R.C.T. Slade, J.R. Varcoe, Comparison of PVDF- and FEP-based radiation-grafted alkaline anion-exchange membranes for use in low temperature portable DMFCs, *J. Mater. Chem.*, 12 (2002) 3371–3373.

124. J.R. Varcoe, R.C.T. Slade, E. Lam How Yee, S.D. Poynton, D.J. Driscoll, D.C. Apperley, Poly(ethylene-co-tetrafluoroethylene)-derived radiation-grafted anion-exchange membrane with properties specifically tailored for application in metal-cation-free alkaline polymer electrolyte fuel cells, *Chem. Mater.*, 19 (2007) 2686–2693.

125. J.R. Varcoe, R.C.T. Slade, An electron-beam-grafted ETFE alkaline anion-exchange membrane in metal-cation-free solid-state alkaline fuel cells, *Electrochem. Commun.*, 8 (2006) 839–843.

126. C. Sollogoub, A. Guinault, C. Bonnebat, M. Bennjima, L. Akrour, J.F. Fauvarque, L. Ogier, Formation and characterization of crosslinked membranes for alkaline fuel cells, *J. Membr. Sci.*, 335 (2009) 37–42.

127. E. Agel, J. Bouet, J.F. Fauvarque, Characterization and use of anionic membranes for alkaline fuel cells, *J. Power Sources*, 101 (2001) 267–274.

128. Y. Luo, J. Guo, C. Wang, D. Chu, Quaternized poly(methyl methacrylate-co-butyl acrylate-co-vinylbenzyl chloride) membrane for alkaline fuel cells, *J. Power Sources*, 195 (2010) 3765–3771.

129. J. Zhang, J. Qiao, G. Jiang, L. Liu, Y. Liu, Cross-linked poly(vinyl alcohol)/poly (diallyldimethylammonium chloride) as anion-exchange membrane for fuel cell applications, *J. Power Sources*, 240 (2013) 359–367.

130. M. Mamlouk, K. Scott, Effect of anion functional groups on the conductivity and performance of anion exchange polymer membrane fuel cells, *J. Power Sources*, 211 (2012) 140–146.

131. M. Mamlouk, J.A. Horsfall, C. Williams, K. Scott, Radiation grafted membranes for superior anion exchange polymer membrane fuel cells performance, *Int. J. Hydrogen Energy*, 37 (2012) 11912–11920.

132. S.R. Narayanan, B. Haines, J. Soler, T.I. Valdez, Electrochemical conversion of carbon dioxide to formate in alkaline polymer electrolyte membrane cells, *J. Electrochem. Soc.*, 158 (2011) A167–A173.

133. S. Meenakshi, S.D. Bhat, A.K. Sahu, S. Alwin, P. Sridhar, S. Pitchumani, Natural and synthetic solid polymer hybrid dual network membranes as electrolytes for direct methanol fuel cells, *J. Solid State Electrochem.*, 16 (2012) 1709–1721.

134. Y.G. Wu, C.M. Wu, F. Yu, T.W. Xu, Y.X. Fu, Free-standing anion-exchange PEO-SiO$_2$ hybrid membranes, *J. Membr. Sci.*, 307 (2008) 28–36.

135. T.N. Danks, R.C.T. Slade, J.R. Varcoe, Alkaline anion-exchange radiation-grafted membranes for possible electrochemical application in fuel cells, *J. Mater. Chem.*, 13 (2003) 712–721.

136. J.R. Varcoe, R.C.T. Slade, E.L.H. Yee, S.D. Poynton, D.J. Driscoll, Investigations into the ex situ methanol, ethanol and ethylene glycol permeabilities of alkaline polymer electrolyte membranes, *J. Power Sources*, 173 (2007) 194–199.

137. J. Fang, P.K. Shen, Quaternized poly(phthalazinon ether sulfone ketone) membrane for anion exchange membrane fuel cells, *J. Membr. Sci.*, 285 (2006) 317–322.

138. J.S. Park, S.H. Park, S.D. Yim, Y.G. Yoon, W.Y. Lee, C.S. Kim, Performance of solid alkaline fuel cells employing anion-exchange membranes, *J. Power Sources*, 178 (2008) 620–626.

139. L. Wu, T.W. Xu, D. Wu, X. Zheng, Preparation and characterization of CPPO/BPPO blend membranes for potential application in alkaline direct methanol fuel cell, *J. Membr. Sci.*, 310 (2008) 577–585.

140. C.-C. Yang, Study of alkaline nanocomposite polymer electrolytes based on PVA–ZrO$_2$–KOH, *Mater. Sci. Eng. B*, 131 (2006) 256–262.

141. S.B. Sang, J.F. Zhang, Q.M. Wu, Y.G. Liao, Influences of Bentonite on conductivity of composite solid alkaline polymer electrolyte PVA-Bentonite-KOH-H$_2$O, *Electrochim. Acta*, 52 (2007) 7315–7321.

142. A.B. Yuan, J. Zhao, Composite alkaline polymer electrolytes and its application to nickel-metal hydride batteries, *Electrochim. Acta*, 51 (2006) 2454–2462.

143. C. Manea, M. Mulder, Characterization of polymer blends of polyethersulfone/sulfonated polysulfone and polyethersulfone/sulfonated polyetheretherketone for direct methanol fuel cell applications, *J. Membr. Sci.*, 206 (2002) 443–453.

144. J.S. Park, J.W. Park, E. Ruckenstein, Thermal and dynamic mechanical analysis of PVA/MC blend hydrogels, *Polymer*, 42 (2001) 4271–4280.

145. X. Andrieu, J.F. Fauvarque, A. Goux, T. Hamaide, R. Mhamdi, T. Vicedo, Solid polymer electrolytes based on statistical poly (ethylene oxide-propylene oxide) copolymers, *Electrochim. Acta*, 40 (1995) 2295–2299.

146. J.L. Qiao, J. Fu, L.L. Liu, J. Zhang, J. Xie, G. Li, Synthesis and properties of chemically cross-linked poly(vinyl alcohol)-poly (acrylamide-co-diallyldimethylammonium chloride) (PVA-PAADDA) for anion-exchange membranes, *Solid State Ionics*, 214 (2012) 6–12.

147. J.L. Qiao, T. Okada, H. Ono, High molecular weight PVA-modified PVA/PAMPS proton-conducting membranes with increased stability and their application in DMFCs, *Solid State Ionics*, 180 (2009) 1318–1323.

148. S.R. Samms, S. Wasmus, R.F. Savinell, Thermal stability of proton conducting acid doped polybenzimidazole in simulated fuel cell environments, *J. Electrochem. Soc.*, 143 (1996) 1225–1232.

149. J.C. Lassegues, B. Desbat, O. Trinquet, F. Cruege, C. Poinsignon, From model solid-state protonic conductors to new polymer electrolytes, *Solid State Ionics*, 35 (1989) 17–25.

150. T.S. Olson, S. Pylypenko, P. Atanassov, K. Asazawa, K. Yamada, H. Tanaka, Anion-exchange membrane fuel cells: Dual-site mechanism of oxygen reduction reaction in alkaline media on cobalt-polypyrrole electrocatalysts, *J. Phys. Chem. C*, 114 (2010) 5049–5059.

151. H. Bunazawa, Y. Yamazaki, Influence of anion ionomer content and silver cathode catalyst on the performance of alkaline membrane electrode assemblies (MEAs) for direct methanol fuel cells (DMFCs), *J. Power Sources*, 182 (2008) 48–51.

152. J.R. Stevens, W. Wieczorek, D. Raducha, K.R. Jeffrey, Proton conducting gel H$_3$PO$_4$ electrolytes, *Solid State Ionics*, 97 (1997) 347–358.

153. D. Raducha, W. Wieczorek, Z. Florjanczyk, J.R. Stevens, Nonaqueous H$_3$PO$_4$-doped gel electrolytes, *J. Phys. Chem.*, 100 (1996) 20126–20133.

154. R. Tanaka, H. Yamamoto, A. Shono, K. Kubo, M. Sakurai, Proton conducting behavior in non-crosslinked and crosslinked polyethylenimine with excess phosphoric acid, *Electrochim. Acta*, 45 (2000) 1385–1389.

11 Aromatic Polymer Electrolyte Membranes as Hydroxide Conductors

Zhaoxia Hu and Shouwen Chen

CONTENTS

11.1 INTRODUCTION

The energy and environmental issues are challenging the human society greatly in the postindustrial age now, in the areas from economic development strategy, high-tech industry, to daily lifestyles. Since the 1960s, the interests and investments in fuel cell technology have grown quickly and branched into five major types, including alkaline fuel cell (AFC), phosphoric acid fuel cell (PAFC), polymer electrolyte membrane fuel cell (PEMFC), molten carbonate fuel cell (MCFC), and solid oxide fuel cell (SOFC), for the diversified fuel choices, high energy efficiency, and environmental friendship.[1] AFC using alkali solution (e.g., KOH) as the electrolyte is

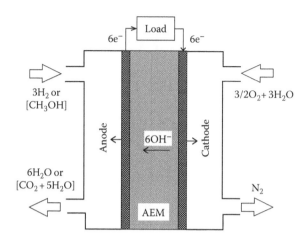

FIGURE 11.1   Construction and working principle of AEM fuel cells.

the pioneer type used in real sense primarily developed by Francis Thomas Bacon, and its application can be dated back to the NASA Apollo program and other demonstration projects. However, it fell behind the PEMFC with proton exchange membranes due to the cost, reliability, operability, and security issues associated with the fuel leakage problem. Therefore, the alternation of polymer electrolyte membrane (PEM) as the hydroxide conductor, named as alkaline polymer electrolyte membrane, instead of alkali solutions drew attentions in the past few years.[2–5] Figure 11.1 illustrates the construction and working principle of an AEM fuel cell.

11.2   POLYMER ELECTROLYTE MEMBRANE

During the last 50 years, PEMs have been drawing considerable attention as an important branch of ion exchange membranes and successfully applied for environmental and energetic associated applications due to the good ion conducting property and some advantages of polymer materials, such as lightweight, mechanical flexibility, as well as film-forming ability. Therefore, they have evolved from laboratory scale to industrial production with significant technical and commercial impacts.

Usually, PEMs are classified into acid PEMs (or cation exchange membrane [CAM]) and alkaline PEMs (or anion exchange membrane [AEM]) depending on the type of cationic or anionic electrolyte they conduct.

From the electrochemical prospective, a good PEM should meet the primary requirements summarized as follows[4,6]:

1. Good ion conduction property at complicated and changeable surrounding conditions, especially for the parameters of temperature and humidity. A minimum ion conductivity of 10^{-3} S/cm at room temperature is necessary.
2. Good adherence to the electrode materials and other parts.
3. Good mechanical and thermal stability during manufacturing and operation.

4. As thin as possible (50–80 μm) but in good strength–resistance–cost balance in dry and wet states.
5. Low electron transport and fuel/oxidizer permeability across the membrane.
6. Low cost.

11.3 AEM

11.3.1 Membrane Preparation

According to the connection way of cationic charged groups to the membrane matrix or their chemical structure, the AEMs can be classified into two types, homogeneous and heterogeneous; the cationic charged groups are chemically bonded to the membrane matrix for the former type while physically mixed with the membrane matrix for the latter one.

To prepare a homogeneous AEM, several approaches are reported available to introduce the cationic charged groups into the membrane matrix. Based on the starting materials, these approaches can be divided into three catalogues[4,7]:

1. Ionic copolymers are prepared from functionalized monomer containing groups that are or could be converted into cationic charged groups and other nonfunctionalized monomers and subsequently used to prepare polymer membranes.
2. Polymer precursors containing moiety that could be converted into halomethyl groups are prepared first and subsequently used to form a membrane followed by functionalization reaction.
3. Start with polymer precursors or polymer blends by introducing cationic charged moieties, followed by the dissolving of polymer and casting it into a membrane.

On the other hand, a heterogeneous AEM refers to the anion exchange material embedded in an inert compound and is usually prepared by ion solvating or hybrid method.[4]

11.3.2 Aromatic Polymer Electrolyte Membranes for AEM Fuel Cells

A good AEM should fulfill stringent mechanical, thermal, and chemical properties as mentioned in Section 11.2. Historically, the first AEM material was developed by researchers from the Tokuyama Soda Company.[8] They introduced quaternary ammonium groups to the divinylbenzene-cross-linked polychloropropene polymer matrix via trimethylamine. Since then, several membrane-associated companies explored various kinds of AEMs and pushed them to commercial market; most of them were based on cross-linked polystyrene, polyvinyl alcohol, low (or high)-density polyethylene, and other aliphatic polymers through irradiation-grafting method. The primary objective of developing these materials was for applications in the fields such as electrodialysis, desalination, selective electrode, and waste acid recovery.[3] However, they showed performance in AEM fuel cells far below practical demands in regard of ion conductivity, power output, and life-span while comparing with its CAM counterpart of Nafion. Over the past few decades, a large number of accomplishments achieved in CAM alternatives have intrigued researchers' interests in the exploration of novel AEM materials.

Given this background, wholly aromatic polymers are considered to be one of the most promising materials to prepare high-performance AEMs due to their availability, diversified structural selection, and anticipated mechanical, thermal, and chemical stability under fuel cell operation conditions as well as cost cutting. Aromatic AEMs are classified according to the chemical structure irrespective of the preparation procedure. This section mainly summarized the various aromatic polymer materials developed and used in AEM fuel cells, including their chemical structure, characterization, performance, modification, and solving-required problems.

11.3.2.1 Polysulfones

Polysulfone (PSU) is a family of thermoplastic polymers; they are classified as PSU, poly(aryl sulfone), and poly(ether sulfone) (PES) by the polymer backbone structure. They are well known for their toughness and stability at high temperatures (−100°C to 150°C), high oxidative stability, and dimensional stability. Hence, it is easy to get the thin membrane with reproducible properties, which have been widely used in many fields like hemodialysis, wastewater treatment, gas separation, and especially PEM fuel cell applications.

AEMs based on PSUs (Udel 1700, Amoco) were initially investigated by Zschocke and Quellmalz back to 1985. They introduced the anion exchange groups by three steps, including chloromethylation (Friedel–Crafts alkylation), quaternization (Menshutkin reaction), and alkalization (Figure 11.2). The obtained quaternary PSU (QPSU) AEMs were chemically stable by soaking in NaOH (40 wt.%) at 70°C–80°C for 300 h and were applied in the electrodialysis process.[9]

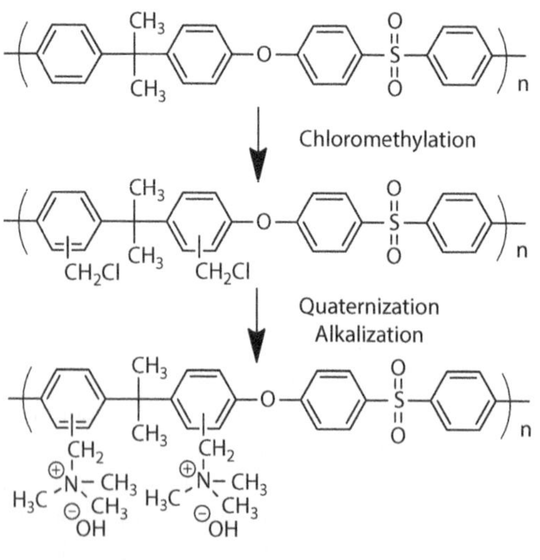

FIGURE 11.2   Synthesis route of PSU-based ionomers by Zschocke and Quellmalz.

Afterward, using commercial PSUs, researches utilized various amination reagents (aliphatic diamines and 1-methylimidazole were chosen) during the quaternization procedure to study the influence on ion conductivity and stability in alkaline solution.[10–12] According to Stamatialis' research, the degradation of the membranes depends on the alkyl chain length of the diamine units; the membranes containing the *N,N,N,N*-tetramethylhexanediammonium (TMHDA) are the most stable under 2 M NaOH solutions at 40°C (Figure 11.3).[10] Lu et al. published a preliminary fuel cell performance study based on this type of AEMs free from nonprecious metal catalysts (Ni, 5 mg/cm^2; Ag, 1 mg/cm^2); the power output reached 50 mW/cm^2 at 0.55 V at 60°C in a H_2/O_2 AEM fuel cell.[13] Using 1,4-diazabicyclo-[2,2,2]-octane (DABCO) as the amination reagent, Wang et al. immersed the quaternized PSU membrane (QDPSU) in 11 M phosphoric acid at room temperature for 3 days and prepared the PA/QDPSU membrane. H_2/O_2 fuel cell tests gave a high power density of 400 mW/cm^2 at 150°C and atmospheric pressure.[14]

Others also prepared AEMs through similar three-step procedures based on lab-synthesized PSUs in the literatures.[15] The polymer backbone structures were not identical to commercial PSUs by changing the diphenol comonomers. Also, various amination reagents besides common trimethylamine were used, such as 1,1,2,3,4-pentamethylguanidine (PMG), DABCO, and morpholine.[14,16]

In 2010, Hickner reported a series of AEMs prepared via the bromination, amination, quaternization, and alkalization of tetramethyl bisphenol A–based PSUs (Figure 11.4). The introduced benzylmethyl moieties could circumvent the polymer postmodification by chloromethylation and control the ion content and distribution more quantitatively along the polymer backbone. Therefore, the ion conductivity and other properties of the AEMs could be tuned over a wide range, and structure–property relationship be clarified more clearly.[17]

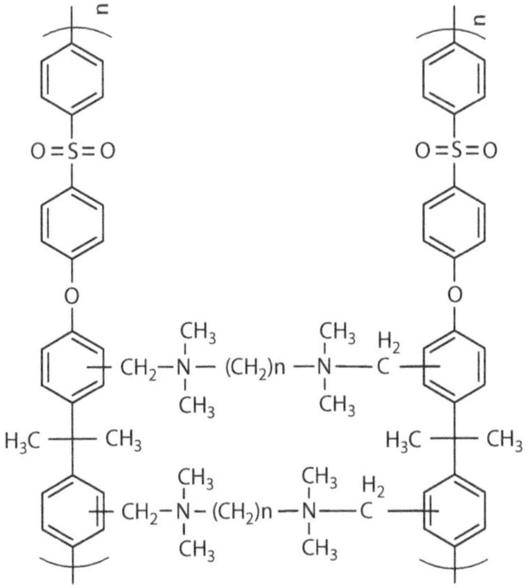

FIGURE 11.3 Chemical structure of cross-linked PSU-based ionomers.

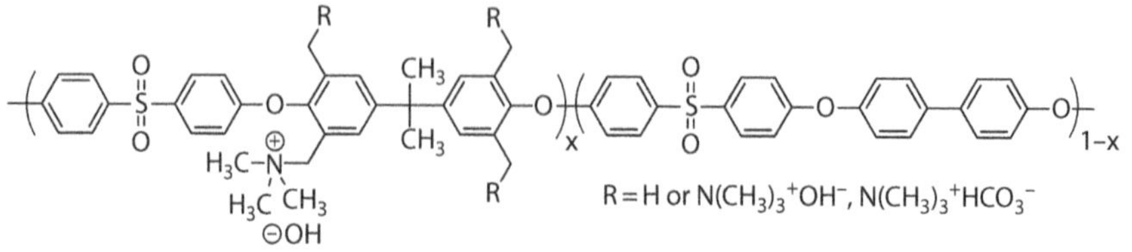

FIGURE 11.4 Chemical structure of PSU ionomers developed by Hickner et al.

Later, Liu prepared self-cross-linked AEM from the brominated polymer pre-cursors based on the strategy of forming membranes with hydrophilic/hydrophobic structure. Lab-synthesized PMG was used as the amination reagent (Figure 11.5). The as-prepared membranes exhibited very low methanol permeability with the minimum of 1.02×10^{-9} cm^{-1} s^{-1} at 30°C.[18]

Borrowing ideas from the side-chain-type proton exchange membranes that pro-vided better ion conductivity and stability, Guiver designed a series of AEMs with *side-chain-type* structures of pendant functional groups from commercial PSUs by functionalization with tertiary amines via lithiation chemistry (Figure 11.6).[19] The obtained AEMs showed considerably lower water uptakes (<20%), low dimen-sional swelling in water, and ion conductivity above 10 mS/cm at room temperature. Although the authors claimed that the long-term stability (800 h) under alkali condi-tion was insufficient due to the pendant substituent, it was still comparable to other reported results.

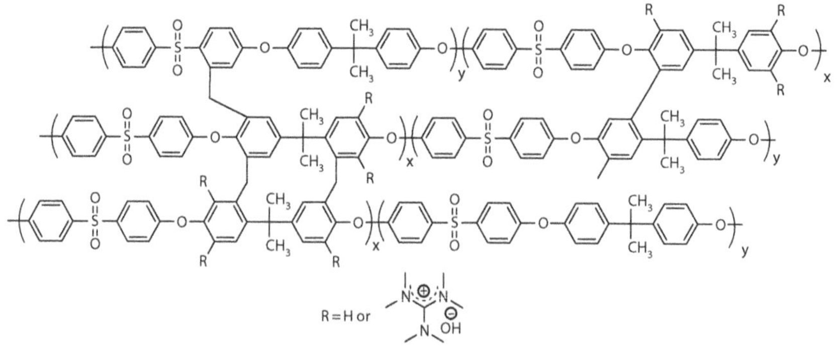

FIGURE 11.5 Chemical structure of self-cross-linked PSU ionomers by PMG.

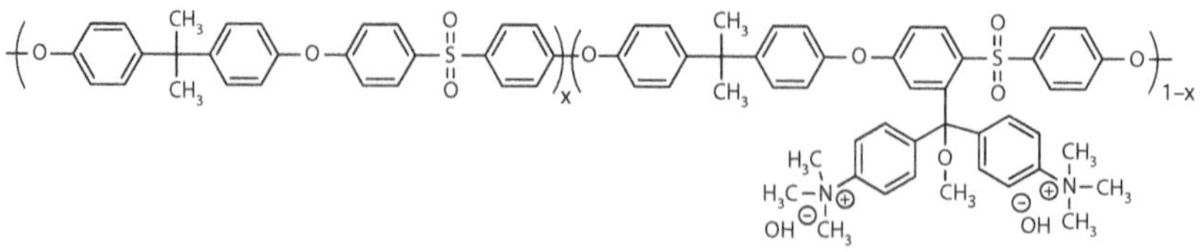

FIGURE 11.6 Chemical structure of side-chain-type PSU ionomers.

11.3.2.2 Poly(Arylene Ether)s

Poly(arylene ether) materials contain a large family of polymers, such as poly (arylene ether ether ketone)s (PEEK), poly(arylene ether ketone)s (PAEK), poly(arylene ether sulfone)s (PAES), and their derivatives. Most of them are high-performance engineering polymer materials; their synthesis has been extensively reported. They are very attractive as PEM alternatives due to their outstanding oxidative and hydrolytic stability, mechanical toughness, and structural diversity. Among the developed aromatic polymer-based AEMs, poly(arylene ether)–derived ones are the most studied and reported in literatures up to date. Most of them are developed according to the proton exchange membranes. Introduction of active proton exchange sites to the poly(arylene ether)s has been achieved by either a polymer postmodification approach or a direct polymerization of functionalized monomers. The same strategies were developed for the introduction of anion exchange sites to the poly(arylene ether)s.

He and colleagues prepared AEMs from commercial PEEK powders by chloromethylation in sulfuric acid, quaternization from trimethylamine, and alkalization (Figure 11.7). The obtained AEMs possess ion exchange capacity (IEC) ranging from 0.43 to 1.35 mmol/g, appropriate water uptake (≤145%), and moderate swelling ratio (≤27%). The membranes with IEC of 0.95 mmol/g showed ionic conductivity of 12 mS/cm at 30°C.[20] They also used 1-methylimidazole as the amination reagent and got imidazolium-functionalized AEMs (PEEK-ImOHs). The prepared PEEK-ImOH membranes with IEC of 1.56–2.24 mmol/g had high ionic conductivity, acceptable water swelling ratio, great tensile strength, surprising flexibility, and high thermal stability (i.e., PEEK-ImOH with IEC of 2.03 mmol/g exhibited ionic conductivity of 52 mS/cm at 20°C, swelling ratio of 51% at 60°C, tensile strength of 78 MPa, elongation-to-break of 168%, and decomposition temperature of 193°C). In a single methanol/O_2 fuel cell test, the open-circuit voltage (OCV) is 0.84 V, and the peak power density is 31 mW/cm².[21] Shah and coworkers also did similar work.[22]

The commercial cardo poly(ether ketones) (PEK-C) were selected by Liu et al. to prepare AEMs by usual three-step postmodification method (Figure 11.8). The final membranes showed ionic conductivity varied from 1.6 to 5.1 mS/cm over the temperature range of 20°C–60°C. Although its ionic conductivity was quite lower compared with other PEEK-based AEMs, the methanol permeability was less than 10^{-9} cm²/s at 30°C in 4 M methanol solutions.[23] Except this way, Zhang and colleagues successfully introduced benzyl chloromethyl groups to the PEK-C matrix via plasma graft polymerization. This approach enables a well preservation in the structure of

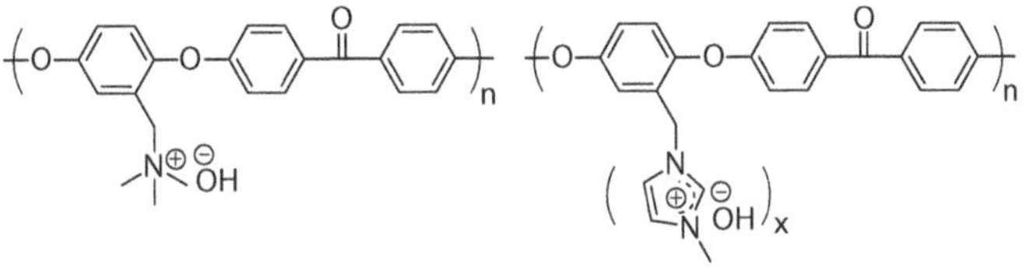

FIGURE 11.7 Chemical structure of commercial PEEK ionomers.

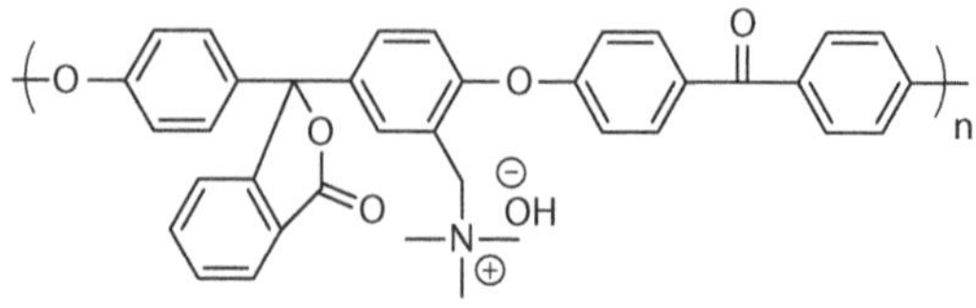

FIGURE 11.8 Chemical structure of commercial cardo poly(ether ketone) ionomers.

functional groups as well as the formation of highly cross-linked structure in the membrane. After amination and alkalization, the resulting membranes exhibited fairly good alkaline and oxidative stability, retaining 86% of the initial ionic conductivity in 6 M KOH solutions and 94% of the initial weight in 3 wt% peroxide solutions for 120 and 262 h, respectively, at 60°C. Ionic conductivity was 8.3 mS/cm, this can be overcome by increasing the quaternary benzyltrimethyl ammonium group content in the polymer matrix in the future, and the ethanol permeability was less than 6.6×10^{-11} cm^2/s at 20°C.[24]

Recently, Zhang et al. also published their work on PEEK-based AEMs; the PEEK copolymers were synthesized by direct step-growth copolymerization from a new monomer containing two tertiary amine groups on 4,4′-dihydroxydiphenyl ether with 4,4′-difluorobenzophenone and 2,2′6,6′-tetramethyl-4,4′-biphenol, instead of the common three-step procedure of chloromethylation–amination–alkalization (Figure 11.9a). The process could precisely control the amount and position of the quaternary ammonium groups along the polymer skeleton and avoid the use of carcinogenic chloromethylation reagent, such as chloromethyl methyl ether, unlike the polymer postmodification approach. However, the use of another toxic chemical of methyl iodide is necessary.

Furthermore, they conducted partial amination of the homopolymers; the residue tertiary amine groups were reacted with p-xylene dichloride to produce a cross-linked membrane (Figure 11.9b). The resulting cross-linked AEMs exhibited ionic conductivity above 10 mS/cm at room temperature, good mechanical properties, and reduced water uptake relative to the pristine uncross-linked ones.[25]

Several reports were also published concerning side-chain-type PEEK-based AEMs as shown in Figure 11.10. New monomers containing pendant methyl groups were first synthesized and used to prepare PEEK copolymers. The obtained copolymers were subsequently bromomethylated by a radical reaction in the presence of NBS/BPO, then converted to quaternary benzyltrimethyl ammonium groups by amination and alkalization. At moderate IEC, they could achieve fairly high ionic conductivities up to 30 mS/cm (0.90–1.35 mmol/g, at 80°C), good dimensional stability, as well as low methanol permeability lower than 2.0×10^{-7} cm^2/s.[26,27]

Fang and Shen reported novel AEMs derived from poly(phthalazinone ether sulfone ketone)s through the polymer postmodification approach as shown in Figure 11.11a. The ionic conductivity of the obtained membranes is as high as 140 mS/cm in 2 mol/L KOH solution at room temperature and thermal stability below 150°C.[28] Others also developed this kind of materials and applied for vanadium redox flow battery applications.[29]

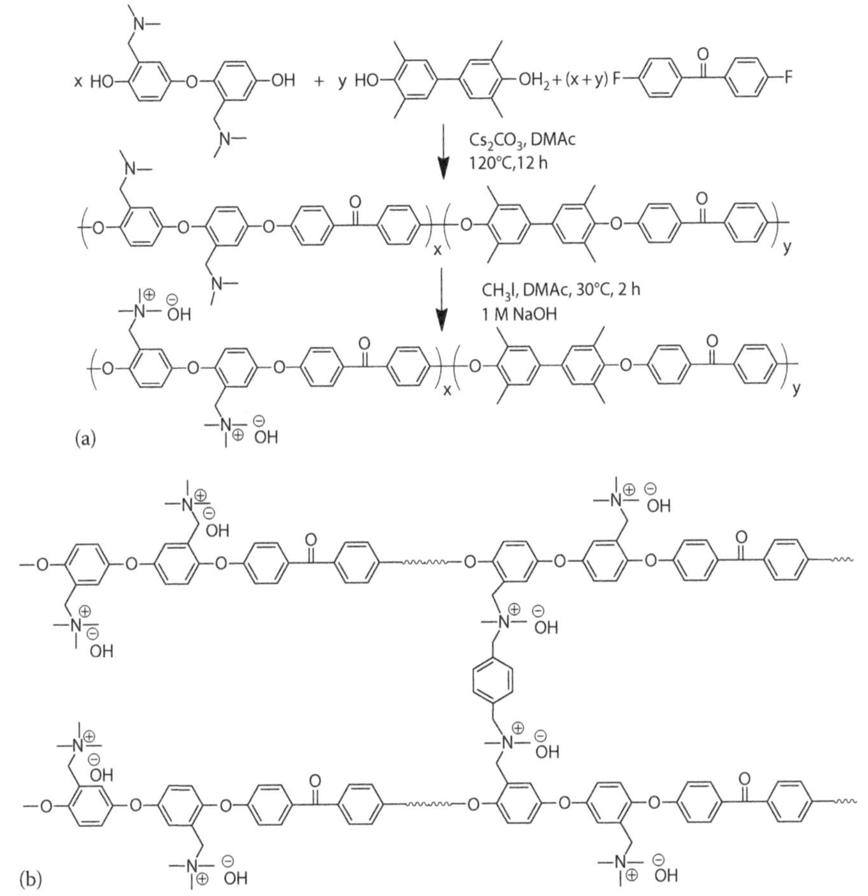

FIGURE 11.9 (a) Synthetic route of PEEK ionomers by direct polymerization. (b) Chemical structure of self-cross-linked PEEK ionomers.

Chen and Hickner designed a PAEK-based AEMs with clustered ionic groups and IEC ranging from 1.12 to 2.88 mmol/g bearing quaternary benzylmethyl ammonium groups (Figure 11.11b).[30] The chemical structure was confirmed by NMR, FTIR, and XPS. SAXS measurements indicated that the densely functionalized quaternary ammonium groups in the solution-cast membranes formed distinct ion clusters, showing significantly improved ionic conductivity compared to those without obvious ion clustering morphology. According to their analysis, it was more likely to form better ionic domain morphology in the low IEC membranes than the high IEC ones.

Based on the exciting results gained from sulfonated poly(arylene ether)s and poly(arylene imide)s containing fluorenyl groups as the proton exchange membranes, which demonstrated the effectiveness of the rigid and bulky fluorenyl groups on membrane performance, especially on the proton conductivity and

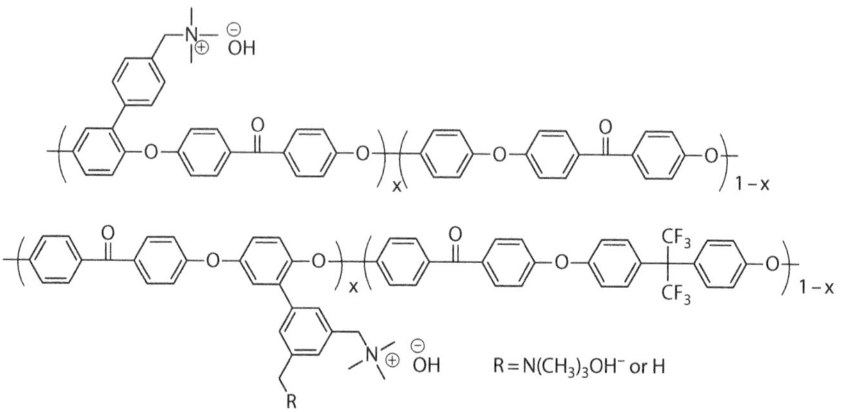

FIGURE 11.10 Chemical structure of side-chain-type PEEK ionomers.

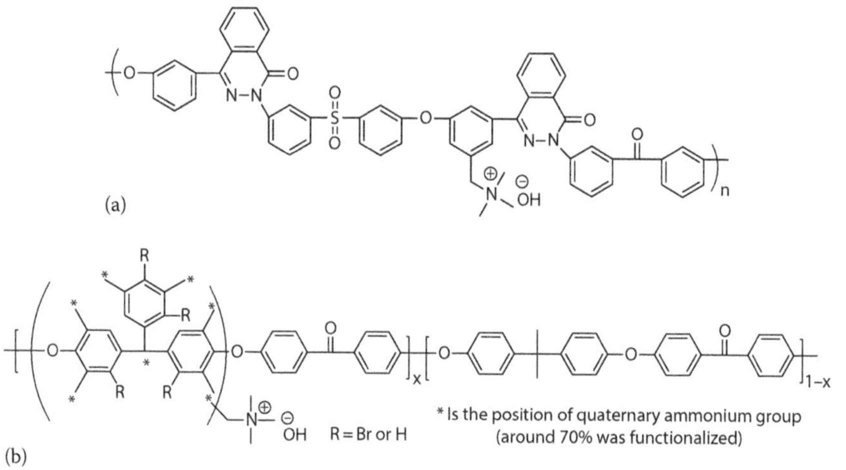

FIGURE 11.11 (a) Chemical structures of poly(phthalazinone ether sulfone ketone) ionomers. (b) Chemical structure of PAEK ionomers.

chemical stability, Watanabe and coworkers introduced fluorenyl groups into the poly(arylene ether)–based AEMs.[31] Figure 11.12 shows the chemical structure of the developed poly(arylene ether sulfone ketone)s containing quaternary benzyltrimethyl ammonio-substituted fluorenyl groups. Due to the random sulfone-*co*-ketone structure with fluorenyl groups, high IEC up to 2.54 mmol/g was achieved for the obtained AEMs. As they anticipated, highly conductive AEMs with ionic conductivity up to 50 mS/cm at 30°C and 78 mS/cm at 60°C, as well as high thermal and chemical stability, were obtained.

Except the idea of introducing fluorenyl groups, they also borrowed ideas from the well-accepted viewpoint that block structure is favorable for the formation of

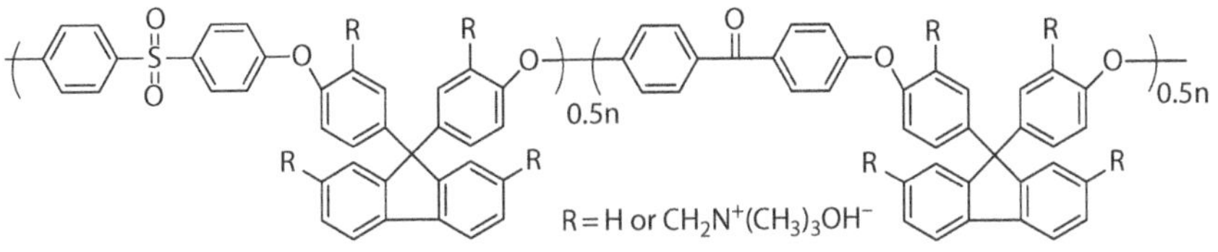

FIGURE 11.12 Chemical structure of fluorenyl-containing poly(arylene ether sulfone ketone) ionomers.

unique phase-separated morphology with well-connected ion transporting pathways, that is, the strategy of utilizing a sequential hydrophilic/hydrophobic structure with dense ionic groups in the hydrophilic groups is also expected to be useful to enhance ionic conductivity and stability for AEMs. Hence, they designed a novel anion conductive block poly(arylene ether)s with hydrophilic blocks containing fluorenyl segments (Figure 11.13a), where the existence of multiple reactive sites in the fluorenyl phenyl rings for chloromethylation functionalization is favorable to the formation of dense ionic groups.[32] STEM observation confirmed the formation of hydrophilic/hydrophobic phase-separation morphology and interconnected ion transporting pathway in the obtained membranes. The ionomer membranes showed considerably higher ionic conductivity, up to 144 mS/cm at 80°C than those of the reported AEMs up to date, and retained high conductivity in hot water at 80°C for 5000 h. In a noble metal-free direct hydrazine fuel cell test operated at 80°C, the maximum power density of 297 mW/cm² (at 826 mA/cm²) was achieved. These results confirmed their concept that the utilization of a multiblock copolymer structure with highly ionized hydrophilic block is effective for improving the ionic conductivity without sacrificing other desirable properties such as stability of the AEMs.

Also using the fluorenyl groups, Hickner and coworkers synthesized a novel monomer by introducing two tertiary amine groups into the backbone phenyl rings of the fluorine first and subsequently prepared a series of AEMs by direct polymerization as shown in Figure 11.13b.[33] They studied the influence of counterions on the water uptake and ionic conductivity in detail to give a comprehensive view of the transport properties of the obtained AEMs. Other researchers also reported similar work on poly(arylene ether)–based AEMs containing fluorenyl groups.[34-36]

A large number of other types of poly(arylene ether)–derived AEMs with different chemical structures were also reported, typically by the modification of polymer skeleton and quaternary ammonio-substituted structure. For example, as shown in Figure 11.14a, Zhang and coworkers chose a type of aromatic dihydric phenol monomer with rigid and bulky structure, 2.5-triptycenediol (TPD), as the comonomer to prepare the poly(arylene ether) copolymers.[37] They excepted that after functionalization, combing with the structural rigidity, nonplanarity, bulkiness, π-electron richness of iptycenes, and high degrees of internal free volume, the polymers containing quaternary ammonio-substituted triptycene groups may also render good water-holding capability, high ionic conductivity, and good dimensional stability even at high IEC level. As expected, the obtained AEMs displayed rather low water uptake, membrane swelling, but good ionic conductivity at high IEC level,

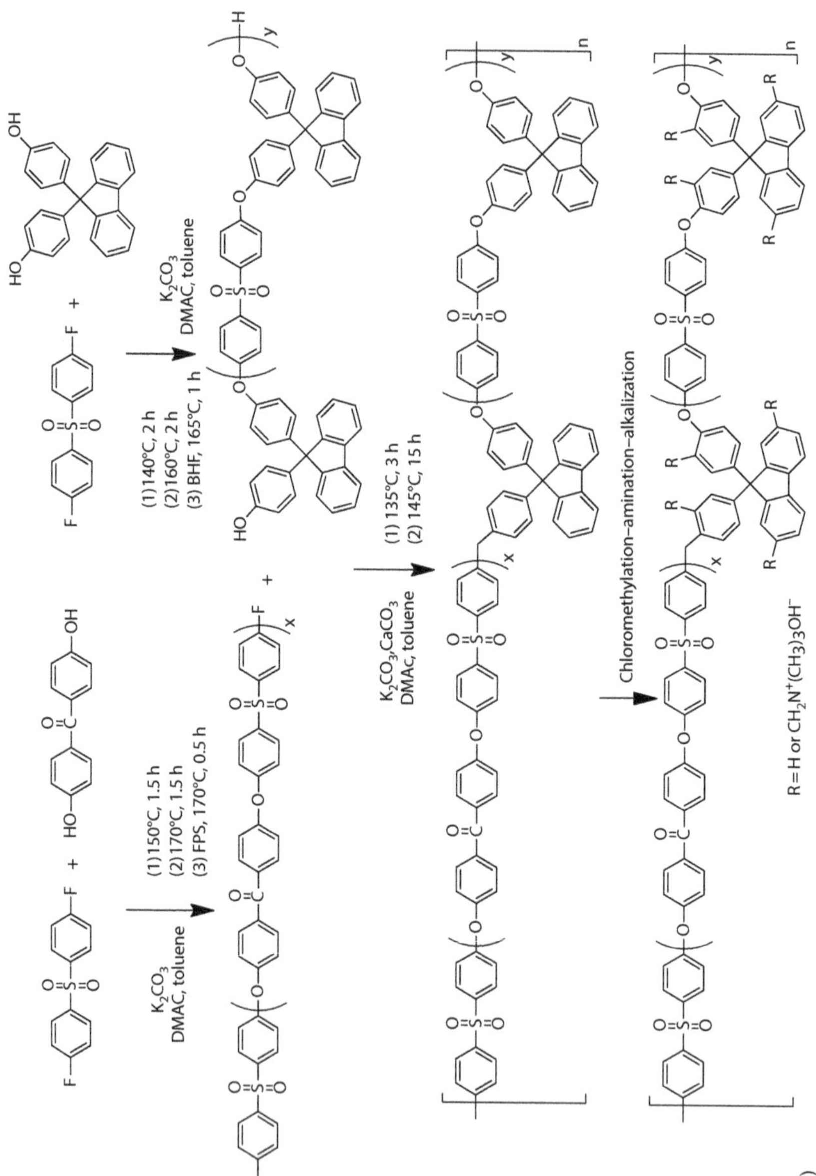

FIGURE 11.13 (a) Synthetic route of multiblock poly(arylene ether) ionomers.

(Continued)

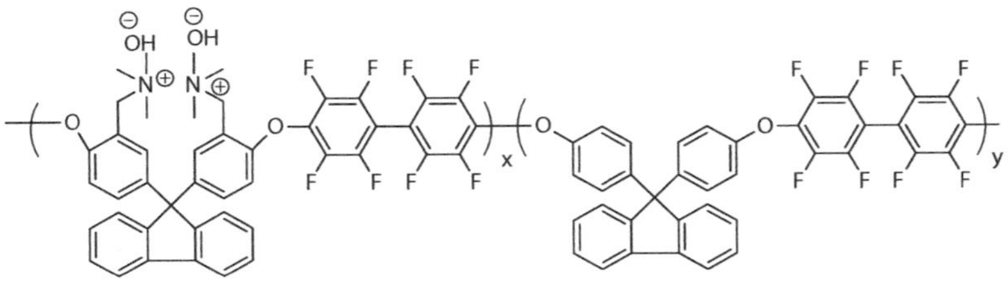

FIGURE 11.13 (Continued)   (b) Chemical structure of poly(fluorenyl ether) ionomers.

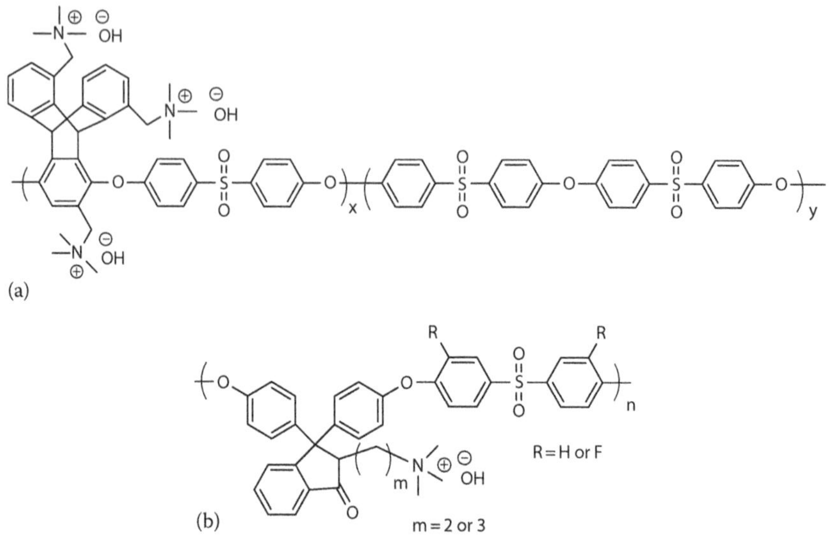

FIGURE 11.14 (a) Chemical structure of poly(arylene ether) ionomers with triptycene groups. (b) Chemical structure of side-chain-type poly(aryl ether sulfone) ionomers.

that is, the membrane with IEC of 1.97 mmol/g exhibited water uptake only 21% and swelling ratio of 11% but ionic conductivity of 29 mS/cm at room temperature. For alkaline stability, the membranes showed no significant change in 4 M NaOH solution at 25°C for 30 days.

They also published their works on novel cardo poly(aryl ether sulfone)–based AEMs containing pendant quaternary ammonium groups on aliphatic side chains by direct polymerization using functionalized monomer (Figure 11.14b). TEM observation revealed that the obtained membranes exhibited a distinct phase-separated morphology comprised interconnect ionic clusters in the size of 1–2 nm. The resulting membranes with IEC of 1.44 mmol/g displayed ionic conductivities varied from 30 to 41 mS/cm at 20°C–60°C.

According to the similar consideration, several monomers with rigid and bulky structure were also selected for the preparation of novel poly(arylene ether) ionomers as shown in Figure 11.15.[38–43] For instance, the membranes derived from *bis*(4-hydroxylphenyl)diphenylmethane and 4,4'-diphenylsulfone/4,4'-diphenylketone (1:1) with IEC of 1.04 mmol/g displayed ionic conductivity of 75 mS/cm at 80°C and strong tensile strength (29.2 MPa) at 25°C and kept 90% of mechanical properties and 82% of ionic conductivity after being treated with 1 M NaOH solution at 60°C for 170 h.[38]

As for the structure of quaternary ammonio-substitutes, it can be realized by the selection of amination reagent or molecular design. For instance, Guiver and Kim et al. reported a kind of quaternary phenylguanidinium-functionalized PAES ionomers recently. As illustrated in Figure 11.16, the bulky guanidinium cations were incorporated into polymers using activated aromatic fluorophenyl-amine reaction, followed by methylation.[44] The prepared ionomers were used as electrode ionomers

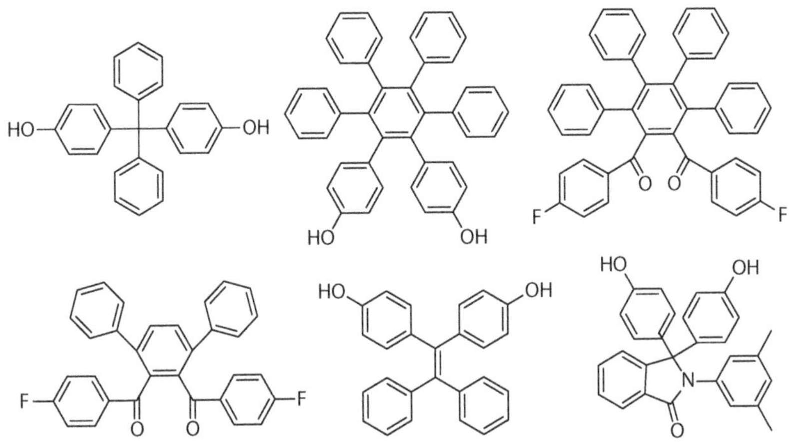

FIGURE 11.15 Chemical structure of selected rigid and bulky monomers for the preparation of poly(arylene ether) ionomers.

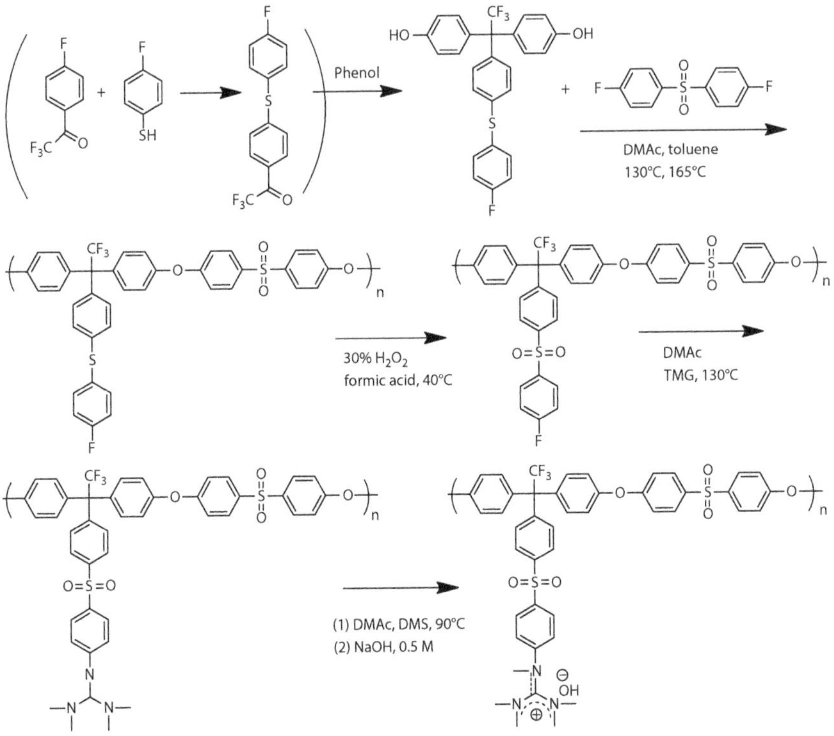

FIGURE 11.16 Synthetic route of phenylguanidinium-functionalized PAES ionomers.

in an AEM fuel cell test; it drew significant currents with platinum catalysts under H_2/O_2 conditions (ca. the peak power density at 80°C is ~200 mW/cm^2) and exhibited superior stability to traditional benzyltrimethyl ammonium–functionalized control ionomers.

11.3.2.3 Polybenzimidazoles

Polybenzimidazole (PBI) is a high-performance polymer containing imidazole groups as an integral part of the polymer chain, with a very high melt point and exceptional thermal and chemical stability. This material has drawn much attention as a candidate for proton exchange membrane, especially considering its success use in high-temperature fuel cell applications as PBI/phosphoric acid blend membranes. As a highly hygroscopic polymer, it can also be doped with aqueous alkali solutions to give a material with high ion conductivity in the 10^{-2} S/cm range; this could meet the basic requirement for practical AEM fuel cell applications. Savadogo and colleagues prepared PBI/KOH (NaOH and LiOH)–based AEMs (Figure 11.17a), which showed ion conductivity in the range of 5×10^{-5} to 10^{-1} S/cm. They studied the effect of soaking time and temperature on the alkali-doping level; the membrane of PBI/KOH exhibited ion conductivity of 9×10^{-2} S/cm, three to four times higher than the corresponding PBI/H_2SO_4 and PBI/H_3PO_4 membranes. PEMFCs based on these alkali-doped PBI membranes exhibited the same performance as those based on Nafion 117.[45] Currently, the PBI-based AEMs are still under active investigation. Sun and coworkers developed KOH-doped PBI membranes. The prepared membranes

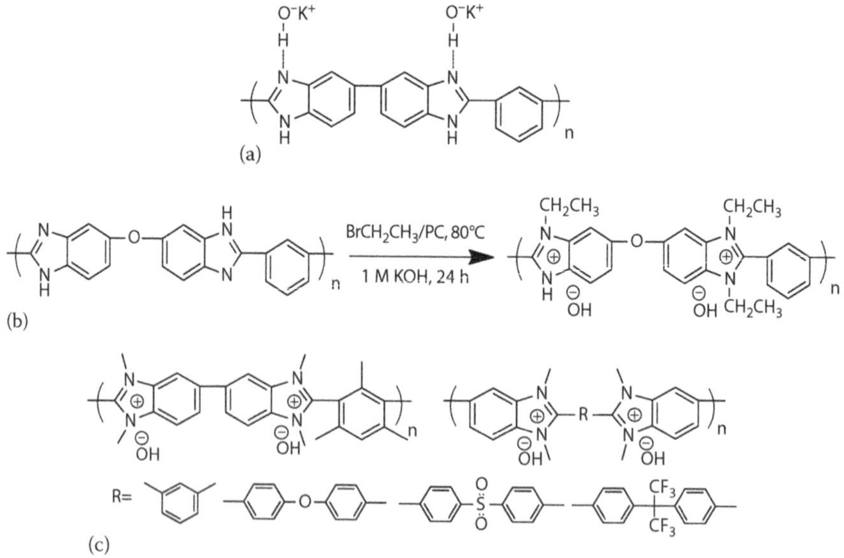

FIGURE 11.17   (a) Chemical structure of alkali-doped PBI ionomers. (b) Chemical structure of CH$_3$CH$_2$-grafting PBI ionomers. (c) Chemical structure of structural modified PBI ionomers.

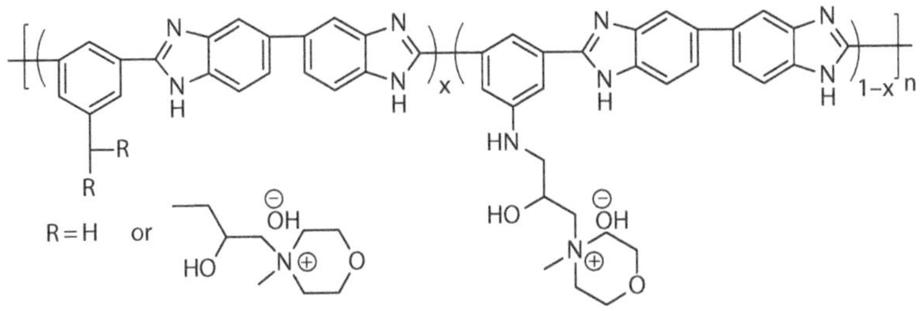

FIGURE 11.18   Chemical structure of PBI ionomers developed by Fang et al.

possessed satisfying thermal stability and comparable mechanical strength with acid-doped PBI ones, and in an alkaline direct methanol fuel cell operated at 90°C, the peak power density reached 31 mW/cm^2.[46]

Furthermore, they chemically grafted the groups of $-CH_2CH_3$ onto the PBI backbone by a green and facile route (Figure 11.17b); the obtained AEM exhibited ionic conductivity and ethanol permeability of 22 mS/cm and 5.24×10^{-8} cm^2/s, respectively. By a single CH_3CH_2OH/air cell test, it delivered a peak power density of about 11 mW/cm^2 at 13°C.[47] A handful of reports on alkali-doped structural-modified PBI were also published recently.[48–50]

Except the alkali-doping method, introducing cationic ion groups onto the PBI backbones is another way to gain PBI-based AEMs. On one hand, by permethylation of the benzimidazolium ring of the PBI derivatives, AEMs could be obtained by OH$^-$ exchange. Holdcroft and colleagues designed a novel PBI by steric crowding around the benzimidazolium reactive C2 position (Figure 11.17c). Although the resulting membranes possess ionic conductivity slightly lower than other types (13.2 mS/cm at 21°C, 1.0 mmol/g), it creates a new opportunity to prepare new PBI-based AEMs with unique or specialized properties.[51] Similar works were reported by others.[52,53]

On the other hand, pendant cationic ionic groups can be directly bounded to the PBI backbones. For example, Fang and coworkers partially grafted 4-methyl-4-glycidylmorpholin-4-ium chloride onto PBI with pendant amino groups (NH$_2$-PBI) and formed cross-linking structure by the reaction between remaining unreacted pendant amino hydrogen atom of NH$_2$-PBI with the epoxy groups of bisphenol A diglycidyl ether (Figure 11.18). The highest ionic conductivity reached 56 mS/cm at 80°C in water. The polymer backbones showed extremely good chemical stability, but the quaternary ammonium groups were unstable in 6.0 M NaOH solutions at 60°C.[54] Nijmeijer also grafted functionalized ion liquids onto the preprepared porous PBI membranes. Although the ionic conductivity only can reach up to 6.62 mS/cm at 20°C, the authors claimed that the membrane properties can be easily tailored toward specific applications by choosing the proper chemistry.[55]

11.3.2.4   Poly(*p*-Phenylene Oxide)s

Poly(*p*-phenylene oxide)s or poly(2,6-dimethyl-1,4-phenylene oxide) (PPO) is also an attractive polymer for the fabrication of membranes for its outstanding film-forming property, as well as thermal and chemical stability. Proton exchange membranes have been successfully developed from PPO by postsulfonation method. It also attracted researchers' notice for the existence of two methyl groups attached to the polymer backbone, which can be brominated easily without using highly carcinogenic chloromethylation reagent. Scientists in GE first applied patents referring to AEMs from PPO with IEC of 3.80 mmol/g.

In 1995, Tsuchida et al. reported their work on the preparation of quaternary ammonium PPO ionomers.[56] The allyl group of the PPO polymer was converted into an iodopropyl group by hydrozirconation with ZrCp$_2$HCl (Schwarz's reagent) followed by the reaction with iodine. Finally, quaternary ammonium PPO ionomers were obtained by quaternization through Menshutkin reaction with a series of trialkylamines as illustrated in Figure 11.19a. The resulting quaternary ammonium PPO ionomers exhibited high thermal stability. However, they reported no electrochemical properties.

Xu and his coworkers did extensive works on PPO-based AEMs (Figure 11.19b). They systematically studied the influence of bromination conditions, membrane preparation procedure, amination conditions, and reagents on the AEM properties. They found that the membrane properties are significantly affected by the bromination processes: IEC and water content are closely related to the benzyl substitution but hardly to the aryl substitution,[57] while properly higher content aryl substitutions could improve the membrane's dimensional stability.[58] Therefore, the properly balanced substitution position is crucial to comply with AEM fuel cell requirements. Various amination reagents except trimethylamine, including triethylamine, pyridine,[59] 4-vinylpyridine,[60] diamines,[61] guanidinium,[62] *tris*(2,4,6-trimethoxyphenyl) phosphine,[63] and benzimidazolium,[64] were used for the purpose of forming cross-linking structure or to offer more detailed information about structure–property relationship and find a better suitable structure.

They also prepared blend AEMs (aminated by triethylamine) derived from chloroacetylated PPO and bromomethylated PPO, which resultantly formed cross-linking structure between them by heat treatment (Figure 11.19c). The ionic conductivity of the as-prepared mechanically enhanced membranes reached 22–32 mS/cm in water at 25°C with low methanol permeability (1.35×10^{-7}–1.46×10^{-7} cm^2/s).[65] Others also reported similar work in the literatures.[66,67]

Recently, Li and his coworkers published several inspiring works on PPO-based AEMs. They successfully incorporated clicked 1,2,3-triazoles into PPO-based AEMs through Cu^I-catalyzed *click chemistry* (Figure 11.20). This structure smartly provided more sites to form efficient and continuous hydrogen-bond networks between the water/hydroxide and the triazole for anion transport. Consequently, these membranes showed an impressive enhancement of ionic conductivity, which was several times higher than the typical trimethylamine-aminated PPO-based AEMs both under fully and partially hydrated conditions as well as chemical stability. The peak power density achieved 188.7 mW/cm^2 at 50°C for the clicked AEMs.[68] Instead of traditional benzyltrimethyl ammonium moieties, they introduced the long alkyl side chains to the nitrogen-centered cation to form a comb-shaped structure (Figure 11.20). The resultant AEMs

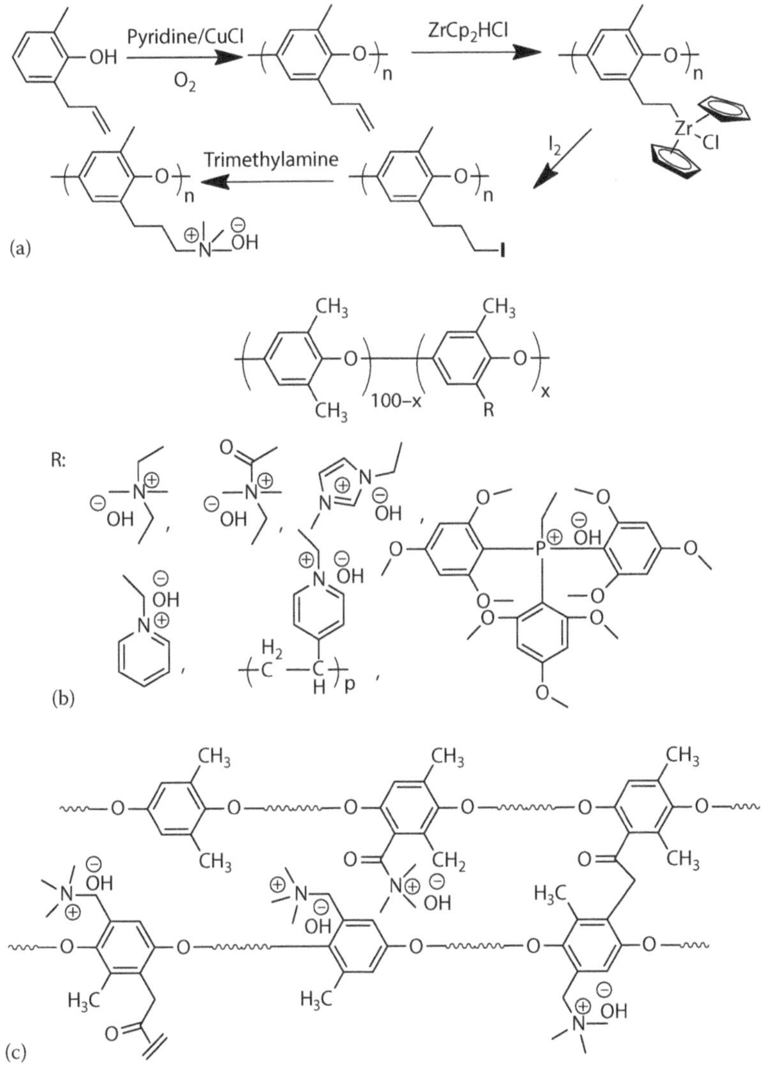

(a)

R:

(b)

(c)

FIGURE 11.19 (a) Synthesis of quaternary ammonium PPO ionomers via a zirconium complex. (b) Chemical structure of PPO ionomers. (c) Chemical structure of self-cross-linked PPO ionomers.

with benzyldimethylhexadecyl ammonium cations retained their high ionic conductivities in 1 M NaOH solutions at 80°C for 2000 h.[69]

11.3.2.5 Other Aromatic Polymers

Other high-performance aromatic polymers have been explored as the AEM precursors in addition to the polymer mentioned earlier. Almost all the conception of

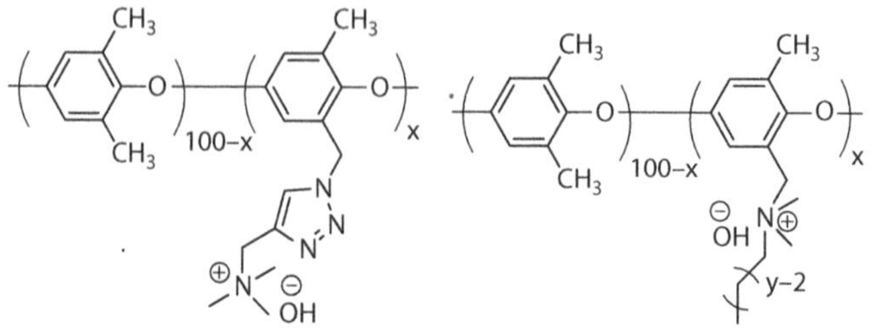

FIGURE 11.20 Chemical structure of side-chain-type PPO ionomers.

structure design came from the proton exchange membrane counterparts by replacing sulfonic acid groups with quaternary ammonium or other cationic groups.

Hibbs and colleagues published a type of AEM based on a polyphenylene backbone prepared by a Diels–Alder reaction (Figure 11.21a). The obtained polymers were transformed to cationic polymers by converting the benzylic methyl groups to bromomethyl groups through a radical reaction, followed by quaternization with trimethylamine in the solid state.[70] The formed membranes (0.40–1.57 mmol/g) were robust and creasable, exhibiting water uptakes of 23%–122% and ion conductivity of 1–50 mS/cm; both were greater than the PSU-based AEMs with similar IECs. The authors explained that the irregularities in the linearity of the polyphenylene backbone and bulky side groups could prevent efficient packing of the polymer. They also did not exhibit any changes in appearance and flexibility after immersing in a stirred 4 M NaOH solution at 60°C for 28 days. They further studied the alkaline stability of these polyphenylene-based AEMs with various cations. The results indicated that the AEMs with resonance-stabilized cations (benzylic guanidinium and imidazolium groups) and with hexane-1-one-6-trimethylammonium side chains showed poor alkaline stability (over 50% loss of ion conductivity after 1 day), while those with hexane-6-trimethylammonium side chains and trimethylammonium cations showed better alkaline stability (5% and 33% loss of ion conductivity after 14 days).[71]

Chen and colleagues reported a novel poly(ether imide)–based AEM material (Figure 11.21b).[72,73] The as-prepared membranes displayed ion conductivity in the range of 2.28–3.51 mS/cm at room temperature and were stable in 1 M KOH solution up to 80°C without losing membrane integrity, but not in concentrated KOH solutions at elevated temperatures.

Kim and Tak proposed AEMs derived from copolyimides (Figure 11.21c); the membranes were stable up to 260°C.[74] However, their ion conductivities were very low due to the low IECs of 0.18–0.33 mmol/g. Besides, the five-membered imide rings of the polymer backbone are sensitive to hydrolysis and could undergo polymer degradation under alkaline conditions inevitably. Given their low methanol permeability and enhanced hydrolytic stability of six-membered imide ring sulfonated polyimide for CAM applications, its AEM derivatives might be promising in low–medium temperature AEM fuel cell applications.

Recently, AEMs derived from poly(aryl ether oxadiazole) ionomers have also drawn attentions from researchers (Figure 11.22). Whether cross-linked or not, the

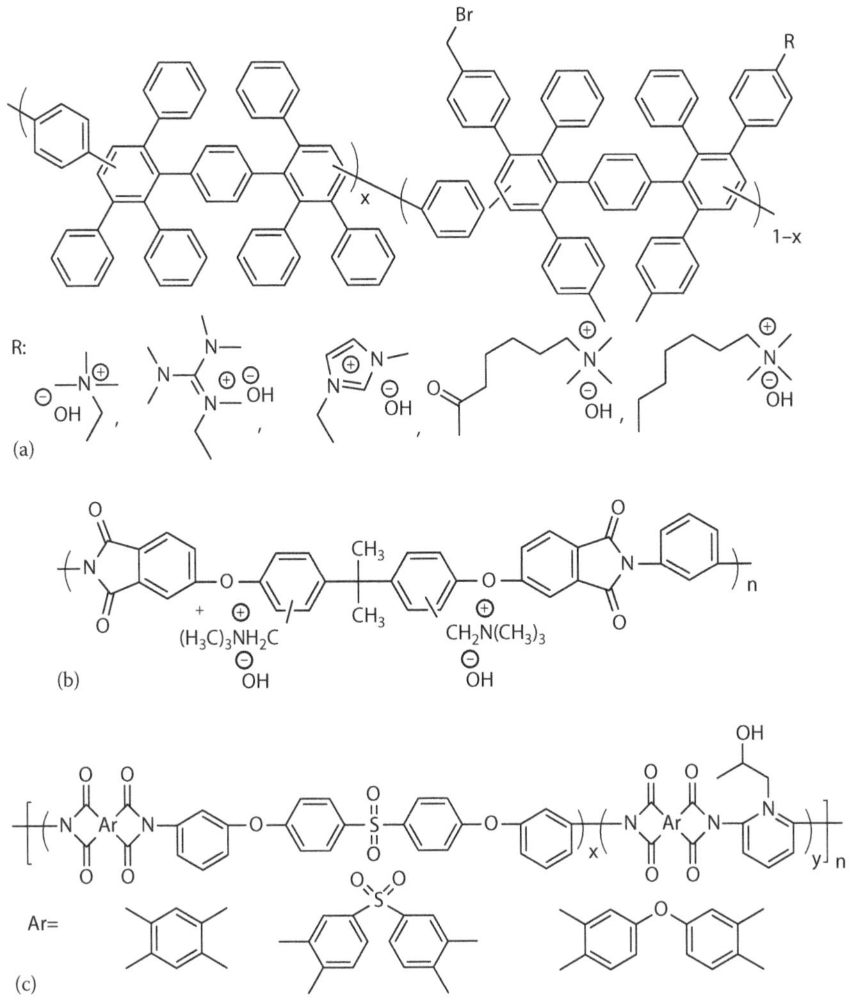

FIGURE 11.21 (a) Chemical structure of poly(phenylene) ionomers. (b) Chemical structure of side-chain-type PPO ionomers. (c) Chemical structures of polyimide ionomers.

prepared membranes generally possess satisfactory ionic conductivity (~10^{-2} S/cm) and good mechanical and thermal stability.[75–78]

Polyarylene sulfonium membranes developed by Tsuchida and colleagues were also investigated as AEM materials (Figure 11.23). The advantage of this kind of material is that it does not need the halomethylation and amination steps to gain positive charged groups along the polymer skeleton. However, there are very few information concerning its chemical, mechanical, and electrochemical properties in alkali conditions.[79]

More systematic fundamental properties, stability, as well as fuel cell performance remain to be investigated for these aforementioned materials. Nevertheless,

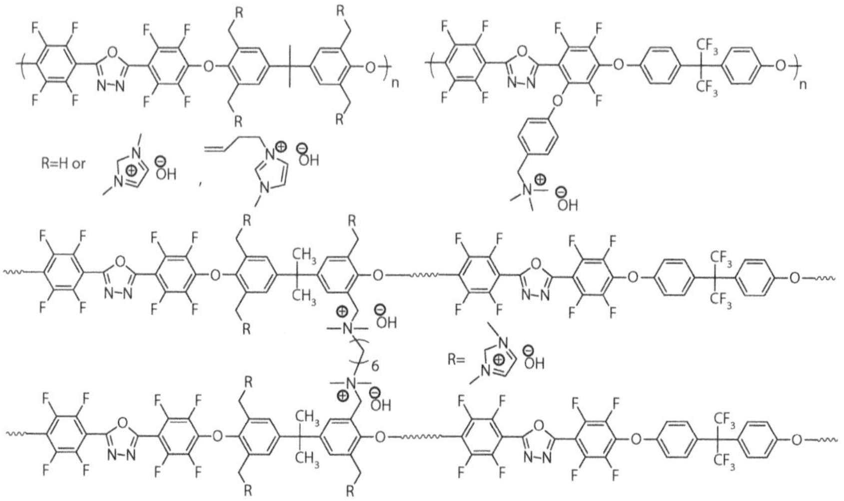

FIGURE 11.22 Chemical structure of poly(aryl ether oxadiazole) ionomers.

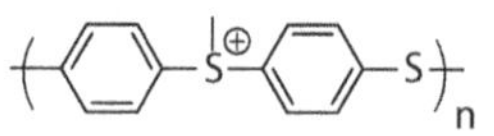

FIGURE 11.23 Chemical structure of polyarylene sulfonium ionomers.

these attempts have given some insight and directions for novel polymeric designs. Moreover, the results will enlighten the researchers on the exploration of high-performance aromatic polymer-based AEM materials in the future.

11.3.3 MEMBRANE MODIFICATION

At the current stage of developing novel aromatic AEM materials, there are still a number of research challenges that must be overcome before they could be successfully applied to practical AEM fuel cell applications. One primary limitation is the chemical stability of both the quaternary ammonium groups and the polymer skeleton under high pH condition. It will lead to deterioration of membrane integrity and reduction of ionic conductivity, resulting in the loss of cell efficiency and performance during practical operation. Besides, AEMs with high ionic conductivity are necessary to guarantee the high cell performances at high current densities. Therefore, several strategies have been carried out to enhance the chemical stability and ionic conductivity on the basis of those strategies proven to be effective in the case of proton exchange membrane materials, including cross-linking, composite (or hybridation), and blend modification.

11.3.3.1 Cross-Linking Modification

Cross-linking is a common choice to improve the mechanical strength and stability of the membrane materials. In Section 11.3.2, the cross-linking treatment of PSU,

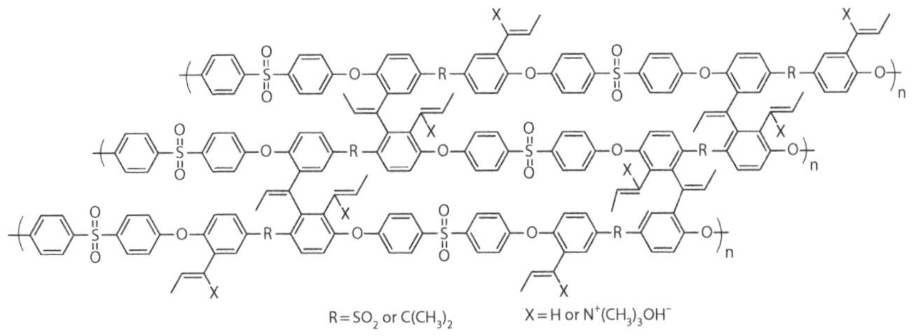

FIGURE 11.24 Chemical structure of self-cross-linked poly(aryl ether sulfone) ionomers.

poly(arylene ether), and PPO-based AEMs has been proven to be very effective to enhance the membrane mechanical stability and suppress the fuel permeability without sacrificing the ionic conductivity. Without using any catalyst or a separate cross-linker, Na et al. prepared a novel self-cross-linked AEMs derived from poly(aryl ether sulfone) with high alkaline stability, excellent dimensional stability, admirable fuel resistance, and high thermal stability (Figure 11.24). The cross-linking reaction occurred between the bromomethyl groups and active hydrogen atoms of aromatic rings through Friedel–Crafts reaction. The obtained self-cross-linked membranes exhibited ionic conductivities about 20 mS/cm and methanol permeability around 2×10^{-9} cm^2/s at 20°C. After immersing in 1 M NaOH solutions for 30 days, negligible decrease in ionic conductivity was observed under a wide temperature range of 20°C–80°C.[80] Similar method was also applied to quaternary ammonium PES and poly(ether ether ketone) AEMs.[81,82]

11.3.3.2 Inorganic–Organic Composite (Hybrid) Modification

Inorganic–organic composite materials have triggered increasing interests owning to the combination of characteristics of both components and the remarkable changes in properties such as mechanical, thermal, electrical, and magnetic arising from the synergism of the properties between the pure organic polymers and inorganic materials.[7] A huge number of reports have been published concerning this modification method for proton exchange membranes.[83]

For the modification of AEMs, common nano-sized inorganic fillers, such as ZrO_2, SiO_2, TiO_2, and multiwalled carbon nanotubes (CNTs), were selected preferentially. As reported, the introduction of inorganic filler will induce crystalline character of the polymer matrix, which leads to enhancement in thermal resistance and mechanical strength as well as fuel crossover suppression. For example, Li prepared a quaternized PAES/nano-ZrO_2 composite AEMs by a simple solution casting method. The introduction of nano-ZrO_2 improves the physicochemical properties in most aspects of the resulting composite AEMs, especially for the membrane containing 7.5% nano-ZrO_2. With the nano-ZrO_2 content above 7.5%, the membranes display ionic conductivity above 41.4 mS/cm at 80°C.[84] Similar results were reported in a quaternized PSU/nano-TiO_2 composite system.[85] For a single fuel cell test using quaternized PSU/nano-ZrO_2 (10%) composite membrane as the polyelectrolyte, the maximum power density reached 250 mW/cm^2 at 60°C.[86] Also by solution casting method, Lue and coworkers prepared a new type of PBI/functionalized multiwalled CNT (<1%) nanocomposite membrane. AEMs were obtained by doping the PBI/CNT composites with KOH solutions. The CNT-containing composites exhibited higher fractional free volumes and higher water diffusivities. In a single fuel cell test (2 M methanol in 6 M KOH/humidi-fied oxygen), the system achieved a maximum power density of 104.7 mW/cm^2 at 90°C, indicating its potential in high-temperature AEM fuel cell applications.[87] PPO/nano-SiO$_2$ hybrid membranes were prepared through a sol-gel reaction with monophenyl triethoxysilane and tetraethoxysilane, followed by heat treatment at 120°C–140°C. The obtained membranes (2.01–2.27 mmol/g) displayed ionic con-ductivity of 12–35 mS/cm in the temperature range of 30°C–90°C. Beginning-of-life fuel cell testing of a membrane yielded an acceptable H$_2$/O$_2$ peak power density of 32 mW/cm^2.[88]

11.3.3.3　Blend Modification

According to the description by Couture et al.,[3] a blend AEM refers to the blend-ing of a water-soluble polymer, which acts as a matrix, and a hydroxide salt, while in some cases, one or more plasticizers can be added to the blend. So far, most of the blend AEM materials are based on nonaromatic polymer blends, such as poly(vinyl alcohol) and poly(ethylene oxide), and blends with potassium hydroxide or 1-ethyl-3-methylimidazolium hydroxide. Only a few reports are based on the aro-matic polymer blends; most of them are PBI/KOH blend membranes as mentioned in Section 11.3.2.3.

From the viewpoint of polymer science, a polymer blend (or polymer mixture) is a member of a class of materials in which at least two polymers are blended together to create a new material with different physical properties, and it can be broadly divided into immiscible polymer blends, compatible polymer blends, and miscible polymer blends. This modification method is widely used for proton exchange mem-branes to reinforce the mechanical strength especially. In fact, the OH$^-$ mobility in an AEM is much lower than the H$^+$ mobility in a proton exchange membrane.[19] Therefore, a thin membrane with no compromise on mechanical and chemical dura-bility is a solution to obtain high ionic conductivity. To date, polytetrafluoroethylene (PTFE)-based blend AEMs are the most studied, for it can offer great mechanical strength, thermal and chemical stability, and low cost.[89–91] For example, a PTFE quaternary DABCO PSU (QDPSU) blend membrane (PTFE-QDPSU) prepared by impregnation of QDPSU solution in PTFE support showed a good mechanical strength (32 MPa), a water uptake of 61% ± 3%, and a 17% ± 2% swelling degree, which is 30% and 25% lower than the pristine QDPSU membrane, respectively. The ionic conductivity reached 51 mS/cm at 55°C in water. SEM observation revealed the dense and homogenous structure. Preliminary date of a single membrane elec-trode assembly (MEA) gave power output of 146 and 103 mW/cm^2 with oxygen and air, respectively.[89] Peak power density of 315 mW/cm^2 was achieved at 50°C in a PTFE quaternary ammonium PSU blend system; in addition, the mechanical strength increased by five times, and swelling ratio reduced by 50% compared with the unmodified membranes.[91]

11.3.4   Membrane Characterization and Measurements

11.3.4.1   Fundamental Physicochemical Properties

The role of an AEM is to conduct hydroxyl ions from cathode to anode at matchable rates to the foreign current, as well as the separation of fuels and oxidants. In addition, its integration with the catalytic electrodes forms the heart, MEA, of the AEM fuel cell system. The basic requirements for an AEM are summarized in Section 11.2. To evaluate the performance of a developed AEM material, the fundamental physicochemical properties, including IEC, water absorbing and dimensional swelling, mechanical and thermal properties, membrane morphology, and methanol permeability, are usually investigated. The physical properties of some selected aromatic AEMs are summarized in Table 11.1.

11.3.4.1.1   Ion Exchange Capacity

For an AEM, IEC is expressed as milliequivalent of cation ionic groups per 1 g of dry polymer (meq/g or mmol/g). It is the basic index that represents the amount of cation ionic groups embedded in the polymer matrix and intercorrelated with other properties, especially the membrane swelling and ion conducting behaviors. For the AEMs prepared through direct polymerization approach, the theoretical IEC values are usually estimated by the feed ratio of starting materials directly. As for the AEMs prepared by postmodification approach, it is almost impossible to estimate the accurate IEC values due to the uncertainty of halomethylation degree, which strongly depends on the reaction condition and polymer type. Experimentally, researchers often determine the actual IEC values by acid–base titration method or Mohr titration method.[17,61,64] In some cases, they also gain IEC values through the ^{1}H NMR spectrum of halomethylated polymers, where the integration of peaks assigned to typical hydrogens is used.[68,69] So far, IEC values of the developed aromatic AEMs are controlled in the wide range of 0.1–3.0 mmol/g, especially in the moderate range of 1.0–2.0 mmol/g.

11.3.4.1.2   Water Absorbing and Dimensional Swelling

For ion conducting process, whether proton or hydroxyl ions, water molecules are necessary for their smooth and rapid transport across the membrane irrespective of the mechanism that the ions are subjected to. Due to the hygroscopic nature of the attached quaternary ammonium groups, the membranes are inclined to absorb water in partially and/or fully hydrated state. Generally, the amount of water that membrane absorbs (also called as water uptake, %) can be simply determined by the weight difference between water-swollen and dry membrane sample after being exposed to the given moisture-laden air or immersed in water at a given temperature.

Water absorbing capacity of the AEMs depends on the number of ion exchange groups (IEC value) and the nature of the polymer matrix of the membrane. In the majority of situations, the former factor is more important; larger IEC value corresponds with the larger water uptakes and vice versa. For the AEMs derived from different types of aromatic ionomers, they exhibit rather different level of water uptakes. Therefore, to better understand the difference of the water absorbing ability

TABLE 11.1

Physical Properties of the Selected Aromatic AEMs

AEM	IEC^a (mmol/g)	WU^b (%)	λ^b (H$_2$O/QA)	SR^b (%)	σ (OH$^-$)^c (mS/cm)	$P_M^{b,d}$ (×10^{-7}cm^2/s)	Power Density (mW/cm^2)	Refs.
PSU/diamines	0.53–1.46	15–32		7.5–25.5	16–21		8.1–30.1 (H$_2$/air, 60°C)	[10,11]
PSU/MIm	1.39–2.46	8.5–93					6–16 (H$_2$/O$_2$, 60°C)	[12]
PSU/DABCO-PA (DS58/DS106)					64/120 (150°C)		323/400 (H$_2$/O$_2$, 150°C)	[14]
PES/morpholine	0.76	6.56	4		16, 79^e			[16]
PSU/TMA	1.48–2.40	55–166	15–44		6.48–27.3 (HCO$_3^-$)	3.7–41		[17]
PSU/PMG (cross-linked)	0.66–1.21	25–87		30–92	13–60	0.01–0.096		[18]
PSU/TMA	1.01–1.74	8.7–18.4	4.8–5.9	3–4	15–38	0.3–1.2		[19]
PEEK/TMA	0.43–1.18	9–30	10–15	3–13	5.6–13			[20]
	1.35	153	63	37	17			
PEEK/MIm	1.56–2.03	40–159	14–43.5	17–41	15–52 / 9–27^f	1.3–6.9	31 (methanol/O$_2$, r.t.)	[21]
PEEK/TMA	0.70–1.15	10.2–28.7	8.1–13.9	16–20	12–20	0.34–0.54	48.1 (H$_2$/air, 50°C)	[22]
PEK-C/TMA	0.11	3.33	16.8		1.6, 5.06^g	0.01		[23]
PEK-C-grafted-VBC/TMA	0.67	18.1	15	13	8.3	6.6 × 10^{-4} (ethanol)		[24]
PAEK/TMA	2.28–2.91	39–145	9.5–27.7	20–198	52–65	14–150		[25]
PAEK/TMA (cross-linked)	3.35–3.50	37–87	6.1–13.8	17–50	5–11			
PEEK/TMA (side-chain-type)	0.62–0.90	5.3–11.9	4.7–7.3	2.4–5.0	5–11	0.23–1.43		[26]
PEEK/TMA (side-chain-type)	0.97–1.35	16.8–69.3	9.6–28.5	4.4–9.9 (4.7–11)^f	5.4–12.3	0.52–1.94		[27]

(Continued)

TABLE 11.1 (Continued)
Physical Properties of the Selected Aromatic AEMs

AEM	IEC[a] (mmol/g)	WU[b] (%)	λ[b] (H_2O/QA)	SR[b] (%)	σ(OH^-)[c] (mS/cm)	P_M[b,d] ($\times 10^{-7}$ cm²/s)	Power Density (mW/cm²)	Refs.
Poly(phthalazinone ether ketone)/TMA	0.70–2.04	12.9–52.3	10.2–14.2					[29]
PAEK/TMA	1.12–2.88	8–41	4.0–7.9	3–33	1.7–12 (Br^-)			[30]
Poly(arylene ether sulfone ketone)/TMA	0.68–2.54	22–160	18–35		0.3–50			[31]
Block poly(arylene ether)/TMA	0.39–2.05	19–112	11.2–36.6		2.4–126[g]		297 (hydrazine/O_2, 80°C) 161 (hydrazine/air, 80°C)	[32]
Poly(fluorenyl ether)/TMA	0.6–2.0	22–81	20.4–22.5		8–30			[33]
PAES/TMA	1.21–1.73	20.6–68.7	9.5–22.1	26.8–58.8 (23.5–48.6)[f]	14.6–30.5[e]			[34]
Poly(fluorenylene ether sulfone)/MIm	1.55–1.96	11.7–35.2	4.2–10	5.2–11.5[e]	15.7–28.2[e]			[35]
Block PAES/TMA	1.20	29.1	13.4		21			[36]
Block PAES/MIm	1.45	44.2	17		30			
PAES/TMA	1.34–2.61	8–94	3.3–20	5–64	12–72			[37]
Poly(arylene ether)/TMA	0.21–2.38	5–19	3.1–13.2		0.37–25			[38]
Poly(tetraphenyl ether sulfone)/MIm	1.94–2.41	26.1–46.8	7.5–10.8	17.1–25.9[e]	26.9–38.4[e]			[39]
PAES/TMA	1.59	42	15		17.9[e]			
Block PAES/TMA	1.14–1.89	16–53	7.8–15.6		11.4–29.3[e]			[40]
Poly(dibenzoylbenzene ether sulfone)/TMA	1.14–1.69	19–64	9.3–21	19.3–50.9[e]	13.9–28.7[e]			[41]

(Continued)

TABLE 11.1 (Continued)
Physical Properties of the Selected Aromatic AEMs

AEM	IEC[a] (mmol/g)	WU[b] (%)	λ[b] (H₂O/QA)	SR[b] (%)	σ (OH⁻)[c] (mS/cm)	P_M[b,d] (×10⁻⁷cm²/s)	Power Density (mW/cm²)	Refs.
PTPPES-QAH	1.16–1.80	22.1–75.6	10.6–23.3	24.9–68.4[e]	22.7–51.2		90–100 (H₂/O₂, 80°C)	[42]
Cardo block PAES/MIm	1.16–1.30	10.4–13.2	5.0–5.6	2.3–2.9	32–35	0.67–0.92		[43]
PBI/KOH					18.4	2.60	31 (methanol-KOH/O₂, 90°C)	[46]
PBI/KOH					22	0.524 (ethanol)	11 (ethanol/air, 13°C)	[47]
ABPBI/NaOH		9.1–19.2			14–23			[48]
Cross-linked PBI/KOH					~10[f]			[49]
PBI/KOH							80 (ethylene glycol/O₂, 60°C)	[50]
Polybenzimidazolium hydroxide	1.0–2.0	82–162	22–45		13.2–9.6			[51]
Cross-linked PBI/grafting-MGMC	1.43–2.12	28–85[g]	10.9–22.3[g]	6.7–34 (6.7–33)[f,g]	0.6–39[g]		3.1 (H₂/O₂, r.t.)	[54]
PPO/4-vinylpyridine	0.3–1.1	6–27.5	11.1–13.9					[60]
PPO/PMG	0.37–2.69	6.1–101	6.1–21	2.5–37.5	71		16 (H₂/O₂, 50°C)	[62]
PPO/tris(2,4,6-trimethoxyphenyl) phosphine	0.49–1.40	10–38[h]		<7[i]	10–17			[63]
PPO/methyl benzimidazole	0.63–2.21	6.8–32	4.1–8.0	2.6–12.7	10–37		13 (H₂/O₂, 35°C)	[64]
Blend CPPO/ BPPO/TMA	1.9–3.3	44.3–280			22–32	1.35–1.46		[65]
PPO/TMA	1.74	24		4.5	16.4[g]		19.5 (H₂/O₂, 70°C)	[66]
Cross-linked PPO/ 2-benzyl-1,1,3,3- tetramethyl guanidine	0.80	20	13.9	<15	10			[67]

(Continued)

TABLE 11.1 (Continued)
Physical Properties of the Selected Aromatic AEMs

AEM	IEC[a] (mmol/g)	WU[b] (%)	λ[b] (H_2O/QA)	SR[b] (%)	σ (OH^-)[c] (mS/cm)	P_M[b,d] ($\times 10^{-7}\,cm^2/s$)	Power Density (mW/cm^2)	Refs.
PPO-clicked 1,2,3-triazole/TMA	0.99–1.50 (¹H NMR)	22.4–32	11.8–12.6		27.8–62		188.7 (H₂/O₂, 50°C)	[68]
PPO/TMA	1.37–2.03	20–59	8.0–16	5–13	6–16			[69]
	3	130	27	180	24			
PPO/dimethylalkyl amine (comb-shaped)	1.08–2.39	8–16	3.5–4.2	2–4	6–28		67–145 (H₂/O₂, 50°C)	
PP(ATMPP)/TMA	2.75	150	30	200	43			[70]
	0.93–1.04	42–48	25–26		4–7			
PP(ADAPP-ATMPP)/TMA	1.39–1.57	97–122	39–45		20–50			
	0.40–1.15	23–78	31–37		7–30			
PP(TMPP)/PMG	1.51	66	24		10(Cl⁻)			[71]
PP(TMPP)/MIm	1.79	59	18		9.7(Cl⁻)			
PP(KC6PP)/TMA	1.46	91	35		13(Cl⁻)			
PP(C6PP)/TMA	1.74	126	40		17.4(Cl⁻)			
Poly(ether-imide)/TMA	0.98	43	24		2.28			[72]
Poly(aryl ether oxadiazole)/MIm	1.14–1.79	13.5–53.9	6.6–16.7	2.0–11.4	4.5–15.9			[75]
Cross-linked poly(aryl ether oxadiazole)/MIm	2.03	102.2	28	25.1	20.2			[77]
		5.9–9.9		1.7–3.9	7.9–19.4			
Poly(aryl ether oxadiazole)/MIm	1.32/1.80	9/64	3.8/19.8	2/13	5.4/17			[78]
Cross-linked poly(aryl ether oxadiazole)/MIm	1.36/1.80	8/51	3.3/15.7	2/6	6.5/16.9			

(Continued)

TABLE 11.1 (Continued)
Physical Properties of the Selected Aromatic AEMs

AEM	IEC[a] (mmol/g)	WU[b] (%)	λ[b] (H$_2$O/QA)	SR[b] (%)	σ(OH$^-$)[c] (mS/cm)	P_M[b,d] ($\times 10^{-7}$cm^2/s)	Power Density (mW/cm^2)	Refs.
Self-cross-linked poly(ether sulfone)/TMA	0.79–1.08	9.4–13.3	6.6–6.8		6.9–8.5			[81]
PEEK/TMA	0.90–1.43	24.6–32.1	12.5–15.2	11.9–15.3	12–17	4.9–6.1		[82]
Cross-linked PEEK/TMA	1.01–1.18	8.2–13.5	4.5–6.4	4.7–6.2	13–14	2.1–2.9		
PAES/TMA-nano ZrO$_2$ (0%–20%)	1.46–1.82	10.3–28.1	3.9–8.6	11.4–7.4[g]	9.5–23.1			[84]
PSU/TMA-ZrO$_2$ (0%–10%)	0.69–0.92	5.2–19	4.2–11		7.3–15.1		85–140 (H$_2$/O$_2$, r.t.) 170–250 (H$_2$/O$_2$, r.t.)	[86]
PBI/KOH–carbon nanotube	2.0–2.3	17.5–21.5			23–35	1.85–2.59 × 10^{-3}	104.7 (methanol- KOH/ O$_2$, 90°C)	[87]
PPO/TMA-SiO$_2$ (0%–13.9%)		32–64			7.9–9.7		32 (H$_2$/O$_2$, 50°C)	[88]
PSU/DABCO- PTFE		61		17	51 (55°C)		146 (H$_2$/O$_2$, 50°C) 103 (H$_2$/air, 50°C)	[89]
Polyetherimides/TMA-PTFE	0.35–0.58	105–138		7–9	3.8–11.9			[90]
PSU/TMA-PTFE	1.27	54.7		14.3	16–17		252–315 (H$_2$/O$_2$, 50°C)	[91]

a   Titrated value.
b   At 20°C–30°C.
c   In-plane direction.
d   Methanol permeability.
e   At 80°C.
f   Through-plane direction.
g   At 60°C.
h   At 40°C.
i   At 70°C.

for the AEMs derived from different types of aromatic ionomers, the parameter of hydration number (λ, the average number of water molecules absorbed per ionic groups) is used.

Water absorbing accompanies the membrane dimensional swelling both in the membrane in-plane and through-plane direction inevitably. An appropriate level of membrane swelling is necessary for the integration and formation of triple-phase interface between the membrane and catalytic electrode of MEA. However, excessive membrane swelling causes reduction in mechanical strength or the delamination of membrane and catalyst layer. As listed in Table 11.1, the swelling ratio differs widely, depending on the IEC level, polymer skeleton structure, and type of attached anion ions. It seems that the cross-linked, blend, and inorganic nanoparticle composite membranes exhibit smaller swelling ratio than those unmodified ones. The dimensional swelling (or swelling ratio, SR, %) is usually expressed as the size change percentage of the membrane sample (strip or circular shape) before and after being exposed to the given moisture-laden air or soaked in water at a given temperature as the same as the determination of water uptakes. In some cases, the volumetric change of the membrane sample is also used.

11.3.4.1.3 Mechanical and Thermal Properties

An AEM has to exhibit sufficient mechanical toughness and stability in order to fulfill its separator function during MEA fabrication and long-term operation under stable or unsteady conditions. The mechanical strength of an AEM is usually characterized by Young's modulus, stress, and strain by using a universal testing instrument. Most of the developed aromatic AEMs possess Young's modulus of ~1.5 GPa, maximum stress at the break of tens of MPa, and elongation at break ~100%, which can meet the requirement for practical fuel cell applications.

Thermal property of an AEM is characterized by thermogravimetric analysis (TGA) or differential scanning calorimetry (DSC). Typically, in a TGA curve for an ionomer, the weight losses present three stages. The first stage is derived from the loss of water absorbed in the polymer matrix; the second stage is attributed to the decomposition of ionic groups bound to the polymer skeleton; and the last stage is ascribed to the degradation of polymer skeleton itself. The second weight loss stage is of the most concern because it is closely related to the permissible (or maximum) operation temperature for the AEM in practical fuel cell operation. The decomposition temperature of quaternary benzyltrimethyl ammonium groups is in the range of 140°C–250°C,[23,36,40,41,84] while 150°C–325°C for the decomposition of quaternary imidazolium ammonium groups[35,36,75] and about 270°C–350°C for the quaternary guanidinium ammonium groups.[62] So far, the fuel cell tests using the developed aromatic AEMs are performed at the temperature ranging from room temperature to 90°C irrespective of the used fuel as listed in Table 11.1, except one phosphoric acid–doped QPSU (aminated by DABCO) membrane that was conducted at 150°C.[14]

11.3.4.1.4 Morphology

Morphology of the AEM material is often characterized by small-angle x-ray scattering (SAXS),[69] energy-dispersive x-ray (EDX),[60,84,89] scanning electron microscopy (SEM),[55,59,65] transmission electron microscopy (TEM),[67,86] and atomic force microscopy (AFM).[35,36,39–41,63] The intensity of the ionomer peak in SAXS spectra can be used to analyze the order of the ionic domains inside the polymer matrix. EDX analysis can give proportional percentages of the distribution of the element uniformity on the surface of the AEM material, especially for those composite materials.[85] SEM and TEM are generally used to observe the membrane morphology in the micrometer or nanometer scale, providing quick and clear images of the membrane microstructures. AFM is often used to characterize the surface morphology of the prepared AEMs.

11.3.4.1.5   Methanol Permeability

For an AEM fabricated in a fuel cell system that used methanol solution as the fuel, the methanol crossover should be considered, and the parameter of methanol permeability is often used. Majority of the developed aromatic AEMs exhibited one to three orders of magnitude lower methanol permeability than that for Nafion series membranes, in the range of 10^{-9}–10^{-7} cm²/s, as shown in Table 11.1. Combing with the opposite direction of methanol diffusion and ion transport across the membrane, the rather low methanol permeability can ensure the stable performance of the fuel cells using methanol solutions as fuels.

In most cases, the methanol permeability (P_M) is determined by ex situ method. The apparatus often contains two tightly closed membrane-separated diffusion half-cells. One compartment is filled with methanol solution with a given concentration, and water is placed in the other compartment under continuous stirring to ensure uniformity. The concentration of the methanol is measured at a certain interval, usually by the gas chromatograph.

11.3.4.1.6   Ionic Conductivity

Ionic conductivity of an AEM reflects the capability of an anionic ion (i.e., OH^-, HCO_3^-, Cl^-, Br^-) moving from one site to another, which is one of its most critical property. The minimum requirement of the ionic conductivity (in OH^- form) for an AEM is 10 mS/cm to apply for practical fuel cell operation. Due to the difference in ion mobility or dissociation (OH^- vs. H^+) between quaternary ammonium functional groups and sulfonic acid groups, the developed AEMs generally possess lower ionic conductivity than the proton exchange membranes with the same polymer skeleton at similar IEC levels.[19] To improve the ionic conductivity of an AEM, various strategies have been adopted: (1) implementing alternative cation chemistries such as quaternary imidazolium, pentamethylguanidinium, and phosphonium over the commonly employed quaternary trimethylammonium and (2) altering the physical network of the AEM, for example, polymer structure design, block copolymer formation, inorganic–organic hybridation, blending, grafting, or pendant structure formation. A decade ago, the most reported ionic conductivity is at the level of 10^{-3}–10^{-2} S/cm, but today, it is not surprising to learn that ionic conductivities exceed 10^{-1} S/cm according to the literatures. Examples can be found in Sections 11.3.2 and 11.3.3, and the ionic conductivity data are listed in Table 11.1.

Ionic conductivity is usually obtained through electrochemical impedance spectroscopy (EIS) technology by setting the membrane sample in deionized water or alkali solutions. Most of the values obtained are referring to the ionic conductivity in the membrane in-plane direction according to the measurements; only a few data in the membrane through-plane direction were reported.[21,49] Besides, unlike that for the proton exchange membranes, the ionic conductivity under various relative humidity is hardly reported so far.

It should also be noted that the ionic conductivity values given in the literatures are based on hydroxyl ions in most cases. According to the results provided by Hickner and his coworkers, the ionic conductivity decreased with the time during exposure to ambient air containing carbon dioxide due to the conversion from OH^- form to HCO_3^- or CO_3^{2-} form, because both the bicarbonate and carbonate anions are bulkier and possess poorer water solvation and lower ionic mobility when compared to the hydroxyl ion.[17,92] Therefore, extreme care is needed during the measurement to prevent exposure of the AEM to atmospheric carbon dioxide, for example, setting the test cell in nitrogen atmosphere or bubbling nitrogen gas into the system.

11.3.4.1.7 Fuel Cell Performance

The ultimate objective for the development of an AEM is to fabricate a fuel cell.

Fuel cell performance is usually evaluated by the OCV, I–V polarization curves, peak power density, and the corresponding current density. Fuel cell characterization using the aromatic AEMs has been mentioned in the work of several research groups. The reported fuel cell performance characteristics span a substantial range from 3.1 to 400 mW/cm^2. The operation condition also differed largely, that is, the cell temperature varied from room temperature to 150°C, the fuels selected diversified from hydrogen, methanol, ethanol, ethylene glycol to hydrazine as summarized in Table 11.1, and the electrode catalysts utilized included noble metal and nonprecious metal. Therefore, the direct comparison of fuel cell data based on different types of AEMs is not always straightforward and can be misleading. For example, He reported the fuel cell performance using an imidazolium-functionalized poly(ether ether ketone) ionomer (PEEK-ImOH) as the membrane electrolyte and electrode ionomer at the same time.[21] The catalyst layer for both anode and cathode contained 0.5 mg Pt/cm^2. The MEA was assembled in a single cell fixture for the fuel cell test. And I–V polarization curves were obtained at room temperature with anode feed of 10 mL/min 2 M methanol solution and cathode feed of 20 mL/min pure O_2. The methanol/O_2 fuel cell with the PEEK-ImOH membrane (2.03 mmol/g) exhibited OCV of 0.84 V and a peak power density of 31 mW/cm^2. Using block quaternary benzyltrimethyl ammonium poly(arylene ether) membrane as the electrolyte, Ni powder and Co with polypyrrole dispersed on carbon as the anode and cathode catalyst, respectively, and hydrazine as the fuel; Watanabe reported a reasonable high fuel cell performance with OCV of 0.71 V and maximum power density of 161 mW/cm^2 at a current density of 446 mA/cm^2 with air and OCV of 0.76 V and maximum power density of 297 mW/cm^2 at a current density of 826 mA/cm^2 with O_2.[32] Apparently, the reported fuel cell performance data are much lower than those using aromatic proton conducting ionomers. There is an urging demand toward improving the power density of AEM fuel cells.

11.3.4.2 Chemical Stability

Usually, the chemical stability issue of an AEM involves the degradation of polymer backbone and decomposition of side-chain functional groups. In order to discern how much these two factors contribute to the stability problem of the developed AEM materials, the membranes are subjected to accelerating aging test by immersing in alkali solution or hot water for a certain time at a given temperature. KOH or NaOH solutions (1.0–6.0 M) are usually chosen as the alkali condition; the test is often carried out at room temperature to 90°C for 24–2000 h. A harsher condition was conducted on a PEEK-based AEM, in which the membrane sample was treated with 10 M NaOH solution at 100°C, but the test only continued for 24 h.[22]

Chemical stability of the AEMs in alkali solutions is called alkaline stability and is commonly assessed by the change in ionic conductivity and IEC before and after the test. This is a direct representation of the loss of the ionic groups inside the membrane matrix, mainly caused by the degradation or elimination of the quaternary ionic groups as mentioned earlier. In some cases, the changes in mass and mechanical properties of the membrane were also used as the criterion.[69]

Under heated or hydrothermal conditions, the possible degradation mechanism for AEMs has been discussed in the literatures.[93,94] Hydrolytic stability of the AEMs is performed by treating the membrane in hot water for a specific period. Only a few reports are concerned with the hydrolytic stability so far. For Watanabe research groups, they performed the hydrolytic stability test in hot water at 80°C for 500 h; the chemical structures before and after the hydrothermal test were analyzed by the use of ^{1}H NMR spectra. A long-term hydrothermal test was also performed for 5000 h, and the ionic conductivity was used to assess the stability. According to their results, the major degradation mode of the AEM (based on quaternary PES) under the hydrothermal condition is most likely to be the elimination of a tertiary amine to form benzyl alcohol and its further decomposition via dehydration.[32]

As well accepted, formation of radicals, such as HO·, H·, and HOO·, in the PEM of the fuel cell is responsible for the chain scission of the polymer electrolyte, for the free radicals attack mainly on the polymer backbone rather than on the vicinity of hydrophilic domains.[22,95] Oxidative stability of the AEMs is generally evaluated in Fenton's reagent at a given temperature for a specific period. The loss in mass, IEC, and ionic conductivity after the treatment is commonly used to assess the AEM oxidative stability.

It has been well known that the quaternary ammonium-functionalized polymers (and hence the corresponding membranes) are generally unstable in an alkaline solution at elevated temperatures due to the nucleophilic attack of OH⁻ anions through Hoffmann degradation or nucleophilic substitution as well as ylide formation as demonstrated in Figure 11.25. To weaken the behavior, researchers have carried out various strategies: (1) synthesizing β-hydrogen-absent quaternary ammoniums or avoiding a coplanar arrangement of β-hydrogen and nitrogen to circumvent the Hoffmann elimination reaction. However, the nucleophilic substitution due to the attack of α-carbon by OH⁻ cannot be averted for most of the alkyl ammonium-functionalized polymers. (2) Introducing heterocyclic system to the quaternary

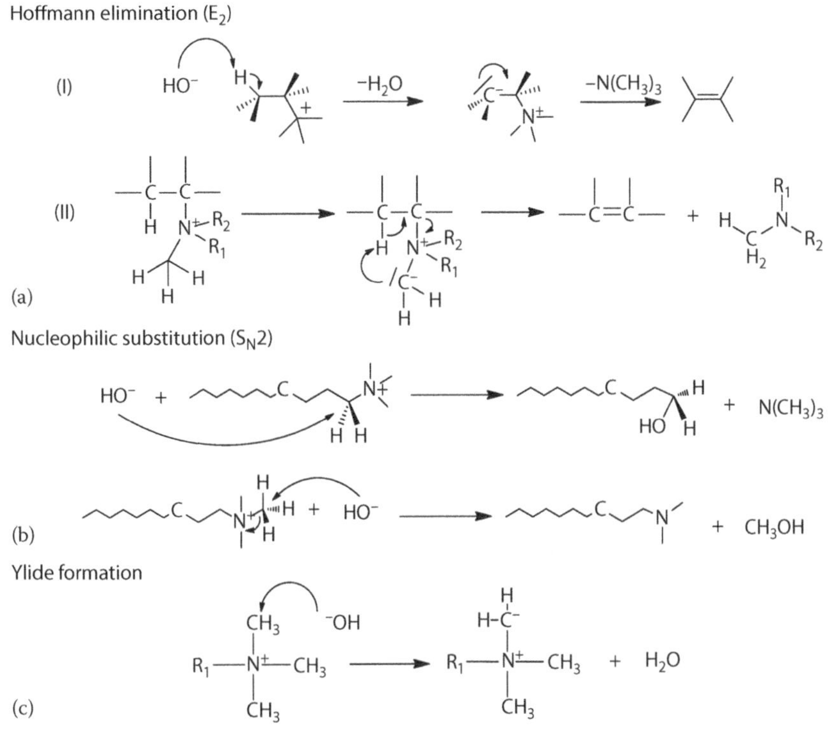

Hoffmann elimination (E$_2$)

(I)

(II)

(a)

Nucleophilic substitution (S$_N$2)

(b)

Ylide formation

(c)

FIGURE 11.25 Degradation mechanism of a quaternary ammonium group by the (a) Hoffmann elimination, (b) nucleophilic substitution, and (c) ylide formation. (From Couture, G. et al., *Prog. Polym. Sci.*, 36(11), 1521, 2011; Merle, G. et al., *J. Membr. Sci.*, 377(1–2), 1, 2011; Chempath, S. et al., *J. Phys. Chem. C*, 112(9), 3179, 2008; Chempath, S. et al., *J. Phys. Chem. C*, 114(27), 11977, 2010; Cope, A.C. and Mehta, A.S., *J. Am. Chem. Soc.*, 85(13), 1949, 1963.)

ammonio-substituted groups, such as replacing benzyltrimethyl-functionalized cations with bulky cations of imidazolium-functionalized cations, benzyl and phenyl guanidinium-functionalized cations, or phosphonium-functionalized cations, because the presence of the conjugated π-bonds or charge delocalization of the cations could stabilize the α-carbon–nitrogen bond. Examples can be found in Sections 11.3.2 and 11.3.3.

11.4 SUMMARY AND PERSPECTIVES

AEM fuel cell using an AEM as the solid electrolyte provides the potential of producing high power densities without cost noble-metal catalyst. The rapid progress in the development of novel AEM, as a key component of the fuel cell system, has pushed this technology to a research focus in fuel cell and green energy device–related field in the past few years.

In this chapter, quaternary ammonium aromatic AEMs have been reviewed as promising hydroxide conductor for PEM fuel cell applications. As discussed earlier, except the good comprehensive performance and diversified chemical structure selection of aromatic polymers, the significant advancements and rich experience gained from the synthesis, preparation, property evaluation methods, and some of the lessons learned during the development of aromatic proton conducting ionomers can be employed to the anion conducting ionomers conveniently. This chapter summarizes the major alternative AEM materials derived from aromatic ionomers so far, including PSUs, poly(arylene ether)s, PBIs, and poly(phenylene oxide)s. Most of them share the same polymer skeleton structure to the cationic ionomers. A number of innovative material structure designs have also been adopted. These developed materials exhibited good thermal and mechanical properties, acceptable water absorbing and swelling behavior, as well as reasonable high ionic conductivity (about 10^{-2} S/cm).

However, there is still more work to be done in the anion conducting ionomers to realize the large-scale commercialization of this technology. First, the AEMs with higher ionic conductivity must be developed to ensure the better fuel cell performance, especially at high current densities. The nature of hydroxyl ion and quaternary ammonium groups cannot provide ionic conduction capacity as high as the proton ion and sulfonic acid groups. To date, most of the developed aromatic AEMs possess ionic conductivity of $\sim 10^{-2}$ S/cm in the fully hydrated state. However, this is not sufficient while taking the partially hydrated atmosphere in real fuel cells at elevated temperatures into consideration. Similar to the dilemma faced in the proton conducting ionomers, by simply improving the IEC level of the anion conducting ionomers is not rational to get high ionic conductivity. Polymer architectural design including cation chemistry modification, block copolymer formation, and pendant structure formation, as well as inorganic nanoparticle composite, seems to be more feasible choices. In addition, a lot of work must be done for better understanding of structure–property relationship of the aromatic ionomers. Also, new synthesis strategies, the development of a variety of new materials, and new methods to analyze their properties are necessary. Second, a more fundamentally daunting challenge is to enhance the AEM chemical (in)stability in alkaline media, because it will lead to the decrease of ionic conductivity, to the deterioration of mechanical strength, and to the loss of efficiency of the cell system. Although the stability evaluated in ex situ hydrothermal or alkali media has been largely reported, it is still questionable whether it can exactly represent their actual behavior in a real fuel cell. Third, the fuel cell performance investigation of the aromatic AEMs is still insufficient. Currently, long-term fuel cell tests for the developed aromatic AEMs are no more than 1000 h, and most of them only provide the primary cell performance for tens to several hundreds of hours. Furthermore, the membrane is not a stand-alone component. It must be combined with the catalysts and electrodes in the MEA. The interfacial properties, its expanding applications as catalyst binder, and the selection of fuels as well as operation conditions should also be considered and thoroughly studied in an application-specific fuel cell system.

It goes without saying that the commercialization of AEM fuel cell systems requires more progress and a breakthrough in membrane performance and production technology. It is a highly complex task and requires extensive cooperation and endeavor of researchers. Nevertheless, the primary works done to date on the structure–property relationship of the aromatic AEMs may be of great importance and reference in the design of high-performance AEMs in the future.

ACKNOWLEDGMENTS

The authors appreciate the support of this research by the National Natural Science Foundation of China (21006052, 21276128), Basic Research Program of Jiangsu Province of China (BK20141398), and the Fundamental Research Funds for the Central Universities (30920130121014).

REFERENCES

1. Appleby, A. J.; Foulkes, F. R., *Fuel Cell Handbook* (7th edn.). EG&G Technical Services, Inc.: Morgantown, WV, 2004.
2. Varcoe, J. R.; Slade, R. C. T., Prospects for alkaline anion-exchange membranes in low temperature fuel cells. *Fuel Cells* 2005, *5*(2), 187–200.
3. Couture, G.; Alaaeddine, A.; Boschet, F.; Ameduri, B., Polymeric materials as anion-exchange membranes for alkaline fuel cells. *Progress in Polymer Science* 2011, *36*(11), 1521–1557.
4. Merle, G.; Wessling, M.; Nijmeijer, K., Anion exchange membranes for alkaline fuel cells: A review. *Journal of Membrane Science* 2011, *377*(1–2), 1–35.
5. Hickner, M. A.; Herring, A. M.; Coughlin, E. B., Anion exchange membranes: Current status and moving forward. *Journal of Polymer Science Part B: Polymer Physics* 2013, *51*(24), 1727–1735.
6. Scherer, G. G., *Fuel Cells I* (Advances in Polymer Science). Springer Verlag: Berlin, Germany, 2008.
7. Xu, T., Ion exchange membranes: State of their development and perspective. *Journal of Membrane Science* 2005, *263*(1–2), 1–29.
8. Yamamoto, M.; Toi, K., Electrodialytic method. JP Patent 61192311, 1986.
9. Zschocke, P.; Quellmaz, D., Novel ion exchange membranes based on an aromatic poly-ethersulfone. *Journal of Membrane Science* 1985, *22*(2–3), 325–332.
10. Komkova, E.; Stamatialis, D.; Strathmann, H.; Wessling, M., Anion-exchange membranes containing diamines: Preparation and stability in alkaline solution. *Journal of Membrane Science* 2004, *244*(1–2), 25–34.
11. Park, J.-S.; Park, S.-H.; Yim, S.-D.; Yoon, Y.-G.; Lee, W.-Y.; Kim, C.-S., Performance of solid alkaline fuel cells employing anion-exchange membranes. *Journal of Power Sources* 2008, *178*(2), 620–626.
12. Zhang, F.; Zhang, H.; Qu, C., Imidazolium functionalized polysulfone anion exchange membrane for fuel cell application. *Journal of Materials Chemistry* 2011, *21*(34), 12744.
13. Lu, S.; Pan, J.; Huang, A., Zhuang, L.; Lu, J., Alkaline polymer electrolyte fuel cells completely free from noble metal catalysts. *Proceedings of the National Academy of Sciences of the Unites States of America* 2008, *105*, 20611–20614.
14. Wang, X.; Xu, C.; Golding, B. T.; Sadeghi, M.; Cao, Y.; Scott, K., A novel phosphoric acid loaded quaternary 1,4-diazabicyclo-[2.2.2]-octane polysulfone membrane for intermediate temperature fuel cells. *International Journal of Hydrogen Energy* 2011, *36*(14), 8550–8556.

15. Abuin, G. C.; Nonjola, P.; Franceschini, E. A.; Izraelevitch, F. H.; Mathe, M. K.; Corti, H. R., Characterization of an anionic-exchange membranes for direct methanol alkaline fuel cells. *International Journal of Hydrogen Energy* 2010, *35*(11), 5849–5854.

16. Hahn, S.-J.; Won, M.; Kim, T.-H., A morpholinium-functionalized poly(ether sulfone) as a novel anion exchange membrane for alkaline fuel cell. *Polymer Bulletin* 2013, *70*(12), 3373–3385.

17. Yan, J.; Hickner, M. A., Anion exchange membranes by bromination of benzylmethyl-containing poly(sulfone)s. *Macromolecules* 2010, *43*(5), 2349–2356.

18. Zhao, C. H.; Gong, Y.; Liu, Q. L.; Zhang, Q. G.; Zhu, A. M., Self-crosslinked anion exchange membranes by bromination of benzylmethyl-containing poly(sulfone)s for direct methanol fuel cells. *International Journal of Hydrogen Energy* 2012, *37*(15), 11383–11393.

19. Li, N.; Zhang, Q.; Wang, C.; Lee, Y. M.; Guiver, M. D., Phenyltrimethylammonium functionalized polysulfone anion exchange membranes. *Macromolecules* 2012, *45*(5), 2411–2419.

20. Yan, X.; He, G.; Gu, S.; Wu, X.; Du, L.; Zhang, H., Quaternized poly(ether ether ketone) hydroxide exchange membranes for fuel cells. *Journal of Membrane Science* 2011, *375*(1–2), 204–211.

21. Yan, X.; Gu, S.; He, G.; Wu, X.; Benziger, J., Imidazolium-functionalized poly(ether ether ketone) as membrane and electrode ionomer for low-temperature alkaline membrane direct methanol fuel cell. *Journal of Power Sources* 2014, *250*, 90–97.

22. Jasti, A.; Prakash, S.; Shahi, V. K., Stable and hydroxide ion conductive membranes for fuel cell applications: Chloromethylation and amination of poly(ether ether ketone). *Journal of Membrane Science* 2013, *428*, 470–479.

23. Xiong, Y.; Liu, Q. L.; Zeng, Q. H., Quaternized cardo polyetherketone anion exchange membrane for direct methanol alkaline fuel cells. *Journal of Power Sources* 2009, *193*(2), 541–546.

24. Hu, J.; Zhang, C.; Jiang, L.; Fang, S.; Zhang, X.; Wang, X.; Meng, Y., Plasma graft-polymerization for synthesis of highly stable hydroxide exchange membrane. *Journal of Power Sources* 2014, *248*, 831–838.

25. Wang, J.; Wang, J.; Zhang, S., Synthesis and characterization of cross-linked poly(arylene ether ketone) containing pendant quaternary ammonium groups for anion-exchange membranes. *Journal of Membrane Science* 2012, *415–416*, 205–212.

26. Xu, S.; Zhang, G.; Zhang, Y.; Zhao, C.; Ma, W.; Sun, H.; Zhang, N.; Zhang, L.; Jiang, H.; Na, H., Synthesis and properties of a novel side-chain-type hydroxide exchange membrane for direct methanol fuel cells (DMFCs). *Journal of Power Sources* 2012, *209*, 228–235.

27. Shen, K.; Pang, J.; Feng, S.; Wang, Y.; Jiang, Z., Synthesis and properties of a novel poly(aryl ether ketone)s with quaternary ammonium pendant groups for anion exchange membranes. *Journal of Membrane Science* 2013, *440*, 20–28.

28. Fang, J.; Shen, P. K., Quaternized poly(phthalazinon ether sulfone ketone) membrane for anion exchange membrane fuel cells. *Journal of Membrane Science* 2006, *285*(1–2), 317–322.

29. Zhang, S.; Yin, C.; Xing, D.; Yang, D.; Jian, X., Preparation of chloromethylated/quaternized poly(phthalazinone ether ketone) anion exchange membrane materials for vanadium redox flow battery applications. *Journal of Membrane Science* 2010, *363* (1–2), 243–249.

30. Chen, D.; Hickner, M. A., Ion clustering in quaternary ammonium functionalized benzylmethyl containing poly(arylene ether ketone)s. *Macromolecules* 2013, *46*(23), 9270–9278.

31. Tanaka, M.; Koike, M.; Miyatake, K.; Watanabe, M., Anion conductive aromatic iono-mers containing fluorenyl groups. *Macromolecules* 2010, *43*(6), 2657–2659.

32. Tanaka, M.; Fukasawa, K.; Nishino, E.; Yamaguchi, S.; Yamada, K.; Tanaka, H.; Bae, B.; Miyatake, K.; Watanabe, M., Anion conductive block poly(arylene ether)s synthesis, properties, and application in alkaline fuel cells. *Journal of the American Chemical Society* 2011, 133, 10646–10654.

33. Chen, D.; Hickner, M. A.; Wang, S.; Pan, J.; Xiao, M.; Meng, Y., Synthesis and characterization of quaternary ammonium functionalized fluorene-containing cardo polymers for potential anion exchange membrane water electrolyzer applications. *International Journal of Hydrogen Energy* 2012, *37*(21), 16168–16176.

34. Seo, D. W.; Hossain, M. A.; Lee, D. H.; Lim, Y. D.; Lee, S. H.; Lee, H. C.; Hong, T. W.; Kim, W. G., Anion conductive poly(arylene ether sulfone)s containing tetra-quaternary ammonium hydroxide on fluorenyl group for alkaline fuel cell application. *Electrochimica Acta* 2012, *86*, 360–365.

35. Hossain, M. A.; Lim, Y.; Lee, S.; Jang, H.; Choi, S.; Hong, T.; Jin, L.; Kim, W. G., Synthesis and characterization of tetra-imidazolium hydroxides poly(fluorenylene ether sulfone) anion exchange membranes. *Reactive and Functional Polymers* 2013, *73*(9), 1299–1305.

36. Rao, A. H. N.; Thankamony, R. L.; Kim, H.-J.; Nam, S.; Kim, T.-H., Imidazolium-functionalized poly(arylene ether sulfone) block copolymer as an anion exchange membrane for alkaline fuel cell. *Polymer* 2013, *54*(1), 111–119.

37. Zhao, Z.; Gong, F.; Zhang, S.; Li, S., Poly(arylene ether sulfone)s ionomers containing quaternized triptycene groups for alkaline fuel cell. *Journal of Power Sources* 2012, *218*, 368–374.

38. Li, X.; Yu, Y.; Liu, Q.; Meng, Y., Synthesis and properties of anion conductive ionomers containing tetraphenyl methane moieties. *ACS Applied Materials & Interfaces* 2012, 4, 3627–3635.

39. Hossain, M. A.; Lim, Y.; Lee, S.; Jang, H.; Choi, S.; Jeon, Y.; Lee, S.; Ju, H.; Kim, W. G., Anion conductive aromatic membrane of poly(tetra phenyl ether sulfone) containing hexa-imidazolium hydroxides for alkaline fuel cell application. *Solid State Ionics* 2013, *262*(1), 754–760.

40. Hossain, M. A.; Lim, Y.; Lee, S.; Jang, H.; Choi, S.; Jeon, Y.; Lim, J.; Kim, W. G., Comparison of alkaline fuel cell membranes of random & block poly(arylene ether sulfone) copolymers containing tetra quaternary ammonium hydroxides. *International Journal of Hydrogen Energy* 2014, 39(6), 2731–2739.

41. Hossain, M. A.; Lim, Y.-D.; Jang, H.-H.; Jeon, Y.-T.; Lim, J.-S.; Lee, S.-H.; Kim, W.-G.; Jeon, H.-S., Anion conductive aromatic ionomers containing a 1,2-dibenzoylbenzene moiety for alkaline fuel cell applications. *Electronic Materials Letters* 2013, 9(6), 797–799.

42. Seo, D. W.; Lim, Y. D.; Hossain, M. A.; Lee, S. H.; Lee, H. C.; Jang, H. H.; Choi, S. Y.; Kim, W. G., Anion conductive poly(tetraphenyl phthalazine ether sulfone) containing tetra quaternary ammonium hydroxide for alkaline fuel cell application. *International Journal of Hydrogen Energy* 2013, *38*(1), 579–587.

43. Rao, A. H. N.; Kim, H.-J.; Nam, S.; Kim, T.-H., Cardo poly(arylene ether sulfone) block copolymers with pendant imidazolium side chains as novel anion exchange membranes for direct methanol alkaline fuel cell. *Polymer* 2013, *54*(26), 6918–6928.

44. Kim, D. S.; Labouriau, A.; Guiver, M. D.; Kim, Y. S., Guanidinium-functionalized anion exchange polymer electrolytes via activated fluorophenyl-amine reaction. *Chemistry of Materials* 2011, *23*(17), 3795–3797.

45. Xing, B.; Savadogo, O., Hydrogen/oxygen polymer electrolyte membrane fuel cells (PEMFCs) based on alkaline-doped polybenzimidazole (PBI). *Electrochemistry Communications* 2000, *2*(10), 697–702.

46. Hou, H.; Sun, G.; He, R.; Sun, B.; Jin, W.; Liu, H.; Xin, Q., Alkali doped polybenzimidazole membrane for alkaline direct methanol fuel cell. *International Journal of Hydrogen Energy* 2008, 33(23), 7172–7126.

47. Hou, H.; Wang, S.; Liu, H.; Sun, L.; Jin, W.; Jing, M.; Jiang, L.; Sun, G., Synthesis and characterization of a new anion exchange membrane by a green and facile route. *International Journal of Hydrogen Energy* 2011, *36*(18), 11955–11960.

48. Luo, H.; Vaivars, G.; Agboola, B.; Mu, S.; Mathe, M., Anion exchange membrane based on alkali doped poly(2,5-benzimidazole) for fuel cell. *Solid State Ionics* 2012, *208*, 52–55.

49. Aili, D.; Hansen, M. K.; Renzaho, R. F.; Li, Q.; Christensen, E.; Jensen, J. O.; Bjerrum, N. J., Heterogeneous anion conducting membranes based on linear and crosslinked KOH doped polybenzimidazole for alkaline water electrolysis. *Journal of Membrane Science* 2013, *447*, 424–432.

50. An, L.; Zeng, L.; Zhao, T. S., An alkaline direct ethylene glycol fuel cell with an alkali-doped polybenzimidazole membrane. *International Journal of Hydrogen Energy* 2013, *38*(25), 10602–10606.

51. Thomas, O. D.; Soo, K. J. W. Y.; Peckham, T. J.; Kulkarni, M. P.; Holdcroft, S., A stable hydroxide-conducting polymer. *Journal of the American Chemical Society* 2012, *134*(26), 10753–10756.

52. Lee, H.-J.; Choi, J.; Han, J. Y.; Kim, H.-J.; Sung, Y.-E.; Kim, H.; Henkensmeier, D.; Ae Cho, E.; Jang, J. H.; Yoo, S. J., Synthesis and characterization of poly(benzimidazolium) membranes for anion exchange membrane fuel cells. *Polymer Bulletin* 2013, *70*(9), 2619–2631.

53. Henkensmeier, D.; Cho, H.; Brela, M.; Michalak, A.; Dyck, A.; Germer, W.; Duong, N. M. H.; Jang, J. H.; Kim, H.-J.; Woo, N.-S.; Lim, T.-H., Anion conducting polymers based on ether linked polybenzimidazole (PBI-OO). *International Journal of Hydrogen Energy* 2013, 39(6), 2842–2853.

54. Xia, Z.; Yuan, S.; Jiang, G.; Guo, X.; Fang, J.; Liu, L.; Qiao, J.; Yin, J., Polybenzimidazoles with pendant quaternary ammonium groups as potential anion exchange membranes for fuel cells. *Journal of Membrane Science* 2012, *390–391*, 152–159.

55. Geraldine Merle, A. C.; de Ven, E. V.; Nijmeijer, K., An easy method for the preparation of anion exchange membranes: Graft-polymerization of ionic liquids in porous supports. *Journal of Applied Polymer Science* 2013, 129, 1143–1150.

56. Tsuchida, E.; Yamamoto, K.; Miyatake, K.; Hara, H., Synthesis of a new series of polycations based on poly[oxy-2-[3-(trialkylammonio)propyl]-6-methyl-1,4-phenylene salts]. *Macromolecules* 1995, *28*(23), 7917–7923.

57. Tongwen, X.; Weihua, Y., Fundamental studies of a new series of anion exchange membranes: Membrane preparation and characterization. *Journal of Membrane Science* 2001, *190*, 159–166.

58. Tongwen, X.; Zha, F. F., Fundamental studies on a new series of anion exchange membranes: Effect of simultaneous amination-crosslinking processes on membranes ion-exchange capacity and dimensional stability. *Journal of Membrane Science* 2002, *199*, 203–210.

59. Li, Y.; Xu, T.; Gong, M., Fundamental studies of a new series of anion exchange membranes: Membranes prepared from bromomethylated poly(2,6-dimethyl-1,4-phenylene oxide) (BPPO) and pyridine. *Journal of Membrane Science* 2006, *279*(1–2), 200–208.

60. Li, Y.; Xu, T., Fundamental studies of a new series of anion exchange membranes: Membranes prepared from bromomethylated poly(2,6-dimethyl-1,4-phenylene oxide) and 4-vinylpyridine. *Journal of Applied Polymer Science* 2009, *114*(5), 3016–3025.

61. Wu, L.; Xu, T.; Yang, W., Fundamental studies of a new series of anion exchange membranes: Membranes prepared through chloroacetylation of poly(2,6-dimethyl-1,4-phenylene oxide) (PPO) followed by quaternary amination. *Journal of Membrane Science* 2006, *286*(1–2), 185–192.

62. Lin, X.; Wu, L.; Liu, Y.; Ong, A. L.; Poynton, S. D.; Varcoe, J. R.; Xu, T., Alkali resistant and conductive guanidinium-based anion-exchange membranes for alkaline polymer electrolyte fuel cells. *Journal of Power Sources* 2012, *217*, 373–380.

63. Jiang, L.; Lin, X.; Ran, J.; Li, C.; Wu, L.; Xu, T., Synthesis and properties of quaternary phosphonium-based anion exchange membrane for fuel cells. *Chinese Journal of Chemistry* 2012, *30*(9), 2241–2246.

64. Lin, X.; Liang, X.; Poynton, S. D.; Varcoe, J. R.; Ong, A. L.; Ran, J.; Li, Y.; Li, Q.; Xu, T., Novel alkaline anion exchange membranes containing pendant benzimidazolium groups for alkaline fuel cells. *Journal of Membrane Science* 2013, *443*, 193–200.

65. Wu, L.; Xu, T.; Wu, D.; Zheng, X., Preparation and characterization of CPPO/BPPO blend membranes for potential application in alkaline direct methanol fuel cell. *Journal of Membrane Science* 2008, *310*(1–2), 577–585.

66. Ong, A. L.; Saad, S.; Lan, R.; Goodfellow, R. J.; Tao, S., Anionic membrane and ionomer based on poly(2,6-dimethyl-1,4-phenylene oxide) for alkaline membrane fuel cells. *Journal of Power Sources* 2011, *196*(20), 8272–8279.

67. Liu, L.; Li, Q.; Dai, J.; Wang, H.; Jin, B.; Bai, R., A facile strategy for the synthesis of guanidinium-functionalized polymer as alkaline anion exchange membrane with improved alkaline stability. *Journal of Membrane Science* 2014, *453*, 52–60.

68. Li, N.; Guiver, M. D.; Binder, W. H., Towards high conductivity in anion-exchange membranes for alkaline fuel cells. *ChemSusChem* 2013, *6*(8), 1376–1383.

69. Li, N.; Leng, Y.; Hickner, M. A.; Wang, C. Y., Highly stable, anion conductive, comb-shaped copolymers for alkaline fuel cells. *Journal of the American Chemical Society* 2013, *135*(27), 10124–10133.

70. Hibbs, M. R.; Fujimoto, C. H.; Cornelius, C. J., Synthesis and characterization of poly(phenylene)-based anion exchange membranes for alkaline fuel cells. *Macromolecules* 2009, *42*(21), 8316–8321.

71. Hibbs, M. R., Alkaline stability of poly(phenylene)-based anion exchange membranes with various cations. *Journal of Polymer Science Part B: Polymer Physics* 2013, *51*(24), 1736–1742.

72. Wang, G.; Weng, Y.; Chu, D.; Xie, D.; Chen, R., Preparation of alkaline anion exchange membranes based on functional poly(ether-imide) polymers for potential fuel cell applications. *Journal of Membrane Science* 2009, *326*(1), 4–8.

73. Wang, G.; Weng, Y.; Zhao, J.; Chen, R.; Xie, D., Preparation of a functional poly(ether imide) membrane for potential alkaline fuel cell applications: Chloromethylation. *Journal of Applied Polymer Science* 2009, *112*(2), 721–727.

74. Kim, I. C.; Tak, T. M., Synthesis of soluble anion-exchange copolyimides and nanofiltration membrane performances. *Macromolecules* 2000, *33*(7), 2391–2396.

75. Hu, Q.; Shang, Y.; Wang, Y.; Xu, M.; Wang, S.; Xie, X.; Liu, Y.; Zhang, H.; Wang, J.; Mao, Z., Preparation and characterization of fluorinated poly(aryl ether oxadiazole)s anion exchange membranes based on imidazolium salts. *International Journal of Hydrogen Energy* 2012, *37*(17), 12659–12665.

76. Liu, G.; Shang, Y.; Xie, X.; Wang, S.; Wang, J.; Wang, Y.; Mao, Z., Synthesis and characterization of anion exchange membranes for alkaline direct methanol fuel cells. *International Journal of Hydrogen Energy* 2012, *37*(1), 848–853.

77. Li, C.; Wang, S.; Wang, W.; Xie, X.; Lv, Y.; Deng, C., A cross-linked fluorinated poly (aryl ether oxadiazole)s using a thermal cross-linking for anion exchange membranes. *International Journal of Hydrogen Energy* 2013, *38*(25), 11038–11044.

78. Wang, W.; Wang, S.; Li, W.; Xie, X.; lv, Y., Synthesis and characterization of a fluorinated cross-linked anion exchange membrane. *International Journal of Hydrogen Energy* 2013, *38*(25), 11045–11052.

79. Dewi, E. L.; Oyaizu, K.; Nishide, H.; Tsuchida, E., Cationic polysulfonium membrane as separator in zinc–air cell. *Journal of Power Sources* 2003, *115*(1), 149–152.

80. Ni, J.; Zhao, C.; Zhang, G.; Zhang, Y.; Wang, J.; Ma, W.; Liu, Z.; Na, H., Novel self-crosslinked poly(aryl ether sulfone) for high alkaline stable and fuel resistant alkaline anion exchange membranes. *Chemical Communications* 2011, *47*(31), 8943–8945.

81. Sun, H.; Zhang, G.; Liu, Z.; Zhang, N.; Zhang, L.; Ma, W.; Zhao, C.; Qi, D.; Li, G.; Na, H., Self-crosslinked alkaline electrolyte membranes based on quaternary ammonium poly(ether sulfone) for high-performance alkaline fuel cells. *International Journal of Hydrogen Energy* 2012, *37*(12), 9873–9881.

82. Xu, S.; Zhang, G.; Zhang, Y.; Zhao, C.; Zhang, L.; Li, M.; Wang, J.; Zhang, N.; Na, H., Cross-linked hydroxide conductive membranes with side chains for direct methanol fuel cell applications. *Journal of Materials Chemistry* 2012, *22*(26), 13295.

83. Tripathi, B. P.; Shahi, V. K., Organic–inorganic nanocomposite polymer electrolyte membranes for fuel cell applications. *Progress in Polymer Science* 2011, *36*(7), 945–979.

84. Li, X.; Yu, Y.; Meng, Y., Novel quaternized poly(arylene ether sulfone)/Nano-ZrO(2) composite anion exchange membranes for alkaline fuel cells. *ACS Applied Materials & Interfaces* 2013, *5*(4), 1414–1422.

85. Nonjola, P. T.; Mathe, M. K.; Modibedi, R. M., Chemical modification of polysulfone: Composite anionic exchange membrane with TiO_2 nano-particles. *International Journal of Hydrogen Energy* 2013, *38*(12), 5115–5121.

86. Vinodh, R.; Purushothaman, M.; Sangeetha, D., Novel quaternized polysulfone/ZrO_2 composite membranes for solid alkaline fuel cell applications. *International Journal of Hydrogen Energy* 2011, *36*(12), 7291–7302.

87. Wu, J.-F.; Lo, C.-F.; Li, L.-Y.; Li, H.-Y.; Chang, C.-M.; Liao, K.-S.; Hu, C.-C.; Liu, Y.-L.; Lue, S. J., Thermally stable polybenzimidazole/carbon nano-tube composites for alkaline direct methanol fuel cell applications. *Journal of Power Sources* 2014, *246*, 39–48.

88. Wu, Y.; Wu, C.; Varcoe, J. R.; Poynton, S. D.; Xu, T.; Fu, Y., Novel silica/poly(2,6-dimethyl-1,4-phenylene oxide) hybrid anion-exchange membranes for alkaline fuel cells: Effect of silica content and the single cell performance. *Journal of Power Sources* 2010, *195*(10), 3069–3076.

89. Wang, X.; Li, M.; Golding, B. T.; Sadeghi, M.; Cao, Y.; Yu, E. H.; Scott, K., A polytetrafluoroethylene-quaternary 1,4-diazabicyclo-[2.2.2]-octane polysulfone composite membrane for alkaline anion exchange membrane fuel cells. *International Journal of Hydrogen Energy* 2011, *36*(16), 10022–10026.

90. Yan, X.; Wang, Y.; He, G.; Hu, Z.; Wu, X.; Du, L., Hydroxide exchange composite membrane based on soluble quaternized polyetherimide for potential applications in fuel cells. *International Journal of Hydrogen Energy* 2013, *38*(19), 7964–7972.

91. Zhao, Y.; Pan, J.; Yu, H.; Yang, D.; Li, J.; Zhuang, L.; Shao, Z.; Yi, B., Quaternary ammonia polysulfone-PTFE composite alkaline anion exchange membrane for fuel cells application. *International Journal of Hydrogen Energy* 2013, *38*(4), 1983–1987.

92. Arges, C. G.; Wang, L.; Parrondo, J.; Ramani, V., Best practices for investigating anion exchange membrane suitability for alkaline electrochemical devices: Case study using quaternary ammonium poly(2,6-dimethyl 1,4-phenylene)oxide anion exchange membranes. *Journal of the Electrochemical Society* 2013, *160*(11), F1258–F1274.

93. Chempath, S.; Einsla, B. R.; Pratt, L. R.; Macomber, C. S.; Boncella, J. M.; Rau, J. A.; Pivovar, B. S., Mechanism of tetraalkylammonium headgroup degradation in alkaline fuel cell membranes. *The Journal of Physical Chemistry C* 2008, *112*(9), 3179–3182.

94. Chempath, S.; Boncella, J. M.; Pratt, L. R.; Henson, N.; Pivovar, B. S., Density functional theory study of degradation of tetraalkylammonium hydroxides. *The Journal of Physical Chemistry C* 2010, *114*(27), 11977–11983.

95. Gubler, L.; Dockheer, S. M.; Koppenol, W. H., Radical (HO·, H· and HOO·) formation and ionomer degradation in polymer electrolyte fuel cells. *Journal of the Electrochemical Society* 2011, *158*(7), B755–B769.
96. Cope, A. C.; Mehta, A. S., Mechanism of the Hofmann elimination reaction: An ylide intermediate in the pyrolysis of a highly branched quaternary hydroxide. *Journal of the American Chemical Society* 1963, *85*(13), 1949–1952.

12 Theoretical Modeling of Polymer Electrolyte Membranes

Su Zhou, Fenglai Pei, and Jeffrey De Lile

CONTENTS

12.1   INTRODUCTION

As an alternative power convertor, proton exchange membrane fuel cell (PEMFC) has been paid more and more attention in many applications. Using hydrogen and air (oxygen) as reactants, the device directly converts chemical energy into electrical energy through electrode reactions. The core component of PEMFC is membrane electrode assembly (commonly used as MEA in bibliography), which puts two carbon fiber papers with spraying Nafion® solution and Pt catalyst on both sides of polymer electrolyte membrane (PEM). The catalyst is close to the PEM.

This chapter will focus on the modeling of MEA and its polymer electrolyte membrane. First, 3D modeling of PEMFC and its MEA will be discussed, and an example will be put forward. Then, dynamic modeling of PEM will be introduced. Further, this chapter will move on to the fault-embedded modeling of PEM. As an extension, application of membranes in other cases will be recommended, such as in lithium battery, vanadium redox flow battery (VRFB), chlor-alkali electrolysis, water electrolysis, and solar cell. Finally, several typical examples will be given, including Pt and Pt alloy simulation with density functional theory (DFT), water formation and Pt adsorption on carbon reactive force field (ReaxFF) simulation, and coarse-grained simulations.

12.2   3D MODELING OF PEM

12.2.1   Introduction on Dimension of PEMFC and MEA

According to dimension considered, PEMFC mathematic models can be divided into four types, that is, 0D model, 1D model, 2D model, and 3D model.

Zero-dimensional model is a concentrated parameter model and can be simulated conveniently. Thus it is commonly applied in the cases where only dynamic behaviors of a PEMFC are dealt with.

While in 1D model, Darcy and Schlögl equations are usually chosen in order to describe the momentum transfer in multiporous layer. As to the formation of ion and chemical components transferred in the membrane, Nernst–Planck and Stefan–Maxwell equation are simplified and applied. Butler–Volmer equation is responsible to the electrode reaction dynamics and current density. The pressure and voltage gradient in membrane is taken as constant. While modeling, reaction mechanisms and conservation equations vertical to flux can be established. In the meantime, parameters in other directions are assumed as invariable.

Two-dimensional model usually considers reaction and transfer both parallel and vertical to flux. It is used to explain pivotal effects in the reactant transfer and also to simulate the quantities and variations of reactants and products. Two-dimensional model includes two structures, parallel and vertical to the flux. These structures have their own advantages and drawbacks. The former considers flow and transfer in gas and liquid phase in manifolds but not in gas diffusion layer (GDL) and catalyst layer behind the ribs. The latter mainly deals with the zone corresponding to the ribs, regardless of the flow and mass transfer of either gas or liquid.

Compared to the previous models, 3D model is one step further, which can monitor important variables, such as temperature, pressure, and humidity, within a stack in 3D and reflect mass transfer and reaction condition. When building a 3D model, powerful commercial computing software plays an important role.

Recently, more and more papers [1–5] started to pay attention to modeling of stacks. Different from 0D models, parameter distributed models consider also spatial effects. The correlations between each single piece are obviously significant since it somehow changes the character of their companions. Among them, several novel approaches of FEM simulation have attracted much of our sight. Experts from Japan, South Korea, Germany, and the United States are focused to the panorama of PEMFCs according to delicate experiments. This is large—partly due to PEMFC applications. It also surely accelerated R&D on PEMFCs in the past decades. The combination of various disciplines can be noticed and promoted by the phenomenon. Advanced modeling methods, for instance, genetic algorithm and optimum algorithm, are established to describe and simulate PEMFCs or PEMs more precisely. At the same time, parameter identification on the base of accurate model is urgently demanded by scientists and engineers. All the aforementioned cases reflect the significance of proper and reasonable models, especially 3D ones. Thanks to the development of the large computation tools, much of the complex and onerous computation can be realized. Fluent and Matrix laboratories are commonly applied in R&D activities.

PEMFC is a highly integrated system, and its function is governed by various disciplines including electrochemistry, mechanical engineering, and polymer chemistry. The MEA can be regarded as PEMFC's heart and soul, its most common form [6] is shown in Figure 12.1. An MEA includes seven components (layers), that is, a PEM, anode and cathode catalyst layers, two GDLs, and two sets of sealing gaskets. An MEA is definitely something more than just a collection of these components. One trifle change in one single layer will obviously influence the performance of the others. In the following content, details of establishing a model in terms of each part in MEA will be introduced and discussed.

12.2.2 MODELING OF MEA

PEM is a significant part in the structure of MEA. Electrons and protons are generated on the surface of the catalyst layer. The electrons travel through the external circuit to a load and do electrical work on the load. The hydrogen protons moving through the electrolyte to cathode are combined with oxygen to form water [1]. To function effectively, the electrolyte has to satisfy the following conditions:

- High ionic conductivity and low electronic conductivity
- Adequate barrier to the reactants
- Chemically and mechanically stable
- Ease of manufacturability and broad availability
- Relatively low cost

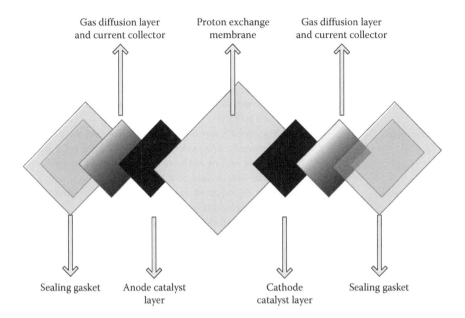

FIGURE 12.1   Layout of MEA.

TABLE 12.1
Governing Equations to Establish the Model of a PEM

Model Features	Equations and Description
Dimension	1, 2, or 3
Model of operation	Steady state or dynamic
Phases	Gas, liquid, or biphase
Mass transport	Nernst equation, drag coefficient, or Stefan–Maxwell equation
Iron transport	Ohm's law
Membrane swelling	Empirical or thermodynamic models
Energy balance	Isothermal or full energy balance

Table 12.1 describes the governing equations related to PEM modeling, and Table 12.2 lists the possible types of models.

12.2.2.1   Modeling of Gas Diffusion Layers

The GDL is located between the catalyst layer and the bipolar plates. In a PEMFC, an MEA consists of a membrane, catalyst layers, and GDLs. The GDLs are the outermost layers and provide electrical contacts between electrodes and bipolar plates. In the meantime, GDLs offer path for water in the electrodes and enable

TABLE 12.2
Types of PEM Models

Type of Model	Description
Microscopic and physical models	Providing a fundamental understanding of many membrane processes such as diffusion and conduction in the membrane on a pore level
Diffusive models	Regarding the membrane as a single phase and assuming that the membrane is vapor equilibrated, where the water and protons dissolve and move by diffusion under the condition of dilute and concentrated solution
Hydraulic models	Assuming the membrane layer has two parts, which are the membrane and liquid water
Hydraulic–diffusive models	Considering both diffusion and pressure-driven flow
Combined models	Combining all the aforementioned types and being suitable for the cases when membrane is either saturated or dehydrated with water

water generated in reaction to exit the electrode surface. This can be concluded as the following:

- Electronic conductivity
- Mechanical support for the proton exchange membrane
- Porous media for the catalyst to cling
- Reactant access to the catalyst layers
- Product removal

The equations used to model GDLs and the related description can be summarized in Table 12.3.

Apparently, comparing to PEM modeling, the governing equations of GDL modeling is less complex. There are abundant porous media models introduced in the literature, where the quantity of the GDL model treats these layers rigorously.

TABLE 12.3
Equations of Modeling GDLs

Model Character	Equations and Description
Dimension	1, 2, or 3
Mode of operation	Steady state or dynamic
Phase	Gas, liquid, or biphase
Mass transport	Nernst–Plank equation, drag coefficient, or Stefan–Maxwell equation
Ion transport	Ohm's law
Energy balance	Isothermal or full energy balance

TABLE 12.4
Types of GDL Models

Types of Model	Description
Liquid phase models	Assuming only liquid phase flow
Gas phase models	Assuming only gas flow
Biphase flow models	Describing how the gas and liquid interact in a porous media

TABLE 12.5
Variables Used in GDL Modeling

Variables	Equation
Overall liquid water flux	Mass balance
Overall membrane water flux	Mass balance
Gas phase component flux	Mass balance
Gas phase component partial pressure	Stefan–Maxwell equation
Liquid pressure	Darcy's law
Electronic phase current density	Ohm's law
Electronic phase potential	Difference balance
Temperature	Energy balance
Total gas pressure	Darcy's law
Liquid saturation	Saturation relation

The flow through the substrate and the interaction of the solid substrate with the molecules are two main aspects in the modeling of GDLs [7]. Fick's diffusion law and Stefan–Maxwell equation can be used in the modeling process. The interaction of the gas and solid, known as the dusty gas model, is due to the application of the kinetic theory to the interaction of the gas–gas and the solid–gas molecule. These models take either gas or liquid phase, or sometimes both. The following table illustrates the types of GDL models described in the literature (Table 12.4).

Besides gas and liquid transportation, GDLs have other properties such as electronic conductivity and evaporation/condensation ability of gas/liquid. The corresponding variables used in GDL modeling are shown in Table 12.5.

12.2.2.2 Modeling of Catalyst Layers

In PEMFCs, electrode layers are the place where electrochemical reactions take place, made up of catalyst and GDL. At the anode, the hydrogen is broken into protons and electrons. The electrons firstly travel to the carbon cloth, flow field plate followed by the contact, and then to the load. The protons travel through the polymer exchange membrane to the cathode. At the cathode catalyst layer, oxygen combines with protons and electrons, and then water is produced. The catalyst layer must get very effective when breaking molecules into protons and electrons with large

TABLE 12.6
Catalyst Layer Modeling Methods

Model Characteristic	Equations and Description
Dimension	1, 2, or 3
Model of operation	Steady state or dynamic
Phases	Gas, liquid, or biphase
Kinetics	Tafel-type expressions, Butler–Volmer equation, or complex kinetic equations
Mass transport	Nernst–Planck equation, drag coefficient, or Stefan–Maxwell equation
Ion transport	Ohm's law
Membrane swelling	Empirical or thermodynamic models
Energy balance	Isothermal or full energy balance

surface area. The catalyst layers are the most complex and the thinnest, due to the multiple phases, porosity, and electrochemical reactions [8]. It is still difficult to find an optional solution to choose a low-cost and effective catalyst.

Some approaches in modeling catalyst layers are illustrated in Table 12.6. The approach depends on the other part of the modeled fuel cell. The catalyst layers can be modeled both microscopically and macroscopically. The microscopic ones consist of pore-level models and quantum models. The quantum models cope with detailed reaction mechanism, elementary transfer reactions, and transition states.

12.2.2.3 Modeling of Flow Field Plates

Single cells must be placed in the stack to distribute fuel and oxidant, and getting together the current to power up the demanded loads (devices). There is no need to consider bipolar plates for a single-cell modeling. However, in the stack, bipolar plates become necessary and have to be modeled. The roles of bipolar plates are manifold. They distribute fuel and oxidant within the single cell, separate the individual cells in the stack, collect current, transport water, humidify gases, and keep the cell's temperature at a suitable level [9]. In order to simultaneously perform these functions, specific plate materials and designs are required. The structures of flow field include straight, serpentine, parallel, and pin type.

This part of modeling is specifically correlated to the following topics:

- Flow field plate materials
- Flow field design
- Channel shape, dimension, and spacing
- Pressure drop in flow channels
- Heat transfer from plate channels to gas

12.2.2.4 Types of PEM Models

Many fuel cell models are single phased. The membrane system consists of three main components including membrane, protons, and water. Thus, naturally speaking, there are three main transport properties. The hydrogen or oxygen crossover is not considered in different models because of its complex effect on the mass transportation and overall cell performance. In this section, several types of models shown in Table 12.2 will be discussed and analyzed.

12.2.2.4.1 Microscopic and Physical Models

Numerous models are based on statistical mechanics, molecular dynamics, and other types of macroscopic phenomena. These models are valuable because they provide a fundamental understanding of behaviors of related species and of conduction through different proton–water complexes. Almost all microscopic models treat the membrane as a two-phase system. Although these models provide valuable information, they are usually too complex to be integrated into an overall fuel cell model. How the membrane structure changes as a function of water content is still under investigation and unclear.

12.2.2.4.2 Diffusive Models

Diffusive membrane models treat the membrane as a single phased one. Typically, this type of system has no true pores, with fluctuating collapsed channels. Indeed, the system is treated as a single, homogeneous phase where protons dissolve and move by diffusion. The proton movement obeys Ohm's law $i = -k\nabla\Phi$, where k is the ionic conductivity of the membrane. The process can be seen as a resistance, which is used in a polarization equation in a 0D model.

12.2.2.4.3 Dilute Solution Theory

The Nernst–Planck equation is

$$N_i = -z_i \frac{F}{RT} D_i C_i \frac{d\Phi_m}{dx} - D_i \frac{dC_i}{dx} + vC_i \qquad (12.1)$$

where
  N_i is the superficial flux of species i
  z_i is the charge number of species i
  C_i is the concentration of species i
  D_i is the diffusion coefficient of species i
  Φ_m is the electrical potential in the membrane
  v is the velocity of H_2O

The first term is a migration term—representing the motion of charged species that result from a potential gradient. The migration flux is related to the potential gradient by a charge number, z_i; concentration, c_i; and mobility, u_i. The second term relates the diffusive flux to the concentration gradient. The last term is a convective term and represents the motion of the species as the bulk motion of the solvent.

While in the single phase systems, the solvent is assumed to be the membrane. Dilution solution theory only considers the interactions between each dissolved species and the solvent. The motion of each charged species is described by the transportation properties, which are the mobility and the diffusion coefficient. They are related by the Nernst–Einstein equation:

$$D_i = RTu_i \tag{12.2}$$

If the solution species are very dilute, then the interactions can be neglected— and just the material balance can be used. If water movement in the membrane is considered, the Nernst–Planck equation will also be needed. As the protons move across the membrane, they induce a flow of water in the same direction. This electroosmotic flow is a result of the proton–water interaction and is not a dilute solution effect because the membrane is regarded as the solvent [10]. This obeys the equation

$$N_w = \xi \frac{i_2}{F} - D_w \nabla c_w \tag{12.3}$$

where ξ is the electroosmotic coefficient. Most of the single-phase models use equation related to Ohm's law.

12.2.2.4.4   Concentrated Solution Theory

Concentration solution theory can be used when an electrolyte is modeled with three species. This model is interpreted as the binary interactions between all of the species. The equations for the three species system are

$$i_2 = -\frac{\kappa \xi}{F} \nabla \mu_{w,2} - \kappa \nabla \Phi_2 \tag{12.4}$$

and

$$N_{w,2} = \xi \frac{i_2}{F} - \alpha_w \nabla \mu_{w,2} \tag{12.5}$$

Here, μ_w represents the chemical potential of water, and α_w is the transport coefficient of water. The equation for the membrane is ignored, because it is dependent on the other two of Gibbs–Duhem of equations. From many models in the literatures, these equations were used in a Stefan–Maxwell framework.

12.2.2.4.5   Hydraulic Models

Many models assume the membrane system to be two phased. This is accomplished by assuming the membrane has pores that are filled with liquid water. The two phases are water and membrane. The additional degree of freedom allows the inclusion of a pressure gradient in the water because of a possibly unknown stress relation between the membrane and fluid at every point in the membrane. Most of these models assume the water to be pure. The water content of membrane is assumed to stay constant while the pores are filled and the membrane has been pre-treated appropriately. The first model ever to describe the membrane in the manner was introduced by Bernardi and Verbrugge. They use Nernst–Planck equation to describe the movement of protons. The velocity of the water is given by Schlogl's equation:

$$v_{w,2} = -\left(\frac{k}{\mu}\right)\nabla p_L - \left(\frac{k_\Phi}{\mu}\right)z_f c_f F \nabla \Phi_2 \qquad (12.6)$$

where k and k_Φ are effective hydraulic and electrokinetic permeability, respectively.

12.2.2.4.6 Combination Models
In order to describe diffusive and hydraulic behavior in consistent manner, the char-acters of the two types must agree with experiment data. For example, a membrane with low water content is expected to be controlled by diffusion, where an uptake isotherm needs to be used. This is due to the fact that the membrane matrix interacts significantly with the water due to binding and solvating on the sulfonic acid sites. A hydraulic pressure in this system may not be defined. On the other hand, when the membrane is saturated, transportation still occurs. This transportation must be due to a hydraulic-pressure gradient since oversaturated activities are nonphysical. A model that combines the concepts from both diffusive and hydraulic models would most accurately describe the membrane system. The two types of models are seen as fully operating at the limits of water concentration and must somehow be averaged between the limits. As mentioned, the hydraulic/diffusive models follow the same pattern, but Schroeder's paradox and its effects on the transportation properties are not considered.

12.2.3   PEM MODELING EXAMPLES

In order to model the electrolyte properly, the transportation of mass, charge, and energy must be included in the model. Contact resistance between electrode and electrolyte can also be significant, which should be incorporated into the model.

12.2.3.1   Mass and Species Conservation
For both water and protons, the mass conservation equation can be represented as

$$\frac{\delta c_i}{\delta x} = -\frac{\delta}{\delta t} N_i \qquad (12.7)$$

where
   i is H_2O or H^+
   c_i is the molar concentration
   N_i is the molar flux due to electroosmotic driving forces and convection

In a diluted solution, N_i is given by Nernst–Planck equation

$$N_i = J_i + c_i u^m \tag{12.8}$$

and u^m is the mixture velocity and J_i is the diffusive flux.

In PEMFCs, two important fluxes or material balances are the proton flux and the water flux. The membrane needs to stay hydrated in order to conduct with hydrogen; therefore, the water profile must be calculated in the electrolyte. In Nafion membrane, two types of water fluxes are present: back diffusion and electroosmotic drag [10]. Both of the fluxes are accounted by the following equation:

$$J_{H_2O}^M = -D_{cH_2O,T} \frac{\delta c_{H_2O}^m}{\delta x} + n_{drag} \frac{i_x}{F} \tag{12.9}$$

where

n_{drag} is the measured drag coefficient
i_x is the proton current in the x-direction
F is the Faraday's constant
λ_{H_2O/SO_3} is the water content $\{/\}$
ρ_{dry}^m is the dry membrane density (kg/m$_3$)
$D_{cH_2O,T}$ is the diffusion coefficient
M^m is the membrane molecular mass (kg/mol)

The water content is not constant in this equation. The resistance of the electrolyte can be estimated using the water content, which can be described as

$$n_{drag} = 2.5 \frac{\lambda_{H_2O/SO_3}}{22} \tag{12.10}$$

$$\lambda_{H_2O/SO_3} = \frac{c_{H_2O}^m}{(\rho_{dry}^m / M^m) - b c_{H_2O}^m} \tag{12.11}$$

where b is the membrane extension coefficient in x-direction and it is determined by experiment. The value of $b=0.0126$ is typically used. The diffusion coefficient $D_{cH_2O,T}$ includes a correction for the temperature and water content. It is expressed in a fixed coordinate system with the dry membrane as

$$D_{cH_2O,T} = D' \left\{ e^{2416((1/303)-(1/T))} \right\} \lambda_{H_2O/SO_3} \frac{1}{a} \frac{1}{17.81 - 78.9a + 108a^2} \tag{12.12}$$

where

a is the activity of water
$D'\{m^2/s\}$ is the diffusion coefficient moving with the swelling of the membrane at constant temperature

$e^{2416((1/303)-(1/T))}$ is added to the aforementioned equation to ensure that even the water contents are below 1.23, $D_{cH_2O,T}$ will not result in negative value. D' is written as

$$D' = 2.64227e^{-13}, \lambda_{H_2O/SO_3}, \quad \lambda_{H_2O/SO_3} \leq 1.23$$

$$D' = 7.75e^{-11}\lambda_{H_2O/SO_3} - 9.5e^{-11} 1.23 < \lambda_{H_2O/SO_3} \leq 6$$

$$D' = 2.5625e^{-11}\lambda_{H_2O/SO_3} + 2.1625e^{-10} 6 < \lambda_{H_2O/SO_3} \leq 14$$

The total molar flux for water can be expressed as

$$N_{H_2O} = J_{H_2O} + \left(c_{H_2O}^m u^m \right) \tag{12.13}$$

where the mixture velocity u^m is given by the momentum equation.

Now, due to the assumption of electroneutrality and the homogeneous distribution of charge sites, the mass conservation of protons is simplified to

$$\frac{\delta c_{H^+}}{\delta x} = 0, \frac{\delta c_{H^+}}{\delta t} = 0 \tag{12.14}$$

Therefore, as soon as a current exists, the membrane will be charged and the proton concentration remains constant. The charge of protons equals to the fixed charges. The diffusive molar flux for the protons can be written as

$$J_{H^+} = -\frac{F}{RT} D_{H^+} c_{H^+} \frac{\delta \phi_m}{\delta x} \tag{12.15}$$

where
  ϕ_m is the membrane proton potential
  D_{H^+} is the proton diffusivity

Combining this diffusive flux with the convective flux results in the total molar flux of hydrogen protons:

$$N_{H^+} = J_{H^+} + \left(c_{H^+}^m u^m \right) \tag{12.16}$$

12.2.3.2 Momentum Equation

For the mixture (water and protons), the momentum equation is assumed to take the form of generalized Darcy relation:

$$u^m = -\frac{Kk_r^g}{\mu}\left[\frac{\partial p}{\partial x} - \rho g \cos\theta\right] \tag{12.17}$$

where
  u^m is the mixture velocity
  K is the absolute permeability of the porous medium
  k_r^g is the relative permeability
  g is the gravity
  θ is the angle between the x-axis (the direction of flow) and the direction of gravity

The mixture density and dynamic viscosity are written as

$$\rho = M_{H^+}c_{H^+} + M_{H^2O}c_{H^2O} \tag{12.18}$$

$$\mu = \frac{M_{H^+}c_{H^+}}{\rho}\mu_{H^+} + \frac{M_{H^2O}c_{H^2O}}{\rho}\mu_{H_2O} \tag{12.19}$$

12.2.3.3 Conservation of Energy Equation

Energy is transported by conduction and convection within three phases of the membrane (polymer and liquid/gas). The effects of ohmic losses within the membrane are taken into account by an additional source term in the energy balance equation. The energy conservation is given by

$$\rho c_p \frac{\partial T}{\partial t} = \lambda_m \frac{\partial^2 T}{\partial x^2} - Mc_p N \frac{\partial T}{\partial x} + R_m \tag{12.20}$$

where

$$\rho c_p = \rho_m^{dry} c_{pm} + \rho_{H_2O}^m c_{p,H_2O}^m + \rho_{H^+} c_{pH^+} \tag{12.21}$$

$$\rho_{H^+} = M_{H^+}c_{H^+}\rho_{H_2O}^m = M_{H_2O}c_{H_2O}^m \tag{12.22}$$

$$Mc_p N = M_{H^+}c_{p,H^+}N_{H^+} + M_{H_2O}c_{p,H_2O}^m N_{H_2O} \tag{12.23}$$

The transient energy effect associated with mass storage within the hydrated membrane is neglected due to the fact that dry membrane mass does not change. In addition, it is several orders of magnitude larger than water that hydrates the membrane.

Substituting the expressions of N_{H_2O} and N_{H^+}, an expanded expression for $Mc_p N$ can be obtained. The source term R_m is given by $R_m = i^2/\sigma_m$, where σ_m is the conductivity of membrane, written as a function of the temperature and water content:

$$\sigma_m = \sigma_{m303}e^{1268\left((1/303)-(1/T)\right)} \tag{12.24}$$

The conductivity of the membrane σ_{m303} at 303 K is given by

$$\sigma_{m303} = 100\left(0.005139\lambda_{H_2O/SO_3} - 0.00326\right), \lambda_{H_2O/SO_3} > 1 \qquad (12.25)$$

12.2.3.4 Ion Transportation and Interface Water Activity Relation

The equation for the proton potential is derived from Ohm's law. It represents the proton flux divided by the membrane conductivity. The electroneutrality assumption allows the total molar proton flux to be related directly to the current density and velocity that represents the convective flux of protons. This results in the following equation:

$$\frac{\partial \Phi_m}{\partial x} = -\frac{i}{\sigma_m} + \frac{F}{\sigma_m} c_{H^+} u^m \qquad (12.26)$$

At the membrane interfaces, the water vapor activity is given by

$$a = \frac{RT}{p_{sat(T)}} c_{H_2O}^g + 2s, \quad a \in [0,...,3] \qquad (12.27)$$

where
$c_{H_2O}^g$ is the water vapor concentration
s is the saturation ratio

An assumption is made that s is zero for activities less than 1, meaning that no liquid water is present in the membrane pores until the activity exceeds 1. The maximal value of the first term is 1. Therefore, a maximum saturation ratio of 1 results in an activity of 3.

12.2.4 BRIEF INTRODUCTION OF CASES

According to the earlier content, we used several types of computational software to solve numerous research issues. Admittedly, the established models cannot be universally applied on each specific problem. Some selected works (results) made by our group in the past few years will be introduced in the following.

ANSYS Fluent software contains the broad physical modeling capabilities required to model flow, turbulence, heat transfer, and reactions for industrial applications, ranging from air flow over an aircraft wing to combustion in a furnace, bubble columns to oil platforms, blood flow to semiconductor manufacturing, and clean room design to wastewater treatment plants. In addition, special models have the ability to modeling in-cylinder combustion, aeroacoustics, turbomachinery, and multiphase systems. GAMBIT is the abbreviation of the six words—Geometry and Mesh Building Intelligence Toolkit. It is built to help analyze and design meshed computing fluid dynamics.

A particular case of PEMFC stack is shown as follows, whose model can be typically used to analyze the basic element of the membrane, for instance, the current

FIGURE 12.2 A picture of fuel cell in Fluent.

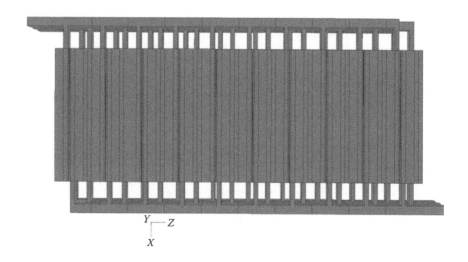

FIGURE 12.3 A picture of stake model in Fluent.

flux and water content. A 3D model in Fluent, which includes 10 single cells, is shown in Figures 12.2 and 12.3.

The dimension of the fuel cell is $2.4 \times 2.88 \times 125$ mm in x-, y-, and z-directions, respectively. This model represents a repeating channel of one larger counterflow PEMFC. The cross-sectional area of the MEA is 300 mm² (2.4 mm × 125 mm).

Aside from the picture, geometrical and operating parameters of the stack (which contains 10 pieces of fuel cells) are shown in Table 12.7.

TABLE 12.7

Geometrical and Operating Parameters

Quantity	Value
Gas channel depth	2.0 mm
Gas channel and shoulder width	2.0 mm
GDL thickness	0.3 mm
Catalyst layer thickness	0.01 mm
Number of fuel cell in the stack	10
Porosity of GDL/catalyst layer	0.8/0.6
Thermal conductivity of membrane	2 W/(m·K)
Thermal conductivity of catalyst layers	8 W/(m·K)
Thermal conductivity of GDL	8 W/(m·K)
Thermal conductivity of bipolar pressure	220 W/(m·K)
Permeability of GDL	10^{-12} m^2
Electronic conductivity in the GDL/land	5,000/1,000,000/Ωm
Anode/cathode inlet pressure	1.0/1.0 atm
Stoichiometry	3.0/3.0
Inlet temperature of gas flow	353 K
Inlet temperature of coolant	333 K
Inlet humidity at both sides	40%
Open circuit voltage	9.5 V
Stack output voltage	6 V
Reference current	1.8 A

Here, we consider three species, namely, H_2, O_2, and H_2O. Therefore, the second value in each list is the calculated oxygen mass flow rate in kilograms per second. In addition, a negative number indicates that there is flow leaving from that boundary. In the last line, the net oxygen consumption is 8.1×10^{-8} kg/s. The molecular weight of oxygen is 32 kg/kmol. Also, since the valence of a diatomic oxygen molecule is 4, there are 4 kmol of electrons released per kilo mole of oxygen. Finally, Faraday's constant is 9.7×10^7 C/kmol-electrons. Thus, the total release of electrons (which is equivalent to the current in amperes) is

$$i = \frac{\dot{m}vF}{M} = \frac{(8.1 \times 10^{-8}) \times (4.0) \times (9.7 \times 10^7)}{32} A = 0.98 \text{ A} \qquad (12.28)$$

The current flux of both the cathode and anode under certain boundary condition is shown in Figures 12.4 and 12.5. These two graphs show that on the corner of each fuel cell, current density is more concentrated, while the position of the high occupied density is different.

The water content of the membrane is shown in Figure 12.6, which is much different from the current density. It is affected by the distance from the cathode. The closer the position to the cathode, the more water content presented. This can be

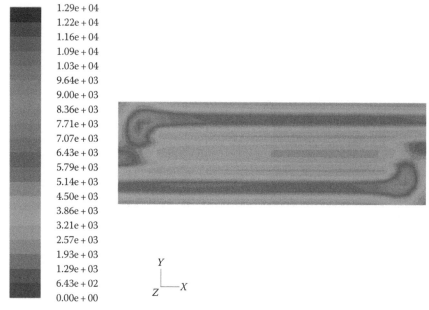

1.29e + 04
1.22e + 04
1.16e + 04
1.09e + 04
1.03e + 04
9.64e + 03
9.00e + 03
8.36e + 03
7.71e + 03
7.07e + 03
6.43e + 03
5.79e + 03
5.14e + 03
4.50e + 03
3.86e + 03
3.21e + 03
2.57e + 03
1.93e + 03
1.29e + 03
6.43e + 02
0.00e + 00

Contours of current flux density magnitude (A/m^2)

FIGURE 12.4   Current flux simulation result on the anode in Fluent.

postulated from the chemical reaction. In contrast to the cathode, the water in anode is more concentrated near the bipolar plate. This leads to dehydration of the membrane at the anode.

12.3   DYNAMICAL MODELING OF PEMs

12.3.1   INTRODUCTION

A system model is used to describe, imitate, or abstract a system. Its modeling follows the concept of identifying certain characteristics or behavior of the system, entirely depending on R&D purposes. Therefore, different models related to a certain system will be set up for corresponding purposes.

As previously described, the membrane system is assumed to have three main components: membrane, protons, and water. In addition, the types of fuel cell membrane models stated in literature include microscopic and physical models, diffusive models, hydraulic models, hydraulic–diffusive models, and combination models [10].

Models can be divided into static and dynamic. This is known as an object at any moment and how this information changes dynamically with events [11]. This representation can provide a different perspective of the same system. Steady-state model shows the structure of the system to be developed. Steady state is just one initial condition for a dynamic model: always get steady-state model for free in dynamic study.

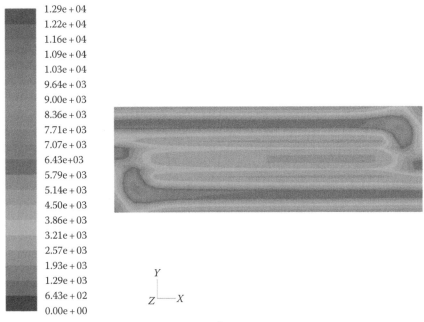

1.29e + 04
1.22e + 04
1.16e + 04
1.09e + 04
1.03e + 04
9.64e + 03
9.00e + 03
8.36e + 03
7.71e + 03
7.07e + 03
6.43e+03
5.79e + 03
5.14e + 03
4.50e + 03
3.86e + 03
3.21e + 03
2.57e + 03
1.93e + 03
1.29e + 03
6.43e + 02
0.00e + 00

Contours of current flux density magnitude (A/m^2)

FIGURE 12.5   Current flux simulation result on the cathode in Fluent.

It optimizes plant for different feed concentrations and loads. It identifies which equipment must be revamped to increase the capacity. The dynamic model is used to express and model the behavior of the system over time, such as describing the system transition from one state to another. Dynamic model describes the operating time and the sequence of system characteristics related to the change of events, the sequence of events, the environment, and the organization of events. When a rapid change with time is measured, the relationship between the input and the output is called dynamic characteristics that can be expressed by differential equations. Dynamic models are sensitivity studies to identify necessary processing measurements. Sensitivity study is to define appropriate control structure (matching of sensors to control elements) and tuning control parameters. It is the validation of the control system by predicting plant response to a variety of load and set point changes. Dynamic simulations are used for (1) identifying experiments to determine transfer function models for control design with the goal to regulate the outputs and (2) validating the controller with the test in a wide variety of scenarios before actual implementation in plant.

As an important part of the PEMFC, PEM is used to separate the cathode and anode reaction gas, while the proton is conducting from the anode to the cathode. Good performance of the membrane is a significant property of the film that largely depends on the water content. Proper control of the water distribution can significantly improve the fuel cell performance. Many researchers regard water transportation phenomena in PEMFC as a research focus. A water diffusion model [12,13] of

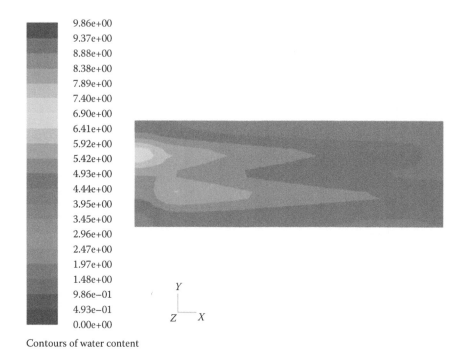

Contours of water content

FIGURE 12.6 Water content distribution of the membrane.

Nafion membrane was proposed, assuming the water vapor and the liquid water are balanced in the film. The water content was then calculated according to the activities on the film surface. Hsing and Futerku [14] created a 2D finite element model in the absence of external humidification. On the assumption that the cathode side is fully wet while liquid and vapor water on the anode side reaches equilibrium, the water content of anode/membrane surface can be calculated. Kulikovsky [15] established a 3D water transfer model. In this model, the diffusion coefficient of water and electroosmotic drag coefficient of the PEM is considered to be in a nonlinear relationship with the water content of the membrane. The aforementioned description is limited to literature steady behavior. Yan et al. [16] studied the dynamic process of passing water obtained under different conditions to reach steady state and the water content of the film. Wang Yun and Wang Chao-Yang [17] pointed out that the dynamic response is associated with the electric double-layer charge and discharge, gas/mass transfer, and the hydration of the film. The hydration of the film takes about 10s to reach equilibrium. The gas mass transfer takes about 0.01–0.15s, and the electric double-layer charge and discharge is much faster. Loo et al. [18] found that when the current step increases, due to the redistribution of water in the membrane lead to a time delay, voltage comes down. The response time is 50s. When flooding occurs, the voltage increment drops. The response time is extended to 150–200s. The study found the main impact of different operating conditions and the detail of the increase level down without changing the overall behavior. Qu et al. [19] show that it takes 20vs for PEMFC's potential load to reach a new steady state. The redistribution of water in the PEMFC depends on electroosmotic drag of water in PEM and back diffusion homeostasis. The dynamic process of PEMFCs depends on passing water in PEM. Shen and Li [20] established a dynamic model of PEM water transfer that investigates water transfer in the membrane's influence of overshoot phenomenon and its impact on the ohmic polarization. The authors draw the conclusion that the current density step change causes membrane water content overshoot phenomenon. Shan-Hai et al. [21] considered electromigration, stress migration, and proliferation of poor water transfer to systematically study the effect of cell temperature, humidification, and water transportation and distribution in the membrane. Currently, Zhou et al. [22] established a half mechanism and semiempirical model that reflects the distribution of stack temperature. It considers the difference between the stack response gas concentration, the water concentration and the distribution of the single-cell voltage resistance, and the effect of different conditions of the stack temperature distribution. Zhou et al. [23] deeply introduced a PEM model applied in a PEMFC engine simulation and described the current state of the domestic PEMFC engines with technical indicators. Zhou et al. [24] built a suitable model on PEMFC humidity control. It simulates and analyzes PEMFC cathode inlet humidity on the maximum output power and current density cutoff. The model simulation results show that it can predict the PEMFC performance with various air relative humidity at cathode inlet. Chen and Zhou [25] established a semiempirical PEMFC model to describe the internal and output characteristics (current, voltage, outlet flow rate, pressure, temperature, etc.) of their dynamic functions. Based on that, they have found a parameter identification method to reduce the work of traditional empirical parameter adjustment.

12.3.2 MATHEMATICAL MODELS OF PEMS

Researchers and engineers did a lot of works on the water transfer in membrane. Water transfer in membrane has two main aspects. First, because of electric drag effect, water passes through the electrolyte to the cathode with protons. Thus, electromigration of water is related with the current density and the number of proton hydration. Second, if the water content of the cathode side is high, the water diffusion reversely occurs from the cathode to the anode. The quantity of water is proportional to the diffusion coefficient and membrane concentration gradient. It is inversely proportional to the thickness of the film. Based on this, different dynamic models related with water transmission in film are established, for example, the model of Shen and Li [20], describing the water molar quantity produced as follows:

$$f_{H_2O} = M_{H_2O} \frac{i}{2F} \tag{12.29}$$

The water quality of electromigration in proton exchange membrane is

$$\omega_h = \frac{i}{F} n_d \tag{12.30}$$

The water quality of concentration diffusion in proton exchange membrane is

$$\omega_d = -D\frac{dc_W}{dx} \tag{12.31}$$

The water transfer amount in membrane during battery runtime is

$$\frac{dc_W}{dt} = -\frac{dQ}{dx} = \frac{d}{dx}\left(D\frac{dc_W}{dx} - \frac{i}{F}n_d\right) \tag{12.32}$$

with

$$n_d = \frac{2.5\lambda}{22} \tag{12.33}$$

and

$$D = 1.25\times10^{-6}\left[\exp\left(\frac{1}{303} - \frac{1}{T}\right)\right] \tag{12.34}$$

$$c_W = \lambda c_S \tag{12.35}$$

Therefore, we have

$$\frac{dc_W}{dt} = -\frac{dQ}{dx} = \frac{d}{dx}\left(D\frac{dc_W}{dx} - \frac{i}{F}n_d\right) \tag{12.36}$$

Electric drag coefficient n_d is a function of water concentration

$$n_d = \frac{2.5\lambda}{22} \tag{12.37}$$

where

$$D = 1.25\times10^{-6}\left[\exp\left(\frac{1}{303} - \frac{1}{T}\right)\right], \quad c_W = \lambda c_S \tag{12.38}$$

The dynamic model created by Ge states that when water transfers in PEM, a random microunit is taken, and since there is no potential gradient in x- and z-direction, the water flux of electromigration is generally calculated as

$$N_{w,y,e} = n_d\frac{I}{F} \tag{12.39}$$

$$N_{w,x,e} = N_{w,z,e} = 0 \tag{12.40}$$

Since there is no pressure gradient in x- and z-direction, the water flux of pressure migration is then

$$N_{w,y,p} = -\frac{k_p}{\mu} c \frac{dp}{dy} \tag{12.41}$$

$$N_{w,x,p} = N_{w,z,p} = 0 \tag{12.42}$$

Water diffusion flux is

$$J_{w,y} = -D_m \frac{\partial e}{\partial y} \tag{12.43}$$

$$J_{w,x} = -D_m \frac{\partial e}{\partial x} \tag{12.44}$$

$$J_{w,z} = 0 \tag{12.45}$$

For $y \ll x$, $J_{w,x} \ll J_{w,y}$. Therefore, the diffusion of x-direction is negligible. The total flux of water passed in the film is

$$N_{w,y} = N_{w,y,e} + N_{w,y,p} + J_{w,y} \tag{12.46}$$

where

$$n_d = 2.5\lambda/22, \quad c = \lambda c_f$$

$$D_m = 10^{-10} \exp\left[2416\left(\frac{1}{303} - \frac{1}{T}\right)\right](2.563 - 0.33\lambda + 0.0264\lambda^2 - 0.000671\lambda^3)$$

A pressure change in membrane is

$$\frac{dp}{dy} = \frac{p_c - p_a}{\delta_m} \tag{12.47}$$

Total water flux in membrane can be obtained by:

$$N_{w,y} = \frac{2.5\lambda}{22} \frac{I}{F} - \frac{k_p}{\mu} \lambda c_f \frac{p_c - p_a}{\delta_m} - D_m c_f \frac{d\lambda}{dy} \tag{12.48}$$

12.3.3 DYNAMIC MODELING INSTANCES

In our group, Chen Hairong did a lot of works in the study of dynamic characteristics of PEMs. She established the relevant dynamic characteristics of proton exchange membrane by semi-semiempirical method mechanism. In the past, the proton exchange membrane water content is usually calculated by the empirical fitting formula that only considers the influence of water pressure changes in airway to PEM water content. The dynamic characteristic of water content in the membrane was not considered. Therefore, to additionally count the water pressure, temperature, and other factors in airway that affect the membrane properties, PEM water content will be introduced in the dynamic model of the membrane humidity control. According to the balance of water content in membrane, the PEM water content λ_m is satisfied in the relationship

$$X_m \rho_m \frac{\partial \lambda_m}{\partial t} = -\frac{\partial j(H_2O)}{\partial y} \tag{12.49}$$

Equation 12.49 points at both ends in y-direction to obtain the dynamic equation on the water content of the PEM (12.50):

$$X_m \rho_m \frac{d\lambda_m}{dt} = -\frac{1}{\delta_m}\left[j(H_2O)\big|_{\delta_m} - j(H_2O)\big|_0 \right] \tag{12.50}$$

where
  X_m is the ion-exchange capacity of proton exchange membrane
  ρ_m is the density of the proton exchange membrane, kg/m³
  δ_m is the thickness of Nafion 112, m
  $j(H_2O)$ is the density of water flux, mol/(s·m²)

Referring to the description of Nebrand's 1D model of proton exchange membrane, the ion-exchange capacity X_m and the density of proton exchange membrane in a wet state ρ_m can be expressed as a function of water content λ_m. The relationship is expressed as follows:

$$X_m = \frac{1}{1 + \lambda_m M(H_2O)X_m^{dry}} X_m^{dry} \tag{12.51}$$

$$\rho_m = \rho(H_2O)\rho_m^{dry} \frac{1 + \lambda_m M(H_2O)X_m^{dry}}{\rho(H_2O) + \lambda_m M(H_2O)X_m^{dry}\lambda_m^{dry}} \tag{12.52}$$

where X_m^{dry}, ρ_m^{dry}, respectively, are the ion-exchange capacity of proton exchange membrane and the density in the dry state. $M(H_2O)$, $\rho(H_2O)$, respectively, are the molar mass and density of water.

According to the balance of charge, the proton exchange membrane can be considered partially neutral. Thus, among the whole range of film thickness, the flux density of proton is constant and described as

$$0 = -\frac{\partial j(H^+)}{\partial y} \Leftrightarrow j(H^+) = \text{const}, \quad \forall y \tag{12.53}$$

According to the law of mass transfer, the formula of flux density H_2O, $j(H_2O)$ and the flux density H^+, $j(H^+)$ are related through Equations 12.49 and 12.52:

$$j(H_2O) = -\frac{t_w k}{1+F^2}\frac{\partial \mu_m(H^+)}{\partial y} + \frac{D_w}{RT}c_m(H_2O)\frac{\partial \mu_m(H_2O)}{\partial y} \qquad (12.54)$$

$$j(H^+) = -\frac{k}{l^2+F^2}\frac{\partial \mu_m(H^+)}{\partial y} + \frac{t_w k}{1+F^2}\frac{\partial \mu_m(H_2O)}{\partial y} \qquad (12.55)$$

where
  t_w is the water molecule transfer coefficient and $t_w = 2.5\lambda_m/22$
  D_w is the diffusion coefficient of water molecules in the membrane, m²/s
  k is the conductivity, s/m
  $c_m(H_2O)$ is the molar concentration of water in the proton exchange membrane

Assuming the material in the proton exchange membrane is in uniformly mixed state, the electrochemical gradient of the component H^+, H_2O in Equations 12.53 and 12.54 can be calculated as

$$\frac{\partial \mu_{m,\alpha}}{\partial y} = RT\frac{\partial}{\partial y}\ln x_{m,\alpha} + l_\alpha F\frac{\partial \phi}{\partial y}, \quad \alpha = \{H^+, H_2O\} \qquad (12.56)$$

where $x_{m,\alpha}$ is the PEM molar fraction of α.

Formula (12.50) is related to the boundary conditions of water equilibrium in proton exchange membrane. It uses the principle of the adsorption isotherm when the membrane is under standard temperature. The water content of the formula is obtained by linear interpolation, that is, the boundary of the water content in the proton exchange membrane:

$$\lambda_m\big|_0 = f(p_{ac}(H_2O),T), \lambda_m\big|_{\delta_m} = f(p_{cc}(H_2O),T) \qquad (12.57)$$

According to Formula (12.35), Faraday's law and the proton equilibrium boundary condition, $j(H^+)|_0$ and $j(H^+)\big|_{\delta_m}$, can be calculated by battery current density in the following relationship:

$$j(H^+)\big|_0 = j(H^+)\big|_{\delta_m} = \frac{i_{cell}}{FL(H^+)} \qquad (12.58)$$

Simultaneously, the solutions of Formula (12.54), (12.55), and (12.57) grant us $\partial\phi/\partial y$ with the value of $\partial\phi/\partial y|_0$ and $\partial\phi/\partial y\big|_{\delta_m}$ across the membrane. The approximate expression is

$$\frac{\partial \phi}{\partial y}\Big|_0 = \frac{\phi_m - \phi_{am}}{0.5\delta_m}, \quad \frac{\partial \phi}{\partial y}\Big|_{\delta_m} = \frac{\phi_{cm} - \phi_m}{0.5\delta_m} \qquad (12.59)$$

The voltage loss $\Delta\phi_m$ in proton exchange membrane is solved by Equation 12.58 as

$$\Delta\phi_m = \varphi_{am} - \varphi_{mc} \qquad (12.60)$$

In conclusion, PEM is a core part of PEMFCs. The membrane resistance and water content of the membrane is crucial to the fuel cell performance. This model describes the dynamic process of the membrane variables (e.g., the flux density of H_2O, the flux density of H^+ and water content).

12.4 FAULT-EMBEDDED MODELING OF PEMs

12.4.1 INTRODUCTION

Fault analysis is a critical issue for improving operability, flexibility, and modularization of commercial PEMFCs. Since membrane is the key component in PEMFCs, the faults in membrane have a big influence on its performance. For PEM fault diagnosis R&D activities, a fault-embedded model can be used as a powerful tool.

Two major membrane faults, that is, water flooding and membrane dehydration, will be discussed. Several models with these two faults will be overviewed in this section. In Sections 12.4.2 and 12.4.1, their respective causes and consequences on PEMFC performance will be discussed. Finally, two fault-embedded models will be introduced for instances in Section 12.4.3.

Water in fuel cell is imported through humidified gas streams from both anode and cathode and additionally produced at the cathode catalyst layer. For water transportation, two types of water flux are present in membrane. One is the back diffusion from cathode to anode. The other one is the electroosmotic drag in reverse direction. Therefore, inappropriate water management can result in water accumulation or even flooding in the cathode or membrane dehydration and has an impact on fuel cell performance.

In recent years, much more attention is attracted to model development and simulation work concerning water management in PEMFCs and corresponding loss mechanisms. The complexity of the applied physics, the modeled components, and dimension of the modeling domains are very different from each other and have their own study focus.

Bernardi [26] is the first person to propose a 1D model to study water management and to identify the humidification conditions that induce either the dehydration or excessive flooding of the membrane.

Liu et al. [27] has developed a 2D partial flooding model that considered pore size distribution of GDL to explain flooding. Two different kinds of pores, hydrophobic pores and hydrophilic pores, were dealt with. For hydrophilic pores, capillary condensation of liquid water will occur before saturation, and water will condense first in smaller pores and then in bigger ones. On the contrary, for hydrophobic pores, water condensation occurs in some extent of oversaturation and first in bigger pores. The fault of water flooding is embedded in this model, assuming that GDL is composed of hydrophilic pores and hydrophobic pores with different diameters. At every local position, they have the same property of pore diameter distribution. The GDL is partially flooded to different extent along the channel.

Gerteisen et al. [28] have summarized the PEMFC modeling development and introduced a 1D, two-phase, transient model including GDL, catalyst layer, and membrane, under the assumption that GDL, catalyst layer, and membrane are spatially resolved in 1D with an agglomerate approach for the structure of catalyst layer. The faults of water flooding and membrane dehydration are embedded by the assumption that the saturation due to the continuous capillary pressure and immobile saturation due to the mixed wettability of the GDL structure are discontinued. In order to allow dehydration of the ionomer on the anode side, the water content is not a constant but follows the Cauchy boundary condition. The model is validated by voltammetry experiments, and the simulated current responses are compared with the measured ones from chronoamperometric experiments.

The models mentioned earlier are limited to single cell. McKay et al. [29] has developed a two-phase isothermal 1D model of reactant and water dynamics. It is validated using a multicell stack. The lumped parameter model depends on six tunable parameters associated with the estimation of voltage, the membrane water vapor transportation, and the accumulation of liquid water in the gas channels. The water flooding fault is embedded in this model by the assumption of liquid water layer of uniform thickness at the GDL channel interface. This water layer spreads across the GDL surface as the liquid water volume in the channel increases, thus, reducing the surface area. This increases the calculated current density that will reduce the cell voltage at a fixed total stack current.

Fault-embedded modeling is essential in the fault analysis and diagnosis. In addition, it is critical to improve the operability and flexibility of commercial equipment. In this section, we will focus on the fault-embedded modeling and discuss the corresponding causes and impacts of the often-appearing PEM faults, that is, water flooding and membrane dehydration.

12.4.2 TYPICAL MEMBRANE FAULTS AND THEIR RESPECTIVE CAUSES

12.4.2.1 Flooding

Excessive water in cathode can obstruct the porous media of oxygen diffusion. It forces oxygen to dissolve and to diffuse through water to reach the active sites. This phenomenon is called water flooding.

12.4.2.2 Level of Air Humidity

The impact of air humidity on water flooding has been observed from many experiments. If the air humidity is larger than the proper value, liquid water can accumulate in cathode and obstruct the diffusion of oxygen.

The relative humidity is defined as the ratio of the two pressures described as

$$RH = \frac{p_w}{p_{sat}} \quad (12.61)$$

where
p_w is the partial pressure
p_{sat} is the saturated vapor pressure of water

The water-saturated air is regarded as fully humidified when $p_w = p_{sat}$.

The experiments carried out by Karimi et al. [30] show that portions of GDL begin to flood, when the inlet relative humidity RH_{in} exceeds 68% and GDL is fully flooded when $RH_{in}=73\%$ ($J=5000$ Am^{-2}, $T=353$ K, $P_{out}=1$ atm). The experiment is based on the fuel cell stack composed of 31 cells with two oxidant inlets at the endpoints of the inlet manifold.

12.4.2.3 Current Density

More and more simulation studies and experiments are focused on the relationship between current density and water flooding. Although the consequences vary with the properties of fuel cells, measuring methods, and model assumptions, they have a similar trend that current density results in water production as well as water removal in electrode. The water flood is dependent upon the combined influence of these two impacts.

The experiments conducted by Karimi indicate that a larger amount of oxidant stream needs to flow in the bipolar flow fields with increase in stack operating current density, which results in significantly low pressures. The low pressure improves water transportation and thus reduces flood.

Baschuk and Li [31] have simulated numerous of water floods under different operating conditions. The trends of water flood with the increase in current density vary with operating temperature and pressure.

12.4.2.4 Operating Pressure and Temperature

The flood decreases with higher temperature and increases with higher pressure.

A higher temperature will not only activate the reactions on the catalyst surface but also accelerate the species transportation. The experiments by Baschuk and Li also indicated that higher temperature increases the current density at which flooding occurs in 1 and 3 atm flooding schemes.

By increasing the stack pressure, the partial pressure of the produced water is augmented, and water tends to condense in the GDL pores, which aggravates water flood.

12.4.2.5 Membrane Dehydration

There must be sufficient water content in the proton exchange membrane due to the fact that proton conductivity is directly proportional to the water content in the membrane. If there is no sufficient water, the membrane will dehydrate and the resistance to proton transfer will sharply increase.

12.4.2.6 Level of Air Humidity

The humidification conditions have a big impact on the water content of membrane. Ciureanu [32] took an experiment on Nafion under various types of humidification: cathodic (ChAd), anodic (CdAh), anodic and cathodic (ChAh), and no humidification at all (CdAd). It is indicated that the resistance of membrane is small and relatively insensitive to the anodic humidification when the fuel cell stack is cathodic humidified (ChAd and ChAh). On the contrary, the resistance is high and strongly dependent on current density and anodic humidification when the stack is not cathodic humidified (CdAh and CdAd).

It is noticed that higher airflow would obviously reduce humidity as temperature increases. Therefore, insufficient humidification and temperature as well as inappropriate airflow can result in membrane dehydration.

12.4.2.7   Water Flux

If the membrane is already dry, the dehydration will get worse with increase in current density. This can be explained by following hypothesis and equations. The assumption from Büchi and Scherer [33] states: the osmotic drag is independent of the membrane thickness and proportional to the current density that induces a water concentration difference between anode and cathode.

The water flux from cathode to anode is driven by either diffusion or convection. In the case of solution diffusion approach, the diffusional flux of water according to Fick's law is

$$j_{H_2O} = -D_{H_2O} \frac{c_{H_2O}^c - c_{H_2O}^a}{\Delta x} \qquad (12.62)$$

where

$c_{H_2O}^a$ and $c_{H_2O}^c$ are the membrane water content at anode and cathode, respectively
Δx is the membrane thickness
D_{H_2O} is the water diffusion coefficient in the membrane

In the case of convection approach, the essential property influencing the back transportation of water is the permeability of K_{H_2O}. The flux equation may be written as

$$j_{H_2O} = -cK_{H_2O} \frac{p_{H_2O}^c - p_{H_2O}^a}{\Delta x} \qquad (12.63)$$

where $p_{H_2O}^c$ and $p_{H_2O}^a$ are the capillary pressures at cathode and anode side of the membrane, depending on pore size determined by the local water content.

The flux of water back to the anode is directly proportional to D_{H_2O} and K_{H_2O}, which is strongly dependent on water content λ and indirectly proportional to Δx. If the membrane is dry and water content is lower than the normal value, the coefficients D_{H_2O} and K_{H_2O} can decrease as well. With the increase in current density, more water is dragged from anode to cathode, while the back transport water is limited by lower coefficients. As a consequence, the dehydration of membrane, especially at the anode side, will get even worse.

12.4.2.8   Operating Temperature

At temperatures of over approximately 60°C, the air will always dry out the electrodes faster than the water produced by the H_2/O_2 reaction.

High temperature will dry the fuel cell because p_{sat} increases very rapidly at high temperatures. The saturated vapor pressure for a range of temperatures is given in Table 12.8. The air ($RH = 70\%$), which might be only moderately dried at ambient temperature, can be sharply done so when heated to about 60°C.

TABLE 12.8

Saturated Vapor Pressure of Water at Selected Temperatures

T/°C	P_{sat}/kPa
15	1.705
20	2.338
30	4.246
40	7.383
50	12.35
60	19.94
70	31.19
80	47.39
90	70.13

For instance, when the air at 30°C with 70% relative humidity, water has the following partial pressure expression:

$$p_w = 0.7 \times p_{sat} = 0.7 \times 4.246 = 2.972 \text{ kPa} \qquad (12.64)$$

If this air is heated to 60°C at constant pressure without adding water, p_w stays the same. Thus, the new relative humidity is then

$$RH = \frac{p_w}{p_{sat}} = \frac{2.972}{19.94} = 14.9\% \qquad (12.65)$$

This is very dry for membrane since it is not only strongly relied on the high water content but also prone to rapid drying out.

12.4.3 PEM PERFORMANCE DEGRADATION CAUSED BY MEMBRANE FAULTS

Water flood and membrane dehydration are two main PEM performance limitations. Rama et al. [35] summarized five top events reflecting PEMFC performance degradation: (1) activation losses, (2) mass transportation losses, (3) ohmic losses, (4) fuel efficiency losses, and (5) catastrophic cell failure. Here, the activation, mass transportation, and ohmic losses will be discussed in details since they are closely related to membrane failures.

12.4.3.1 Mass Transportation Losses

When fuel cell begins to operate and electrical power begins to output, the electrochemical reaction can lead to the depletion of reaction in catalyst layer. This depletion will affect the performance of fuel cell through mass transportation or concentration losses. The two major mass transportation effects considered in fuel cell modeling are: (1) convective mass transfer, which occurs in flow channels due to hydrodynamic transportation and the relatively large-sized channels ($\sim$1 mm to 1 cm), and (2) diffusional mass transfer that occurs in the electrodes because of tiny pore sizes. The mass transportation losses therefore consist of two parts: the resistance to the convective mass transfer and the resistance to the diffusional mass transfer. Water flooding in cathode mainly affects the diffusional mass transfer.

When liquid water accumulates in cathode electrode, it can hinder the transportation of the reactant species by blocking the pores in the porous GDL. In addition, the active sites can be covered by liquid water in catalyst layer creating a barrier through which oxygen would have to diffuse.

The total mass transportation resistance can be derived using Fick's law. Thus, the rate of convective mass transfer can be written as

$$\dot{m} = A_{elec} h_m (C_0 - C_S) \tag{12.66}$$

where

A_{elec} is the electrode surface area
h_m is the mass transfer coefficient,
C_0 is the concentration at which the reactant is supplied to the flow channel
C_S is the concentration at electrode surface

The rate of diffusional transportation through the GDL at steady state is

$$\dot{m} = A_{elec} D^{eff} \left(\frac{C_S - C_i}{\delta} \right) \tag{12.67}$$

where

C_i is the reactant concentration at the GDL/catalyst interface
δ is the GDL thickness
D^{eff} is the effective diffusion coefficient for the porous GDL, which is dependent upon the bulk diffusion coefficient D and the pore structure

Assuming the pore size is uniform and GDL is free from water flood, then

$$D^{eff} = D\Phi^{3/2} \tag{12.68}$$

where Φ is the electrode porosity.

The total resistance to the transportation of reactant to the reaction sites can be expressed by combining Equations 12.66 and 12.67:

$$\dot{m} = \frac{C_0 - C_i}{(1/h_m A_{elec}) + (\delta/A_{elec} D^{eff})} \tag{12.69}$$

where

$1/h_m A_{elec}$ is the resistance to the convective mass transfer
$\delta/A_{elec} D^{eff}$ is the resistance to the diffusional mass transfer through the GDL

In the paper from Baschuk and Li, they found that liquid water can be partially occupied in the void region of electrode. In other words, some of the void regions in the electrode are flooded. Thus, the effective diffusion coefficient D^{eff} will be smaller when electrode is flooded with large resistance. Small amount of water present in the electrode backing can influence the performance at high current densities.

When catalyst layer is flooded, the active sites are covered by liquid water. Baschuk and Li employed $R_{56}R_{78}R_{89}$ to denote the resistance of oxygen in a partially flooded catalyst layer and derived the equivalent diffusion coefficient D^{eff} in catalyst layer as

$$\frac{\delta_c}{D^{eff}} = \left(\frac{RT}{H_{O_2}} R_{56} + R_{78} + R_{89} \right) \tag{12.70}$$

where

δ_c is the catalyst layer thickness

H_{O_2} is Henry's constant for oxygen gas dissolution in liquid water

The resistance of oxygen in catalyst layer $R_{56}R_{78}R_{89}$ varies with different degree of water flood, which indeed influence the diffusion coefficient. However, this effect on the cell voltage is relatively small. Due to the fact that the resistance caused by membrane fraction in the void region of the catalyst layer is so high, the added resistance caused by liquid water is nearly negligible.

In conclusion, mass transportation losses will rise due to the decreased effective diffusion coefficient in both the gas diffusion and the catalyst layer.

12.4.3.2 Ohmic Resistance

In fuel cell, electrons and ions are two main charged particles. Both electronic and ionic losses should be considered in ohmic resistance. The electronic losses between the bipolar, cooling, and contact plates are caused by the degree of contact due to compression of the fuel cell stack. The ionic charge losses that occur in the fuel cell membrane when H^+ ions travel through the electrolyte are more complicated than the electronic losses.

Severe dehydration of the membrane can result in significant high ohmic resistance. First, this is because the fuel cell ohmic over potential is mainly caused by ionic resistance in the electrolyte. Second, the conductivity of membrane, a critical factor of the ionic resistance, is a very strong function of its own water content.

Membrane water content was initially measured by Zawodzinski as a function of water activity for Nafion 117.

Water vapor activity is expressed as

$$a_{water\text{-}vap} = \frac{p_w}{p_{sat}} \tag{12.71}$$

where

p_w denotes the partial pressure of water vapor in the system

p_{sat} represents saturated water vapor pressure for the system at operating temperature

The relationship between water activity on the membrane surface and water content can be described as

$$\lambda_{30} = 0.043 + 17.18 a_{water\text{-}vap} - 39.85(a_{water\text{-}vap})^2 + 36(a_{water\text{-}vap})^3 \qquad (12.72)$$

$$\lambda_{80} = 0.3 + 10.8 a_{water\text{-}vap} - 16(a_{water\text{-}vap})^2 + 14.1(a_{water\text{-}vap})^3 \qquad (12.73)$$

The ionic conductivity σ (S/cm) is correlated to water content and temperature as

$$\sigma = (0.005139\lambda - 0.00326)\exp\left[1268\left(\frac{1}{303} - \frac{1}{T}\right)\right] \qquad (12.74)$$

Since the conductivity is proportional to resistance, the resistance of membrane changes with water saturation and thickness. The total resistance of a membrane R_m is found by integrating the local resistance over the membrane thickness:

$$R_m = \int_0^m \frac{dz}{\sigma[\lambda(z)]} \qquad (12.75)$$

At constant temperature, the water content will go up with the increase of water vapor activity. The conductivity of membrane can also get better to deliver low ionic losses for fuel cell. In contrast, inadequate water can lead to dehydration and low conductivity of membrane.

Büchi and Scherer [33] has also observed the current dependence of membrane resistance in different Nafion membranes. The result is that the thicker the membrane in the fuel cell, the more distinct the increase of the resistance with current density. In short, it indicated that for multiple membrane cells, the increase in resistance is confined to anode.

12.4.3.3 Activation Resistance

Activation resistance is caused by the slow reactions taking place on the surface of the electrodes. A proportion of the generated voltage is lost in driving the chemical reaction that transfers the electrons from or to the electrode.

Ciureanu [32] found that the ohmic drop, which is caused by dehydrated cathode and anode, cannot entirely account for the observed voltage decaying. The additional resistance may be caused by the dehydration effect of Nafion in the catalyst layer. The presence of polymer in catalyst layer is important since it provides continued proton paths from membrane to the active sites. Dehydration can produce interruption to some of those paths, which is equivalent to partially disappearance of the active sites. Thus, we can assign this contribution to an increase in activation resistance that is closely related to the dehydration of Nafion in catalyst layer.

12.4.3.4 Other Influences

In addition to the mentioned mass transportation limits, nonhomogeneous current density, ineffective heat removal, and membrane swelling are also possible consequences of water flood. Furthermore, the delamination of fuel cell components in thermal cycling associated with freeze/thaw processes can be resulted from excessive water.

The dehydration of membrane may result in temporary performance losses in addition to permanent material damage [36].

12.4.4 CASES OF FAULT-EMBEDDED MODELING

In this section, two papers are introduced as a fault-embedded model instance. The first model describes a 60-cell fuel cell stack. The relationship between the voltage and parameters is precisely presented, such as temperature and water content. Therefore, the fault can be embedded into the system model by changing the value of these parameters with step or ramp input signals. On the other hand, the second paper has a different idea. The model contains six tunable parameters, including two water-related parameters that influence the degree of liquid water accumulation in the gas channel. The rest of the four parameters are identified by experiments and determined the calculated voltage. Therefore, the different degree of flooding can be simulated.

The model from the first paper is not specially designed for a certain fault, so it can be used to examine different faults. However, the model needs to be tolerable for the changing input of the related parameters. Its disadvantage is the inaccuracy and the over ideal input, such as step signal. In contrast, the second model is originally designed for fault analysis—flooding. Thus, its response voltage or current density is more accurate. However, it is not suitable for other applications.

12.4.4.1 Case 1

Based on the model of single cell presented by Chen and detailed mentioned in Section 12.3.3, Su et al. [22] have tried to extend it to a 60-cell fuel cell stack and embed some faults. The modeling process is divided into four parts—temperature distribution model, modeling of mass transfer, modeling of flow volume pressure in the distribution manifolds, and voltage calculation. Due to the relatively comprehensive modeling of temperature, she analyzes the effect of extremely high temperature on the fuel cell. Another embedded fault is the membrane degradation, which is achieved by changing the value of water content.

In order to deal with the modeling of large fuel cell stack, Li separates the stack into 15 modules that each contains 4 single cells. It is shown in Figure 12.7.

All the assumptions of this model are summarized:

- Heat exchange between cooling channel control volume and the air/gas channel control volume and the air is neglected.
- Uniform temperature distribution in control volume.
- Coolant and coolant flow channel is equivalent to a control volume.
- Heat convection in coolant flow channel is converted to thermal conduction.
- Sectional areas of different control volume are assumed to be the same.
- The difference of coolant inlet temperature and flow rate is neglected.

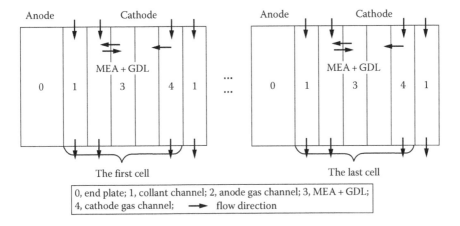

FIGURE 12.7 Structure of fuel cell stack.

The first part is the temperature distribution model. Four processes are taken into consideration. They are heat changes caused by flowing of species, heat exchange between control volumes and air, and heat exchange between control volumes and the heat generated by electrochemical reactions. According to the conservation of energy, the dynamic equation of temperature can be derived. The second part is the modeling of mass transfer. The conservation of mass is applied in the gas channel at anode/cathode as well as the MEA control volume. The third part describes the flow volume pressure in the distribution manifolds. The last part calculates the fuel cell stack voltage.

Based on the model mentioned earlier, the author sets some faults using the step and ramp signal.

The fault of membrane dehydration can be induced by inappropriate temperature. Therefore, a temperature ramp signal is put into the model. From 0 to 60 s, the fuel cell stack works under the normal condition of 60°C. From 60 to 80 s, the temperature rises linearly to 107°C (180 K). The voltage degradation is distinct, shown in Figure 12.8. In addition, the water content in membrane has a critical effect on the fuel cell performance. The fault occurs when the water content decreases to 0.1% of the initial value at 60 s, shown in Figure 12.9.

12.4.4.2 Case 2

McKay et al. [29] has developed a two-phase isothermal 1D gas and water dynamics model. The fault-embedded modeling methodology used is efficient for predicting temporal fuel cell voltage behavior in flooding.

The fault embedded in this model is water flooding in anode. It is modeled by the assumption that liquid water at the GDL–channel interface forms a layer of uniform thickness and decreases the anode active area. In order to simulate different operating conditions including flooding and nonflooding, they employed six tunable physical parameters—the parameter used for scaling the *stack-level* membrane

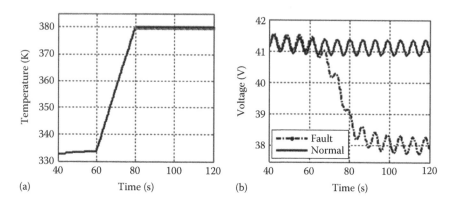

FIGURE 12.8   The input signal of temperature (a) and the voltage response (b).

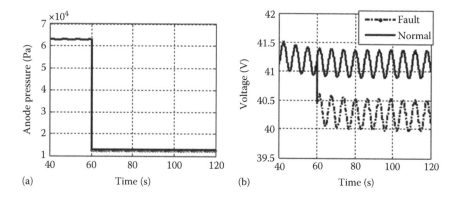

FIGURE 12.9   The input signal of pressure (a) and the voltage response (b).

back diffusion and the thickness of the liquid water and the scaling factor used in the equation of voltage losses K_1, K_2, K_3, K_4. These tunable parameters are identified by testing a 1.4 kW (24 cell, 300 cm²) stack with pressure-regulated pure hydrogen under a range of operating conditions. Given a set of values for α_w and t_{wl}, the voltage parameters are identified using linear least square fit to minimize the difference between the measured average cell voltage $\bar{v}$ and the modeled cell voltage $\hat{v}$.

The model is operated under relatively low current density and based on certain assumptions. All the assumptions of this model are summarized:

- The volume of liquid water within the GDL does not restrict the volume occupied by the gases.
- The internal cell structure (gas channel, GDL, and membrane) is assumed to be isothermal and equal to the coolant outlet temperature.
- The gas channels are treated as homogeneous and lumped parameter.
- The only mechanism for removing liquid water from the gas channels is evaporation.

- All gases behave ideally and H_2, O_2, and N_2 molecules do not cross through the membrane.
- The convective transportation of gas due to bulk flow is neglected.

The modeling process can be divided into four parts.

The first part is modeling of liquid water capillary transportation that describes the liquid water dynamics in the GDL. When the GDL pores are filled with liquid water, the capillary pressure is increased that causes water flow to adjacent pores. As a result, the liquid water flows through the GDL and injects into the channel. The equation of the liquid water dynamics arises from mass flow W_1 and molar evaporation rate R_{evap} [29]. The calculation of W_1 and R_{evap} is also interpreted in details. The notion of the reduced liquid water saturation S is a function of the immobile saturation s_{im}. S is a function of the immobile saturation s_{im}. When $s \leq s_{im}$, the liquid water path becomes discontinuous and liquid water capillary interrupts. The value of S then turns to zero.

The second part is the modeling of gas species diffusion. The diffusion of gas species in the GDL is a function of the concentration gradient. In addition, diffusion in both anode and cathode is considered as binary diffusion. This is because water vapor and hydrogen/oxygen diffusion exist simultaneously. It should be noted that the effective diffusivity of the gas constituents in the GDL$\langle D_j \rangle$ is closely related to porosity ε, effective diffusivity D_j, and liquid water saturation s.

The third part plays a very important role in modeling. It is all about the modeling of boundary conditions of membrane, the cathode channel, and the anode channel. First, the author separates each GDL into three discrete volumes, shown in Figure 12.10. Spatial discretization of the GDL yields 18 coupled ordinary differential equations (ODEs). They describe the gas constituent concentrations and liquid water saturation to approximate the solution of the original PDEs.

The consumed fluxes of hydrogen and oxygen at the electrode are shown in Equation 12.76 [29]. Another critical value is the molar flux of water vapor through the membrane $N_{v,mb}$ that contains the key tunable parameter α_w:

$$N_{v,mb} = n_d \frac{i}{F} - \alpha_w D_w \frac{c_{v,ca,mb} - c_{v,an,mb}}{t_{mb}} \tag{12.76}$$

where
n_d is the electroosmotic drag coefficient
i is the nominal fuel cell current density
F is the Faraday constant
D_w is the membrane water vapor diffusion coefficient
$c_{v,ca,mb}$ and $c_{v,an,mb}$ are the water vapor concentration in the electrode at the membrane surface
t_{mb} is the membrane thickness

It is important to note that only back diffusion and electroosmotic drag are taken into consideration since this function is assumed to neglect convective water

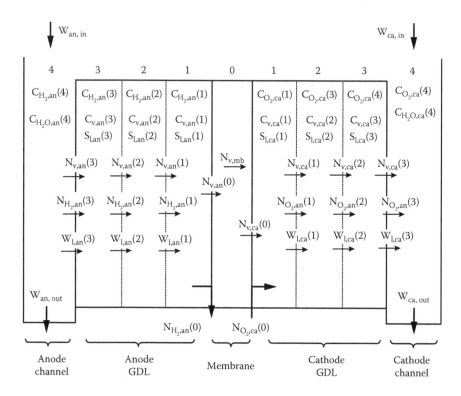

FIGURE 12.10 Discretization of GDL. The solid arrow is used to indicate the direction of the assumed mass flow rate, while the dashed arrow is used to indicate periodic mass flow rate.

transportation due to the small water pressure gradient at this relatively low current density.

As the GDL is separated into three sections, the water activity in the first GDL section ($L = 1$) is assumed to be equal to the membrane water activity at the membrane–GDL interface. Then, the average value of the water activity at anode and cathode is considered to be the lumped membrane water activity. The diffusion and electroosmotic drag can be then derived by using the lumped membrane water activity.

The boundary conditions at the anode channel and the cathode channel are similar to each other. The inputs are the experimentally inlet conditions, including the dry air mass flow rate, temperature, total gas pressure, and humidity. The outlet total pressure is also used for the calculation of the mass flow rate leaving the anode channel.

The last part is the estimated output voltage. In this section, the voltage equation is presented as a mapping from the apparent current density, reactant concentrations, temperatures, and membrane humidity conditions.

When water accumulates in the anode channel to a certain level, it will cause flooding and occupying part of the fuel cell active area. This aspect will consequently increase the apparent current density as:

$$i_{app} = \frac{I(A)}{10,000\,A_{app}(m^2)} \tag{12.77}$$

$$A_{app} = A_{fc} - \frac{2m_{l,an}(L+1)}{n_{cells}\rho_l t_{wl}} \tag{12.78}$$

where
   A_{app} is the apparent fuel cell area
   n_{cells} is the number of cells in the stack
   t_{wl} is the thickness of liquid water

The notion of apparent current density is the simplification of the flooding phenomena. The introduction of this parameter can simulate the experiment to observe dynamic voltage behavior of a multicell stack under the operation of both flooding and nonflooding.

The other four tunable parameters are presented in the equation of activation and ohmic voltage losses:

$$U_{act} = K_1 \frac{RT}{F} \ln\left(\frac{i_{app} + i_{loss}}{i_0}\right) \tag{12.79}$$

$$i_0 = K_2 \left(\frac{p_{O_2,ca}(1)}{p_0}\right)^{K_3} \exp\left[-\frac{E_C}{RT}\left(1 - \frac{T}{T_0}\right)\right] \tag{12.80}$$

$$U_{ohmic} = K_4 \left[\frac{t_{mb}}{b_{11}\lambda_{mb} - b_{12}} e^{-1268((1/303)-(1/T))}\right] i_{app} \tag{12.81}$$

In these equations, i_{loss} is the loss current density due to hydrogen crossover, i_0 is the exchange current density, E_C is the activation energy for oxygen reduction on Pt, T_0 is the reference temperature, and b_{11}/b_{12} are the experimentally obtained parameters.

Assuming that the concentration voltage loss at high current density is neglected, the calculated voltage is then

$$\hat{v} = E - K_1 \frac{RT}{F}\left(\ln\left(\frac{i_{app} - i_{loss}}{i_0}\right) + \frac{E_C}{R}\left(\frac{1}{T} - \frac{1}{T_0}\right)\right) + \ln(K_2)K_1\frac{RT}{F}$$

$$+ K_3 K_1 \frac{RT}{F} \ln\left(\frac{p_{O_2,ca}(1)}{p_0}\right) - K_4\left[\frac{t_{mb}}{b_{11}\lambda_{mb} - b_{12}} e^{-1268((1/303)-(1/T))}\right] i_{app} \tag{12.82}$$

In the following section, the parameter identification and model calibration are presented. Least square method is applied to minimize the difference between the measured average cell voltage and the estimated cell voltage. Thus, the voltage parameters are identified with a given set of α_w and t_{wl}. In order to model calibration, a portion of experimentally measured data is recorded under operating condition of flooding and nonflooding. The parameters are identified through least square method over $\alpha_w \in [7,12]$ and $t_{wl} \in [0.12, 0.14 \text{ mm}]$. The identified parameters are shown in Table 12.9.

Finally, the model calibration results are obtained and illustrated in Figure 12.11.

The calibration result illustrates the apparent current density that fluctuates with the action of anode purge. When the anode is flooded, the apparent active area reduces, and the apparent current density increases. Following the purge, the liquid water is removed, and the anode gas channel is released from flooding that decreases

TABLE 12.9

Experimentally Identified Parameter Values

Parameter	Tuned Value
K_1	1.00
K_2	1.24
K_3	2.05
K_4	3.40
α_w	10.0
t_{wl}	0.14

Source: McKay, D.A. et al., *J. Power Sources*, 178(1), 207, 2008.

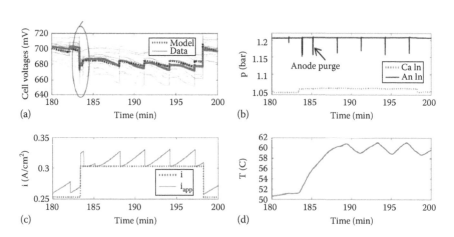

(a)

(b)

(c)

(d)

FIGURE 12.11 Model calibration results. The first subplot (a) shows the 24 individual cell voltages in thin faint lines with the average cell voltage in a thick solid line. The model estimated average cell voltage is illustrated in a thick dotted line. The second subplot (b) shows the nominal and apparent current densities. The third subplot (c) is the anode and cathode inlet total pressures. The fourth subplot (d) is the temperature of the water coolant leaving the stack. (From McKay, D.A. et al., *J. Power Sources*, 178(1), 207, 2008.)

the apparent current density. The first plot shows the comparison of measured averaged cell voltages and the calculated cell voltages. The calculated voltage gradually decays with increasing apparent current density. Indeed, this is consistent with the experiment results.

In summary, this paper focuses on the fault of water flooding at anode and presents a fault-embedded model. This methodology, which employs tunable parameters to simulate the fault effects, is inspiring.

12.5 APPLICATIONS OF MEMBRANES

12.5.1 APPLICATIONS OF PEM FOR FUEL CELLS

Fuel cell, as a new innovative chemical power source, has the following advantages: high energy conversion efficiency, cleanliness, and low noise. Nowadays, fuel cell is divided into alkaline fuel cell (AFC), phosphoric acid fuel cell (PAFC), molten carbonate fuel cell (MCFC), solid electrolyte fuel cell (SOFC), and PEMFC. Some oxyhydrogen fuel cells use the asbestos membrane as electrolyte, which has better performance than ion-exchange membrane (IEM). However, this membrane uses a large number of rare high-cost precious metals. Thus, the application is restricted. MCFC uses mixture of alkali metal carbonate $LiAlO_2$ as membrane. Similarly, the cost of MCFC is high, which at the same time is still in experiment stage. PEMFC currently grows extremely fast and eventually will become the electric car core power generation device. The performance of proton exchange membrane (PEM) greatly affects the performance of PEMFC. We will focus on PEMFC's application and development in the next section.

12.5.1.1 Physical Description for Proton Exchange Membrane

Proton exchange membrane (PEM) is the key component of PEMFC, which directly affects the fuel cell performance and life span. The function of PEM is to charge carrier with protons, to separate the reactant gases, and to work as electronic insulator to stop electrons from passing through the membrane. In order to achieve high efficiency in fuel cell applications, the membrane must possess the following properties [37]:

- High proton conductivity to support high currents with minimum resistive losses
- Zero electronic conductivity
- Adequate mechanical strength and stability
- Chemical and electrochemical stability under operating conditions
- Moisture control in stack
- Extremely low fuel or oxygen bypass to maximize efficiency
- Production costs compatible with intended application

Now, a perfluorosulfonic acid membrane produced by DuPont Company named "Nafion®" is commonly used in PEMFC industry. It has various excellent properties such as chemical stability, electrochemical properties, and thermostability.

Nafion membrane is not only used in PEMFC but also in membrane auxiliary electrolysis technology and desalination. It consists of a polytetrafluoroethylene backbone and regular-spaced long perfluorovinylether pendant side chains terminated by
a sulfonate ion group. We use EW (equivalent weight, the quality of resin per mole of
the ion-exchange group) to represent the content of sulfonate ions. These ions cause
polymer to absorb large amount of water. The proton conductivity of perfluorosulfonic acid membrane is associated with the acid concentration of the membrane,
which is represented by the value of EW. Lower EW usually can achieve higher fuel
cell performance. Figure 12.12 illustrates the comparison of single-cell performance
between Nafion and Aciplex®. The two membranes have identical thickness but significant different EW. Thus, the single-cell performances of the two are similar,
where Aciplex has a slightly lower EW than Nafion. In conclusion, Aciplex has better
performance at high current density. Figure 12.13 shows the performance of single
cell between Nafion 1135 and Flemion® SH80. The results show that the thicknesses
of the two are similar, but the EW of Flemion is much higher. Indeed, Flemion has
better performance at high current density than Nafion.

The level of hydration and the thickness of membrane affect PEM performance
in terms of fuel cell application stability [39]. The PEM performance is related to its
proton conductivity, when the conductivity is also related to the membrane humidity.
Therefore, the higher membrane humidity is, the higher proton conductivity will be
achieved. When relative humidity of Nafion 117 is less than 35%, membrane electrical conductivity will be significantly reduced. In short, Nafion 117 membrane has
almost become an insulator.

On the other hand, reducing membrane thickness can avoid water drag or
water crossover. With lower membrane resistance (an enhancement in membrane

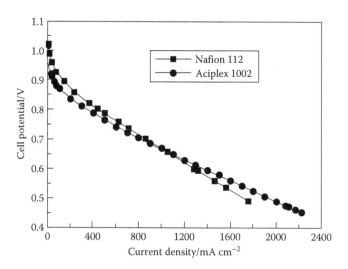

FIGURE 12.12   The comparison of single-cell performance between Nafion membrane and
Aciplex membrane. (From Yi Baolian, *Fuel Cell—Principal, Technology, Application[M]*,
The Chemical Industrial Press, Beijing, China, 2003.)

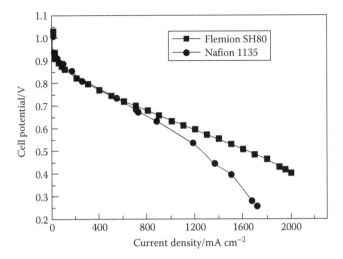

FIGURE 12.13 The comparison of single-cell performance between Nafion 1135 membrane and Flemion SH80 membrane. (From Yi Baolian, *Fuel Cell—Principal, Technology, Application[M]*, The Chemical Industrial Press, Beijing, China, 2003.)

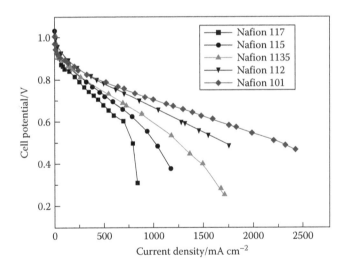

FIGURE 12.14 The comparison of performance of PEMFC assembled with different thickness of membranes. (From Yi Baolian, *Fuel Cell—Principal, Technology, Application[M]*, The Chemical Industrial Press, Beijing, China, 2003.)

conductivity), there will be lower cost and fewer cases of rapid hydration. As a result, reducing membrane thickness is an improvement in fuel cell performance. Figure 12.14 shows that the thickness of a membrane not only influences the performance of PEMFC, but also fuel cell's limited current density. The thinner the membrane is, the higher the limited current density of fuel cell is [40]. However, there is

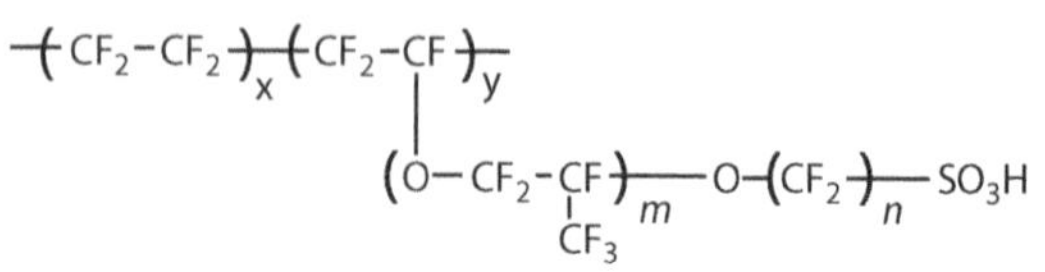

$$\left(CF_2-CF_2 \right)_x \left(CF_2-CF \right)_y$$

$$\left(O-CF_2-CF \right)_m \hspace{-1mm} O \hspace{-1mm} \left(CF_2 \right)_n \hspace{-1mm} SO_3H$$
$$\qquad\qquad CF_3$$

Nafion 117	$m \geq 1, n = 2, x = 5\text{--}13.5, y = 1000$
Flemion	$m = 0, 1; n = 1\text{--}5$
Aciplex	$m = 0, 3; n = 2\text{--}5, x = 1.5\text{--}14$
Dow membrane	$m = 0, n = 2, x = 3.6\text{--}10$

FIGURE 12.15 Chemical structures of perfluorinated polymer electrolyte membranes. (From Rikukawa, M. and Sanui, K., *Progr. Polymer Sci. (Oxford)*, 25, 1463, 2000.)

a limit in reducing membrane thickness due to the difficulties in durability and fuel bypass.

Nafion, Dow (Dow Chemical Company), Flemion (Asahi Glass Company), and Aciplex (Asahi Chemical Company) are all perfluorinated sulfonic acid membrane. Figure 12.15 illustrates their chemical structures [41]. Dow has lower EW (the quality of resin per mole of the ion-exchange group) than Nafion, which induces a higher electrical conductivity. However, Dow is difficult to be produced and its cost is much higher. In fact, the cost of Nafion membrane is as high as up to $500–$800 per square meter. In short, the scientists are studying partially perfluorinated or nonfluorinated PEM to lower the cost in order to find more suitable fuel cell membrane.

12.5.1.2 Development of PEM for Fuel Cells

Although Nafion membrane has good performance, it is too expensive. As a result, to find PEM with better performance and cheaper price has become an important research topic in the future.

One option is to use Nafion and PTFE composite membrane that is reinforced with microporous medium. Here, the transmission channel is provided by Nafion. This composition can improve the mechanical strength and stability of the membrane. It can also minimize the membrane thickness to save the material cost. At the same time, this structure can improve the distribution of water and the proton conductivity of the membrane [42–44]. Gore and Associates Company has developed such product named Gore-Select™. The shrinkage of Gore-Select is a quarter of Nafion 117 to better establish dimensional stability. Comparing to Nafion 117, its mechanical strength enhancement is obvious [45–48].

Another option is to use partially perfluorinated membrane. Ballard Company develops such product called BAM3 that has good thermal stability, chemical stability, and mechanical properties, low EW ($EW = 407$), and high water content. The thickness of hydration BAM-407 membrane is 140 μm where water content is 87 wt%. In contrast, Nafion 117 membrane has 190 μm of hydration thickness and 19 wt% of water content. The solubility coefficient of oxygen in BAM-407 is 4 times less than that in Nafion 117, but the diffusion coefficient of oxygen in BAM-407 is 4 times larger than Nafion 117 [49,50].

The preparation complexity of perfluorosulfonated membrane and the high cost have restricted PEMFC from commercialization. Many researchers are dedicated to the development of nonfluorinated PEM. The American company Dais has developed styrene/ethylene–butylene/styrene triblock polymer [51]. This membrane is especially suitable for small power PEMFC working at room temperature. The lifetime of the membrane is up to 4000 h. Baglio did some experiments to test the performance comparison of portable direct methanol fuel cell mini-stacks between a low-cost nonfluorinated polymer electrolyte and Nafion membrane. He found that at room temperature, a single-cell nonfluorinated membrane can achieve maximum power density of about 18 mW/cm^2. As a comparison, the value was 31 mW/cm^2 for Nafion 117 membrane. Despite the lower performance, the nonfluorinated membrane showed good characteristics for application in portable DMFCs especially regarded to the perspectives of significant cost reduction [52].

We can also add inorganic acid to the resin to form the composite membrane. This membrane has a good electrical conductivity at high temperature where the proton hardly carries water molecules. This simplifies the water management. More details can be observed from literatures.

12.5.2 Applications of Membrane for Lithium Battery

The application of lithium batteries is very broad. It has been used in spacecraft, torpedoes, rockets, planes, cars, and other advanced technologies. In order to ensure effective work of lithium batteries or lithium-ion batteries, the membrane plays an important role. The main effect of membrane in the lithium battery is to separate anode and cathode, so that the electrons cannot go through the circuit inside the batteries where the ions can move freely.

12.5.2.1 Physical Properties of Membrane for the Lithium Battery

Polyethylene (PE) and polypropylene (PP) have high porosity, low resistance, high tensile strength, good acid and alkali resistance, good elasticity, etc. Thus, the commercial membranes of lithium or lithium-ion battery are mainly made of PE or PP. The membranes of lithium batteries can be produced by dry method and wet method according to the different preparation technology. Their membrane micropore formation mechanisms are different [53,54].

The basic characteristic of the membrane applied to lithium or lithium-ion battery can be summarized as follows:

- Good electrical insulation
- Good permeability to electrolyte ions with low resistance
- Good chemical stability and electrochemical stability to electrolyte
- Good wettability to electrolyte
- Good mechanical strength with the minimum thickness
- Lower cost

Membrane does not participate in any reaction within the battery. However, the structure and characteristics of the membrane might directly impact on the performance of battery. The characteristics of membrane are mainly referred to the appearance, thickness, quantities, tightness, resistance, dry and wet tensile strength, porosity, pore size, the fluid absorption rate, the ability of keeping the electrolyte, expansion rate, etc. When choosing the membrane for Li or Li-ion batteries, we need to pay extra attention to the following points: (1) thickness, (2) porosity, (3) pore size, (4) permeability, (5) mechanical strength (tensile strength and puncture strength), (6) wettability, (7) shrinkage, (8) thermal shutdown, (9) meltdown or high temperature integrity, (10) open pore, (11) electronic bridging, (12) uniform pore, (13) cost, etc. [55]. Some important physical properties are described as follows:

1. Thickness

   In batteries with the same size, the membrane is thicker. The number of winding layer is less, and the capacity of battery will reduce correspondingly. For thicker membrane, the resistance of puncture will be slightly higher. The security of battery will also be higher. Under the condition of the same porosity, the thicker the membrane is, the poorer permeability and the higher internal resistance will be. Generally speaking, the thickness of membrane directly influences the security, capacity, and internal resistance of the battery.

2. Porosity

   Porosity is the ratio of the volume of pore to the total volume of membrane. Larger porosity can facilitate the lithium ions getting through. In contrast, large porosity will decrease the mechanical strength. Currently, the commercial membrane's porosity is between 40% and 60%.

3. Pore size

   The pore size influences the permeability rate of membrane. Large pore will cause microcircuit shortage. It directly affects the internal resistance and short circuiting rate of battery.

4. Permeability (Gurley No.)

   The time it takes a certain gas volume to pass through one square inch of membrane under a certain constant pressure is the permeability (Gurley No.). It is proportional to the internal resistance of battery, that is, the higher Gurley No., the higher internal resistance. However, we must realize that there is no direct relationship between the internal resistance and the permeability. The current commercial membrane's permeability is approximately 200–800 s/100 mL.

5. Puncture strength

   Puncture strength is the ability to resist external force from puncturing the membrane. It affects the short circuit rate and battery safety. For winding membrane battery, higher puncture strength is required. The common requirements of puncture strength for wet method should be greater than 30 g/20 μm.

6. Thermal shutdown

   The membrane would lead to close porosity which indeed leads to open circuit when the temperature reaches the thermal shutdown. It can prevent the internal temperature from rising further due to the large internal current.

Therefore, this feature can provide additional safety to lithium batteries. Of course, there is a close relationship between thermal shutdown and the melting point of material. The different microstructures also have certain influence on the thermal shutdown.

12.5.2.2 Development of the Membrane for Lithium or Lithium-Ion Battery

At present, a lot of commercial membranes are made of polyolefin, such as PE or PP. The lithium-ion battery membrane is usually 10–40 μm thick. The pore size is 30–200 nm. The thermal shutdown of PE and PP membrane is about 90–130°C and 165°C. When the battery temperature is higher than the melting point of the membrane, the membrane will melt, which leads to plate contact and causes short circuit. This can even cause intensive exothermic reaction between the plate and the electrolyte, which even causes the battery to explode. Thus, it is extremely important to develop more safe and advanced membrane for lithium battery.

In recent years, membrane manufacturers have developed several kinds of advanced products. This includes three layers of membrane (PP/PE/PP) developed by Celgard company (Figure 12.16) [56]. The shutdown of the PE layer in the middle of this membrane is around 130°C. The PP layer attached on the PE layer has a shutdown of about 165°C. Thus, it can keep mechanical strength and safety under 165°C. However, this kind of membrane porosity is too low and has poor wettability of electrolyte. Therefore, the thermal stability is limited to be under 165°C.

The Tonen Chemical Nasu company develops a new membrane in cooperation with ExxonMobil's affiliate company. The thermal stability of this membrane can reach 185°C, which maintains a good mechanical strength. The German company

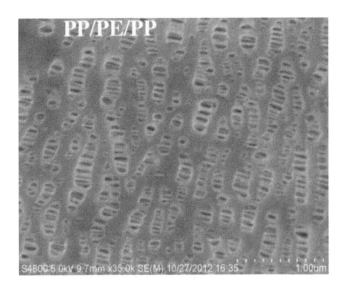

FIGURE 12.16   The three layers of membrane (PP/PE/PP) developed by Celgard company. (From Xiong, M. et al., *Carbohydr. Polymer*, 101, 1140, 2014.)

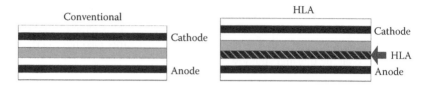

FIGURE 12.17 Heat resistance membrane developed by MBI.

Degussa uses polyethylene glycol terephthalate (PET) nonwoven fabric material as the base material, whose surface is evenly coated with nanoceramic particles. Its thermal stability is up to 260°C, which has fairly good wettability to the electrolyte. However, its mechanical strength is poor.

Matsushita Battery Industrial (MBI) develops the technology of heat-resistant layer for batteries that significantly improves safety. It adds a metal oxide layer to the middle of a plate and a diaphragm (Figure 12.17). The metal oxide layer can prevent battery from explosion when the plate is in short circuit caused by the diaphragm overheating shrinkage.

Due to the fact that the affinity of PE and PP membrane with electrolyte is poor, a lot of researchers are focused on it. Cheng and Sun [57] coat PP membrane (Celgard 2400 single-layer membrane) with PE mixed with nanometer silicon dioxide to improve the wettability of the membrane. Miao et al. [58] coat a three-layer composite microporous membrane (PP/PE/PP) with polyvinylidene fluoride (PVDF) surface processing (Figure 12.18). This technology reduces the membrane thickness and battery volume.

Above all, compounding the different diaphragm or adding the chemicals on the diaphragm is the membrane designing direction for lithium or lithium-ion batteries in the future.

12.5.3 APPLICATIONS OF MEMBRANE FOR VANADIUM REDOX FLOW BATTERY

Vanadium redox flow battery (VRFB) by Skyllas-Kazacos et al. has been extensively studied as a promising energy storage system emerging in response to the increasing global implementation of renewable energy technologies [59,60]. IEM plays the central role in this battery. For VRFB, IEM is used to prevent the crossover of vanadium ions. Meanwhile, it allows the transportation of ions to complete the conducting circuit.

The ideal membrane for VRFB should possess the following properties:

- Low permeation rates of vanadium ions to minimize self-discharging
- High conductivity
- Good chemical stability
- Competitive cost

In present, the available commercial membranes cannot meet all these requirements simultaneously. For instance, the Selemion CMV membrane shows lowest chemical stability, while Nafion 112 and New Selemion are excellent in vanadium

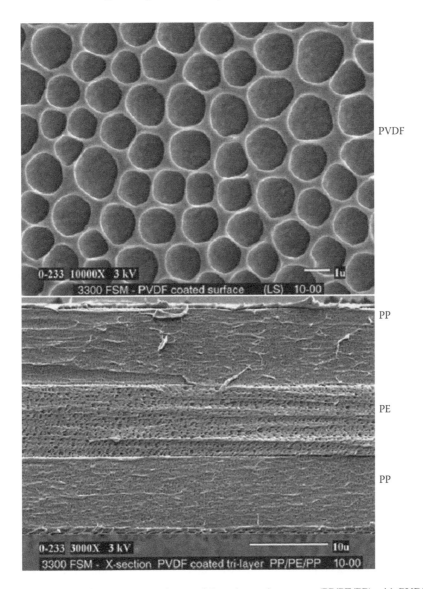

FIGURE 12.18 Microporous membrane of three layered structure (PP/PE/PP) with PVDF. (From Miao, R. et al., *Power Sources*, 184, 420, 2008.)

redox battery. The composite membrane prepared by cross-linking Daramic and divinylbenzene (DVB) presents good chemical stability similar to Selemion anion exchange membrane (AMV). Sulfonation of the composite membrane and AMV membrane did not improve their stability in the vanadium solution. Nafion membrane (DuPont, USA) suffers from high permeability of vanadium ions and high price [61]. Many researches are devoted to the development of new IEMs for VRFB such as nonperfluorinated aromatic membranes, composite membranes, and grafted membranes.

For nonperfluorinated aromatic membranes, there are sulfonated poly(ether ether ketone) (SPEEK), sulfonated poly(arylene thioether ketone) (SPTK), and sulfonated poly(arylene thioether ketone ketone) (SPTKK). VRFB single cells with SPEEK, SPTK, or SPTKK show very high energy efficiency, longer duration, and lower cost comparing to those of Nafion at much higher columbic efficiency [62,63]. The VRFB assembled with the SPFEK/3% SiO_2 and SPFEK/9% SiO_2 membranes exhibits higher columbic efficiency and average discharge voltage comparing to the ones assembled with the SPFEK membrane under all the tested current densities [64]. After the comparison to Nafion 112 and Nafion 117 membrane, quaternized poly(phthalazinone ether sulfone) anion exchange membranes exhibit higher energy efficiency with better vanadium selectivity [65].

Nafion/sulfonated poly(ether ether ketone) (SPEEK)-layered composite membrane (N/S membrane) consists of a thin layer of recast Nafion membrane and a layer of SPEEK membrane to reduce the cost in VRFB system while keeping the same chemical stability. The VRFB single cell with N/S membrane exhibited higher columbic efficiency and lower voltage efficiency comparing to VRFB single-cell Nafion membrane [66]. More composite membranes for VRFB are available in literatures [67–70].

12.5.4 APPLICATION OF MEMBRANES FOR CHLOR-ALKALI ELECTROLYSIS

12.5.4.1 Introduction

The chlor-alkali process is an industrial process of using sodium chloride electrolysis to produce chlorine and sodium hydroxide (caustic soda). Besides the two main products, hydrogen is also produced through this process.

Diaphragm is the first electrolysis cell went into operation in the late 1880s. A diaphragm made by asbestos fiber is used in this cell to separate anode from cathode. However, due to its structure, the diaphragm can only prevent H_2 and Cl_2 from contacting. It cannot prevent ions in the solutions such as OH^- and Cl^- traveling through, which leads to the low current efficiency and the poor quality of produced caustic soda. The OH^- travel rate from cathode compartment to anode compartment depend mainly on the flow rate of sodium chloride traveling through the diaphragm, which is determined by the static pressure difference between the anode and cathode compartment [71]. Therefore, to prevent the OH^-, the height of brine in the anode compartment is maintained above the level in the cathode side. In contrast, this cannot prevent the OH^- at an expected degree.

To solve this problem, a membrane that allows the passage of Na^+ cations while at the same time prevents the Cl^+ and OH^+ from crossing replaces the diaphragm and the membrane cell in 1970s.

Among all the IEM, the perfluorinated membranes developed for the usage in chlor-alkali cells exhibit super selectivity and high thermal and chemical stability.

12.5.4.2 Properties of Perfluorinated Membranes

12.5.4.2.1 Basic Structure

Based on the cluster network model [72], the perfluorinated membranes undergo phase separation on a molecular scale when swollen by contact of water. Clusters are formed when sodium ions (Na^+) separated from the fixed ironic sites joints the aqueous water separated from the fluorocarbon matrix. The ions and the sorbed solutions are all in clusters. A cluster's diameter varies from 3 to 5 nm and contains approximately 70 ion-exchange sites. The clusters are connected by short narrow channels with diameter of 1 nm estimated from hydraulic permeability data. The channels are formed by fixed ionic sites hydrated and embedded with water phase.

The three commercial perfluorinated membranes that have been extensively studied and used are

1. Nafion (perfluorosulfonate membrane produced by DuPont)
2. Flemion (perfluorocarboxylate membrane produced by Asahi Glass)
3. Aciplex (bilayered perfluorosulfonate and carboxylate membrane produced by Asahi Kasei)

Both perfluorosulfonate membranes and perfluorocarboxylate membranes are cation exchange membrane (cation Na^+) and are based on perfluorocarbon polymer. However, the fixed group of perfluorosulfonate membranes is sulfonate pendant ($-SO_3-$) while perfluorocarboxylate is carboxylate pendant ($-COO-$). This structure difference leads to different physicochemical properties.

12.5.4.2.2 Water Uptake

The water uptake of perfluorinated membranes depends on their chemical structure such as hydration of the fixed ionic groups and the counter ions. It also depends on the nature of the electrolyte and the pretreatment procedure [73]. The hydration of the fixed ionic groups of perfluorosulfonate membranes is stronger than that in perfluorocarboxylate. Therefore, the water absorption in perfluorosulfonate is higher compared to perfluorocarboxylate.

12.5.4.2.3 Ion-Exchange Capacity

The ion-exchange capacity is usually expressed in millimoles associated with the fixed ionic groups per gram of the polymer. It depends on the EW, equivalent weight of membrane.

$$EC(\text{mg/eqg}) = \frac{1000}{EW(\text{g/eqg})} \quad (EC \text{ in mg/eqg dry resin}) \quad (12.83)$$

Perfluorosulfonate membranes have lower EW than perfluorocarboxylate, which result in higher ionic exchange capacity under the same conditions (the electrolyte solutions are same).

12.5.4.2.4   Selectivity

The selectivity of membranes is the ability to repel anions and allow only cations to transport. This ability is resulted from the repulsion to the anions by the fixed ionic groups and the co-ions of the same electrical charge. The extend selectivity is determined mainly by the repulsion of fixed ionic groups in the short narrow channels that are described in its structure.

The concentration of fixed ionic groups (in mol/L) can be described as [74]

$$\overline{C_f} = \frac{EC \cdot f_i}{V_w \cdot f_w} = \frac{1000 d_e}{EW \cdot W_e} \left(\frac{f_i}{f_w} \right) \tag{12.84}$$

where
  W_e, d_e, and V_w refer to the weight, density, and volume of the electrolyte absorbed
    by the membrane
  f_i is the fraction of the ion groups in the clusters
  f_w is the fraction of the electrolyte solution in the clusters
  EC is the ion-exchange capacity
  EW is the equivalent weight of the membrane

The calculated values [5] demonstrate that $\overline{C_f}$ increases with rising EW when the electrolyte solutions are the same in anode and cathode. Researchers have also found that when the water content in a membrane increases, $\overline{C_f}$ decreases [75].

Perfluorosulfonate membranes have lower EW than perfluorocarboxylate membranes and contain more water due to the stronger hydration of fixed ionic groups resulted in lower selectivity compared to perfluorocarboxylate under the same conditions (the electrolyte solutions are same).

Good selectivity cannot only block the migration of OH⁻ to get high current efficiency but also hinder the transportation of Cl⁻ to achieve high quality of caustic soda.

12.5.4.2.5   Conductivity

The conductivity is a strong function of cation's nature and increases with rising water content until it reaches a plateau [76]. Indeed, the conductivity of perfluorosulfonate membranes is higher than perfluorocarboxylate membranes.

12.5.4.2.6   Diffusivity

NaOH and NaCl solutions have lower self-diffusion coefficient values for Na^+ in perfluorosulfonate membranes [77,78] than in perfluorosulfonate membranes. This behavior reflects the high activation energy of perfluorocarboxylate membranes.

12.5.4.2.7   HCl Used in Anode for OH⁻ Neutralization

The migration of OH⁻ from cathode to anode will significantly decrease the current efficiency, thus increasing energy consumption. To reduce the OH⁻, HCl is often added to the brine in the anode. However, the method can be applied only when the membrane's pKa is below the solution's pH. The perfluorosulfonate membranes have small pKa due to the acid property of fixed ionic groups and can be kept stable when HCl is added in the anode while perfluorocarboxylate is in contrast.

To combine the advantages of the two types, bilayer membranes with the sulfonic acid side facing the anode and the carboxylic acid side facing the cathode were developed. Table 12.10 shows the comparison of the three types perfluorinated membranes.

12.5.4.3 Further Development of Membrane

The development of membranes in chlor-alkali is similar to that in PEMFC and in water electrolysis. The resistance of a membrane is being reduced its durability and physical and chemical stability improved when operating under critical situation, and avoid the swell of membranes.

However, the biggest difference is that the lifetime of the membranes used in chlor-alkali strongly depends on the purity of the saturated sodium chloride brine. The impurity elements like Mg^{2+} or Ca^{2+} will travel across the IEM and react with OH^- to form barriers. Therefore, the development is also focused on the antipollution.

This is achieved by applying sacrificial core material [79]. The so-called sacrificial core material is a fiber woven into the membrane by pretreatment, and it will be dissolved in operation. After dissolution the formative hollow become the channels of water and ion. Therefore, the sediment of $Mg(OH)_2$ and $Ca(OH)_2$ may not block the membrane, which increases the antipollution ability. On the other hand, the resistance decreases. Figure 12.19 shows the sacrificial core material used in Nafion and Aciplex.

12.5.5 APPLICATION OF MEMBRANES FOR WATER ELECTROLYSIS

12.5.5.1 Introduction

Water electrolysis is a method to generate hydrogen. The water electrolysis process can be treated as a superposition of concurrent or sequential electrochemical reaction occurring in the vicinity of electrodes (half reactions) whose overall effect is to split water molecule under the influence of a direct electric current and separate gaseous products (hydrogen and oxygen) [80].

Among all the water electrolysis methods, alkaline electrolysis is a matured technology for hydrogen production up to the megawatt range, which is used at a commercial level to the most extent [81]. Alkaline electrolysis uses a diaphragm made by asbestos fiber to keep the product hydrogen and oxygen gas apart from each other while at the same time permeable to the hydroxide ions and water molecules. However, due to the structure drawbacks of the diaphragm, alkaline electrolysis has three major problems: the low partial load range, limited current density, and low operating pressure [81]. To solve this problem, the diaphragm is replaced by a proton exchange membrane (PEM) to introduce water electrolysis. A Nafion membrane, which is a perfluorosulfonic acid polymer manufactured by DuPont, is the most commonly used membrane in PEM water electrolysis cell due to its excellent chemical stability, mechanical strength, thermal stability, and proton conductivity.

TABLE 12.10
Properties of the Three Types of Perfluorinated Membranes

Membrane Type	pKa	Hydration	Water Content	Current Efficiency (%)	Conductivity	Chemical Stability	pH in Anode
Perfluorosulfonate	<1	High	High	75	High	Super	>1
Perfluorocarboxylate	2–3	Low	Low	96	Low	Good	>3
Bilayer	2–3/<1	Low/high	Low/high	96	High	Good	>1

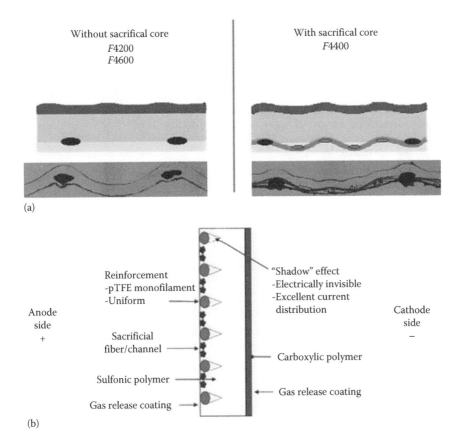

FIGURE 12.19 Sacrificial core material used in Aciplex (a) and Nafion (b). (From Cheng, D., *Chlor-Alkali Industry*, 45(12), December 2009.)

12.5.5.2 Properties of Membranes under Water Electrolysis Condition

The properties of Nafion membranes have been studied extensively to analyze PEMFC's operation [82]. The research methods and results can be well applied to the properties of Nafion membranes under water electrolysis.

PEMs are also used as electrolytes in PEMFCs whose electrolyzer has similar configuration to PEMFC. However, the hydration state of the membrane differs between fuel cell operation and electrolysis operation. During PEMFC operation, the membrane is humidified and equilibrated with water vapor, whereas during electrolysis operation, Nafion membrane is exposed to liquid phase of water and fully hydrated during water electrolysis. This leads to the big differences related to water uptake, swelling ratio, and proton conductivity between PEM water electrolysis and PEMFC.

12.5.5.2.1 Water Uptake

Different from PEMFC whose membrane is equilibrated with water vapor, water uptake decreases when the temperature increases [83–85]. The water uptake of membranes in PEM water electrolysis is equilibrated with liquid water and strongly depends on the pretreatment of the membranes [86]. The water uptake dried at room temperature is roughly twice of that dried at an elevated temperature of 105°C. It is also independent of immerse temperature. The water uptake of a membrane without a pretreatment of vacuum drying is relatively high and remains constant up to approximately 100°C. Water uptake by Nafion membrane from liquid water is higher than that from saturated water vapor (relative humidity 100%) [84,86,88,89].

12.5.5.2.2   Swelling Ratio
Due to water uptake, a membrane swells in both directions, namely, in plane (length/width) and through plane (thickness). Hiroshi Ito in his article has given detail data on swelling ratio of different Nafion membrane [89].

12.5.5.2.3   Conductivity
Proton conductivity of a Nafion membrane depends on water content and temperature. Zawodzinski et al. [87] measured the conductivity of a Nafion 117 membrane as a function of water content at 30°C and reported that the conductivity increases linearly with water content when the water uptake coefficient is from 2 to 22 (fully humidified). They also measured the temperature dependence of the conductivity on a Nafion 117 membrane immersed in liquid water in the temperature range from 25°C to 90°C. They indeed found the relationships between conductivity, water content and temperature.

12.5.5.3   Further Development of Membranes
The development of membranes in PEM electrolysis is mainly focused on two aspects: finding alternatives of Nafion while maintaining the ion-exchange characteristics and durability and reducing cross-permeation phenomenon of the membranes to work at high pressure.

Nafion membrane has many excellent properties that we have described earlier, but on the other hand, it has many disadvantages, especially its high price.

Research groups have been concentrating their efforts to make less expensive proton exchange membranes, who also focus on improving their ion-exchange characteristics and durability for PEM electrolysis. Masson et al. developed nonfluorine membranes by radiation grafting of styrene groups on a PE matrix, followed by chemical sulfonation of the resulting polymer [90]. Linkous et al. and Wei et al. evaluated different types of engineering polymers and identified a few options that could withstand the conditions found in PEM water electrolyzers [91,92]. From all these, polybenzimidazoles (PBI), poly(ether ether ketones) (PEEK), poly(ether sulfones) (PES), and sulfonated poly phenyl quinoxaline (SPPQ) were selected for sulfonation into membranes used for PEM electrolysis. Jang et al. chose SPEEK and sulfonated block copolymer of polysulfone (PSF) and poly(phenylene sulfide sulfone) (SPSfco-PPSS) to make SPEs for water electrolysis [93]. SPEEK polymer is considered to have high strength, and it is an easy membrane formation material. SPEEK also has a fairly high degree of sulfonation, which enables high proton conductivity. However, SPEEK membranes tend to swell excessively or even dissolve at elevated temperature. It needs other polymer structures to be reinforced. These alternative membranes have shown rather low current densities and low durability compared to standard Nafion membranes [94].

Another development is focused on high operating pressure. Due to the low gas cross rate of membrane, PEM electrolyzer allows a compact system design in which high operation pressures in both anode and cathode are achieved. The increasing pressure minimizes the expansion and dehydration of the membrane, keeping the integrity of the catalytic layer [83]. In addition to that, the high pressure has the advantage of delivering hydrogen that requires less energy for further compression and hydrogen storage. It also helps to remove the water products from the electrode.

However, during electrolysis operation at such high pressure, numerous difficulties needed to be overcome. The most critical problem is the cross-permeation phenomenon that occurs across the PEM during electrolysis. In this phenomenon, hydrogen and oxygen produced at both sides of the electrode permeate through the PEM and then mix together at the respective electrode compartment. This mixing increases the danger of gas explosions and increases the concentration of impurity gases. The main goal of a recent R&D project called GenHyPEM was to develop low-cost and high-pressure PEM electrolyzers [95].

12.5.6 APPLICATION OF MEMBRANES IN SOLAR CELL

A solar cell (also called a photovoltaic cell) is an electrical device that converts the energy of light directly into electricity by the photovoltaic effect. It is a form of photoelectric cell exposed to light generating and supporting electric current without being attached to any external voltage source. However, it requires an external load for power consumption. A typical solar cell is made up of five parts: cover glass, antireflective coating with contacts, N-type silicon, P-type silicon, and back contact. The structure is shown in Figure 12.20 [96].

A membrane is used to form the back contact that covers the entire back surface of the solar cell. The purpose of the membrane is to improve the mechanical strength

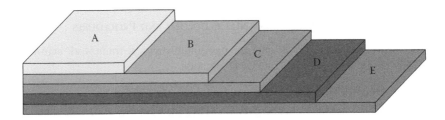

FIGURE 12.20 Standard silicon solar cell cross section. A, cover glass; B, antireflective coating with contacts; C, N-type silicon; D, P-type silicon; E, back contact. (From Characterization of Photovoltaic Devices by Spectroscopic Ellipsometry Using Equipment From Horiba Scientific, Horiba Scientific—Thin Films Division, Date Added: April 20, 2007 | Updated: June 11, 2013.)

of solar panels, to prevent moisture penetration into the sealing layer and impact on the battery life and power generation efficiency. The membranes are required to have good heat aging resistance, high temperature resistance, hydrolysis resistance, and corrosion resistance, as well as the ability to withstand light irradiation.

One of the most commonly used back-contact membranes is TPT [97]. T represents polytetrafluoroethylene membrane (PVF membrane, most commonly used Tedlar®). It has the following functions: solar cell sealing, anti-aging, UV resisting and water proofing. P represents polyester film PET that obstructs the water moisture, which is electrically insulated. TPT consists of three thin membranes. The two external membranes are all T membranes, while the internal is P membrane.

Other commonly used back contacts are TPE, FPF, and FPE, in which E represents ethylene vinyl acetate (EVA) membrane and F refers to fluorocarbon coatings (FC).

12.6 TYPICAL CASE ANALYSIS

As the name indicated, a polymer electrolyte membrane fuel cell's major function largely depends on the membrane electrode assembly. It is composed of catalyst, catalyst support, and ionomer membrane. Each component stated earlier can be theoretically analyzed with either finite element based computational fluid dynamic methods, molecular dynamic methods based on quantum mechanics, Newtonian mechanics or semi-empirical methods based on hybrid of classical and quantum world. In this section, a concise overview of the main simulation methods including computational fluid dynamics (CFD) is discussed. Typical material simulation and analysis methods are surveyed. Their capabilities are illustrated, and the pros and cons of each modeling technique is highlighted. In addition, multiscale coarse-grained simulation is used to model ionomer-induced carbon corrosion as an example of degradation model is presented. Furthermore, surface segregation of noble metals in Pt alloys, dual paths for oxygen reduction reaction (ORR), and solvation effects on ORR have been modeled with DFT are discussed. Platinum–carbon aggregate mismatch-induced microstructure changes modeled with ReaxFF, coarse-grained models of self-assembly of ionomer, and solvent dielectric property–induced microstructure formation during fabrication of MEA are succinctly described.

12.6.1 OVERVIEW OF SIMULATION METHODOLOGY AND PROCEDURES

In this section, we concisely discuss the basic mathematical framework behind the DFT, ReaxFF, coarse graining, and CFD methods.

12.6.1.1 Density Functional Theory

The core of DFT is the Hohenberg–Kohn (HK) theorem [98–100] that states that for a given ground-state density $n_0(r)$, the corresponding ground-state wave function $\psi_0(r_1, r_2, r_N)$ can be calculated. Thus, ψ_0 is a functional of n_0 and consequently all the ground-state observables are functions of n_0. Therefore, if ψ_0 can be calculated from n_0 and vice versa, both functions are equivalent and contain exactly the same information. The basic functional form of HK theorem was reintroduced by Levy [102] and Lieb [103] stating that for a given ground-state density $n_0(r)$, energy can be expressed as

$$E_{v0}[n] = \min \left\langle \Psi \middle| T + U + V \middle| \Psi \right\rangle \tag{12.85}$$

where
  E_{v0} denotes the ground-state energy in potential $V(r)$
  $T, U,$ and V represent kinetic, coulomb, and potential energy operators, respectively

The ground-state wave function ψ_0 reproduces $n_0(r)$ and minimizes the energy. For an arbitrary density $n(r)$,

$$E_v[n] = \min \left\langle \Psi \middle| T + U + V \middle| \Psi \right\rangle \tag{12.86}$$

if n is a density different from the ground-state density n_0 in potential $V(r)$, then the n produced by Ψ is different from the ground-state wave function ψ_0. According to the variational principle, the minimum value obtained from $E_v[n]$ is minimized by the ground-state density n_0 that is denoted as E_{v0}. The total energy functional can be written as [100,102,103]

$$E_V\left[n\right] = \min \left\langle \Psi \middle| T + U \middle| \Psi \right\rangle + \int d^3 rn(r)v(r) = F\left[n\right] + V\left[n\right] \tag{12.87}$$

where
  $F[n]$ is the energy function
  $V[n]$ is the external potential dependent on density

The universality of the aforementioned internal energy function $F[n]$ allows us to define the ground-state wave function ψ_0 to an N-particle system that delivers the minimum of $F[n]$ and reproduces n_0. Kohn and Sham (KS) [101] made this more practical by using noninteractive electrons that are known as Kohn–Sham electrons. They put forward a mathematical assumption [99,101–103]:

$$n(r) = \sum |\Psi_i(r)|^2 \tag{12.88}$$

$$n(r) = n_s(r) = \sum f_i |\Phi_i(r)|^2 \tag{12.89}$$

The usual way of solving such problem is to start with an initial guess for $n(r)$ to calculate the corresponding $v_s(r)$ (noninteracting potential) and then solve the differential equation for Φ_i (wave function of molecular orbitals). From this method, one can calculate a new density and use Equation 12.89 to start again. The process is repeated until it converges. When the solution converges to ground-state density n_0, the total energy can be calculated from the following equation [101,104]:

$$E_0 = \sum \frac{\varepsilon_i q^2}{2} \frac{\int d^3 r \int d^3 r' n_0(r) n_0(r')}{|r - r'|} \int d^3 r V_{xc}(r) n_0(r) + E_{xc}\left[n_0\right] \tag{12.90}$$

As you can see, E_{xc} exchange–correlation energy and V_{xc} exchange–correlation potential bear all the assumptions in KS DFT theory because the functional form of this is not known and approximations have to be made to get useful results. There are two main approximations found in the literature, known as local density approximation (LDA) and generalized gradient approximation (GGA) [99,101,102,104]. The first approximation assumed that $E_{xc}[n]$ is a sum of contributions from each point in real space, depending only on the density at each point. Since E_{xc} is assumed to be universal, it could be taken the same as that of the homogeneous electron gas of a given density. GGA assumes that E_{xc} is a sum of contributions from each point in real space depending only on the density and its gradient at each point. Thus, a heterogeneous electron gas is considered [104,105].

12.6.1.2   Reactive Force Field Method

Computer experiments particularly use quantum chemical approaches that provide accurate result with intense computational cost. Classical or semiempirical methods on the other hand are able to simulate thousands or up to millions of atoms of a system with pairwise Lennard-Jones (LJ)-type potentials [104–107]. Thus, LJ-type potentials are very accurate for inert gas systems [108], whereas they are unable to describe reactions or they do so by predetermined reactive sites within the molecules of the reactive system [109]. van Duin and coworkers [109–115] developed bond-order-dependent reactive force field technique is called ReaxFF as a solution to the aforementioned problems. Therefore, ReaxFF force field is intended to simulate reactions. They are successfully implemented to study hydrocarbon combustion [112,115,116] that is based on C–H–O combustion parameters, fuel cell [110,111], metal oxides [117–122], proteins [123,124], phosphates [125,126], and catalyst surface reactions and nanotubes [110–113] based on ReaxFF water parameters [127]. Bond order is the number of chemical bonds between a pair of atoms that depends only on the number and relative positions of other atoms that they interact with [127]. Parameterization of ReaxFFs is achieved using experimental and quantum mechanical data. Therefore, ReaxFF calculations are fairly accurate and robust. The total energy of the molecule is calculated as the combination of bonded and nonbonded interaction energies.

$$E_{ReaxFF} = E_{self} + E_{bond} + E_{over} + E_{under} + E_{lp} + E_{conj} + E_{H-bond}$$

$$+ E_{val} + E_{tor} + E_{other} + E_{waals} + E_{Coul} \tag{12.91}$$

Self-energy term accounts for the difference in energy of charge states of an atom, bonded interactions, overcoordination penalty, undercoordination stability, lone pair, conjugation energy, hydrogen bond energy, valence angle, torsion angle contributions, and other energy terms (C_2 species, triple bonds, etc.) are considered as bonded energies in a molecule. Nonbonded energies are van der Waals and Coulomb interactions, respectively [112,115]. Energies accounted for three and four body interactions are represented by Econj in above equation. Except for the last two terms in the previous equation, all other terms disappear during bond dissociation. Electronegativity equalization method (EEM), fast approach for charge calculation based on DFT and Sanderson's Electronegativity Equalization Principle, are used as charge equilibration methods in ReaxFF [128]. The fundamental assumption of ReaxFF is that: bond order can be directly obtained from interatomic distance (r_{ij}) of a pair of atoms and their environment [115]. Bond order (BO'_{ij}) can be calculated from the following equations for sigma, pi, and double-pi bonds, respectively.

$$BO'_{ij} = Bo'^{\sigma}_{ij} + Bo'^{\pi}_{ij} + Bo'^{\pi\pi}_{ij} = \exp\left[p^1_{bo} \left(\frac{r_{ij}}{r^{\sigma}_0} \right)^{p^2_{bo}} \right] + \exp\left[p^3_{bo} \left(\frac{r_{ij}}{r^{\sigma}_0} \right)^{p^4_{bo}} \right]$$

$$+ \exp\left[p^5_{bo} \left(\frac{r_{ij}}{r^{\sigma}_0} \right)^{p^6_{bo}} \right] \tag{12.92}$$

Full description of all the separate equations for each energy term can be found in Refs. [114] and [115]. The disadvantage of this method is the limited transferability from system to system and the difficulty to parameterize for a given system without extensive experience.

12.6.1.3 Coarse-Graining Methods

Computational cost associated with detailed atomistic models motivates to reduce the number of particles by mapping certain number of atoms/atomic groups into single bead (as shown in Figure 12.21) using the many-body potential of mean force (W) [129–131]. This reduces the number of force calculations and memory requirements.

$$W(R) = -k_B T \ln z(R) \tag{12.93}$$

$W(R)$ is the total Boltzmann weight on volume element of atomistic configurations (r) that are mapped to a specific Coarse-Grained (CG) configuration (R), known as

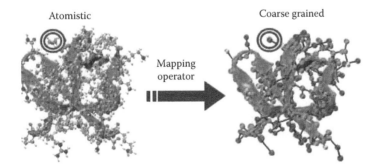

Atomistic   Coarse grained

Mapping operator

FIGURE 12.21 The mapping of an atomistic structure to a coarse-grained structure. Note the circled group in atomistic structure coarse grained into single bead. (Reprinted with the permission from Noid, W.G., Perspectives: Coarse-grained models for biomolecular systems, *J. Chem. Phys.*, 139(090901), 1–25, 2013. Copyright 2013, American Institute of Physics.)

k_B is known as Boltzmann constant and T is the temperature. The natural logarithm of CG configuration–dependent function [130] is illustrated as

$$z(R) = \int dr \exp\left[\frac{-u(r)}{k_R T}\right] \delta(M(r) - R) \qquad (12.94)$$

where u is the potential of the atomistic model and M is the mapping operator defined as

$$M_I(r) = \sum_i c_{Ii} r_I = R_I \qquad (12.95)$$

The CG configuration Cartesian coordinates of site I are determined as a linear combination of atomic Cartesian coordinates (r_I) with positive, constant coefficients that often correspond to the center of mass for the associated atomic group [130]. However, $W(R)$ is either too complex for numerical simulations or impossible to find a solution for most systems. A number of different simplified approximations for CG interactions (top–bottom, based on thermodynamic/macroscopic properties; bottom-up, based on detail atomistic model; and knowledge based, based on experimental evidence) pave the way to efficient and accurate CG simulations [130,131].

12.6.1.4 Computational Fluid Dynamic Method

The fluid flow, heat, and mass transfer in fuel cell are commonly expressed by Navier–Stokes equation, energy equation, mass conservation equation, the associated chemical reaction equations, the charge conservation equation for the current density distribution, and their derivatives [132]. The theoretical maximum electrochemical work that a fuel cell can do may be determined by the Nernst equation as follows:

$$E = E_0 - \frac{RT}{Fn_e} \ln(\prod_k P_k^m) \qquad (12.96)$$

where

E_0 is the potential under standard reference pressure ($P = 1$ atm)
R is the universal gas constant
T is the temperature
n_e is the number of electrons transferred in reaction
F is the Faraday constant
P_k^m is the partial pressure of the species k in the mixture with m stoichiometric coefficient
E is the theoretical potential of the fuel cell [133]

Navier-Stokes equation and the continuity equation for the fluid flow via isotropic porous medium are expressed in tensor notation and in Cartesian coordinates as

$$\frac{1}{\varepsilon}\frac{\partial(\rho u^i)}{\partial t} + \frac{1}{\varepsilon^2}\frac{\partial(\rho u^i u^j)}{\partial x^j} = -\frac{\partial p}{\partial x^i} + \frac{\partial}{\partial x^j}\left[\frac{\mu}{\varepsilon}\left(\frac{\partial u^i}{\partial x^j} + \frac{\partial u^j}{\partial x^i}\right)\right] + S_u \qquad (12.97)$$

where

ε is the porosity of the material

u^i, u^j are the fluid velocity components

x^i, x^j are the components of Cartesian coordinate system

p, ρ, μ are the pressure, density, and molecular viscosity of fluid flow, respectively

S_u is the resistance force on the fluid flow due to porous matrix

The left-hand side of the equation represents the rate of change of momentum per unit volume with time caused by convection. The right-hand side represents forces exerted on the fluid flow due to pressure difference, viscous shear force, and porous matrix [132]. Besides these convective transportation of the reactive species, diffusion occurs in fluid flow that can be described from species transportation equation with conservation law of chemical species:

$$\frac{\partial(\rho Y_k)}{\partial t} + \frac{1}{\varepsilon}\frac{\partial(\rho u^i Y_k)}{\partial x^i} = \frac{\partial}{\partial x^i}\left[\rho D_k \frac{\partial Y_k}{\partial x^i}\right] + S_k \qquad (12.98)$$

where Y_k, S_k, and D_k are the mass fraction, the production or consumption, and the diffusivity of species k, respectively [132]. The Butler–Volmer equation describes the charge transfer process in electrodes. The proton exchange membrane (PEM) fuel cell anode and cathode reactions can be expressed as follows:

$$j_a = i_{0,a}^{ref}\sqrt{\frac{C_{H_2}}{C_{H_2}^{ref}}}\left[\exp\left(\frac{\alpha_a F}{RT}\eta_{act,a}\right) - \exp\left(\frac{\alpha_c F}{RT}\eta_{act,c}\right)\right]$$

$$j_c = i_{0,c}^{ref}\sqrt{\frac{C_{O_2}}{C_{O_2}^{ref}}}\left[\exp\left(\frac{\alpha_a F}{RT}\eta_{act,a}\right) - \exp\left(\frac{\alpha_c F}{RT}\eta_{act,c}\right)\right] \qquad (12.99)$$

where i_0^{ref}, j, α, C, and η denote reference current density, charge-transfer current density, anodic transfer coefficient, concentration of the chemical species at the reaction site, and active overpotential, respectively [132,133]. The subscripts a and c denote anode and cathode of the fuel cell. The reference current density has to be obtained experimentally as it depends on the catalyst material and operational conditions [132]. Tafel equation is a good approximation to the aforementioned equations at extremely small overpotentials [133].

12.6.2 PT AND PT ALLOY SIMULATION WITH DFT

Here, we provide brief overview of recent findings on quantum mechanical studies of ORR, hydrogen oxidation, solvation effect on various transition metal surfaces, and potential energy surfaces of reaction steps linked to active sites on platinum and Pt alloy nanoparticles.

12.6.2.1 Active Sites on Pt and Dual Path for Oxygen Reduction Reaction

The platinum and its alloys are the pacemakers of PEMFC. Better understanding of their structure and functional relationship eventually contributes to significant improvements in this technology. There are recent advancements of simulating catalysts in PEMFC, with the help of ab initio and DFT methods to figure out exact locations of oxygen and hydrogen atoms bind to catalyst and possible paths for ORR reaction, shed light on ways to improve catalyst reactions, and design suitable heterogeneous catalyst for fuel cells. The most promising results can be found from applications of interstitial electron model proposed by McAdon and Goddard III [134–136]. It is used to calculate the best model for adsorption of oxygen with variety of sizes and shapes in face-centered-cubic (fcc) metals such as platinum. Pt12 cluster is the simplest and the most useful single-plane model to find accurate bond energies for the fcc, bridge, and on-top site of the Pt(111) surface except hexagonal closed pack (hcp) site [134,136]. These four active sites are defined in the literature of Jacob and Goddard III [137] as follows:

- hcp threefold sites, where the adsorbate is above a threefold position and second surface layer underneath it, contains Pt atoms.
- fcc threefold sites, although adsorbate is above a threefold position, no Pt atom on the second layer.
- On-top site, where there is a Pt atom directly under the adsorbate atom.
- Bridge site, two neighboring surface Pt atoms are bridged by adsorbate atom.

For all the four active sites, $Pt_{9,10,9}$ cluster is the best that contains 9 Pt atoms on the top layer, 10 atoms on the second layer, and again 9 atoms on the third Pt layer since it allows interstitial bonding orbitals (IBO) with tetrahedra to terminate at the surface [137]. However, the authors strongly advised to select proper model before the simulation begins. If one uses a slab model with periodic boundary conditions, the system then will appear semi-infinite characters. Therefore, care must be taken to mimic bulk-like behavior in its center. Large vacuum spacing is necessary to avoid interaction between neighboring adsorbate and virtual copies of the unit cells. If one uses slab model with finite system representation, care should be given to study appropriate cluster size and shape to reproduce accurate material properties and energies prior to simulation. The authors found minimum 20 atoms with at least three layers for platinum. Other metals such as aluminum and copper require 100 atoms and 56 atoms with three layers respectively to reproduce accurate binding energies [137].

Electrochemical ORR can be expressed as

$$O_2 + 2e^- + 2H^+ \rightarrow H_2O_2 \quad E^0 = 0.67 \text{ V}$$

This produces two electron transfers with low energy, low efficiency, and corrosive hydrogen peroxide–generating process. That is a highly unwanted by-product that brings numerous problems to fuel cells.

$$O_2 + 4e^- + 4H^+ \rightarrow H_2O \quad E^0 = 1.23 \text{ V}$$

This reaction is the most desirable, high energy–producing, and complete four-electron reduction of oxygen to water. Although it seems simple, the exact potential energy surface of this reaction with all the intermediates and transition states are not known. This also emphasizes the importance of molecular scale directly observes the kinetics of reactions to thoroughly understand the surface phenomenon. Computational modeling approaches were performed by distinguished scientists who came across two distinct reaction pathways in the Pt catalyst [137–139]. They are known as O_2 dissociation and OOH formation (O_2 nondissociation) pathways.

The gas-phase O_2 dissociation pathways are facilitated by adsorbed highly mobile H atoms because they have undergone barrier-less dissociation on Pt(111) surface and have high vibration frequency of 2456 cm^{-1} [140,141]. Molecular oxygen absorbed on the fcc (0.31 eV), bridge site (0.49 eV), and tilted sites (0.06 eV) with binding energies are presented in brackets [137]. Migration of atomic oxygen from tilted site to fcc site due to low dissociation barrier (0.24eV) is an important observation has been made in these oxygen adsorption simulations. As a result most of the atomic oxygen end up at the fcc site. As the simulation goes on, structure changes of adsorbed oxygen molecules dissociating from the bridge site to the tilted sites dramatically reduce overall dissociation barrier at bridge site from 1.33 to 0.65 eV [137]. Adsorbed hydrogen atoms and adsorbed oxygen atoms bonded to form adsorbed OH. Finally, adsorbed hydrogen and adsorbed OH formed adsorbed water on Pt surface, which was eventually desorbed to H_2O gas. Breaking Pt–H bond (2.73 eV), breaking Pt–OH bond (2.06 eV) and formation of lone-pair orbital bond in oxygen (0.60 eV), Pt–H_2O bond breaking to desorb the water (0.60 eV), and formation of O–H bond (5.25 eV) determine the formation of gas-phase water [137,142]. Formation of adsorbed OH out of adsorbed H and O is the rate-determining step (RDS) with the highest activation barrier of 1.37 eV [137,138,143–145]. Son et al. [146] were able to calculate H_3O^+ barriers in vacuum at zero potential and discovered ORR reactions involving hydronium ions without reaction barriers. The summary of the O_2 dissociation path can be schematically shown as

$$O_2 \longrightarrow 2O \xrightarrow[\text{RDS}]{2H} 2OH \xrightarrow{2H} 2H_2O$$

Cyclic OOH and noncyclic OOH formations have been observed in OOH formation pathway. However, cyclic structure soon collapses to noncyclic OOH that dissociates to adsorbed oxygen and hydroxyl at hcp sites [137–139]. Then, adsorbed H and O atoms produced adsorbed OH, and eventually, adsorbed H and OH combine together to form adsorbed water molecule. This mechanism is illustrated as

$$O_2 \xrightarrow{H} OOH \xrightarrow{RDS} O + OH \xrightarrow{H} 2OH \xrightarrow{2H} 2H_2O$$

The rate-limiting step of this reaction is dissociation of OOH into O^{ad} and H^{ad} atoms, but some researchers suggested peroxide formation is the rate-limiting step. It is a possible alternative as it has 0.47 eV activation barrier than 0.74 eV OOH dissociation barrier [137,142,147]. Similar to DFT calculation, experimental evidence has been reported for hydrogen-activated oxygen-dissociated OOH formation on Pt–Ru nanoparticles [148].

12.6.2.2 Reactivity and Surface Segregation of Base Metals on Pt Alloys

Quantum mechanical simulations further revealed major differences of water formation reaction in metals. Certain metals such as Pd, Pt, Ag and Au preferred OOH formation mechanism while Ni, Cu, Ir and Rh preferred the oxygen dissociative mechanism [149]. In addition, oxygen strongly binding metals (Rh, Ir, Cu, and Ni) allowed oxygen molecules to dissociate to surface oxygen atoms, but further reduction is kinetically hindered. Weakly binding metals (Ag and Au) allowed partial reduction of oxygen molecules to H_2O_2 because O–O bonds cannot break effectively on such metal surfaces and Pt and Pd metals selectively reduced oxygen to water completely by effectively breaking O–O bonds and forming O–H bonds [149].

Using the Gaussian basis function with Perdew–Burke–Ernzerhof (PBE) exchange–correlation functional [150] DFT-GGA [151,152] in SeqQuest code [153], Yu et al. [141] calculated Y_3X-type alloys. Here, Y is Pt, Pd, and Rh, and X is Re, Os, Mo, Ru, Ir, Tc, Rh, Co, Ta, Nb, and Ni, with the hypothesis that surface layer consists of 100% noble metals when the second layer consists of 50% electropositive metal that decreases the critical barriers of ORR in fuel cell systems. Large positive segregation energies indicated that the top layer is fully covered with pure noble metals and the second layer is partially (50%) covered with transition electropositive metals. Therefore, such values for segregation energy are considered as a favorable alloy catalyst for ORR reaction. Yu et al. [141] concluded that the best alternative for Pt alloy is Pd_3W alloy. It is six times less expensive than Pt, which at the same time has a lower barrier for direct OH formation and water formation. It also has showed activities comparable to pure Pt. This conclusion was further justified by Sarkar et al. [154] when they carried out an experiment using Pd–W as a catalyst for ORR using $Pd_{95}W_5$ nanoparticles. Okube and coworkers [155] published topologically sensitive surface segregation of Au–Pd alloy system. Local environmental changes of Pd in hydrogen adsorption region could be interpreted either due to formation of PdH or as extensive transfer of Pd to the surface maximizing interaction of Pd with hydrogen [155]. X-ray adsorption fine structure refinement experiments gave evidence for the absence of PdH; thus, the observed behavior strongly suggested the surface segregation of Pd. The gold surface segregation in the absence of hydrogen is also observed. They also reported three adsorbed hydrogen atoms stabilized three Pd atoms segregated on surface, forming trimmer-like ring structure. DFT simulation studies further revealed trios of hydrogen drawing trios of Pd from subsurface to surface. They found that the adsorption of hydrogen on the surface changes the equilibrium surface structure of Au–Pd alloy system and induced the catalyst atoms to take this new structure or stayed in metastable state such as edge or kink due to kinetic hindrance. Desorption of the product rearranged the surface to original equilibrium state. However, if this readjustment is not kinetically favorable, it remains in metastable state. Therefore, this may leave little or no evidence of the true catalytic structure under reaction conditions. Similar DFT and experimental combined studies by Voelker et al. [156] found that O_2 induced surface segregation of gold in Au–Cu alloy system. After exposure to O_2, they observed surface enrichment with Cu leading to an inversion of surface metals that depends on the circumstance [156]. Those research results suggested that high sensitive surface equilibrium states under reaction conditions of bimetallic alloys should be thoroughly investigated before applying it in fuel cells. Nevertheless, these information are extremely valuable in catalyst designing purposes for specific tasks.

12.6.2.3 Solvation Effects in ORR

Operating fuel cell involves solvent environment, and it is important to study water formation reaction under solvent conditions to understand the solvation effects on overall reaction kinetics. Adding water molecules explicitly or entire water bilayer to unit cells of simulating catalyst is considered as explicit solvent model. This method, however, does not resemble fully saturated system, and water molecules are introduced during the simulation that directly participates in the ORR [148]. Poisson–Boltzmann implicit solvent model is a method to resemble solvent as a continuum rather than individual molecules in explicit models [160,162]. This method is computationally inexhaustive and accurate enough to reproduce reliable results for the atomistic energy calculations.

Gas-phase reaction kinetics of ORR obtained from DFT determines the formation of OH^{ad} with the highest barrier (1.37 eV) [139,157], whereas experimental evidence shows O_2^{ad} dissociation as the rate-limiting step [148,158]. This difference is caused by the solvation of reactants and products that changes the overall kinetics of reactions dramatically. The Poisson–Boltzmann implicit solvent model gave a reasonable picture of the solvated systems and led to a consistent model for estimating the contribution of solvent all along the reaction surface [159,160]. During the simulation, Sha et al. [141,157] have taken solvent-accessible surface as the boundary condition for the six-layered continuum solvent with spherical radius of 1.4 Å and a dielectric constant of 78. Solvation reduced the oxygen dissociation barrier from 0.58 to 0.27 eV while decreasing the hydroxyl formation barrier up to 0.59 eV from 1.37 eV at gas phase [161]. Solvation energy for OH is about 0.54 eV at the first layer and increasing up to 0.59 eV at the second layer. Water-binding energy goes up to 0.58 eV [161]. The reaction barriers for O_2 dissociation, OH formation and H_2O formation changed by −0.31 eV, 0.36 eV and 0.12 eV respectively, for hydrated systems under O_2 dissociation mechanism. In OOH mechanism, OOH formation changed by −0.03 eV, OOH dissociation by −0.17 eV, OH formation by 0.36 eV, and H_2O formation by 0.12 eV [161–164]. The aforementioned overall changes of RDS under solvation conditions suggested either different mechanism govern in Pt(111) surface in wet condition or ORR may involve hydronium induced mechanisms. Further DFT studies of water with the influence of transition-metal interactions [159–165] and experimental evidences [147] shed light on possible different mechanisms in fuel cell systems.

$$O^{ad} + H_2O^{ad} \rightarrow 2OH^{ad} \left[E_a = 0.50 \text{ eV (Pt)}, 0.49 \text{ eV (Pd)} \right]$$

$$OH^{ad} + H^{ad} \rightarrow H_2O^{ad} \left[E_a = 0.24 \text{ eV (Pt)}, 0.70 \text{ eV (Pd)} \right]$$

These two reactions observed from DFT studies gave new direction to improve ORR by reducing the barrier for O^{ad} hydration [166]. Activation energies for Pt and Pd suggested the importance of Pd as ORR catalyst as it has a low barrier for adsorbed oxygen hydration reaction. Similar models of different transition elements found that Co has the least barrier 0.04 eV followed by Ni with 0.20 eV, which already proved to enhance activity as Pt–Ni and Pt–Co alloys through lower activation barriers for O^{ad} hydration reaction [166].

12.6.2.4 Anode Catalyst Layer

Oxidation of hydrogen, which is exactly the opposite of the Volmer reaction [167], required 32 eV that decomposed into hydration of proton (22 eV), metal (9 eV), electrode, and bulk of electrolyte (~1 eV) contributions [168]. On pure Pt(111) surface, DFT calculation (B3LYP with GGA) resulted in similar adsorption energies for different active sites. The values are top site 2.66 eV, bridge site 2.61 eV, fcc site 2.62 eV, and hcp site 2.49 eV, respectively [137]. Jacob and Goddard III [137] concluded that adsorbed hydrogen in pure Pt has a very low barrier to move around on the surface. Therefore, hydrogen is highly mobile on all four active sites. On the other hand, hydrogen on Pt–Ni alloy can be localized or mobile depending on the kinetic energy. Hydrogen with 0.26 eV or greater kinetic energies is highly mobile and with less than 0.26 eV localized to hcp-Pt$_3$ sites [137].

Using DFT in Dacapo codè [169], Santos et al. [168] formulated a method to calculate activation energy of a system with the effect of electronic structure. They have used three metals Cd(0001), Au(111), and Pt(111) to demonstrate the validity of the method by selecting noncatalyst, medium catalyst, and good catalyst for hydrogen oxidation reaction. Furthermore, they reported dissociation and activation barriers for hydrogen on Cd and Au that are 0.72 eV (1.52 eV) and 0.38 eV (0.68 eV), respectively, and values within the brackets represent activation energies. The source of the hydrogen should be pure enough to run this anodic reaction up to the standard conditions because there are plenty of evidences for CO poisoning that caused the malfunction of the anode, drastically affecting the processes of hydrogen oxidation [170]. This was modeled by Springer et al. [171] and suggested that the rate of hydrogen dissociative adsorption changed on catalyst active sites. Further, they have reported the possibility to increase the tolerance for CO poisoning by increasing second transition metal concentration that was thought to be an introduction of more d-band electrons that weaken the C–O bond.

12.6.3 WATER FORMATION AND PT ADSORPTION ON CARBON REAXFF SIMULATION

In this particular ReaxFF method, more detail information is initially obtained from quantum mechanics that is used to parameterize ReaxFF for the system of

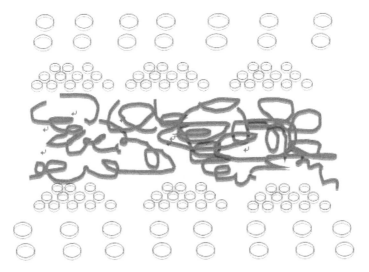

FIGURE 12.22 ReaxFF mesoscale model cartoon that simulates a whole fuel cell consists of Nafion membrane (light gray), anode (top, dark gray) and cathode (bottom, dark gray particles), and carbon support (black) particles.

interest either in engineering studies, macromolecular system, or biological drug design. Bond order–dependent force fields at the beginning are used only on covalent systems such as hydrocarbons to describe bond properties and reactions. However, ReaxFF became popular in the scientific modeling community, and thus, it applies with metallic and ionic bond parameterization* [172]. Goddard III and coworkers [110] introduced multiparadigm multiscale simulation using ReaxFF to simulate the entire fuel cell. In this simulation, they have used negative net charged fixed in the middle of the cathode catalyst and positive net charged fixed in the middle of the anode catalyst to study electric field effect on adsorbates on catalyst surfaces because such a nonequilibrium system cannot be modeled using standard DFT codes. The authors modeled a complete fuel cell cathode nanoparticle with 87-Pt atoms cluster (20 Å in diameter) with graphite layer as carbon support and Nafion membrane consisted of sulfates. Figure 12.22 is a cartoon representing complete fuel cell simulation with ReaxFF. It shows carbon in black color while Pt particles in dark gray and polymer electrolyte membrane in light gray. The completely wet cathode surface facilitates access of protons and high oxygen diffusivity at hydrophobic regions of Nafion membrane were observed in this simulation [110]. Current and temperature effect was studied using the same model by achieving steady-state condition for cathode and anode. The steady-state was achieved by adding molecular oxygen at a regular rate between the top two layers of the graphite but excluding the footprint of the catalyst and turning off the interactions of the O_2 with the top graphite layer. Then an H^+ is added to water at a regular rate and water is removed,

---

* Parameterization for thiol–gold systems can be found in appendix A of Ref. [171] and Al/MoO$_3$ system in Ref. [170].

near the graphite but away from the catalyst layer, at a regular rate. That process was carried out for various temperatures to simulate the thermal effects on catalyst. Buehler et al. [111] studied the water formation reaction at Pt(111) surface using ReaxFF at finite temperatures and found a smooth controlled reaction at 600 K with Pt catalyst. Without Pt catalyst, it is either uncontrollable or does not happen due to the higher activation energy barrier. Production of water molecules as a function of temperature expresses the common view that water production rate is rising in fuel cell systems. They have used NVT (number of particles, volume, and temperature constant) ensemble with 0.25 fs time steps to calculate the results. Arrhenius' plot was generated and fitted the data to calculate activation barrier, which was found to be ~12 kcal/mol. The method they used is simple and fast to produce the reaction barrier rather than running another DFT calculation for it is an advantage.

Carbon support is one of the important components in catalyst layer. The carbon has been used in the fuel cell experiments due to its inert nature in acidic and basic conditions in the fuel cell environment, high conductivity with Pt cluster and high surface to volume ratio [174]. Sanz-Navorro et al. [175] successfully applied ReaxFF method to calculate 300–800 carbon atom platelets with Pt clusters and found the mismatch between carbon platelets' edge atomic structure and the adsorbed face of the Pt(100) cluster leads to a desorption of a few Pt atoms from the cluster and restructuring of it. Three different 10-layer graphite lattices and 4-layer thin graphite lattice have been built with 1.42 Å between carbon atoms and an interlayer distance of 3.41 Å to resemble the exact distances in graphite. Pt-fcc structure was distorted due to adsorption on the carbon substrate with mean bond length of isolated cluster of 2.847 Å. For the adsorbed cluster, the value is 2.888 Å that eventually gains 0.014 strain to a Pt–Pt bond [175]. Adsorption energies of Pt adatoms are varied between −167 kcal/mol (twofold site) and −251 kcal/mol (fourfold site) depending on the binding site, which were calculated with another force field called BrennerFF. BrennerFF was found to be very different from ReaxFF that preferred twofold sites. Nevertheless, Zhang et al. [176] found experimental evidence for Pt adsorption on fourfold sites in the graphite interlayer compatible with ReaxFF predictions. The bond strain effect is cascading to neighboring Pt atoms and carbon atoms to relax the structure by migrating atoms to voids available due to the initial mismatch of structures causing restructuring of the Pt cluster. Similar result was experimentally observed in Ag clusters deposited on graphite [177].

12.6.4 COARSE-GRAINED SIMULATIONS

To obtain reasonable results from modeling, at least one mole of a substance equivalent to 10^{23} atoms should be used. Even with the contemporary computer technologies, this is still a challenging task to perform. Therefore, upscaling of models to grasp the broad picture of the physical nature of a system with simplified atomistic details paves the way to coarse-graining methods. Nafion is one of the most commonly used materials as cation exchange membrane in electrochemical devices. Walther Grot of E. I. DuPont discovered it in the late 1960s, in which it is produced by copolymerization of perfluorinated vinyl ether comonomers (4-methyl-3, 6-dioxa-7-octene-1-sulfonyl fluoride) with tetrafluoroethylene (Teflon) [178].

Therefore, Nafion is composed of poly-tetrafluoroethylene backbone with sulfo-nated groups at the terminals. The acidity and ionic functionality of the membrane is determined by chemical conversion of sulfonyl fluoride groups (SO_2F) to sulfonic acid (SO_3H) and the relative amount of comonomers specified during the polymer-ization [179]. Nafion backbone is highly hydrophobic, whereas sulfonated terminals are hydrophilic in nature. Thus, due to its complicated structure, coarse-grained simulation technique is highly appropriate to use in order to simplify and model ionomer.

12.6.4.1 Carbon Corrosion

To widespread commercialization of PEMFC, as automotive application, 3000–5000 h of operation is demanded. However, presently, its maximum durability under automotive-operating conditions is roughly 1000 h [180]. The physiochemical prop-erties of ionomer, microstructure of Pt alloy catalyst, and carbon corrosion ultimately lead to degradation of MEA [181]. Although there is no severe corrosion occurred in nonzero currents, much has been changed after PEMFC power cycle of start-up and shutdown operation for automotive applications [182,183]. Corrosion leads to perfor-mance degradation by reducing volume and connectivity of catalyst layer.

Franco et al. [184] introduced multiscale/multiphysics model for PEMFC phe-nomena that is designed to capture atomistic (elementary kinetic processes) and macroscopic data (electrochemical observables) using ab initio and surface science data with reasonable computational efforts [181,185,186]. Malek and Franco [181] studied carbon corrosion with special emphasis to water coverage and ionomer morphology impact applying latter model with multiscale coarse-grained methods. Three water molecules and hydronium ion are represented by polar beads, each with 0.47 nm in radius and 0.43 nm^3 in volume. Sulfonic acid groups are represented as coarse-grained spherical charge beads with elementary positive charge. On the other hand, backbone of Nafion is represented as coarse-grained chain of 40 nonpolar beads with elementary negative charge that stands for four monomer units of back-bone [181]. Platinum nanoparticle is coarse grained with 5:1 mapping in cuboctahe-dral morphology. However, for carbon model, coarse-grained potential parameters are obtained from atomistic level interactions. In this endeavor, Martini force field method [187] was used as initial guess, and Boltzmann inversion method [188] was used to improve the interaction parameters as the following:

$$F_{i,m}^{\rightarrow CG} = \sum_{i \neq j} f_{ij}^{CG}(|R_{ij,m}^{\rightarrow CG}|) \frac{R_{ij,m}^{\rightarrow CG}}{|R_{ij,m}^{\rightarrow CG}|} \tag{12.100}$$

where coarse-grained position, force, and interaction parameter are represented as $R_{ij,m}^{\rightarrow CG}$, $F_{i,m}^{\rightarrow CG}$, and f_{ij}^{CG}, respectively. Atomistic fullerene system of C540 coarse grained with 9:1 mapping of carbon atoms forms 0.47 nm radius beads [189]. This carbon particle resembles the properties of Vulcan. During the initial stage of sim-ulation, ionomer/carbon and Pt/carbon ratios are kept as 0:9 and 1:1, respectively [181]. They have removed carbon from simulation environment at different carbon loss percentages during the simulation. A random number generator based on a given reaction probability distribution was used as random procedure to remove carbon to resemble corrosion. All the simulations were carried out in modified GROMACS package [190] with VMD v1.8.1 visualization molecular dynamic commercial package [191]. Detailed simulation methodology can be found in Ref. [181]. In the absence of Pt nanoparticles, external surface of ionomer is mostly hydrophilic, and sulfonic acid groups are formed in highly ordered array pointing away from carbon surface. In the presence of Pt, ionomer is more clustered and less connected. Increasing carbon corrosion will decrease ionomer coverage. It will be stabilized after 5% of carbon is lost. Water coverage increases with up to 40% carbon loss and then decreases with rising corrosion. This would mainly due to hydrophilic ionomer side chains and Pt particles compete for the water. Thus, after 40% carbon loss, most of the Pt–C surface covered by ionomer reduces the exposure of Pt. As a result, water coverage tends to reduce [181].

12.6.4.2 Solvent Polarity and Microstructure Formation

Malek et al. [192] further developed coarse-grained molecular dynamics (CGMD) simulation (Figure 12.23) to find the microstructure formation in catalyst layers during fabrication through equilibration with various C–Pt cluster sizes. Predefined sizes have been assigned to C–Pt particles, hydronium molecules, Nafion, and water. These were represented as different colored spherical beads, and then renormalized interaction energy parameters between beads have been specified. This sort of simplification with experimental information is very useful in CG such that molecular volume of Nafion ionomer of 0.305 nm^3 can be used to assign the length scales

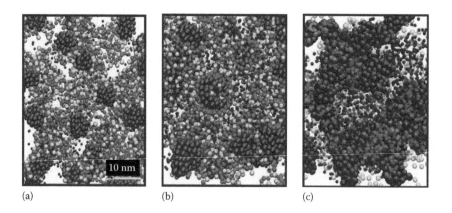

(a)   (b)   (c)

FIGURE 12.23 Multilevel coarse graining of carbon and Pt–C particle aggregates. Snapshots along the trajectory during microstructure formation at (a) $t = 100$ ps and (b) $t = 40$ ns and final equilibrated configuration (c) $t = 320$ ns. At (c), configuration Nafion side chains are represented by transparent green beads, but backbone beads are removed. This structure also represents separate hydrophilic and hydrophobic regions. The carbon particles are black spheres, side chains are transparent light gray beads, and hydronium ions are dark gray bead. (Reprinted with permission from Malek, K. et al., Self-organization in catalyst layers of polymer electrolyte fuel cells, *J. Phys. Chem. C*, 111, 2007, 13627–13634. Copyright 2007 American Chemical Society.)

of predefined beads and give more reliable results from models [192]. Implicit solvent approach was carried out to use three different continuum dielectric mediums, namely, water, isopropyl alcohol, and hexane, with permittivity of 80, 20, and 2, respectively. Carbon–Pt clusters form agglomerates, and Nafion is segregated into hydrophobic and hydrophilic regions. It is then attached to carbon aggregates. However, it has been observed that agglomerate sizes of C–Pt depend on permittivity. This phenomenon is vividly seen in Figure 12.23 (c-equilibrium configuration) as carbon and Nafion backbone segregated to form a hydrophobic region while hydronium ions and Nafion side chains made a hydrophilic region. The higher the dielectric constant on the medium, the size of aggregate is reduced from (2 hexane) 33 nm to (80 water) 15 nm. Similarly, primary pore sizes are reduced with high permittivity of medium, indicating the charged carbon clusters produce smaller agglomerates [192]. This was a clear indication of solvent effect in aggregate sizes.

REFERENCES

1. Martín, I.S., Ursúa, A., and Sanchis, P. Modeling of PEM fuel cell performance: steady-state and dynamic experiment validation. *Polymer Electrolyte Membrane Fuel Cells*, 7, 670–700, February 10, 2014.
2. Uday, K.C., Abbott, T.E., and Das, S.K. PEM fuel cell modeling using differential evolution. *Energy*, 40, 387–399, 2012.
3. Li, Q., Chen, W., Wang, Y., Liu, S., and Jia, J. Parameter identification for PEM fuel cell mechanism model based on effective informed adaptive particle swarm optimization. *IEEE Transactions on Industrial Electronics*, 58(6), 2410–2419, June 2011.
4. Jung, J.-H., Ahmed, S., and Enjeti, P. PEM fuel cell stack model development for real-time simulation applications. *IEEE Transactions on Industrial Electronics*, 58(9), 4217–4231, September 2011.
5. Basu, S. *Recently Trends in Fuel Cell Science and Technology*. Springer, New York, 2007.
6. Dai, C. et al. Seeker optimization algorithm for global optimization: A case study on optimal modelling of proton exchange membrane fuel cell (PEMFC). *International Journal of Electrical Power and Energy Systems*, 33(3), 369–376, March 2011.
7. Grimm, M. Modeling gas flow in PEMFC channels: Part I—Flow pattern transitions and pressure drop in a simulated ex situ channel with uniform water injection through the GDL. *International Journal of Hydrogen Energy*, 37(17), 12489–12503, September 2012.
8. Khorasani, A.N. Molecular modeling of proton and water distribution in catalyst layer pores of polymer electrolyte fuel cells. Abstract MA2012-011059.
9. He, S., Mench, M.M., and Tadigadapa, S. Thin film temperature sensor for real-time measurement of electrolyte temperature in a polymer electrolyte fuel cell. *Sensors Actuators A*, 12, 170–177, 2006.
10. Spiegel, C. PEM fuel cell modeling and simulation using MATLAB. Academic Press, Amsterdam, the Netherlands, 2011.
11. Barton, P.I. Dynamic modeling and simulation. Department of Chemical Engineering Massachusetts Institute of Technology, Cambridge, MA, 1997.
12. Springer, T.E., Zawodzinski, T.A., and Gottesfeld, S. Polymer electrolyte fuel cell model. *Journal of the Electrochemical Society*, 38, 2334–2342, 1991.
13. Springer, T.E., Wilson, M.S., and Gottefeld, S. Modeling and experimental diagnostics in polymer electrolyte fuel cells. *Journal of the Electrochemical Society*, 140, 3513–3526, 1993.

14. Futerko, P. and Hsing, I.-M. Thermodynamics of water vapor uptake in perfluorosulfonic acid membranes. *Journal of the Electrochemical Society*, 146, 2049–2053, 1999.
15. Kulikovsky, A.A. Quasi-3D modeling of water transport in polymer electrolyte fuel cells. *Journal of the Electrochemical Society*, 150, A1432–A1439, 2003.
16. Wei-Mon, Y., Hsin-Sen, C., Jian-Yao, C., Chyi-Yeou, S., and Falin, C. Transient analysis of water transport in PEM fuel cells. *Journal of Power Sources*, 162, 1147–1156, 2006.
17. Yun, W. and Chao-Yang, W. Transient analysis of polymer electrolyte fuel cells. *Electrochimica Acta*, 50, 1307–1315, 2005.
18. Loo, K.H., Wong, K.H., Tan, S.C., Lai, Y.M., and Tse, C.K. Characterization of the dynamic response of proton exchange membrane fuel cells: A numerical study. *International Journal of Hydrogen Energy*, 35, 11861–11877, 2010.
19. Shu-Guo, Q., Xiao-Jin, L., Chang-Chun, K., Zhi-Gang, S., and Bao-Lian, Y. Experimental and modeling study on water dynamic transport of the proton exchange membrane fuel cell under transient air flow and load change. *Journal of Power Sources*, 195, 6629–6636, 2010.
20. Shen, J. and Li, Y. Dynamic characteristics of proton exchange membrane fuel cell internal water transfer. *Chemical Industry and Engineering Progress*. In Chinese, 31, 100–103, 2012.
21. Ge, S., Yi, B., and Xu, H. Model of water transport for proton-exchange membrane fuel cell (PEMFC). *Journal of Chemical Engineering*. In Chinese, 50, 39–48, 1999.
22. Zhou, S., Li, Z., Zhai, S., and Chen, F. Modeling study and dynamic analysis under special working conditions for a PEMFC stack. *Acta Energiae Solaris Sinica*. In Chinese, 32, 1123–1128, 2011.
23. Zhou, S., Ji, G., Ma, T., Chen, F., and Zhang, T. The state of the art of PEMFC for automotive application. *Automotive Engineering*. In Chinese, 32, 749–760, 2010.
24. Zhou, S., Li, J., Ji, G., and Gao, K. Analysis and simulation of the model for PEMFC cathode inlet relative humidity control. In Chinese, 7642–7645, 2013.
25. Chen, H. and Zhou, S. Modeling and simulation of a PEMFC based on the dynamic characteristics of membrane. *Journal of Qingdao Technological University*. In Chinese, 25, 1–11, 2010.
26. Bernardi, D.M. Water-balance calculations for solid-polymer-electrolyte fuel cells. *Journal of the Electrochemical Society*, 137(11), 3344, 1990.
27. Liu, Z., Mao, Z., and Wang, C. A two dimensional partial flooding model for PEMFC. *Journal of Power Sources*, 158(2), 1229–1239, 2006.
28. Gerteisen, D., Heilmann, T., and Ziegler, C. Modeling the phenomena of dehydration and flooding of a polymer electrolyte membrane fuel cell. *Journal of Power Sources*, 187(1), 165–181, 2009.
29. McKay, D.A., Siegel, J.B., Ott, W., and Stefanopoulou, A.G. Parameterization and prediction of temporal fuel cell voltage behavior during flooding and drying conditions. *Journal of Power Sources*, 178(1), 207–222, 2008.
30. Karimi, G., Jafarpour, F., and Li, X. Characterization of flooding and two-phase flow in polymer electrolyte membrane fuel cell stacks. *Journal of Power Sources*, 187(1), 156–164, 2009.
31. Baschuk, J.J. and Li, X. Modelling of polymer electrolyte membrane fuel cells with variable degrees of water flooding. *Journal of Power Sources*, 86(1–2), 181–196, 2000.
32. Ciureanu, M. Effects of Nafion. *Journal of Applied Electrochemistry*, 34(7), 705–714, 2004.
33. Büchi, F.N. and Scherer, G.G. Investigation of the transversal water profile in Nafion membranes in polymer electrolyte fuel cells. *Journal of the Electrochemical Society*, 148(3), A183, 2001.

34. Larminie, J. and Dicks, A. *Fuel Cell Systems Explained*, 2nd edn. Wiley, New York.
35. Rama, P., Chen, R., and Andrews, J. *Failure Analysis of Polymer Electrolyte Fuel Cells*, SAE International, Detroit, MI, 2008.
36. Bazylak, A. Liquid water visualization in PEM fuel cells: A review. *International Journal of Hydrogen Energy*, 34(9), 3845–3857, 2009.
37. Smitha, B., Sridhar, S., and Khan, A.A. Solid polymer electrolyte membranes for fuel cell applications—A review. *Journal of Membrane Science*,259, 10–26, 2005.
38. Baolian, Y. *Fuel Cell—Principal, Technology, Application*. The Chemical Industrial Press, Beijing, China, 2003.
39. Appleby, A.J. and Foulkes, F.R. *Fuel Cell Handbook*. Van Nostrand Reinhold, New York, 1989.
40. Du, X.Z., Yu, J.R., and Yu, B.L. Performances of proton exchange membrane fuel cells with alternate membranes. *Physical Chemistry Chemical Physics*, 2001(3), 3175–3179, 2001.
41. Rikukawa, M. and Sanui, K. Proton-conducting polymer electrolyte membranes based on hydrocarbon polymers. *Progress in Polymer Science (Oxford)*, 25, 1463–1502, 2000.
42. Penner, R.M. and Martin, C.R. Ion transporting composite membranes. *Journal of the Electrochemical Society*,132, 514–515, 1985.
43. Liu, C. and Martin, C.R. Ion transporting composite membranes. *Journal of the Electrochemical Society*, 137, 510–515, 1990.
44. Verbrugge, M.W., Hill, R.F., and Schneider, E.W. Composite membranes for fuel-cell applications. *AICHE Journal*, 38(1), 93–100, 1992.
45. Sang-Yeoul, A., Lee, Y.C., Hab, H.Y., Seong-Ahn, H., and In-Hwan, O. Properties of the reinforced composite membranes formed by melt soluble ion conducting polymer resins for PEMFCs. *Electrochimica Acta*, 50, 571–575, 2004.
46. Hontsu, S., Nakamori, M., and Kata, N. Fabrication and characterization of PFSI/PTFE composite proton exchange membranes. *Journal of the Application of Physics*, 37, 1169, 1998.
47. Timmons, C.L., and Hess, D.W. Electrochemical cleaning of post-plasma etch fluoro-carbon residues using reductive radical anion chemistry. *Electrochemical and Solid-State Letters*, 7, 302–305, 2004.
48. Tang, H.L., Pan, M., and Jiang, S.P. Fabrication and characterization of PFSI/ePTFE composite proton exchange membranes of polymer electrolyte fuel cells. *Electrochemistry Acta*, 52, 5304–5311, 2007.
49. Basura, V.I., Beattie, P.D., and Holdcroft, S. Solid-state electrochemical oxygen reduction at Pt/Nafion 117 and Pt/BAM3GTM407 interfaces. *Journal of Electroanalytical Chemistry*, 458, 1–5, 1998.
50. Basura, V.I., Beattie, P.D., and Holdcroft, S. Temperature and pressure dependence of O_2 reduction at Pt/Nafion117 and Pt/BAMTM407 interfaces. *Journal of Electroanalytical Chemistry*, 468, 180–192, 1999.
51. Ehrenberg, S.G., Serpico, J.M., Sheikh-Ali, B.M., Tangredi, T.N., Zador, E., and Wnek, G.E. Hydrocarbon PEM/electrode assemblies for low-cost fuel cells: Development, performance and market opportunities. *Workshop on Fuel Cell. Proceedings of the Second International Symposium on New Materials for Fuel-Cell and Modern Battery Systems II*, Montréal, Québec, Canada, 1997, pp. 828–835.
52. Baglio, V., Stassi, A., Modica, E., Antonucci, V., Arico, A.S., Caracino, P., Ballabio, O., Colombo, M., and Kopnin, E. Performance comparison of portable direct methanol fuel cell mini-stacks based on a low-cost fluorine-free polymer electrolyte and Nafion membrane. *Electrochimica Acta*, 55(20), 6022–6027, 2010.
53. Wu, D. and Liu, C. The research and development of Lithium-ion battery membrane and development. *Journal of New Material*, 9, 48–53, 2006.

54. Yi, Y., Hu, X., and Gao, K. The research and development status of Lithium-ion battery membrane. *Battery*, 35(6), 468–470, 2005.
55. Huang, K., Wang, Z., and Liu, S. *The Principle and Key Technology of Lithium-Ion Battery*. Chemical Industry, Beijing, China, 2007.
56. Xiong, M., Tang, H., Wang, Y., and Pan, M. Ethylcellulose-coated polyolefin separators for lithium-ion batteries with improved safety performance. *Carbohydrate Polymers*, 101, 1140–1146, 2014.
57. Cheng, H. and Sun, W. The application of microporous filter membrane for the lithium-ion battery. *Plastic*, 33(2), 39–43, 2003.
58. Miao, R., Liu, B., Zhu, Z., Liu, Y., Li, J., Wang, X., and Li, Q. PVDF-HFP based porous polymer electrolyte membranes for lithium-ion batteries. *Journal of Power Sources*, 184, 420–426, 2008.
59. Sum, E., Rychcik, M., and Skyllas-Kazacos, M. Investigation of the V(V)/V(IV) system for use in the positive half-cell of a redox battery. *Journal of Power Sources*, 16, 85–95, 1985.
60. Sum, E. and Skyllas-Kazacos, M. A study of the V(II)/V(III) redox couple for redox flow cell applications. *Journal of Power Sources*, 15, 179–190, 1985.
61. Mohammadi, T. and Kazacos, M.S. Evaluation of the chemical stability of some membranes in vanadium solution. *Journal of Applied Electrochemistry*, 27, 153–160, 1997.
62. Chen, D.Y., Wang, S.J., Xiao, M., and Meng, Y.Z. Synthesis and characterization of novel sulfonated poly(arylene thioether) ionomers for vanadium redox flow battery applications. *Energy and Environmental Science*, 3, 622–628, 2010.
63. Mai, Z.S., Zhang, H.M., Li, X.F., Bi, C., and Dai, H. Sulfonated poly(tetramethydiphenyl ether ether ketone) membranes for vanadium redox flow battery application. *Journal of Power Sources*, 196, 482–487, 2011.
64. Chen, D., Wang, S., Xiao, M., Han, D., and Meng, Y. Sulfonated poly (fluorenyl ether ketone) membrane with embedded silica rich layer and enhanced proton selectivity for vanadium redox flow battery. *Journal of Power Sources*, 195, 7701–7708, 2010.
65. Jian, X.G., Xing, D.B., Zhang, S.H., Yin, C.X., and Zhang, B.G. Effect of amination agent on the properties of quaternized poly(phthalazinone ether sulfone) anion exchange membrane for vanadium redox flow battery application. *Journal of Membrane Science*, 354, 68–73, 2010.
66. Luo, Q., Zhang, H., Chen, J., You, D., Sun, C., and Zhang, Y. Preparation and characterization of Nafion/SPEEK layered composite membrane and its application in vanadium redox flow battery. *Journal of Membrane Science*, 325, 553–558, 2008.
67. Teng, X., Zhao, Y., Xi, J., Wu, Z., Qiu, X., and Chen, L. Nafion/organic silica modified TiO$_2$ composite membrane for vanadium redox flow battery via in situ solgel reactions. *Journal of Membrane Science*, 341, 149–154, 2009.
68. Tian, B., Yan, C.W., and Wang, F.H. Proton conducting composite membrane from Daramic/Nafion for vanadium redox flow battery. *Journal of Membrane Science*, 234, 51–54, 2004.
69. Xi, J., Wu, Z., Qiu, X., and Chen, L. Nafion/SiO$_2$ hybrid membrane for vanadium redox flow battery. *Journal of Power Sources*, 166, 531–536, 2007.
70. Zeng, J., Jiang, C., Wang, Y., Chen, J., Zhu, S., Zhao, B., and Wang, R. Studies on polypyrrole modified nafion membrane for vanadium redox flow battery. *Electrochemistry Communications*, 10, 372–375, 2008.
71. Fang, D. and Jiang, L. *Chlor-Alkali Technology*. Chemical Industry Press, Beijing, China, 1990, pp. 227–229 (in Chinese).
72. Mauritz, K.A. and Moore, R.B. State of understanding of Nafion. Department of Polymer Science, The University of Southern Mississippi, Hattiesburg, MS, Received July 19, 2004.

73. Davis, T.A., Genders, J.D., and Fletcher, D. *A First Course in Ion Permeable Membranes*. The Electrochemical Consultancy, Alresford, U.K.

74. O'Brien, T.F. and Bommaraju, T.V. *Handbook of Chlor-Alkali Technology. Volume 1: Fundamentals*. Springer, Boston, MA, 2005, p. 306.

75. Hanai, T. *Membranes and Ions*. Kagaku Dojin, Kyoto, Japan, p. 161.

76. O'Brien, T.F. and Bommaraju, T.V. *Handbook of Chlor-Alkali Technology. Volume 1: Fundamentals*. Springer, Boston, MA, 2005, p. 321.

77. Rodmacq, B., Coey, J.M., Escoubes, M., Roche, E. Water absorption in neutralized nation membranes. In T. Zwardowski, H.L. Yeager, and B. O'Dell (eds.), *Water in Polymers*, ACS Symposium Series, Washington, DC, Vol. 127, 1980, p. 487; *Journal of the Electrochemical Society*, 129, 328.

78. Yeager, H.L., Sodium ion diffusion and migration in perfluorinated ionomer membranes. In E.B. Yeager, B. Schumm, Jr., K. Mauritz, K. Abbey, D. Blankenship, and J. Akridge (eds.), *Membranes and Ionic and Electronic Conducting Polymers, PV 83-3*, The Electrochemical Society, Pennington, NJ, 1982, p. 134.

79. Cheng, D. The membrane application for the past 10 years in chlor-alkali. *Chlor-Alkali Industry*, 45(12), 15–21, 2009.

80. Naterer, G.F. *Hydrogen Production from Nuclear Energy*. Springer, London, U.K., 2013.

81. Ursua, A., Gandia, L.M., and Sanchis, P. Hydrogen production from water electrolysis: current status and future trends. *Proceedings of the IEEE*, 100(2), 410e26, February 2012.

82. Carmoa, M., Fritza, D.L., Mergela, J., and Stoltena, D. A comprehensive review on PEM water electrolysis. *International Journal of Hydrogen Energy*, 38(12), 4901–4934, 2013.

83. Doyle, M. and Rajendran, G. Perfluorinated membranes. In W. Vielstich, A. Lamm, and H.A. Gasteiger (eds.), *Handbook of Fuel Cells*, Vol. 3. John Wiley & Sons, Chichester, U.K., 2004, Chapter 30.

84. Rieke, P.C. and Vanderborgh, N.E. Temperature dependence of water content and proton conductivity in poly-perfluorosulfonic acid membranes. *Journal of Membrane Science*, 32(2–3), 313–328, 1987.

85. Hinatsu, J.T., Mizuhata, M., and Takenaka, H. Water uptake of perfluorosulfonic acid membranes from liquid water and water vapor. *Journal of the Electrochemical Society*, 141, 1493–1498, 1994.

86. Broka, K. and Ekdunge, P. Oxygen and hydrogen permeation properties and water uptake of Nafion 117 membrane and recast film for PEM fuel cell. *Journal of Applied Electrochemistry*, 27(2), 117–123, 1997.

87. Zawodzinski, T.Z., Derouin, C., Radzinski, S., Sherman, R.J., Smith, V.T., and Springer, T.E. Water uptake by and transport through Nafion 117 membranes. *Journal of the Electrochemical Society*, 140, 1041–1047, 1993.

88. Schroeder, P. Uber erstarrungs und quellungserscheinungenvon gelatin. *Zeitschrift für Physikalische Chemie*, 45, 75e128, 1903 [in German]; Weber, A.Z. and Newman, J. Transport in polymer-electrolyte membranes I. Physical model. *Journal of the Electrochemical Society*, 150, A1008e15, 2003.

89. Itoa, H., Maedaa, T., Nakanoa, A., and Takenakaba, H. Properties of Nafion membranes under PEM water electrolysis conditions. *International Journal of Hydrogen Energy*, 36(17), 10527–10540, 2011.

90. Masson, J.P., Molina, R., Roth, E., Gaussens, G., and Lemaire, F. Obtention and evaluation of polyethylene-based solid polymer electrolyte membranes for hydrogen-production. *International Journal of Hydrogen Energy*, 7(2), 167–171, 1982.

91. Guoqiang, W., Li, X., Chengde, H., and Yuxin, W. SPE water electrolysis with SPEEK/PES blend membrane. *International Journal of Hydrogen Energy*, 35(15), 7778–7783, August 2010.

92. Linkous, C.A., Anderson, H.R., Kopitzke, R.W., and Nelson, G.L. Development of new proton exchange membrane electrolytes for water electrolysis at higher temperatures. *International Journal of Hydrogen Energy*, 23(7), 525–529, 1998.

93. Jang, I.Y., Kweon, O.H., Kim, K.E., Hwang, G.J., Moon, S.B., and Kang, A.S. Application of poly-sulfone (PSf) and polyether ether ketone(PEEK) tungstophosphoric acid (TPA) composite membranes for water electrolysis. *Journal of Membrane Science*, 322(1), 154–161, 2008.

94. Goni-Urtiaga, A., Presvytes, D., and Scott, K. Solid acids as electrolyte materials for proton exchange membrane (PEM) electrolysis: Review. *International Journal of Hydrogen Energy*, 37(4), 3358–3372, February 2012.

95. Millet, P. GenHyPEM: An EC-supported STREP program on high pressure PEM water electrolysis. Laboratoire de Physico-Chimie de l'Etat Solide, Institut de Chimie Moléculaire et des Matériaux d'Orsay, Paris, France.

96. Characterization of Photovoltaic Devices by Spectroscopic Ellipsometry Using Equipment From Horiba Scientific, Horiba Scientific—Thin Films Division, Date Added: April 20, 2007 | Updated: June 11, 2013.

97. Xiaojian, L. and Hui-Li, Y. Study on the application of fluoroplastic film for solar power and fuel cell. Beijing University of Chemical Technology, Beijing Key Laboratory of Polymer Processing, Beijing, China.

98. Steinhauser, A.O. *Computational Multi Scale Modeling of Fluid and Solid-Theory and Application*, 2nd edn. Springer, Berlin, Germany, 2008.

99. Capelle, K. *A Birds Eye view of Density Functional Theory*, 5th edn. Brazilian journal of Physics, 36(4a), 1318–1343, 2006. http://arxiv.org/archive/cond-mat/0211443.

100. Hohenberg, P. and Kohn, W. Inhomogeneous electron gas. *Physical Review*, 136, B864–B871, 1964.

101. Kohn, W. and Sham, L. Self-consistent equations including exchange and correlation effects. *Physical Review*, 140, A1133–A1138, 1965.

102. Levy, M. Electron densities in search of Hamiltonians. *Physical Review A*, 26, 1200–1208, 1982.

103. Lieb, E.H. Density functionals for Coulomb systems. *Journal of Quantum Chemistry*, 24, 243–277, 1983.

104. Frankel, D. and Smit, B. *Understanding Molecular Simulations-Algorithms to Applications*, 2nd edn. New York Press Scientific Publications, San Diego, CA, 2002.

105. Nakano, A., Kalia, R.K., Nomura, K.I., Sharma, A., Vashishta, P., Shimojo, F., van Duin, A.C.T., Goddard III, W.A., Biswas, R., and Srivastava, D. A divide and conquer/cellular decomposition framework for million-to-billion atom simulations of chemical reactions. *Computational Materials Science*, 38, 642–652, 2007.

106. Buehler, M. and Grossman, J. Materials Engineering Lectures, 2011. Massachusetts Institute of Technology: MIT OpenCouseWare, http://ocw.mit.edu (December 24, 2012). License: Creative Commons BY-NC-SA.

107. Lennard-Jones, J.E. The electronic structure of some diatomic molecules. *Transactions of the Faraday Society*, 25, 668–686, 1927.

108. Rahman, A. Correlations in the motion of atoms in liquid argon. *Journal of Physical Review*, 136, A405–A411, 1964.

109. Russo, M.F. Jr. and van Duin, A.C.T. Atomistic-scale simulations of chemical reactions: Bridging from quantum chemistry to engineering. *Nuclear Instruments and Methods in Physics Research B*, 269, 1549–1554, 2011.

110. Goddard III, W.A., Merinov, B., van Duin, A.C.T., Jacob, T., Blanco, M., Molinero, V., Jang, S.S., and Jang, Y.H. Multi-paradigm multi-scale simulations for fuel cell catalysts and membranes. *Molecular Simulation*, 32, 251–268, 2006.

111. Buehler, M., van Duin, A.C.T., Timo, J., Jang, Y.H., Berinov, B., and Goddard III, W.A. Formation of water at a Pt(111) surface: A study using reactive force fields (ReaxFF). *MRS Proceedings*, Boston, MA, 2005, p. 900. doi:10.1557/PROC-0900-O03-09.

112. Chenoweth, K., van Duin, A.C.T., and Goddard III, W.A. ReaxFF reactive force field for molecular dynamics simulations of hydrocarbon oxidation. *Journal of Physical Chemistry A*, 112, 1040–1053, 2008.

113. Goddard III, W.A., van Duin, A.C.T., Chenoweth, K., Cheng, M.J., Pudar, S., Oxgaard, J., Merinov, B., Jang, Y.H., and Persson, P. Development of the ReaxFF reactive force field for mechanistic studies of catalytic selective oxidation processes on BiMoOx. *Topics in Catalysis*, 38, 93–103, 2006.

114. Nielson, K.D., van Duin, A.C.T., Oxgaard, J., Deng, W.Q., and Goddard III, W.A. Development of the ReaxFF reactive force field for describing transition metal catalyzed reactions, with application to the initial stages of the catalytic formation of carbon nanotubes. *Journal of Physical Chemistry A*, 109, 493–499, 2005.

115. van Duin, A.C.T., Dasgupta, S., Lorant, F., and Goddard III, W.A. ReaxFF: A reactive force field for hydrocarbons. *Journal of Physical Chemistry A*, 105, 9396–9409, 2001.

116. Chenoweth, K., van Duin, A.C.T., Dasgupta, S., and Goddard III, W.A. Initiation mechanisms and kinetics of pyrolysis and combustion of JP-10 hydrocarbon jet fuel. *Journal of Physical Chemistry A*, 113, 1740–1746, 2009.

117. Russo, M., Li, R., Mench, M., and van Duin, A.C.T. Molecular dynamic simulation of aluminum–water reactions using the ReaxFF reactive force field. *International Journal of Hydrogen Energy*, 36, 5828–5835, 2011.

118. Raymand, D., van Duin, A.C.T., Goddard, W.A., Hermansson, K., and Spangberg, D. Hydroxylation structure and proton transfer reactivity at the zinc oxide–water interface. *Journal of Physical Chemistry C*, 115, 8573–8579, 2011.

119. van Duin, A.C.T., Bryantsev, V.S., Diallo, M.S., Goddard, W.A., Rahaman, O., Doren, D.J., Raymand, D., and Hermansson, K. Development and validation of a ReaxFF reactive force field for Cu cation/water interactions and copper metal/metal oxide/metal hydroxide condensed phases. *Journal of Physical Chemistry A*, 114, 9507–9514, 2010.

120. Raymand, D., van Duin, A.C.T., Spangberg, D., Goddard, W.A., and Hermansson, K. Water adsorption on stepped ZnO surfaces from MD simulation. *Surface Science*, 604, 741–752, 2010.

121. Fogarty, J.C., Aktulga, H.M., Grama, A.Y., van Duin, A.C.T., and Pandit, S.A. A reactive molecular dynamic simulation of the silica–water interface. *Journal of Chemical Physics*, 132(174704), 1–10, 2010.

122. Aryanpour, M., van Duin, A.C.T., and Kubicki, J.D. Development of a reactive force field for iron-oxyhydroxide systems. *Journal of Physical Chemistry A*, 114, 6298–6307, 2010.

123. Rahaman, O., van Duin, A.C.T., Goddard, W.A., and Doren, D.J. Development of a ReaxFF reactive force field for glycine and application to solvent effect and tautomerization. *Journal of Physical Chemistry B*, 115, 249–261, 2011.

124. Rahaman, O., van Duin, A.C.T., Bryantsev, V.S., Mueller, J.E., Solares, S.D., Goddard, W.A., and Doren, D.J. Development of a ReaxFF reactive force field for aqueous chloride and copper chloride. *Journal of Physical Chemistry A*, 114, 3556–3568, 2010.

125. Abolfath, R.M., van Duin, A.C.T., Biswas, P., and Brabec, T. Reactive molecular dynamic study on the first steps of DNA damage by free hydroxyl radicals. *Journal of Physical Chemistry A*, 115, 11045–11049, 2011.

126. Zhu, R., Janetzko, F., Zhang, Y., van Duin, A.C.T., Goddard, W.A., and Salahub, D.R. Characterization of the active site of yeast RNA polymerase 11 by DFT and ReaxFF calculations. *Theoretical Chemistry Accounts*, 120, 479–489, 2008.

127. Shin, Y.K., Shan, T.S., Liang, T., Noordhoek, M.J., Sinnott, S.B., van Duin, A.C.T., and Phillpot, S.R. Variable charge many-body interatomic potentials. *MRS Bulletin*, 37, 504–512, 2012.

128. Mortier, W.J., Ghosh, S.K., and Shankar, S. Electronegativity-equalization method for the calculation of atomic charges in molecules. *Journal of the American Chemical Society*, 108, 4315–4320, 1986.

129. Kirkwood, J.G. Statistical mechanics of fluid mixtures. *Journal of Chemical Physics*, 3, 300–313, 1935.

130. Noid, W.G. Perspectives: Coarse-grained models for biomolecular systems. *Journal of Chemical Physics*, 139(090901), 1–25, 2013.

131. Noid, W.G., Chu, J.W., Ayton, G.S., Krishna, V., Izvekowv, S., Voth, G.S., Das, A., and Anderson, H.C. The multiscale coarse-graining method. I. A rigorous bridge between atomistic and coarse-grained models. *Journal of Chemical Physics*, 128(244144), 1–11, 2008.

132. Ma, L., Ingham, D.B., Pourkashanian, M., and Carcadea, E. Review of the computational fluid dynamics modeling of fuel cells. *Journal of Fuel Cell Science and Technology*, 2, 246–257, 2005.

133. Um, S., Wang, C.Y., and Chenb, K.S. Computational fluid dynamics modeling of proton exchange membrane fuel cells. *Journal of the Electrochemical Society*, 147(12), 4485–4493, 2000.

134. McAdon, M.H. and Goddard III, W.A. New concepts of metallic bonding based on valence-bond ideas. *Physical Review Letters*, 55, 2563–2567, 1985.

135. Kua, J. and Goddard III, W.A. Chemisorption of organics on platinum. 1. The interstitial electron model. *Journal of Physical Chemistry B*, 102(47), 9481–9491, 1998.

136. Li, M. and Goddard III, W.A. Interstitial-electron model for lattice dynamics in fcc metals. *Physical Review B*, 40, 12155–12163, 1989.

137. Jacob, T. and Goddard III, W.A. Water formation on Pt and Pt-based alloys: A theoretical description of a catalytic reaction. *ChemPhysChem*, 7, 992–1005, 2006.

138. Kitchin, J.R., Norskov, J.K., Barteau, M.A., and Chen, J.G. Modification of the surface electronics and chemical properties of Pt(111) by subsurface 3D transition metals. *Journal of Chemical Physics*, 120, 10240–10246, 2004.

139. Janik, M.J., Taylor, C.D., and Neurock, M. First-principles analysis of the initial electroreduction steps of oxygen over Pt(111) *Journal of the Electrochemical Society*, 156, B126–B135, 2009.

140. Jacob, T. and Goddard III, W.A. Adsorption of atomic H and O on the (111) surface of Pt3Ni alloys. *Journal of Physical Chemistry B*, 108, 8311–8323, 2004.

141. Yu, H.T., Sha, Y., Merinov, V., and Goddard III, W.A. Improved Non-Pt alloys for the oxygen reduction reaction at fuel cell cathodes predicted from quantum mechanics. *Journal of Physical Chemistry C*, 114, 11527–11533, 2010.

142. Jacob, T., Richard, P.M., and Goddard III, W.A. Chemisorption of atomic oxygen on Pt(111) from DFT studies of Pt-clusters. *Journal of Physical Chemistry B*, 107, 9465–9476, 2003.

143. Jacob, T., Fritzsche, S., Sepp, W.D., Fricke, B., and Anton, J. Cluster size convergent full relativistic density-functional calculations of single atom adsorption. *Physics Letters A*, 300, 71–75, 2002.

144. Anderson, A.B. and Albu, T.V. Catalytic effect of platinum on oxygen reduction an ab initio model including electrode potential dependence. *Journal of the Electrochemical Society*, 147, 4229–4238, 2000.

145. Ogasawara, H., Naslund, L.A., McNaughton, J., Anniyev, T., and Nilsson, A. Double role of water in the fuel cell oxygen reduction reaction. *ECS Transactions*, 16(2), 1385–1394, 2008.

146. Son, D.N., Nakanishi, H., David, M.Y., and Kasai, H. Oxygen reduction on Pt(111) cathode of fuel cells. *Journal of the Physical Society of Japan*, 78, 114601–114611, 2009.

147. Sachs, C., Hildebrand, M., Voelkening, S., Wintterlin, J., and Ertl, G. Reaction fronts in the oxidation of hydrogen on Pt(111): Scanning tunneling microscopy experiments and reaction–diffusion modeling. *Journal of Chemical Physics*, 116, 5759–5773, 2002.

148. Alayoglu, S., Nilekar, A.U., Mavrikakis, M., and Eichhorn, B. Ru–Pt core–shell nanoparticles for preferential oxidation of carbon monoxide in hydrogen. *Nature Materials*, 7, 333–338, 2008.

149. Denise, F.C., Nilekar, A.U., Xu, Y., and Mavrikakis, M. Partial and complete reduction of O_2 by hydrogen on transition metal surfaces. *Surface Science*, 604, 1565–1575, 2010.

150. Feibelman, P.J. Force and total-energy calculations for a spatially compact adsorbate on an extended, metallic crystal surface. *Physical Review B*, 35, 2626–2646, 1987.

151. Perdew, J.P., Burke, K., and Ernzerhof, M. Generalized gradient approximation made simple. *Physical Review Letters*, 77, 3865–3868, 1996.

152. Ceperly, M.D. and Alder, B.J. Ground state of the electron gas by a stochastic method. *Physical Review Letters*, 45, 566–569, 1980.

153. Schultz, P. SeqQuest, electronic structure code. Sandia National Laboratory, Albuquerque, NM, http://dft.sandia.gov/Quest/. (December 2013).

154. Sarkar, A., Murugan, A.V., and Manthiram, A. Low cost Pd–W nanoalloy electrocatalysts for oxygen reduction reaction in fuel cells. *Journal of Materials Chemistry*, 19, 159–165, 2009.

155. Okube, M., Petrykin, V., Mueller, J.E., Fantauzzi, D., Krtil, P., and Jacob, T. Topologically sensitive surface segregations of Au-Pd alloys in electrocatalytic hydrogen evolution. *ChemElectroChem*, 1, 207–212, 2014.

156. Voelker, E., Williams, F.J., Calvo, E.J., Jacob, T., and Schiffrin, D.J. O_2 induced surface segregation in Au-Cu alloys studied by angle resolved XPS and DFT modelling. *Physical Chemistry Chemical Physics*, 14, 7448–7455, 2012.

157. Sha, Y., Yu, T.H., Liu, Y., Boris, V.M., and Goddard III, W.A. Theoretical study of solvent effects on the platinum catalyzed oxygen reduction reaction. *Journal of Physical Chemistry Letters*, 1, 856–861, 2010.

158. Hammer, B. and Norskov, J.K. Electronic factors determining the reactivity of metal surfaces. *Surface Science*, 343, 211–220, 1995.

159. Ogasawara, H., Brena, B., Nordlund, D., Nyberg, M., Pelmenschikov, A., Pettersson, L., and Nilsson, A. Structure and bonding of water on Pt(111). *Physical Review Letters*, 89(276102), 1–4, 2002.

160. Tannor, D.J., Marten, B., Murphy, R., Friesner, R.A., Stikoff, D., Nicholls, A., Ringnalda, M., Goddard III, W.A., and Honig, B. Accurate first principles calculation of molecular charge distributions and solvation energies from ab initio quantum mechanics and continuum dielectric theory. *Journal of the American Chemical Society*, 116, 11875–11882, 2003.

161. Cai, Y. and Anderson, A.B. The reversible hydrogen electrode: Potential-dependent activation energies over platinum from quantum theory. *Journal of Physical Chemistry B*, 108, 9829–9833, 2004.

162. Tomasi, J. and Persico, M. Molecular interactions in solution: An overview of methods based on continuous distributions of the solvent. *Chemical Reviews*, 94, 2027–2094, 1994.

163. Karlberg, G.S., Olsson, F.E., Persson, M., and Wahnstrom, G. Energetics, vibrational spectrum, and scanning tunneling microscopy images for the intermediate in water production reaction on Pt(111) from density functional calculations. *Journal of Chemical Physics*, 119(9), 4865–4872, 2003.

164. Zimbitas, G., Gallagher, M.E., Darling, G.R., and Hodgson, A. Wetting of mixed OH/ H_2O layers on Pt(111). *Journal of Chemical Physics*, 128(7), 074701-1–074701-12, 2008.

165. Sha, Y., Boris, V.M., Shirvanina, P., and Goddard III, W.A. Oxygen hydration mechanism for the oxygen reduction reaction at Pt and Pd fuel cell catalysts. *Journal of Physical Chemistry Letters*, 2, 572–576, 2011.

166. Meng, S., Wang, E.G., and Gao, S.W. Water adsorption on metal surfaces: A general picture from density functional theory studies. *Physical Review B*, 69, 195404-1–195404-13, 2004.

167. Volmer, M. and Erdey-Gruz, T. Zuer Theorie der Wasserstoffueberspannung. *Zeitschrift für Physikalische Chemie A*, 150A, 203–213, 1930.

168. Santos, E., Poetting, K., and Schmickler, W. On the catalysis of the hydrogen oxidation. *Faraday Discussions*, 140, 209–218, 2009.

169. Hammer, B., Hansen, L.B., and Norskov, J.K. Improved adsorption energetics within density-functional theory using revised Perdew-Burke-Ernzerhof functionals. *Physical Review B: Condensed Matter and Materials Physics*, 59, 7413–7421, 1999. www.fysik.dtu.dk/campos.

170. Gastiger, A.H., Kocha, S.S., Sompalli, B., and Wagner, F.T. Activity benchmarks and requirements for Pt, Pt-alloy, and non-Pt oxygen reduction catalysts for PEMFCs. *Applied Catalysis B: Environmental*, 56, 9–35, 2005.

171. Springer, T.E., Rockward, T., Zawodzinski, T.A., and Gottesfeld, S. Model for polymer electrolyte fuel cell operation on reformate feed: Effects of CO, H_2 dilution, and high fuel utilization. *Journal of the Electrochemical Society*, 148, A11–A23, 2001.

172. Song, W.X. and Zhao, S.J. Development of the ReaxFF reactive force field for aluminum molybdenum alloy. *Journal of Materials Research*, 28, 1155–1164, 2013.

173. Jaervi, T.T., Kuronen, A., Hakala, M., Nordlund, K., van Duin, A.C.T., Goddard III, W.A., and Jacob, T. Development of a ReaxFF description for gold. *European Physical Journal B*, 66, 75–79, 2008.

174. Che, G.L., Lakshmi, B.B., Fisher, E.R., and Martin, C.R. Carbon nanotubule membranes for electrochemical energy storage and production. *Nature*, 393, 346–349, 1998.

175. Sanz-Navorro, C.F., Astrand, P.O., Chen, D., Ronning, M., van Duin, A.C.T., Jacob, T., and Goddard III, W.A. Molecular dynamics simulations of the interactions between platinum clusters and carbon platelets. *Journal of Physical Chemistry A*, 112, 1392–1402, 2008.

176. Zhang, Y., Toebes, M.L., van der Eerden, A., O'Gradi, W.E., de Jong, K.P., and Koningsberger, D.C. Metal particle size and structure of the metal–support interface of carbon-supported platinum catalysts as determined with EXAFS spectroscopy. *Journal of Physical Chemistry B*, 108, 18509–18519, 2004.

177. Goldby, I.M., Kuipers, L., von Issendorff, B., Palmer, R.E. Diffusion and aggregation of size-selected silver clusters on a graphite surface. *Applied Physics Letters*, 69, 2819–2821, 1996.

178. Connolly, D.J. and Gresham, W.F. Fluorocarbon vinyl ether polymers, US Patent 3,282,875 (November 1, 1966).

179. Curtain, D.E., Lousenberg, R.D., Henry, T.J., Tangeman, P.C., and Tisack, M.E. Advanced materials for improved PEMFC performance and life. *Journal of Power Sources*, 131, 41–48, 2004.

180. Borup, R. et al. Scientific aspects of polymer electrolyte fuel cell durability and degradation. *Chemical Reviews*, 107, 3904–3951, 2007.

181. Malek, K. and Franco, A.A. Microstructure-based modeling of aging mechanisms in catalyst layers of polymer electrolyte fuel cells. *Journal of Physical Chemistry B*, 115, 8088–8101, 2011.

182. Tang, H., Qi, Z., Ramani, M., and Elter, J.F. PEM fuel cell cathode carbon corrosion due to the formation of air/fuel boundary at the anode. *Journal of Power Sources*, 158, 1306–1312, 2006.
183. Perry, M.L., Patterson, T.W., and Reiser, C. Systems strategies to mitigate carbon corrosion in fuel cells. *ECS Transactions*, 3(1), 783–795, 2006.
184. Franco, A.A., Schott, P., Jallut, C., and Maschke, B. A multi-scale dynamic mechanistic model for the transient analysis of PEFCs. *Fuel Cell*, 7(2), 99–117, 2007.
185. Gabriel, M., Genovese, L., Krosnicki, G., Lemaire, O., Deustch, T., and Franco, A.A. Metallofullerenes as fuel cell electro catalysts: A theoretical investigation of adsorbates on $C_{59}Pt$. *Physical Chemistry Chemical Physics*, 12, 9406–9412, 2010.
186. Ferreira de Morais, R., Loffreda, D., Sautet, P., and Franco, A.A. Towards a multiscale modeling methodology for the prediction of the electro-activity of PEM fuel cell catalysts. *ECS Transactions*, 25, 167–173, 2010.
187. Marrink, S.J. and Tieleman, D.P. Perspective on the Martini model. *Chemical Society Reviews*, 42, 6801–6822, 2013.
188. Bayramoglu, B. and Faller, R. Modeling of polystyrene under confinement: Exploring the limits of iterative Boltzmann inversion. *Macromolecules*, 46, 7957–7976, 2013.
189. Izvekov, S. and Violi, A. A coarse-grained molecular dynamics study of carbon nanoparticles aggregation. *Journal of Chemical Theory and Computation*, 2(3), 504–512, 2006.
190. Lindahl, E., Hess, B., and van der Spoel, D. GROMACS 3.0 a package for molecular simulation and trajectory analysis. *Journal of Molecular Modeling*, 7(8), 306–317, 2001.
191. Humphrey, W., Dalke, A., and Schulten, K. VMD: Visual molecular dynamics. *Journal of Molecular Graphics*, 14(1), 33–38, 1996.
192. Malek, K., Eikerling, M., Wang, Q., Navessin, T., and Liu, Z. Self-organization in catalyst layers of polymer electrolyte fuel cells. *Journal of Physical Chemistry C*, 111, 13627–13634, 2007.

Index